QUANTUM MECHANICS:
Principles & Applications

Prof. S. Devanarayanan, Ph.D.; D.Sc.,

Dip(Uppsala)

Second Edition

Original Edition, 2005, by SCITECH Publishers, Chennai, India
Second Print 2010
ISBN 81 88429 619

ISBN-13: 978-1539180913

ISBN-10: 1539180913

October 2016

CreateSpace Independent Publishing

(Amazon.com)

Dedication

This book is dedicated to my revered

Parents

and

the Deity of Ambalapuzha Temple, Kerala

1.
    || *Aswaanana Mahaa Budhe Sarva Saastra Visaaradha* ||
    || *Jnaananda Mayam Devam Nirmala Spatikaakruthim*
    || *Aadharam Sarva Vidyaanam Hayagrivam Upaasmahe* ||

(I prostrate myself before **Hayagriva**, the God who is the Personification of knowledge and happiness, who is very pure, and who is the basis of all learning).

Sage Agasthya

2. When asked by Hans Reichenbach how he found his theory of relativity, Albert Einstein's answer was that '*it was because he was so strongly convinced of the harmony of the Universe*'.

WMO- Tech. Note 162

FOREWARD

Since the dawn of Quantum Mechanics, a steady stream of books is appearing in the subject. Some of the earlier books in the field are classics, like Dirac's book. Others serve the interests of research workers or postgraduate students.

The author of this book was my colleague at the Department of Physics, Kerala University for several years. He has made significant contributions in different branches of Physics such as Solid State Physics, Atmospheric Physics, Raman Spectroscopy, *etc.*, and besides one of the authors of a monograph on Thermal Expansion. The present book has grown out of the course of lectures that he has been delivering for the M.Sc. and M.Phil. students for the past two decades. Starting from basics and axiomatic foundation, the book leads to twenty-six chapters to all essential aspects of quantum mechanics up to relativistic theory, fields and Feynman diagrams.

A noteworthy feature of the book is the large number of problems and worked out examples. The importance of solving problems in the training and evaluation processes needs hardly to be stressed. Students generally have a tendency to answer essay type questions only; however, if problems are made compulsory, this will force them to *think*, imparting at the same time training in the application of basic principles to different physical problems.

I hope that the book will be received well by teachers and students alike.

Sd/-

Dr. K.S. Viswanathan, F.A.Sc.

(Raman Centenary Medal Awardee)

Chennai

AUTHOR'S PREFACE

This book has evolved from the lecture notes prepared by the author in teaching a two-semester course in Quantum Mechanics to students of post-graduate and M.Phil Degree courses in Physics for over 15 years at the University of Kerala. The present work, therefore, essentially covers syllabi for Quantum Mechanics for different PG level courses in various Universities.

A study of basic Quantum Mechanics has necessarily been a compulsory (core) part not only for any Graduate Degree programme in Physics but also for any graduate or modern interdisciplinary courses involving chemistry and biology like biotechnology, biophysics, molecular biology as well as for engineering courses. A sound knowledge of basic quantum Mechanics is indispensable for a majority of scientists engaged in research.

Comprised of 26 chapters, the book is addressed to PG students in Physics who have had introductory physics and mathematics through calculus. The emphasis is on a thorough understanding of the principles of quantum theory with particular attention given to the application of these principles in the modern branches of atomic physics, molecular physics, spectroscopy and of engineering. These first 10 Chapters of this book, suitable for the M.Sc. First year students, deals with pre-quantum phenomena, old quantum theory, wave mechanics, applications to problems yielding exact solutions, where the problems discussed are (1) particle in a box (electrons in a metal), (2) particles in a potential well (tunneling / radioactive decay), (3) linear harmonic oscillator (vibrating molecule), both the traditional approach and the operator method, (4) rotating molecule, and (5) hydrogen atom. The subject matter is suitable for one Semester course consisting 48 credits.

Advanced Quantum Mechanics, described in 16 Chapters, is designed to the reader to understand Matrix mechanics, Angular momentum, Addition of momenta, Spin, quantum mechanical problems which require approximate methods to yield solutions, scattering theory, radiation theory, two- and many-body systems, relativistic quantum mechanics, elementary quantum field theory, and related topics. These Chapters are addressed to students in Second year M.Sc. Course in Physics, and M.Phil. In other words these 16 Chapters can be suitable for the second one semester course consisting of 48 credits. These are essential for atomic, molecular and nuclear Spectroscopy, many-body problems, sub-atomic physics and solid-state physics.

An attempt has been made to include lengthy mathematical arguments possible in the text wherever possible, and the reader needs only to minimize consultation with other books. Problems, of varying degrees of difficulty, have been included as examples at the right places with detailed solutions so as to help the students to understand the principles involved and get well prepared for good performance in examinations, both qualifying and competitive. The detailed discussions provided in the text are almost sure to answer any objective type questions from quantum mechanics. The examples with solutions in the book are meant to be an integral part of the text..

The present work is essentially a text at the GRADUATE & PG level courses in Quantum mechanics at any University, and suitable either as single paper in Quantum mechanics in the Annual Examination Scheme, or as two Semesters, each of 48 credits, in the Semester System I hope it will serve as a general reference book for scientists and engineering who are engaged in research and development work involving NEW PHYSICS.

The *outstanding features* of this book are:
(1) There are a maximum number of conceptually difficult Examples, but calculation wise easy, followed by solutions in each section to enable students to understand and use the book as a self-guide for study and assist teachers to efficient instructions in their classes. No other book so far published has this feature.
(2) All effort was taken by the author to high light every possible aspect of physical concepts, physical principles, mathematical principles, as subheadings to enable readers from omission of these while understanding the subject, and to simplify preparation for qualifying / competitive examinations.
(3) It is probably the first student-friendly textbook in quantum mechanics.
(4) The content and presentation of the subject quantum mechanics is kept in an expert teacher's style.

For the physical scientist and engineer, mathematics is a tool, and is second in importance to understanding the physical ideas. For your convenience, some of the most useful mathematical relations are listed in an appendix at the end of the book. A bibliography listing more than sixty books on quantum mechanics and related topics is included in the end of the book.

It is a pleasure to acknowledge my personal and intellectual debt to (late) Professor Dr. V.S. Venkatasubramanian, Indian Institute of Science, for his intelligent discussions and lectures on the subject during 1963 -1968, and Professor Dr. K.S. Viswanathan, earlier Head, Department of Physics, University of Kerala, for assigning the author to teach Quantum mechanics for several years at the University of Kerala, which really enabled me acquire enough source materials in and exposition to quantum mechanics. The influence of Professor (late) C.V. Raman, N.L. and Professor (late) R.S Krishnan on inculcating in the author the spirit of dedication and hard work cannot be forgotten. The author expresses his sincere thanks to his wife Mrs. Chitra for her full cooperation and all kind of support throughout during the process of preparation of the material, without which it could not have become a reality. Both Chitra and I had acquired enough expertise after the co-authoring the Monograph on THERMAL EXPANSION OF CRYSTALS (Pergamon, Oxford, 1979). Last but not the least the help, at various stages in the preparation of the final version of the manuscript, from both my son Mr. Ajith Shankar and daughter Ms. Aparna Gayathri, is recorded herewith.

Thanks to M/s SCITECH Publications, Chennai for readily accepting the manuscript for publication. Finally, I am sure that there may be found mistakes crept on various counts in the text as a result of human error, which may kindly brought to my notice (sdevanarayanan@yahoo.com) for future use.

October 2nd, 2004
S. Devanarayanan

PREFACE

of Second Edition

Mathematics pervades us, and it has shaped our understanding of the World in innumerable ways. According to the mathematician Ian Stewart there are seventeen mathematical equations that changed the world. This list includes the Schrodinger equation of Quantum Mechanics.

A study of basic Quantum Mechanics has necessarily been a compulsory (core) part not only for any Graduate Degree programme in Physics but also for any graduate or modern interdisciplinary courses involving chemistry and biology like biotechnology, biophysics, molecular biology as well as for engineering courses. A sound knowledge of basic quantum Mechanics is indispensable for a majority of scientists engaged in research. It is not surprising that the text of my book "*Quantum Mechanics – Principles and Applications*" (SCITECH, Chennai, India, 2005, 2010)" (*ISBN* : 81 88429 619) has shown signs of ageing and needs rectification, and few suggestions and criticisms was received in last ten years. This Second edition has evolved from the earlier text. I know the role of the Nobel Prize winner in Physics Enrico Fermi, I have learnt a lot on this topic of quantum mechanics in 1968 through my personal copy "Notes on Quantum Mechanics" by E Fermi.

The second edition incorporates only minor changes in contents in Chapter 4:
(1) New titles are given to most of the Chapters with few change in contents
(2) Review Questions with answers are included at the end of each Chapter. This will be useful to students and question paper setters.

I am sure that there may be found mistakes crept on various counts in the text, which may kindly be brought to my notice for future use, through chsd1976@gmail.com.

S. Devanarayanan

January 15, 2016

LIST OF CONTENTS

Pages

Dedication

Foreword

Author's Preface

Preface of second edition

Chapter 1: **A Crisis in Physics**: Experimental developments during the first quarter of the 20th Century   1

1. Introduction: 2. A crisis in physics: 3. Theory of photons and Planck's radiation law: 4. Discovery of the photon and Photons as particles: 5. Nuclear atom- Bohr's theory of H-atom 6. Sommerfeld model of atom 7. Vector Model of the Atom: 8 Success of the old quantum theory: Review Questions.

Chapter 2: **Concepts in Wave Mechanics** – Wave-Particle Duality.   40

1. Introduction: 2. Further Developments 3. Diffraction of electrons; 4. . Wave packet description of material particles: 5. Particle spectrum; 6. Conclusions: 7. Wave packet Description, 8. Heisenberg Uncertainty principle 11. Gedanken Experiments, Review Questions.

Chapter 3: **Quantum Mechanics - 1**: Wave Mechanical Approach   68

1. Introduction: 2. Electron and the wave function: 3. Dirac delta function: 4. Analogy between optics and mechanics: 5. Schrodinger wave equation: 6. Deduction of the basic equation - non-relativistic. 7. Salient features, Review Questions.

Chapter 4: **Quantum Mechanics - 2:** Postulatory Approach to Wave Mechanics: 87

1. Introduction: 2. Postulates in Quantum Mechanics: Postulate 1 to Postulate 7: 3. Deduction of Schrodinger equation using mathematical operators: 4. conservation of probability density 5. Hydro dynamical wave theory, 6. Commutation Rules: 7. Rigorous method for the Uncertainty principle: 8 Separation of time-dependent Wave function: 9. Parity. 10. Spectrum of the Hamiltonian: 14. Closure property. Review Questions.

Chapter 5: **Exact Solutions: Applications 1 to 3**: Particle in a Box, -Tunneling – Electron in a Periodic Lattice   130

1. Introduction, 2. Exact Solutions; 3. Particle in a box: 4. Transmission and Reflection at a Barrier, 5. Finite potential Barrier, 6. Alpha decay, 7. Kronig-Penney Model Potential barrier, Review Questions.

Chapter 6. **Exact Solutions: Applications 4** - Linear Oscillator – Traditional Approach.   158

1. Introduction: 2. Linear Oscillator. 3. Quantum mechanical Oscillator. 4. Conclusions, Review Questions.

Chapter 7. Exact **Solutions: Quantum mechanical Oscillator**-II: Operator method: 175

1. Preliminaries: 2. Energy spectrum of the oscillator, 3. Number operator, 4. Explicit form 5. Second Quantization, Review Questions

Chapter 8. **Exact Solutions: Applications 5** - Spherically Symmetric System- 193

1. Introduction: 2. The dumb bell model of the diatomic molecule, 3. Wave function, $\psi(\mathbf{r})$ for the rigid rotor: 4. The polar factor of $\psi(r)$: 5. Rotational Spectra: 6. Properties of Polar Functions, 7. Rigid rotator Wave functions, Review Questions.

Chapter 9. **Exact Solutions: Applications 6** - Wave Mechanics of the Hydrogen Atom   209

1. Introduction 2. Wave-mechanics of H-atom: 3. The Schrodinger Equation: 4. Radial wave equation: 5. Discrete Energy Spectrum: 6. Properties of Radial functions: 7. Total wave function 8. Interpretation of Quantum numbers: 9. Electron Probability Density: 10. Atomic Orbitals: 11. Chemical Bonding: 12. Quantum Virial Theorem. Review Questions.

Chapter 10. **Paradoxes in Quantum mechanics**   248

1. Introduction 2. Paradoxes in quantum mechanics. 3. EPR Argument.- The Bell Inequality. 4. Quantum Information.

Chapter 11. **Quantum Mechanics – 3**: Mathematical Formalism – Matrix Mechanics: 259

1. Introduction: 2. Mathematical preliminaries 3. Hilbert space, Matrices, 4. Dirac's bracket notation, 5. Representation of kets, bras and operators in matrix mechanics, 6. Unitary transformation in quantum mechanics; 7. Types of pictures (equation of motion), Review Questions.

Chapter 12: **Types of Pictures (Equations of Motion)**: S - , H- & I- Pictures   319

1. Introduction: 2. The Evolution operator, 3. The S-picture; 4. The H-picture: 5. Dirac representation ` 6. Comparison of the S-, H-, and Interaction pictures, Review Questions.

Chapter 13: **Harmonic Oscillator Revisited**- Matrix Formulation: 341

1. Introduction, 2. Oscillator pure states, 3. Oscillator in a mixed state, 4) q-Deformed Oscillator, Review Questions.

Chapter 14: **Invariance Principles and Conserved quantities in Quantum Mechanics**: 355

1. Introduction, 2. Coordinate and state vector Transformations, 3. Spatial Translation and Conservation of momentum, 4. Temporal Translations and energy conservation, 5. Space Inversion and parity conservation, 6. Spatial Rotation and angular momentum conservation, 7. Time Reversal Invariance, 8. Conservation laws and constants of motion, Review Questions.

Chapter 15: **Angular Momentum – Properties**: 380

1. Introduction. 2. Properties of ℓ-vectors 3. Shift Operators: ℓ_+ and ℓ_- : 4. Angular momenta in spherical coordinates: 5. Spectra of operators ℓ_z and ℓ^2: 6. Simultaneous eigen states of ℓ^2 and ℓ_z, Review Questions.

Chapter 16: **Angular Momentum – Matrices**. 399

1. Introduction 2. Unambiguous determination of angular momentum matrices, Review Questions.

Chapter 17: **Intrinsic Angular Momentum - Spin-$\frac{1}{2}$ particles**: 408

1. Introduction – 2. Description of Particles with spin – $\frac{1}{2}$. 3. Spinors; Spin-up and down states, Review Questions.

Chapter 18: **Addition of Angular Momenta**: Total Angular Momentum; CG Coefficients; 419
1. Introduction – 2. Vector model for combining angular momenta. 3.. Addition of two angular momenta, general case 4. Clebsch-Gordan coefficients; 5. The Wigner-Eckart theorem, Review Questions.

Chapter 19: **Perturbation Theory of Stationary State**: 438

1. Introduction, 2. Time Independent Perturbation; 3.Noin-degenerate case. 4. II order perturbation; 5. Normal state of Helium atom; 6. Degenerate case; 7. Stark effect in hydrogen, Review Questions.

Chapter 20: **Variational and WKB Approximation Methods**· 468

1. Introduction. 2 The Ritz method of Variation 3. Applied to spectrum of helium-4. The JWKB method, Review Questions.

Chapter 21. **Fine Structure of Hydrogen** (Application of Perturbation Theory to Real Hydrogen) and Zeeman Effect: 482

1. Introduction; 2. Perturbation calculations; 3. Second order perturbation correction; 4. Total energy Shift, 5. Zeeman Effect in Hydrogen; 6.The 21-*cm* line, Review Questions.

Chapter 22: **Quantum Mechanics: -4**: Scattering Theory (Collision or Reaction

Theory): 504

1. Introduction, 2. Scattering fundamentals; Potential scattering, Scattering amplitude, Integral formula, differential cross section, 3. Scattering at low energies and method of partial waves, Optical theorem, s-wave scattering, scattering by attractive square potential, 4. Elastic Scattering at higher energies, The Born approximation, Standard form, Review Questions.

Chapter 23:   **Identical Particles (Many Body Systems):**   528

1. Introduction, 2. Symmetry and Ant-symmetry, Central Field Approximation, Slater determinant, 3. The Helium atom; 4. Para- and Ortho-helium. 5. Many-electron atoms, 6. The Hartree -Fock SCF method, 7. Correlation energy; 8. Feynman Diagrams, Review Questions.

Chapter 24.   **Time Dependent Perturbation Theory**   563

1. Introduction 2. Perturbative expansion, 3. Constant Perturbation 4. Transitions to continuous spectra -The Fermi Golden Rule ,5. Einstein Transition Probabilities 6. Lifetime and energy uncertainty- 7. The Periodic (Harmonic) Perturbation 8. Multi-pole Radiations, Review Questions.

Chapter 25. **Quantum Mechanics – 5** : Relativistic Quantum Mechanics: 587
1. Introduction- Preliminaries on Lorentz transformation, Metric Tensors, $g_{\mu\nu}$ and $g^{\mu\nu}$, 4-vectors;- 2. Single Particle Relativistic Equations, Klein-Gordon equation, Its solutions, Difficulties;- 3. Hamiltonian of charged particle in EM Field, 4.. Dirac equation for a Free particle, 5. Hole Theory, Negative energy states, Potential Catastrophe, its interpretation, Eigen spinors,- 6. Dirac Particle in an EM Field, Spin-magnetic properties,-7. The Dirac Particle in a Central Potential,   8. Kepler problem in Dirac theory, 9. Beyond the Fine Structure corrections, The Lamb Shift,   10. Hyperfine Structure. 11. Conclusions. Review Questions.

Chapter 26.   **Quantum Mechanics – 6 : Interacting Fields**; Quantum Field Theory, Quantum Electrodynamics: 639

1. Introduction, 2. Quantum Field Theory – General Formalism, 3. Second Quantization, 4. Interacting Fields; 5. The Quantum Field; 6. Quantum Electrodynamics, Quantization of Gauge field, 7. Renormalization; 8. Feynman Diagrams, 9. Renormalization and theory of the Lamb shift, 10. Colour and Quantum Chromodynamics, 11. Toward a Theory of Everything, Review Questions.

Appendix A   Table of Physical Constants   690

Bibliography   691

About the Author   694

---ooo0ooo---

Chapter 1

A Crisis in Physics: Experiments and Bohr Theory of H-Atom

Chapter 1

A Crisis in Physics: Experiments and Bohr Theory of H-Atom

No language which lends itself to visualizability can describe the quantum jumps

- Max Born

1. INTRODUCTION:

Mechanics is the branch of Physics dealing with the effects of forces on the motions of physical bodies, which for the most part were explicable in terms of what one now calls classical theoretical physics. In the classical picture the world is composed of distinct elements, each possessing a definite position and velocity. The motions of mechanical objects were believed to be explicable in terms of what we now call classical physics. Newtonian Mechanics (1687) explained the motion of mechanical objects on both celestial and terrestrial scales. Classical Mechanics is a computational scheme based on Newton's Laws of Motion., in both classical and terrestrial scales. Application of this theory to

(a) Molecular motions produced results in the kinetic theory of gases.
(b) Discovery of the electron in 1897 by JJ Thomson consisted in showing that it behaved like a Newtonian particle, and
(c) The wave nature of light had been strongly suggested by the diffraction experiments of Thomas Young in 1803, and was put on a firmer foundation in 1864 by James Clerk Maxwell's discovery of the connection between optical and electrical phenomena.

Taking all this into account it is not surprising that at the end of the nineteenth century many physicists thought that all interesting questions had been asked and that finding the right answers would be merely a matter of time. The wave theory could explain interference, diffraction and polarization (also reflection, refraction and superposition) but not the other experimental results to be described below.

1.1 A CRISIS IN PHYSICS:

The end of the 19thCentury and the beginning of the 20thCentury witnessed a **crisis in physics**. This was because several experiments indicated that the *interaction of the electromagnetic radiation with matter* was not entirely in accordance with the laws of electromagnetism (Due to Andre Marie Ampere, Hans Christian Oersted, Laplace, Michael Faraday, Joseph Henry, J C Maxwell and many others). These Laws are synthesized in Maxwell's Equations for the EM Field. But a series of experimental results required **concepts totally incompatible** *with Classical Physics*. The background of the crisis above, and to expose the new concepts, which led to the development of a new mechanics, called <u>Wave Mechanics</u>, *i.e.*, of the 'quantum theory', will make the transition from the classical to quantum theory less mysterious for you. The **new concepts** are:

a) The *particle properties of radiation*,
b) The *wave properties of matter*, and
c) The *quantization of physical quantities*.

These will emerge in the following phenomena from several DECISIVE EXPERIMENTS performed during 1900 -1930:

i) Black body radiation - Planck hypothesis (Max Planck, 1900),
ii) The photoelectric effect - photon hypothesis (Albert Einstein, 1905),
iii) The Bohr's theory of the Hydrogen atom – optical line spectra (Niels Bohr's atom model, 1913).
iv) The Compton Effect - particle nature of radiation, and later the Raman Effect,
v) Electron diffraction, or, X-ray spectra, and

All these experiments could not be explained by means of the *wave theory of light*.

1.2 WAVE-PARTICLE DUALITY:

Irregular and non-repeating acceleration without oscillation is a source of low frequency radiation. A metal heated to incandescence emits light in a similar way from random oscillations of electrons in this metal. This type of radiation is called *thermal radiation*.

1.2.1 THEORY OF PHOTONS

1.2.1.1 *Wien's theory*:
The most fundamental of all of equations in wave theory is given by
$$c = \nu \lambda, \quad (1.3.1)$$
relates c, the velocity of light, ν frequency of the wave having wavelength, λ (See Fig 1.1).

In 1894, Wilhelm Wien, using the very general arguments of Classical Physics showed that the energy density $u(\lambda, T)$ of radiation,
$$u(\lambda, T) = \lambda^{-5} f(\lambda, T) \quad (1.3.2)$$
where $f(\lambda, T)$ = a function of wave length, λ and temperature, T, inside the cavity of a blackbody, leading to *Wien's Displacement Law*:
$$\lambda_{max} = b/T \quad (1.3.3)$$
where $b = 2.90 \times 10^{-3} m\,K$. This is an empirical law.

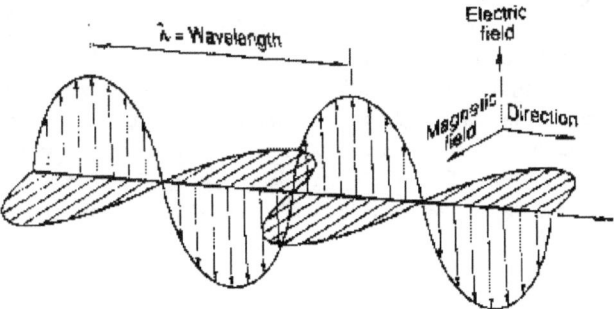

Fig. 1.1 EM Wave structure.

1.2.1.2 *Rayleigh-Jeans law:* (R-J Law).

When the cavity is in thermal equilibrium at temperature T each of the standing waves must be assigned an average kinetic energy $k_B T$, where k_B = Boltzmann constant = $1.3805 \times 10^{-23} JK^{-1}$, deriving the Law (by Lord Rayleigh):

$$u(\nu, T) = \rho(\nu)\, \bar{E}(\nu) = 8\pi \frac{\nu^2}{c^3} k_B T \qquad (1.3.4)$$

1.2.1.3 *Lummer and Pringsheim's experiment*:

In 1899 measurements by Otto Lummer and E. Pringsheim resulted in intensity $I(\nu)$ versus ν for different temperatures showed some simple characteristics:

The distribution of frequencies is a function of temperature of the *blackbody*,

i)   The total amount of radiation emitted $\int I(\nu)\, d\nu$ increases with increasing T,

ii)   The position of the peak maximum shifts toward higher ν with increasing equilibrium temperature.

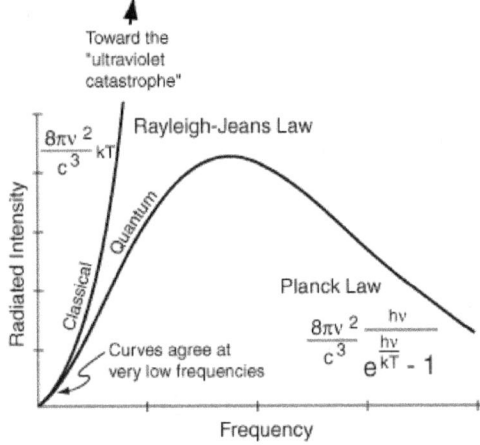

Fig. 1.2 Radiation distribution and '*Ultra Violet Catastrophe*'

1.2.1.4 The Ultra-Violet Catastrophe:

A shortcoming of the R-J Law, which attempted to describe the radiance of a blackbody at various frequencies of the EM spectrum, is its failure at the UV end of this spectrum of thermal distribution; and the Wien's Law failed at the red end of the spectrum, as shown by the *Lummer-Pringsheim experiment*. It was clearly wrong because as the frequency increased, the radiance increased without bound; something quite not observed; this was dubbed the "Ultraviolet Catastrophe", in thermal radiation (Fig. 1.2 and Fig. 1.3). It was later reconciled and explained by the introduction of the Planck radiation Law.

1.3. PLANCK'S RADIATION DISTRIBUTION LAW:

In 1900, Max Planck found a formula for an ingenious interpretation between the high frequency Wien Law and the low frequency R-J Law. He made use of the *Maxwell Boltzmann Distribution*, viz.,

$$N = N_0 e^{-\varepsilon/k_B T} \qquad (1.4.1)$$

and combined with his quantum hypothesis, viz.,

$$\varepsilon = n\hbar\omega = nh\nu, \text{ where } n = 0,1,2,\ldots \qquad (1.4.2)$$

Planck constant $h = 6.6256 \times 10^{-34} J-s$

$\hbar = (h/2\pi) = 1.054 \times 10^{-34} J-s$, and

$\omega = 2\pi\nu$ = the angular frequency of the oscillation.

Plank's distribution Law gives the expression for the distribution of the maximum intensity of radiation in the spectrum of the blackbody. It states:

$$du(\nu,T) = \left[\frac{8\pi h \nu^3}{(e^{h\nu/k_B T}-1)}\right]\frac{1}{c^3}d\nu \qquad (1.4.3)$$

The *Planck's hypothesis* was a revolutionary break from classical EMTheory based on Maxwell's Equations. The Wien's Law and the R-J Law were found to be the *short wave* and *long wave* limits, respectively, of the Planck Law, expressed by equations (1.4.3), (1.4.4) or (1.4.5).

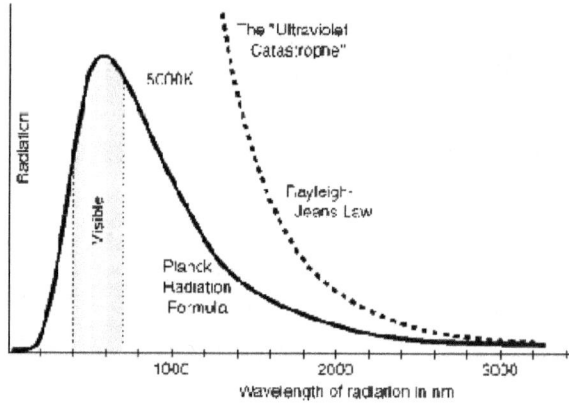

Fig. 1.3. Planck radiation distribution at T = 500 K.

Energy per unit volume per unit frequency

$$S_\nu = \left[\frac{8\pi h \nu^3}{(e^{h\nu/k_BT}-1)}\right]\frac{1}{c^3} \qquad (1.4.4)$$

Energy per unit volume per unit wavelength

$$S_\lambda = \left[\frac{8\pi h c}{(e^{hc/\lambda k_BT}-1)}\right]\frac{1}{\lambda^5} \qquad (1.4.5)$$

The Planck Law is found to confirm the Lummer-Pringsheims's experimental results. <u>It should be noted that most radiating bodies – the Sun, an incandescent filament lamp, and a hot gas – do not behave like black bodies.</u>

	# modes peer unit frequency per unit volume	Probability of occupying modes	Average energy mode
CLASSICAL	$8\pi(\nu^2/c^3)$	Equal for all modes	k_BT
QUANTUM	$8\pi(\nu^2/c^3)$	Quantized modes: require $h\nu$ energy to excite upper modes, less pprobable	$\left[\frac{h\nu}{(e^{h\nu/k_BT}-1)}\right]$

Fig. 1.4. Comparison of classical and quantum pictures of Black body radiation.

Worked out Example 1.1.
The star Sirius (assumed to be a blackbody) appears as blue. It is emitting a spectrum with $\lambda_{max} = 260.0\ nm$. Estimate the surface temperature of Sirius.

Solution:
   $\boxed{Step\#1}$ Being a blackbody use the empirical Wien's Law, $\lambda_{max} = b/T$, where $b = 2.90 \times 10^{-3}\ m\ K$; $\lambda_{max} = 260.0\ nm$.
   $\boxed{Step\#2}$ $\lambda_{max} = (b = 2.90 \times 10^{-3}\ m\ K)/T = 260.0\ nm$.
   $\boxed{Step\#3}$ $T = (b = 2.90 \times 10^{-3}\ m\ K)/(260.0 \times 10^{-9}\ m) \sim 11{,}000\ K$.

Worked out Example 1.2.
Find the Wien's constant from the Planck radiation distribution law.

Solution: $\boxed{Step\#1}$ The Planck Law in terms of λ is $du(\lambda,T) = \left[\frac{8\pi h c}{(e^{hc/\lambda k_BT}-1)}\right]\frac{1}{\lambda^5}d\lambda$

. $\boxed{Step\#2}$ Set $x = hc/\lambda k_BT$; $du/d\lambda$ gives $x = e^{-x} + \frac{x}{5} = 1$.

$\boxed{Step\#3}$ On iteration, one gets $x_{max} = 4.965 = 1 hc/\lambda_{max}k_BT$;

i.e., $\lambda_{max}T = hc/[(4.965)k_B] = b$, the Wien's constant.

Worked out Example 1.3.
A pendulum of mass 1g on a 1m string oscillates with angular amplitude of 5° and frequency 0.5Hz, having total observed energy of oscillation $E = 3.7 \times 10^{-5}\ J$. Apply Planck's hypothesis and comment on the result.

Solution:
   $\boxed{Step\#1}$ The minimum increment by which this oscillator energy that may change is given by the Planck's hypothesis $\varepsilon = n\hbar\omega \equiv nh\nu$;

$\boxed{Step\ \#2}$ $\varepsilon = (6.625 \times 10^{-34}\ J-s) \cdot (0.5\ s^{-1}) = (3.3 \times 10^{-34}\ J)\ E\ /\ (3.7 \times 10^{-5}\ J)$

$\therefore \varepsilon = 9 \times 10^{-29}\ E$ (To the appropriate number of significant figures).

$\boxed{Remarks}$ Such a change is quite undetectable. Thus quantization of the available oscillation energies could be curiously a small effect in almost all its macroscopic manifestations and yet emerge as the mechanism by which UV Catastrophe in thermal radiation is averted.

1.3.1 SPECIFIC HEAT OF SOLIDS:

As early as 1872, there were observed departures from
i) Equipartition Law, and
ii) DULONG-PETIT Law, *viz.* product of (Mol. wt. or At. wt.) and (Specific heat) of a solid = constant).

The unqualified success of the Planck's distribution law gave an explanation to this problem of specific heats of solids. No further discussion on this topic will be made here.

1.4 DISCOVERY OF THE PHOTON (Photon hypothesis)/ PHOTOELECTRIC EFFECT:

1.4.1 Background:

Albert Michelson (1881) and Edward Morley (1887) showed that Maxwell's *aether* does not exist, but Maxwell's empirically based equations in electromagnetism remained an excellent fit to observations. The ejection of electrons from a metal by the action of EM wave is called the photoelectric effect. In 1887, Heinrich Hertz discovered this when he was engaged in experimenting with EM waves to confirm the existence of radio energy, as predicted by Maxwell's Theory (1884). He succeeded in creating and detecting EM waves. As transmitter he used a spark gap connected to a resonating circuit, which determined the frequency, ν and also acted as the antenna.

1.4.2 PHOTOELECTRIC EFFECT:

Then further experiments established the following:
i) When polished metal plates are irradiated they may emit photoelectrons (Fig. 1.5).

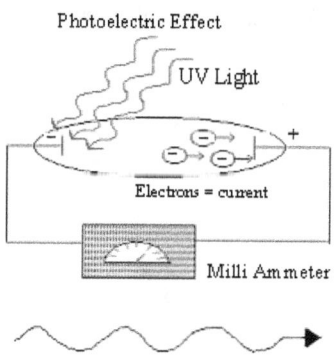

Photon = Wave-particle of Light

Fig. 1.5 Photoelectric experimental schematic diagram.

ii) Emission of photoelectrons by metals depends on the wavelength λ of the light. For each substance there is a minimum or **Threshold frequency**, v_0 of EM radiation such that there is no photoelectric effect if the light falling on it is of frequency, $v < v_0$.

iii) The photoelectric current, i_p, is proportional to the intensity of the light (Fig. 1.6).

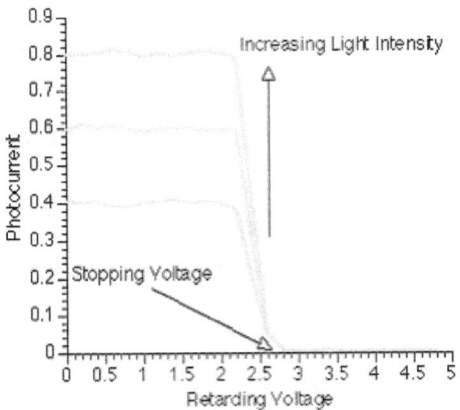

Fig. 1.6 Plot of Photo-current versus retarding voltage, with increasing light intensity.

Iv) The energy of the photoelectrons is independent of the intensity of the light source, but varies linearly with the frequency v of incident light (Fig. 1.7).

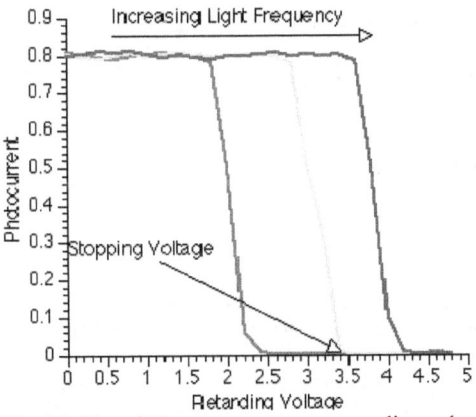

Fig. 1.7 Plot of Photo-current versus retarding voltage, with increasing light frequency..

1.4.3 Threshold Equation:

Robert A. Millikan carried out a complete investigation of the characteristics of the photoelectric effect. If φ_0 = energy required by a photoelectron to escape from given metal, electronic kinetic energy is related to the frequency of light, v.

Kinetic energy of an electron = photon energy − work function.

$$\tfrac{1}{2} m v^2 = h\nu - \varphi_0 \tag{1.5.1}$$

$$h\nu - \varphi_0 = eV_0 \tag{1.5.2}$$

$$\boxed{\tfrac{1}{2} m v^2 = h\nu - \varphi_0 = eV_0} \tag{1.5.3}$$

is called the **Threshold Equation**.
V_0 is the *stopping potential*.

1.4.4   Einstein's Explanation:

Albert Einstein (1905) gave this explanation for the dependence of photoelectron emission on the frequency of the radiation, for which he was awarded the Nobel Prize in Physics in 1921.

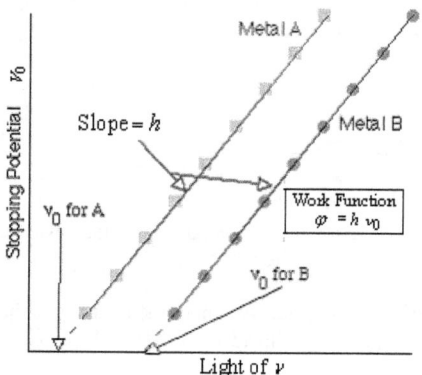

Fig. 1.8 Plot of stopping potential versus light frequency.

Thus Einstein adopted the Planck's hypothesis and applied it to the EM radiation. The family of straight lines in the plot

$$\boxed{eV_0 = h(\nu - \nu_0)} \tag{1.5.4}$$

is depicted in Fig.1.8. Thus the energy associated with the quantum is transferred as a unit to the electron - quite paradoxical!!

1.4.5   Conclusions

(1) The success of the 'photon hypothesis' is to explain the photoelectric effect,
(2) According to Einstein energy is quantized and each photon carries an amount of energy $h\nu$ (and not $nh\nu$), $\therefore E = h\omega$
(3) The dual character of EM radiation became established. It sometimes behaves like a wave motion and sometimes like a stream of corpuscular quanta. The wave-particle duality of EM radiation is hereby established.

Worked out Example 1.4

When UV radiation of 100 nm is falling on silver surface, photoelectrons emitted can be completely retarded with a potential of 7.7 V. Find the work function of silver.
Solution:

$\boxed{Step\#1}$ $h\nu = [(3 \times 10^8\, ms^{-1})/\ (100 \times 10^{-9}\, m)]\cdot[(6.625 \times 10^{-34}\, J-s)/\ (1.6 \times 10^{-19}\, J)/eV)]$

= 12.4 eV.

$\boxed{Step\#2}$ $\varphi_0 = h\nu - eV_0 = (12.4 - 7.7)\, eV = \hbar\nu_0$;

$\boxed{Step\#3}$ $\nu_0 = 4.7\, eV / h = (4.7\, eV \times 1.6 \times 10^{-19}\, J) / (6.625 \times 10^{-34}\, J\text{-}s)$

$\nu_0 = 1 \times 10^{15}\, Hz$.

1.5. PHOTONS AS PARTICLES (X-RAYS and the COMPTON EFFECT):

A.H. Compton (1923) performed another experiment leading to the same paradoxical position, as by Photoelectric Effect. The experiment provided the most direct evidence for the <u>particle nature of radiation</u>. According to the classical EM theory, when an EM wave of frequency ν is incident on free charged particles, such as electrons, the charges absorb EM radiation and are forced to oscillate with frequency ν. These oscillating charges in turn reradiate EM radiation of the same frequency ν. This leads to the prediction of intensity, I, of radiation observed at an angle ϑ,

(i) I varies as $(1+\cos^2 \vartheta)$, and

(ii) I does not depend on the incident ($\lambda_{inc} \equiv \lambda$ and $\lambda_{sc} \equiv \lambda'$), i.e. $\lambda_{sc} = \lambda_{inc}$.

These hold good for the visible and longer λ. This is **Coherent** *scattering*. On the other hand, Compton found that the radiation of shorter λ, like X-rays, scattered fails to obey the classical theory. The scattered X-rays are found to consist of two frequencies: one which is the same as that of the incident frequency and the other of frequency, $\nu' < \nu$ (or $\lambda' > \lambda$) shifted relative to the λ by an amount depending on the scattered angle ϑ. The incident wavelength λ is called *unmodified wavelength* while λ' is called the *modified* wavelength. This is called **incoherent** *scattering*. The correct explanation of the presence of λ' in the X-ray scattering was given by Compton, by adopting the Planck's Quantum hypothesis and applying the Laws of the conservation of momentum and energy. Thus the incident radiation is treated as a beam of photons of energy $E = nh\nu$ propagating with velocity c and elastically scattering off electrons. The collision between the incident X-ray and a free electron at rest is shown schematically in Fig 1.9.

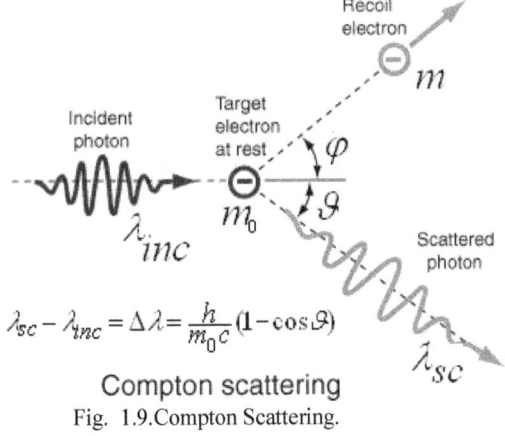

Compton scattering

Fig. 1.9. Compton Scattering.

Maxwell's Equation gives the relation between energy and momentum for a *plane EM wave*. $E = cp$, for mass less particles. By analogy with relativistic particle kinematics (to be dealt with in Chapter 25),

$$E = c\sqrt{(m_0^2 c^2 + p^2)} \qquad (1.6.1)$$

where m_0 = rest mass of the particle, p = linear momentum of the particle,
Velocity at this momentum is given by

$$v = \frac{dE}{dp} = \frac{pc^2}{E} = \frac{pc^2}{c\sqrt{(m_0^2 c^2 + p^2)}} \qquad (1.6.2)$$

For photons, $\quad \frac{dE}{dp} = c$ (always), $\qquad (1.6.3)$

$m_0 = 0$, $p = h\nu/c$, and $\varepsilon = h\nu$; $m = h\nu/c^2$.
A mathematical analysis of the collision (Fig 1.9) leads to

$$h\nu' = \frac{h\nu}{[1 + (h\nu/m_0 c^2)(1 - \cos\vartheta)]}.$$

But $\lambda'\nu' = \lambda\nu = c$.

∴ $\quad \Delta\lambda = (\lambda' - \lambda) = h(1 - \cos\vartheta)/m_0 c \qquad (1.6.4)$

Thus the modified wavelength $\lambda' > \lambda$.

$$\Delta\lambda = \lambda_C (1 - \cos\vartheta) = 2\lambda_C \sin^2\frac{\vartheta}{2} \qquad (1.6.5)$$

where $\boxed{\lambda_C = \frac{h}{m_0 c} = 0.02426 \, \overset{o}{A}}$ (approx.). $\qquad (1.6.6)$

λ_C is called the <u>Compton wavelength</u> of the electron.

$$\lambda' = \lambda + \lambda_C (1 - \cos\vartheta) \qquad (1.6.7)$$

These results do agree with the measured values of λ' at various angles ϑ. On the basis of this interpretation of the Compton Effect together with the previous discussion of black body radiation (Section 1.4) and the photoelectric effect (Section 1.5), one may conclude that a photon is a quantum of EM radiation of energy and momentum emitted / absorbed in a single process by a charged particle. The EM interaction is, therefore, considered as an exchange of photon, the photons transfer energy and momentum from one charge to the other.

Worked out Example 1.5

A metallic target is exposed to X-rays of energy 100 keV. Find the energy of the X-rays scattered under the Compton Effect at an angle of $30°$ to the incident, the energy of the recoiling electron and the angle of the recoil electron coming out.

Solution: $\boxed{Step\#1}$ $(\lambda' - \lambda) = h(1-\cos\vartheta)/m_0 c = hc(\frac{1}{E'} - \frac{1}{E}) = (\frac{1}{E'} - \frac{1}{100\,keV})$

$= (1 - \cos 30°)/m_0 c^2 = (1 - 0.866)/510\,keV = 1/3810\,keV$; ∴ $E' = 97.5\,keV$;

$\boxed{Step\#2}$ Kinetic energy $= (E - E') = 2.5\,keV$.

$\boxed{Step\#3}$ Momentum, $(p')^2 = p^2 + p_e^2 - 2pp_e \cos\varphi$; $E = pc$, $E' = p'c$;

$\boxed{Step\#4}$ $\cos\varphi = [E^2 - (E')^2 + p_e^2] / 2pp_e c$;

$= [E^2 - (E')^2 + T^2(1 + 2E_0/T)] / 2ET[1 + E_0/T]^{1/2}$; $\varphi = 73°$.

1.6 THE NUCLEAR ATOM

1.6.1. Background on an Atomic View:

Isaac Newton was the first to resolve white light into separate colours by dispersion with a glass prism. In 1752 Th.Melvill showed emission lines from light emitted by incandescent gases. G.Kirchhoff in 1861 began the study of atomic spectra, while R.W. Bunsen worked on spectra of alkali metals. James Clerk Maxwell (1864) proposed the EMTheory The existence of a spectrum composed of well-defined frequencies was a problem that puzzled physicists at the end of the 19th Century and the beginning of the 20th Century. In fact a low-pressure discharge through hydrogen emits an unusually simple spectrum consisting of a *series* of lines that converge toward a limit in the near ultraviolet. J.J. Balmer (1885) represented all these lines of the *series* by the so-called **Balmer formula** bearing his name, viz. frequency,

$$\boxed{v = cR\left(\frac{1}{2^2} - \frac{1}{n^2}\right)}, n = 3, 4, 5, \dots \qquad (1.7.1)$$

R is called the *Rydberg Constant*, after Johannes R. Rydberg (1889J, the distinguished Swedish spectroscopist. Similarly, in 1906 Lyman, and later Paschen discovered similar series represented by formula similar to the *Balmer formula*. It was in 1897 that J J Thomson discovered the electrons.

1.6.2 Jellium Model of Atom:

The assumption that electrons are constituents of all atoms was a reasonable inference from such experiments as the photoelectric effect, ionization of gases in discharge tubes, *etc*. J.J. Thomson's e/m experiment turned out the result that the hydrogen atom to be 1836 times as heavy as the electron, if one assumed the same magnitude of charge for each. Thomson proposed an atomic model in which electrons were embedded in a massive matrix of positive charge filling a volume of roughly one atomic diameter ($\sim 1 \overset{o}{A}$).

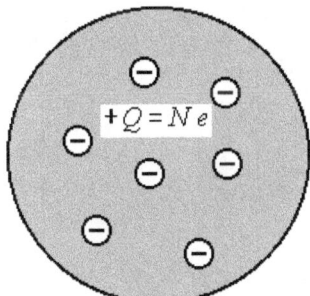

Fig. 1.10 Thomson model of atom in early 1900's

This **jellium** (or *plum-pudding*) model of the atom (Fig. 1.10) could not explain H. Geiger and E. Marsden's large angle scattering results by α-rays.

1.6.3 Rutherford's Nuclear Atom:

In order to account for the unexpected large angle scattering of α-particles, Ernst Rutherford in (1911) proposed the *nuclear atom model*. Rutherford is therefore, credited the discovery of the nucleus. All of the positive charge and essentially all of the atomic mass were concentrated in a **nucleus** of dimension much less than that of the atomic dimensions. The application of the Newton's Laws of motion and the Coulomb's Law – both pillars of classical physics-gave results in accord with experimental data for atoms. However, the Rutherford model of the atom is not in accord with the EM theory-another pillar of classical physics-which predicts that centripetally accelerated electric charges radiate energy continuously in the form of EM wave. If this were the case, the energy of the dynamic system would decrease continuously and planetary charge would spiral into the nucleus after a normal lifetime of ~10^{-8} s. But this is not so because most elementary atoms have infinite life times. Thus the dynamics of the atomic electrons requires a more sophisticated approach. The nuclear atom model –

i) Settled the problem associated with α-scattering,
ii) Failed to explain the stability of atoms and
iii) Atoms do not radiate unless excited and when the radiation does occur its spectrum consists of discrete frequencies rather than a continuum of frequencies. Amazingly it turned out that a single empirical relation by Rydberg gave all the wavelengths of atomic hydrogen.

With a fresh PhD degree in his pocket, Niels Hendrik David Bohr journeyed from Denmark to England in 1911 and came under the influence of two giants of atomic science namely J J Thomson of Cambridge and E Rutherford in Manchester. Back in Copenhagen in1912 he worked on the details of a new model of atom.

1.7.1. BOHR'S THEORY OF HYDROGEN ATOM:

In 1913 Niels Bohr working in Rutherford laboratory advanced a new and revolutionary idea to solve the problems left out by the nuclear atom model. Bohr used the *photon hypothesis* as analyzed in section 1.4 and extended *Planck's quantum hypothesis, viz.,*

$$\boxed{E = \hbar \omega}.\tag{1.7.2}$$

1.7.2.1. Bohr advanced a series of POSTULATES as follows:
a) Rutherford's nuclear model of the atom was adopted; on this basis the hydrogen atom should consist of a singly positively charged ($+e$) massive nucleus and an electron ($-e$) of light mass outside the nucleus (Fig 1.11).

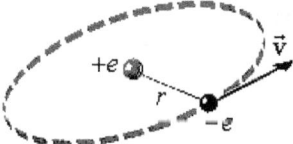

Fig 1.11 Bohr Hydrogen Atom Model

b) It assumed the *Coulomb's Law* of force and *Newton's Laws* of motion to be applicable in the atomic domain.
c) The path of the electron around the nucleus should be a *conic section*.
d) The conic section is a circle of radius r with the nucleus at the center of the circle)
e) POSTULATE 1. This relates to the *mechanics* of the atom (the idea of the *stationary state*). Only the electron orbits are *allowed* (or *permissible*) for which the angular momentum (L) of the electron is an integral multiple of $\hbar$,

$$\boxed{L = \frac{n h}{2\pi} = n \hbar, \quad n = 0, 1, 2, 3, \ldots, \text{an integer}..} \quad (1.7.3)$$

and that no energy is radiated while the accelerated electron remains in any of the *permissible orbits* And that no energy is radiated while the accelerated electron remains in any of the permissible orbits,

f) An electron moving in one of the stable orbits does not radiate,

g) P0STULATE 2: relates to the *electrodynamics* of the atom (idea of quantum jump). An electron can make

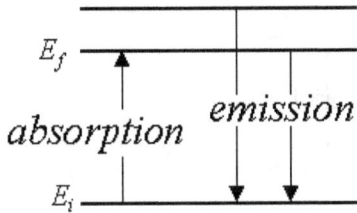

Fig 1.12 Bohr's Second Postulate

discontinuous transition from one allowed orbit to another causing a change in the energy of the atom which appears as emission or absorption of radiant energy, according to the scheme:

$$\boxed{E_i - E_f = h \nu} \quad (1.7.4)$$

where E_i and E_f denote the initial and final values of energy of the atom and ν is the frequency of the radiation emitted /absorbed.

1.7.2.2 ORBITS given by Classical Physics
Consider a one-electron atom such as hydrogen.

Let $Z e$ = Nuclear charge, (Atomic number, $Z = 1$, for hydrogen)

e = electronic charge $e = 1.6021 \times 10^{-19} C$,

M = nuclear mass,

m = electronic mass, $m_e = 9.1094 \times 10^{-31}$ kg.

$$\varepsilon_0 = 1/c^2 \mu_o, \quad (1.7.5)$$

$\varepsilon_0 = 8.8542 \times 10^{-12} \, F \, m^{-1}$ is permittivity of free space

$k = (1/4\pi \varepsilon_0) = 8.9875 \times 10^9 \, F^{-1}m$, the Coulomb Constant,

$$k = (1/4\pi \varepsilon_0) = 8.9875 \times 10^9 \, Nm^2 C^{-2}.$$

Centrifugal force, $\quad \vec{F}_C = \dfrac{m v^2}{r}$ \hfill (1.7.6)

Coulomb force, $\quad \vec{F} = \dfrac{k q_1 q_2}{r}$ \hfill (1.7.7)

$$\vec{F}_{EM} = \dfrac{1}{4\pi \varepsilon_0} \dfrac{-Z e^2}{r^2} \qquad (1.7.8)$$

In equilibrium, $\quad \vec{F}_C = \vec{F}_{EM}$

i.e., $\quad \dfrac{m v^2}{r} = \dfrac{1}{4\pi \varepsilon_0} \dfrac{-Z e^2}{r^2}$ \hfill (1.7.9)

* Note: Hereafter, k will not be included for convenience.

1.7.2.3.   QUANTUM ORBITS: Quantum number to Distinguish Electrons
By postulate L, angular momentum of the electron in orbit with radius r_n is

$$L = m v_n r_n \qquad (1.7.10)$$

$$m v_n r_n = \dfrac{Z e^2}{v_n} = \dfrac{n h}{2\pi} = n = 0, 1, 2, 3, \ldots \qquad (1.7.11)$$

$$\therefore \quad v_n = \dfrac{2\pi Z e^2}{n h} \qquad (1.7.12)$$

$$\boxed{r_n = \dfrac{n^2 h^2}{4\pi^2 m Z e^2} = \dfrac{n^2 h^2}{m Z e^2}} \qquad (1.7.13)$$

$\therefore$ Total energy, E = potential energy, V + kinetic energy, T,
$$E = V + T \qquad (1.7.14)$$

1.7.2.4.   QUANTIZED (DISCRETE) ENERGY LEVELS:
$$E_n = \dfrac{m v_n^2}{2} + \dfrac{-Z e^2}{r_n}$$

The system being a conservative one, and $\vec{F} = -\dfrac{dV}{dr}$, where $V = -\dfrac{Z e^2}{r}$

$$E_n = \dfrac{+Z e^2}{2 r_n} + \dfrac{-Z e^2}{r_n} = \dfrac{-Z e^2}{2 r_n}$$

$$= \dfrac{-Z e^2}{2}\left(1 / \dfrac{n^2 h^2}{4\pi^2 m Z e^2}\right)$$

i.e., $\quad \boxed{E_n = \dfrac{-2\pi^2 m Z^2 e^4}{n^2 h^2}} \qquad (1.7.15)$

This expression indicates that the *quantization of energy* has arisen as a result of assuming that the electron's angular momentum is quantized (postulate 1). The quantity n is referred to as the QUANTUM NUMBER of the nth orbit. The existence of *discrete energy* states for electrons in atoms (Fig 1.13) is given by this result.

1.7.3. CALCULATION of some of the quantities that emerge from the Bohr Theory. They are:
$$m c^2 = 0.51\ MeV\ ;\ h/mc = 3.9 \times 10^{-11}\ cm;\ h/mc^2 = 1.3 \times 10^{-21}\ s$$

1.7.3.1. FINE STRUCTURE CONSTANT, α

Defining a dimensionless quantity called the *fine structure constant*, denoted by the symbol α and defined as
$$\alpha = ke^2/\hbar c = 7.2973 \times 10^{-3} = 1/137.036 \tag{1.7.16}$$

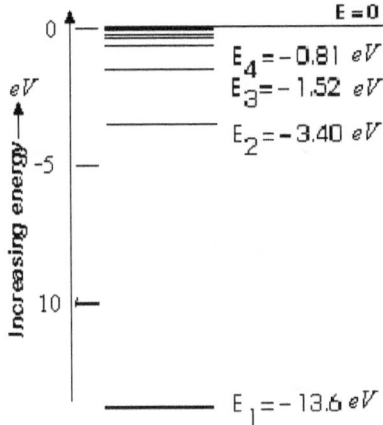

Fig 1.13   Discrete energy levels of Hydrogen atom

1.7.3.2 RADIUS OF THE ATOM, r_n

The radius of the orbit of an electron becomes from (1.7.13)
$$r_n = \frac{n^2 \hbar^2}{m Z e^2}$$

i.e.,
$$r_n = \frac{n^2}{Z} \cdot \frac{\hbar c}{e^2} \cdot \frac{\hbar}{m c} = \frac{n^2}{Z \alpha} \cdot \frac{\hbar}{m c}$$

This yields
$$r_n = \frac{137 n^2}{Z} \cdot (3.9 \times 10^{-11}\ cm)$$

whereby
$$\boxed{r_n = \frac{n^2}{Z} \cdot (0.0529167\ nm)} \tag{1.7.17}$$

1.7.3.3 BOHR ATOMIC RADIUS constant, a_o

The radius of the lowest Bohr orbit, denoted by a_o, is when $n = 1$.
So, for hydrogen atom, $Z = 1$, and from equation (1.7.17) Bohr first radius
$$a_o \equiv r_1 = \frac{1}{Z} \cdot (0.0529167\ nm) \xrightarrow{Z=1} (0.0529167\ nm).$$
$$a_o = (0.0529167\ nm) \quad \text{(For hydrogen atom).} \tag{1.7.18}$$

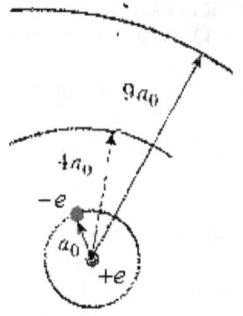

Fig. 1.14   Hydrogen atom radii, $r_n = n^2 a_o$.

For the nth orbit of hydrogen atom (Fig 1.14),

$$\boxed{r_n = n^2 a_o} \quad (1.7.19)$$

1.7.3.4 The BINDING ENERGY, E_n of the electron in an atom:

From Equation (1.7.15)

$E_n = \dfrac{-2\pi^2 m Z^2 e^4}{n^2 h^2}$.   This can be rearranged in other forms:

$$E_n = \left(-\frac{1}{2}\frac{m c^2}{n^2}\right)\left(\frac{Z e^4}{h^2 c^2}\right) \quad (1.7.20)$$

$$E_n = -13.6\left(\frac{Z^2}{n^2}\right) eV \quad (1.7.21)$$

The energy of an electron in the lowest orbit

$$E_1 = -13.6\, Z^2\ eV \quad (1.7.22)$$

The **negative sign** indicates that the electron is *bound to the nucleus*.

$$E_n = E_1 / n^2 \quad (1.7.23)$$

1.7.3.5 SPECTRUM of Hydrogen theoretically derived

Hydrogen, the simplest of all the elements, has been investigated most extensively both experimentally and theoretically. As long ago as 1885, Balmer succeeded in obtaining an empirical simple relationship among the wave numbers of the lines in the visible region of the hydrogen spectrum. Balmer's equation expressed in modern notation is

$$\bar{v} = \frac{1}{\lambda} = \frac{v}{c} = R\left(\frac{1}{2^2} - \frac{1}{n^2}\right); \quad n = 3, 4, 5, \ldots \quad (1.7.24)$$

where $\bar{v} = \frac{1}{\lambda} = \frac{v}{c}$ = Wave number of a spectral line, and R is the Rydberg Constant.

Postulate 2, viz, Equation (1.7.4), can be used to calculate the energies and the frequencies of the possible transitions. If the electron is in the initial state, with the energy E_i and makes a transition to a final state E_f, the energy $h\nu$ of the photon emitted / absorbed (Fig. 1.12) is given by.

$$h\nu = E_i - E_f$$

where $E_n = E_1/n^2$

$$v = \frac{E_i - E_f}{h} = \frac{E_1}{h}\left(\frac{1}{n_i^2} - \frac{1}{n_f^2}\right) \quad (1.7.25)$$

Wave number, $\bar{v} = \frac{1}{\lambda} = \frac{v}{c}$,

$$\bar{v} = \frac{E_1}{h}\left(\frac{1}{n_i^2} - \frac{1}{n_f^2}\right) = R\left(\frac{1}{n_i^2} - \frac{1}{n_f^2}\right) \quad (1.7.26)$$

where R is the Rydberg constant given by

$$R = \left(\frac{E_1}{h}\right) = \left(+\frac{1}{2} m c^2 \alpha^2 Z^2 / hc\right) = [m e^4 / 4\pi h^3 c] \quad (1.7.27)$$

1.7.4.1 ENERGY LEVEL DIAGRAM:
The energy levels of the hydrogen atom are schematically shown in Fig 1.1

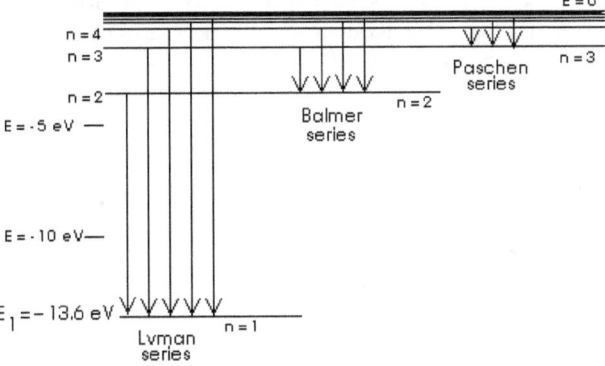

Fig 1.15 Energy level Diagram

1.7.4.2 LINE SPECTRAL SERIES:
It is known from Equation (1.7.26)

$$\bar{v} = R\left(\frac{1}{n_i^2} - \frac{1}{n_f^2}\right).$$

For a given value of n_i, the set of transitions from $n_2 = n_1 + 1, n_1 + 2, n_1 + 3$, etc. constitutes a SERIES of lines, and these *Series* bear the names of their discoverers or principal investigators:

$n_1 = 1, n_2 = 2, 3, 4,$, etc	LYMAN series; Ultra Violet
$n_2 = 2, n_3 = 3, 4, 5,...$ etc.	BALMER series; Visible
$n_3 = 3, n_4 = 4, 5, 6,$, etc	PASCHEN series; Infra Red
$n_4 = 4, n_5 = 5, 6, 7,$, etc	BRACKETT series; Far Infra Red
$n_5 = 5, n_6 = 6, 7, 8,$, etc	PFUND series; Far Infra Red
$n_6 = 6, n_7 = 7, 8, 9,$, etc	HUMPHREY'S series; Far Infra Red

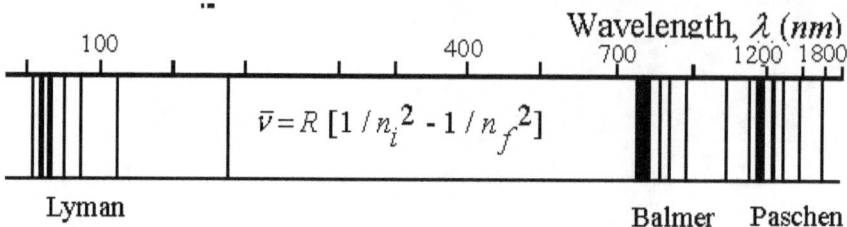

Fig.1.16 Energy level diagram and Spectral series for Hydrogen atom.

The spectral Series are illustrated in the spectrum of Fig 1.16.

1.7.4.3. SERIES LIMIT, $\bar{v}_\infty$

The limit of each Series $\bar{v}_\infty$ is the wave number of the transition that just succeeds in ionizing the atom. This limit corresponds to $n_2 = \infty$ in each Series (Fig. 1.17), and therefore,

$$\bar{v}_\infty = R_H = \frac{1}{n_1^2} \qquad (1.7.28)$$

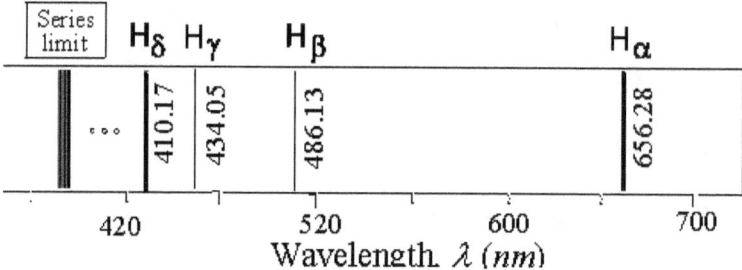

Fig. 1.17 Series Limit: Hydrogen spectrum

1.7.5 COMPARISON WITH OBSERVATIONS:

For hydrogen, the Rydberg constant R_H is

$$R_H = \left(+\tfrac{1}{2} m\, c^2 \alpha^2 Z^2 / hc\right) = [m\, e^4 / 4\pi\, h^3\, c] = 109740\ cm^{-1}. \qquad (1.7.29)$$

Experimental measurement gave $R_H = 109677.5\ cm^{-1}$. Hence there is a very good agreement between theory and experiment. This is also the same as the empirical relation given by Johannes R. Rydberg (1889) for the Balmer Series. ($n_i = 2$, $n_f = n$).

From the *Bohr Theory of the hydrogen atom*, there emerged.
a) Bohr's <u>Correspondence Principle</u>, and
b) The <u>quantization of both angular momentum and energy</u>.

1.7.6 ENERGIES of EXCITATION and IONIZATION:

The **energy-level diagram** directly represents two important quantities characteristic of the atom, *viz.*, the '*energies of excitation* and *ionization*.'

1.7.6.1 IONIZATION ENERGY, I

The *Ionization energy*, I, of the atom is the minimum energy required to ionize it from its ground state, the state with $n_i = 1$. This is the energy difference between the ground state ($n = 1$) and the convergence limit ($n = \infty$). In this situation the electron is in the outermost possible orbit and is said to ionize the atom. This is the *ionization energy*, and is seen to be

$$I = hc R_H \quad (1.7.30)$$
$$= 1312 \; kJ \; mol^{-1} = 13.60 \; eV.$$

1.7.6.2 EXCITATION ENERGY:

From the diagram it is inferred that the excitation of the hydrogen atom requires a minimum the energy needed to lift the atom from the ground state ($n = 1$) to the lowest excited state $n = 2$, for the Lyman Series. This energy difference is called the *excitation energy*, and is computed from Bohr's Theory as

$$(E_2 - E_1) = 16.31 \times 10^{-19} J = 10.19 \; eV. \quad (1.7.31)$$
$$(E_\infty - E_1) = 21.76 \times 10^{-19} J = 13.58 \; eV \quad (1.7.32)$$

Worked out Example 1.6

Determine the shortest and longest wavelengths of the Lyman's series of hydrogen. Given, $R_H = 109740 \; cm^{-1}$.

Solution: $\boxed{Step\#1}$ Lyman series $\bar{v} = R [1/n_i^2 - 1/n_f^2]$; where $n_i = 1$, $n_f = 2, 3, 4,$, etc (1.7.31)

$\boxed{Step\#2}$ $R_H = 109740 \; cm^{-1}$; λ attains maximum value, λ_{Max} when $n_f = 2$;

$\boxed{Step\#3}$ $(1/\lambda_{Max}) = (R_H = 109740 \; cm^{-1}) \cdot [1/(n_i = 1)^2 - 1/(n_f = 2)^2]$,

whence $\lambda_{Max} = 121.5 \; nm$.

$\boxed{Step\#4}$ Similarly, $(1/\lambda_{Min}) = (R_H = 109740 \; cm^{-1}) \cdot [1/(n_i = 1)^2 - 1/(n_f = \infty)^2]$,

whence $\lambda_{Min} = 91.2 \; nm$.

Worked out Example 1.7

In a transition to a state of excitation energy 10.19 eV, a hydrogen atom emits a 489.0 nm photon. Determine the binding energy of the initial state. $h = 6.6256 \times 10^{-34} \; J-s$, $c = 2.997925 \times 10^8 \; ms^{-1}$

Solution: $\boxed{Step\#1}$ The energy of the emitted photon.

$hv = hc/\lambda = (h = 6.6256 \times 10^{-34} J-s) \cdot (c = 2.997 \times 10^8 \; ms^{-1}) / (489.0 \; nm)$
$= 2.54 \; eV.$

$\boxed{Step\#2}$ The corresponding excited state has energy

$E_n - E_1 = +10.19 \; eV = -13.6 \; eV + 10.19 \; eV = -3.41 \; eV.$

$\boxed{Step\#3}$ $E_{BE} - E_n = hv$; $E_{BE} - E_n = E_f - (-3.41 \; eV) = 2.54 \; eV$.; whence $E_{BE} = -0.87 \; eV$ The binding energy of the electron $= 0.87 \; eV$.

1.7.7 RITZ COMBINATION RULE

It has been known for a long time (1905) that many combinations exist between the observed frequencies of the various lines emitted from the same atom. When the frequency of the first line of the Lyman series is denoted by L_1 and other frequencies of the Lyman series and Balmer series correspondingly, the empirical relation holds

$$\boxed{L_1 + B_1 = L_2} \tag{1.7.32}$$

This relation is called the *Combination Rule* (W. Ritz, 1905).

1.7.8 BohrCorrespondence Principle

This principle by Bohr (1923), in essence, states that results of classical physics should be contained as limiting cases (of quantum physics), where to avert the masses and dimensions of the system considered are made to approach to those of classical systems and h can be neglected of quantum mechanical results, *i.e.*, for large values of quantum number, n a quantum equation would go to the corresponding classical equation. This is illustrated in Fig 1.15. This principle, however, should be viewed as a guideline rather than a quantitative law of physics. (Comment on results of Example 1.17).

1.7.8.1 The Correspondence Principle illustrated in a Bohr atom:

According to W. Ritz's combination principle, and equation (1.7.25)

$$v = \frac{E_i - E_f}{h} = \frac{E_1}{h}\left(\frac{1}{n_i^2} - \frac{1}{n_f^2}\right)$$

where $v = v_{if}$ is the frequency of the radiation or spectral line.

i.e., $\quad = -cR(n_f^2 - n_i^2)/n_i^2 n_f^2$

$\quad\quad = cR(2n-1)/n^2(n-1)^2$, for $n_f = n-1$, and $n_i = n$.

When n becomes large, $n_i = n_f = n$, $n_i - n_f = \Delta n$ and $v_{if} = cR\, 2\Delta n/n^3$

$$\therefore\ v_{if} = [c2\Delta n/n^3(n-1)^2]\left(\frac{-me^4}{4\pi\hbar^3 c}\right) = \left(\frac{me^4}{2\pi n^3 \hbar^3}\right)\cdot\Delta n \tag{1.7.33}$$

giving in the case of a Bohr atom,

$$v_{if} = \left(\frac{1}{2\pi}\right)\sqrt{\frac{e^2}{m\, r_n^3}}\cdot\Delta n \xrightarrow{\Delta n = 1,\ n\to\infty} \left(\frac{1}{2\pi}\right)\sqrt{\frac{e^2}{m\, r_n^3}}. \tag{1.7.34}$$

Therefore, as $n \to \infty$, and since $(mv^2/r) = (Ze^2/r^2)$,

$$v_{QM} = \left(\frac{1}{2\pi}\right)\sqrt{\frac{e^2}{m\, r^3}} = \frac{v}{2\pi r} = \frac{\omega}{2\pi} = v_{CL} \tag{1.7.35}$$

Because classically, an electron moving in a circular orbit with velocity v, with angular frequency $\omega = \frac{v}{r}$, radiates with the frequency of motion, v_{CL}.

1.7.9. The SELECTION RULE

The result from Equation (1.7.35) is not true for the $(n+2) \to n$ transition. Thus employing the Bohr's Correspondence Principle, the *Selection Rule*, $\Delta n = 1$ for a transition is readily obtained.

1.7.10. BOHR–SOMMERFELD RELATIVISTIC MODEL OF H-ATOM
(Relative Motion of the Hydrogen Nucleus
or *Correction to the Finite Mass of the Nucleus*)

The extra-ordinary success with which Bohr's simple hydrogen atom model not only explains quantitatively the Balmer series but also predicts the existence of other spectral series encourages one to proceed with further refinements of the theory. Bohr's theory *failed* to account for the spectrum of any atom having more than one electron (Z =1). A more general form of the atom was tried by N. Bohr & Arnold Sommerfeld, called the *relativistic elliptical atom model*.

So far it was assumed that the nucleus remained fixed at the centre of the circular orbits of the atom. This is true only if the mass of the nucleus is infinitely large. If the nuclear mass M were finite both the nucleus and the electron in the hydrogen atom would rotate about their common centre of mass.

Let M and V are the mass and velocity of the nucleus. Since the atom as such has internal energy as well as momentum, conservation of energy and momentum give us,

Total kinetic energy of the atom $T = \frac{1}{2}[M V^2 + m v^2]$.

Total momentum of the atom $P = (MV + mv) = 0$; i.e., $p = mv = MV$.

$$\therefore T = \frac{p^2}{2}[1/M + 1/m] = (p^2/2\mu)$$

where $\mu = \frac{mM}{m+M}$ (1.7.36)

μ is called the *Reduced Mass* of the atom (See Chapter 8 for derivation). The corrected expression for the energy of the electron in the atom is given by Equation (1.7.37):

$$E_n = -\frac{2\pi^2 \mu Z^2 e^4}{n^2 h^2}, \quad n = 1, 2, 3, 4, ...$$ (1.7.37)

1.8.1. APPLICATION to the Helium Atom

It is known that helium (He), $Z = 2$, gas emits a much more complicated spectrum than that of atomic hydrogen. However, under very violent conditions in an electric discharge, there appears an additional spectrum with a simple structure resembling the one of hydrogen. This simple spectrum of He is attributed to *ionized* helium atoms; He^+, while the complicated, more easily excited spectrum is attributed to *neutral* atomic helium.

In the visible range the He-ion spectrum shows a series given by the formula

$$v = 4 c R \left(\frac{1}{4^2} - \frac{1}{n_f^2}\right); \quad n = 5, 6, 7, ...$$ (1.8.1)

It is seen that every 2nd line (for $n = 6, 8, 10, ..$) coincides with a line of the Balmer series of hydrogen, provided R here is taken to be the same as that of hydrogen. But the helium series has additional lines (for $n = 5, 7, 9, ...$) one of them between any two successive lines of the Balmer series.

What prediction does Bohr theory make regarding the He spectrum?

$$v = 4 c R \left(\frac{1}{n_i^2} - \frac{1}{n_f^2}\right)$$ (1.8.2)

He^+ ion and neutral H atom are identical except that Bohr's final equation (1.8.2) contains the factor Z^2, which predicts the factor 4 in equation (1.8.2), whereas for hydrogen the factor is 1. This prediction agrees with the observed fact and thus strengthens confidence in Bohr's theory. It has been found that for lithium Li^{2+} ion whose emission spectrum is described by

$$v = 9 c R \left(\frac{1}{n_i^2} - \frac{1}{n_f^2}\right) \qquad (1.8.3)$$

where the factor $9 = z^2 = 3^2$, where the atomic number Z of Li is 3. These results give a strong confirmation of the General Law, viz.

> The Atomic number Z = The Nuclear charge, in multiple units of e

Hence by applying Bohr Theory to the simplest spectra emitted by the *light* elements, this important law is corroborated.

1.8.2 Spectrum of Heavy Hydroen, called Deuterium, D:
It can be shown that the wavelength difference between the red Balmer line of hydrogen, λ_H and deuterium, λ_D (in terms of λ_H) as $(\lambda_H - \lambda_D)$ = a quantity agreeing with the observation. It is to be commented that the *Displaced Balmer Line* computed as above observed '*heavy hydrogen*'!

Worked out Example 1.22

Calculate the ratio of the mass of the proton to that of the electron using the knowledge of the Rydberg constants for hydrogen and ionized helium. Given: $M_{He} = 3.9717 M_H$, from mass spectrographic data. $R_H = 109677.8 \ cm^{-1}$, $R_{He} = 109722.4 \ cm^{-1}$.

Solution: $\boxed{Step\#1}$ $R_H = [\mu_H \ e^4 / 4\pi \ h^3 \ c] = 109677.8 \ cm^{-1}$

$\boxed{Step\#2}$ $R_{He} = [\mu_{He} \ e^4 / 4\pi \ h^3 \ c] = 109722.4 \ cm^{-1}$; $\mu_H = M_H m/(M_H + m)$;
$\mu_{He} = M_{He} m/(M_{He} + m)$;

$\boxed{Step\#3}$ $R_H / R_{He} = \mu_H / \mu_{He} = (1 + m/M_H)/(1 + m/3.9717 M_H)$; $(M_H/m) = 1840$.
This is in excellent agreement with values determined by other methods.

1.9 CHARACTERISTIC X-ray spectra

Henry Gwyn-Jeffreys Moseley (1913) made a systematic investigation of the characteristic X-ray spectra of the elements (Fig. 1.18).

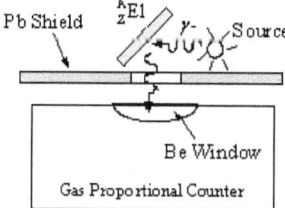

Fig. 1.18   HG.J Moseley's experiment.

Plotting the square root of the frequency, $\sqrt{v}$ of one of the lines, say K_α line, against the atomic number, Z of the element $^A_Z El$ emitting the line, Moseley obtained the *Moseley*

Diagram ($\sqrt{v}$ *versus Z* graph), consisted of doublet straight lines, which to a good approximation, is given by

$$v = C(Z-a)^2 \qquad (1.9.1)$$

where C and a are constants. Relation (1.9.1) is known as <u>Moseley's Law</u>. (Fig 1.19). For the K_α line,

$$C = \tfrac{3}{4} cR, \text{ and } a \approx 1 \qquad (1.9.2)$$

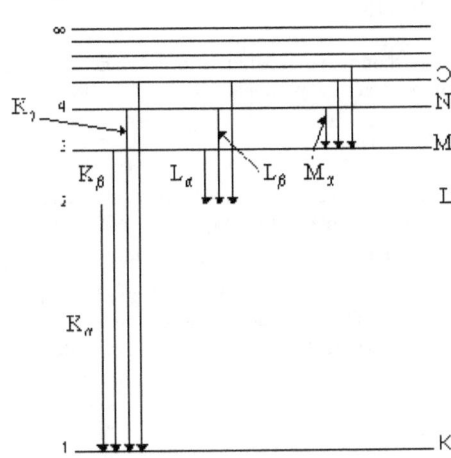

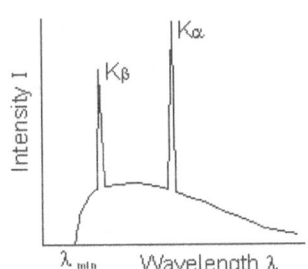

X ray emission spectrum
Fig 1.19

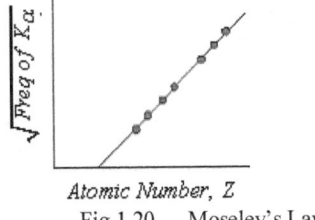

Fig 1.20   Moseley's Law

Moseley's Law places the element in the proper location in the Periodic Table. Moseley's work can best be shown by rewriting equation (1.8.3) for the frequency of the K_α line to read

$$v_{K\alpha} = cR(Z-1)^2 \left(\frac{1}{1^2} - \frac{1}{2^2}\right) \quad (1.9.3)$$

The interpretation of this equation on the Bohr Theory, not to be described here, was quite satisfactory.

1.10 FURTHER DEVELOPMENTS

In order to arrive at a more thorough evaluation of Bohr's Theory, it is necessary to report on the further comparison of experiment and theory, as follows.

1.10.1 The FRANCK - HERTZ *Experiment*

From the early spectroscopic work it is clear that atoms emitted radiation at discrete frequencies; from Bohr's model, the frequency v of the radiation is related to the change of energy levels through $E_i - E_f = hv$, it is then to be expected that transfer of energy to atomic electrons by any mechanism should always be in discrete amounts. One such mechanism of energy transfer is through inelastic scattering of low-energy electrons. J. Franck and G. Hertz (1914) set out to demonstrate these considerations.

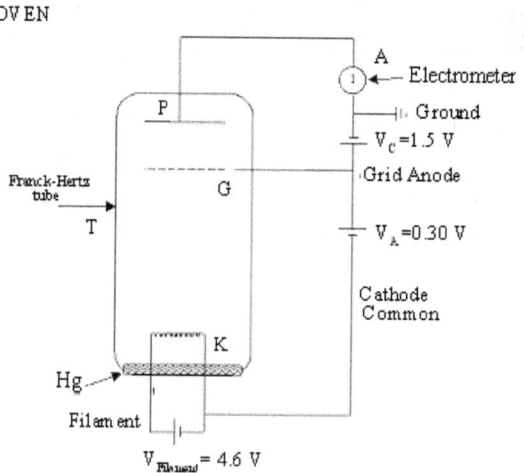

Fig. 1.15 Frank-Hertz Experimental Set up.

Experimentally.
i) It is possible to excite atoms with low-energy electron bombardment.
ii) The energy transferred from electrons in the atoms always had discrete values, and
iii) The values so obtained for the energy levels were in agreement with spectroscopic results.

Thus the existence of atomic energy levels put forward by Bohr can be proved directly. The experimental set up is shown Fig. 1.15.

The Frank-Hertz tube T, usually a vacuum tube, say a tetrode, is filled with the vapor of the experimental substance, say Mercury vapor at low pressure. The filament when heated supplies electrons, which are accelerated by the potential V_0 applied between the grid G_2 and the cathode K. The grid G_1 helps in minimizing space charge effects. The plate P is maintained at a potential slightly negative with respect to the grid G_2 to enable in making the dips in the plate current more prominent. The current I_P is read by the nano-ammeter or galvanometer A.

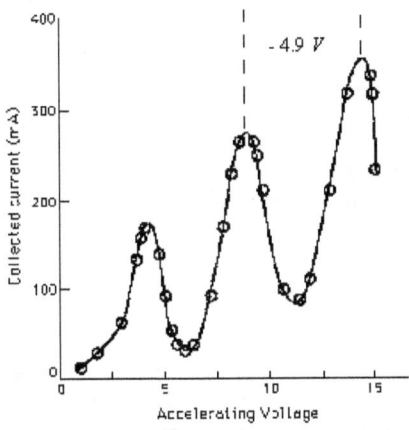

Franck-Hertz Data for Mercury

Fig. 1.16 Franck-Hertz experimental data for Mercury

The current is measured as a function of V_0. The electrons satisfy the equation

$$\tfrac{1}{2}mv^2 = eV_0 \qquad (1.10.1)$$

The retarding potential V_r between G_2 and P prevents electrons having negligible kinetic energy from contribution to the I_P. This is due to loss of energy to electrons during collisions with the mercury atoms. Such a drop in I_P was found to occur at $4.9\ eV$. This is the energy voltage corresponding to the 253.6 nm line of mercury appearing in the emission spectrum of mercury vapor.

$$\tfrac{1}{2}mv^2 = eV_0 = h\nu = \tfrac{hc}{\lambda} \qquad (1.10.2)$$

Photon energy corresponding to 253.6 nm line is $4.86\ eV$ This behavior, of drop in current and appearance of new emission lines, is repeated at multiples of $4.89\ V$, as shown in Fig.1.16.

These voltages corresponding to the peaks in I_p are called *exciting potentials*. This experimental set up can lead to a plot of the energy amplitude spectrum curve by measuring point by point. If a CRO is included in the experimental circuit, the display of the plot can be obtained on the screen.

1.10.2   FAILURE OF BOHR'S THEORY to explain spectral structure:

Though Bohr's Theory of hydrogen atom was successful in predicting the spectrum of hydrogen; there remained an unexplained '***fine structure***' or '*splitting*' of the spectral lines. This splitting of around 100 *ppm* (hundred parts per million) could not be detected in early spectrometers, which had low resolving power.

1.11 SOMMERFELD'S RELATIVISTIC MODEL of hydrogen atom:

1.11.1 QUANTIZATION RULE:

According to Arnold Sommerfeld, *elliptical* as well as *circular orbits* can be allowed; the velocity of an electron in an orbit of large *eccentricity* could become relativistic. W. Wilson and A. Sommerfeld independently (1915, 1916) discovered that a method of quantifying the action integrals of Classical Mechanics. A *phase point* represents a system in *phase space* and each generalized coordinate q_i, and its conjugate momentum p_i, must be periodic in time, which together form the 2N coordinates in phase space. The action integral $\oint$ takes over a cycle of the motion is quantized. *i.e.* action integral J is expressed by the **Wilson–Sommerfeld quantization Rule**,

$$\boxed{J_i = \oint p_i \cdot dq_i = n_i h}. \qquad (1.11.1)$$

where n_i is an integer. This is the first generalized postulate for the determination of the permissible orbits.

In the case of circular orbitrary motion there is only one coordinate, which is a periodic function of time, *viz.* the angle, Φ that the radius vector r makes with the x- axis. In the case of elliptic motion both φ and length, r of the vector, $\vec{r}$, vary periodically as shown in Fig.1.17. The elliptic orbits will, therefore, be determined by the two quantum conditions:

$$J_\varphi = \oint p_\varphi \cdot d\varphi = n_\varphi h \qquad (1.11.2)$$

$$J_r = \oint p_r \cdot dr = n_r h \qquad (1.11.3)$$

where n_φ and n_r are both integers (quantum numbers) and

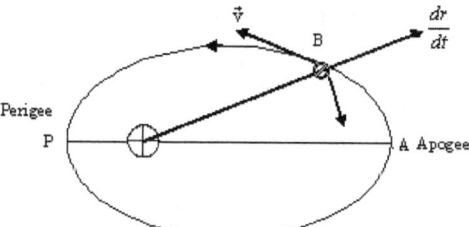

Fig. 1.17 Elliptical orbit

angular momentum $\quad p_\varphi = m r^2 \frac{d\varphi}{dt} \qquad (1.11.4)$

and radial momentum, $\quad p_r = m \frac{dr}{dt} \qquad (1.11.5)$

if a = semi-major axis, b = semi-minor axis and ε = eccentricity

$$\varepsilon = \sqrt{1 - \frac{b^2}{a^2}} \qquad (1.11.6)$$

$$J_r = \oint p_r \cdot dr = p_\varphi \{\oint [\tfrac{1}{r^2}(\tfrac{dr}{d\varphi})^2 d\varphi]\} = p_\varphi \Im \qquad (1.11.7)$$

where $\Im$ is the integral within the symbols, viz.,

$$\oint \tfrac{1}{r^2}(\tfrac{dr}{d\varphi})^2 d\varphi = [\tfrac{2\pi}{\sqrt{1-\varepsilon^2}} - 2\pi] = 2\pi \qquad (1.11.8)$$

$$J_r = p_\varphi \Im = [\tfrac{2\pi}{\sqrt{1-\varepsilon^2}}]p_\varphi - 2\pi p_\varphi = n_r h \qquad (1.11.9)$$

$$J_\varphi = \oint p_\varphi \cdot d\varphi = 2\pi p_\varphi = n_\varphi h \qquad (1.11.10)$$

This leads to *eccentricity*, ε, of the orbit

$$\varepsilon = \sqrt{1 - (\tfrac{n_r}{n})^2} \qquad (1.11.11)$$

where n is the **Principal Quantum Number**

$$n = (n_r + n_\varphi) \qquad (1.11.12)$$

It will be easy to show that the **total energy**, E_n

$$E_n = E_{n\varphi} + E_{n_r} = -\tfrac{2\pi^2 \mu Z^2 e^4}{n^2 h^2}, \quad n = 1, 2, 3, 4, \ldots \qquad (1.11.13)$$

which is the identical expression for E of a Bohr atom.

1.11.2 DEGENERACY of orbits predicted:

The introduction of the elliptical orbits has, however, brought the concepts of **multiplicity** of orbits and *degeneracy* of the nth orbit. It was at this point that Sommerfeld introduced a relativistic correction for the mass of the electron. So long as the atom is simply one object orbiting another, it is a direct prediction of Coulomb Law that the

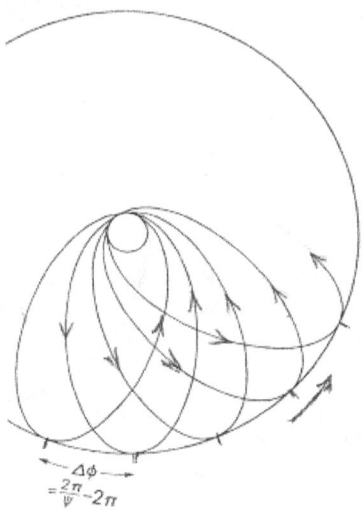

Fig. 1.18 Sommerfeld showed that the path of an electron is a rosette.

same path in space is retraced indefinitely. But if anything interferes with the simple interaction, the orbit will *precess*; the points of *apogee* and *perigee* progressively creep around in a circular fashion. Sommerfeld showed that the path of an electron is a rosette, *i.e.*, a PRECESSING ELLIPSE.(Fig. 1.18). With this refinement the energy expression becomes

$$E_n = \left[-\frac{2\pi^2 \mu Z^2 e^4}{n^2 h^2} \right] \left(1 + (\frac{Z^2 \alpha^2}{n})(\frac{1}{n_\varphi} - \frac{3}{4n}) \right), \quad n = 1, 2, 3, 4, \ldots \ldots (1.11.14)$$

In this way the relativistic equation correctly accounted for the fine structure splitting of the lines of the hydrogen spectrum. In this theory, however, there are no restrictions on the value of the principal quantum number n. Only those transitions are observed for which n_φ changes by unity

$$\Delta n_\varphi = n_{\varphi i} - n_{\varphi f} = \pm 1 \qquad (1.11.15)$$

This is called the **Selection Rule** for n_φ for *allowed transitions*.

1.12   VECTOR MODEL of the Atom:

1.12.1   PRELIMINARIES

1) The electron was assumed to be moving in a circular orbit with angular velocity, ω. Angular velocity is a vector quantity and is, therefore, represented by vector along the axis of rotation, perpendicular to the plane of the orbit. Because of this angular velocity, the electron has a definite *angular momentum*, $L \equiv p_\varphi$, which, on Bohr's theory, was a whole multiple of $h/2\pi \equiv \hbar$.

$$L = \Im \omega = n_\varphi h / 2\pi \equiv n_\varphi \hbar \qquad (1.12.1)$$

2) In the *vector atom model*, the quantum number n_φ is replaced by the quantum number ℓ, *i.e.*, $n_\varphi = \ell$ and the electron is assigned the *orbital angular momentum*,

$$\boxed{L \equiv p_\varphi = \ell h / 2\pi \equiv \ell \hbar} \qquad (1.12.2)$$

3) Being a vector quantity it is expressed as a vector along the axis of rotation. The quantity $h/2\pi \equiv \hbar$ is widely used as the *Unit of Angular Momentum*.

1.12.2   MULTI - ELECTRON ATOMS

Despite the accuracy with which the quantum theory, which is elegant and simple, accounts for the properties of the hydrogen atom, it cannot completely describe it without the including *electron spin* and the *Pauli Exclusion Principle*.

1.12.2.1 Neutral Atom

A *neutral atom* consists of a nucleus of positive charge, $+Ze$ and Z electrons. The *atomic state* is described by the quantum numbers of the individual electrons. The energy, E_n of the atom is decided by the principal quantum number, n, while the ground state energy, E_1 of a neutral atom is

$$E_1 = -13.6 \, Z^2 \, eV \qquad (1.12.3)$$

$$E_n = E_1 / n^2 \qquad (1.12.4)$$

1.12.2.2 Electron Spin and the Four Quantum Numbers

a) One of the two most conspicuous *shortcomings* of the quantum theory is the explanation for the experimental fact that many spectral lines are formed of two or more very closely spaced separate lines. An example is the **fine structure** of the first line, 656.3 nm, of the Balmer Series of hydrogen.

b) Secondly, the **Zeeman Effect** could not be accounted for. An electron in an atom has, because of orbital motion around the nucleus, in addition to the Principal Quantum Number, n, an Orbital Momentum Quantum Number, ℓ, such that $\ell < n$, and $\ell = 0, 1, 2, 3, ..., (n-1)$.

c) In *Normal Zeeman Effect*, an atom located in the presence of a magnetic field of flux density, B gains magnetic energy, V_ℓ as a result of each electron ℓ-level splits into ($2\ell + 1$) magnetic sub-levels denoted by *Orbital magnetic Quantum Number*, m_ℓ, which can take on values $m_\ell = \pm\ell,\ \pm(\ell-1),\ \pm(\ell-2), ...etc$

$$V_\ell = m_\ell\ (eh/4\pi m)\ B \tag{1.12.5}$$

2) To explain the features in *Anomalous Zeeman Effect*, S.A. Goudsmit and G.E. Uhlenbeck (1925) proposed that electron possesses an *intrinsic spin angular momentum*, S independent of its orbital angular momentum, L. The magnitude of the spin angular momentum, S due to an electron given by the *Intrinsic Electron Spin* quantum number, $s = \frac{1}{2}$ the electron.

3) As in the case of orbital momentum, *spatial quantization* of the electron spin momentum is given by the *Spin Magnetic quantum number*, m_s. $m_s = \pm\frac{1}{2}$. The fine structure doubling of spectral lines may be explained on the basis that the electron acquires interaction energy, V_m due to the interaction between the orbital magnetic moment and spin magnetic moment of the electron(s) in the atom.

$$V_m = -\mu\ B\ Cos\theta \tag{1.12.4}$$

where θ is the angle between magnetic moment, $\mu = eS/m$ and magnetic field of flux density, B.

1.12.3 Bohr's AufbauPrinzip

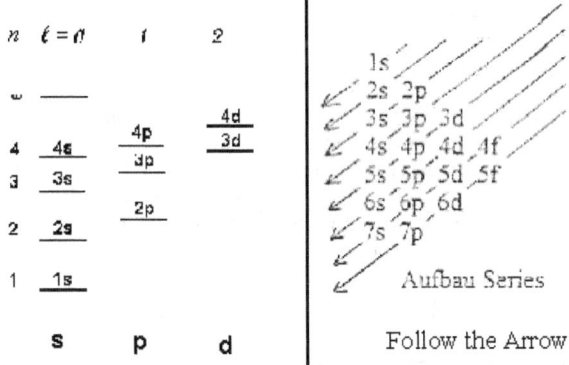

Fig 1.19 A mnemonic diagram for the calculation of the ground state configuration of ions.

According to this Principle, the electrons are added progressively to the various orbitals in their order of increasing energies starting with the orbital of lowest energy; *i.e.* 1s; 2s; 2p; 3s; 3p; 4s; 3d; 4p; 5s; 4d; 5p; 6s; 4f; 5d; 6p; 7s; *etc.*
This is illustrated for easy memory in the scheme in Fig 1.19.

1.12.4 Pauli's Exclusion Principle

In 1925, Wolfgang Pauli discovered the fundamental principle that

> No two electrons in an atom can possess the same set of four quantum numbers,

viz., n, ℓ, m_ℓ, and m_s.

The Exclusion Principle can also be stated thus:
"No two electrons in an atom can exist in the same quantum state".

> "A quantum state of an electron in an atom is unique and is specified by the four quantum numbers", *viz.*, n, ℓ, m_ℓ, and m_s

1.12.5 TOTAL ANGULAR MOMENTUM, j

In many cases, for example in the *alkali elements*, the motion of a single electron produces the changes in the atomic configuration giving rise to the optical spectrum. Total angular momentum of a single electron, $j\, h/2\pi$ = the vector sum of the orbital angular momentum, $\ell\, h/2\pi$ and spin angular momentum, $s\, h/2\pi$, of the single electron. The vector $\vec{j}$, representing the total angular momentum, is defined by

$$\vec{j} = \vec{\ell} + \vec{s} \tag{1.12.5}$$

Since $s = \tfrac{1}{2}$, always, $\vec{j}$ can have only two values for a given value of $\vec{\ell}$, *viz.*, $(\ell \pm \tfrac{1}{2})$, except when $\ell = 0$.

In the addition of the two vectors, $\vec{\ell}$ and $\vec{s}$, to form the vector $\vec{j}$, the magnitude of $\vec{\ell}$ is taken as $\sqrt{\ell(\ell+1)}$ and that of $\vec{s}$ is taken as $\sqrt{s(s+1)}$. The angle, $(\vec{s}, \vec{\ell})$ between the vectors $\vec{\ell}$ and $\vec{s}$ can be obtained and it is given by

$$\cos(\vec{s}, \vec{\ell}) = \frac{j(j+1) - \ell(\ell+1) - s(s+1)}{2\sqrt{\ell(\ell+1)}\vec{j}\sqrt{\ell(\ell+1)}\vec{j}} \tag{1.12.6}$$

1.12.6 Electronic Configuration

1.12.6.1 Atomic Shell

The several electrons in an atom interact directly with one another, as revealed by the Exclusion Principle in governing their behavior. Each electron is supposed to exist in a constant mean force field. All the electrons that have the same principal quantum number, n are, on the average, approximately at the same distance from the nucleus. It is conventional that each electron is said to occupy the same **Group, Energy Level, or Atomic Shell**. These shells are given symbols (Roman capital letters) as follows:

Quantum number, n	1	2	3	4	5,	.	.
Shell Symbol	K	L	M	N	O		

(1.12.7)

1.12.6.2 Subshell

In a complex atom electrons in a particular shell increase in energy with increasing ℓ. Electrons that share a certain value of ℓ in a shell are said to occupy the same ***Subgroup, Sublevel or Subshell***.

1.12.7 SPECTRAL NOTATION:

It is customary to specify angular momentum states of electrons in an atom by a letter, with s corresponding to $\ell = 0$, p to $\ell = 1$, etc., according to scheme given below:

Quantum number, ℓ	0	1	2	3	4	5	6,	etc to $(n-1)$
Spectral Symbol	s	p	d	f	g	h	i	etc.

(1.12.8)

In this notation a state in which $n = 2$, $\ell = 0$, is a **2s state**, one in which $n = 4$, $\ell = 2$, is a **4d state**.

1.12.8 Electron Configuration of Atoms

The occupancy of the various subshells in an atom is usually expressed with help of the *notation* as follows: Each subshell is identified by its principal quantum number n followed by the letter corresponding to its orbital quantum number, ℓ. A superscript after the letter indicates the number of electrons in that subshell given by $\sum 2(2\ell+1)$.

For example, the *electron configuration* of sodium atom ($Z = 11$), in the normal state, is written as

$$_{11}Na: \boxed{1s^2 2s^2 2s^6 3s^1} \qquad (1.12.9)$$

which means that
i) There are two 1s ($n = 1$, $\ell = 0$) electrons,
ii) Two 2s electrons ($n = 2$, $\ell = 0$),
iii) Six 2p electrons, and
iv) One 3s electron.
v) A subshell is filled or completed when the sum of the vectors, $\sum m_\ell = 0$.
vi) Also the sum of the vectors, $\sum m_s = 0$, for completed subgroup. Hence, in order *to determine the angular momentum of an atom, only those electrons, which are external to the closed shells, need be considered.*

1.12.8.1 Spectral Symbol
Capital letters are used to represent the total angular momentum of an atom, according to the following scheme:

Quantum number, L	0	1	2	3	4	5	6,	etc to $(n-1)$
Spectral Symbol	S	P	D	F	G	H		etc.

(1.12.10)

1.12.8.2 Notation of Total Angular Momentum of an Atom
The value of the total angular momentum of an atom, J, is written as a subscript at the lower right of the letter representing the particular L value of the atomic state. The number of possible values of J for a given value of L is written as a superscript at the upper left of the letter representing the L value. Thus

$$^2P_{1/2}, \, ^2P_{3/2}, \text{ read } \textit{"doublet P one half"}, \text{ etc.} \qquad (1.12.11)$$

or

$$^3P_2, \, ^3P_1, \, ^3P_0, \text{ read } \textit{"triplet P two"}, \text{ and so on.} \qquad (1.12.12)$$

1.12.8.3 Multiplicity Symbol
The *superscript* is an indication of the **Multiplicity** of the terms of the atomic configuration.

1.12.9 DISPACEMENT LAW

According to the **DisplacementLaw**, any singly charged ion has the same type of spectrum as the neutral atom of the preceding element in the Periodic Table, but shifted to higher frequencies. One gets an "*isoelectronic sequence*" a row in the Periodic Table, which are reduced to the same number of external electrons.

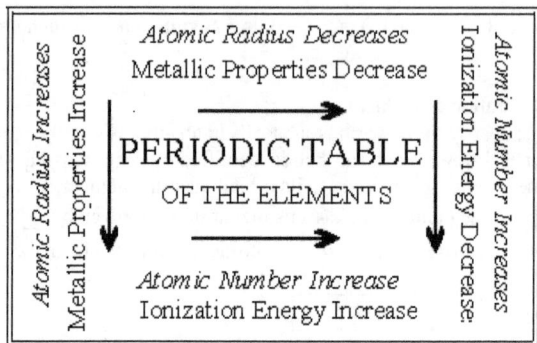

Fig 1.20   Periodic Table of the Elements

1.12.10 Periodic Table of the Elements

The great success of the Pauli Principle is that it explains many aspects of the Periodic Table of Elements (Fig. 1.20 and Fig. 1.21) on the basis of quantum numbers, which originally were introduced for an entirely different purpose, *i.e.* for the interpretation of spectra.

1A									Noble
1 H $1s^1$	2A			3A	4A	5A	6A	7A	2 He $1s^2$
3 Li $1s^2 2s^1$	4 Be $1s^2 2s^2$		$1s^2$	5 B $2s^2 2p^1$	6 C $2s^2 2p^2$	7 N $2s^2 2p^3$	8 O $2s^2 2p^4$	9 F $2s^2 2p^5$	10 Ne $2s^2 2p^6$
11 Na [Ne]$3s^1$	12 Mg [Ne]$3s^2$	1B	2B	13 Al [Ne] $3s^2 3p^1$	14 Si $3s^2 3p^2$	15 P $3s^2 3p^3$	16 S $3s^2 3p^4$	17 Cl $3s^2 3p^5$	18 Ar $3s^2 3p^6$
19 K [Ar]$4s^1$		29 Cu [Ar]$3d^{10}$ $4s^1$	30 Zn $4s^2$	31 Ga $4s^2 4p^1$	32 Ge $4s^2 4p^2$	33 As $4s^2 4p^3$	34 Se $4s^2 4p^4$	35 Br $4s^2 4p^5$	36 Kr $4s^2 4p^6$
37 Rb [Kr]$5s^1$		47 Ag [Kr]$4d^{10}$ $5s^1$	48 Cd $5s^2$	49 In $5s^2 5p^1$	50 Sn $5s^2 5p^2$	51 Sb $5s^2 5p^3$	52 Te $5s^2 5p^4$	53 I $5s^2 5p^5$	54 Xe $5s^2 5p^6$
55 Cs [Xe]$6s^1$		79 Au [Xe] $4f^{14} 5d^{10}$ $6s^1$	80 Hg $6s^2$	81 Tl $6s^2 6p^1$	82 Pb $6s^2 6p^2$	83 Bi $6s^2 6p^3$	84 Po $6s^2 6p^4$	85 At $6s^2 6p^5$	86 Rn $6s^2 6p^6$

Fig. 1.22 Abridged Periodic Table of Elements

1.13.1 **SUCCESS** of the OLD QUANTUM THEORY:

1.13.1.1 The **success of Bohr's theory and Sommerfeld's generalization** can be summarized by listing the observations that are predicted by the theory:
1.13.1.2 The line spectrum of hydrogen atoms
1.13.1.3 The diameter of the hydrogen atom
1.13.1.4 The line spectra of helium ions and certain ions of the next heavier elements
1.13.1.5 The small wavelength shift explained by the relative motion of the nucleus and the centre of mass
1.13.1.6 The fine structure of lines explained by the elliptical orbit treated with the relativity correction.
1.13.1.7 The Stark Effect.

1.13.2 **SHORTCOMINGS** of THE OLD QUANTUM THEORY

The combination of what we have come to call the '*old quantum theory*' lies in the unification by Wilson and Sommerfeld of the Planck and Bohr quantization principles as special cases of the quantization of the action integral. Its short comings are:

1.13.2.1 A spectrum shows that some spectral lines appear to produce a denser image on the film detector than others. Bohr Theory is silent over this intensity of lines, time as a variable does not appear in the Bohr Theory, which is clearly incomplete in this account. Bohr Theory is unable to deal with complex atoms.
1.13.2.2 The Coulomb law worked out but the classical theory of radiation by accelerated charges did not except the limit of large value of the principal quantum number n for which electrons become unbound and hence no longer described by the theory.
1.13.2.3 The spectra of atoms having more than one valence electron have to be explained.
1.13.2.4 The fine structure observed in many situations.
1.13.2.5 The effect of magnetic field on atomic spectra.
1.13.2.6 The logical reason for assuming different quantum numbers.

One cannot doubt that the quantum theory represents a great step forward toward the truth. But its validity is limited, because of its shortcomings.
Subsequently the theory was supplemented by the hypothesis of the electron spin. Old Quantum theory when greatly refined, and later the work of W.E. Lamb, Jr. and R.C. Rutherford (1947) led to further refinement of the fundamental assumptions.

1.13.3 The Exotic Quantum World

We are living in a quantum universe, not a classical one. Most people however are still living in the classical Newtonian world which expired at the beginning of the twentieth century. Our world is very subtle and much more mysterious than the "building blocks" view of the universe would indicate. Many people lead their lives at the macroscopic level as if quantum reality were only true for atoms and somehow not true for larger things like people and their immediate environment, but the correspondence principle by which the quantum world is supposed to fade into the classical world never works out for a host of reasons.

Chaotic, self-critical and certain other processes may inflate quantum effects in unforeseen ways to the macroscopic level. The physics underlying conscious interaction with the physical world may likewise depend both on quantum effects, criticality and chaos in its functioning. The entire universe itself may be a self-consistent interconnected whole which has emerged from a single quantum wave function, therefore it is non-classical in its entire description. For these reasons it is necessary for us to understand how the quantum works and how it may differ from our classical view of order, solidity, determinism and mechanism.

Today the combined work of these three men is known as the Old Quantum Theory. The old quantum theory, resulting in Bohr's orbital model of the atom could point to certain real successes: Derivation of the Balmer formula, quantum numbers and selection rules for energy states in an atom, explanation of the Periodic Table and the Pauli Exclusion Principle. The old quantum theory relied heavily on the Newtonian Mechanics, but sought to supplement it with supplementary conditions. But what about the particle / wave character of the electron? (de Broglie, 1923) In the early 1920's it was clear that the quantum theory as it then existed was unsatisfactory.

In the mid 1920's two distinct and seemingly independent versions of a new quantum theory were presented:
i) Matrix mechanics (W. Heisenberg, 1925)
ii) Wave Mechanics (E. Schrödinger, 1926)
iii) Soon after their discovery these two formulations were shown to be equivalent, forming the basis of present-day quantum theory.

REVIEW QUESTIONS

R.Q. 1.1 Consider a Radio Station operating on a frequency of 98 MHz, and it radiates a power of 200 kW. Find how many quanta of energy are emitted per second (To an appropriate number of significant figures). (Answer: $n = 3 \times 10^{30} E$).

R.Q. 1.2 Find the wavelength and frequency of a 1.0 keV photon. (Answer: $\lambda = 1$ nm, $\nu = 2.4 \times 10^{17}$ Hz.

R.Q. 1.3 What is the momentum of a 12.0 MeV photon? (Answer: $p = 6 \times 10^{-21} kg\ ms^{-1}$).

R.Q. 1.4 What is the maximum wavelength of the photon that will separate a molecule whose binding energy is 15 eV? (Answer: $\lambda = 80$ nm).

R.Q. 1.5 Find the equivalent energy does a photon have if its momentum is equal to that of a 3 MeV electron. (Answer: Using $E^2 = E_0^2 + p_0^2 + c^2$, $E_{photon} = p_{photon}\ c = 3 MeV$).

R.Q. 1.6 Find the average number of quanta that fall on the Earth per $m^2 s^{-1}$ from the Sun. Given: the Solar constant, $S_\odot = 1.4\ kW\ m^{-2}$; Sun's surface temperature $T_\odot = 6000 K$; and Wien's constant $b = 2.90 \times 10^{-9}\ m\ K$. (Answer: $n = 3 \times 10^{-34}\ m^{-2} s^{-1}$).

R.Q. 1.7 Light of wavelength 589.3 nm is incident on Potassium surface. The stopping potential for the electrons is 0.36 eV. Calculate the maximum energy of the photoelectron as well as the threshold frequency. Given: $c = 2.997925 \times 10^8\ ms^{-1}$. (Answer: $\varphi_0 = 1.79\ eV$, $\nu_0 = 4.3 \times 10^{14}\ Hz$).

R.Q. 1.8 Light, having a wavelength of 500.0 nm, falls on a metal surface characterized by photoelectric work function of 1.90 eV. Find the energy of the photon in eV, the kinetic energy of the most energetic photoelectron in eV and in $Joules$, and the stopping potential. (Answer: $\varepsilon = 2.4\ eV$, $E = 0.57 eV, = 9.11 \times 10^{-20} J$; $V_0 = 0.57 eV$).

R.Q. 1.9 The work function, φ_0 of cesium and tungsten are, respectively, 1.8 eV and 4.5 eV. Which metals yield photoelectrons bombarded with light of wavelength 500.0 nm? Calculate the stopping potential in *Volts*. (Answers: φ_0 (W) > E = 2.48 eV, so no photoelectrons; but φ_0 (Cs) < E = 2.48 eV, so photoelectrons; $V_0 = 0.68\ V$).

R.Q. 1.10 Calculate the stopping potential for photoelectrons ejected from a metallic surface by visible 600.0 nm photons strike at it. Given, the stopping potential for the metal by 375 nm photons is 1.870 V.
(Answer: $V_0 = 0.628\ V$).

R.Q. 1.11 A free, stationary, electron is being hit by an x-ray photon 0.0500 nm. A photon scatters at 90°. Determine the momenta of the incident X-ray, the scattered X-ray and the electron. (Answers: $\lambda' = 0.0524$ nm, $p = h/\lambda = 1.33 \times 10^{-23}$ kg m s^{-1}; $p' = h/\lambda' = 1.26 \times 10^{-23}$ kg m s^{-1}; $\varphi = 43.6^\circ$; $p_e = \beta mv = 1.83 \times 10^{-23}$ kg m s^{-1} at $\varphi = 43.6^\circ$ to incident photon).

R.Q. 1.12 Determine the Rydberg constant for hydrogen, which emits largest wavelength in Lyman series, is 121.5 nm. (Answer: $R = 1.097 \times 10^4$ cm^{-1})..

R.Q. 1.13 How many different photons can be emitted by hydrogen atoms that undergo transitions to the ground state from the $n = 5$ level? (Answer: The number of photons emitted for $n_i = 1$ and $n_f = n$ is given by $N = n(n-1)/2$; $N = 5(5-1)/2 = 10$).

R.Q. 1.14 Determine the wavelength of the Lyman H_α line

R.Q. 1.15 Determine the wavelength of the Balmer H_α line

R.Q. 1.16 If the frequency of X-ray K_α emitted from the element with atomic number 31 is v, find the frequency of K_α X-ray emitted from the element with atomic number 51. (Answer: $v_{K\alpha} = (5/3)^2 \, v$).

R.Q. 1.17 Find the wavelength of the Copper Cu K_α line, given $C = (1/4) \, c \, R$ and $a \approx 1$

(Answer: $\lambda_{K_\alpha} = 1.5418$ Å).

R.Q. 1.18 Given the magnetic quantum number mℓ of an electron in a particular excited state of the hydrogen atom is 5, Write the possible choice of quantum numbers n, ℓ, m_ℓ, m_{ss} [Answer: respectively, 8, 6, 5, 1/2].

R.Q. 1.19 According to Pauli principle $2n^2$ electrons can occupy the major shell with principal quantum number n. Out of 50 electrons in the major shell n = 5, what is the number of electrons having the magnetic quantum number $m_\ell = 1$ [Answer: 10].

R.Q. 1.20 In the first Bohr orbit of a hydrogen atom the total energy of the electron is -21.6×10^{-19} J. Determine the potential energy [Answer: -43.52×10^{-19} J).

R.Q. 1.21 UV Catastrophe is observed under what context?

R.Q. 1.22 What do you mean by the term quantization of physical quantities? Which was the first quantity to be quantized and under what circumstance? Give in order of time the various quantization of quantities when the Bohr theory of atom was formed.

R.Q. 1.23 What is the minimum angle the angular momentum vector may make with the z-axis in the case $\ell = 3$. [Answer: $\vartheta = \cos^{-1}(3/\sqrt{12}) = 30^\circ$].

R.Q. 1.24 The atoms in a sample of hydrogen are all in their ground states. Now a beam of electrons of KE of 12.5 eV collides with the sample. What wavelengths will be emitted? [Answer: $\lambda = 103$ nm (Lyman), 656 nm (Balmer), and 122 nm (Lyman).]

R.Q. 1.25 Sodium metal has a work function of 2.0 eV. Calculate the threshold frequency for which electrons are emitted. [Answer: 4.8×10^{14} Hz].

R.Q. 1.26 In the Photoelectric Effect, the maximum kinetic energy of electrons emitted from a metal is 1.6×10^{-19} J. When the frequency of the emitted radiation is 7.5×10^{14} Hz, what is the minimum frequency of radiation for which electrons will be ejected from the metal? [Answer: 5.2×10^{14} Hz]

R.Q. 1.27 The intensity versus wavelength distribution curves at two temperatures shows that the radiation emitted by two light bulbs at temperature of 1500 K and 2000 K. What are the peak wavelengths λ_{max} of the radiation from the light bulbs? Do the data support Wien's Law, $\lambda_{max} T$ = constant? [Answer: For 1500 K, $\lambda_{max} = 1 \mu m$, for 2000 K, $\lambda_{max} = 0.75 \mu m$].

R.Q. 1.28 Given Wien's Law $\lambda_{max} T = 2.9 \times 10^{-3}$ $m K$, what is the peak wavelength of the 2.7 K cosmic microwave background blackbody radiation? In what part of the EM spectrum does it fall? [Answer: 10^{-3} m .in the micro-wave region].

R.Q. 129 Light of wavelength 589.3 nm is incident on Potassium surface. The stopping potential for the electrons is 0.36 eV. Calculate the maximum energy of the photoelectron as well as the threshold frequency.[Answer: 4.3×10^{14} Hz]

R.Q. 1.30 Using the Bohr theory, arrive at the energy levels of H-atom.

R.Q. 1.31 How many different photons can be emitted by hydrogen atoms that undergo transitions to the ground state from the n = 5 level? [Answer: 10].

R.Q. 1.32 Calculate the ratio of the mass of the proton to that of the electron using the knowledge of the Rydberg constants for hydrogen and ionized helium. Given: $M_{He} = 3.9717$ M_H, from mass spectrographic data [Answer: $M_H /m = 1840$]

R.Q. 1.33 Find the reduced mass of HF in terms of mass unit $u = 1.66 \times 10^{-27}$ kg.

[Answer: (19/20) u]

R.Q. 1.34 De Broglie's relationship gives the wavelength of a hydrogen atom moving with a velocity of 10^3 m/s as [Answer: 3.96×10^{-10} m].

R.Q. 1.35 Find the equivalent energy a photon has if its momentum is that of a 3 MeV electron.

[Answer: 3 MeV].

R.Q. 1.36 The Lyman emission (2p → 1s) of hydrogen, a major component of the solar spectrum, occurs at a wavenumber ($\bar{v}$) of 82259 cm^{-1} Use this to determine the value of R (in cm^{-1}).

R.Q. 1.37 The shortest wavelength emission lines of the Balmer series (np → 2s) and (nd → 2p) that have been observed in the laboratory have $n = 18$, but from interstellar space a Balmer line with $n = 42$ has been detected. What is the wavenumber of this line?

R.Q. 1.38 Derive the relation that predicts the frequencies of the line spectrum of hydrogen.

[Answer: $v = (2\pi me^4 / h^3)(1/n_i^2 - 1/n_f^2)$

R.Q. 1.39 Derive the energy of an electron in the H-atom using Bohr's formulation.].

R.Q. 1.40 Using the formula for quantized orbits of electron in a H-atom, show that the ground state radius is 0.529×10^{-8} cm.

R.Q. 1.41 Derive an expression for the frequencies at which light is absorbed by hydrogen atom in its ground state., using the atomic model of Bohr. Why should these frequencies be slightly different for positronium?

Chapter 2

Concepts in Wave Mechanics – Wave-Particle Duality

Chapter 2

Concepts in Wave Mechanics – Wave-Particle Duality

Anyone who has not been shocked by quantum physics has not understood it - Niels Bohr

2.1 INTRODUCTION:

For more than a century the old controversy between the *Corpuscular Theory* and the *Wave Theory* of light seemed to have been decided, since the phenomena of diffraction and interference gave clear evidence of the wave nature Thomas Young, 1803; A. Fresnel, 1815). Much later, however, these were followed by the EM Theory (J.C. Maxwell, 1864) and its experimental verification (H. K.Hertz, 1889) – had firmly established the wave nature of EM Radiation. Then came new experimental facts – spectrum of Blackbody Radiation (Otto Lummer &E. Pringsheim, 1900; Max Planck, 1901), the Photoelectric Effect (A. Einstein, 1905), continuous X-ray spectrum, the Compton Effect (A.H. Compton, 1923), *etc.* in which some sort of particle nature to EM Radiation was assigned. The Photoelectric Effect and the Compton Effect furnished proof of the *quantum nature of light* and so revived the Corpuscular Theory – This *dual nature* of EM Radiation continued to be generally accepted from 1905 to 1924. These two theories appeared to be incompatible.

Therefore there was speculation in 1923 that the converse might also be true. Classical Physics had already the Maupertuis's *Least Action Principle* (less metaphysically the *Hamilton's Principle*) in mechanics and the *Fermat's Principle* (1657) in optics. In this Chapter these two principles will be examined.:

2.1.1 The Maupertuis' Least Action Principle

To express the *Lest Action Principle* in Mechanics mathematically consider: Time integral of *Lagrangian* $\hat{L}$ and action S are related by

$$S = \int_{t_1}^{t_2} \hat{L}\, dt = extremum \text{ (minimum in most cases).} \tag{2.1.1}$$

Principle of stationary particle means

$$\delta S = \int_{t_1}^{t_2} \hat{L}\, dt = 0 \tag{2.1.2}$$

for the motion of a system between any two instants t_1 and t_2. The Lagrangian is given by

$$\boxed{\hat{L}(q,q',t) = T + V} \tag{2.1.3}$$
$$\text{(Lagrangian)} \quad \text{(KE)} \quad \text{(PE)}$$

2.1.2. Fermat's Principle:

It is the *Principle of Least Time* or of the *Shortest Optical Path* in geometrical Optics. It is expressed as

$$\boxed{\int_P^Q \frac{1}{w} \, ds = Minimum} \tag{2.1.4}$$

$$\delta \int_P^Q \frac{1}{w} \, ds = 0. \tag{2.1.5}$$

It asserts that the time taken for light to travel with speed w, from position P to Q of an actual ray, is shorter than the path of any other curve arbitrarily chosen, adjacent to the path, *i.e.*,

$$\frac{1}{c} \int n \, ds = c \int dt = 0 \tag{2.1.6}$$

n is the refractive index of the medium and c is the velocity of light.
The calculus version is

$$\frac{d}{ds} \int n(s) \, ds = 0 \tag{2.1.7}$$

2.1.3. De BROGLIE HYPOTHESIS:

In 1924, Louis Victor de Broglie, guided by the analogy of Fermat's Principle in optics and the least action principle in mechanics, was led to suggest that the *dual nature* of radiation (**wave-particle**) should have its counterpart in a *dual nature* (**particle-wave**) of matter. After looking deeply into the *Special Theory of Relativity* and the *PhotonHypothesis*, he suggested a more fundamental relation between waves and particle. According to the Special Theory of Relativity for a photon, its energy $E(=\varepsilon)$

$$\boxed{\varepsilon = h\nu} \tag{2.1.8}$$

where ν is the frequency of the photon, is

$$E = c\sqrt{m_0^2 c^2 + p^2} \tag{2.1.9}$$

with $m_0 = 0$, for a photon, its momentum p is

$$E = pc,$$
and $\quad E = h\nu$

This means $\boxed{p = h\frac{\nu}{c} = \frac{h}{\lambda}} \tag{2.1.10}$

It can be seen that the left side of the equality symbol of both the equations (2.1.8) and (2.1.10),
i.e., E and p are characteristics of the particle nature, and the right side of the ' = ' symbol of these equations are characteristics of waves, as further clearly illustrated in equation (2.1.11).

$$\boxed{\begin{array}{ccc} \swarrow \varepsilon & = h & \swarrow \nu \\ \text{(Particle nature)} & & \text{(Wave characteristic)} \\ \nwarrow p & = h & \frac{1}{\lambda} \end{array}} \tag{2.1.11}$$

These two sets of quantities – particle and wave – are connected through the Planck's constant h. Drawing upon an intuitive expectation that nature is symmetric, de Broglie asserted that

$$\boxed{\lambda = h/p} \quad (2.1.12)$$

is completely a general formula that applies (or governs) to or extended to material particles (electrons, protons, α-particles, etc) as well as photons. The difference is that
a) Photon has $m_o = 0$, and $v = c$,
b) Matter has $m_o \neq 0$, and $v < c$.

Thus the momentum of a particle is $p = mv$ and consequently, the **deBroglie relation** becomes

$$\lambda = h/p = \frac{h}{mv} \quad (2.1.13)$$

where $\quad m = m_o / \sqrt{1 - \frac{v^2}{c^2}} \quad (2.1.14)$

Now kinetic energy

$$T = \tfrac{1}{2} mv^2 = \frac{p^2}{2m}, \quad (2.1.15)$$

or $\quad p = \sqrt{2mT}. \quad (2.1.16)$

Therefore, the de Broglie relation (2.1.13) becomes

$$\lambda = h/p = \frac{h}{mv} = \frac{h}{\sqrt{2mT}}$$

i.e., $\quad \lambda = \sqrt{\frac{150}{V \text{ (in kV)}}} \overset{o}{A}. \quad (2.1.17)$

Thus the de Broglie wavelength $\lambda = h/p$ is associated with every particle, and the frequency associated with the monochromatic field $v = E/h$

2.1.4. Fundamental Wave-Particle Relationships (Useful Forms of the Planck and De Broglie Relations)

Introducing the wave number k and angular frequency ω

$$k = 2\pi/\lambda, \quad (2.1.18)$$
$$\omega = 2\pi v,$$

one can write in more symmetric or useful forms the PLANCK RELATION as

$$\boxed{E = \hbar \omega}. \quad (2.1.19)$$

and DE BROGLIE RELATION as

$$\boxed{p = \hbar k}. \quad (2.1.20)$$

where $\hbar = (h/2\pi) = 1.054 \times 10^{-34} J-s$.

If this is correct, one may expect that whenever the motion of a particle is disturbed in such a way that the field associated with it can not propagate freely, interference and diffraction phenomena should be observed, as in the case for elastic and EM waves. **This is indeed what happens!**

Since $\quad v = c\lambda$,
Equation (2.1.10) can be written as
$$p = hv/c \quad (2.1.21)$$

2.1.5. The Appropriate Relations are $p = \frac{h}{\lambda}$ and $\varepsilon = hv$

It is known that the momentum of a photon is *always* expressed as $p = h/\lambda$, rather than $p = hv/c$. The two expressions are *equivalent* for a photon. However, in the case of a massive particle with a wave nature, the two expressions are **NOT** equivalent. It is found that the expression (2.1.10) is correct while expression (2.1.21) is not. There is for all phenomena a fundamental relationship between p and λ, not between p and v. There is a fundamental relationship complementary to expression (2.1.10), between energy ε and frequency v (not between ε and λ), so energy is expressed as $\varepsilon = h\,v$ (2.1.8), rather than $\varepsilon = h\,c\,/\lambda$; although the two are again equivalent in the case of a photon.

Worked out Example 2.1

Calculate the de Broglie wavelength of an electron moving with energy of 54 eV. Given, $m_e = 9.109 \times 10^{-31}\ kg$, $h = 6.63 \times 10^{-34}\ J-s$. $1\ eV = 1.6021 \times 10^{-19}\ J$.

Solution: Step#1

$T = 54\ eV;\ m_e = 9.109 \times 10^{-31}\ kg;\ h = 6.63 \times 10^{-34}\ J-s;$
$1\ eV = 1.6021 \times 10^{-19}\ J;$

$$\lambda = \frac{h}{\sqrt{2\,m\,T}} = \frac{(6.63 \times 10^{-34} J-s)\cdot}{\sqrt{2\,(9.109 \times 10^{-31}\ kg)\cdot(54\ eV)\cdot(1.6021 \times 10^{-19} J/eV)}} = 0.167\ nm.$$

Worked out Example 2.2

What is the energy of a photon of wavelength 0.386 nm? Given, $h = 6.63 \times 10^{-34}\ J-s$, $1\ eV = 1.6021 \times 10^{-19}\ J$. $c = 2.997 \times 10^8\ ms^{-1}$.

Solution: $E = hc/\lambda = \dfrac{(6.63 \times 10^{-34} J-s)\cdot(2.997 \times 10^8\ ms^{-1})}{\sqrt{(0.386 \times 10^{-9}\ m)\cdot(1.6021 \times 10^{-19}\ J/eV)}} = 3.21\ keV$.

2.1.4. QUANTITATIVE EXPLANATION OF THE QUANTUM CONDITION of Bohr's Postulate 1:

It is useful to examine the de Broglie's waves in the hydrogen atom. The theoretical implications of the de Broglie *wavelength of matter* are interesting. De Broglie *hypothesis* provides one with more fundamental basis of quantization of electron orbits in the Bohr atom. From equations (1.7.13) and (1.7.17) orbit of hydrogen atom has radius

$$r_n = \frac{n^2\,\hbar^2}{m\,Z\,e^2} = \frac{n^2}{Z}\cdot(0.0529\ nm) \qquad (2.1.22)$$

i.e., $\qquad r_n = n^2\,a_o$.
where $\qquad a_o = (0.0529\ nm)$.

The circumference of the first orbit $= 2\pi r_{n=1} = 2\pi\,a_o = \dfrac{2\pi\,\hbar^2}{m\,e^2}$. $\qquad (2.1.23)$

Wavelength of electron in the fist orbit $= \lambda_1 = h/p_1 = \dfrac{2\pi\,\hbar}{m\,v_1}$ $\qquad (2.1.24)$

Applying Bohr's postulate (1.7.3), for $n = 1$,
$L_1 = m\,v_1\,r_1 = m\,v_1\,a_o = \hbar$. $\qquad (2.1.25)$

whence wavelength, $\qquad \lambda_1 = \dfrac{2\pi\,\hbar}{\hbar/a_o} = 2\pi a_o$. $\qquad (2.1.26)$

Therefore, circumference of the first orbit = de Broglie wavelength of the associated electron in $n = 1$.

Why is it so? The answer lies in the *interference* property of deBroglie waves in an atom. Constructive and destructive interference will take place and the wave will re-enforce or rapidly damp to zero amplitude as in Fig 2.1. It indicates the formation of stationary (or standing) waves for the various orbits, $2\pi r = n\lambda$, $n = 1, 2, 3, ...$ It can be seen that fractional number of λ_n cannot persist because destructive interference will occur. Bending an acoustical resonator into a circle, a doughnut with no obstructing walls inside will be the case for the $n = 1$ orbit.

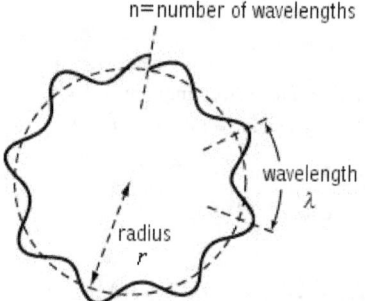

n must be an integer for a standing wave

Fig. 2.1 Bohr's postulate on quantum condition merging with the de Broglie concept.

This distinguishes the *allowed stable* Bohr orbit or an orbit, which fails to satisfy quantum theory, *i.e.* orbital *standing waves* according to de Broglie. This first successful application of deBroglie's "λ – MOMENTUM" condition, explains in a simple and straightforward way the puzzling A*ngular Momentum Quantization Rule* of Bohr. The merging of Bohr's and de Broglie's concepts provided much of the impetus that led within a short period of time to the development of the modern theory of atomic structure and quantum mechanical terms.

However, this discussion should not be interpreted as derivation of Bohr's postulate, for ssentially what I have done is replace Bohr's postulate with deBroglie's postulate, assuming electrons in states are represented as stationary waves.

2.2. FURTHER DEVELOPMENTS

This generalization proposed by de Broglie had revolutionary consequences for it implied a *wave nature of matter*. A quarter of century after Planck had introduced a new quantum constant h into the description of nature, there was the major developments in a short span of time, as given below:

1) 1923: Discovery of electron diffraction- wave nature of matter (C.J. Davisson & L.H. Germer),
2) 1924: Wave nature of matter was predicted (Louis deBroglie),
3) 1925: Electron spin was suggested (G.E. Uhlenbeck and S.A. Goudsmit),
4) 1925: Te Exclusive Principle (W. Pauli),
5) 1925: Mathematical theory of Quantum Mechanics (W. Heisenberg, M. Born, Jordan),
6) 1926: Wave theory of Quantum Mechanics (E. Schrödinger),
7) 1926: Probability interpretation (M. Born),
8) 1927: Wave nature of matter verified experimentally (G.P. Thomson, C.J. Davisson, L.H. Germer),
9) 1927: The Uncertainty Principle (W. Heisenberg),
10) 1928: Relativistic Quantum theory of electron (P.A.M. Dirac),
11) 1930: A more abstract form of Quantum Mechanics (P.A.M. Dirac),

12) 1930: Particle nature of radiation verified using visible radiation, and phonons instead of electrons, like the Compton Effect (C.V. Raman),
13) 1947: Lamb Shift (W.E. Lamb and R.C. Retherford).
14) 1947 - 49: Quantum Electrodynamics (Richard P. Feynman, Julian Schwinger, Sin-Itiro Tomonoga).
15) 1948: Renormalization.
16) 1973: Quantum Chromo-dynamics (QCD)- a theory for quark interactions (Greenberg).and early ideas about G U Ts, Neutral currents.

2.2.1 ELECTRON DIFFRACTION Experiment by Davisson& Germer:

In a systematic investigation of "secondary electrons" produced by electron impact on metals, in the Bell Telephone Laboratories C.J. Davisson and L.H. Germer explored the scattering of a ell-defined beam of electrons impinging on a single crystal of a metal and independently by G.P. Thomson (son of J.J. Thomson) in the experiment using a narrow pencil of electrons incident on a thin crystalline film of gold (analog of Kikuchi's Laue analysis) in 1927, confirming de Broglie's hypothesis by demonstrating that electrons exhibit diffraction when they are scattered by crystals in gold foils. The transmission of electrons through a thin metal foil showed, instead of blurred image (if electrons were particles), gave circular diffraction patterns. Since the wavelengths associated with electrons of 10 – 100 eV correspond to the atomic spacing of most of the crystals it was conjectured that such electrons ought to be diffracted by crystals in the same manner as William H. Bragg's X-ray diffraction. Fig.2.2 shows schematically the experimental arrangement.

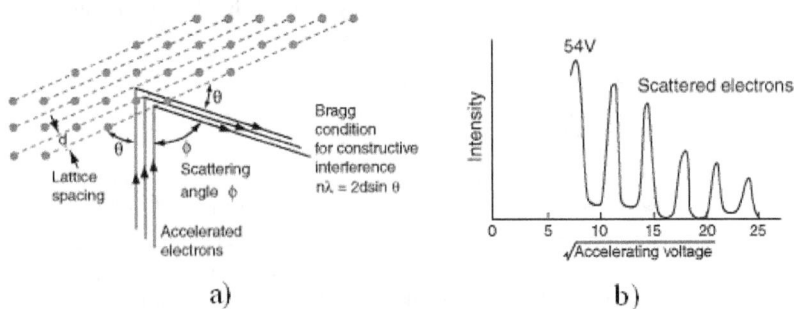

Fig. 2.2 Davisson and Germer, electron diffraction theory.

$$\underset{\substack{Electron \\ wavelength}}{\frac{1}{\lambda}} = \underset{\substack{Bragg \\ law}}{\frac{n}{2d\sin\theta}} = \underset{\substack{deBroglie \\ relationship}}{\frac{p}{h}} = \frac{\sqrt{2mE}}{h} = \underset{\substack{Acceleration \\ through\ voltage\ V}}{\frac{\sqrt{2meV}}{h}} \quad (2.1.1)$$

These scientists were involved in measuring the (scattering) angular distribution of intensity I of scattered electrons, arising from the striking (primary incident) fast (54 eV) electrons on the target of Nickel metal. The electrons are accelerated through a potential difference, p.d. V, of 54 V in the electron gun. These electrons fall on the target at normal incidence and the intensity of the scattered electrons as a function of the angle φ. It was plotted in the form of a polar diagram, Fig. 2.3, at any angle φ,

$I \propto |r|.$

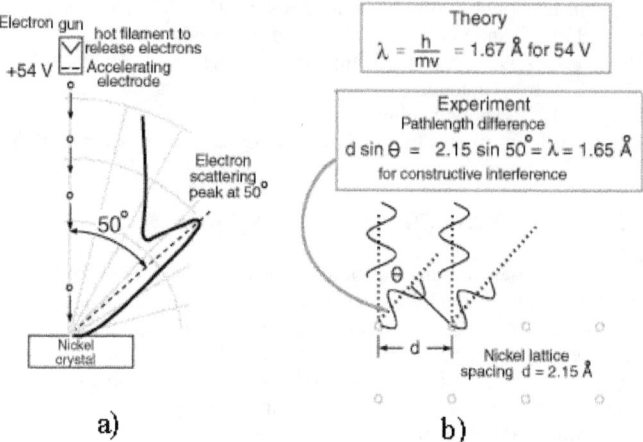

Fig. 2.3 Details of the electron diffraction experiment.

This is so classically, *i.e.* if there is no wave nature of the electrons, the intensity of the electrons would decrease monotonically with scattering angle φ with no large fraction coming at any particular φ. An explosion was found to break the vacuum and the target got oxidized. The experiment was conducted with a Nickel target (after cleaning the oxidized layer of the surface by heating it in a high temperature oven, *i.e.* by lengthy heat treatment of the sample), the results were quite different.

$N(\varphi = 0)$ = 0
$N(\varphi = 35°)$ = peak
$N(\varphi = 50°)$ = peak

The strong peak at $\varphi = 35°$ is expected from either particle or wave theory. But the peak at a scattering angle $\varphi = 50°$ when the p.d. was V = 54 V what had happened was that the heating of the metallic sample resulted in orienting the micro-crystals in the sample so as to become a large single crystal. The de Broglie waves of the electron beam satisfied the Bragg Law, causing a diffraction peak.

As $\theta + \frac{\varphi}{2} = 90°$ for $\varphi = 50°$ means $\theta = 65°$.

For nickel, $2d \sin \theta = n\lambda$ means 2 (0.91 Å) Sin 65° = 1 λ, gives . $\lambda = 1.65$ Å.

This calculation is according to Bragg's Law.

According to de Broglie's hypothesis, as per equation (2.1.17)

$\lambda = h/p$; $\lambda = \sqrt{\frac{150}{V \text{ (in kV)}}}$ Å.; whence $\lambda = \sqrt{\frac{150}{0.54}}$ Å $= 1.67$ Å.

This value quite agrees with that obtained from Bragg's Law, within the experimental error. Hence the de Broglie's hypothesis of the wave nature for the electron, and generally moving bodies, is confirmed and firmly established.

The neutrons have proved to be the best particles for demonstrating the phenomena of diffraction and interference. As a bench mark for $\lambda - p$ relationship, one may note that a neutron moving at 9,000 *mph* or $4 \times 10^4 \, ms^{-1}$ – about half the speed of an orbiting astronaut has a wavelength, $\lambda = 1 \, \text{Å}$.

Finally, an *electron in motion* is to be cognized as an **electron wave**, and a neutron as a 'neutron wave', or in general, matter in motion can be referred to as a matter wave or *particle wave*.

2.2.2   Is the Wave-Particle Duality Real?

From what have been described earlier, it is now known that electrons and photons both appear sometimes as waves and sometimes as particles. Does each exist as part wave and part particle? Or are they both capable of transforming back and forth between these two descriptions? These are clear by realizing that when one makes a wave or a particle classification one has forced a classical description on entities that are essentially non-classical. Electrons and photons both do not obey the rules of classical mechanics – their behavior is described correctly only by Quantum Mechanics. Ambiguity will naturally arise when one uses classical ideas to describe quantum entities.

The only meaningful way to discuss the behaviour of the object studied is in terms of the results of *measurements*. Therefore, as an electron (or a photon) appears as a wave (or as a particle) depends on the nature of the measurement that is made. These two characters of an entity, like an electron or a photon, therefore, lies in the eye of the beholder.

2.3   WAVE PACKET DESCRIPTION OF MATERIAL PARTICLES

2.3.1   WAVE FUNCTION, denoted by the Greek symbol, *psi*, Ψ

Human sensory experience tells one that the objects one touches and sees have a well-defined shape and size and, therefore, are LOCALIZED in space. One tends to extrapolate and think of the fundamental (microscopic) particles (electrons, protons, neutrons, etc) as characterized by a spherical shape, having a radius, as well as mass and charge. This extrapolation is beyond direct sensory experience. Experimenters have shown that this picture of the basic constituents of matter is erroneous. The dynamical behaviour of elementary particles requires that one can associate with each particle – matter field, in the same but reverse way that one associates a photon with an EM field. Further electrons and photons do not obey the rules of classical bodies. It is thus appropriate to look into the question: how to describe 'PARTICLE RADIATION' and 'MATTER WAVES? What kind of wave phenomenon is involved in the *electron waves* of deBroglie?

i)   In **EM waves**: the varying physical quantity in space and time is the EM field,
ii) In **acoustic waves**: pressure varies in space and time,
iii) In **matter waves**: What is it whose variations constitute deBroglie waves?
iv) The varying quantity characterizing deBroglie waves is called matter field amplitude, which for convenience, is referred to as the wave function, denoted by the Greek symbol, psi, Ψ.

2.3.2 THE PRINCIPLE OF LINEAR SUPERPOSITION:

In order to account for interference effects physicists assume the validity of the *Principle of Superposition*. This law of fields sometimes taken for granted is so important that an explicit notice needs to be taken of it. In optics:

2.3.2.1 With COHERENT light, accordingly, the amplitudes A_1 and A_2 of the light waves involved at a point are added *vectorially* and the resultant amplitude is squared to yield the time average of the intensity, I, at the point:

$$I = (A_1 + A_2)^2$$
$$I = (A_1^2 + A_2^2 + 2A_1 A_2 = I_1 + I_2 + 2A_1 A_2 \qquad (2.3.1)$$

where $2A_1 A_2$ is called the *interference term*.

2.3.2.2 With INCOHERENT light, the intensity, I, of the resultant superposed wave
$$I = I_1 + I_2 \qquad (2.3.2)$$
and the interference term $= 2A_1 A_2 = 0$.

From our knowledge of fields and waves, it is known that a particle localized within s certain region of space, Δz, in the z-direction, should correspond to a matter field amplitude (wave function, Ψ)*large* in that region or *very small* outside that region.
Note: Elsewhere in the world the idea of superposition is not always valid:
For example,
1) If a *litre* of water is added to one *litre* of alcohol the total volume of the liquids turns out to be LESS THAN 2 l*itres*!
2) One neutron mass + one proton mass is GREATER THAN the mass of a resulting deuteron!

2.3.3 SUPERPOSITION OF PLANE WAVES – TIME DEPENEDENCE - Conventional Approach

Consider two plane harmonic waves propagating along the z-axis (Fig. 2.4) is described by equatons:

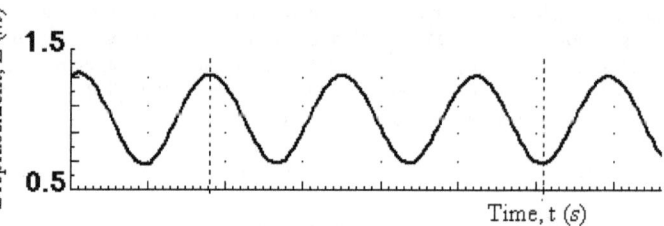

Fig. 2.4Plane wave.

$$u_1(z,t) = A_1 [e^{-i(k_1 z - \omega_1 t)}], \qquad (2.3.3)$$
$$u_2(z,t) = A_2 [e^{-i(k_2 z - \omega_2 t)}] \qquad (2.3.4)$$

where the angular frequencies $\omega_1 \approx \omega_2$;

and defining $k_1 = k_o - \frac{\Delta k}{2}$, $k_2 = k_o + \frac{\Delta k}{2}$,

$$\omega_o \equiv <\omega> = (\omega_1 + \omega_2)/2 \tag{2.3.5}$$

The wave vectors / propagation constants $k_1 \approx k_2$, and

$$k_o \equiv <k> = (k_1 + k_2)/2 \tag{2.3.6}$$

The amplitudes $A_1 = A_2 = A_o$.

2.3.3.1 WAVE and GROUP velocities:

By superposing these *two component waves* one gets the **compound wave** (Fig. 2.5)

$$U(z,t) = u_1(z,t) + u_2(z,t)$$
$$= 2A_o \left\{ [e^{-i(k_1 z - \omega_1 t)}] + [e^{-i(k_2 z - \omega_2 t)}] \right\}$$
$$= 2A_o \left\{ e^{-i(k_o z - \omega_o t)} \right\} \cdot [e^{i\theta} + e^{-i\theta}]$$
$$= [2A_o Cos\theta] \left\{ e^{-i(k_o z - \omega_o t)} \right\} \tag{2.3.7}$$

where a) $[2A_o Cos\theta] = 2A_o Cos(\Delta k_o z - \Delta \omega_o t)$. (2.3.8)
= Real part of the AMPLITUDE WAVE, and

b) $\left\{ e^{-i(k_o z - \omega_o t)} \right\}$ = the PHASE WAVE. (2.3.9)

2.3.3.2 The COMPOUND WAVE:

The compound wave, shown in Fig. 2.5, is written as

$$\boxed{\begin{aligned} U(z,t) &= 2A_o Cos(\Delta k\, z - \Delta \omega\, t)\, Cos(k_o z - \omega_o t) \\ &= \sum \frac{1}{\sqrt{2\pi}} g(\omega)\, e^{i\omega_o t} \Delta \omega \end{aligned}} \tag{2.3.10}$$

The details are clearly shown in Fig. 2.5.

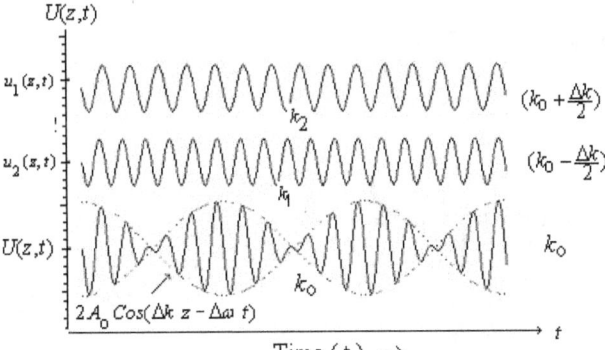

Fig 2.5 Combination of two plane waves, in phase and interfere constructively, and destructively. The dashed-line curve is the compound wave..

2.3.3.3 FEATURES of the Compound Wave

2.3.3.3.1 PHASE VELOCITY, $\bar{w}$

The second cosine function in real part of $U(z,t)$ is the original wave, having the

phase (wave front) $= (k_0 z - \omega_0 t)$,
and for a constant phase point
$$(k_0 z - \omega_0 t) = \text{constant}.$$
When differentiated this becomes
$$(k_0 dz - \omega_0 dt) = 0$$
Defining the PHASE VELOCITY, $\vec{w} \equiv \vec{v}_p \equiv \vec{v}_\varphi$ as

$$\boxed{\vec{w} \equiv \vec{v}_p \equiv \vec{v}_\varphi = \frac{dz}{dt} = \frac{\omega_0}{k_0}} \qquad (2.3.11)$$

2.3.3.3.2 The AMPLITUDE of the real part of U:
The amplitude of the Real part of $U(z,t)$ is $2A_0 \cos(\Delta k\, z - \Delta\omega\, t)$. This amplitude itself is a function of ω and t giving a wave.

2.3.3.3.3 The AMPLITUDE of the compound wave has a value [$2A_0$], which is twice the value of either of the individual component wave.

2.3.3.3.4 The AMPLITUDE is a *modulation* envelope (Fig. 2.5 and Fig 2.6).
It ravels at a velocity different from the *phase velocity*, $\vec{w}$. A given constant value of the magnitude can be followed only if the phase
$$(\Delta k\, z - \Delta\omega\, t) = \text{invariant with time, } t.$$
This means $\quad d(\Delta k\, z - \Delta\omega\, t) = 0$,
or, $\quad\quad (\Delta k\, dz - \Delta\omega\, dt) = 0$,
resulting in the definition of the a new velocity,

$$\frac{dz}{dt} = \frac{\Delta\omega}{\Delta k}, \qquad (2.3.12)$$

for a particular ω.

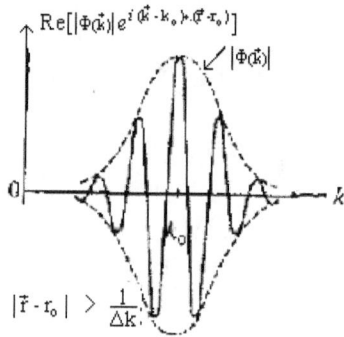

Fig. 2.6 A de Broglie pilot wave / Modulation Amplitude Envelope.

2.3.3.3.5 GROUP VELOCITY, $\vec{u}$
Defining the GROUP VELOCITY, $\vec{u}$:

$$\vec{u} \equiv \vec{v}_g = Lt_{\Delta k \to 0}(\tfrac{\Delta\omega}{\Delta k}) = Lt_{\Delta k \to 0}(\tfrac{\omega-\omega_0}{k-k_0})_{k \to k_0} = \tfrac{d\omega}{dk} \qquad (2.3.13)$$

$$\boxed{\vec{u} \equiv \vec{v}_g = \tfrac{d\omega}{dk}}. \qquad (2.3.14)$$

Provided that the ω-s and k-s are scattered only slightly about their mean values ω_0 and k_0, then one can use $<\omega>$ and $<k>$, and the result will be applicable to any number of harmonic waves – WAVE PACKET, or WAVE GROUP.
$\bar{u}$ depends on the way in which the frequency varies with k. Therefore,

$$\boxed{\bar{u} \equiv \bar{v}_g = \left(\frac{\omega_0}{k_0}\right) \rightarrow \frac{<\omega>}{<k>}}. \qquad (2.3.15)$$

2.3.4 FOURIER SYNTHESIS – An Alternative Approach to Superposition - Wave Train

In the case of a plane harmonic traveling wave, the phase velocity is not the velocity one observes when one analyses a wave motion. In a WAVE TRAIN (*i.e. continuous harmonic wave* or an *infinite linear plane wave*) of infinite length the wave has a wavelength, λ and a frequency, ν or angular frequency, ω. But a wave of this nature is not adequate to transmit a signal, because <u>a signal implies something that begins at a certain time and ends at a certain later time</u>. In other words, **a signal means a PULSE and not a WAVE TRAIN**. A wave pulse has no constant magnitude along the direction of propagation, viz. the z-axis. This requires a *Fourier analysis* of the pulse. When it is done as seen in Section 2.3.3, the PULSE actually is a **wave packet** containing several component waves.

In order to establish a complete description of the motion of a particle by the motion of a wave, one must
a) Find a suitable representation of a single particle, and
b) Establish the kinematical equivalence of a ray and a particle trajectory.

To satisfy condition a), *i.e.*, one has to find a suitable representation of a single particle. In Section 2.3.4, it is explained that the superposition of a group of plane waves of nearly the same wave length $\lambda_0 = \frac{2\pi}{k_0}$ and moving forward with frequency ω_0 with phase velocity $w = \omega_0/k_0$, that interfere destructively everywhere except in a small region results in a *wave packet*. In the one-dimension (1-D) case a wave packet is represented by the compound wave (2.3.10), viz.,

$$U(z,t) = \sum_{n=1\,to\,\infty} \frac{1}{\sqrt{2\pi}} g(\omega) e^{i\,\omega_n t} \Delta\omega \qquad (2.3.16)$$

where the factor $1/\sqrt{2\pi}$ appears for symmetry reasons.
The most general superposition such waves may be represented, in Fourier analysis, by a **Fourier integral**.
Assume the form at $z=0$ and $t=0$ of $\Psi(z,t)$ is

$$\Psi(0,0) = \frac{1}{\sqrt{2\pi}} \int_{(k_0-\Delta k/2)}^{(k_0+\Delta k/2)} \Phi((k_z)\,d(k_z) \qquad (2.3.17)$$

This is a wave PULSE. In evaluating the integral, z and t are treated as fixed parameters, while $\omega = \omega(k)$. At a later time Δt and distance Δz, the form of $\Psi(z,t)$ is

$$\Psi(\Delta z, \Delta t) = \frac{1}{\sqrt{2\pi}} \int_{(k_0-\Delta k/2)}^{(k_0+\Delta k/2)} \Phi((k_z)\{e^{i\,(k_z\,\Delta z - \Delta\omega\,t)}\}\,d(k_z) \qquad (2.3.18)$$

This can be written as

$$\Psi(\Delta z, \Delta t) = \frac{1}{\sqrt{2\pi}} \{e^{i\,(k_0\,\Delta z - \omega_0\,\Delta t)}\} \int_{(k_0-\Delta k/2)}^{(k_0+\Delta k/2)} \Phi((k_z)\{e^{i\,[(k-k_0)\,\Delta z - (\omega-\omega_0)\,\Delta t]}\}\,d(k_z) \qquad (2.3.19)$$

where the phase wave term = $\{e^{i(k_o \Delta z - \omega_o \Delta t)}\}$ (2.3.20)

and the amplitude wave term = $\dfrac{1}{\sqrt{2\pi}} \displaystyle\int_{(k_o - \Delta k/2)}^{(k_o + \Delta k/2)} \Phi((k_z)) \{e^{i[(k - k_o)\Delta z - (\omega - \omega_o)\Delta t]}\} d(k_z)$

(2.3.21)

Note: The behavior in time of such a wave packet depends on how $\omega(k)$ depends on k, i.e., the LAW OF DISPERSION.

In the amplitude term

$[(k - k_o)\Delta z - (\omega - \omega_o)\Delta t] = 0$. (2.3.22)

means $\Delta z / \Delta t = (\omega - \omega_o)/(k - k_o)$ (2.3.23)

i.e., z is allowed to vary holding t fast, so that

$z = t \dfrac{d\omega}{dk}\big|_{k \to k_o}$ (2.3.24)

Expanding $\omega(k)$ in a *Taylor's Series* about k_o, one gets

$\omega(k) = \omega_o + (k - k_o)\dfrac{d\omega}{dk}\big|_{\omega \to \omega_o}$ + Negligible higher order terms

$= \omega_o + (k - k_o)\dfrac{d\omega}{dk}\big|_{\omega \to \omega_o}$ (2.3.25)

Therefore, $\dfrac{\Delta z}{\Delta t} = \dfrac{(\omega - \omega_o)}{(k - k_o)} = \left(\dfrac{d\omega}{dk}\right)_{(\omega = \omega_o)}$ (2.3.26)

This is called the *group velocity* of the centre of the *wave group*.

The phase term is $e^{i(k_o \Delta z - \omega_o \Delta t)}$, which has a constant phase point,

and it gives $\dfrac{\Delta z}{\Delta t} = \dfrac{\omega_o}{k_o}$ (2.3.27)

This is defined as the PHASE or WAVE-FRONT VELOCITY denoted by

$\vec{w} \equiv \vec{v}_\varphi = \left(\dfrac{\omega_o}{k_o}\right)$

$\Psi(z,t) = \dfrac{1}{\sqrt{2\pi}} \int \Phi(k_z) \{e^{i(k_z z - \omega t)}\} dk_z$ (2.3.28)

where the amplitude function

$\Phi(k_z) = \dfrac{1}{\sqrt{2\pi}} \int \Psi(z) \{e^{-i(k_z z - \omega t)}\} dz$ (2.3.29)

$\Phi(k_z)$ is called the FOURIER INTEGRAL TRANSFORM of $\Psi(z,t)$.

$\Psi(z,t)$ is confined to spatial region V_z in *normal space* as shown in Fig 2.5.

If $\dfrac{\partial \Phi(k_z)}{\partial t} = 0$, is satisfied by $\Psi(z,t)$, then

$\int \left\{ \dfrac{\partial \Psi(z)}{\partial t} + i\omega \Psi(z) \cdot [e^{-i(k_z z - \omega t)}] \right\} dz = 0$ (2.3.30)

When $\Phi(k_z)$ is an oscillatory function of k,

with wave length, $\lambda_k = \dfrac{dk}{d(k_z z - \omega t)} = [z - t\left(\dfrac{d\omega}{dk}\right)]^{-1}$ (2.3.31)

Such a wave packet is constructed so that it is concentrated in a volume V_k in k-space. The dimension of $V_k \propto (V_z)^{-1}$. V_k is concentrated in the region around k_o, where k_o is some mean vector. λ_k = small in the region $(k_o - \Delta k/2)$ to $(k_o + \Delta k/2)$, and where $z = t\left(\dfrac{d\omega}{dk}\right)$ at $k < (k_o \pm \Delta k/2)$, $\Psi(z)$ = maximum, at $k = k_o$; $\Psi(z) = 0$, if k and k_o are very different.

Condition 2, viz., the KINEMATICAL EQUIVALENCE OF RAY AND PARTICLE TRAJECTORY, can be accomplished if $\vec{u} \equiv \vec{v}$.

In other words, if the wave group is to represent a particle, we require that $\vec{u} = v$.
Now $\quad \vec{u} = \vec{v} = p/m$

Since $E = \dfrac{p^2}{2m} + V$, and $p = \hbar k$,

∴ $\quad dE = 2p\,dp / 2m + 0$

$dE = p\,dp/m + 0 = v\,dp$

∴ $\quad v = \dfrac{dE}{dp} = \dfrac{1}{\hbar}\dfrac{dE}{dk} = \dfrac{d(E/\hbar)}{dk} = \dfrac{d\omega}{dk}$

The group velocity of the wave is given by

$$\vec{u} = \dfrac{d\omega}{dk}$$

Since it is required that $\vec{u} = \vec{v}$,

$$\vec{u} = \dfrac{d\omega}{dk} = \dfrac{d(E/\hbar)}{dk},$$

or, $\quad \omega = E/\hbar$

and $\quad E = \hbar\omega$

This establishes the fact that it is reasonable to consider describing the motion of a particle by use of a LOCALIZED wave, WAVE PACKET, if one requires that $E = \hbar\omega$. This is just Planck's quantization of energy condition!

2.3.5.1 GENERALIZING Expressions (2.3.28) and (2.3.29) in Three-Dimensions (3-D):

Displacement vector $\quad \vec{r} \equiv r\,(x, y, z)$,

Wave vector $\quad \vec{k} \equiv k\,(k_x, k_y, k_z)$,

Volume element $\quad d\tau = dx\,dy\,dz$, and $\hfill (2.3.32)$

$$d^3k \equiv d\vec{k} = dk_x dk_x dk_x \hfill (2.3.33)$$

The <u>Fourier Transform Pair</u>, in multiple dimensions, are:

$$\Psi(z,t) \equiv \Psi(\vec{r},t) = \dfrac{1}{\sqrt{2\pi}} \int \Phi(\vec{k})\,\{e^{i\,(\vec{k}\cdot\vec{r} - \omega t)}\}\,d\vec{k} \hfill (2.3.34)$$

$$\Phi(\vec{k},t) = \dfrac{1}{\sqrt{2\pi}} \int \Psi(\vec{r}')\,\{e^{-i\,(\vec{k}\cdot\vec{r} - \omega t)}\}\,d\tau' \hfill (2.335)$$

Note: It is the Real Part of $\Psi(\vec{r},t)$ is the linear superposition of waves giving the *wave packet*, as in Fig 2.7.

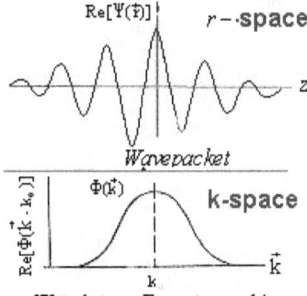

Fig 2.7 Variation of the function with respect to k.

2.3.5.2 PARTICLE SPECTRUM (Particles as function of mass):

In Quantum Mechanics matter consist of various microscopic entities and their mass spectrum is given in Table 2.1 and the classical bodies are also listed for information.

Table 2.1   Particle Spectrum.

Mass (kg)	Matter	Energy	wave	particle	EM spectrum (Hz)	
10^{30}	stars	highest	yes		10^{20}	γ-rays
10^{20}	Planets		yes			
10^{10}		higher	yes			
100	Automobiles	"	yes		10^{18}	X-rays
	Golf balls		yes			
					10^{14}	Visible
10^{-10}	Dust grains	low		yes		
10^{-20}					10^{12}	
	Molecules, Atoms	lower	yes	↑ ↓	10^{10}	Microwave
10^{-30}	Electrons	lower		yes	10^{8}	Radio waves
10^{-40}	Neutrinos	lowest		yes	10^{6}	

2.3.5.3 Conclusions

1) The wave velocity, $\bar{w}$, of the matter waves has no physical significance as $\bar{w}$ is the phase velocity of the individual superposing waves of the wave packet such that $\bar{w} = c^2/v$ meaning $\bar{w} > cc$

2) The group velocity of the center of the group – the velocity of the wave packet – is the same as that of the particle, viz. v. $\bar{u} = v$. This means the packet can guide the motion of the particle and also localize it. Thus the deBoglie wave group associated with a moving body travels with the same velocity as the particle That is it is possible to have a wave motion which has the particle characteristic of being in a small region of space and which will move with the particle velocity, v.

3) A de Broglie wave cannot be represented simply by a *simple infinite plane harmonic wave* equation, viz.,

$$\Psi(z,t) = \Psi_o Sin(kz - \omega t),$$
$$\Psi(z,t) = \sum \psi_i \{e^{i(k_i \cdot z - \omega_i t)}\},$$
$$k_i = (k_o + \Delta k/2)$$

within the limits $k_i = (k_o \pm \Delta k/2)$. Here the range Δk needed for the summation depends on the degree of localization of the particle in space. In general, a matter wave is represented a superposition of continuum of (phase) plane waves, viz.

$$\Psi(z,t) = \int \Phi(\vec{k}_z) \{e^{i(\vec{k}_z z - \omega t)}\} dk_i;$$

where $\Phi(\vec{k}_z)$ serves as a *weighing factor* for the wave function it multiplies to give $\Psi(z,t)$, i.e., it tells how much of that particular wave function contributes to the packet at z and time t. A plot of $\Phi(\vec{k}_z)$ *versus k* is known as the **Spectrum of states**.

$$\Psi(z,t) = \frac{1}{\sqrt{2\pi}} \int_{(k_0 - \Delta k/2)}^{(k_0 + \Delta k/2)} \Phi((k_z)) \{e^{i[(kz - \omega t)]}\} d(k_z)$$

is a more general form of the expression for a wave packet. Here the amplitude function, $\Phi((k_z))$ represents all the component phase (i.e., a train) of waves forming the integral the same phase at the same space – time, z_o and t_o, with the approximate wavelength $\lambda_k = 2\pi/k_0$, was moving forward individually with a phase velocity $\bar{w} = \omega_0/k_0$, but having a variable amplitude such that the whole disturbance is confined to a short interval of the z-axis, whose centre z_o moves forward with the group velocity, $u = v$, particle velocity in the classical limit.

Thus MATTER WAVES CANNOT BE REPRESENTED BY A SINGLE REAL FUNCTION, LIKE EM WAVES, THEY BY NATURE HAVE TWO INEXTRICABLY RELATED PARTS.

Worked out Example 2.3

The wave function\ $\Psi(z,t) = A\{e^{i[(1.58 \times 10^{-12} m^{-1}) \cdot z - (7.91 \times 10^{14} s^{-1}) t]}}$ represents a certain free particle. Find the momentum, energy, velocity and mass of the particle. Given, $\hbar = (h/2\pi) = 1.054 \times 10^{-34} J-s$,

Solution:

The fundamental wave-particle relationships are, $E = \hbar\omega$, and $p = \hbar k$.

Momentum, $p = \hbar k = (1.58 \times 10^{-12} m^{-1}) \cdot (1.054 \times 10^{-34} J-s) = 1.67 \times 10^{-22} kgms^{-1}$.;

Energy, $E = \hbar\omega = (1.054 \times 10^{-34} J-s) \cdot (7.91 \times 10^{14} s^{-1}) = 8.35 \times 10^{-18} J$.

$E = p^2/2m + (V = 0)$.

Mass, $m = \frac{p^2}{2E} = \frac{(1.67 \times 10^{-22} kgms^{-1})^2}{2(8.35 \times 10^{-18} J)} = 1.67 \times 10^{-27} kg$

Phase velocity, $v = \frac{\omega}{k} = \frac{(7.91 \times 10^{14} s^{-1})}{(1.58 \times 10^{12} m^{-1})} = 5 \times 10^4 ms^{-1}$.

2.4 THE BEHAVIOUR IN TIME OF A WAVE PACKET AND THE LAW OF DISPERSION

2.4.1 Normal Dispersion - Relation between u, v, w:

If we consider a group of waves of slightly different wavelengths, λ, traveling in a dispersive medium, i.e. one in which the velocity $\bar{w}$ of individual component of wave is a function λ, i.e., $\bar{w} = \bar{w}(\lambda)$. This is because of the change of relative phase of the components in time the wave packet is no longer propagated without a change in shape. By definition w = (ω/k). All the phase waves have different velocities, w, and the question of u is more complicated. If c = velocity of light in free space (vacuum), and n is the refractive index of the medium, $\omega = k\bar{w} =$ a *non-linear* function of k.

i.e., $\omega \equiv \omega(k) = ck/n$.

Group velocity by definition $\bar{u} = \frac{d\omega}{dk}$

$$\bar{u} = \frac{d(c/n)}{dk} = \frac{c}{n} \cdot 1 + ck \, (-\frac{1}{n^2}) \frac{dn}{dk} = w - (\frac{ck}{n^2}) \frac{dn}{dk} \qquad (2.4.1)$$

$n = c/w, \quad dn = -(\frac{c}{w^2}) dw$

if λ_0 = wave length denoting the interior or average one of the group,

$$dk = \frac{-2\pi}{\lambda_0^2} d\lambda = \frac{k_0}{\lambda_0} d\lambda = \frac{dn}{dk}\left[\frac{-c}{w^2} dw\right] / \left(\frac{k_0}{\lambda_0} d\lambda\right) = \frac{n^2}{c} \cdot \frac{\lambda_0}{k_0} \cdot \frac{dw}{d\lambda},$$

$$u = w(\lambda) - \lambda_0 \cdot \frac{dw}{d\lambda} = \text{(resolving power)}. \ \Delta w. \qquad (2.4.2)$$

Further, $\qquad dn = \frac{n^2}{c} dw$

$$u = w - \frac{\lambda_0}{d\lambda} dw = w - \frac{\lambda_0}{d\lambda}\left(\frac{c}{n^2}\right) dn = w + \left(\frac{c}{n^2}\right) \lambda_0 \frac{dn}{d\lambda} \qquad (2.4.3)$$

$$u = w + \left(\frac{c}{n^2}\right) \cdot \Delta n \ \text{(Resolving Power)} \qquad (2.4.4)$$

Worked out Example 2.4

Show that $u^{-1} = w^{-1} - c^{-1} \frac{dn}{d\lambda}$ $u = w + k \, (dw/dk)$.

Solution: $u = w - \lambda_0 \frac{dw}{d\lambda}$; $\omega = kw$; $u = \frac{d\omega}{dk} = w \cdot 1 + k \frac{dw}{dk}$;

$k = \frac{2\pi}{\lambda}$; $dk = \left(-\frac{2\pi}{\lambda^2}\right) d\lambda$; $\frac{k}{dk} = \left(-\frac{\lambda}{d\lambda}\right)$;

$$\bar{u} = \bar{w} + k \frac{d\bar{w}}{dk} = \bar{w} - \lambda \frac{d\bar{w}}{d\lambda} \qquad (2.4.5)$$

There are thus several equivalent expressions for the relation between $\bar{u}$ and $\bar{w}$. It is worthy to note that as the interval Δk is increased the speed in w of the harmonic components of the wave packet, in a dispersive medium, becomes more marked; the wave packet is deformed rapidly, and $\bar{u}$ as the velocity of the whole loses its physical significance. For normal dispersion, $\bar{u} < \bar{w} < c$.

2.4.2   Anomalous Dispersion:

For anomalous dispersion, which occurs for EM waves in conducting media and of extreme UV radiation in glass, the relation is $\bar{u} > \bar{w} > c$.

2.5 THE DEGREE OF LOCALIZATION OF A PARTICLE REPRESENTED by a wave packet:

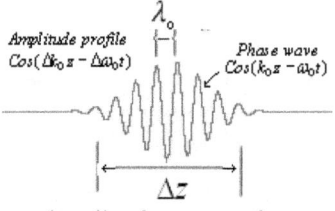

a localized wave packet

Fig 2.8 A localized wavepacket

It was seen that a particle in motion is has associated wave packet. It is still to be investigated as to how much the wave packet is extended over a region of space as it should not conflict with the small particle located at a specific position.

2.5.1 TIME-FREQUENCY domain:

Consider the superposition of n phase waves (phasors) in 1-D.,

$$u_1(z,t) = A_o \iota e^{i(k_1 \cdot z - \omega_1 t)} \quad (2.5.1)$$

$$u_2(z,t) = A_o \iota e^{i(k_2 \cdot z - \omega_2 t)}$$

$$u_3(z,t) = A_o \iota e^{i(k_3 \cdot z - \omega_3 t)} \quad (2.5.2)$$

where the angular frequencies phasors

$$\omega_2 - \omega_1 = \omega_3 - \omega_2 = \cdots = \omega_n - \omega_{n-1} = \partial \omega \quad (2.5.3)$$

$$\partial \omega = \frac{\omega_n - \omega_1}{n-1}. \quad (2.5.4)$$

If all the phasors are all parallel at time t = 0, they will be spread uniformly at time t. The angle between the adjacent phasors will be $\partial \theta$.

$$\partial \theta = t \partial \omega = \frac{\omega_n - \omega_1}{n-1} t. \quad (2.5.5)$$

At $t = 0$, amplitude will be maximum. Let $\partial \omega$ = very small, so that the amplitude does not change very much during on revolution. As the phasors fan out,

$$\partial \theta_{t=0} = (\partial \omega)(t=0) = \frac{\omega_n - \omega_1}{n-1}(t=0) = 0. \quad (2.5.6)$$

From $t = 0$, the magnitude of the resultant decreases to zero. The first zero occurs when the fan covers the entire angle 2π, and the angle between two adjacent phasors is $(2\pi / n)$. If the time for the first zero is specified as t_1, at $t = t_1$: amplitude = 0

$$\partial \theta_{t=1} = (\partial \omega)(t=t_1) = \frac{\omega_n - \omega_1}{n-1}(t_1 = \frac{2\pi}{n}). \quad (2.5.7)$$

$$\therefore \quad t_1 = \frac{2\pi}{n(\omega_n - \omega_1)}(n-1). \quad (2.5.8)$$

Similarly the second maximum will occur at t_2. At $t = t_2$;

$$\partial \theta_{t=2} = (\partial \omega)(t=t_2) = \frac{\omega_n - \omega_1}{n-1} t_2 = 2\pi.$$

$$t_2 = \frac{2\pi}{(\omega_n - \omega_1)}(n-1). \quad (2.5.9)$$

The envelope will repeat thereafter the shape it took between $t = 0$ and $t = t_1$, so that the resultant sum of the phasors as a function of time will be as in Fig 2.5.

2.5.2. CO-ORDINATE-MOMENTUM DOMAIN

Wave groups are regions of large amplitudes. Let z is varied at fixed time. Following analysis similar to the one adopted for the time-frequency domain, one can arrive at the results below: It was seen earlier that there were too many wave groups to reasonably represent a particle. There should be only one group to represent a particle. This achieved by the condition $t_2 - 0 = t_2$ to be infinite. This can be done if

$t_2 = \frac{2\pi}{(\omega_n - \omega_1)}(n-1) =$ infinite, i.e., $n \to \infty$. This does not mean $\omega_n - \omega_1$ must be infinite.

The SIZE of the group in time $2t_1$, is called the TIME DURATION, Δt, of the group.

$\Delta t = 2t_1$, for $n \to \infty$,

$$t_1(n \to \infty) = \frac{2\pi}{(\omega_n - \omega_1)}. \quad (2.5.10)$$

If the FREQUENCY SPREAD is $\Delta \omega = \omega_n - \omega$
The TEMPORAL SIZE of the group

$$\Delta t = 2t_1 = 4\pi / \Delta \omega \quad (2.5.11)$$

In the same way the SPATIAL EXTENT (or size), Δz, of the group is

$$\Delta z \approx \lambda = 4\pi / \Delta k \quad (2.5.12)$$

The two conditions of temporal and spatial size estimated cannot be fulfilled: be cause of limited values of Δt and Δk. So one could construct a wave group having particle-like characteristic, extending over a very limited region of space. Each of the phasors has a phase velocity $\bar{w}$, and the wave group has $\bar{u} = v$. If u and v differed, i.e., $\bar{u} \neq v$, the particle would soon be in a region where the amplitude of the wave becomes negligible and the wave would not give a useful representation of the position of the particle.

2.6 HEISENBERG'S UNCERTAINTY PRINCIPLE:

It is possible to imagine configurations of waves that are much *localized* (Fig 2.8). It was seen in this Chapter that wave groups can be achieved by superposing phase waves of different frequencies in a spatial wave, so that they interfere with one another completely outside of a given spatial region. This involves Fourier integration. On the other hand, a moving body is a matter wave rather than a <u>localized entity</u> suggests that there is a fundamental limit to the accuracy with which a measurement on the particle properties can be made. The idea of localizability of waves is extremely important for this.
As discussed in Section 2.3.3, consider two phase waves, equations (2.3.3) and (2.3.4), propagating along the z-axis is described by

$$u_1(z,t) = A_1 [e^{-i(k_1 z - \omega_1 t)}], \quad (2.6.1)$$

$$u_2(z,t) = A_2 [e^{-i(k_2 z - \omega_2 t)}] \quad (2.6.2)$$

where the various quantities have been defined as before.
By superposing these *two component waves* one gets the **compound wave,** equation (2.3.7) as

$$U(z,t) = u_1(z,t) + u_2(z,t) = [2A_0 Cos\theta] \left\{ e^{-i(k_0 z - \omega_0 t)} \right\} \quad (2.6.3)$$

The compound wave, shown in Fig. 2.5, is written as equation (2.3.10)

$$U(z,t) = 2A_0 Cos(\Delta k\ z - \Delta \omega\ t)\ Cos(k_0 z - \omega_0 t) \quad (2.6.4)$$

Let the two phase waves differ by $\Delta \lambda = 10\%$ in λ.
Five cycles from the point of maximum reinforcement, the phasors $u_1(z,t)$ and $u_2(z,t)$ interfere destructively and five cycles further on, they again reinforce. This alternation of constructive and destructive interference produces partial localization of the wave, $U(z,t)$, a bunching together of the wave into regions each about 10 wavelengths in extent). In general from what was discussed in Section 2.5, the inherent <u>Uncertainty</u>, Δz in the position of the group is of the same order of the magnitude λ in space, *i.e.* modulation wavelength $\lambda \approx \Delta z = 4\pi / \Delta k$. Because the phasors that constitute the group are a combination of waves

specified by ($k_o + \Delta k$) and ($k_o - \Delta k$), the best measurements of k_o will have still an inherent uncertainty of dominant k_o of about $2\Delta k$.

But $\quad \Delta z = 4\pi / \Delta k$, $\hspace{5cm}$ (2.6.5)

Or $\quad \Delta z \cdot \Delta k = 4\pi$. $\hspace{5cm}$ (2.6.6)

This is called the **Reciprocity Relation** between the 'sizes' or 'widths' of the wave group in z- and k-spaces. *i.e.*, Δk is related to *the uncertainty in the position* z of a wave group. This relation is for the spatial part of the wave packet. Here the average value

$\quad k_o = (k_1 + k_2)/2$, $\hspace{5cm}$ (2.6.7)

and it is related to an average wavelength λ_o by

$\quad \lambda_o = 2\pi / \Delta k_o = 2\pi /(h/p_o) = p_o / h$. $\hspace{3cm}$ (2.6.8)

Therefore an *uncertainty* Δk in k_o, of the phase waves associated with the particle, results in an *uncertainty*, Δp, in the particle's momentum, p_o, such that

$\quad \Delta p = 2h / \Delta z$, $\hspace{5cm}$ (2.6.9)

i.e., $\quad \Delta z \cdot \Delta p = 2h$. $\hspace{5cm}$ (2.6.10)

This is the *lowest limit of accuracy*, and so it varies with the choice of the wave function $U(z,t)$. A more general relation may be written as

$\quad \Delta z \cdot \Delta p_z \geq \hbar / 2$. $\hspace{5cm}$ (2.6.11)

The sign ≥ is used because Δz and Δp of the above equation are *irreducible minima or spreads* that are consequence of the wave nature of moving particles. The inability to observe the wave and particle aspects of a moving body at the same time illustrates the Principle of Cmplementariy, enunciated first by Niels Bohr. Thus light waves, sound waves, gravitational waves, and deBroglie waves manifest themselves as photons, phonons, gravitons, and massive matter, respectively. The two aspects are never contradictory, they always appear separately. The *particle aspects* appear under situations where the criteria of geometrical optics obtain at the reduced apertures *diffraction effects* appear, and the *wave aspects* take over.

The equation (2.6.11), viz., $\Delta z \cdot \Delta p_z \geq \hbar / 2$, is one of the <u>co-ordinate-momentum Uncertainty Principle</u> obtained first by Werner Heisenberg in 1927. The Uncertainty Principle may be stated as follows:

"*It is impossible to measure simultaneously and precisely both position and momentum of a particle*".

A more realistic method of approach than given above (to be dealt with later) yields the more widely used Uncertainty Relation:

$\quad \Delta z \cdot \Delta p_z \geq \hbar / 2$

$\quad \Delta x \cdot \Delta p_x \geq \hbar / 2$ $\hspace{5cm}$ (2.6.12)

$\quad \Delta y \cdot \Delta p_y \geq \hbar / 2$

for 3–D *spatial part* of a wave packet.

Thus in quantum physics, *position and momentum*, just like particle behavior and wave aspects of a system, are *complimentary properties* of the system, and the theory does not admit the possibility of an experiment in which both could be established simultaneously. The smallness of $\hbar$ guaranties that only for a microscopic system the usual notions of classical physics fail. The uncertainty principle can also be formulated in terms of other *conjugate variables*. If the energy involved in a measurement is in the form of EM waves, the limited time available restricts the accuracy with which the frequency of the waves can be determined. ν, the frequency of the waves = Number of waves counted/ time interval,

$$\Delta \nu = 1/\Delta t .\qquad (2.6.13)$$

The corresponding energy change,

$$\hbar \Delta E = h \Delta \nu = h/\Delta t ,$$

or $\quad \Delta E \cdot \Delta t \geq h .$

A more *realistic* calculation changes this relation to

$$\Delta E \cdot \Delta t \geq \hbar/2 \qquad (2.6.14)$$

in the *temporal part* of a wave packet.
This is called the *Time-Energy Uncertainty Relation*.

Worked out Example 2.5

Calculate the minimum energy of an electron when it is trapped in a crystal lattice vacancy. Assume that the electron can move in a spherical volume of radius, a_o. Given,

$\hbar = 1.054 \times 10^{-34} J-s$, $m_e = 9.11 \times 10^{-31}$ kg , $1 eV = 1.602 \times 10^{-19} J$

Solution:

$\Delta E \cdot \Delta t \geq \hbar/2$,and $\Delta z \cdot \Delta p_z \geq \hbar/2$; $p_{min} = \Delta p_z \geq \hbar/\Delta z = tp_{min} = \hbar/a_o$,

$E_{min} = p^2/2m = \hbar^2/m\, a_o^2 = (1.054 \times 10^{-34} J-s)^2 /8 \,(9.11 \times 10^{-31}$ kg$)(4 \times 10^{-10} m)^2$

$= (0.097 \times 10^{-19} J)/(1.602 \times 10^{-19} J/eV) = 0.061 eV$ ..

Worked out Example 2.6.

Suppose that the momentum of an electron can be measured to an accuracy of one part in thousand. Determine the minimum uncertainty in the position of the electron if it is moving with a speed of 1.8×10^8 m/s. Given, $\hbar = 1.054 \times 10^{-34} J-s$, $m_e = 9.11 \times 10^{-31}$ kg .

Solution:

Relativistically, mass of an electron is $m = m_o / \sqrt{1-v^2/c^2}$,

$\Delta z \geq \hbar/2\Delta p_z = \hbar/(10^{-3} mv) = [\hbar\sqrt{1-v^2/c^2}]/(10^{-3} v\, m_o)$

$= (1.054 \times 10^{-34} J-s)^2 / \,[(9.11 \times 10^{-31}$ kg$)\cdot(1.8 \times 10^8\, ms^{-1})\cdot(10^{-3})]$

$\Delta z = 5.1 \times 10^{-10} m = 0.51\, nm.$

2.6.1 GEDANKEN EXPERIMENTS (Proof that wave-particle duality acts to prohibit a violation of the Uncertainty Principle).

A number of *Gedanken* (or 'thought') experiments can show in detail how the wave-particle duality acts to conspire to prohibit a violation of the Principle of Indeterminacy.

2.6.2 Heisenberg Gamma Ray Microscope

As suggested by N. Bohr, consider the experimental set up (Fig. 2.9) whose purpose is to measure the position of an electron. The electrons are in beam having well-defined momentum p_z and moving in the negative Z-direction. When light of wavelength λ is scattered from the electron during the observation process through the microscope(along the y-axis) the momentum of the electron, which is being measured will be affected because the incident light itself carries momentum, $p_y = \hbar/\lambda$. Consider the experiment with a single photon which is shone along the +Z-axis. Due to Compton Effect a photon scattered by electron (when light reflected from a particle passes) recoils through the objective lens of the microscope, a diffraction pattern is produced at the location of the eye (or photographic film). Thus a fuzzy pattern, rather than a precise point will be observed with normally intense light consisting of many photons.

From Physical Optics, diffraction theory of light shows that the diameter, D, of the central disc of the diffraction patterns given approximately by

$$D = \lambda / \sin\alpha \qquad . \tag{2.7.1}$$

where λ = wavelength of the photon,
2α = angle subtended at the particle by the objective lens (Fig 2.9).

$$\sin\alpha = d/f , \tag{2.7.2}$$

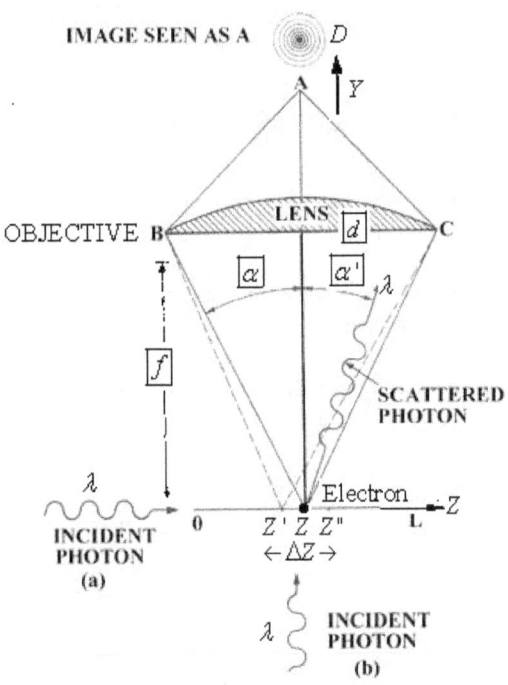

Fig. 2.9 Heisenberg microscope

where d = diameter of the aperture of the objective. Since the single photon would have arrived at somewhere in the central disc of the diffraction pattern and f is the focal length of the lens and represents the distance of object and the lens.

Hence the *uncertainty in the position* of the electron can be taken as

$$\Delta z = \lambda / \sin\alpha \qquad . \tag{2.7.3}$$

$\Delta z \to$ small, when i) $\lambda \to$ small, and ii) $\sin\alpha \to$ large.

So the uncertainty in the position of the electron can be made as small as desired by using a sufficiently small λ.

In the scattering process some of the momentum of the photon will be transferred to the electron. If the momentum of the scattered photon were known exactly, it would be a relatively easy matter to work backwards to determine how the original momentum of the electron was affected. However, it is only known to the observer is that the scattered photon entered the objective lens somewhere, its Z-component of momentum could have a magnitude anywhere between 0 and p_y and $\text{Sin}\alpha$, where $p_y = h/\lambda$, is the momentum of the photon. Hence the Z- component of momentum of the electron will be given by (Fig 2.9).

$$\Delta p_y = p_y \, \text{Sin}\alpha = (h/\lambda) \cdot (d/2y)$$

Δp_y can be made as small as desired by making λ = sufficiently large, but then Δz becomes correspondingly larger.

So, $\quad \Delta z \cdot \Delta p_y = (\lambda/\text{Sin}\alpha) \cdot (h/\lambda) \cdot (\text{Sin}\alpha)$

$$= h \; \Delta z \cdot \Delta p_y = h, \qquad (2.7.4)$$

which result is in agreement with the Uncertainty Principle.

Thus it is the measurement process that introduces the *indeterminacy*, $\Delta z \cdot \Delta p_y$. On the contrary, <u>the indeterminacy is inherent in the nature of a moving body</u>.

2.7.2 The SINGLE-SLIT Diffraction Electron Wave Experiment:

Suppose that the x-co-ordinate of an electron is to be determined by observing whether or not it passes through a slit, of width b, the electron traveling along the Z-axis (Fig. 2.10). In this way, the electrons are localized in the X-direction to within a distance b, i.e., one cannot say exactly where the electron crosses the slit. Thus the position of the electron is uncertain by the width of the slit, and one may write $\Delta x = b$. When a monochromatic wave of wavelength λ passes through a slit of width b, a diffraction pattern will be produced in the screen. The location of the first point of zero intensity is found from the theory of diffraction to be

$$\text{Sin}\beta = d = \lambda/b.$$

Because of the associated de Broglie wave / wave packet, whose, $\lambda = h/p$, the electron will be diffracted as it passes through the slit (*i.e.* the ends of the slit disturbs the electron field and results in a corresponding change in the motion of the electron), giving a diffraction pattern on the screen, instead of a spot. This means the electron will acquire some unknown momentum in the X-direction. The chances are that the electron will most probably strike at the screen somewhere within the central region of the diffraction pattern. Thus the magnitude of the momentum of the electron fallsbetween $p \, \text{Sin}\beta$ and 0, where p is the initial momentum,

i.e., $\quad \Delta p_x = p \, \text{Sin}\beta = p(\lambda/b) = (h/\lambda)(\lambda/b) = h/b. \qquad (2.7.5)$

Δp_x can be chosen as small as desired by making λ = sufficiently larger. Taking the product of the two uncertainties,

$$\Delta z \Delta p_x = b(h/b) = h. \qquad (2.7.6)$$

This relation is in agreement with the Uncertainty Principle.

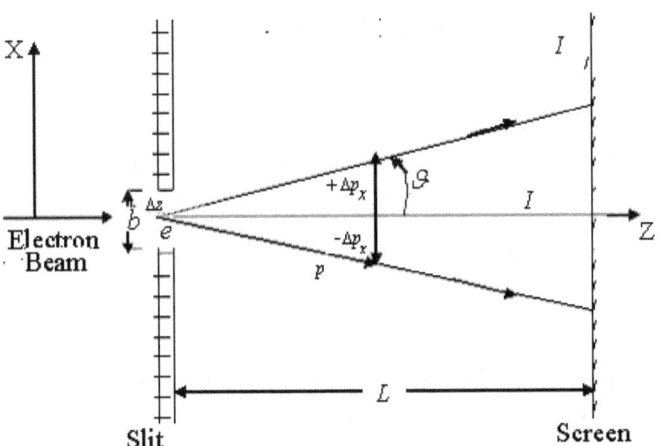

Fig. 2.10   Single-slit diffraction of an electron matter wave.

In analogy to the n-D *Configuration Space* in the Lagrange formulation one can define a 2n-D *Cartesian space* formed by *generalized co-ordinates* of position q_i and momentum p_i. To describe the dynamical state of a particle in motion in a geometrical way, by its phase point, whose position and momentum at each instant, a representative space called *Phase Space* is used. For 1-D motion of a body the phase space has 2-D, with the abscissa corresponding to the Z- and the ordinate to the p_x. In Classical Mechanics the state of a body is represented in phase space by point (z. p_x). The particle moves the representative point and describes a curve in phase space. In Quantum Mechanics the picture is quite different. By dividing the phase space into cells, of sides Δz and Δp_x so that $\Delta z \cdot \Delta p_x \geq h/2$, as time goes on, the path of the representative point falls within a '*ribbon-like path*'.

REVIEW QUESTIONS

R.Q. 2.1　　Where in the EM spectrum does the wavelength of a 10 eV electron appear? Given: $h = 6.63 \times 10^{-34} J-s$, $1 eV = 1.6021 \times 10^{-19} J$, $c = 2.997 \times 10^8 ms^{-1}$, $m_e = 0.5110 \ MeV$.

R.Q. 2.2　　What is the ratio of the Compton wavelength and de Broglie wavelength of a particle moving with relativistic speed v? (Answer: Solution: $\lambda_C / \lambda = \sqrt{(E/E_0)^2 - 1}$).

R.Q. 2.3　　What is the deBroglie wavelength of a student of mass 70 kg moving with velocity of 2 m/s? [Answer: $\lambda = 4.7 \times 10^{-16} \ m$].

R.Q. 2.4　　De Broglie's relationship gives the wavelength of a hydrogen atom moving with a velocity of 10^3 m/s as [Answer: $3.96 \times 10^{-10} \ m$].

R.Q. 2.5　　Find the equivalent energy a photon has if its momentum is that of a 3 MeV electron. [Answer: 3 MeV].

R.Q. 2.6　　What is the energy of a proton of wavelength 0.5 fm? Given, $h = 6.63 \times 10^{-34} J-s$, $1 eV = 1.6021 \times 10^{-19} J$, $c = 2.997 \times 10^8 ms^{-1}$ $M_p = 938.2 \ MeV$.(Answer: $E = E_0 + T$; $E = 2650 \ MeV$; $T = 1710 \ MeV$).

R.Q. 2.7　　Calculate the deBroglie wavelength associated with a) a Golf ball of mass 50 g moving with a speed of 20 m/s., b) a free proton moving with a speed of 2200 m/s, and c) a free electron moving with kinetic energy of 10 keV. (Answer: a) $\lambda(ball) = 6 \times 10^{-23}$ nm; b) $\lambda(proton) = 0.19 \ nm$, c) $\lambda(electron) = 0.39 \ nm$).

R.Q. 2.8　　Find the relation between particle velocity v and wave velocity $\bar{w}$ in the case of matter waves. (Answer: $\bar{w} = v/2$, as V = 0, for a free particle $\bar{w} = c^2/v \neq v$).

R.Q. 2.9　　Using the energy-momentum expression $E = c\sqrt{m_e^2 c^2 + p^2}$ show that a) the phase velocity of the electron of deBroglie wave length, $\lambda = h/p$, is $\bar{w} > c$, and b) the group velocity of the electron corresponding to $\lambda = h/p$, is $\bar{u} = v$. and c) $\bar{u} = \bar{w}$.

R.Q. 2.10　　Show that $\frac{1}{u} = \left(\frac{\partial}{\partial v} \frac{1}{\lambda}\right)_{\lambda = \lambda_0}$.

R.Q. 2.11　　Show that $u^{-1} = w^{-1} - c^{-1} \frac{dn}{d\lambda}$

R.Q. 2.12　　Find relations between $\bar{u}, \bar{w}, \&\ c$ for Standing waves (Answers: $\bar{u} = 0, w < c$).

R.Q. 2.13　　Derive the de Broglie wavelength by using $\bar{u} = \bar{w} - \lambda(d\bar{w}/d\lambda)$.

R.Q. 2.14 Find out the wave and group velocities of an electron of wavelength 0.015 nm. Given, $h = 6.63 \times 10^{-34} J-s$, $m_e = 9.11 \times 10^{-31}$ kg $c = 2.997 \times 10^8 ms^{-1}$ (Answers: $v = 4.8 \times 10^7 ms^{-1}$, $u = v = 4.8 \times 10^7 ms^{-1}$;.in the relativistic range. $w = c^2/v = 1.852 \times 10^9 ms^{-1}$).

R.Q. 2.15 Find the de Broglie wavelength of a particle of mass m moving with RMS speed of a Maxwellian distribution at temperatures T. (Answer: $\lambda = h/\sqrt{3m k_B T}$).

R.Q. 2.16 Show that if $\Psi(z) = [\pi^{1/4} \sigma_z^{-1/2}]\{e^{-z^2/2\sigma_z^2}\}$ is a Gaussian distribution function, then its Fourier inverse transform is also Gaussian. $\sigma_z = $ r m s deviation (standard deviation) from $<z> = 0$.

R.Q. 2.17 Show from de Broglie hypothesis and the expression that $w = c^2/v$, that i) the phase velocity w in free space
of the de Broglie waves associated with a moving particle of rest mass m_o is given by the relation $w = c\sqrt{1 + (m_e c \lambda/h)^2}$. Ii) Which λ-s have the greater w, the long ones or shorter? Iii) Will each exceed c? Does the equation indicate that there is dispersion of de Broglie waves in free space?

R.Q. 2.18 In emitting a photon, an atom radiates for approximately 10^{-9} sec. What is the uncertainty in energy of the photon? Given, $c = 2.997 \times 10^8 ms^{-1}$ (Answer: $\Delta E = 4 \times 10^{-6} eV$, $\Delta v = 1.6 \times 10^8 Hz$).

R.Q. 2.19 A body of mass 1 g falls through a height of 1 cm. If all the energy acquired in the fall were converted to light of 600 nm wavelength, how many phonons would be emitted? Given: $m_e = 0.5110\ MeV$, $c = 2.997 \times 10^8 ms^{-1}$, $h = 6.63 \times 10^{-34} J-s$, $g = 9.8\ ms^{-2}$.

R.Q. 2.20 What is the uncertainty in the location of a photon of wavelength 300.0 nm if this λ is known to an accuracy of one part per million? Given, $c = 2.997 \times 10^8 ms^{-1}$ (Answer: $\Delta p = 4.13 \times 10^{-6} eV/c$, $\Delta z = 47.8 nm$).

R.Q. 2.21 What is the minimum uncertainty in the energy state of an atom if the electron remains in this state for 10^{-8} s.? Given, $c = 2.997 \times 10^8 ms^{-1}$ (Answer: : $\Delta E = 0.3 \times 10^{-7} eV$).

R.Q. 2.22 The width of a spectral line of wavelength 400.0 nm is measured as to 10^{-3} nm. What is the average time that the atomic system remains in the corresponding energy state? Given, $c = 2.997 \times 10^8 ms^{-1}$ (Answer: $\Delta t = 8\ ns$).

R.Q. 2.23 Examine if electrons are present within atomic nuclei. Given the radius of atom in nucleus is 1 f m. Given, $c = 2.997 \times 10^8 ms^{-1}$, $h = 6.63 \times 10^{-34} J-s$, $m_e = 0.5110\ MeV$ (Answer: $\Delta p_x \geq 0.556 \times 10^{-10} kg-ms^{-1}$,.$T == 10.2\ MeV$., i.e., electrons cannot be present within atomic nuclei).

R.Q. 2.24   It is desired to measure the position and momentum of an electron by observing it through a microscope. Analyze the observation process to show that results consistent with the principle of indeterminacy are obtained.

R.Q. 2.25   The position of a particle is measured by passing it through a slit of width b. Find the corresponding uncertainty introduced in the momentum of the particle. (Answer: h/b).

Chapter 3

QUANTUM MECHANICS - I :
WAVE MECHANICAL APPROACH

Chapter 3

QUANTUM MECHANICS - I : WAVE MECHANICAL APPROACH

"The important thing in science is not so much to obtain new facts as to discover new ways of thinking about them".   ~William Lawrence Bragg

3.1   INTRODUCTION

Quantum Mechanics provides an understanding of all of the phenomena discussed in Chapter 1. It is indispensable for the understanding of atoms, molecules, atomic nuclei, and aggregates of these. In this Chapter the study of Quantum Mechanics will be through Erwin Schrödinger formulation by deducting the Schrödinger Equation and the appropriate interpretation of its solutions. It can in no way be derived from Classical Physics, since it requires new concepts, which are unknown to Classical Physics.

In analogy to the "n-D" *configuration space* in the *Lagrange formulation*, one can define a "2n-D" *Cartesian space* formed by generalized co-ordinates of position q_i and momentum p_i. To describe the dynamical state of a particle in motion in a geometrical way, by its phase point, whose position and momentum at each instant, a representative space called *Phase space* is used. For 1-D motion of a body the phase space has 2-D, with the abscissa corresponding to the z- and the ordinate to the p_x, (as described in Fig 3.1). In Classical Mechanics the state of a body is represented in phase space by point (z, p_x). The particle moves the representative point and describes a curve in phase space. In quantum mechanics the picture is quite different. By dividing the phase space into cells, of sides Δz and Δp_x so that $\Delta z \Delta p_x \geq \hbar/2$, as time goes on the path of the representative point falls within a *ribbon-like path*.

The Old Quantum Theory resulting in Bohr's model of the hydrogen atom and its modifications by A. Sommerfeld could point in certain real successes of the hydrogen spectrum, i.e. the derivation of the Balmer formula and selection rules for the energy states in an atom, explanation of the Periodic Table of Elements, and the Exclusion Principle.

But how could one think of an electron in the atom, as a particle or as a wave? However, now both these concepts regarding an atomic electron must continue. By this one is closer to the underlying essence of quantum principle. After the lapse of the first quarter of the 20th Century, there appeared three distinct and independent developments on a complete new quantum theory were published, which were shown to be equivalent. They were
(1) the first mathematical formulation of the so-called MATRIX MECHANICS by Werner Heisenberg in 1925,
(2) the next – WAVE MECHANICS by Erwin Schrödinger in 1925, and
(3) the QUANTUM MECHANICS by Paul Dirac in 1926.

Both Heisenberg and Pauli had attended a series of lectures by N. Bohr in 1922. After discussions, Heisenberg hated the imaginary electron orbits in the *atom model*.

After meeting Max Born, Heisenberg completely rejected the idea of the little solar system model of atom and treated the atom as virtual oscillator that could produce all the frequencies of the spectrum, which was transformed, with matrix methods, by Born as matrix mechanics, and W. Pauli supported it.

3.2 ELECTRON AND THE WAVE FUNCTION, $\Psi(z)$

The talented Erwin Schrödinger in Zurich did not take well with the new mathematical theory on quantum mechanics from W. Heisenberg and despised it because it is devoid of pictures and is full of mathematical jugglery. He had been thinking of about waves. An atomic electron is essentially confined to a small region of space with dimensions of the order of 1 nm. Its associated '*matter field*' may be expressed in terms of standing waves *localized* in this region. This matter field is designated by the Greek symbol, *psi*, $\Psi(z)$. This is the variable quantity characterizing matter waves (de Broglie waves) and it is currently known as the '**wave function**'. $\Psi(z,t)$ = wave function of a particle with its location z in space at time t. Once again, a meaningful interpretation of the wave function $\Psi(z,t)$ can be obtained by drawing an analogy with the amplitude of the electric field, $E(z, t)$, of the EM radiation, i.e. by replacing the $y(z, t)$ by $E(z, t)$ in the plane wave equation

$$y(z,t) = y_0 Cos(kz - \omega t), \qquad (3.2.1)$$

to get $\quad E(z,t) = E_0 Cos(kz - \omega t). \qquad (3.2.2)$

3.2.1. PROBABILITY INTERPRETATION for EM radiation:

Consider a beam of *monochromatic* radiation,
$$E(z,t) = E_0 e^{i(kz - \omega t)} \qquad (3.2.3)$$
which is incident at right angle to a screen. The intensity, I, of illumination (the energy per unit area per unit time) is
$$I = \varepsilon_0 E^2 c \ J \ m^{-2} s^{-1} \qquad (3.2.4)$$
where ε_0 = permittivity of free space
E = the magnitude of the field (instantaneous on the screen.
Consider N photons, each of energy,
$$\varepsilon = \hbar\omega, \qquad (3.2.5)$$
falls per $m^2 s^{-1}$, on the screen. According to the photon picture,
$$I(\varepsilon) = N\hbar\omega = \varepsilon_0 E^2 c \qquad (3.2.6)$$
from which
$$N = (\varepsilon_0 c / h \omega) E^2. \qquad (3.2.7)$$

So in terms of quantum interpretation, a quantity which is undergoing oscillation, *viz.*, E is not probabilistic, but the E^2 gives the probability of finding a photon at a given place. This is an interesting result. The left hand side of the equation represents the particle model of EM radiation, whereas the right hand side of the equation presents the wave model. Now suppose that the intensity of EM radiation falling on the screen is recorded, and then it is not possible to predict the exact position and time of arrival of each photon on the screen. While there will be only a random distribution of the photons on the screen, the average number of photons arriving per unit time is constant and predictable.

This situation is similar to the one in Statistical Mechanics concerning the *kinetic theory of gases*, where one can only calculate the overall average effects of the system rather than the individual particle behavior. That means the concepts of *probability* are to be used. Since as per equation (3.2.7)

$$N = (\varepsilon_o c / \hbar \omega) E^2 .$$

$E^2 \propto$ Probability of observing the photon, specified by its magnitude E, and as per equation (3.2.3)

$$E(z,t) = E_o e^{i(kz - \omega t)}$$

Worked out Example 3.1

Determine the photon flux associated with a beam of monochromatic light of wavelength 300.0 nm and of intensity 3×10^{-14} Wm^{-2}. Given, $\hbar = 1.054 \times 10^{-34} J-s$, $c = 2.997 \times 10^8 ms^{-1}$.

Solution:

$$\varepsilon = \hbar \omega = 2\pi \hbar c / \lambda = 2\pi \frac{(1.054 \times 10^{-34} J-s) \cdot (2.997 \times 10^8 ms^{-1})}{300.0 \times 10^{-9} m} = (6.63 \times 10^{-19} J) / \gamma ,$$

The number of γ photons, $N = I / \hbar \omega = \frac{(3 \times 10^{-14} W m^{-2})}{[(6.63 \times 10^{-19} J)/\gamma]} = 4 \times 10^{14} \gamma m^{-2}$

3.2.2 PROBABILISTIC INTERPRETATION of $\Psi(z,t)$ by Born:

In an analogous way in Quantum Mechanics one can define a wave amplitude function, or a SPATIAL WAVE FUNCTION, or simply, a **wavefunction**, for an electron, $\Psi(z,t)$, such that its modulus squared, i.e. $|\Psi(z,t)|^2$ is proportional to the probability of finding an electron at position z at time t.
$\Psi(z,t)$ is a COMPLEX wave function written as

$$\boxed{\Psi(z,t) = \frac{1}{\sqrt{2\pi}} \int \Phi(k_z) e^{i(kz - \omega t)} dk_z} \qquad (3.2.9)$$

This complex quantity is NOT directly OBSERVABLE, i.e. NOT MEASURABLE, yet loosely speaking can be visualized as describing an electronic 'cloud'. That is, however, $\Psi(z,t)$ has no direct physical significance; it cannot be interpreted in terms of an experiment. On the other hand, $E(z,t)$ describing an electric field of an EM wave has physical meaning.

Max Born (1926) found the correct answer – one can only make probability statements. For continuous distribution,

$$\Psi(x,y,z,t) \equiv \Psi(\vec{r},t) . \qquad (3.2.10)$$

Is a quantity such that the product $|\Psi(\vec{r},t)|^2 d\tau$ is the probability of observing a (non-relativistic) particle in a volume element $d\tau$ ($= dx\, dy\, dz$), in the configuration space of the quantum system, at time t. If the system is stationary – i.e. if it is independent of time – one may write the *probability of observing a particle as proportional* to

$$\boxed{|\Psi(\vec{r})|^2 d\tau = \Psi(\vec{r})^* \cdot \Psi(\vec{r}) d\tau}, \qquad (3.2.11)$$

where $\Psi(\bar{r})^*$ is the complex conjugate of $\Psi(\bar{r})$. $\Psi(\bar{r})^*$ is obtained by replacing $i\ (\equiv \sqrt{-1})$ by $-i$ in $\Psi(\bar{r})$.

3.2.2.1 Coordinate Probability, $\rho(\bar{r},t)$

In the problem of Example 3.1, the *co-ordinate probability* per unit length = co-ordinate probability density, $\rho(z)$, of finding the particle at z is

$$\rho(z) = \frac{|\Psi(z)|^2 dz}{dz} = |\Psi(z)|^2 \tag{3.2.12}$$

For a wave motion in space, the co-ordinate probability density $\rho(\bar{r},t)$ of observing the particle within a volume element $d\tau$, around $\bar{r}$, at time t

$$\rho(\bar{r},t) = |\Psi(\bar{r},t)|^2 \tag{3.2.13}$$

3.2.2.2 Momentum Probability, $\rho(p,t)$

Similar to $\rho(\bar{r},t)$, there is momentum probability density,

$$\rho(\bar{p},t) = |\Phi(\bar{p},t)|^2 \tag{3.2.14}$$

But there is one fundamental difference between $\Psi(\bar{r})$, describing a wave packet (matter wave), and $E(r)$, describing the electric field of an EM wave: $E(r)$ and $|E(\bar{r})|^2$ both have physical meaning, while $\Psi(\bar{r})$ itself has NO PHYSICAL MEANING, $|\Psi(\bar{r})|^2$ has the meaning of total co-ordinate probability

$$P(\bar{r},t) = \iiint_{entire\ space} |\Psi(\bar{r},t)|^2\, d\tau = \iiint_{entire\ space} \Psi(\bar{r},t)^* \cdot \Psi(\bar{r},t)\, d\tau \tag{3.2.15}$$

where the triple integration is carried out over the entire volume, τ in question. This is the probability of finding the particle 'anywhere' in the space in question. The integral (3.2.15) must be finite in order to represent a real particle. Because of the physical significance of $P(\bar{r},t)$ and $P(p,t)$, both of these quantities are such that

$$P(\bar{r},t) \geq 0,\ \text{and}\ P(p,t) \geq 0.$$

It is convenient, now, to define the co-ordinate probability density for the particle as

$$\rho(\bar{r},t) = \left[\iiint_{entire\ space} |\Psi(\bar{r},t)|^2\, d\tau\right] / \left(\iiint_{entire\ space} d\tau\right) \tag{3.2.16}$$

$$\rho(\bar{r},t) = \left[\iiint_{entire\ space} |\Psi(\bar{r},t)|^2\, d\tau\right] / (Total\ Volume)$$

3.2.2 MATHEMATICAL REQUIREMENTS on $\Psi(\bar{r},t)$

The probabilistic interpretation of $\Psi(r, t)$ by Born implies upon $\Psi(\bar{r},t)$ a certain mathematical criterion to be satisfied. Now an electron in an atom should be somewhere in space, and therefore, the probability of finding it anywhere in the space in question must be real and positive, *i.e.* **a certainty**, or 1. Hence when only one particle exists in a system, $\Psi(\vec{r},t)$ is so scaled that

$$\iiint \rho(\vec{r},t)\, d\tau = 1 \quad \text{(For continuous distribution)} \qquad (3.2.17)$$

i.e., the integral must converge.

Or, $\quad \sum \rho(\vec{r},t) = 1 \quad$ (For discrete distribution) $\qquad (3.2.18)$

Worked out Example 3.2
If a fair dice is thrown, the outcomes are 1, 2, 3, 4, 5, or 6 (The six sides), forming a discrete distribution. Find the mean μ of the results on x.

Solution:
The probability p_i of measuring the value of x to be x_i is:

$p_1 = \frac{1}{6}, \qquad p_2 = p_3 = p_4 = p_5 = p_6 = \frac{1}{6}.$

For discrete distribution,

$$\sum p(\vec{r},t) = 6(\tfrac{1}{6}) = 1,$$

$$P_{ii} = \sum x_i\, p_i\, \vec{r},t) = 1(\tfrac{1}{6}) + 1(\tfrac{1}{6}) + 2(\tfrac{1}{6}) + 3(\tfrac{1}{6}) + 4(\tfrac{1}{6}) + 5(\tfrac{1}{6}) + 6(\tfrac{1}{6}) = (\tfrac{7}{2})$$

This means μ is not any one of the x_i which are actual measurements 1, 2, 3, 4, 5, or 6.

3.2.3 Quadratically Integrablility or Square Integrability or NORMALIZATION of $\Psi(\vec{r},t)$

$$P(\vec{r},t) = K\,|\Psi(\vec{r},t)|^2\, d\tau = K\Psi(\vec{r},t)^* \cdot \Psi(\vec{r},t)\, d\tau$$

where K = a constant or a function of time.

Now, $\quad \iiint \Psi(\vec{r},t)^* \cdot \Psi(\vec{r},t)\, d\tau = \iiint \Psi(\vec{r},t) \cdot \Psi(\vec{r},t)^*\, d\tau = 1 \qquad (3.2.19)$

If the integral converges to a finite value,

$$\iiint \Psi(\vec{r},t) \cdot \Psi(\vec{r},t)^*\, d\tau = 1/K \qquad (3.2.20)$$

then $\Psi(\vec{r},t)$ is said to be **Quadratically (Square) Integrable**. A majority of the mathematical functions do not satisfy this condition, and for them this hypothesis is not applicable since K must vanish. These *non-quadratically integrable functions* never correspond to actual experimental physical situations.

For physically acceptable wave functions

$$\iiint_{\text{Entire space}} |\hat{\Psi}(\vec{r},t)|^2\, d\tau = \iiint_{\text{Entire space}} \hat{\Psi}(\vec{r},t)^* \cdot \hat{\Psi}(\vec{r},t)\, d\tau = 1$$

or, $\quad \boxed{\iiint_{\text{Entire space}} |\hat{\Psi}(\vec{r},t)|^2\, d\tau = 1} \qquad (3.2.21)$

This means the $\Psi(\vec{r},t)$ has the dimensions of $\sqrt[2]{(\text{length})^{-3}}$. Evaluation of integral (3.2.21) enables evaluation of K, the *constant of proportionality*. When the above relation (3.2.21) is true the wave function $\hat{\Psi}(\vec{r},t)$ is said to be NORMALIZED to unity. The expression (3.2.21) is also called as the **Normalization Relation** or <u>Born's Condition</u>. Born's condition imposes severe limitation on the possible forms of wave function $\Psi(\vec{r},t)$, since it is not always possible to satisfy this condition for an arbitrary function, or this normalization relation is obeyed by not all forms of $\Psi(\vec{r},t)$.

If $\quad \iiint \Psi(\vec{r},t) \cdot \Psi(\vec{r},t)^* \, d\tau = g$, constant,

then the new wave function, $\hat{\Psi}(\vec{r},t)$, ('hat' over Ψ) known as the NORMALIZED WAVE FUNCTION.

$$\hat{\Psi}(\vec{r},t) = \Psi(\vec{r},t)/\sqrt{g} . \qquad (3.2.22)$$

$\hat{\Psi}(\vec{r},t)$ represents the same physical situation as $\Psi(\vec{r},t)$, and in addition

$$\iiint_{Entire\ space} \hat{\Psi}(\vec{r},t) \cdot \hat{\Psi}(\vec{r},t)^* \, d\tau = 1. \qquad (3.2.23)$$

If $\sqrt{g}$ = a constant in time means the integral exists in the case of monochromatic wave functions.

Worked out Example 3.3

Given that $\int_{-\infty}^{+\infty} e^{-au^2} \, du = \sqrt{\frac{\pi}{a}}$, find the normalization constant for the normal (Gaussian) wave packet $\Phi(k,t) = A \, e^{-k^2/2\sigma^2}$, where σ = standard deviation of the distribution and is a measure of the spread of the wave packet.

Solution:

The probability that an observation will lie between two values k_1 and k_2 is represented by the area under the curve between the ordinates k_1 and k_2, given by $\int_{k_1}^{k_2} dk \, e^{k^2/2\sigma^2}$.

Writing $k/\sigma = t$, $dk = \sigma dt$. Putting $(k_1-k_o)/\sigma = t_1$, $(k_2-k_o)/\sigma = t_2$; $A\sigma \int_{t_1}^{t_2} dt \, e^{t^2/2}$ = the difference between two integrals = $A\sigma\sqrt{\pi/2}$.

Note that $\Phi(\vec{k})$ is normalized means $\int dk \, \Phi^*(\vec{k}) \cdot \Phi(\vec{k}) = 1$, and so $[A\sigma\sqrt{\pi/2}]^2 = 1$, whence $A = \frac{1}{\sigma}\sqrt{2/\pi}$.

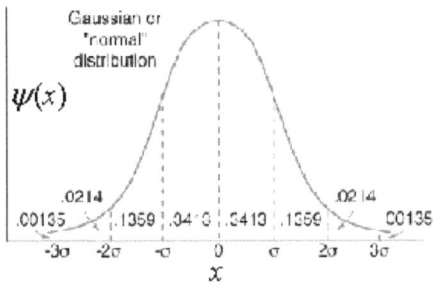

Fig 3.1   Gaussian distribution function $\psi(x) = A \, e^{-\beta x^2}$, where A and β are constants

3.3   DIRAC DELTA FUNCTION, $\delta(x)$

Kronecker's delta is

$\delta_{lm} = 1, \text{ for } l = m, \quad \delta_{lm} = 0, l \neq m$.

It holds good for the case of functions of a discrete variable. On the other hand, **Dirac delta** function, $\delta(x)$, plays the role for functions of a *continuous variable*, as Kronecker δ_{lm} for discrete variable functions. The need for $\delta(x)$ function arises naturally from the Fourier series.

Suppose that $\varphi(k_z)$ is substituted into $\psi(z)$ where

$$\psi(z) = \frac{1}{\sqrt{2\pi}} \int \varphi(k_z) e^{i(k_z z)} dk_z \qquad (3.3.1)$$

$$\varphi(k_z) = \frac{1}{\sqrt{2\pi}} \int \psi(z) e^{i(k_z z)} dz \qquad (3.3.2)$$

$$\frac{1}{\sqrt{2\pi}} \frac{1}{\sqrt{2\pi}} \int [\int e^{i(k_z z)} dk_z] \psi(z') e^{-i(k_z z')} dz' \qquad (3.3.3)$$

Suppose now that one interchanges, without question, the order of integrations,

$$\psi(z) = \int dz' \, \psi(z') [\frac{1}{2\pi} \int \{e^{i[k_z (z-z')]}\}] \qquad (3.3.4)$$

For this to be true, define the 1-D delta function

$$\delta(z) \equiv \delta(z - z') \equiv \delta(z, z') \qquad (3.3.5)$$

As the quantity obtained later as Equation (3.3.15) is

$$\delta(z - z') = \frac{1}{2\pi} \int_{-\infty}^{+\infty} dk_z \, e^{i(z-z')k}$$

which must be a peculiar kind of function. Strictly speaking the *impulse symbol* is no function at all. Such a function was understood by PAM Dirac and is commonly known as **Dirac δ-function**. It is not a function in the real mathematical sense; but is rather a 'distribution' or 'generalized function', a delta-function is often defined by the following properties:

3.3.1 Some PROPERTIES of delta-function

1) $\delta(z - z') = \begin{cases} = 0, & z \neq a \\ = 1, & b < z < c \end{cases}$; (3.3.6)

$$\int_{b<z<c} \delta(z - z') dz = 1 \qquad (3.3.7)$$

This integral is not a meaningful quantity until some convention for interpreting it is declared. The range of integration is infinitesimally small. Or,

2) $\delta(\tau) d\tau \begin{cases} = 0, & -\infty < \tau < 0 \\ = 1, & 0 < \tau < +\infty \end{cases}$; (3.3.7)

3) $\delta(z) = \delta(-z')$; (3.3.8)
$z \, \delta(z) = 0$; (3.3.9)
$z \, \delta'(z) = -\delta(z)$); (3.3.10)
$\delta'(-z) = -\delta'(z)$; (3.3.11)
$\delta(az) = \frac{1}{|a|} \delta(z)$; (3.3.12)

4) Since the $\delta(z) = 0$ everywhere but at the singular point, the only contribution to an integral occurs at this point. Its most important property called SHIFTING PROPERTY, is

$$\boxed{\int f(x)\delta(z-a)dz = f(a)}. \tag{3.3.13}$$

In Quantum Mechanics this means,

$$\int_{-\infty}^{+\infty} \psi(z')\delta(z-z')dz' = \psi(z). \tag{3.3.14}$$

with $\psi(z')$ as a smooth function in the range of values the argument of δ-function takes.

(5) The δ-function may be thought of as an infinitely high, infinitely thin spike – *i.e.* a thin spike function, in the description of an impulse force, or charge density for a point charge – having unit area but a non-zero amplitude at only one point, the point of singularity, where the amplitude becomes infinity. It is written as

$$\boxed{\delta(z-z') = \frac{1}{2\pi}\int_{-\infty < z' < +\infty} dk_z \, e^{i(z-z')k_z}} \tag{3.3.15}$$

The δ-function is NOT quadratically integrable, and it has the type of normalization introduced for continuous spectrum of eigen functions.

3.3.2 δ-function NORMALIZATION:

Since infinite plane waves are not 'well-behaved' wave functions in the sense that they do not vanish at $z = \pm\infty$, special procedures must be used in order to normalize them. One such procedure is the δ-*function normalization*. Thus consider the function

$$\psi(z) = A e^{ikz}$$

$$\int \psi^*(z)\psi(z)dz = |A|^2 \int e^{ikz}\{e^{-ik'z}\}dz;$$

$$= |A|^2 \int \{e^{iz(k-k')}\}dz;$$

$$= |A|^2 \, 2\pi \, \delta(k-k');$$

Defining, $\int \psi^*(z)\psi(z)dz = \delta(k-k')$ \hfill (3.3.16)

for δ-function normalization, $|A| = \frac{1}{\sqrt{2\pi}}$

for normalization of a plane wave function to the δ-function.

It is now known that the integration over a continuum of states is identical with the Fourier transform of $\psi(z)$ and $\varphi(k)$ with the time factor included.

$\psi(z)$ and $\varphi(k)$ are Fourier transforms of each other.

Consider a wave packet whose SPECTRUM OF k-VALUES to $\varphi(k)$ is a δ-function in k-space, *i.e.*, $\quad \varphi(k) = \delta(k-k')$, \hfill (3.3.17)

then $\quad \psi(z) = FT[\varphi(k)] = \frac{1}{2\pi}\int dk \, \varphi(k)\{e^{izk}\} = \frac{1}{2\pi}\int dk \, \delta(k-k')\{e^{izk}\}$

$$\psi(z) = \frac{1}{2\pi}\int dk \, \{e^{izk}\} \tag{3.3.18}$$

which is a *plane wave*

It is worthy to note that the infinite plane wave and δ-function are Fourier transforms of each other.

$$FT\{\delta(z-a)\} = e^{ikz}, \text{ infinite plane wave} \tag{3.3.19}$$

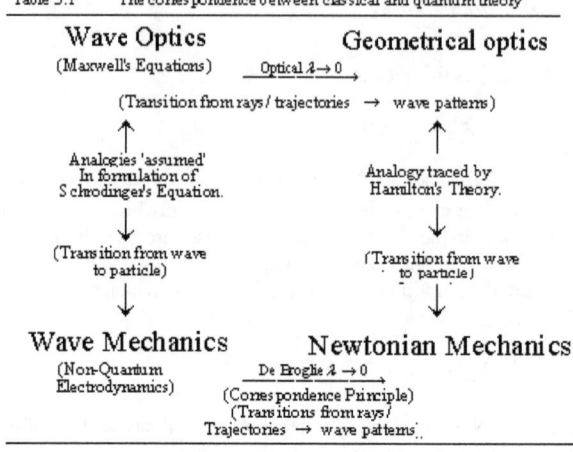

Table 3.1   The correspondence between classical and quantum theory

Similarly if a plane wave $f(x)$ consists of a sum of individual phase waves,
$$f(x) = \sum \delta(x - x_i),$$
then $f(x) = \sum (\text{Individul Transforms}) = \sum e^{i k x_i}$.

3.4.   ANALOGY BETWEEN OPTICS AND MEACHANICS:

The correspondence between classical and quantum theory can be presented in the way shown in Table 3.1
The Schrödinger's wave Equation does in fact give a correct limiting transition. This equation is, as it were, a fourth member in the above correspondence.

3.5   Schrödinger's Wave Equation -THE BASIC EQUATION in Quantum Mechanics - DEDUCTION:

The concept of matter waves expressed in the de Broglie relation, $p = \hbar k$ has suggested the hypothesis that an atomic particle defined by the complex wave function $\Psi(\bar{r},t)$, the well-known differential equation of waves, either traveling along a string or stationary, will be helpful in describing the dynamical state of atoms. This partial differential equation is a powerful simplification. This will describe the progress of the particle in space and time. Such a partial differential equation, whose solutions will correspond to the motion of the particle, describes the dependence of the amplitude on both displacement coordinate (x) and time (t). Since the concept of *wave packet* is not a result of previous physical theories, such a wave equation can not be derived, like Newton's Laws of motion are not derived, but must be postulated against experimental results. Erwin Schrödinger described the wave packet associated with a matter wave by $\Psi(\bar{r},t)$, and that he system does not change with time t. The **postulates** are:

1) The Laws of conventional Optics are a consequence of the linearity of Maxwell's Equations, which permits the superposition of different waved- Huygen's construction.

Therefore, the *fundamental wave equation* must be linear too in $\Psi(\vec{r})$, which is an assumption (insuring the validity of) known as the Principle of Superposition of states.
2) The differential equation must be of the first order with respect to time (*i.e.* first order in $\partial / \partial t$), and
3) The wave equation must be consistent with the de Broglie's hypothesis as well as with the Correspondence Principle.

3.5.1 Deduction of FREE-TIME EQUATION (TISE) for FREE particles:

There are several ways to make the transition from Classical Mechanics to Quantum Mechanics. One approach to the problem will be to exploit the close resemblance between EM waves and Matter waves, and to get a wave equation that describes the behaviour of a *wave function*. The usual basis of the mathematical theory of wave motion is furnished by a **'differential equation'**. The usual partial differential equation of a plane monochromatic wave is, in 1 - D motion,

$$\boxed{\frac{\partial^2 W(z,t)}{\partial z^2} = \frac{1}{w^2} \frac{\partial^2 W(z,t)}{\partial t^2}}. \tag{3.5.1}$$

where $W(z, t)$ is the amplitude, at z and normal to z and time t, of the EM wave, described by

$$W(z,t) = W_0 e^{i(kz - \omega t)} \tag{3.5.2}$$

where the phase velocity $w = c$, the velocity of light in vacuum for the EM waves. Schrödinger proposed that the de Broglie waves associated with an electron must obey this equation. Further the fundamental relations [equations (2.1.19) and (2.1.20)], *viz.* Einstein and de Broglie relation

$$E = \hbar \omega \tag{3.5.3}$$
$$p = \hbar k \tag{3.5.4}$$

where $\quad k = 2\pi / \lambda \tag{3.5.5}$
$$\omega = 2\pi \nu \tag{3.5.6}$$

are both assumed to be valid.
The phase velocity, $w =$ not physically significant for matter waves, and velocity of the electron
$\quad\quad u =$ group velocity of the matter wave.
$\quad\quad w = \omega / k = p / 2m \rightarrow u \neq$ constant
But w is a function of p.
The differential equation of the motion of an electron of mass, m, is, therefore,

$$\boxed{(\partial^2 / \partial z^2)\psi(z,t) = \frac{1}{u^2} (\partial^2 / \partial t^2)\psi(z,t)}. \tag{3.5.7}$$

A solution to this equation representing a standing (infinite plane) waveform is possible only if $u =$ constant. By the *method of separation of variables*, $\psi(z,t)$ may be expressed in the form

$$\boxed{\psi(z,t) = \psi(z) \cdot f(t)} \tag{3.5.8}$$

in which $\psi(z)$ is a function of co-ordinates only and $f(t)$ is a function of time only. For *standing wave* motion it is well known that

$$f(t) = e^{-i\omega t} \tag{3.5.9}$$

In 1-D case,
$$\psi(z,t) = \psi(z) \cdot e^{-i\omega t} \tag{3.5.10}$$

$$(\partial^2/\partial z^2)\,\psi(z,t) = \frac{1}{u^2}(\partial^2/\partial t^2)\,\psi(z,t)$$

becomes $(\partial^2/\partial t^2)\,\psi(z,t) = (\partial^2/\partial t^2)[\psi(z)\cdot e^{-i\omega t}] = (\partial/\partial t)(\partial/\partial t)[\psi(z)\cdot e^{-i\omega t}]$

$$= (-i\omega)(\partial/\partial t)[\psi(z)\cdot e^{-i\omega t}] = (-i\omega)(-i\omega)[\psi(z,t)]$$

$$(-i\omega)^2[\psi(z,t)] = -\omega^2[\psi(z)\cdot e^{-i\omega t}]$$

$$(\partial^2/\partial z^2)\,\psi(z,t) = e^{-i\omega t}](\partial^2/\partial z^2)\,\psi(z) = (\partial^2/\partial z^2)\,\psi(z)\cdot f(t)$$

Combining these two equations,

$$(\partial^2/\partial z^2)\,\psi(z,t) = (-\omega^2)(\partial^2/\partial t^2)\,\psi(z,t).$$

$$(\partial^2/\partial z^2)\,\psi(z,t) = \frac{1}{u^2}(\partial^2/\partial t^2)\,\psi(z,t)$$

$$\therefore (\partial^2/\partial z^2)\,\psi(z)\cdot f(t) = \frac{1}{u^2}(-\omega^2)\psi(z)\cdot f(t)$$

$$(d^2/dz^2)\,\psi(z) = (-\omega^2/u^2)\,\psi(z) = (-k^2)\,\psi(z) \quad (3.5.11)$$

$$(d^2/dz^2)\,\psi(z) = (-4\pi^2/\lambda^2)\,\psi(z) \quad (3.5.12)$$

$$\therefore (d^2/dz^2)\,\psi(z) = -(mv/\hbar)^2\,\psi(z) \quad (3.5.13)$$

It will be noted that the time variable, t, has been eliminated from this equation, which consequently represents the variation of the amplitude function $\psi(z)$ with z at a definite instant for a particle in a *potential-free* region.

3.5.2 Deduction of FREE-TIME (TISE) EQUATION for BOUND particles:

For a particle in an interacting system (i.e. for a bound particle), it is acted upon by a force, $\vec{F}$, which can be expressed as the derivative of a scalar potential function, $V(z)$, the at very low velocities, i.e. $u \ll c$, the force field,

$$\vec{F} = -(\partial/\partial z)\,V(z). \quad (3.5.14)$$

Total energy of the particle, E

$$E = T + V = \tfrac{1}{2}mv^2 + V(z) \quad (3.5.15)$$

$$E = \tfrac{1}{2m}p^2 + V(z), \quad (3.5.16)$$

Or, $\quad p^2 = 2m\,[E - V(z)] \quad (3.517)$

$$p^2 = \{\hbar(2\pi/\lambda)\}^2 = 2m\,[E - V(z)]$$

$$\lambda^2 = \hbar^2/\{2m\,[E - V(z)]\}. \quad (3.5.18)$$

Assuming the relation (3.5.12) and substituting for λ^2 from the relation (3.5.18),

$$(d^2/dz^2)\,\psi(z) = \{(-4\pi^2/\hbar^2)/2m\,[E - V(z)]\}\,\psi(z)$$

On re-arranging terms,

$$\boxed{\frac{d^2}{dt^2}\psi(z) + \frac{2m}{\hbar^2}[E - V(z)]\,\psi(z) = 0} \quad (3.5.19)$$

3.5.3 STEADY STATE FORM of Schrödinger's Equation (TISE):

In general, using the *Laplacian operator* ∇^2 defined as the generalization to 3-D of the differential operator $(\partial^2/\partial z^2)$, equation (3.5.19) becomes

$$\boxed{\nabla^2 \psi(x,y,z) + \frac{2m}{\hbar^2}[E - V(x,y,z)]\,\psi(x,y,z) = 0} \quad \text{(TISE)} \quad (3.5.20)$$

This is the FIRST of Schrödinger's Equation, known as <u>Schrödinger's AMPLITUDE Equation</u> (TISE), by means of which most of the applications of Quantum Mechanics are made. This equation describes the dynamics at stationary conditions, and is also known by steady state form (or **time-independent form**, or *free-time form*) of the Schrödinger Equation (TISE). This is applicable to mono-energetic or *monochromatic* $\psi(x, y, z)$-s only. In this case, the probability of finding the particle at r is given by equation (3.2.15)

$$P(r) = P(r, t) = \iiint |\hat{\Psi}(\vec{r},t)|^2 \, d\tau = \iiint \hat{\Psi}(\vec{r},t)^* \cdot \hat{\Psi}(\vec{r},t) \, d\tau$$

Consider $\quad \psi(z) = A\, e^{-\beta z^2}$

Normalization of $\psi(z)$ gives $A = (2\beta/\pi)^{1/4}$

Now, $\quad \varphi(k_z) = \frac{1}{\sqrt{2\pi}} \int \psi(z)\, e^{i(k_z z)}\, dz$

$\qquad = \frac{1}{\sqrt{2\pi}} \int e^{i(k_z z)}\, dk\, A\, e^{-\beta(k_z - k')^2}$

$\qquad = \frac{A}{\sqrt{2\pi}} \int e^{i(k_z z)}\, e^{-\beta z^2}\, dz$

$\qquad = \frac{A}{\sqrt{2\pi}} \int e^{(i k_z z - \beta z^2)}\, dz$

$\qquad = \frac{A}{\sqrt{2\pi}} \left[\int e^{-\beta z^2}\, \cos kz\, dz + i \int e^{-\beta z^2}\, \sin kz\, dz \right]$

$\qquad = \frac{A}{\sqrt{2\pi}} \left[\int e^{-\beta z^2}\, \cos kz\, dz + 0 \right]$

$\therefore \iiint |\hat{\Psi}(\vec{r},t)|^2 \, d\tau = \iiint \hat{\Psi}(\vec{r},t)^* \cdot \hat{\Psi}(\vec{r},t) \, d\tau = \text{Constant in time.} \quad (3.2.15)$

3.6 DEDUCTION of the BASIC EQUATION (**TDSE**) of Non-Relativistic Quantum Mechanics:

In the case of wave packets where *polychromatic* wave functions are used, one needs a general differential equation.

3.6.1 For FREE particles:
The constraints used are:
(i) Total energy of the particle
$\qquad E = \frac{1}{2m} p^2 + V(z), \quad$ where $V(z) = 0.$
(ii) Einstein relation (2.1.19)
$\qquad E = \hbar \omega$
(iii) and de Broglie relation (2.1.20)
$\qquad p = \hbar k$

where $k = 2\pi/\lambda$, and $\omega = 2\pi\nu$,

$\qquad \therefore E = \frac{1}{2m} p^2 = = (\hbar k)^2 / 2m = \hbar \omega. \qquad (3.6.1)$

$\omega / k^2 = \hbar / 2m = $ constant, independent of p.

(iv) the free particle can be represented by a complex harmonic function or r and t, i.e., by a function of $(\vec{k} \cdot \vec{r} - \omega t)$, as $\vec{w} = \omega / k$.

$$\psi(\vec{r},t) = \psi_0\, e^{i(\vec{k}\cdot\vec{r} - \omega t)} \tag{3.5.20}$$

which can be written as

$$\psi(\vec{r},t) = \psi_0\, e^{-(i/\hbar)(E t - \vec{p}\cdot\vec{r})} \tag{3.6.2}$$

This exponential form is very important when derivatives are to be found. Let E and p characterizing the particle are both constants. Differentiating with respect to time, t once gives

$$(\partial/\partial t)\psi(\vec{r},t) = (-i/\hbar) E \psi(\vec{r},t) \tag{3.6.3}$$

this means E = a constant, or

$$E\, \psi(\vec{r},t) = (-\hbar/i)\,(\partial/\partial t)\,\psi(\vec{r},t) \tag{3.6.4}$$

Thus the differential operator,

$$\{(-\hbar/i)\,(\partial/\partial t)\} = [i\hbar\,(\partial/\partial t)]$$

when operating on $\psi(\vec{r},t)$, is equivalent to the total energy, E. This means E (or its equivalent ω is eliminated from the Schrödinger's first equation (TISE), equation (3.5.20), by differentiation of $\psi(\vec{r},t)$.

$$\boxed{\hat{E} \rightarrow \{(-\hbar/i)\,(\partial/\partial t)\} = [i\hbar\,(\partial/\partial t)]}\,. \tag{3.6.5}$$

Differentiating twice with respect to displacement, r, the function $\psi(\vec{r},t)$, equation (3.6.2) becomes

i.e., $\quad \psi(\vec{r},t) = \psi_0\, e^{-(i/\hbar)E}\, e^{(i/\hbar)(xp_x + yp_y + zp_z)}$. $\tag{3.6.6}$

$$[(\partial/\partial x) + (\partial/\partial y) + \partial/\partial z)]\,\psi(x,y,z,t) = \hat{\nabla}\psi(x,y,z,t)$$
$$= (i/\hbar)[p_x + p_y + p_z]\psi(\vec{r},t)$$
$$[p_x]\,\psi(x,y,z,t) = [-i\hbar\,(\partial/\partial x)]\,\psi(\vec{r},t) = [-i\hbar\,\nabla_x]\,\psi(\vec{r},t) \tag{3.6.7}$$

This relation is interpreted as meaning that the momentum p_x] may be replaced by the differential operator, $[-i\hbar\,\nabla_x]$, which operates on $\psi(\vec{r},t)$. Therefore,

$$\boxed{\hat{p}_x \rightarrow [-i\hbar\,(\partial/\partial x)] = [-i\hbar\,\hat{\nabla}_x]} \tag{3.6.8}$$

From (3.6.7), $\hat{\nabla}_z\, \psi(z,t) = (i/\hbar)\hat{p}_z\, \psi(z,t)$ $\tag{3.6.9}$

Operating it with $\hat{\nabla}_z$

$$\hat{\nabla}_z\,\{\hat{\nabla}_z \psi(z,t)\} = (i/\hbar)\hat{p}_z\{(i/\hbar)\hat{p}_z\, \psi(z,t)\}$$
$$\hat{\nabla}_z^2\, \psi(z,t) = (-\hat{p}_z^2/\hbar^2)\,\psi(z,t)$$
$$\hat{p}_z^2\, \psi(z,t) = -\hbar^2 \hat{\nabla}_z^2\, \psi(z,t) \tag{3.6.10}$$

This relationship is interpreted thus: $\hat{p}_z^2$ can be replaced by the differential operator, $-\hbar^2\hat{\nabla}_z^2$, which generates on $\psi(\vec{r},t)$.

$$\boxed{\hat{p}_z^2 \rightarrow -\hbar^2\hat{\nabla}_z^2}\,. \tag{3.6.11}$$

So from $E = \hat{p}_z^2/2m$,

$$E\,\psi(z,t) = [\hat{p}_z^2/2m]\,\psi(z,t) \tag{3.6.12}$$

$$\{(-\hbar/i)\,(\partial/\partial t)\}\,\psi(z,t) = [-\hbar^2\hat{\nabla}_z^2/2m]\,\psi(z,t) \tag{3.6.13}$$

or from $E = \hat{p}_z^2/2m$,

$$\{[\hat{p}_z^2/2m] - E\}\psi(z,t) = 0 \tag{3.6.14}$$

i.e., $\qquad \{(1/\sqrt{2\pi\hbar})^3[\hat{p}_z^2/2m] - E\}\varphi(p)e^{-(i/\hbar)E\,t - p\bullet\vec{r}} = 0 \tag{3.6.15}$

3.6.2 For the BOUND particle

In the case of a free particle, at non-relativistic speeds,
$$\vec{u} \ll c \tag{3.6.16}$$
the force field is given by
$$\vec{F} = -\hat{\nabla}V(r) \tag{3.6.17}$$
and the total energy
$$E = T + V(r) + m_o c^2 \tag{3.6.18}$$
(which can be ignored, since in atomic physics $m_o c^2 < 100 eV$).

$$E = T + V(r) = \hat{p}^2/2m + V(r) \tag{3.6.19}$$

$$= (\hbar k)^2/2m + V(r) = [-\hbar^2\hat{\nabla}_z^2/2m] + V(r) \tag{3.6.20}$$

Applying. $E \rightarrow (-\hbar/i)\,(\partial/\partial t) \tag{3.6.21}$

$$(-\hbar/i)\,(\partial/\partial t)\,\psi(\vec{r},t) = \{[-\hbar^2\hat{\nabla}_z^2/2m] + V(r)\}\,\psi(\vec{r},t) \tag{3.6.22}$$

so one gets

$$\boxed{(i\hbar)\,(\partial/\partial t)\,\psi(\vec{r},t) = [(-i\hbar\hat{\nabla})^2/2m]\,\psi(\vec{r},t) + V(r)\}\psi(\vec{r},t)} \tag{3.6.23}$$

The layman finds such a law as
$$(\partial/\partial t) = k(\partial^2/\partial z^2) \tag{3.6.24}$$
much less simple than *it oozes* (or it diffuses or it flows) of which it is a mathematical statement.
JBS Haldane, the physicist, reverses this judgment, and his statement is certainly the more fruitful of the two as far as the prediction is concerned.
Equation (3.6.23) is a partial differential equation of the first order with respect to time, t. This is the famous *TIME DEPENDENT Schrödinger Equation* (**TDSE**), or, QM Wave Equation in its GENERAL FORM for matter waves.
This equation may written, using
$$E\psi(\vec{r},t) = (i\hbar)\,(\partial/\partial t)\,\psi(\vec{r},t), \tag{3.6.25}$$

$$\boxed{\hat{\nabla}^2\,\psi(\vec{r},t) + \tfrac{2m}{\hbar^2}[(i\hbar)\,(\partial/\partial t) - V(r)]\psi(\vec{r},t) = 0} \quad \textbf{(TDSE)} \tag{3.6.26}$$

This is the SECOND of Schrödinger's Equations, (**TDSE**). which is another form of the general wave equation.
Defining the quantum mechanical equivalent of the total energy E in Newtonian Mechanics as the corresponding operator called Hamiltonian, $\hat{H}$,

$$\hat{H} = T + V(r) = \hat{p}^2/2m + V(r) \tag{3.6.27}$$

$$\hat{H} = (-i\hbar)^2 \ (\partial/\partial r)(\partial/\partial r)/2m \ + \ V(r) \qquad (3.6.28)$$

since $\quad \hat{p}^2 = \hat{p} \cdot \hat{p} \qquad\qquad\qquad\qquad\qquad\qquad\qquad\quad (3.6.29)$

$$\hat{H} = [-\hbar^2 \hat{\nabla}^2 / 2m] + V(r) \qquad\qquad\qquad\qquad\quad (3.6.30)$$

Written in the full form,
$$\hat{H}(\vec{r},\vec{p}) = [-\hbar^2/2m][\partial^2/\partial x^2 + \partial^2/\partial y^2 + \partial^2/\partial z^2] + V(x,y,z)$$

i.e., $\quad \hat{H}(\vec{r},\vec{p}) = [-\hbar^2 \hat{\nabla}^2 / 2m] + V(x,y,z) \qquad\qquad (3.6.31)$

3.6.3 THREE NEW QUANTUM MECHANICAL OPERATORS:

From the discussions carried out in section 3.6.1. there appeared three new operators, which are: the Hamiltonian (energy) operator,

$$\hat{H}(\vec{r},\vec{p}) = [-\hbar^2 \hat{\nabla}^2 / 2m] + V(x,y,z) \qquad\qquad (3.6.31)$$

1) The operator corresponding to the total energy, $\hat{E}$,
$$\hat{E} \rightarrow (i\hbar)\ (\partial/\partial t) \qquad\qquad\qquad\qquad\qquad\qquad (3.6.5)$$

2) The linear momentum operator, $\hat{p}$
$$\hat{p}_x \rightarrow [-i\hbar\ (\partial/\partial x)] = [-i\hbar\ \hat{\nabla}_x] \qquad\qquad\qquad (3.6.8)$$
$$\hat{p}^2 = \hat{p} \cdot \hat{p} \ \rightarrow -\hbar^2 \hat{\nabla}_z^2 \qquad\qquad\qquad\qquad\qquad (3.6.11)$$

3.6.4 OPERATOR FORM of the Schrödinger's equation (SE)

The Schrödinger equation may be written in the operator form as

$$\boxed{\hat{H}(\vec{r},\vec{p})\ \psi(\vec{r},t) = E\ \psi(\vec{r},t)}. \qquad\qquad (3.6.32)$$

Or, $\quad \boxed{\hat{H}(\vec{r},\vec{p})\ \psi(\vec{r},t) = (i\hbar)\ (\partial/\partial t)\ \psi(\vec{r},t)} \qquad (3.6.33)$

3.7. SALIENT FEATURES of Schrödinger's Equation:

Schrödinger extended hypothetically the application of the wave equation (3.5.12)

$$(d^2/dz^2)\ \psi(z) + 4\pi^2/\lambda^2)\ \psi(z) = 0 \qquad\qquad (3.7.1)$$

to the vast realm of atomic physics by introducing the bold hypothesis that equation (3.7.1) describes the motion of a particle, e.g., an electron, when the wavelength, λ, is identified with the de Broglie wavelength $\lambda = h/mv$ of the particle. This introduces the momentum mv of the particle, which depends on the kinetic energy (T) and potential energy (V), both depending on the specific system to which the equation is applied (revolving electron, vibrating atom, etc.). ψ is the wave function associated with the particle in terms of spatial coordinate/s at an instance of time t.

REVIEW QUESTIONS

R.Q. 3.1　　How many photons emanate per second from a laser source with 10 mW, 633.0 nm? $c = 2.997 \times 10^8 \, ms^{-1}$, $\hbar = 1.054 \times 10^{-34} \, J-s$. (Answer: $3.18 \times 10^{14} \gamma \, s^{-1}$).

R.Q. 3.2　　"The choice of $\Psi(r)$ rather than $\Psi^*(r)$ as the wave function in wave mechanics is purely a matter of convention, since the same physical results would be obtained by replacing $\Psi(r)$ with $\Psi^*(r)$". Illustrate the validity of the statement.

R.Q. 3.3　　By means of Parseval's formula, show that if the spatial wave function $\Psi(\vec{r},t)$ is normalized to unity, so is the momentum wave function $\Phi(\vec{p},t)$.

R.Q. 3.4　　Given that $\int_{-\infty}^{+\infty} e^{-au^2} du = \sqrt{\frac{\pi}{a}}$, find the normalization constant for the normal (Gaussian) wave packet $\Phi(k,t) = A \, e^{-k^2/2\sigma^2}$, where σ = standard deviation of the distribution and is a measure of the spread of the wave packet. (Answer: $A = \sqrt{(2/\pi)}(1/\sigma)$)

R.Q. 3.5　　An amplitude function is a square pulse $g(k) = \begin{cases} 1/\sqrt{\varepsilon}, & -\varepsilon/2 \leq k \leq +\varepsilon/2 \\ 0, & |k| > \varepsilon/2 \end{cases}$
Find the Fourier transform of $g(k)$. (Answer: $(\sqrt{\varepsilon}/[2\pi])$ sinc $(x \, \varepsilon/2)$).

R.Q. 3.6　　Find $\varphi(k)$, the Fourier transform of the Normal (Gaussian) distribution function $\psi(x) = A \, e^{-\beta x^2}$, where A and β are constants. (Answer: $g(k) = \sqrt{2\sigma_k} \, e^{-(k-k')^2/2\sigma^2}$, $\sigma_k = (\Delta x/2) = (1/\sqrt{2\beta})$)

R.Q. 3.7　　Find the momentum representation of the 2s-state of hydrogen, viz. $\psi(r) = \frac{1}{\sqrt{32\pi}} (r^{-2}) e^{-r/2}$, with $(\hbar^2/me^2)$ as unit of r.

R.Q. 3.7　　Consider the state of a system represented by the function $\psi(z) = N \, e^{-(z/2\Delta)^2} e^{i k_o z}$ Derive from this, the state function in the momentum co-ordinates, including the normalization constant. Given the standard integral $\int_{-\infty}^{+\infty} e^{-ax^2 + bx} dx = \sqrt{\frac{\pi}{a}} \, e^{b^2/4a}$ (Answer: $\varphi(k) = \frac{N\Delta}{\sqrt{2}} e^{i \Delta^2 (k_o - k)^2}$. $\varphi(k)$ is Gaussian, centers at $k = k_o$; and width $\Delta k = (1/2\Delta)$.

R.Q. 3.8　　Normalize $u_1(x) = A_1 \, e^{-ax^2}$ and $u_2(x) = A_2 \, e^{-ax^2}$ over the interval $-\infty \leq x \leq +\infty$. Are these two functions orthogonal over this interval? Are these two functions orthogonal over the interval $0 \leq x \leq +\infty$?

R.Q. 3.9 a) Find the Fourier transform of $f(x) = 1, |x| < a$, and $= 0, |x| > a$ is $[2\sin(as)]/s$, using the result $\int \{e^{-\beta p^2} \cos kp\} \, dp = \sqrt{\pi/4\beta} \, e^{-p^2/4\beta}$, and b) Obtain the Fourier transform of $e^{-\beta p^2}$, and hence derive Fourier transform of $\delta(x)$, using $\delta(x) = Lt_{n \to \infty} (\sqrt{n/2\pi} \, e^{-nx^2})$.

R.Q. 3.10 Find the normalization constant, N, for the Gaussian wave packet $\psi(x) = N e^{-(x-x')^2 / 2\sigma^2}$. Obtain the Fourier transform of $\psi(x)$ and verify that it is normalized.

R.Q. 3.11 $f(x) = \delta(z + a/2') + \delta(z - a/2)$, find $FT[f(x)]$. (Answer: $2\cos(ka/2)$).

R.Q. 3.12 Consider a particle in an infinitely deep box, which extends from 0 to L.

1) Write down the wavefunctions and energies

Assume that the particle is in the lowest level of the box. At time $t=0$, the box is suddenly expanded to twice its original width, and now extends from 0 to 2L. (At $t=0$, the system can be described by the wavefunction from 1).)

2) Write the wavefunction as a superposition of the eigenstates of the new box

(Hint: Use the following formula:

$$\int_0^a \sin(bx)\sin(cx)dx = -\frac{1}{2}[\frac{\sin\{(b+c)a\}}{(b+c)}]\{\frac{\sin\{(b-c)a\}}{(b-c)}\})$$

3) Describe what will happen to this system as a function of time – will the particle start moving? (Will the wavefunction be time-dependent?).

4) Calculate at what times the wavepacket will come back to its original state.

Answer: 2) (n odd)

R.Q. 3.13 Consider a hydrogen atom. At time $t = 0$, the wavefunction is the following superposition of energy eigenfunctions $\psi_{n\ell m}(r)$
$$\Psi(r, t=0) = N\{2 \psi_{100}(r) - \psi_{210}(r) + \psi_{311}(r)\}$$

a) Normalize the wavefunction b) Is the wavefunction of even or odd parity? c) What are the probabilities of finding the system in the ground state? The state (200) ? Or (311)? Another eigenstate? d) What is the expectation value of the total energy? Of the Operators $\hat{L}^2$, and $\hat{L}_z$? (Answer:)

R.Q. 3.14 Classically-forbidden region in hydrogen. Classically, any region of space where the kinetic energy of a particle is negative is forbidden.

a) Show that the classically forbidden region for the ground state of hydrogen is $r > 2a_0$, where a_0 is the Bohr radius.

b) Calculate the probability of finding the electron in this region. (Answer: 0.238).

&&&&&&&&&&&

Chapter 4

Quantum Mechanics - 2:
Postulatory Approach to Wave Mechanics

Chapter 4

Quantum Mechanics - 2: Postulatory Approach to Wave Mechanics

"The saddest aspect of life right now is that science gathers knowledge faster than society gathers wisdom".   ~Isaac Asimov

4.1   INTRODUCTION

It is known from Chapter 3 that Schrödinger's wave Equation, like Newton's Laws, cannot be derived. There are two distinctive approaches to obtain it –

(1) The plausibility argument, and

(2) The *postulatory approach*.

It is fundamental to Quantum Mechanics, in contrast to Classical Mechanics, that all physical quantities of a system cannot be measured simultaneously to great precision, even in theory. It is, therefore, useful to discuss what can be measured involving unfamiliar language, and the second unfamiliar aspect results from the difference between the mathematics used in Quantum Mechanics and that used in Classical Mechanics – quantum mechanical equations may be written in terms of certain operators.

The *wave nature* of matter and the *Correspondence Principle* have led to certain fundamental relations regarded as postulates of Quantum Mechanics. The basic postulates can be stated in different ways. The different sets of postulates differ widely in form and in emphasis, though it can be understood that the various sets are mathematically and physically equivalent.

4.2   POSTULATES IN QUANTUM MECHANICS

A set of postulates in quantum mechanics is presented as follows:

4.2.1   POSTULATE # 1

"WAVE FUNCTIONS $\Psi(\vec{r}, t)$ and $\Phi(\vec{k}, t)$ DESCRIBE PHYSICAL SYSTEMS"

Existence of two wave functions $\Psi(\vec{r},t)$ and $\Phi(\vec{k},t)$ and their meaning are described in this postulate.

4.2.1.1 On $\Psi(\vec{r},t)$ and $\Phi(\vec{k},t)$:

For a physical system consisting of a particle moving in a non-conservative field of force (produced by an externally applied potential, V(r)), that there is an associated wave function $\Psi(\vec{r},t)$ in configuration/ co-ordinate or real space, that the wave function $\Psi(\vec{r},t)$ determines everything that can be known about the system in quantum mechanics. In a similar way, $\Phi(k,t)$ determines every thing that can be known about the system in the momentum/ Fourier or k-space. In other words, $\Psi(r,t)$ and $\Phi(k,t)$ are a pair of wave functions associated with the particle in the **dual space**.

$\Psi(\vec{r},t)$ must be **well behaved**, or a *class Type A* function or a physically admissible; because the application of Fourier theorem is justified only to functions whose integrals involved are convergent. This requires

(a) FINITE function everywhere ('Finite' means $\Psi(\vec{r},t)$ at the boundary points, to give energy values in conformity with the experimental results),

(b) SINGLE-VALUEDFUNCTION [i.e. $\Psi(\vec{r},t)$ has unique value], and

(c) FUNCTION CONTINUITY: Function $\Psi(\vec{r},t)$ must be CONTINUOUS everywhere of the 'configuration space' (x, y, z) of the system, in a region or bound space, under consideration. This is true only if the following so-called BOUNDARY CONDITIONS are satisfied:

(I) Amplitude continuity: Every $\Psi(\vec{r},t)$ must be continuous function of space,

(ii) Slope and curvature continuities: Derivatives of $\Psi(\vec{r},t)$, viz. $\Psi'(\vec{r},t)$, (i.e. the slope of $\Psi(\vec{r},t)$), and $\Psi''(\vec{r},t)$ (i.e. curvature of $\Psi(\vec{r},t)$), with respect to spatial co-ordinates, r, must also be continuous functions of r, for all r, i.e. everywhere, except where the potential $V(r)$ is finite. $\Psi(\vec{r},t)$ with $+\Psi''(\vec{r},t)$ looks like concave downward, and one with $-\Psi''(\vec{r},t)$ looks like convex upward. $\Psi''(\vec{r},t)$ is proportional to the amplitude of $\Psi(\vec{r},t)$. $\Psi(\vec{r},t)$ must be 'twice differentiable'.

(iii) $\Psi(\vec{r},t)$ must VANISH at infinity, i.e. $\Psi(\vec{r},t) \to 0$, as $x, y, z \to \pm\infty$.

(iv) Born's Normalization condition, equation (3.2.23): In addition to these, $\Psi(\vec{r},t)$ must satisfy Born's Normalization condition, that $\Psi(\vec{r},t) = \hat{\Psi}(\vec{r},t)$

$$\boxed{\int_{All\ space} \hat{\Psi}(\vec{r},t) \cdot \hat{\Psi}^*(\vec{r},t)\, d\tau = 1} \qquad (4.2.1)$$

$$\int_{All\ space} |\hat{\Psi}(\vec{r},t)|^2\ d\tau = 1 \qquad (4.2.2)$$

This means the PROBABILITY of observing a particle in a volume element, $d\tau$, is

$$\hat{\Psi}(\vec{r},t)\cdot\hat{\Psi}^*(\vec{r},t)\ d\tau = \hat{\Psi}(\vec{r},t)|^2\ d\tau$$

in configuration space, and $\Psi(r, t)$ must be square integrable –

$$\iiint |\hat{\Psi}(\vec{r},t)|^2\ d\tau\ \text{should converge,}$$

i.e., it exists in all space, *i.e.* $\Psi(\vec{r},t)$ must be *square integrable* and twice differentiable function of x, y, z.

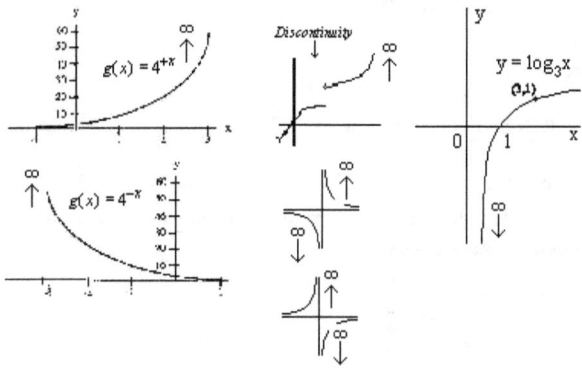

Fig 4.1 Functions which do not satisfy one of the conditions for *well behaviour*

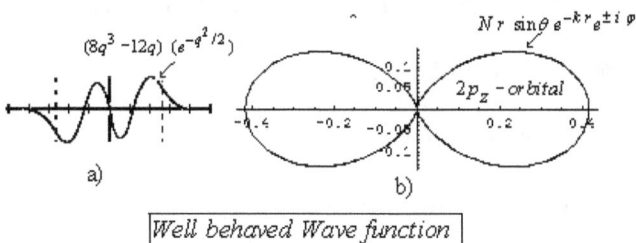

Fig 4.2 Two typical Well behaved functions

(v) The meaning $\Psi(\vec{r},t)$ is expressed as

$$\rho(\vec{r},t)\,dx\,dy\,dz = |\hat{\Psi}(\vec{r},t)|^2 dx\,dy\,dz \qquad (4.2.3)$$

Some typical functions which are NOT physically acceptable are shown in Fig 4.1.

Examples of *Well behaved* wave functions (Fig 4.2) and typical mathematical expressions are given below.

1) $A\,Cos\theta\,e^{-i\pi(kz-t)}$

2) $\hat{u}(x,y) = Sin\,k_x x \cdot Sin\,k_y y$

3) $N\,e^{-q^2/2}$,; $A\,(2q^3 - 3q)\,e^{-q^2/2}$

4) $\psi(\vartheta,\varphi) = N\,(\sin\theta\,\cos\theta\,e^{+i\varphi})$

5) $N r\,\sin\theta\,e^{-k r} e^{\pm i\varphi}$

4.2.1.2 RELATION between $\Psi(\vec{r},t)$ and $\Phi(\vec{k},t)$

The motion of a particle, described by a wave function in Cartesian space $\Psi(r, t)$, is also equally described by the wave function $\Phi(\vec{p}, t)$ in *momentum space* or $\Phi(\vec{k}, t)$ in *Fourier space*. If

$$\boxed{\hat{\Psi}(\vec{r},t) = \left(\frac{1}{\sqrt{2\pi}}\right)^3 \iiint \Phi(\vec{k})\,e^{i(\vec{k}\cdot\vec{r}-\omega t)}\,dk} \qquad (4.2.4)$$

$$\boxed{\Phi(\vec{k},t) = \left(\frac{1}{\sqrt{2\pi}}\right)^3 \iiint \hat{\Psi}(\vec{r})\,e^{-i(\vec{k}\cdot\vec{r}-\omega t)}\,d\tau} \qquad (4.2.5)$$

Here also $\Phi(\vec{k}, t)$ must satisfy all the conditions stated (i) to (v),

i.e., $\Phi(\vec{k}, t)$ must be <u>well behaved</u>, meaning

$$\iiint \hat{\Phi}(\vec{k},t) \cdot \hat{\Phi}^*(\vec{k},t)\,dk = 1 \qquad (4.2.6)$$

$$\rho(k,t)\,d^3k = |\Phi(\vec{k},t)|^2 d^3k \qquad (4.2.7)$$

4.2.2 POSTULATE # 2

TO EVERY OBSERVABLE, q THERE CORRESPONDS AN OPERATOR, $\hat{Q}$

and that when measurements are made on a system the only *observable* results q and the eigen value q of $\hat{Q}$

An *observable* is physically measurable quantity, which is usually familiar from Classical Mechanics.

Examples of observable quantities are:

Momentum component $\hat{p}_x$ and corresponding quantum mechanical operator,

$$\hat{p}_x \to [-i\hbar\, (\partial/\partial x)] = [-i\hbar\, \hat{\nabla}_x]$$

Momentum, $\hat{p}_z^2 \to -\hbar^2 \hat{\nabla}_z^2$ and

Energy, $\hat{E} \to (i\hbar)\, (\partial/\partial t)$

4.2.2.1 OPERATOR defined

Just like a function is something but a rule by which, given a number, one gets another number; an operator is a symbol for an instruction to carry out a certain mathematical transformation on the function following it. Thus let an operator, $\hat{\xi}$, is defined such that when it operates on a function *f(x)* results in multiplying *f(x)* by the independent variable *x*. Symbolically one writes this as

$$\hat{\xi} f(x) = x f(x) \qquad (4.2.8)$$

Similarly differentiation, with respect to independent variable *x*, denoted by operator symbol $\hat{\delta}$, when operated on a function *f(x)* gives

$$\hat{\delta} f(x) = f'(x).$$

4.2.2.2 Operator ALGEBRA:

1) <u>Distributive Law</u>

Summation of tow operators ξ and δ is

$$(\hat{\xi} + \hat{\delta}) f(x) = x f(x) + \hat{\delta} f(x) \qquad (4.2.9)$$

2) <u>Associative Law</u>

The product of two operators stated as

$$(\hat{\xi}\, \hat{\delta}) f(x) = \hat{\xi}\, [\hat{\delta} f(x)] \neq \hat{\delta}[\hat{\xi} f'(x)]$$

i.e., $\hat{\xi}\, [\hat{\delta}] \neq \hat{\delta}[\hat{\xi}\,]$ (4.2.10)

3) <u>Commuttivity</u>

If the product of two operators stated as

$$(\hat{\xi}\,\hat{\delta})f(x) = \hat{\xi}\,[\hat{\delta}f(x)] = \hat{\delta}[\hat{\xi}f'(x)],$$

i.e., $\quad \hat{\xi}\,[\hat{\delta}] = \hat{\delta}[\hat{\xi}\,]$, $\hspace{4cm}$ (4.2.11)

and the two operators $\hat{\xi}$ & $\hat{\delta}$ are said to *commute* each other. This is bracket is called the '*commutator bracket*'. The operators commute if their commutator bracket, expressed as

$$[\hat{\xi},\hat{\delta}] = 0, \quad \text{or} \quad (\hat{\xi}\,\hat{\delta} - \hat{\delta}\,\hat{\xi}) = 0 \hspace{2cm} (4.2.12)$$

An illustration: As per definitions of $\hat{\xi}$ and $\hat{\delta}$,

$$\hat{\xi}f(x) = x f(x), \text{ and } \hat{\delta}f(x) = f'(x).$$

$$(\hat{\xi}\,\hat{\delta})f(x) = \hat{\xi}\,[\hat{\delta}f(x)] = \hat{\xi}f''(x) = x f''(x), \text{ and}$$

$$(\hat{\delta}\,\hat{\xi})f(x) = \hat{\delta}\,[\hat{\xi}f(x)] = \hat{\delta}[x f(x)] = f(x).1 + x f'(x)$$

$$\therefore \hat{\xi}\,\hat{\delta} \neq \hat{\delta}\,\hat{\xi}$$

4) Operators are not limited to functions of one variable;

$$\hat{\delta}[f(x,y)] = (\partial/\partial x) f(x).$$

5) Vector operator, $\hat{\nabla}^2$:

Most frequently used vector operator in quantum mechanics is

$$\hat{\nabla}^2 = \partial^2/\partial x^2 + \partial^2/\partial y^2 + \partial^2/\partial z^2. \hspace{3cm} (4,2.13)$$

Thus $\quad \hat{\nabla}^2[f(x,y,z)] = \partial^2 f/\partial x^2 + \partial^2 f/\partial y^2 + \partial^2 f/\partial z^2$

6) Continuous Functions

Quantum mechanical functions are continuous functions in the complete range of the variables, and which gives a finite result when squared and integrated over the entire range of the variables. Quantum mechanical functions themselves are called QUANTUM MECHANICAL VARIABLES, and the 'quantum mechanical operators are called QUANTUM MECHANICAL OBSERVABLES.

4.2.3 EIGEN FUNCTIONS:

Consider $\hat{\xi}f(x) = x f(x)$, and $\hat{\xi}$ is replaced by $\hat{Q}$, and *f(x)* is a function such that

$$\hat{Q}f(x) = q f(x), \hspace{5cm} (4.2.14)$$

where q = a constant. If this relation is obeyed then such variables, *i.e.* of functions (which are finite, continuous and single-valued) are called EIGEN FUNCTIONS (or, 'proper or 'characteristic' functions) of the operator $\hat{Q}$. The various possible values of the quantity (constant) q are called the EIGEN VALUES (or *characteristic values*) of the operator $\hat{Q}$, *i.e.* the values that a physical quantity can take in Quantum Mechanics. The equation (4.2.14) viz. $\hat{Q} f(x) = q f(x)$, itself is known as the Eigen value Equation.

For example,

$$\hat{\delta} e^{kx} = k e^{kx},$$

where e^{kx} is single-valued, continuous and finite in the range $-\pi < x < +\pi$. Here k is either real or complex.

4.2.4 LINEAR OPERATOR:

In the preceding example, every value of k is an *eigen value* of $\hat{\delta}$. The operator $\hat{\delta}$ is then called a *linear operator*.

Examples of linear operators: $\ddot{x}$, $\hat{\nabla}^2$, $\int dx$, $(a+d/dx)$, $[\partial/\partial t]$. Performing certain operations with the above linear operators — multiplication by a number, performing addition or taking product may obtain additional linear operators.

Examples: $c\hat{\nabla}^2$, $[\hat{\nabla}^2 + (\partial/\partial t)]$, *etc.*

Let the operator $\hat{P}$ is defined such that when it operates on *f(x)*, viz.

$$\hat{P}[f(x)] = \{f(x)\}^2,$$

then $\quad \hat{P}[f(x)+g(x)] = \{f(x)\}^2 + 2f(x)g(x) + \{g(x)\}^2$

$$= \hat{P}[f(x)] + 2f(x)g(x) + \hat{P}[g(x)]$$

$$\hat{P}[f(x)+g(x)] \neq \hat{P}[f(x)] + \hat{P}[g(x)]$$

$\hat{P}$ is NOT a linear operator.

The general condition for an operator $\hat{P}$ to be LINEAR is to satisfy the relation

$$\boxed{\hat{P}[bf(x)+c g(x)] \neq b \hat{P}[f(x)] + c \hat{P}[g(x)]} \qquad (4.2.15)$$

Examples of *anti-linear* operator are: $y(d/dx)$, $Sin(..)$, $Cos(...)$, *etc.*

4.2.5 DYNAMIC VARIABLES

Quantum Mechanics is based on the premise that the observables (*i.e.*, physical quantities measurable in a laboratory or otherwise, such as position, energy, linear momentum, angular momentum, *etc*), which are associated with elementary particles, can be associated with linear operators. The choice of these operators is arbitrary in Quantum Mechanics; but it must satisfy the condition, *viz.* when it operates on a wave function it gives the observable quantity times the wave function. Thus if $\hat{Q}$ is an operator corresponding to the observable q, then

$$\hat{Q}\,\psi_n = q_n\,\psi_n \tag{4.2.16}$$

In Quantum Mechanics, the *eigen values* are constants q, which are **dynamic varaiables**, *i.e.*, Observable quantities. Thus the TISE (time-independent Schrödinger Equation),

$$\hat{H}\,\psi_n = E_n\,\psi_n, \tag{4.2.17}$$

is an <u>Eigen value Equation</u>. Therefore, E_n is said to be a *conserved quantity* since it is an eigen value. To solve a problem in Quantum Mechanics means to find the eigen values and eigen functions by Solving the related Schrödinger Equation.

A few of the quantum mechanical operators corresponding to different classical variables are listed in Table 4.1.

Table 4.1. Dynamical Variables and corresponding Operators

Classical quantity (Dynamical variable)		QM operators Representative space	
		Co-ordinate space	Momentum space
Position (x, y, z)	q	$\hat{q}$	$(i\hbar\dfrac{\partial}{\partial p_q}) = (i\hbar\nabla_q)$
	q^2	$\hat{q}^2$	$-\hbar^2(\partial^2/\partial p_q^2) = (-\hbar^2\nabla_q^2)$
Momentum	p_q	$-i\hbar\,(\partial/\partial q)$	$\hat{p}_q$
	p	$-i\hbar\,(\partial/\partial q)$	$\hat{p}$
	p^2	$(-\hbar^2\nabla_q^2)$	$\hat{p}_q^2$
Energy	$\bar{p}$	$i\hbar\,(\partial/\partial t)$	$i\hbar\,(\partial/\partial t)$
Angular Momentum	$\hat{l}_x$	$\hat{l}_x = -i\hbar\,(y\,\partial/\partial z - z\,\partial/\partial y)$	
	$\hat{l}_y$	$\hat{l}_y = -i\hbar\,(z\,\partial/\partial x - x\,\partial/\partial z)$	
	$\hat{l}_z$	$\hat{l}_z = -i\hbar\,(x\,\partial/\partial y - y\,\partial/\partial x)$	

4.2.6 POSTULATE #3

If a physical system is described by a wave function Ψ, the *average value* or the <u>expectation value</u> q, of any observable quantity q corresponding to the operator, $\hat{Q}$, of the physical system in the state Ψ given by

$$\boxed{<\hat{Q}> = <q> = \left(\frac{\iiint \Psi^* \hat{Q} \Psi \, d\tau}{\iiint \Psi^* \Psi \, d\tau} \right)} \qquad (4.2.18)$$

$<q>$ may be *real or complex quantity*, depending on the nature of $\hat{Q}$. But experimentally one gets q = real.

If the wave function $\Psi = \hat{\Psi}$ is normalized to unity, then the denominator in the expression above is unity.

i.e., $\quad <\hat{Q}> = <q> = \iiint \hat{\Psi}^* \hat{Q} \hat{\Psi} \, d\tau$

Worked out Example. 4.1

Given the wave function of a quantum particle is $\psi_n(x) = A \, e^{i[\alpha x - \alpha^2 \hbar \, t/2m]}$. Determine the expectation value of its momentum.

Solution: Step # 1 $\quad$ Given, $\psi_n(x) = A \, e^{i[\alpha x - \alpha^2 \hbar \, t/2m]}$; $<\hat{p}_x> = \int \psi_n^*(x) \, \hat{p}_x \, \psi_n(x) dx$

Step # 2 $\quad$ Now $-i\hbar(d/dx)\psi_n(x) = (-A \, i \, \hbar) \, (i\alpha) \, e^{i(\alpha x - \alpha^2 \hbar t/2m)} = \alpha \hbar \psi_n(x)$

Step # 3 $\quad <\hat{p}_x> = \alpha\hbar \int \psi_n^*(x) \, \psi_n(x) dx = \alpha\hbar \; ; <p_x> \, = \alpha \, \hbar$.

4.2.4 POSTULATE # 4

THE OPERATOR ASSOCIATED WITH A PHYSICALLY MEASURABLE QUANTITY IS HERMITIAN

Or, $\quad$ For every dynamical variable there exists a corresponding Hermitian operator

In Quantum Mechanics linear operators may be chosen to correspond to dynamic variables as seen in Postulate 2. The Eigen value equation is (4.2.16), *viz.*,

$$\hat{Q} \psi_n = q_n \psi_n$$

4.2.4.1 ADJOINT operator $\hat{A}^\dagger$

The adjoint operator, $\hat{A}^\dagger$, of the operator $\hat{A}$ is defined by

$$\boxed{\int (\hat{A}^\dagger \varphi)^* \psi \, dx = \int \varphi^* \hat{A}^\dagger \psi \, dx} \qquad (4.2.19)$$

for arbitrary φ and ψ.

4.2.4.2 HERMITIAN operator:

The operator $\mathbb{C}$ is called Hermitian (self-adjoint), if

$$\hat{C}^\dagger = \hat{C}. \qquad (4.2.20)$$

$$\int (\hat{C}^\dagger \varphi)^* \psi \, dx = \int \varphi^* \hat{C}^\dagger \varphi) \, dx, \qquad (4.2.21)$$

means $(\hat{C}\hat{H})^\dagger = \hat{H}^\dagger \hat{C}^\dagger$. $\qquad (4.2.22)$

Further $[\hat{C}\hat{H}, \hat{\Im}] = \hat{C}[\hat{H}, \hat{\Im}] + [\hat{C}, \hat{\Im}]\hat{H}$ $\qquad (4.2.23)$

$$[\hat{C}, \hat{H}]^\dagger = [\hat{H}^\dagger, \hat{C}^\dagger] \qquad (4.2.24)$$

4.2.4.3 HERMICITY CONDITION

The *condition for Hermiticity* of an operator $\hat{\mathbb{F}}$ is that the following equation should always holds good with respect to a pair of 'acceptable' arbitrary functions ψ_1 and ψ_2, defined over a range of configuration space, viz.,

$$\int (\hat{\mathbb{F}}\psi_1)^* \psi_2 \, dx = \int \psi_1^* \hat{\mathbb{F}}\psi_2) \, dx, \qquad (4.2.25)$$

or as $\boxed{\int \psi_1^* (\hat{\mathbb{F}}\psi_2) \, dx = \int \psi_2^* \hat{\mathbb{F}}\psi_1 \, dx = \int \hat{\mathbb{F}}^* \psi_2^* \psi_2 \, dx = \int (\psi_1^* \hat{\mathbb{F}} \psi_2)^* dx}$ $\qquad (4.2.26)$

4.2.4.4 BAKER-HAUSDORFT Identity

$$e^{\hat{A}}\hat{B}e^{-\hat{A}} = \hat{B} + [\hat{A}, \hat{B}] + \frac{1}{2!}[\hat{A},[\hat{A}, \hat{B}] + \cdots \qquad (4.2.27)$$

where $e^{\hat{A}} \equiv \sum (\frac{1}{\nu!}) A^\nu$ is defined by the *power series*.

If $\quad [\hat{A}, \hat{B}], \hat{A}] = 0$,

and $\quad [\hat{A}, \hat{B}], \hat{B}] = 0$,

then $\quad e^{\hat{A}} e^{\hat{B}} = e^{\hat{B}} e^{\hat{A}} + e^{\hat{A}\hat{B}} e^{[\hat{A},\hat{B}]}$. $\qquad (4.2.28)$

and $\quad e^{\hat{A}+\hat{B}} = e^{\hat{A}} e^{\hat{B}} = e^{\hat{A}\hat{B}} e^{-[\hat{A},\hat{B}]/2}$ $\qquad (4.2.29)$

Worked out Example 4.2

Show that the expectation value of an operator must satisfy $<\hat{Q}> = <\hat{Q}>^*$, so as it to be Hermitian.

Solution: Step # 1   Given, consider an operator defined by $\hat{\delta} = \dfrac{d}{dx}$ and a function $f(x) = e^{ikx}$.

Mathematically, an operator $\hat{F}$ to be Hermitian with respect to a pair of functions means that

$$\int (\hat{F}\psi_1)^* \psi_2 \, dx = \int \psi_1^* (\hat{F}\psi_2) \, dx$$

Step # 2 $<\hat{\delta}> = \int (\psi^* \hat{\delta} \psi) \, dx = \int e^{-ikx} (d/dx) \, e^{ikx} \, dx = \int \psi^* (ik) \psi \, dx = ik$

Step # 3 $<\hat{\delta}>^* = \left(\int \psi^* \hat{\delta} \psi \, dx\right)^* = \int \psi \, \hat{\delta}^* \psi^* \, dx$

$= \int e^{ikx} (d/dx) \, e^{-ikx} \, dx = \int \psi (-ik) \psi^* \, dx = -ik$.

Step # 4   i.e. $<\hat{\delta}> \neq <\hat{\delta}>^*$, It is, therefore, evident that δ is not hermitian.

and $\left\langle \hat{\delta} \right\rangle^* = -ik$.

4.2.4.5   PRODUCT of two Hermitian operators:

For two Hermitian operators $\hat{P}$ and $\hat{Q}$, their product $\hat{P}\hat{Q}$ = Hermitian, if $[\hat{P}, \hat{Q}] = 0$

Proof: Let $\hat{P}$ and $\hat{Q}$ are both Hermitian operators

$\hat{P} \psi_n = p_n \psi_n$ and $\hat{Q} \psi_n = q_n \psi_n$.

$\hat{Q}\hat{P} \psi_n = \hat{Q}(\hat{P}\psi_n) = \hat{Q} p_n \psi_n = p_n \hat{Q}\psi_n) = p_n q_n \psi_n$

$= q_n p_n \psi_n = q_n \hat{P}\psi_n = \hat{P} q_n \psi_n = \hat{P}(\hat{Q}\psi_n) = \hat{P}\hat{Q}\psi_n$.

i.e., $[\hat{P}, \hat{Q}] \psi_n = 0$, which means $[\hat{P}, \hat{Q}] = 0$.

On the other hand, if $[\hat{P}, \hat{Q}] = \hat{R} \neq 0$,

$\hat{R}$ is said to be **Skew Hermitian** or *anti-Hermitian*.

It can be shown that

$[\hat{P}, \hat{Q}\hat{H}] = [\hat{P}, \hat{Q}]\hat{H} + \hat{Q}[\hat{P}, \hat{H}]$ (4.2.33)

$[\hat{A}, \hat{B}+\hat{C}] = [\hat{A}, \hat{B}] + [\hat{A}, \hat{C}]$

$[\hat{P}\hat{Q}, \hat{H}] = [\hat{P}, \hat{H}]\hat{Q} + \hat{P}[\hat{Q}, \hat{H}]$

$[\hat{P}, [\hat{Q}, \hat{H}]] + [\hat{Q}, [\hat{H}, \hat{P}]] + [\hat{H}, [\hat{P}, \hat{Q}]] = 0$.

4.2.5 THREE PROPERTIES OF A HERMITIAN OPERATOR

Hermitian operators have three properties that are of importance in physics, both classical and quantum physics. These will be discussed as three theorems as follows.

4.2.5.1 THEOREM 1: PROPERTY # 1 of a Hermitian Operator

Statement   The Eigen values of a Hermitian operator are all Real

Proof: Step 1 From the statement it is known the following:

For a well behave wave function, ψ_n,

i) $\hat{Q}$ is a Hermitian operator, ii) ψ_n = a well-behaved wave function, iii) $\int \psi_n^* \psi_n \, dz \neq 0$, and ii) $\hat{Q}\psi_n = q_n \psi_n$, q_n is eigen value of $\hat{Q}$ for ψ_n., v) Hermiticity condition.

$\int (\hat{F}\psi_1)^* \psi_2 \, dx = \int \psi_1^* \hat{F}\psi_2 \, dx$

Step 2 LHS = $\int (\hat{Q}\psi_n)^* \psi_n \, dx = \int (q_n \psi_n)^* \psi_n \, dx = q_n^* \int \psi_n^* \psi_n \, dx$.

Step 3 RHS = $= \int \psi_1^* \hat{Q}\psi_2 \, dx = \int \psi_n^* q_n \psi_n \, dx = q_n \int \psi_n^* \psi_n \, dx$

Step 4 The Hermiticity condition requires that LHS = RHS.

$\therefore \qquad q_n^* \int \psi_n^* \psi_n \, dx = q_n \int \psi_n^* \psi_n \, dx$.

i.e., $(q_n^* - q_n) \int \psi_n^* \psi_n \, dx = 0$.,

Step 5 But $\int \psi_n^* \psi_n \, dx \neq 0$

This is possible only if $(q_n^* - q_n) = 0$,

Or, $q_n^* = q_n$ \hfill (4.2.34)

which is possible only if q_n = a real quantity.

4.2.6 THEOREM No, 2: PROPERTY # 2 of a Hermitian Operator
or. Ortho-Normality of Wave Functions:

The *eigen functions* of a Hermitian operator are **orthogonal** to one another, if the corresponding eigen values are unequal, *i.e.* distinct

Proof: Step 1 From the statement it is known the following:

i) $\hat{Q}$ is a Hermitian operator, ii) two well behaved wave functions, ψ_m and ψ_n, iii) $\int \psi_n^* \psi_n \, dz \neq 0$, $\int \psi_m^* \psi_m \, dx \neq 0$ and iv) $\hat{Q} \psi_n = q_n \psi_n$, q_n is eigen value of $\hat{Q}$ for ψ_n.; $\hat{Q} \psi_m = q_m \psi_m$, q_m is eigen value of $\hat{Q}$ for ψ_m; q_n & q_m are real; v) Hermiticity condition. $\int (\hat{F}\psi_1)^* \psi_2 \, dx = \int \psi_1^* \hat{F}\psi_2 \, dx$

Step 2 $\text{LHS} = \int (\hat{Q}\psi_m)^* \psi_n \, dx = \int (q_m \psi_m)^* \psi_n \, dx = q_m \int \psi_m^* \psi_n \, dx$

Step 3 $\text{RHS} = \int \psi_m^* \hat{Q} \psi_n) \, dx = \int \psi_m^* q_n \psi_n \, dx = q_n \int \psi_m^* \psi_n \, dx$

Step 4   The Hermiticity condition requires that LHS = RHS.

$$q_m \int \psi_m^* \psi_n \, dx = q_n \int \psi_m^* \psi_n \, dx$$

$$\therefore (q_m - q_n) \int \psi_m^* \psi_n \, dx = 0.$$

Generally, $(q_m - q_n) \neq 0$,

and so this is possible only if

$$\boxed{\int \psi_m^* \psi_n \, dx = 0}. \quad (4.2.35)$$

Equation (4.2.35))is called the OVERLAP INTEGRAL, also known as the **Orthogonality Relation** between two wave functions, ψ_n and ψ_m, of the system. Therefore, the wave functions ψ_n and ψ_m are called a pair of orthogonal wave functions. This integral has the physical meaning as follows: The values of both ψ_n and ψ_m at all points space will add together that the positive values are exactly cancelled out by the negative values.

N.B. *Not all overlap integrals are zero because not all wave functions are orthogonal to one another.* This is important in the theory of chemical bonding (Fig. 4.2). If a good overlap between atoms within a molecule or molecules occurs then there is a good chance of strong bonding.

The condition of orthogonality is the continuum analogue of the vanishing property of a scalar product of two vectors.

4.2.7   Ortho-normality Relation:

The two important relations between the wave functions of a system are

i) the Born's condition of normalization, *viz.*

$$\int \hat{\psi}_n{}^* \hat{\psi}_n \, dz = 1, \text{ and } \int \hat{\psi}_m{}^* \hat{\psi}_m \, dz = 1 \qquad (4.2.1)$$

and (ii) the orthogonal relation, *viz.*

$$\int \psi_m{}^* \psi_n \, dz = 0, \qquad (4.2.35)$$

A general expression comprising these two condition may be expressed by

$$\boxed{\int \hat{\psi}_m{}^* \hat{\psi}_n \, dz = \delta_{mn}} \qquad (42.36)$$

where δ_{mn} the Kroneckar delta.

It is known as the **Ortho-normality Relation**. This relation tells that the wave functions $\hat{\psi}_n$ and $\hat{\psi}_m$ are both normalized and orthogonalized.

4.2.8 EXAMPLES of orthogonal wave functions:

(i): Both the 1s and 2s orbitals of the same atom are orthogonal to each other.

(ii) The 1s orbital of one atom A and a 2p orbital of a neighboring atom B are mutually orthogonal.(Fig 4.2). Here the *s*-orbital and *p*-orbital in this orientation are orthogonal because the integral over the region where their product is positive is cancelled by the integral over the region where their product is negative.

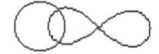
σ overlap can result from
the overlap of an s orbital
with a p orbital

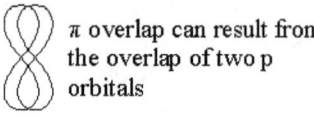
π overlap can result from
the overlap of two p
orbitals

Fig. 4.2 Visual representation of two Overlapping functions.

4.2.9 POSTULATE # 5

PROPERTY # 3 of a Hermitian Operator

A COMPLETE SET of ortho-normal functions. SUPERPOSITION OF STATES

Wave functions ψ_i can be superposed to form a new wave function, $\Psi(\vec{r})$, which is, itself, a physically valid representation of a more possible state of the particle.

This is the fundamental assumption underlying the representation of a particle by a wave packet, where the Fourier integral provides the means for superposing a continuum of states. Accordingly, a set of n functions is linearly independent if the linear equation

$$\sum c_i \, \psi_i \neq 0, \qquad (4.2.37)$$

is satisfied, implying all c_i-s are zero. *i.e.* none of these ψ_i-s can be expressed as a linear combination of the others. c_i-s are <u>expansion coefficients</u> and ψ_i-s are the known <u>discrete states</u>.

$$\hat{Q}\,\psi_i = q_i\,\psi_i. \qquad (4.2.16)$$

This fundamental Principle in Quantum Mechanics has no correspondence in Classical Mechanics. Thus a general (arbitrary) normalized function Ψ can be written as a combination of a *complete set*.

$$\boxed{\Psi(\bar{r}) = \sum c_i \, \hat{\psi}_i(\bar{r}) \neq 0} \qquad (4.2.38)$$

where $\int \hat{\psi}_m^{*} \hat{\psi}_n \, dz = \delta_{mn}.$ (4.2.36)

Such a set of functions $\hat{\psi}_1, \hat{\psi}_2, \hat{\psi}_3, \cdots$ *etc.* is said to constitute a COMPLETE ORTHO-NORMAL SET of a particular operator $\hat{Q}$. The expansion coefficients, c_i are given by

$$\boxed{c_i = \int \hat{\psi}_i^{*}(\bar{r})\,\Psi(\bar{r})\,d\tau} \qquad (4.2.39)$$

4.2.9.1 Compatible Observable:

If a set of functions $\hat{\psi}_i$ is an eigen function of both operators $\hat{R}$ and $\hat{S}$ corresponding to physical observable, the corresponding observables r and s are said to be **compatible** (if not *conjugate* or *complementary*).

If $\qquad \hat{Q}\Psi = \hat{R}\Psi,$ (4.2.40)

the $\hat{Q}$ and $\hat{R}$ are called **equivalent** operators.

4.2.9.2 THEOREM No. 3: DEGENERACY

If an eigen value q of the operator $\hat{Q}$ is <u>degenerate</u>,

$$\boxed{\hat{Q}\left(\sum c_i \, \hat{\psi}_i(\bar{r})\right) = q\left(\sum c_i \, \hat{\psi}_i(\bar{r})\right)} \qquad (4.2.41)$$

For all values of c_i, where every well-behaved wave function $\hat{\psi}_i$ can be approximated by a series, and q_i of the operator $\hat{Q}$ are <u>non-degenerate</u>. If there are n linearly independent solutions for each eigen value the phenomenon is called "*n-fold degeneracy*" The system represented by the eigen value is *n*-fold degenerate.

4.2.9.3 THEOREM No. 4:

> If two observables are *compatible*, their operators *commute*.

Proof: Consider two operators $\hat{R}$ and $\hat{S}$ such that because they are compatible,

$$\hat{R}\,\hat{\psi}_i = r_i\,\hat{\psi}_i, \text{ and } \hat{S}\,\hat{\psi}_i = s_i\,\hat{\psi}_i.$$

$$\hat{R}\hat{S})\,\hat{\psi}_i = \hat{R}\,s_i\,\hat{\psi}_i = s_i\,\hat{R}\,\hat{\psi}_i = s_i\,r_i\,\hat{\psi}_i$$

$$= r_i\,s_i\hat{\psi}_i = r_i\,\hat{S}\,\hat{\psi}_i = \hat{S}\,r_i\psi_i = \hat{S}\,\hat{R}\,\hat{\psi}_i.$$

$$= (\hat{S}\,\hat{R})\,\hat{\psi}_i.$$

$$(\hat{R}\,\hat{S}) - (\hat{S}\,\hat{R}) = 0.$$

Multiplying by c_i and summing up for all values of *i*, one gets

$$(\hat{R}\hat{S} - \hat{S}\hat{R})\sum c_i\,\hat{\psi}_i = (\hat{R}\hat{S} - \hat{S}\hat{R})\Psi = 0.$$

$$[\hat{R},\hat{S}] = (\hat{R}\hat{S} - \hat{S}\hat{R}) = 0.$$

Thus two dynamic variables are said to be compatible, if they can be specified simultaneously with complete accuracy. Otherwise, they are called complementary or conjugate variables.

4.2.9.4 THEOREM No. 5

> If two operators $\hat{Q}$ and $\hat{R}$ *commute* with each other, and either $\hat{Q}$ or $\hat{R}$ has *non - deg enerate eigevalues*, its eigen functions are also eigenstates of the other operator.

Proof: Step 1 Given $\hat{Q}\,\psi_i = q_i\,\psi_i$, and $\hat{R}\,\hat{\psi}_i = r_i\,\hat{\psi}_i$, q_i are distinct,

$$[\hat{Q},\hat{R}] = (\hat{Q}\hat{R} - \hat{R}\hat{Q}) = 0.$$

Step 2 $\quad \hat{Q}\,(\hat{R}\,\hat{\psi}_i) = \hat{R}(\hat{Q}\,\hat{\psi}_i) = (\hat{R}\,q_i\,\hat{\psi}_i) = q_i\,(\hat{R}\,\hat{\psi}_i) = q_i\,(r_i\hat{\psi}_i) = r_i(q_i\hat{\psi}_i)$

$$= r_i(\hat{Q}\,\psi_i) = \hat{R}\,q_i\,\psi_i = (\hat{R}\,\hat{Q})\,\hat{\psi}_i$$

This establishes the $\hat{\psi}_i$ are simultaneously eigen-functions of both $\hat{Q}$ and $\hat{R}$.

4.2.13 THEOREM No.6

> If two operators $\hat{Q}$ and $\hat{R}$ *commute* with each other, there exists a *complete set of eigenstates*, which are *simultaneously* eigenstates of both $\hat{Q}$ and $\hat{R}$.

Proof: Given: $\hat{Q}\, \psi_i = \hat{R}\, \hat{\psi}_i = q\, \hat{\psi}_i$,

q is the *m*th order degenerate eigenvalue of $\hat{Q}$, *i.e.* $q_1 = q_2 = q_3 = \cdots = q_m$,

$$[\hat{Q}, \hat{R}] = (\hat{Q}\hat{R} - \hat{R}\hat{Q}) = 0 .$$

Operating on the equation $\hat{Q}\, \psi_i = \hat{R}\, \hat{\psi}_i = q\, \hat{\psi}_i$ by $\hat{R}$,

$$\hat{R}(\hat{Q}\, \psi_i) = \hat{R}\, q\, \hat{\psi}_i = q(\hat{R}\, \hat{\psi}_i)$$

$$\Psi = \sum c_i \hat{\psi}_i$$

$$\mathbb{R}\Psi = \sum c_i \hat{\psi}_i = (\sum c_i\, q_{ik}\, \hat{\psi}_k) = (\sum r\, c_k\, \hat{\psi}_k)$$

where $(\sum c_i\, q_{ik}) = r\delta_{ik}$.

This leads to $|(q_{ik} - r\, \delta_{ik}| = 0$,

which is the set of m linear equations in the m unknowns c_k, having a non-zero solution for the c_k, provided that the constant r satisfies the above *Characteristic* or *Secular Equation*.

All the Postulates 1 to 5 have concern only for the KINEMATICS (description at a particular time) of a quantum system. Postulate 6 specifies the TIME DEVELOPMENT of the state function of a quantum system.

4.2.14. POSTULATE #6

> THE SCHRODINGER EQUATION DESCRIBES THE BEHAVIOUR OF A WAVE FUNCTION, $\Psi(\vec{r}, t)$ IN SPACE AND TIME

In other words, the Schrödinger Wave Equation (TDSE) determines the development in time of the state function $\Psi(\vec{r}, t)$ (known for its initial form) in an undisturbed system, by the relation between $(\partial/\partial t)\Psi(\vec{r}, t)$ and $\Psi(\vec{r}, t)$, *i.e.* by

$$\left(\frac{\partial^2}{\partial x^2} + \frac{\partial^2}{\partial y^2} + \frac{\partial^2}{\partial z^2}\right)\Psi(\vec{r}, t) + \frac{2m}{\hbar^2}(E - V)\Psi(\vec{r}, t) = 0$$

Also by $\quad\hat{H}(\vec{r},\vec{p})\Psi(\vec{r},t) = \hat{E}\,\Psi(\vec{r},t)$

or, $\quad\hat{H}(\vec{r},\vec{p})\Psi(\vec{r},t) = i\hbar\,(\partial/\partial t)\Psi(\vec{r},t)$. $\quad$ (4.2.42)

This fundamental equation in Quantum Mechanics is known as the WAVE EQUATION.

4.2.15 POSTULATE #7

Two operators $\hat{Q}$ and $\hat{R}$ of quantum theory are such that their commutator brackets are proportional to the corresponding classical *Poisson brackets* of the classically defined functions f and r, according to the expression

$$\boxed{[\hat{Q},\hat{R}] = [\hat{Q}\hat{R} - \hat{R}\hat{Q}] \rightleftharpoons \hbar\{q,r\}} \quad (4.2.43)$$

where $\{q,r\}$ is the classical Poisson Bracket for the variables q and r.

The variables, if any in the Poisson Bracket are to be replaced by operators, observing: -

i) The co-ordinates and momenta must be expressed in Cartesian system of co-ordinates.

ii) The operator must be Hermitian.

These are the limitations and ambiguities.

If F and G are two functions of the canonical variables, q_i and p_i, the Poisson Bracket is defined as

$$\{F,G\} = \sum\left(\frac{\partial F}{\partial q'_i}\frac{\partial G}{\partial p'_i} - \frac{\partial F}{\partial p'_i}\frac{\partial G}{\partial q'_i}\right), \quad (4.2.44)$$

where $\quad q'_i = \frac{\partial H}{\partial p_i}, \quad p'_i = -\frac{\partial H}{\partial q_i} \quad$ (4.2.45)

and $\quad \frac{dF}{dt} = \frac{\partial F}{\partial t} + \{F,H\}$. $\quad$ (4.2.46)

This is the *dynamical; Equation of Motion* of a classical system, where

A feature of Poisson brackets is that they provide a test for recognizing constants of the motion; If

$$\frac{dF}{dt} = -\{F,H\} \quad (4.2.49)$$

then F = a <u>constant of motion</u>.

i.e., if $\{F,H\} = 0$ $\quad$ (4.2.50)

F does not depend on time t explicitly.

Such **constants of motion are known as conserved**.

Poisson brackets provide a powerful tool in formulating quantum theory: -

4.3.1 Time Variation of expectation values, i.e. motion of wave packets. EHRENFEST'S THEOREM:

Now there arises an important question of the classical limit of quantum theory. Because of the Uncertainty Principle, parameters of a system cannot be determined with complete accuracy. Before the advent of Quantum Mechanics, by convention uncertain experiments were repeated an number of times, say n, and to determine the average value of several measurements of, for example, a parameter, x, according to

$$\bar{x} = \frac{1}{n}\sum_1^n x_k ,\qquad(4.3.10)$$

where x_k = result of kth measurement,

If n_k = discrete number of times certain values of x_k occur,

$$\bar{x} = \frac{1}{n}\sum_1^n (x_k \frac{n_k}{n}) = \sum_1^n (x_k n_k) / \sum_1^n (n_k) \qquad(4.3.11)$$

$$\bar{x} = \frac{1}{n}\sum_1^n (n_k x_k) \qquad(4.3.13)$$

Discrete Probability of getting the value x_k is denoted by

$$p_k = n_k / \sum_1^n (n_k) = \frac{n_k}{n} .\qquad(4.3.14)$$

Average value, $\bar{x} = \sum p_k x_k$ \qquad(4.3.15)

= Expectation value of x = <x>.

Extending this to a continuous distribution of probabilities, ρ_x, by writing

$$<x> \ = \int \rho_x x dx = \int x dW ,\qquad(4.3.16)$$

where $dW = \rho_x dx$, and $\int \rho_x dx = 1$.

$$\rho_x = dW / dx \qquad(4.3.17)$$

The procedure used to get the expectation value of a parameter x can be extended to functions if one or more variables.

$$<f(x,y,z)> \ = \iiint \rho(x,y,z) \cdot f(x,y,z)\, dxdydz \qquad(4.3.18)$$

Replacing $f(x, y, z)$ by $\Psi(x, y, z)$ in Quantum Mechanics,

$$\rho(x, y, z) = \iiint \Psi^*(x, y, z) \cdot \Psi(x, y, z) \, dx\,dy\,dz \qquad (4.3.19)$$

$$<x> = \iiint \Psi^*(x) \cdot \Psi(x) \, dx \qquad (4.3.20)$$

$$= \underset{n \to \infty}{Lt} \frac{1}{n} \sum_{1}^{n}(x_k) \qquad (4.3.21)$$

Now consider the expectation value < momentum >.

Experimentally, this is easily defined. On the other hand, theoretically it is required define he expectation value of a QM operator. There is ambiguity as to its definition: Is it

$$<\hat{p}_x> = \int (-i\hbar \, \partial/\partial x)\Psi^* \Psi dx, \qquad (4.3.22)$$

or

$$<\hat{p}_x> = \int \Psi^* (-i\hbar \, \partial/\partial x)\Psi dx \qquad (4.3.23)$$

or

= what else?

To find out **Paul Ehrenfest** has given a procedure. Accordingly,

$$<x(t)> = \int x(t)\Psi^* \Psi dx \qquad (4.3.24)$$

In the motion of a classical particle, if the potential energy changes are negligible over the dimension of the wave packet,

Average value of v_x is $\bar{v}_x = \frac{d<x>}{dt}$

$$\therefore m\bar{v}_x = m\frac{d<x>}{dt} v_x = <\hat{p}_x>$$

4.3.2. To Express $<\hat{p}_x>$ in terms of $(-i\hbar \, \partial/\partial x)$ Operator

Result of measurement on x is such that $x \neq x(t)$.

$$<\frac{d<x>}{dt} = \frac{d}{dt}\int \Psi^*(x) \cdot \Psi(x) \, dx \qquad (4.3.25)$$

$$<\frac{d<x>}{dt} = \int x(t)\{\Psi^*(x) \cdot \frac{d}{dt}\Psi(x) + \Psi(x) \cdot \frac{d}{dt}\Psi^*(x)\} \, dx.$$

But the Schrödinger Equation (TDSE) is

$$i\hbar \, (\partial/\partial t)\Psi(x, t) = [-\frac{\hbar^2 \nabla^2}{2m} + V(x)]\Psi(x, t) \qquad (4.3.26)$$

i.e., $(\partial/\partial t)\,\Psi(x,t) = [\frac{i\hbar\nabla^2}{2m} + V(x)]\,\Psi(x,t)$

and $(\partial/\partial t)\,\Psi^*(x,t) = [\frac{-i\hbar\nabla^2}{2m} + V(x)]\,\Psi^*(x,t)$. (4.3.27)

Choosing arbitrarily the form of the wave function to be

$$\Psi(x,t) = A\,e^{i(kx-\omega t)} = A\,e^{i(px-Et)/\hbar} \qquad (4.3.28)$$

$$\frac{d\langle x\rangle}{dt} = \int x\{\Psi^*\,\frac{i\hbar\nabla^2}{2m}\cdot\Psi - (i/\hbar)V\Psi\} + x\,\Psi\,\frac{-i\hbar\nabla^2}{2m}\cdot\Psi^* + (i/\hbar)V\Psi^*\}dx \quad (4.3.29)$$

Since ∇^2 is Hermitian,

$$\frac{d\langle x\rangle}{dt} = \int [x\Psi^*\{\frac{i\hbar\nabla^2}{2m}\cdot\Psi\} - \Psi^*\{\frac{i\hbar\nabla^2}{2m}\cdot x\Psi\}]dx$$

$$= \{\frac{i\hbar}{2m}\}\int [\Psi^* x\nabla^2\cdot\Psi\} - \nabla^2\Psi\cdot x\Psi]dx$$

because $\nabla^2(x\Psi) = \nabla\cdot\nabla(x\Psi) = \nabla\cdot\{\Psi\nabla x + x\nabla\Psi\}$

$$= \Psi(\nabla\cdot\nabla x) + \nabla x\cdot\nabla\Psi + (\nabla\Psi\cdot\nabla x + x\nabla\cdot\nabla\Psi) \qquad (4.3.30)$$

$$= 0 + 2\nabla x\cdot\nabla\Psi + x\nabla^2\Psi.$$

$$<\frac{d\langle x\rangle}{dt} = \{\frac{i\hbar}{2m}\}\int [\Psi^*\{-2\nabla x\cdot\nabla\Psi\}]\,dx = \{\frac{i\hbar}{2m}\}\int [\Psi^*\{\nabla\Psi\}]\,dx,$$

$$= \{\frac{1}{m}\}\int [\Psi^*(-i\hbar\nabla)\Psi]\,dx = \frac{\langle\hat{p}_x\rangle}{m}$$

$$\therefore\quad \frac{d\langle x\rangle}{dt} = \frac{\langle\hat{p}_x\rangle}{m} \qquad (4.3.31)$$

which is NEWTON'S II LAW!

$$\therefore\quad m\frac{d\langle x\rangle}{dt} = \langle\hat{p}_x\rangle = \int [\Psi^*(-i\hbar\nabla)\Psi]\,dx$$

This equation is known as EHRENFEST'S THEOREM.

Now the question arises $\boxed{\text{why is it that }(-i\hbar\nabla)\text{ is not the factor }(+i\hbar\nabla)\,?}$.

The negative sign comes from an arbitrary choice made in the phase of the wave function, viz.,

$$\Psi(x,t) = A\,e^{i(kx-\omega t)} = A\,e^{i(px-Et)/\hbar}.$$

If instead, the choice

$$\Psi(x,t) = A\, e^{-i(kx-\omega t)} = A\, e^{-i(px-Et)/\hbar} \qquad (4.3.32)$$

were made then the treatment would certainly have led to a positive sign in the Ehrenfest theorem,

i.e., $\quad m\frac{d<x>}{dt} = <\hat{p}_x> = \int [\Psi^*(+i\hbar\nabla)\Psi]\, dx \qquad (4.3.33)$

4.3.3 To Prove $\frac{dp}{dt} = -\nabla V$

Or, find the Conservation Principle for WAVE PACKETS from the Schrödinger Equation:

Consider the time rate of change of $<\hat{p}_x>$

$$\tfrac{d}{dt}<\hat{p}_x> = \tfrac{d}{dt}\int [\Psi^*(-i\hbar\nabla)\Psi]\, dx,$$

$$= -i\hbar\left[\int \frac{\partial\Psi^*}{\partial t}\frac{\partial}{\partial x}dx + \int \Psi^*\frac{\partial}{\partial x}\frac{\partial\Psi}{\partial t}dx\right], \qquad (4.3.34)$$

According to <u>Green's Theorem</u>,

$$\iiint (f\cdot\nabla^2 g)-(g\cdot\nabla^2 f)\, d\tau = \iiint \nabla\, (f\nabla g)-(g\cdot\nabla f)\, d\tau. \qquad (4.3.36)$$

Applying Green's Theorem and the Schrödinger Equation for $(\partial\psi/\partial t)$ and $(\partial\psi^*/\partial t)$, in $<\frac{d\hat{p}_x}{dt}>$,

$$<\tfrac{d\hat{p}_x}{dt}> = (-i\hbar)\int (\tfrac{-i\hbar}{2m})[\nabla^2\Psi^*\nabla\Psi)+\Psi^*\nabla(\tfrac{i\hbar\nabla^2}{2m})\Psi)\, dx$$

$$+\int [\nabla\Psi^*\nabla\Psi - \Psi^*\nabla\Psi^* V\Psi)\, dx, \qquad (4.3.37)$$

where the Schrödinger equation is used.

$$<\tfrac{d\hat{p}_x}{dt}> = (\tfrac{-\hbar^2}{2m})\int [(\nabla^2\Psi^*)\cdot\nabla\Psi - \Psi^*\nabla^2\Psi^*]\, dx + \int [V\Psi^*\nabla\Psi - \Psi^*\nabla V\Psi)\, dx \qquad (4.3.38)$$

According to Green's II identity, the 1st integral vanishes, when taken over a very large surface. $<\tfrac{d\hat{p}_x}{dt}> = \int [\Psi^*(\partial V/\partial x)\Psi]\, dx \qquad (4.3.39)$

$$= -\int [\Psi^*(\nabla V)\Psi]\, dx = -<\nabla_x V> = <F_x> \qquad (4.3.40)$$

This is the QM equivalent of the 1-D *Newton's II Law of Motion*, valid for expectation values. This is also known as **Ehrenfest's Theorem**.

Thus the Ehrenfest Theorem shows the correspondence between the motion of a wave packet and the motion of the associated particle considered in Classical Physics. This treatment also demonstrates that the Newtonian laws of motion, viz. $\frac{dx}{dt} = \{\frac{P}{m}\}$ and $\frac{d<\hat{p}_x>}{dt} = -<\nabla_x V>$, are satisfied exactly by the wave motion of a wave packet described by Ψ, which is a solution of the Schrödinger Equation.

Worked out Example 4.3

Find the products of the uncertainties in x and p_x for the Gaussian wave packet $\psi(x) = N\,e^{-(x-x')^2/2\sigma^2}$. Given,. $\int_{-\infty}^{+\infty} e^{-au^2}\,e^{-bu}\,du = \sqrt{\frac{\pi}{a}}\,e^{b^2/4a}$

Solution: Step # 1 Given,. $\int_{-\infty}^{+\infty} e^{-au^2}\,e^{-bu}\,du = \sqrt{\frac{\pi}{a}}\,e^{b^2/4a}$

Step # 2 Using $\Delta x = [<x^2> - <x>^2]^{1/2}$, $\Delta p_x = [<p_x^2> - <p_x>^2]^{1/2}$, calculate the uncertainties in Δx and Δp_x.

Step # 3 Apply $\int_{-\infty}^{+\infty} e^{-au^2}\,e^{-bu}\,du = \sqrt{\frac{\pi}{a}}\,e^{b^2/4a}$, and show that $\Delta x \Delta p_x \geq \hbar/2$.

Worked out Example 4.4

A particle of mass m moving in a potential $V(r)$ has the Hamiltonian, $\hat{H}(\vec{r},\vec{p}) = [\frac{-\hbar^2 \nabla^2}{2m}] + V(r)$ $m\frac{d<r>}{dt} = m\frac{d<r>}{dt} = <\hat{p}_x>$, and $\frac{d<\hat{p}_x>}{dt} = -<\nabla_x V>$. Use the result that $i\hbar \frac{d<\hat{Q}>}{dt} = -<[\hat{Q}, \hat{H}]>$ and Ehrenfest's Theorem,.

Solution: Step # 1 Given: $i\hbar \frac{d<\hat{x}>}{dt} = -<[\hat{x}, \hat{H}]>$

Step # 3 Now, $<[\hat{x}, \hat{H}]> = \hat{x}\{[\frac{-\hbar^2\nabla^2}{2m}] + V(r)\} - \{[\frac{-\hbar^2\nabla^2}{2m}] + V(r)\}\hat{x}$

$= [\frac{-\hbar^2}{2m}]\{\hat{x}\nabla^2 - \nabla^2\hat{x}\} = [\frac{\hbar^2}{m}]\nabla$

Step # 3 $\therefore \frac{d<r>}{dt} = <\hat{p}_x>/m$ Similarly, $\frac{d<\hat{p}_x>}{dt} = -<\hat{\nabla}_x V>$.

4.4.1 CONSERVATION OF PROBABILITY:

Total probability $|\Psi|^2$ must be conserved: i.e. Normalization of a wave function Ψ is time independent

Proof: TDSE is $\hat{H}(\bar{r},\bar{p})\,\psi(r,t) = E\,\psi(r,t)$

$$\hat{H}(\bar{r},\bar{p})\,\psi(r,t) = (i\hbar\,\partial/\partial t)\,\psi(r,t)$$

i.e., $(i\hbar\,\partial/\partial t)\,\psi(r,t) = \{[\frac{-\hbar^2\nabla^2}{2m}] + V(r)\}\,\psi(r,t)$

and $(-i\hbar\,\partial/\partial t)\,\psi^*(r,t) = \{[\frac{-\hbar^2\nabla^2}{2m}] + V(r)\}\,\psi^*(r,t)$

Pre-multiplying the first equation by $\psi^*(r,t)$ and so by $\psi(r,t)$ the second equation, and subtracting,

$$[\tfrac{-\hbar^2}{2m}]\{\psi^*\nabla^2\psi - \psi\nabla^2\psi^*\} = i\hbar\,[\psi^*(\partial/\partial t)\,\psi + \psi(\partial/\partial t)\,\psi^*]$$

since $\psi^*\nabla\psi = \psi\nabla\psi^*$

$$\nabla^2 = [\partial^2/\partial x^2 + \partial^2/\partial y^2 + \partial^2/\partial z^2]$$

$$\hat{\nabla}^2\psi = \hat{\nabla}\cdot\hat{\nabla}\psi$$

$$\{[\tfrac{-\hbar^2}{2m}]\{\psi^*\hat{\nabla}\cdot\hat{\nabla}\psi - \psi\hat{\nabla}\cdot\hat{\nabla}\psi^*\} = -i\hbar\,[\psi^*(\partial/\partial t)\,\psi + \psi(\partial/\partial t)\,\psi^*]$$

$$[(\partial/\partial t)\,\psi^*\psi + \nabla\cdot(-i\hbar/2m)[\psi\nabla\psi^* - \psi^*\nabla\psi] = 0$$

Assuming $V\psi^*\psi = \psi^*V\psi$

i.e., $[\hat{H},V] = 0$

Integrating, $\iiint [(\partial/\partial t)\,\psi^*\psi]d\tau = -\iiint \nabla\cdot(-i\hbar/2m)[\psi\nabla\psi^* - \psi^*\nabla\psi]d\tau$ (4.4.1)

According to **Green's theorem**, $\iiint \hat{\nabla}\cdot V\,d\tau = \iint \hat{\nabla}V\,dS$

$(\partial/\partial t)\iiint[\psi^*\psi]d\tau = -\nabla(i\hbar/2m)\iint[\psi\nabla\psi^* - \psi^*\nabla\psi]\,\hat{n}\cdot dS$ (4.4.2)

where $\hat{n}$ is the outward *unit vector* to the surface. When the volume involved is the entire space, for a wave packet, *i.e.* surface S is at ∞, where ψ and $\nabla\psi$ vanish. So S also vanishes sufficiently fast as $|x|\to\infty$, that ∇ itself vanishes.

$$\therefore\ (\partial/\partial t)\iiint \psi^*\psi\,d\tau = 0 \qquad (4.4.3)$$

Thus for the state ψ the probability density, $\iiint \psi^* \psi d\tau$, is a constant in time. In other words, the numerical coefficient of ψ that normalizes it must be independent of time in order that the wave function may satisfy the Schrödinger equation.

In other words,

$$(\partial/\partial t)\iiint |\psi|^2 d\tau = 0 \tag{4.4.4}$$

i.e., the total probability is conserved, i.e., $|\psi|^2$ is time-independent. This is in virtue of the physical significance of $|\psi|^2 d\tau$.

4.4.2 HYDRODYNAMICAL WAVE THEORY of Schrödinger and Conservation of Probability Density

Following the procedure leading to equations (4.41) and (4.4.2) are obtained as in Section 4.4. Defining the position probability, ρ, of finding the particle at location r in the elemental volume

$d\tau = dxdydz$ is

$\rho d\tau \equiv \psi^*(\vec{r},t) \, \psi(\vec{r},t) \, d\tau$

$$(\partial/\partial t) \iiint \rho d\tau = -\nabla(i\hbar/2m)\iint [\psi \nabla \psi^* - \psi^* \nabla \psi] \hat{n} \cdot dS \tag{4.4.5}$$

Build up rate / unit time of the probability density in the volume element $d\tau$ between τ_1 and τ_2

$$= -(i\hbar/2m)[\psi \nabla \psi^* - \psi^* \nabla \psi].$$

It is known that the stream of charged particles constitutes the flow of an electric current.

To describe qualitatively the flow of such a current in terms of wave functions, the procedure is as follows:

Defining the PROBABILITY DENSITY OF CURRENT (*flux of probability density or stream density*) by the symbol, $J(\vec{r},t)$ by

$$J(\vec{r},t) = -(i\hbar/2m)[\psi \nabla \psi^* - \psi^* \nabla \psi] \tag{4.4.6}$$

$$= (\hbar/m) \, \text{Im} \, [\psi \nabla \psi^*], \tag{4.4.7}$$

since $[\psi^* \nabla \psi] = [\psi \nabla \psi^*]^*$, and

$$[\psi \nabla \psi^* - \psi^* \nabla \psi] = 2i \, \text{Im} \, [\psi \nabla \psi^*] \tag{4.4.8}$$

One may find also the explicit expression above as

$$J(\vec{r},t) = \text{Re}\,al\;\psi^*(\hbar/im)\nabla\psi \qquad (4.4.9)$$

$$\therefore \quad (\partial/\partial t)\iiint \rho\,d\tau = \nabla(-i\hbar/2m)[\psi\nabla\psi^* - \psi^*\nabla\psi]\,d\tau. \qquad (4.4.10)$$

becomes
$$(\partial/\partial t)\iiint \rho\,d\tau = \iiint[-\nabla\cdot J(\vec{r},t)]\,d\tau \qquad (4.4.11)$$

$$\boxed{(\partial/\partial t)\iiint \rho\,d\tau + \iiint[\hat{\nabla}\cdot J(\vec{r},t)]\,d\tau = 0} \qquad (4.4.12)$$

This equation arises in any theory in which an extensive quantity (like mass, charge and heat energy). This equation is known to satisfy a Law *of Conservation and the Conservation of Probability Density*. It is similar to the **Equation of Continuity in Hydrodynamics**; hence the name the 'Hydro-dynamical Wave Theory of Schrödinger'.

Multiplying the equation (4.4.12) for the conservation of probability density, by the electronic charge, e,

$$\boxed{(\partial/\partial t)\iiint \rho e\,d\tau + \iiint[\hat{\nabla}\cdot e\,J(\vec{r},t)]\,d\tau = 0} \qquad (4.4.13)$$

one then gets the *charge conservation*. This equation is the Continuity Equation in Electro-Magnetism.

This hydro-dynamical wave theory of ψ has strong criticisms, such as a) a wave packet being diffused in its structure must dissipate in course of time, and b) a charged particle like an electron is more or less a permanent entity!

4.5 COMMUTATION RULES

TO FIND THE COMMUTATOR [z, pz] using Poisson bracket:

The rules of commutation of linear operators in Wave Mechanics are derived from the dynamical variables by the use of Poisson Brackets. Consider $\hat{Q}$ and $\hat{R}$ are two dynamical quantities such that, as seen in Postulate VII, *i.e.*, equation (4.2.43)

$$[\hat{Q},\hat{R}] = [\hat{Q}\hat{R} - \hat{R}\hat{Q}] \rightleftharpoons \hbar\{q,r\} \qquad (4.5.1)$$

where $[\hat{Q},\hat{R}]$ is the operator to be associated with the classically computed Poisson Bracket of q and r. The Planck constant is introduced to define the quantum condition. Let q and r are the variables and represent z and p_z - coordinates of a single particle. Then in the *base* z and p_z,

$$\{z, p_z\}_{z, p_z} = \left[(\tfrac{d}{dz} z)(\tfrac{d}{dp} p_z) - (\tfrac{d}{dp} z)(\tfrac{d}{dz} p_z) \right] = 1 \qquad (4.5.2)$$

$$\therefore \quad [\hat{z}, \hat{p}_z] = (\hat{z}\hat{p}_z - \hat{p}_z\hat{z}) = i\hbar \{z, p_z\}_{z, p_z} = i\hbar \qquad (4.5.3)$$

Worked out Example 4.5

Using the QM operator equivalent of z and $\hat{p}_z$, obtain $[\hat{z}, \hat{p}_z]$.

Solution: $\boxed{\text{Step \# 1}}$ Consider a well behaved function $\psi(z)$.

$z \to \hat{z}$; and $p_z \to \hat{p}_z = -i\hbar \, (\partial / \partial z)$.

$\boxed{\text{Step \# 2}} \quad [\hat{z}, \hat{p}_z]\psi(z) = (\hat{z}\hat{p}_z - \hat{p}_z\hat{z})\psi(z)$

$\hat{z}\hat{p}_z \, \psi(z) = z\{-i\hbar \, (\partial / \partial z)\} \, \psi(z) = -i\hbar \, z(\partial / \partial z) \, \psi(z)$

$\boxed{\text{Step \# 3}} \quad \hat{p}_z z \, \psi(z) = \{-i\hbar \, (\partial / \partial z)\} z \, \psi(z) = -i\hbar \, (\partial / \partial z)\{z \, \psi(z)$

$\qquad = (-i\hbar) \, \{(z\partial / \partial z) \, \psi(z) + \psi(z) \cdot 1\} = z \, \hat{p}_z \psi(z) + (-i\hbar)\psi(z)$

$\therefore \quad (\hat{z}\hat{p}_z - \hat{p}_z\hat{z})\psi(z) = i\hbar \psi(z)$

$\therefore \quad [\hat{z}, \hat{p}_z] = (\hat{z}\hat{p}_z - \hat{p}_z\hat{z}) = i\hbar \qquad (4.5.4)$

or, $\quad [\hat{p}_z, \hat{z}] = (\hat{p}_z\hat{z} - \hat{z}\hat{p}_z) = -i\hbar. \qquad (4.5.5)$

Thus, $\hat{z}$ and $\hat{p}_z$ lack commutativity. $\hat{z}$ and $\hat{p}_z$ are conjugate (or complementary) variables. In contrast to ordinary multiplication in Classical Mechanics, multiplication of dynamic variables in Quantum Mechanics is, in general, non-commutative.

4.6 HEISENBERG'S UNCERTAINTY PRINCIPLE:

4.6.1 Definition of UNCERTAINTY:

The difference between a given measurement of z, denoted by z_{exp}, and its expectation value $<z>$ is

$$\Delta z = (z_{exp} - <z>) \qquad (4.6.1)$$

Δz = Deviation of measurement z from the mean value of z, i.e. $<z>$.

By analogy with the standard deviation in Statistics, UNCERTAINTY of z is defined as **Root Mean Square Deviation** of z, *i.e.* the *square root of the variance*, *i.e.* the RMS error, Δz, in the distributions of $|\Psi(z,t)|^2$.

Deviation square $(\delta z)^2 = <(z_{exp} - <z>)^2>$. (4.6.2)

Adopting, $\Delta z = \sqrt{<(z_{exp} - <z>)^2>} = \sqrt{<(z^2 - 2z<z> + <z>^2)>}$

$= \sqrt{(<z^2> - 2<z><z> + <z>^2)} = \sqrt{(<z^2> - <z>^2)}$ (4.6.3)

∴ $\Delta z = \sqrt{(<z^2> - <z>^2)}$ (4.6.4)

Similarly if p_z is the RMS deviation of the distribution in $|\Phi(p_z,t)|^2$,

$\Delta p_z = \sqrt{(<p_z^2> - <p_z>^2)}$. (4.6.5)

4.6.2 Derivation of the PRECISE STATEMENT of the position-momentum UNCERTAINTY Relation –
Conventional Operator Approach

Step # 1 Let $\hat{A}$ and $\hat{B}$ are two real hermitian operators, and

$[\hat{A}, \hat{B}] = i \hat{C}$ (4.7.7)

= Constant, i.e. NOT an operator.

This means the eigen function of $\hat{A}$ is not an eigen function of $\hat{B}$.

$(\hat{A}\hat{B})$ = hermitian, if $[\hat{A}, \hat{B}] = 0$.

Let $\hat{Q}$ be an arbitrary operator defined as

$\hat{Q} = (\hat{A} + i\lambda \hat{B})$ (4.7.8)

where λ = a real number variable.

Step # 2 Let the system be in arbitrary state Ψ, which is not an eigen function of $\hat{A}$ or $\hat{B}$. The conjugate of $\hat{Q}$ is $\hat{Q}^\dagger$.

∴ $\hat{Q}^\dagger = (\hat{A} + i\lambda \hat{B})^\dagger = (\hat{A}^\dagger - i\lambda \hat{B}^\dagger) = (\hat{A} - i\lambda \hat{B})$. (4.7.9)

Since $\hat{A}^\dagger = \hat{A}$, $\hat{B}^\dagger = \hat{B}$

Step # 3   The mean value in a state Ψ of a physical quantity associated with the Hermitian $\hat{Q}\hat{Q}^\dagger$, is

$$<\hat{Q}\hat{Q}^\dagger> = \int \Psi^*(z,t) \cdot \hat{Q}\hat{Q}^\dagger \cdot \Psi(z,t) dz$$

$$= \int (\Psi^* \hat{Q})[\hat{Q}^\dagger \cdot \Psi] dz = \int (\hat{Q}^\dagger \Psi)^* [\hat{Q}^\dagger \Psi] dz = \int |(\hat{Q}^\dagger \Psi)^*|^2 dz \quad (4.7.10)$$

as $<\hat{Q}\hat{Q}^\dagger>$ has to be a real quantity.

i.e., $<\hat{Q}\hat{Q}^\dagger> = \int \{(\hat{A} - i\lambda \hat{B})\Psi\}^* [(\hat{A} - i\lambda \hat{B})\Psi] dz \geq 0$ (4.7.11)

∴ $\int [\Psi^* \{(\hat{A} + i\lambda \hat{B})(\hat{A} - i\lambda \hat{B})\}\Psi] dz \geq 0$

i.e., $<(\hat{A} + i\lambda \hat{B})(\hat{A} - i\lambda \hat{B})> \geq 0$,

or $<(\hat{A} + i\lambda \hat{B})(\hat{A} - i\lambda \hat{B})> =$ positive, always,

or $0 \leq <(\hat{A} + i\lambda \hat{B})(\hat{A} - i\lambda \hat{B})>$ (4.7.12)

$0 \leq <(\hat{A}^2 + \lambda^2 \hat{B}^2 - i\lambda \hat{A}\hat{B} + i\lambda \hat{B}\hat{A}>$

$0 \leq <\hat{A}^2> + \lambda^2 <\hat{B}^2> - i\lambda <\hat{A}\hat{B} - \hat{B}\hat{A}>$ (4.7.13)

Step # 4 $0 \leq <\hat{A}^2> + \lambda^2 <\hat{B}^2> + \lambda \hat{C} = f(\lambda)$ (4.7.14)

where $\hat{C} = -i <\hat{A}\hat{B} - \hat{B}\hat{A}>$ (4.7.15)

$0 \leq f(\lambda)$ $f(\lambda)$ being a function of λ.

This inequality must hold irrespective of the magnitude of λ, where $f(\lambda)$ has no maximum. So it is necessary ask for which value of λ the function $f(\lambda) =$ smallest.

Step # 5   A minimum value for $f(\lambda)$ can be found as follows:

By differentiating with respect to λ, the function $f(\lambda)$, and

setting $\frac{d}{d\lambda} f(\lambda) = 0$, (4.7.14)

$\frac{d}{d\lambda} f(\lambda) - 0 + \hat{C} + 2\lambda <\hat{B}^2> = 0$.

∴ $\lambda_{min} = -\frac{\hat{C}}{2<\hat{B}^2>} = i^2 \frac{\hat{C}}{2<\hat{B}^2>}$.

∴ $\lambda_{min} = i\frac{<\hat{A}\hat{B}-\hat{B}\hat{A}>}{2<\hat{B}^2>} = i\frac{<[\hat{A},\hat{B}]>}{2<\hat{B}^2>}$

∴ $f(\lambda_{min}) = <\hat{A}^2> + \lambda_{min}^2 <\hat{B}^2> + \lambda_{min} \hat{C}$ (4.7.15)

$= <\hat{A}^2> + \left(i\frac{<[\hat{A},\hat{B}]>}{2<\hat{B}^2>}\right)^2 <\hat{B}^2> + \left(i\frac{<[\hat{A},\hat{B}]>}{2<\hat{B}^2>}\right)\hat{C}$

$= <\hat{A}^2> + \left(i\frac{<[\hat{A},\hat{B}]>}{4<\hat{B}^2>}\right) \geq 0.$

∴ $<\hat{A}^2><\hat{B}^2> \geq -<[\hat{A},\hat{B}]>^2/4.$ (4.7.16)

Step # 6   Now introducing an operator that describes the deviation of an individual measurement of an observable quantity described by the operator,

i.e., $\hat{A} \equiv \delta q = (q_{exp} - <q>),$ (4.7.17)

and $\hat{B} \equiv \delta p = (p_{exp} - <p>)$ (4.7.18)

∴ $<\hat{A}^2><\hat{B}^2> \geq -<[\hat{A},\hat{B}]>^2/4,$ (4.7.19)

becomes $<\delta q^2><\delta p^2> \geq -<[q,p]>^2/4.$ (4.7.20)

Step # 7   Defining $\Delta q = \sqrt{(q-<q>)^2} = \sqrt{(<q^2>-<q>^2)} = \sqrt{(<\delta q^2>)}$ (4.7.21)

And similarly, $\Delta p_z = \sqrt{(<p_z^2>-<p_z>^2)} = \sqrt{(<\delta p_z^2>)}$ (4.7.22)

Equation (4.7.20) becomes using

$(\Delta q \cdot \Delta p_q)^2 \geq -<(q'p_q - p_q q')>^2/4 = -(i\hbar)^2 \delta_{qq'}$ (4.7.23)

$\cdot (\Delta q \cdot \Delta p_q)^2 > \hbar^2/4$

∴ $(\Delta q \cdot \Delta p_q) \geq \hbar/2$ (4.7.25)

This is the same as HEISENBERG'S Uncertainty Relation.

The **Principle of Indeterminacy** is thus seen to be a direct consequence of non-commutativity of the operators q and p, i.e. q and p are CANONICALLY CONJUGATE pair of operators.

There are other different ways to arrive at the Uncertainty Relation, for example, look into book by Messiah or by Schwable.

CHWARZ' INEQUALITY states that scalar product

$$(f,g) = \int f^* g \, dz \qquad (4.7.29)$$

$$\int h^* h \, dz \geq 0 \qquad (4.7.30)$$

with $h = \int g^* g \, dz \qquad$ f- $\int$ f*.f dz g. $\qquad (4.7.31)$

It states that for any two complex functions; say f and g, of the real variables which are continuous over a certain domain in configuration space,

$$\left(\int f^* f \, dz\right) \cdot \left(\int g^* g \, dz\right) \geq \left(\int f^* g \, dz\right)^2$$

and obviously, $\left(\int f^* g \, dz\right)^2 \geq \left(\int \text{Im}\, (f^* g) \, dz\right)^2.$

$$= \left(\int \frac{1}{2i} (f^* g - g^* f) \, dz\right) \qquad (4.7.32)$$

If $\qquad f \equiv (p'\Psi) \text{ and } g \equiv (q'\Psi) \qquad (4.7.33)$

$$\therefore (\Delta\hat{p})^2 \cdot (\Delta\hat{q})^2 = \frac{1}{2i} \left\{ \int (p'\Psi)^* \cdot (q'\Psi) \, dz - \int (q'\Psi)^* \cdot (p'\Psi) \, dz \right\}^2$$

$$\geq \left\{ \int \Psi^* \cdot \frac{1}{2i} (p'q' - q'p')\Psi \, dz \right\} \geq \left\{ \int \Psi^* \cdot \frac{1}{2i} [p', q']\Psi \, dz \right\}^2 \qquad (4.7.34)$$

$$\geq \; <\frac{1}{2i} [p', q']>^2 \qquad (4.7.35)$$

Obviously [p', q'] = [p, q] = an operator.

$\therefore \qquad (\Delta\hat{p}) \cdot (\Delta\hat{q}) \geq \left| <\frac{1}{2i} [p, q]> \right| \qquad (4.7.36)$

This is the mathematical expression of HEISENBERG'S UNCERTAINTY PRINCIPLE.

$$(\Delta\hat{p}_z) \cdot (\Delta z) \geq \left| <\frac{1}{2i} [\hat{p}_z, \hat{z}]> \right|$$

$$\geq \left| \frac{1}{2i} [-i\hbar] \right| \geq \hbar/2 \qquad (4.7.37)$$

since $\quad [\hat{z}, \hat{p}_z] = i\hbar.$

$\therefore \qquad (\Delta\hat{p}_z) \cdot (\Delta z) \geq \hbar/2,$

$\qquad (\Delta\hat{p}_x) \cdot (\Delta x) \geq \hbar/2,$

$$(\Delta \hat{p}_y)\cdot(\Delta y) \geq \hbar/2 \qquad (4.7.38)$$

4.6.4 Uncertainty Principle for a WAVE PACKET:

According to de Broglie's fundamental assumption, in Quantum Mechanics, a freely moving particle is associated with an infinitely extended plane wave

$$\Psi_1(z,t) = A\, e^{-i(kz - \omega t)}$$

with energy E and momentum p given by expressions (2.1.16) and (2.1.17), viz.

$$E = \hbar\omega,$$

and $\quad p = \hbar k$.

The wave must be of infinite extent, otherwise it would not have a definite wavelength, but an infinitely extended wave does not determine any definite space points. The wave must be of infinite extent, otherwise it would not have a definite wavelength, but an infinitely extended wave does not determine any definite space points. The uncertainty principle, $(\Delta \hat{p}_z)\cdot(\Delta z) \geq \hbar/2$, tells that any measurement that determines the coordinate z of the particle with certain accuracy (Δz) excludes a measurement of the momentum, $\hat{p}_z$, to a better accuracy $(\Delta \hat{p}_z)$ than determined by the equation $(\Delta \hat{p}_z)\cdot(\Delta z) \geq \hbar/2$.

The Uncertainty Principle has far reaching consequences, viz. it implies that the observer perturbs the system by observing (measuring) it. For instance, by a precise observation of the localization point of the particle, the observer introduces a complete uncertainty in its momentum; the latter is by no means just a lack of knowledge to him, i.e., if the position of the particle is determined within the desired accuracy, i.e. to the observer's will, he interferes at the same time with the further *temporal* development of the quantum system and thus cannot make exact predictions about the momentum which is a variable, complementary (or, *conjugate*) to the spatial variable.

Worked out Example 4.6

Using the time-energy Uncertainty Relation estimate the mass, m_π, of a π-meson, knowing that the effective distance over which nuclear forces act is about 1.4 *fm*. Given, $c = 2.997 \times 10^8\, ms^{-1}$, $\hbar = 1.054 \times 10^{-34}\, J-s$ and $m_e = 9.1094 \times 10^{-31}\, kg$

Solution: Step # 1 $\Delta t = a/c$, where $a = 1.4\, fm$

Step # 2 Since $\Delta t = a/c$, where $a = 1.4\, fm$

∴ $\quad \Delta E = \hbar c/a,\ m_\pi = \Delta E/c^2 = \hbar/ca$

i.e., $m_\pi\ m_\pi = \dfrac{(1.054 \times 10^{-34} J\text{-}s)}{(2.997 \times 10^8\ ms^{-1}).(1.4 \times 10^{-15} m).(9.1094 \times 10^{-31}\ kg/m_e)} \approx 270\ m_e$

4.7 SEPARATION of Stationary WAVE FUNCTION fin ψ(r, t):

Partial differential equations are usually difficult to solve in terms of simple functions except for one very important class of cases: that class for which the solutions happen to be the 'product' of functions of the variables. A *linear partial differential equation* then separates into *ordinary differential equation*.

$$\nabla^2 \Psi(\vec{r},t) + \dfrac{2m}{\hbar^2}(i\hbar\partial/\partial t - V)\Psi(\vec{r},t) = 0 \qquad (4.7.1)$$

If $\qquad V(r, t) = V(r),$ \qquad (4.7.2)

only then *separation of variable technique* works.

Let $\qquad \Psi(\vec{r},t) = \psi(r)\cdot f(t),$ \qquad (4.7.3)

is the product of the SPATIAL function $\psi(r)$ and TEMPORAL function $f(t)$, of the independent variables r and t. That is, time-dependence factors out of these functions.

$$\hat{H}\Psi(\vec{r},t) = \hat{H}\psi(r)\cdot f(t) = [i\hbar\partial/\partial t]\ \psi(r)\cdot f(t) \qquad (4.7.4)$$

$$\dfrac{\hat{H}\ \psi(r)}{\psi(r)} = \dfrac{[i\hbar\ \partial/\partial t] f(t)}{f(t)}\ . \qquad (4.7.5)$$

The left hand side of the equation is a function of r only and the right hand side of the equation is a function of t only, *i.e.* the two are functions of one variable only. Further r and t are two *independent variables*; the above equation is true only if each side can be equal only to a constant, which is denoted by W.

i.e., $\qquad \dfrac{\hat{H}\ \psi(r)}{\psi(r)} = \dfrac{[i\hbar\ \partial/\partial t] f(t)}{f(t)} = W \qquad (4.7.6)$

$$\dfrac{d}{dt} f(t) = (-i/\hbar)\ W\ f(t) \qquad (4.7.7)$$

This means $Log\ f(t) = (-i/\hbar)\ W\ t$

$$\boxed{f(t) = e^{(-i/\hbar)\ W\ t}} \qquad (4.7.8)$$

Also, $\qquad \dfrac{\hat{H}\ \psi(r)}{\psi(r)} = W \qquad (4.7.9)$

i.e., $\qquad \hat{H}\psi(r) = W\psi(r) \qquad (4.7.10)$

and this only reveals that the spatial function $\psi(r)$ is labeled by the energy W, *i.e.*, $\boxed{\psi(r) \text{ is an energy eigen function}}$.

In other words, $\boxed{\hat{H}\, \psi_{W_n}(r) = W_n\, \psi_{W_n}(r)}$ (4.7.11)

where W_n is the *eigen value*, the energy of the nth state. Since the energy W_n remains constant in time t, the solutions are known as **Stationary States**. It is thus seen that the nth particular solution ψ_{W_n} satisfies the TISE, *i.e.*, ψ_{W_n} satisfying this equation corresponds to a state of the particle for which the energy is sharp or well defined or discrete or precisely known and does not change with time. The physically admissible wave function, which is the solution of the general Schrödinger's Equation (TDSE), is the full wave function, $\Psi(\vec{r},t)$, for the state is then

$$\boxed{\Psi(\vec{r},t) = \psi(r)\cdot e^{(-i/\hbar)\,W\,t}}$$ (4.7.12)

Using the Superposition Principle (Section 4.2.9, Postulate 5) a general solution of the TISE for the admissible wave function, may then be expanded in terms of the solutions just obtained for the stationary state equation given below: This is the linear combination of monochromatic wave functions of a special type,

viz., $\qquad \Psi(\vec{r},t) = \sum c_i\, \hat{\psi}_i(r)\, e^{(-i/\hbar)\,W\,t}$. (4.7.13)

This is also because of the linearity of the Schrödinger Equation. The sufficient condition for the equation (4.8.13) to be a valid expansion for a general wave function $\Psi(\vec{r},t)$ is that the functions $\hat{\psi}_i(r)$ form a *complete ortho-normal set* of functions.

4.8 PARITY ($\hat{\Pi}$) of Wave functions

The property of *symmetry or anti-symmetry* of wave functions is called PARITY. Eigen functions of the Hamiltonian, $\hat{H}$, of a system fall into two classes –

i) ODD functions of z

ii) EVEN functions of z.

This property arises because the HAMILTONIAN itself is an EVEN function of z (if V(z), the potential, is symmetric about the origin, *i.e.*, $V(-z) = V(z)$ for a bound system). Therefore, $V(z)$ = even.

$\therefore \qquad \hat{H}(-z) = \hat{H}(+z)$ = even function of x. (4.8.1)

$\therefore \qquad \hat{H}\,\psi(z) = \hat{E}\,\psi(z)$ (4.8.2)

Changing $\quad z \rightarrow (-z)$, everywhere,

$$\hat{H}(-z)\,\psi(-z) = \hat{E}\,\psi(-z)$$

∴ $$\hat{H}(z)\,\psi(-z) = \hat{E}\,\psi(-z). \qquad (4.8.3)$$

Since the eigen functions are non-degenerate (*i.e.*, for each energy there is only one eigen function, $\psi(-z)$ must be the same as $\psi(z)$ up to a normalization factor. For *linear operators* eigen functions are real.

$$\hat{\Pi}\psi(z) = \lambda\,\psi(z).$$

$$\hat{\Pi}\psi(-z) = \lambda\,\psi(-z)$$

$$\hat{\Pi}^2\psi(z) = \lambda\,\hat{\Pi}\,\psi(z) = \hat{\Pi}\,\lambda\,\psi(z) = \lambda^2\psi(z) \qquad (4.8.4)$$

i.e., the eigen function is restored when $z \rightarrow (-z)$ is made twice.

$$\psi(-z) = \lambda\,\psi(z) \qquad (4.8.5)$$

$$\psi(z) = \lambda\,\psi(-z) = \lambda^2$$

$$\lambda^2 = 1,\ \lambda = \pm 1$$

The eigen functions which belong to

(a) EVEN parity

$$\psi_e(z) = +\,\psi_e(-z) \qquad (4.8.6)$$

$\lambda = +1$, they are SYMMETRIC about the origin,

(b) ODD parity

$$\psi_o(z) = -\,\psi_o(-z) \qquad (4.8.7)$$

$\lambda = -1$ they are ANTI-SYMMETRIC about the origin.

Almost all interactions in nature are such that they give even functions of the $\hat{H}$. Only for weak interactions the eigen functions do have a definite parity.

An arbitrary function $\psi(z)$ can always be written as sum of an even function $\psi_e(z)$ and an odd function, $\psi_o(z)$.

$$\psi(z) = \tfrac{1}{2}\{\psi(z) + \psi(-z)\} + \tfrac{1}{2}\{\psi(z) - \psi(-z)\} \qquad (4.8.8)$$

4.9 SPECTRUM OF THE HAMILTONIAN (Eigen value spectrum):

Consider the Hamiltonian operator $\hat{H}$, which is n example of a real linear operator. All the ortho-normal eigen functions $\hat{\psi}_i(z)$ of $\hat{H}$ form a complete set, in the mathematical sense that an arbitrary normalized wave function $\Psi(z)$ can be expanded in terms of such eigen functions $\Psi(z)$ (or $\Psi(z)$ is a state which is a superposition of eigen states of $\hat{H}$) such that

$$\Psi(z,t) = \sum c_i(t)\, \hat{\psi}_i(z) \tag{4.9.1}$$

where the expansion coefficients cj are given by

$$c_i(t) = \int \hat{\psi}_i^*(z) \Psi(z,t) dz \tag{4.9.2}$$

and where

$$\int \hat{\psi}_i^* \hat{\psi}_j dz = \delta_{ij} \tag{4.9.3}$$

4.9.1 Discrete Spectrum

The subscript i classifies a set of all discrete eigen values E_i of E, since

$$\int \hat{\psi}_i^* \hat{\psi}_j dz = \delta_{ij}$$

$$<\hat{H}> = \int \Psi_E^* \hat{H} \Psi_E dz$$

$$= \int \left(\sum c_i(t)\, \hat{\psi}_i(z) \right)^* \hat{E} \left(\sum c_i(t)\, \hat{\psi}_i(z) \right) dz$$

$$= \sum E\, c_i(t)^2 \tag{4.9.4}$$

But by normalization,

$$\int \left(\sum c_i(t)\, \hat{\psi}_i(z) \right)^* \hat{E} \left(\sum c_i(t)\, \hat{\psi}_i(z) \right) dz \tag{4.9.5}$$

$$= \sum \sum \int c_i(t)\, \hat{\psi}_i^*(z)\, c_i(t)\, \hat{\psi}_i(z) dz$$

$$= \sum \sum c_i^* c_i\, \delta_{ij} = \sum c_i^* \left\{ \sum c_i\, \delta_{ij} \right\} = \sum c_i^* \left\{ \sum c_i \right\}$$

$$= \sum ci^* \sum cj = \sum |ci|^2 \tag{4.9.6}$$

where Einstein's *summation convention* is used.

$$\int \Psi^* \Psi\, dz = \int |\Psi|^2\, dz = \sum c_i^2 = 1 \tag{4.9.7}$$

This is called the <u>Completeness Relation,</u> which should be equal to unity. Here the *average value* of the quantity g in the given state Ψ

$$g_{av} = \sum g_i |c_i|^2 \qquad (4.9.8)$$

where c_i, is the expansion coefficient, is determined as follows:

$$\int \hat{\psi}_i^* \Psi \, dz = \int \hat{\psi}_i^* \left(\sum c_i \hat{\psi}_i \right) dz$$

$$= \sum c_i \int \hat{\psi}_i^* \hat{\psi}_i \, dz = \sum c_i \delta_{ij} = c_i \qquad (4.9.9)$$

The subscript *i* classifies a set of all discrete eigen values of *E*. The totality of all eigen values of $\hat{H}$ is termed its <u>energy spectrum</u> (or *eigen value spectrum*, if the operator is not $\hat{H}$). These eigen values form a **discrete set**, in Quantum Mechanics, and so the result is <u>discrete specrum</u> of eigen values of the system. A stationary state of a discrete spectrum always corresponds to a finite motion of the system, *i.e.* one in which neither the system nor any part of it moves off to infinity. (For, $\int |\Psi|^2 \, dz$ = finite, *i.e.* the system is in a bound state.). In general, **discrete eigen values are associated with bound states** (analogous to the *closed orbits* of Classical Mechanics.).

4.9.2 Continuous Spectrum

The assumption that *quadratic integrability* of eigen functions (*i.e.* functions satisfying $\hat{Q} \Psi - q \Psi$, where q = real) restricts that only very few linear operators would have complete set of eigen functions, and many common physical quantities would not qualify as observables. Discrete spectrum of eigen values was discussed in the previous Section. However, the eigen values of $\hat{Q}$, may also be continuous. To know continuous spectrum, it is assumed that the Postulate V (Section 4.2.9) holds good so that

i) $\qquad \Psi(z,t) = \sum c_i(t) \hat{\psi}_i(z) \qquad (4.9.1)$

where the expansion coefficients c_i are given by

$$c_i = \int \hat{\psi}_i^* \Psi \, dz, \qquad (4.9.2)$$

and $\qquad \int \hat{\psi}_i^* \hat{\psi}_i \, dz = \delta_{ij} \qquad (4.9.3)$

ii) $\qquad \hat{Q} \Psi = q \Psi \qquad (4.9.4)$

iii) where $\qquad q$ = real

(iv) Eigen functions, which are NOT *quadratically integrable*, can appear in the expansion of a quadratically integrable function $\hat{\psi}_i$ only with infinitesimal amplitude. Hence these functions are part of the complete set of functions only if they belong to a

CONTINUUM of real eigen values (In classical mechanics generally quantities run through a continuous series of values). In Quantum Mechanics also there are physical quantities (example, coordinates) whose eigen values occupy a continuous range).

The superposition procedure can, therefore, be generalized,

as $\quad \Psi(z,t) = \sum c_i(t) \, \hat{\psi}_i(z)$ \hfill (4.9.5)

where the symbol $\sum$ means for calculation of any physical quantity $\sum c_i(t) \, \hat{\psi}_i(z)$ can be used instead of Ψ even if the sum does not converge to the state it represents, and write

$$\Psi(z,t) \equiv \sum c_i(t) \, \hat{\psi}_i(z) + \int_{E_1}^{E_2} c_E \, \Psi_E \, dE \qquad (4.9.6)$$

with limits of integration E_1 to E_2, the interval within which the continuous set of eigen functions occur. The eigen functions are *Delta normalized and Orthogonalized* in the case of *continuous spectrum*,

viz., $\quad \boxed{\int \Psi_E^*(z') \, \Psi_E(z) \, dz = \delta(z-z')}$, \hfill (4.9.7)

This means eigen functions are subject to **E - normalization** (*i.e.* $\hat{H}$ - normalization). This gives the Orthogonality Relation of the continuum eigen functions. These functions form a continuous spectrum,

i.e., $\quad \int |\Psi|^2 \, dz \to Diverges$. \hfill (4.9.8)

The divergence of this integral, $\int |\Psi|^2 \, dz$, is always due to the fact that

$$|\Psi|^2 \neq 0, \text{ at } \infty \qquad (4.9.9)$$

The stationary states of a continuous spectrum correspond to an infinite motion of the system. (*i.e.* the system is in the free state).or associated with scattering states (corresponding to open classical orbits). Since the eigen values must be real,

NORMALIZATION of $\Psi_E(z')$ is given by

$$\int \delta(E - E') \, dE' = 1 \qquad (4.9.10)$$

$$(E - E') \, \delta(E - E') = 0 \qquad (4.9.11)$$

$$\int [\hat{H} \, \Psi_E^*(z)] \, \Psi_E(z) \, dz = \int \Psi_E^*(z) \, [\hat{H} \, \Psi_E(z)] \, dz \qquad (4.9.12)$$

is the Hermiticity condition to all *physically admissible* eigen functions, whether they are quadratically integrable or not.. For CONTINUOUS SPECTRUM, the *expansion coefficients* are given by

$$c_E = \int \hat{\psi}_E^*(z)\, \Psi_E(z)\, dz \qquad (4.9.13)$$

The continuous spectrum is seen to be as important as the discrete spectrum, and both are needed to make a set of eigen functions COMPLETE.

4.9.3 CLOSURE PROPERTY of Orthonormal set of Eigen Functions:

Another important relation (or condition) which a given set of ortho-normal functions must satisfy, if it is to be complete, can be derived from the identity

$$\Psi(z,t) \equiv \sum c_i(t)\, \hat{\psi}_i(z) \qquad (4.9.14)$$

$$\Psi(z,t) = \sum \hat{\psi}_i(z) + \int \hat{\psi}_i^*(z)\, \Psi(z't)\, dz \qquad (4.9.15)$$

REVIEW QUESTIONS

R.Q. 4..1 Examine the displacement operator $\hat{x}$ for Hermitian, for well behaved wave functions.

R.Q. 4.2 Examine if the momentum operator $\hat{p}_x = -i\hbar(\partial/\partial x)$ is Hermitian (or not) to the class of integrable and twice differentiable functions. Answer: $\hat{p}_x = -i\hbar(\partial/\partial x)$ is Hermitian).

R.Q. 4.3 Define a hermitian operator $\hat{Q}$ of a physical system.

R.Q. 4.4 Prove that for well-behaved wave functions of a physical system having a hermitian operator $\hat{Q}$, its eigenvalues q are all real.

R.Q. 4.5 Prove that the Laplacian operator is Hermitian.

R.Q. 4.6 Prove that the operator $\hat{E} = -i\hbar(\partial/\partial t)$ is Hermitian for well-behaved wave functions.

R..Q. 4.7 Show that the Hamiltonian operator $\hat{H}(z,p_z) = [\frac{-\hbar^2}{2m}\nabla^2 + V(z)]$ is Hermitian in a space of well-behaved functions.

R.Q. 4.8 State which, if any, of the following operators are Hermitian. (i) $\hat{z}\hat{p}_z$, and ii) ($\hat{z}-i\hat{p}_z$). Give reasons for your answer. (Answers: i) $\hat{z}$ and $\hat{p}_z$ are both Hermitian; $(\hat{z}\hat{p}_z) \neq$ Hermitian, because $[\hat{z},\hat{p}_z] \neq 0$. (ii) $(\hat{z}-i\hat{p}_z)$ contains Hermitian $\hat{z}$ and non-Hermitian $i\hat{p}_z$, so $(\hat{z}-i\hat{p}_z) \neq$ Hermitian operator.)

R.Q. 4.9 Given three operators $\hat{A}$, $\hat{B}$ and $\hat{C}$, express the commutator $[\hat{A}\hat{B},\hat{C}]$ in terms of the commutators

$[\hat{A},\hat{C}]$ and $[\hat{B},\hat{C}]$. (Answer: $[\hat{A}\hat{B},\hat{C}] = \hat{A}[\hat{B},\hat{C}]+[\hat{A},\hat{C}]\hat{B}$).

R.Q. 4.10 If $\hat{P}$ and $\hat{Q}$ are two Hermitian operators and if $[\hat{P},\hat{Q}]=i\hat{C}$, prove that $\hat{C}$ is a Hermitian operator. (Answer:

$(\hat{P}\hat{Q}-i\hat{C}) \neq$ Hermitian; i.e. $i\hat{C}$ is not Hermitian. This means $\hat{C}$ is Hermitian).

R.Q. 4.811 Obtain the expectation value of the variable z by means of the momentum wave functions. (Answer: Start using the expressions for the FT pair for $\psi(z)$ and $\varphi(p_z)$;

$<z> = \int dp_z\, \varphi^*(p_z)\, [i\hbar\, \partial/\partial p_z]\, \varphi(p_z)$).

R.Q. 4. 12 Prove hat $[\hat{H},\hat{p}]=0$, for a free particle, i.e. $\hat{H}=p^2/2m$.

R.Q. 4.13 Suppose E_n denotes the energy eigen values of a 1-D system, and $\psi_n(r)$, the corresponding energy eigen functions. If the normalized wave function of the system is given by $\Psi(z,t=0) = \frac{1}{\sqrt{2}}e^{i\alpha_1}\hat{\psi}_1(z) + \frac{1}{\sqrt{3}}e^{i\alpha_2}\hat{\psi}_2(z) + \frac{1}{\sqrt{6}}e^{i\alpha_3}\hat{\psi}_3(z)\alpha_i$, where α_i are constants. (1) Write down the wave function $\Psi(z,t)$ at time t, (2) Find the probability that at time t a measurement of the energy of the system gives the value E_2., (3) Does $<z>$ vary with time? (4) Does $<p_z>$ vary with time? (5) Does $E = <\hat{H}>$ vary with time? (Answers:: 1.

$\psi_n(z,t) = \frac{1}{\sqrt{2}}e^{i\alpha_1}\hat{\psi}_1(z)e^{-iE_1t} + \frac{1}{\sqrt{3}}e^{i\alpha_2}\hat{\psi}_2(z)e^{-iE_2t} + \frac{1}{\sqrt{6}}e^{i\alpha_3}\hat{\psi}_3(z)e^{-iE_3t}$; 2. Probability
= 1/3 ; 3. $<z>$ is time-dependent.; 4. $<p_z>$ is time-dependent; 5. $E = <\hat{H}>$ is time-dependent).

R.Q. 4.14 Evaluate $<x^n>$, for the physical system given by $\psi_n(x) = (\frac{1}{\sigma\sqrt{\pi}})^{1/2}e^{-x^2/2\sigma^2}$.

Given, , $\Gamma(1/2) = \sqrt{\pi}$, $\Gamma(m) = \Gamma(m+1)\Gamma(m-1)$, $\Gamma(m) = \int_0^\infty e^{-w}w^{m-1}dw$ (Answer:

$<x^n> = \sigma^n\Gamma(m-1) \equiv <x^{2s}> = \{\frac{n!}{(n/2)!}\}(\frac{\sigma}{2})^n, n=0,2,4,6,..$)

R.Q. 4.15 Obtain the Hamiltonian operator $\mathcal{H}$ for a single particle in the Cartesian rectangular co-ordinate system.

R.Q. 4.16 Deduce the Schrödinger Equation using the operators equivalent to the variables of a system.

R.Q. 4.17 Arrive at an expression for the QM operator $\hat{H}$ for a single particle in spherical co-ordinates. It is known $\hat{p}_r = -i\hbar \partial/\partial r$, $\hat{p}_\theta = -i\hbar \partial/\partial \theta$, and $\hat{p}_\varphi = -i\hbar \partial/\partial \varphi$

R.Q. 4.18 Compute the probability current density for a beam of particles under free motion (Answer: $J(\vec{r},t) = \frac{p}{m}|A|^2$)

R.Q. 4.19 Show that $J(\vec{r},t) = \frac{p}{m}|A|^2$, when $\Psi(z,t) = A\, e^{-i(p\,z - E\,t)/\hbar}$, and $J(\vec{r},t) = (-i\hbar/2m)\,[\psi\, \nabla \psi^* - \psi^* \nabla \psi]$.

R.Q. 4.20 Find the commutator between the gradient operator and x, with respect to a well behaved wave function $\psi(x)$. (Answer: $[\nabla, x]\,\psi(x) = \psi(x)$).

R.Q. 4.21 Find $[\nabla, x^2]$ (Answer: $[\nabla, x^2] = 2x$

R.Q. 4.22 Find $[\hat{H}, \hat{p}]$ for a freely moving particle, where $\hat{H}(\vec{r},\vec{p}) = [\frac{-\hbar^2 \nabla^2}{2m}]$.

R.Q. 4.23 Find the average value of an operator, say $\hat{p}_z \hat{z}$, in Quantum Mechanics. (Answer: $-i\hbar + <\hat{p}_z \hat{z}>$).

R.Q. 4.24 Show that $<\hat{z}\hat{p}_z> - <\hat{p}_z\hat{z}> = i\hbar$.

R.Q. 4.25 Show that $<\hat{z}\hat{p}_z - \hat{p}_z\hat{z}> = i\hbar$.

R.Q. 4.26 Show that $<\hat{q}'\hat{p}_q - \hat{p}_q\hat{q}'> = (i\hbar)\delta_{qq'}$

R.Q. 4.27 Show that $[\hat{p}_z, \hat{z}^n] = -n\,(i\hbar)\,z^{n-1}, n>1$, and $[\hat{p}_z, A(z)] = -(i\hbar)(\partial A(z)/\partial z)$, when A = A(z) is a differentiable function of z. (Answer: Use $[A, B^n] = \sum B_k\,[A, B]\,B^{n-1-k}$)

R.Q. 4.28 Find the shape of the wave packet for which the Uncertainty Principle attains its theoretical minimum value, so that $(\Delta \hat{p}_z)\cdot(\Delta z) \geq \hbar/2$. Use $<\hat{p}_z> = 0$ and $<\hat{z}> = 0$

R.Q. 4.29 A particle was prepared in a state with wave function $\Psi(z) = N\, e^{-z^2/2\Gamma}$; where $N = (1/\pi\,\Gamma)^{1/4}$. Evaluate δz and δp_z and confirm that the Uncertainty Principle is satisfied. (Answer: $<p_z>^2 = \hbar^2/2\Gamma$; $<z>^2 = \Gamma/2$; $\Delta z \Delta p_z = \sqrt{\{\Gamma/2\}(\hbar^2/2\Gamma)\Gamma/2} = \hbar/2$).

R.Q 4.30   Prove that the Parity operator $\hat{\Pi}$, defined by $\hat{\Pi}\psi(z) = \psi(-z)$, is Hermitian.

R.Q. 4.31   What do you mean by a *physically acceptable* (i.e *well-behaved*) wave function to represent a quantum system?

R.Q. 4.32 $\psi(x) = A e^{-\lambda (x - x_0)^2}$ Evaluate A so that $\psi(x)$ is normalized. A and x_0 are real
[Answer: $A = (2\lambda/\pi)^{1/4}$]

R.Q. 4.33   Deduce the Stationary state Schrödinger Equation (TISE) of a particle in motion a potential field.

R.Q. 4.34   Replace the classical mechanical expression $KE = mv^2/2$, in three dimensions, with its corresponding QM operator. [Answer: $-\frac{\hbar^2}{2m}[(\partial^2/\partial x^2) + (\partial^2/\partial y^2) + (\partial^2/\partial z^2)]$]

R.Q. 4.35   Prove that the operator $E = (-\hbar/i)(\partial/\partial t)$ is Hermitian for well-behaved wave functions.

R.Q. 4.36   What are two important properties of the TDSE?

R.Q. 4.37   Suppose $\psi(z,t) = A(z - z^3) e^{-i E t/\hbar}$. Find V(z) that the Schrödinger Equation is satisfied.

R.Q. 4.38   In a Uncertainty product, one of the factor is an angle, what is the other factor? Explain.

R.Q. 4.39   Show that $u(x) = e^{-x^2/2}$ is an eigen function of the operator $\hat{A} = \left(\frac{\partial^2}{\partial n^2} - x^2\right)$. Find the eigen value.

R.Q. 4.40   Distinguish between Probability density and Probability current density. Write down the equation relating them.

R.Q. 4.41   Stte and explain Ehrenfest's theorem.

&&&&&&&&&&

Chapter 5

EXACT SOLUTIONS – 1 TO 3

Particle in a Box, Tunneling,

and Electron in a Periodic Lattice

Chapter 5

EXACT SOLUTIONS – 1 TO 3

Particle in a Box, Tunneling, and Electron in a Periodic Lattice

"We cannot predict what comes out of a singularity ... It is a disaster for science".

- Stephen Hawking

5.1 INTRODUCTION:

In this Chapter, Quantum Mechanics is applied to simple but very general useful systems The objectives are to construct the Schrödinger Equation of the particle in a potential system under consideration; find the wave functions, the energy levels, probability distributions for the particle, and to apply the results of the various systems to practical problems in modern physics.

At a first glance Quantum Mechanics seems a poor substitute for Newtonian Mechanics, but a closer analysis will show that Quantum mechanics is a well-developed version of Classical Mechanics. It will be seen that Quantum Mechanics is the best effort to date to formulate a mechanics suitable for macroscopic and microscopic systems.

5.2 EXACT SOLUTIONS - Linear Motion

5.2.1 FREE-PARTICLE MOTION – CONTINUOUS SPECTRUM:

For particles that are not subject to force of any kind, i.e. in a potential free region, in one-Dimensional case,

$$V(x) = V_o = 0, \text{ everywhere.} \quad (5.2.1)$$

Let m = mass of the particle, and
p_x = its momentum.

The Hamiltonian of the free-particle is

$$\hat{H} = T + V(x) = \frac{1}{2m} p_x^2 + 0$$

$$= \frac{-\hbar^2}{2m} \left(\frac{\partial^2}{\partial x^2} \right). \quad (5.2.2)$$

The stationary states are represented by $u(x)$, and $u(x)$ must satisfy the Schrödinger Time Independent Equation (TISE) for the free particle.

i.e., $\hat{H} u(x) = E u(x)$ (5.2.3)

$$= \frac{-\hbar^2}{2m} \left(\frac{\partial^2}{\partial x^2} \right) u(x) = E u(x)$$

$$\nabla^2 u(x) + \frac{2m}{\hbar^2} E u(x) = 0.$$

$$\nabla^2 u(x) + k^2 u(x) = 0 \quad (5.2.4)$$

with $k^2 = \frac{2m}{\hbar^2}$. \quad (5.2.5)

The solutions of equation (5.2.4) are:
$$u(x) = N_+ [e^{+i(k_s x)}] + N_- [e^{-i(k_s x)}]. \quad (5.2.6)$$

where N_+ and N_- are normalization constants of the incident and the reflected waves.

Since $\hat{p}_x u(x) = \hbar k \, u(x) = \hbar k \left\{ N_+ [e^{+i(k_s x)}] + N_- [e^{-i(k_s x)}] \right\}$

i.e., $\hat{p}_x u(x) = \hbar k [e^{+i(k_s x)}]$

and $\hat{p}_x u(x) = -\hbar k [e^{-i(k_s x)}]$.

It is difficult to normalize $u(x)$, as some of these have constant amplitudes from $x = \pm \infty$. The eigen values of $\hat{p}_x$ are $\pm \hbar k$. The <u>energy of a free particle</u> is, therefore, given by equation (5.2.5).

$$\boxed{E = \frac{\hbar^2 k^2}{2m}}. \quad (5.2.7)$$

This analysis of free particle motion shows that the energy values form a CONTINUOUS SPECTRUM.

5.3. PARTICLE IN A BOX - Particle in a SQUARE WELL of infinite depth, or a particle enclosed within rigid walls.

This is a simple application of Quantum Mechanics. Consider a particle of mass m, constrained by impenetrable walls/ sides, whose motion is restricted, in the region of a 1-D potential, such as a gas molecule in a box, or a free electron in a piece of metal. (If the interaction between electrons and positive ions is neglected and if the height of the potential barrier is much larger than the k E of the electrons), as shown in diagram. A pure form of this type of system is not observed in nature. A free particle is enclosed in a box whose walls, at $x = 0$ and $x = a$, **cannot be penetrated** by the particle in motion. The boundary conditions (shown Fig. 5.1) are
$$V(x) = 0, \quad 0 < x < a \quad (5.3.1)$$
Since the potential increases abruptly at $x = 0$, and $x = a$,
$$V(x) = V_\infty = \infty, \quad 0 \leq x \geq a. \quad (5.3.2.)$$
This physically means very strong repulsive forces act on the particle at those two points, forcing the particle to reverse its motion. The physical situation is represented by a particle confined to a *Rectangular/ Square well potential*. It is desired to find the wave function and energy of the particle within this potential well.

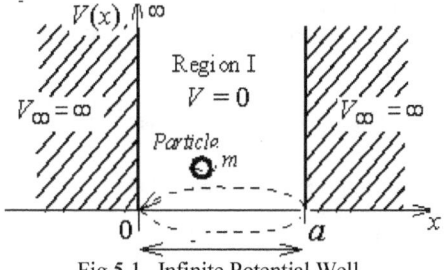

Fig 5.1 Infinite Potential Well

5.3.1 Free Particle Region - described by potential equation (5.3.1) and shown as Region I in Fig 5.1,

$$V(x) = 0, \quad 0 < x < a.$$

Step # 1 Construct the Schrödinger Equation: Let the stationary states are represented by $u(x)$. The Hamiltonian of the free-particle is

$$\hat{H} = T + V(x) = \frac{-\hbar^2}{2m}\left(\frac{\partial^2}{\partial x^2}\right). \quad (5.3.2)$$

the TISE for the free particle is equation (5.2.4)

$$\nabla^2 u(x) + k^2 u(x) = 0. \quad (5.3.3)$$

with $\quad k^2 = 2mE/\hbar^2 \quad (5.3.4)$

This is a differential equation of classical simple harmonic motion.

Step # 2 The general solution (using the superposition form) of equation (5.3.3) are:

$$u(x) = N_+ e^{+i(kx)} + N_- e^{-i(kx)} \quad (5.3.5)$$

which contains motion back and forth. N_+ and N_- are arbitrary constants:

Step # 3 To evaluate N_+ and N_-.

The wave functions have to be finite, and $u(x)$ must be continuous. For well behaved wave functions they have to be subject to the boundary conditions (Postulate 1) $u(x = 0) = 0$, and applying this in equation (5.3.5)

$N_+ \cdot (1) + N_- \cdot (1) = 0$

i.e., $\quad N_+ = -N_- \quad (5.3.6)$

$\therefore \quad u(x) = N_+ [e^{+i(kx)} - e^{-i(kx)}] \quad (5.3.7)$

$u(x) = N_+ [(\cos kx + i \sin kx) - (\cos kx - i \sin kx)]$

i.e., $\quad u(x) = N_+ [2i \sin kx]$

or $\quad u(x) = A \sin kx \quad (5.3.8)$

Step # 4 Again, applying $u(x = a)$ in (5.3.8) gives

$u(x = a) = A \sin ka$

Step # 5 $u(x = a) = 0$ for $u_s(x)$ to be continuous at $x = a$.

i.e., $\quad u(x = a) = A \sin ka = 0$

$A \neq 0$,

otherwise it results in a trivial solution. Thus

$\sin ka = 0$

This is possible only if

$ka = s\pi$, where $s = \pm 1, \pm 2, \pm 3, \ldots \quad (5.3.9)$

and $s = 0$ is not possible since $ka \neq 0$.

Therefore the boundary conditions above are satisfied only for a discrete sequence of values of k,

where $\quad k_n = n\pi/a \quad (5.3.10)$

Step # 6 The de Broglie Relation is $p = \hbar k$, from equation (5.3.10)

$\hbar k_n = n\hbar\pi/a$

are the possible values of momentum for the particle

5.3.2 Energy Quantization (DISCRETE SPECTRUM):

Step # 7 From equation (5.3.4)

$k^2 = 2mE/\hbar^2$

$E_n = \hbar^2(n\pi/a)^2/2m$

$$\boxed{E_n = \frac{\hbar^2\pi^2}{2m}\frac{n^2}{a^2} = \frac{h^2}{8m}\frac{n^2}{a^2}; \quad n = 1, 2, 3, \ldots}$$ \hfill 5.3.11)

This expression defines thus a discrete set of ALLOWED energy eigen values for the particle in motion within the box. n is, therefore, called the QUANTUM NUMBER. The conclusion is, therefore, that the particle cannot have any arbitrary energy, but only as per equation (5.3.11).

5.3.3 Features of the Energy Levels

1) The set of allowed energies - the stationary states - for the particle in the box are known. That is, the energy of the particle is QUANTIZED.
2) There are many QUADRATICALLY SPACED energy levels possible. The situation of only certain energy values are permitted is not a peculiarity of this particular quantum system, but it generally holds whenever Schrödinger's Equation is solved for a potential energy which confines the particle to move in a limited region. Energy quantization is due to the fact that the wave function is determined by the potential energy and boundary conditions.

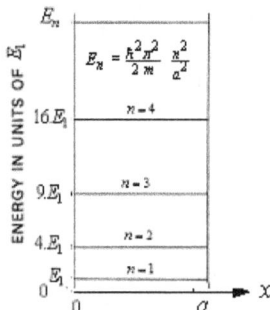

Fig. 5.2 Quadratically spaced Discrete energy levels of a particle in a box

3) All energy levels are *non-degenerate*.
4) ZERO-POINT Energy, E_1. It is quite interesting to note that the minimum energy of a particle in a 1-D box is

$$E_1 = \frac{h^2}{8ma^2} \neq 0$$

This minimum energy is related to the Uncertainty Principle.
As the position uncertainty is $(\Delta x) = 2a$, the width of the well, $(\Delta p_x) = 2p_x$ (back and forth),

$(\Delta p_x)(\Delta x) \geq \hbar/2$,

i.e., $(2p_x)(2a) \geq \hbar/2$

$p_x \geq \hbar/8a$

= the momentum uncertainty

corresponding to (Δx). Since the minimum kinetic energy,

$$E = \frac{p_x^2}{2m} = \frac{(\hbar/8a)^2}{2m}.$$

i.e. $E = \dfrac{h^2}{128\, m\, a^2}$

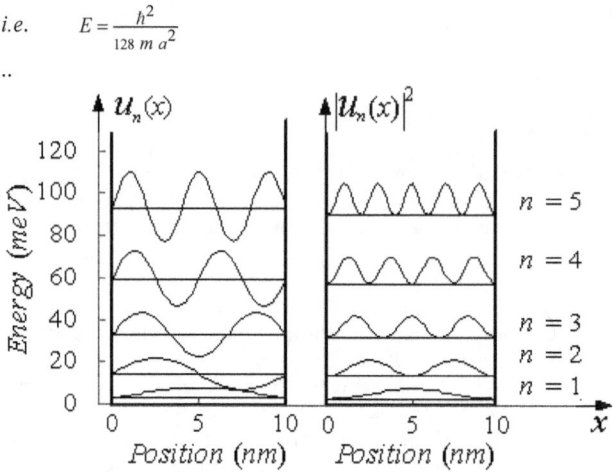

Fig. 5.3 The $u_n(x)$, $|u_n(x)|^2$ and energy levels of a particle in a well.

The existence of a *zero-point energy* is typical of all problems in which a particle is confined to move in a limited region. Thus there is the absence of the QUIESCENT State (state of zero energy). The energy levels of an electron in an infinite well of width 0.39 nm are compared with those of finite depth in Fig. 5.3.

5.3.4 Wave function of the BoundState
The wave function of the system must be of the form (using equations (5.3.8) and (5.3.9):
$$u_n(x) = A\, \operatorname{Sin}(n\pi/a)\, x \qquad (5.3.13)$$
for $0 < x < a$. It has definite parity.
$$u(x) = 0, \text{ for } x < 0, x > a$$
This is illustrated in Fig.5.3. $u_n(x)$ for the first 3 as well as 7th energy levels are represented, together with the position probability distribution function $P_n(x) = |u_n(x)|^2$ for these states are shown in Fig. 5.4. Thus a feature to be stressed about the Schrödinger Equation is that the quantum numbers are not introduced in it as a special assumption, they arise just the same way as that shown in the particle in a box.

5.3.5 Determination of coefficient A - NORMALIZATION of the wave function:
To normalize the wave functions, recollect the Bohr's condition,
$$\int_0^a u_n^*(x)\, u_n(x)\, dx = 1.$$
$$= \int_0^a [A\, \operatorname{Sin}(n\pi x/a)]^*\, [A\, \operatorname{Sin}(n\pi x/a)]\, dx$$
$$= \int_0^a [A^2\, \operatorname{Sin}^2(n\pi x/a)]\, dx$$
$$= [\tfrac{a}{n\pi} A^2] \int_0^{n\pi} [\operatorname{Sin}^2(n\pi x/a)]\, d(\tfrac{n\pi}{a} x)$$
Put $\xi = (n\pi/a)\, x$,
$$= [\tfrac{a}{n\pi} A^2] \int_0^{n\pi} [\operatorname{Sin} \xi]\, d\xi$$
where Standard Integral

$$\int_0^{n\pi} [\operatorname{Sin} \xi] \, d\xi = \xi/2, \tag{5.3.15}$$

$$\int_0^a u_n^*(x) u_n(x) \, dx = [\tfrac{a}{n\pi} A^2][\tfrac{\xi}{2} - \tfrac{1}{4}\operatorname{Sin} 2\xi]_0^{n\pi}$$

$$= [\tfrac{a}{n\pi} A^2](\tfrac{n\pi}{2}) = [\tfrac{a}{2} A^2] = 1,$$

for normalization. $A = \pm\sqrt{2/a}$. \hfill (5.3.16)

So the normalized wave function for the particle inside the well is obtained as

$$\boxed{u_n(x) = \pm\sqrt{2/a} \operatorname{Sin}(n\pi x/a)} \tag{5.3.17}$$

$$\boxed{u_n(x,t) = \pm\sqrt{2/a} \, e^{-i\omega t} \operatorname{Sin}(n\pi x/a); \text{ for } 0 \le x \le a} \tag{5.3.18}$$

Note 1: If one deals with a potential well with boundaries $-a/2 \le x \le +a/2$, one will get two wave functions as solutions:

$$u_n(x) = \pm\sqrt{\tfrac{2}{a}} \operatorname{Sin}(\tfrac{n\pi}{a} x); \text{ for } n = \text{even}$$

and $\qquad u_n(x) = \pm\sqrt{\tfrac{2}{a}} \operatorname{Cos}(\tfrac{n\pi}{a} x); \text{ for } n = \text{Odd}.$

Thus the treatment of a simple method was applied to obtain a 'rough estimate' of the zero-point energy of a system.

Note 2: every increase in number of dimension, wave functions are multiplied, and energies are added.

Worked out Example 5.1

Calculate the energy of an electron confined in an atom, treating it as a particle in a box. Given the radius of the atom is the Bohr radius a_o. Given, $m_e = 9.11 \times 10^{-31}$ kg, $1 \, eV = 1.6021 \times 10^{-19}$ J, $a_o = 0.53$ nm, $h = 1.054 \times 10^{-34}$ J–s

Solution: Step # 1 Given: $h = 6.6256 \times 10^{-34}$ J – s; $m_e = 9.1094 \times 10^{-31}$ kg;

$1 \, eV = 1.6021 \times 10^{-19}$ J $\;\; a_o = \hbar^2/m_e e^2 = 0.529167 \times 10^{-10}$ m; $E_n = \left(\dfrac{\hbar^2}{2m}\right)\left(\dfrac{n\pi}{a_o}\right)^2$,

where; $n=1$

Let the atom is considered as an infinitely deep well with width a_o, containing an electron. Consider the electron in the ground state ($n=1$).

Step # 2 $E_n = \dfrac{n^2 \pi^2 \hbar^2}{2 m_o a^2}$, where $n = 0, 1, 2, 3, \ldots$ \hfill (5.3.11)

$$E_1 = \dfrac{\pi^2 \hbar^2}{2 m_o a^2} = \dfrac{\pi^2 (1.054 \times 10^{-34} J\text{-}s)^2}{2 (9.11 \times 10^{-31} kg)(0.53 \times 10^{-9} m)^2} = \dfrac{(6.05 \times 10^{-18} J)}{(1.6021 \times 10^{-19} J/eV)} = -37 \, eV$$

(the negative sign is required to indicate that the electron is bound within the atom). But the ground state energy of the electron in a H-atom = -13.6 eV.

5.3.6 TRANSMISSION and REFLECTION at a BARRIER or POTENTIAL- Discontinuities, A FINITE PTENTIAL STEP) RECTANGULAR PTENTIAL WALL

The simplest of the bound problem of a quantum system was discussed in Section 5.3. In this section an asymmetric problem will be treated. A classical particle (for example, a bowling ball) with kinetic energy, T, moves toward a potential barrier.

$$V(x<0) < E$$
$$V(x = 0) = E$$
$$V(x>0) > E$$
$$E < V_0$$

The ball rolls up to the slope, and a short time later, at $x = 0$ all the kinetic energy is transformed into potential energy, and
$$V(x = 0) = E. \quad (5.4.1)$$
At this moment the ball reverses its direction of motion and starts to roll back down the slope. It will never be found in region II,
i.e., at $x > 0$, if $V(x>0) > E$. (5.4.2)
which represents the bound system. It will be found that at region I,
$$V(x<0) < E \quad (5.4.3)$$
i.e., one has the free (unbound) system. To investigate the situation as a quantum mechanical system, let a particle with kinetic energy, with total energy, E approaches a potential barrier of height V_0. Depending upon E and V_0, there are two cases:

(i) $E < V_0$,
and (ii) $E > V_0$.

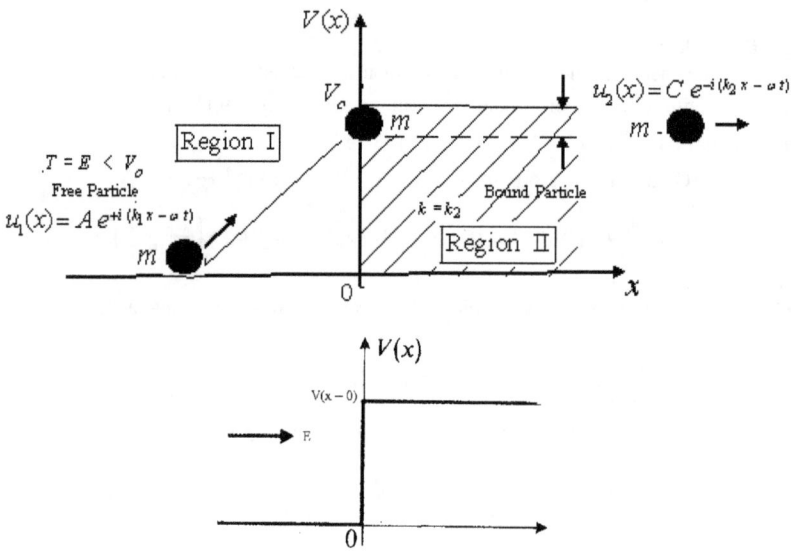

Fig. 5.4 Potential step

5.4.1.1 Case I: $V_0 > E$; $x > 0$; V_0 = constant . (Fig 5.4)

For mathematical convenience, let there is a 'discontinuous transition' from
$$V(x = 0) \rightarrow V(x) = V(0) = V_0.$$
for $x \le 0$, in Region I :

Step # 1 The Schrödinger Equation is

$$\nabla^2 u_1(x) + [2mE/\hbar^2] u_1(x) = 0 \quad (5.4.4)$$
Putting $\quad k_1^2 = 2mE/\hbar^2 \quad (5.4.5)$
$$\nabla^2 u_1(x) + k_1^2 u_1(x) = 0 \quad (5.4.6)$$

Step # 2 The solutions of equation (5.4.5) are:
$$u_1(x) = A e^{-(i k_1 x)} + B e^{+(i k_1 x)} \quad (5.4.7)$$
where A and B are normalization constants of the 'incident' and the 'reflected' waves from the wall. A and B are arbitrary constants.

Case 2: $E > V_0, x > 0$; $V_0 =$ constant in **Region II**:

Step # 3 $\quad \nabla^2 u_2(x) + [(2mE/\hbar^2) - V_0](u_2(x)) = 0.$ (5.4.8)
Putting $\quad k_2^2 = 2m(V_0 - E)/\hbar^2 \quad (5.4.9)$
$$\nabla^2 u_2(x) + k_2^2 u_2(x) = 0 \quad (5.4.10)$$

Step # 4 The general solution of equation (5.4.10) is
$$u_2(x) = C e^{-(k_2 x)} + D e^{+(k_2 x)}. \quad (5.4.11)$$
where C and D are arbitrary constants.

Step # 5 For a physically meaningful solution
$u_2(x) =$ finite, for $x =$ large.
This is as per Postulate 1. This means the second term in equation (5.4.11) violates $u_2(x)$ to be finite. This is possible only if D = 0.

∴ $\quad u_2(x) = C e^{-(k_2 x)}. \quad (5.4.12)$

Case 3: Thus the wave equation is obeyed within the two regions based on finiteness of the wave function.

Step # 6 It is known that a discontinuity of $\nabla u(x) \equiv u'(x)$ or of u(x) would result in letting $u(x) \equiv u''(x) \to \infty$,
and this case (u" suffering a discontinuity) occurs only if E or $V \to \infty$ in the Schrödinger Equation at x = 0 (the boundary of regions I and II). Still let us assume that u(x) and its derivatives shall be continuous everywhere including the point x = 0.
The boundary conditions are:
Amplitude continuity: $\quad u_1(x=0) = u_2(x=0) \quad (5.4.13)$
Slope continuity: $\nabla u(x=0) = u_1'(x=0)$
$$= \nabla u_2(x=0) = u_2'(x=0) \quad (5.4.14)$$

Step # 7 Applying the condition (5.4.13), from (5.4.7) and (5.4.12),
$$u_1(0) = A + B = A + B u_2(0)$$
$$= C^{-k_2 0} = C.$$
That means, $\quad C = A + B \quad (5.4.15)$
Again, from (5.4.7) and (5.4.12), applying condition (5.4.14) gives
$$u_1'(x=0) = -k_2 C$$
but $\quad C = A + B. \quad (5.4.15)$
$B(k_2 - i k_1) = -A(k_2 + i k_1)$
$\alpha = B/A = -(k_2 + i k_1)/(k_2 - i k_1).$
∴ $\quad = -(k_2^2 + k_1^2)/(k_2^2 - i k_1^2). \quad (5.4.16)$
∴ $\quad C = A + B = A + \alpha A.$

$$\alpha = B/A = \{2k_1/(k_2 + ik_1)\} \tag{5.4.17}$$

Equations (5.4.16) and (5.4.17) for the unknowns are not homogeneous. So they have a solution for many values for k_1 and k_2. This means the energy levels E of the particle, given by equations (5.4.5) and (5.4.9), form a CONTINUOUS SPECTRUM in complete conformity with the fact that it moves in an infinite region. Hence

$$u_1(x,t) = u_1(x) e^{-(iEt/\hbar)}.$$
$$= A e^{-i(k_1 x - Et/\hbar)} + B e^{+(ik_1 x - Et/\hbar)}. \tag{5.4.18}$$

$u_1(x)$ is normalized so that A = 1.

$u_1(x,t)$ consists of

(a) a term, $e^{+i(k_1 x - \omega t)}$, representing a wave traveling in the positive x-direction – the incident beam,

(b) and a term, $e^{-i(k_1 x - \omega t)}$, representing a wave traveling in the negative x-direction – reflected beam.

Thus equation (5.4.18) represents the wave function of a particle moving in a 1-D potential of the form $E < V_o$,

(v) The probability density of the particle in the classically forbidden region II is, from equations (5.4.12) and (5.4.17),

$$|u_2(x)|^2 = |C e^{-k_2 x}|^2.$$
$$= |\{2k_1/(k_1 + ik_2)\} A e^{-k_2 x}|^2.$$
$$= \{4k_1^2/(k_1^2 + k_2^2)\} A^2 e^{-2k_2 x}.$$
$$= \{4E A^2/V_o\} e^{-2\sqrt{2m(V_o - E)/\hbar^2} \, x} \tag{5.4.19}$$

(a) At x = 0, $\quad |u_2(x)|^2 = \{4E A^2/V_o\} \cdot 1 \tag{5.4.20}$

(b) At $x = \frac{1}{2}k_2 = \hbar/2\sqrt{2m(V_o - E)}$, $|u_2(x)|^2$ = negligible.

As $V_0 \to \infty$, both the maximum value of the function and the distance at which it becomes negligible, and one obtains the necessary boundary condition,

$$u_2(0) = 0$$
$$u_1(x) = A e^{-i(k_1 x)} + B e^{+(ik_1 x)}. \tag{5.4.7}$$
$$(k_2 \pm ik_1) = \sqrt{\{(k_2^2 + k_1^2)\}} \, e^{\pm i \arctan(k_1/k_2)}$$

If $\quad \delta = \tan^{-1}(k_1/k_2) \tag{5.4.8}$

$$u_1(x) = [A\, 2i\, e^{-i \tan^{-1}(k_1/k_2)}] \, Sin[k_1 x - \tan^{-1}(k_1/k_2)]. \tag{5.4.9}$$

This function $u_1(x)$ has the form of a standing wave $Sin[k_1 x - t\delta]$. But for the unimportant constant,

$[A\, 2i\, e^{-i\delta}]$, $u_1(x)$ is purely real.

Probability current density:

$$\vec{J}(x,t) = (-i\hbar/2m)[\psi \nabla \psi^* - \psi^* \nabla \psi] = 0.$$

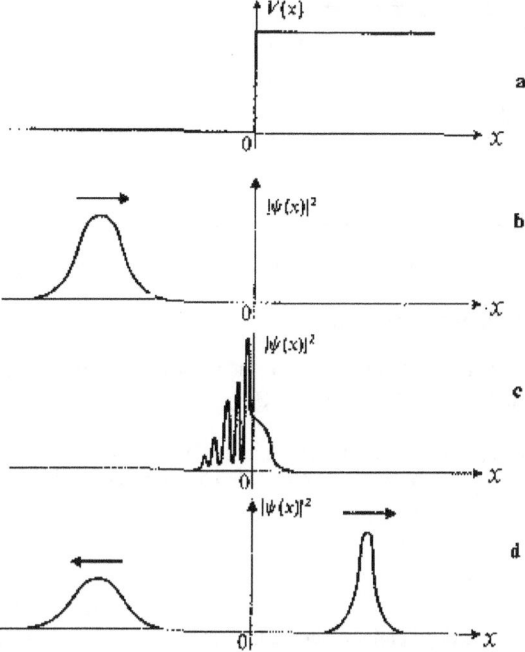

Fig. 5.5 Behavior of a Wave packet at a Potential Step, (a) $E > V_0$;
(b) particle approaching the step; (c) in x<0 region incident and reflected waves interfere. (d) after a time t, reflected and transmitted wave factions.

This is equivalent to the assertion that TOTAL REFLECTION occurs. This result may be arrived also as follows: -
The intensity of the incident and reflected beams in equation (5.4.18) are in the ratio
(Incident amplitude)^2 : (reflected amplitude)^2
$$\equiv |A|^2 : |B|^2$$
$$\equiv 1 : |\alpha|^2.$$
Probability of reflection $\equiv$ Reflectance,
$$R = |B/A|^2 = |\alpha|^2.$$
But $\quad |\alpha|^2 = \alpha^*\alpha$
Reflectance,
$$R = \left\{-(k_2^2 + k_1^2)/(k_2 + i k_1)^2\right\}\left[-(k_2^2 + k_1^2)/(k_2 - i k_1)^2\right]$$
$$= \left\{-(k_2^2 + k_1^2)/(k_2 + i k_1)^2\right\} = 1 \qquad (5.4.21)$$

Intensity of reflected beam = incident intensity. This means the particle suffers 'total internal reflection' as in classical mechanics. This also follows from the fact that the wave function in region II decays exponentially to zero. If T is the transmiitivity of particles, then T + R = 1.

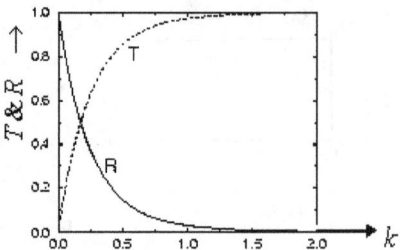
Fig. 5.6 Transmittance (T) and reflectance (R) of particles at a Potential Step

Therefore, T = 0, for the above case.

5.5. **FINITE POTENTIAL BARRIER: or POTENTIAL BARRIER PENETRATION or QUANTUM MECHANICAL TUNNELING:**

It is necessary to compute the transmission and reflection coefficients of a particle incident on a 1-D potential barrier, of Fig 5.7, mathematically expressed as

$$V(x) \begin{cases} = 0, & \text{for } x < a \\ = V_1, & a < x < b \\ = V_2, & x > b \end{cases} \qquad (5.5.1)$$

where $V_2 < E < V_1$,

One may view a general potential function, having the shape shown in Fig 5.7, which can be simplified for mathematical analysis as shown by the dotted line.
Classically, since $E < V_1$, the particle in region I can never penetrate the potential barrier II and appear in region III.
The 1-D Schrödinger Equations for the three regions are:

Region 1 $\qquad \nabla^2 u_1(x) + k_1^2 u_1(x) = 0 \qquad (5.5.3)$

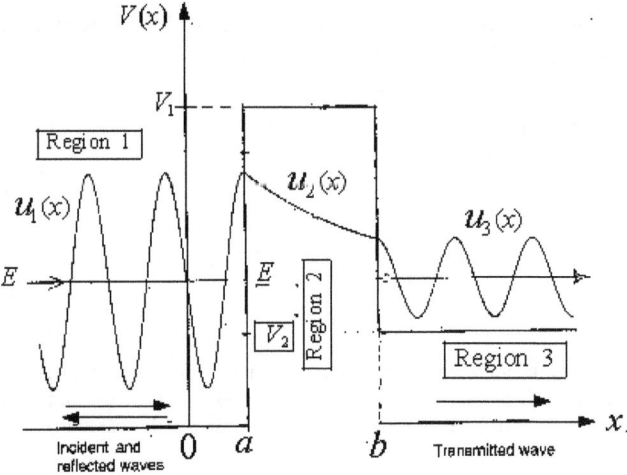
Fig. 5.7 Finite Potential Barrier and Tunnel Effect

| Region 2 | $\nabla^2 u_2(x) + k_2^2 u_2(x) = 0$ | (5.5.4) |

| Region 3 | $\nabla^2 u_3(x) + k_3^2 u_3(x) = 0$ | (5.5.5) |

Putting $\qquad k_1^2 = 2mE/\hbar^2$ (5.5.6)

$$k_2^2 = 2m(V_1 - E)/\hbar^2 \qquad (5.5.7)$$
$$k_3^2 = 2m(E - V_2)/\hbar^2 \qquad (5.5.8)$$

The solutions of equations (5.5.3), (5.5.4) and (5.5.5) are:

$$u_1(x) = A e^{-(ik_1 x)} + B e^{+(ik_1 x)} \qquad (5.4.7)/(5.5.9)$$

where A and B are arbitrary constants.

$$u_2(x) = D e^{-(k_2 x)} + C e^{+(k_2 x)}. \qquad (5.4.11)/(5.5.10)$$

where C and D are arbitrary constants.

$$u_3(x) = 0 + F e^{i(k_3 x)}. \qquad (5.5.11)$$

where F is an arbitrary constant.

Equation (5.5.11) has only one term, and $E e^{-i(k_3 x)}$ is zero, because there is no particle to travel from right to left, in region III. Since there are two boundaries: $x = a$ and $x = b$, the boundary conditions are:

$$u_1(a) = u_2(a) \qquad (5.5.12)a$$
$$u'_1(a) = u'_2(a) \qquad (5.5.12)b$$
$$u_2(b) = u_3(b), \qquad (5.5.12)c$$
and $\quad u'_2(b) = u'_3(b) \qquad (5.5.12)d$

Using condition (5.5.12)a in equations (5.5.9) and (5.5.10),

$$u_1(a) = A e^{-(ik_1 a)} + B e^{+(ik_1 a)}. \qquad (5.5.9)$$
$$= u_2(a) = D e^{-(k_2 a)} + C e^{+(k_2 a)} \qquad (5.5.13)$$

Applying condition (5.512)b in equations (5.5.9) and (5.5.10),

$$A(ik_1) e^{-(ik_1 a)} + B(-ik_1) e^{+(ik_1 a)}.$$
$$= D(-k_2) e^{-(k_2 a)} + C(k_2) e^{+(k_2 a)}. \qquad (5.5.14)$$

Applying condition (5.5.12)c in equations (5.5.14) and (5.5.15),

$$D e^{-(k_2 b)} + C e^{+(k_2 b)} = + F e^{i(k_3 b)} \qquad (5.5.15)$$

Applying condition (5.5.12)d in equations (5.5.14) and (5.5.15),

$$D(-k_2) e^{-(k_2 b)} + C(k_2) e^{+(k_2 b)} = F(ik_3) e^{i(k_3 b)} \qquad (5.5.16)$$

From equations (5.5.13) and (5.5.14) one gets

$$\begin{bmatrix} A \\ B \end{bmatrix} = \frac{1}{2} \begin{pmatrix} [1 + k_2/(ik_1)] e^{+(k_2 a + ik_1 a)} & [1 - k_2/(ik_1)] e^{-(k_2 a + ik_1 a)} \\ [1 - k_2/(ik_1)] e^{+(k_2 a - ik_1 a)} & [1 + k_2/(ik_1)] e^{-(k_2 a - ik_1 a)} \end{pmatrix} \begin{bmatrix} C \\ D \end{bmatrix} \qquad (5.5.17)$$

From equations (5.5.15) and (5.5.16) one gets

$$\begin{bmatrix} C \\ D \end{bmatrix} = \frac{1}{2} \begin{bmatrix} [1 + ik_3/k_2] e^{-k_2 b + ik_3 b} \\ [1 - ik_3/k_2] e^{+k_2 b + ik_3 b} \end{bmatrix} \begin{bmatrix} F \\ 0 \end{bmatrix} \qquad (5.5.18)$$

Eliminating B, C, & D, and writing BARRIER WIDTH, c as

$$c = b - a \tag{5.5.19}$$

$$\frac{F}{A} = \frac{(4\,i\,k_1 k_2)\,e^{\,i(k_1 a - k_2 b)}}{[(k_2+ik_1)(k_2+ik_3)e^{-k_2 c}] - [(k_2-ik_1)(k_2-ik_3)e^{\,k_2 c}]} \tag{5.5.20}$$

5.5.1 TUNNELING EFFECT:

The expression (5.5.20) for (F / A) shows that
$$(F/A) \neq 0 \tag{5.5.21}$$
which indicates that
$$u_2(x) \neq 0, \text{ generally,}$$
i.e., the motion of the particle in region II represented by a non-zero wave function $u_2(x)$, and there is, therefore, a finite probability of the particle penetrating the potential barrier (region II).
This effect is known as the <u>Tunnel Effect</u> (*tunneling phenomenon*), and is significant in the study of
(1) Radio-activity,
(2) Theory of emission from metal surfaces, and
(3) Working of solid-state electronic devices like tunnel diode. The transparency, or, TRANSMISSION COEFFICIENT τ, is defined as

$$\tau = \frac{\varpi_3}{\varpi_1}. \tag{5.5.22}$$

ϖ_3, is Probability of finding the particle in region III
ϖ_1 is Probability of finding the particle in region I at any time, t.

$$\tau = \frac{\varpi_3}{\varpi_1} = \left|\frac{F}{A}\right|^2 \frac{k_3}{k_1} \tag{5.5.23}$$

Equation (5.5.20) can be written as
$$\frac{F}{A} = \frac{1 + (V_1 - V_2)^2 \, Sinh^2 k_2 c}{[4E\,(V_1 - V_2 - E)^{1/2}]}. \tag{5.5.24}$$

From equations (5.5.23) and (5.5.24),

$$\tau \triangleq \left|\frac{F}{A}\right|^2 = G(k_1, k_2, k_3)\,e^{-2 k_2 c}. \tag{5.5.25}$$

$$\tau \triangleq \frac{16 E}{(V_1 - V_2)} \frac{1 - E}{(V_1 - V_2)} \, e^{-2 k_2 c} \tag{5.5.26}$$

even though $\quad E < V_1$.

Or, $\quad \boxed{\varpi_3 = \varpi_1 \, G(k_1, k_2, k_3) \, e^{-2 k_2 c}} \tag{5.5.27}$

Thus τ or ϖ_3 depends on c only, via the factor $e^{-2 k_2 c}$

Reflection coefficient, $\quad R = 1 - \tau \tag{5.5.28}$

5.5.2 APPLICATIONS OF QUANTUM Mechanical Tunnel Effect:

5.5.2.1 Visible Region:

Variable output coupler for lasers. The phenomenon of penetration of an optical barrier is recalled at this context. Barrier penetration is analogous to the following situation: When a visible beam of radiation gets totally reflected at the glass-to-air interface, it actually penetrates into the (rare) air medium. However, in this forbidden region the amplitude decreases exponentially over a distance of the order of a wavelength, λ of the radiation. No radiation escapes permanently, and one observes **'total reflection'**. If one brings another piece of glass to within one λ of the radiation of the reflecting surface, visible radiation can escape into it even though the two reflecting surfaces do not touch each other. Here the air between the two prisms, as shown, in which the long faces of the two 45° prisms are arranged face to face each other.

5.5.2.2 Raman's Experiment:
Sir C.V. Raman performed the first but different experiment using a sharp metallic edge, which was just close to the surface but not in contact with a totally reflecting prism observed that light was reflected from the sharp edge. This shows that light penetrates into the rare medium.

5.5.2.3 Atomic domain
Inversion motion oh nitrogen atom in the ammonia, NH_3 molecule.

5.5.2.4 Nuclear physics
George Gamow (1926) presented theory of emission of α – decay.

5.5.2.5 Solid-state:
Penetration of electrons in a solid state device through a very thin layer of insulating material between two conductors (*eg.* Tunnel diode)

5.5.2.6 Superconductivity
Electrons can tunnel through an insulator in pairs – tunneling from V_2 through the layer to $V_1 (V_2 > V_1)$, electron pair loses an amount of potential energy, $\Delta E = e\ (V_2 - V_1)^2$, which can be in the form of a photon. This is the *Josephson Effect*.

5.6 ALPHA DECAY – Geiger-Nuttal Relation

5.6.1 Paradox of Alpha-Decay:
The radioactive nuclide $^{214}_{84}Po$ emits α – particles with kinetic energy 7.68 *MeV*, and it has been found that no absorption of α – particles by $^{238}_{92}U$ foil occurs, though it may scatter them. But $^{238}_{92}U$ emits α – particles of 4.20 *MeV*, though this energy is less than the depth of the well of 40 *MeV* for $^{238}_{92}U$. This paradox was inexplicable on the arguments base on classical physics. Quantum Mechanics provides a straightforward explanation. In fact the theory of α-decay in nuclear physics, developed independently by G. Gamow and R. Gurney & E. Condon, in 1928, was greeted as an especially striking confirmation of Quantum Mechanics. The explanation of α-decay was one of the first successful applications of the quantum theory.

5.6.2 Geiger-Nuttal Formula: GENERAL POTENTIAL BARRIER-
ASophisticated Approach to the problem

Let the nuclear potential field has the general form
Coulomb potential $V(r) = \frac{2Ze^2}{r}$, for $R_1 \leq r \leq R_2$ (5.6.1)

where nuclear charge $= +Ze$, Charge on an α-particle $= +2e$, The Tunnel Effect discussed in Section 5.5 suggests that, a particle of energy E is capable of escaping from the closed region of the nucleus, because of the probability that the wave function associated with the α-particle can break through the barrier as if there were a hole in it. If ϖ_1, ϖ_2, and ϖ_3 are the probabilities of finding the particle in regions I, II and III, respectively,

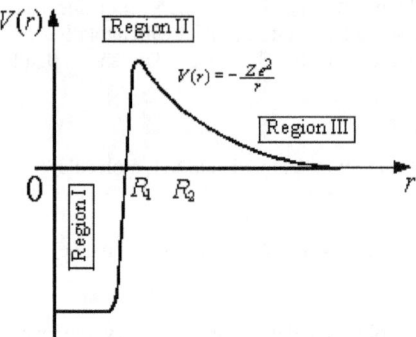

Fig. 5.8 Nuclear Potential

$$\varpi_3 = \varpi_1 \, G(k_1, k_2, k_3) \, e^{-2 k_2 c}. \tag{5.5.27}$$

i.e.,
$$\varpi_3 \propto e^{-2 \sqrt{[2m(V_1-E)]}/\hbar] (R_2-R_1)} \tag{5.6.2}$$

is the probability per collision of the potential barrier. If v = number of encounters to get through the barrier per unit time,

$$v = 1/\varpi_3 \propto e^{2 k_2 c} \tag{5.6.3}$$

If t_o = time between two encounters,

$$t_o \approx 2R_1/v, \tag{5.6.4}$$

where v = velocity with which α-particle is leaving the nucleus
$$v = \sqrt{2E/m}$$

Probability of α-particle emission per unit time,
$$\varpi \approx (v/2R_1 t_o) \approx (t_o/2R_1)\sqrt{2E/m} = 1/\tau$$

This is because if λ = radio-active *decay constant*, the Radio-active Law is

$$dN(t)/dt = \lambda N(t) \Rightarrow N(t) = N_0 e^{-\lambda t} \tag{5.6.5}$$

Similarly, $\varpi = \varpi_0 e^{-\lambda t}$

$$d\varpi_3/dt = -\lambda \, \varpi_1 \propto \varpi_1/\tau_{1/2}, \tag{5.6.6}$$

where $\tau_{1/2}$ = half-life of the radioactive material.

$$(Log 2)/\lambda = 0.693/\lambda. \tag{5.6.7}$$

This is suggested because ϖ_1 refers to the bound particle; while ϖ_3 is for the free (radiated) particle. Since the interest centres on a single particle, 'density' is to be replaced by 'continuity probability'.

Equation of continuity gives
$$\partial \varpi_1/\partial t + \nabla \varpi_1 = 0$$

i.e., $d\varpi_1/dt = -d\varpi_1/dx \propto -\varpi_1$

Using equation (5.4.46),

$$\varpi_3 \triangleq \varpi_1 G(k_1, k_2, k_3) \, e^{-2 k_2 c} . \tag{5.6.8}$$

$$\therefore \quad d\varpi_3 / dt \triangleq (d\varpi_1 / dt) \, G(k_1, k_2, k_3) \, e^{-2 k_2 c}$$

$$\propto \varpi_1 \, e^{-2 k_2 c} \tag{5.6.9}$$

Comparing equations (5.6.5) and (5.6.7),

$$\tau_{1/2} \propto e^{2 k_2 c}$$

or, $\quad \lambda \propto e^{-2 k_2 c}$

so that $\therefore \; Log \tau_{1/2} = 2 k_2 \, c + B (a \text{ Constant})$ (5.6.10)

Or, $\quad Log \lambda = -2 k_2 \, c + B_o (a \text{ Constant})$

Now one can replace

$$k_2 \, c \text{ by } k_2 \, c \equiv \int_{R_1}^{R_2} k_2 \, dr \tag{5.6.11}$$

Equation (5.6.2) becomes

$$\varpi_3 \propto e^{-\int_{R_1}^{R_2} 2 \sqrt{[2m(V_1-E)]/\hbar} \, dr} \tag{5.6.12}$$

$$\propto e^{-\int_{R_1}^{R_2} (2/\hbar) \sqrt{[2m(V_1-E)]} \, dr} . \tag{5.6.13}$$

To evaluate the integral:

$$E = V(R_2) + T = \frac{2 Z e^2}{R_2} \tag{5.6.13}$$

Assuming T as negligibly small compared to

$$E \int_{R_1}^{R_2} k_2 \, dr = -\frac{2}{\hbar} \int_{R_1}^{R_2} \sqrt{2m \, (V_1 - E)} \, dr$$

Note: Substitute for $[V(r) - E] = (\frac{2 Z e^2}{r} - E)$

$$= E(\frac{2 Z e^2}{rE} - 1) = E(\frac{1}{y} - 1)$$

Now let $\quad r = (\frac{2 Z e^2}{E}) y \to y = \frac{E}{2 Z e^2} \, r = E/V(r)$

$$dr = (\frac{2 Z e^2}{E}) \, dy$$

$$\int_{R_1}^{R_2} dr = (\frac{2 Z e^2}{E}) \int_{R_1 E / 2 Z e^2}^{R_2 E / 2 Z e^2} dy$$

$$\varepsilon = R_1 E / 2 Z e^2$$

$$\therefore \quad \int_{R_1}^{R_2} k_2 \, dr = \frac{1}{\hbar} \sqrt{2m E} \, (\frac{2 Z e^2}{E}) \int_{\varepsilon}^{1} \sqrt{\frac{1}{y} - 1} \, dy \tag{5.6.14}$$

This is a standard integral, and it can be evaluated.

Writing, $y = Sin^2 \theta$

$$\int_{\varepsilon}^{1} \sqrt{\frac{1}{y} - 1} \, dy \approx \left(\frac{\pi}{2} - \sqrt{\frac{R_1}{R_2}} - \sqrt{\frac{R_1}{R_2}} \right) \varepsilon$$

$$\therefore \int_{R_1}^{R_2} k_2 \, dr \cong \frac{1}{\hbar} \sqrt{2m/E} \, (\frac{2 Z e^2}{E}) \left(\frac{\pi}{2} R_2 - 2\sqrt{R_1 R_2} \right) \tag{5.6.15}$$

Equation (5.6.15) becomes

$$\int_{R_1}^{R_2} k_2 \, dr \cong -\sqrt{2m/E} \left(\frac{\pi Z e^2}{\hbar}\right)$$

and $\sqrt{E} \propto v$

$$\therefore k_2 c \cong \sqrt{2m/E} \left(\frac{\pi Z e^2}{\hbar}\right) \approx \sqrt{4m(Z/v)} \left(\frac{\pi e^2}{\hbar}\right) \quad (5.6.16)$$

$$2k_2 c \propto (Z/v) = A(Z/v) \quad (5.6.17)$$

Substituting this equation (5.6.17) in equation (5.6.10), results in the equation:

$$\boxed{Log_n \tau_{1/2} = 2 k_2 \, c + B(a\ Constant)} \quad (5.6.10)$$

∴

$$\boxed{Log_n \tau_{1/2} = A(Z/v) + B(a\ Constant)} \quad (5.6.18)$$

where both A and B are constants.

$$\boxed{Log_n \lambda = -A(Z/v) - B} \quad (5.6.19)$$

Putting $R_2 \equiv R$, $R_1 = R_o$, and eliminating $R \equiv R_2$,

∴

$$\int_{R_o}^{R} k_2 \, dr \approx \left(\frac{\sqrt{2}\,(0.98)\,(1727 MeV)}{(197.1 MeV.F)^2}\right)^{1/2} E^{1/2} (3.14)(2.88 MeV\ F) Z E^{-1}$$

giving equation (5.6.19) as

$$\boxed{Log_n \lambda = Log_n \frac{v}{2R_o} + 2.97 Z^{1/2} R_o^{1/2} - 3.95\, Z\, E^{-1/2}} \quad (5.6.20)$$

where E is expressed in MeV.

Or,

$$\boxed{Log_c \lambda = Log_b \frac{vN}{2R_o} + 1.28\, Z^{1/2} R_o^{1/2} - 1.71\, Z\, E^{-1/2}} \quad (5.6.21)$$

$$\boxed{Log_n \tau_{1/2} + 2.97\, Z^{1/2} R_o^{1/2} = 3.95\, Z\, E^{-1/2}} \quad (5.6.22)$$

where N = # of α-decay (=1) within the nucleus at any given instant, t.

Plot $Log_c \lambda$ versus $Z\, E^{-1/2}$ for a number of representative α-particle nuclides from each of the 4 decay series, viz. Thorium, Neptunium, Uranium and Actinium Series having mass number, $4n$, $4n+1$, $4n+2$, and $4n+3$, respectively. It will be seen that
(1) the straight line drawn through the points has a slope of (−1.71) as required by the theory. The position of the line can be used to determine R_o, the radius of the nuclide.
(2) The tremendous range of λ that is represented (corresponding to

$\tau_{1/2} \approx 10^{-6} s$ to 10^{10} yrs) by the slope of the line is not subject to adjustment (This is a span of a range of 23 orders of magnitude). These two points show that the agreement of the barrier penetration theory with observations must be considered as the most satisfactory or remarkable. The position of the line gives

$$Log_b \frac{vN}{2R_o} + 1.28\, Z^{1/2} R_o^{1/2} \approx 55.5$$

Dependence of $Log_b \frac{vN}{2R_o}$ on $\frac{v}{2R_o}$ is quite weak. This means one gets quite accurate values of R_0. Thus, if N = 1, $v \approx 2 \times 10^9\, cms^{-1}$, $R_o \approx 10^{-12}\, cm$ (from Rutherford experiment) $Z \approx 85$, for radio-active α-active nuclide

$$\approx [(55.5 - 21.0)/(1.98 \times 9.2)]^2 = 8.5 \times 10^{-13}\, cm.$$

5.6.3   Particle Tunneling (Electrons in a metal) Fowler-Nordheim Formula
Since this topic is beyond the scope of the present book it will not be discussed here.

5.7 THE KRONIG-PENNEY MODEL OF POTENTIAL FOR N ELECTRON IN A PERIODIC LATTICE
Or Multiple square well potential) or Zone Theory:

5.7.1 TWO ISOLATED IDENTICAL SQUARE WELL POTENTIALS:

If two identical square well potentials are isolated from each other and each has a particle in its ground state, energy of the system is just the sum of the two ground state energies.
From equation (5.311)

$$E_n = \frac{h^2}{8m} \frac{n^2}{a^2}; \quad n = 1, 2, 3, \ldots \quad (5.7.1)$$

where n = 1.

$$E_1 = \frac{h^2}{8m}\left(\frac{1}{a^2} + \frac{1}{b^2}\right)$$

5.7.2 TWO ISOLATED SQUARE WELL POTENTIALS separated by a very narrow barrier:

In this case the two will interact and results are as described below:
(a) Total wave function: is a linear combination of the two single-well ground state functions as shown in, for the symmetric as well as the anti-symmetric cases.
(b) The energies: of the two states are not the same,
(c) Energy Splitting: between the two states occurs. It increases as the width of the barrier decreases. This is so because in the limit as the barrier vanishes, the symmetric wave function corresponds to the ground state of a well having twice the width of the original wells. Hence the energy decreases. As the wells coalesce, the anti-symmetric function corresponds to the first excited state of the well of double width.
(d) The decrease in energy due to increased width of the well is compensated by excitation of the particle to the first state above the ground state. Thus the above model has the following features:

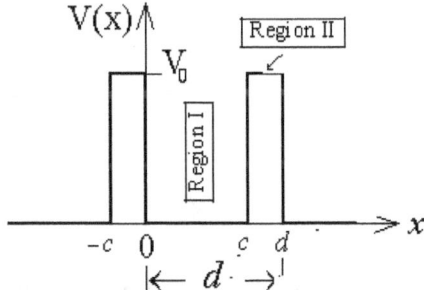

Fig. 5.9 Two isolated Square Wells.

(i) The two wells, which have the same energy levels are large separations, interact more strongly as they are brought together closer; and as this interaction splits each original level into two separate energy levels. The motion of the electron in the 1-D potential is considered. (a) Since the absolute value of the potential has no influence on the motion of the electron, assume the potential to be zero, half way between lattice points.

(ii) The two ends of the 1-D crystal are joined to form a 1-D crystal ring, which take care of the vanishing of the boundary conditions, and provide instead periodic (cyclic) boundary conditions, as shown in. Fig 5.9.

5.7.3 BLOCH THEOREM Discussion

Let $V(x)$ = crystal potential and
d = period of the crystal potential.
As per Postulate 1,
$$V(x + d) = V(x) \quad (5.7.2)$$
Though the shape of the $V(x)$ = same for all wells in the complete lattice,
$$\psi(x+d) \neq \psi(x). \quad (5.7.3)$$
i.e., eigen function need not be periodic with d. The Schrödinger Equation remains the same even if $V(x)$ is replaced with $V(x + d)$,

i.e., $(p_x^2/2m)\,\psi(x)+[V(x)-E]\,\psi(x)$
$$\equiv (p_x^2/2m)\,\psi(x)+[V(x+d)-E]\,\psi(x).$$

But this leaves a constant factor (lag in phase constant) in its eigen functions due to time for the wave to progress from one well to the next.
$$\nabla^2 \psi(x) - E\,\psi(x) \equiv \nabla^2 \psi(x+d) - E\,\psi(x+d)$$
Therefore, $\quad \psi(x+d) = \mu_1\,\psi(x) \quad (5.7.4)$
where μ_1 = a complex constant.
$$\psi(x+2d) = \mu_2\,\psi(x+d) = \mu_2\,\mu_1\,\psi(x)$$
and $\quad \psi(x+Nd) = \mu_N \cdots \cdot \mu_2\,\mu_1\,\psi(x) \quad (5.7.5)$
where N = total number of interacting potentials/ ion in the **crystal ring**.
$$\psi(x+Nd) = \psi(x) \quad (5.7.6)$$
∴ $\quad \mu_N \cdots \cdot \mu_2\,\mu_1 = 1. \quad (5.7.7)$
Because of the symmetry of the crystal ring,
∴ $\quad \mu_N = \cdots = \mu_2 = \mu_1 = \mu \quad (5.7.8)$
∴ $\quad \mu_N \cdots \cdot \mu_2\,\mu_1 = \mu^N = 1 \quad (5.7.9)$
Therefore, $\quad \mu = e^{i2\pi n/N} \quad (5.7.10)$
where $n = 0, 1, 2, 3, …, (N-1)$. Now we have
$$\psi(x+Nd) = \mu^N \psi(x) = e^{i2\pi N/N}\psi(x).$$
$$\psi(x+md) = e^{i2\pi m/N}\psi(x) \quad (5.7.11)$$
or, $\quad \psi(x) = e^{i2\pi nx/Nd}\psi(x). \quad (5.7.12)$
or, $\quad \psi(x,t) = e^{i(k_n x - \omega t)}\psi(x) \quad (5.7.13)$
where $\quad k_n = 2\pi n/Nd \quad (5.7.14)$

The solutions to the Schrödinger Equation, for waves in the 1-D crystal, are given by equation (5.7,13). These solutions give the same behavior in every unit cell except for a lag in phase angle corresponding to the time required for the wave to progress from one cell to the next. The proof that the waves of this type are eigen functions is due to F. Bloch (1928), and equation (5.7.13) is called the **Bloch Theorem**. $\psi(x,t)$ is called <u>Bloch Function</u>.
In this w(x) is periodic so that (since replacing x by $(x + m\,d)$ in equation (5.7.13) to give) periodic function
$$w(x+d) = w(x) \quad (5.7.15)$$

The Bloch Theorem is the central theorem in electron *Band Theory of Solids*. Mathematicians know it as 'Floquet's Theorem'.

5.7.4 THE KRONIG-PENNEY MODEL of an Infinite Lattice:

To go on with the solution of the Schrödinger Equation, the actual shape of the periodic potential, V(x), has to be specified.

5.7.4.1 Infinite Lattice

To simplify things, one assumes the physically impossible, yet mathematically convenient, model of potential in treating the electron energies in solids. L. Kronig and W.G. Penney (1931) made an important generalization of the square well potential, and it is schematically as in Fig 5.10. Here the number of interacting potential wells, *N*, is extremely large so that each of the single-well levels is split into *N* levels spaced so close together that they form nearly continuous energy levels

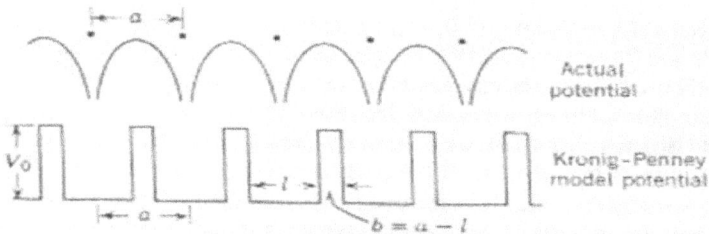

Fig. 5.10 Kronig-Penney Model of Crystal Lattice.

5.7.4.2 BLOCH THEOREM

Bloch has shown that the solution for such a periodic lattice, viz.

$$\psi(x,t) = e^{i(k_n x - \omega t)} w(x) \tag{5.7.13}$$

is in the form of a plane wave, $e^{i(k_n x - \omega t)}$ and a function, *w(x)* having the periodicity of the lattice, where

$$w(x+d) = w(x) \tag{5.7.15}$$

and $\quad k_n = 2\pi n / N$. $\tag{5.7.14}$

$$\nabla^2 \psi(x) + \frac{2m}{\hbar^2}[E - V(x)]\, \psi(x) = 0 \tag{5.7.16}$$

is the Schrödinger Equation with periodic potential having the solution in the form of equation (5.7.13), substituting which one gets

$$(d^2/dx^2)\, w(x) + 2\, i\, k\, (dw/dx) + (2m/\hbar^2)[E - (\hbar^2 k^2 / 2m) - V(x)]\, w(x) = 0 \tag{5.7.17}$$

This is the differential equation in *w(x)*. To solve this equation substitute

$$w(x) = e^{\Omega x}. \tag{5.7.18}$$

where $\quad \Omega = -k \pm \sqrt{2m(E-V)/\hbar^2} \tag{5.7.19}$

Case 1: In region I of Fig 5.9, V(x) = 0.
The general solution of the equation (5.7.17) becomes

$$w_o(x) = A\, e^{i(\alpha - k)x} + B\, e^{-i(\alpha - k)x}. \tag{5.7.20}$$

where $\quad \alpha = \sqrt{2mE/\hbar^2}$. $\tag{5.7.21}$

Case 2: In region II, V(x) = V_o.

The general solution of the equation (5.7.17) becomes
$$w_v(x) = C e^{i(\beta-k)x} + D e^{-i(\beta-k)x} \qquad (5.7.22)$$
where $\beta = \sqrt{2m(E-V_o)/\hbar^2}$ $\qquad (5.7.23)$

Boundary conditions:
$$w_o(x=0) = w_v(x=0)$$
$$w_o(x=c) = w_v(x=c)$$
$$(d/dx)w_o(x=0) = (d/dx)w_v(x=0)$$
$$(d/dx)w_o(x=c) = (d/dx)w_v(x=c) \qquad (5.7.24)$$

Applying the 4 boundary conditions (5.7.24) in equations (5.7.20) and (5.7.22), one will get 4 linear, homogeneous equations which are sufficient to determine the 4 constants A, B, C, D, if the determinant of their coefficients vanishes. The 4th order determinant of transcendental functions of energy E is difficult to solve for E. This, therefore, requires further simplification. For this let the region II is narrowed and at the same time V_o is increased, such that
$$V_o(d-c) = K \text{ (constant)} \qquad (5.7.25)$$
Since $\psi(x)$ is continuous at $x = c$ and at $x = d$, equation (5.7.17) is integrated, and narrowing down the potential well so that $\psi(x)$ = constant, practically as $x \to d$ from c. With the potential well narrowed to this extent, $V_o(d-c) = K$ means V_o is large enough for both k and E to be assumed as negligibly small inside the well. Recalling equation (5.7.17) becomes
$$(d^2/dx^2) w(x) + 2ik(dw/dx) - (2mV_o/\hbar^2) w(x) = 0 \qquad (5.7.26)$$
Integrating equation (5.7.26) over the period, c to d, of the potential well
$$\int_c^d (d^2/dx^2) w(x)dx + 2ik\int_c^d (d/dx)w(x)dx - (2mV_o/\hbar^2)\int_c^d (w(x)dx = 0$$
$$[(d/dx) w]_c^d + [2 i kw]_c^d - [(2mV_o/\hbar^2)wx]_c^d = 0$$
Valued between c and d, this yields
$$[(d/dx) w]_d - [(d/dx) w]_c + [2ik](w_d - w_c) - [(2mV_o/\hbar^2)w(d)K = 0$$
Because of the periodicity of w(x) by equation (5.7.15) and its continuity at $x = 0$,
$$w(d) = w(0) = w(c) \qquad (5.7.27)$$
and the second integrand vanishes.
$$\therefore \qquad (d/dx) w]_o - [(d/dx) w]_c - [(2mV_o/\hbar^2)w(0) = 0 \qquad (5.7.28)$$
where the first two terms are in the region V(x) = 0.
$$\therefore \qquad [(d/dx) w_o]_o - [(d/dx) w_o]_c - [(2mV_o/\hbar^2)w_o(0) = 0 \qquad (5.7.29)$$

5.7.4.3 TRANSCENDENTAL Equation:

Using equation (5.7.20) for $w_o(x)$ and substituting in (5.7.29)
$$A i (\alpha-k)[1 - e^{i(\alpha-k)x}] - B i (\alpha+k)[1 - e^{-i(\alpha-k)x}] - [(2m/\hbar^2) K(A+B) = 0 \quad (5.7.30)$$
Equations (5.7.20) and (5.7.27) yields
$$A + B = A e^{i(\alpha-k)c} + B e^{-i(\alpha+k)c} . \qquad (5.7.31)$$
Equations (5.7.30) and (5.7.31) are two linear homogeneous equations for A and B. These can be solved if the determinant of the coefficients vanishes. *i.e.,*

$$\begin{vmatrix} [1- e^{i(\alpha - k)c}] & [1- e^{-i(\alpha + k)c}] \\ i(\alpha - k)[1- e^{i(\alpha - k)x}] & -i(\alpha - k)[1- e^{i(\alpha - k)x}] \\ -(2m/\hbar^2)K & -(2m/\hbar^2)K \end{vmatrix} = 0 \quad (5.7.32)$$

This gives the **Transcendental equation** for α, viz.,

$$\boxed{Cos(kd) = Cos(\alpha d) + [P/\alpha d]\, Sin(\alpha d)} \, . \quad (5.7.33)$$

where $\quad P = \lim_{c \to d}\left(\tfrac{1}{2}\beta_o^2 b c\right) = \left(m c V_o b / \hbar^2\right).$ (5.7.34)

$$= \tfrac{1}{2}\lambda \quad \alpha = \sqrt{2mE/\hbar^2} \quad (5.7.21)$$

$$= \text{constant}$$

$$\beta_o = \sqrt{2m\,(V_o - E)/\hbar^2}\,. \quad (5.7.23)$$

$$P = \tfrac{1}{2}\lambda$$

if $\quad V(x) = (\hbar^2/2m)(\lambda/d)\sum \delta(x - nd),$

i.e., a series of repulsive δ-function potentials. This means, one can obtain the allowed energy levels of an electron in a periodic potential by the following two interpretations of equation (5.7.33).

5.7.4.4 BAND STRUCTURE:

Interpretation of the right hand side of equation (5.7.33):
$Cos(kd) = Cos(\alpha d) + [P/\alpha d]\, Sin(\alpha d)$.

is the equation which will be satisfied only for those values of E for which the right hand side (RHS) such that $-1 < RHS < +1$, because
$Cos(kd) = \pm 1$

For real values of k, the physically meaningful solutions of equation (5.7.33) must be within these two limits, which are shown in Fig. 5.11, by the heavy horizontal lines along the (αd)-axis. These energies are called ALLOWED BANDS (allowed regions or allowed ZONES) – which an electron in a periodic potential is allowed to take. They are shown shaded in Fig.5.11. In other words, in a crystal only certain energy bands are permitted for an electron. Between the allowed bands are the FORBIDDEN BANDS (forbidden regions or forbidden zones) of energies, which the electron cannot take while it is moving through a periodic potential.

Case (i) $V_o b \to$ large, allowed zones become very narrow and the electron can not move freely. This is what happens in the case of ELECTRONS that are TIGHTLY BOUND to the NUCLEUS.

Case(ii) On the other hand, as $V_o b \to 0$, the allowed zones spread so much that the forbidden zones disappear. This is what happens in the case of VALENCE ELECTRONS IN AN ATOM.

Case (iii) For $V_o b \to 0$, the situation reduces to the case of FREE ELECTRONS.

Interpretation of the left hand side (LHS) of equation (5.7. 33)

5.8.4.5. DISPERSION DIAGRAM; BAND STRUCTURE:

Consider the term $Cos(kd)$, which is the LHS of this equation. This function can assume only those values which correspond to allowed values of E. The heavy lines in Fig. 5.11, show the plot of these allowed values of E *versus* $(k\,d)$. If these allowed values are projected to the right, one gets the ALLOWED bands or ZONES, and between these bands are the forbidden bands. This plot is the DISPERSION diagram of E *versus* $(k\,d)$ for electrons in the Kronig-Penney potential.

The graph of E *versus* k is known as the BAND STRUCTURE. The *dashed-line-parabola* in Fig. 5.11 corresponds to the case of a free electron for which $E_n = \hbar^2 k_n^2 / 2m$. Note that the *heavy lines* depart slightly from the dashed line parabola only in the neighborhood of $\pm n\pi$; where n = an integer. This means that the electron moving in a periodic potential behaves like a free electron for most values of k; except those near $\pm n\pi$. For large values of E, the two are almost similar

5.8.4.6 BRILLOUIN ZONES:

It is important to note that the DISCONTINUITIES in the allowed values of E occur at

$$Cos(kd) = \pm 1 \quad (5.7.34)$$

i.e., $\quad kd = \pm n\pi, n = 0, 1, 2, 3, \ldots$

If one substitutes $\quad k = 2\pi/\lambda$ and $n\lambda = 2d$, $\quad (5.7.35)$

which corresponds to the Bragg condition in diffraction, provided that one substitutes $\theta = \pi/2$, in

$$n\lambda = 2\,d\,Sin\theta$$

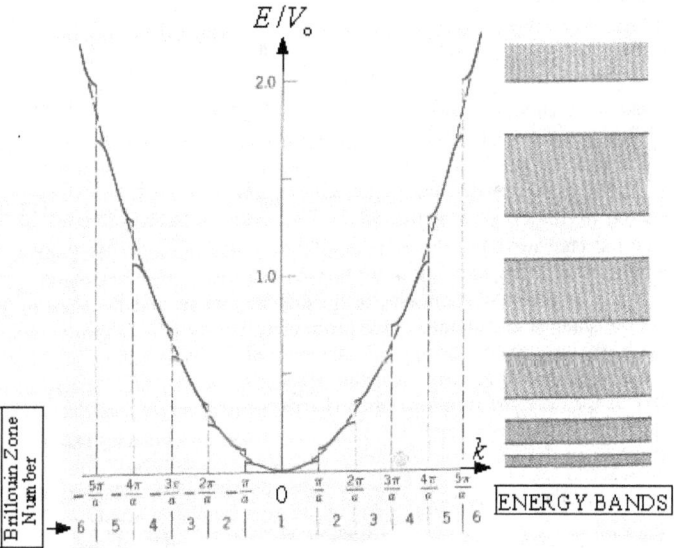

Fig. 5.11 E versus k plot showing the Band structure of a Periodic Lattice,

This implies that the discontinuities represent locations at which the electron waves are reflected backward. The region that contains electrons with momentum such

that $0 < k < \pi/d$, i.e. $n = 1$, in equation (5.7.34), is called the FIRST BRILLOUIN ZONE (I B. Z.).
The SECOND Brillouin Zone is one that contains electron with values of k, such that $\pi/d < k < 2\pi/d$, so also the 3rd, 4th, etc.

Within the B Z, the energy is a continuous function of α with a continuous derivative. A single continuous branch is called an ENERGY BAND. Discontinuities in energy occur only on the zone surfaces. Taking the allowed values of E from

$$\alpha = \sqrt{2mE/\hbar^2} \qquad (5.7.21)$$
$$E_n = \hbar^2(\alpha d)^2 / 2md^2 \qquad (5.7.36)$$

Outside the limits of ± 1, k must be complex with a non-zero imaginary part. The corresponding ranges of E are forbidden. Thus alternate regions of allowed and forbidden energy bands are formed. The grouping of the permitted energy values into these bands is one of the most important and characteristic features of the behavior of electrons in periodic lattices,

i) As $K \to$ large (i.e. higher the potential separating zero potential regions) the energy bands become narrow.

ii) As $K \to \infty$, the crystal tends to become a series of independent square wells, and energy bands go over to the discrete eigenvalues as for a square well

5.8.5 One Dimensional Molecule:
Some aspects of what gives rise to molecules are exhibited by the example of a particle in a double potential well. This problem is not to be described here.

REVIEW QUESTIONS

R.Q. 5.1 Calculate the energy of a proton in a nucleus, by using the theory particle in a box. The dimension of a nucleus is $R = 1 \times 10^{-13}$ m. Given, $h = 1.054 \times 10^{-34}$ J-s, $M_p = 1.67 \times 10^{-27}$ kg, $1\, eV = 1.6021 \times 10^{-19}$ J (Answer: 2 MeV).

R.Q. 5.2 Write the normalized wave functions for a particle in an infinite square well of width 2a. Find the expectation value of x^2 in such a well and that $<x^2>$ approaches the classically expected value for large quantum numbers. (Answers:
$u_n(x) = \pm\sqrt{1/a}\, Sin(n\pi x/2a)$. $<x^2> = (a^2/3)[1-(6/n^2\pi^2)]$
$<x^2> \xrightarrow{n \to large} a^2/3$).

R.Q 5.3 Explain the fact that the energy of the lowest state of a particle in the potential well $V(x) = V_\infty = \infty$, $-a \le x \ge +a$, is different from zero. (Answer: $E_1 \ge h^2/[8m(\Delta x)^2]$, $(\Delta p_x)^2 \ge h^2\pi^2/[4a)^2$; $E_1 > 0$, which is different from zero).

R.Q. 5.4 Determine the expectation values of position x, and position square x^2, for a particle in a box. (Answer: $<x> = [-\frac{2}{\xi^2 a}][\frac{\xi^2}{4} - \frac{1}{4}\xi\, Sin\, 2\xi - \frac{1}{8}Cos2\xi]_0^{n\pi} = 0$;

$<x^2> = [4/(\frac{n\pi}{a})^3 a][-\frac{2}{\xi^2 a}][\frac{\xi^3}{6} - [\frac{\xi^2}{4} - \frac{1}{8}]\, Sin\, 2\xi - \frac{1}{4}(\xi Cos2\xi)]_0^{n\pi/2} = a^2\{[(1/12)-(1/2)]n^2\pi^2\}$).

R.Q. 5.5 Determine the expectation values of momentum p_x and square of momentum, p_x^2, for a particle in a box. of $V(x) = 0$, $|x| < a/2$; $V(x) = \infty$, $|x| \geq a/2$
(Answers: $<p_x> = 0; <p_x^2> = (n^2\pi^2\hbar^2)/(2ma^2)$)

R.Q. 5.6 A particle is in an infinitely deep square potential well of size a. Determine the probability distribution for the first three values of the quantum number for momentum for the normal state of the particle.
(Answer: $|c_k|^2 = \{\pi\hbar^3 a/(p^2 a^2 - \pi^2\hbar^2)^2\} Cos^2(pa/2\hbar)$).

R.Q. 5.7 Consider a 2-D box of length a and width b. Show, by solving the Schrödinger equation, that if a particle is confined to this box, the allowed wave functions and energies are: $u_n(x,y) = \sqrt{\frac{2}{ab}} Sin(\frac{n\pi}{a}x) \cdot Sin(\frac{n\pi}{b}y)$;

and $E_{n_x, n_y} = \frac{\hbar^2\pi^2}{2m}\left(\frac{n_x^2}{a^2} + \frac{n_y^2}{b^2}\right)$..where n_x, n_y are quantum numbers.

R.Q. 5.8 Consider the Heisenberg Uncertainty Principle. Calculate the minimum energy uncertainty of a particle may have when in an infinite potential well of width a, i.e. assume that $\Delta x = a$. How does this value compare with the one obtained in treating the particle in a box? (Answer: $<(\Delta x)^2> = (a/\pi)^2[(\pi^2/3)-2] > (a/\pi)^2$;
$E_1 \geq \hbar^2 /[8m)(a/\pi)^2] \geq ;\hbar^2\pi^2/(8ma)^2$]).

R.Q. 5.9 Determine the pressure exerted on the walls of a rectangular 'potential box' by a particle inside it. (Answer: $P(\perp x)=[\hbar^2\pi^2 n_x^2/(ma)^2]/bc$).

R.Q. 5.10 A particle is trapped in an infinite square well of width '$2a$', such that the wave function is $U_n(x) = C\{Cos(\frac{\pi}{2a}x) + Sin(\frac{3\pi}{a}x) + \frac{1}{4}Cos(\frac{3\pi}{2a}x)\}$ inside the well $U_n(x) = 0$, outside the well. (a) Calculate the coefficient C. (b) If a measurement is made of the total energy, what are the possible results of such a measurement, and what is the probability to measure each of them? (Answers: a) $C = 4\sqrt{1/33a}$;b) $E_1 = -\hbar^2\pi^2/(8ma)^2$, $E_2 = -9\hbar^2\pi^2/(2ma)^2$, $E_3 = -9\hbar^2\pi^2/(8ma)^2$, $P_1 = 16/33$, $P_2 = 16/33$, $P_3 = 1/33$, as $\sum P_i = 1$)

R.Q. 5.11 In a simple model of nucleus, there an Z protons and N neutrons confined to an infinitely deep square potential well. (a) Obtain an expression for the density of the energy levels (i.e. the number of levels / unit energy interval) in this potential, (b) when the nucleus is in the lowest energy state, what is the maximum k E of single nucleon? (c) Show that if the nuclear density is constant, this energy is independent of the number of nucleons. (d) How can the model be modified to take into account of the electric forces between the protons? (Answers: a) $E_{n_x, n_y, n_z} = \frac{\hbar^2\pi^2}{2m}\left(\frac{n_x^2}{a^2} + \frac{n_y^2}{b^2} + \frac{n_z^2}{c^2}\right)$; $dn/dE = \frac{ma^2}{\hbar^2\pi^2 n}$;

b) $E_f = [(3^{2/3}\pi^{4/3}\hbar^2 N^{2/3})/2ma^2]$ for a proton, $E_f = [(3^{2/3}\pi^{4/3}\hbar^2 N^{2/3})/2ma^2]$ for a neutron, A=N+Z, d) One has to enhance the total potential acting on the protons. Thus the energy levels of protons are all elevated, compared to those of neutrons).

R.Q. 5.12 Consider a particle of mass m moving in a 1-D infinite square potential well, $V(x) = 0$, at $|x| < a$; $V(x) = \infty$, $|x| \geq a$. Suppose that at time t = 0, the wave function is given by $u(x, t=0) = A(a^2 - x^2)$. Find the probability that a measurement of energy will give the value, $E_n = [\hbar^2 \pi^2 n^2/(8ma^2)]$ Obtain the average value of energy. (Answers: $P_n = 960/\pi^6 n^6$ for n=odd, $= 0$ for n even; $E = <\hat{H}> = 5\hbar^2/(4ma^2)$).

R.Q. 5.13 What is the lowest possible energy an electron will have when it is trapped in a vacancy in a crystal lattice such that its motion is restricted to a spherical volume of radius, 0.25 nm. (Answer: $E_1 = 0.155 eV$).

R.Q. 5.14 Estimate the kinetic energy of a nucleon in a Carbon nucleus. The diameter of the nucleus is $\sim 2 \times 10^{-15}$ m .(Answer: $\sim 7 MeV$).

R.Q. 5.15 A rectangular box with perfectly reflecting walls has the dimensions $a = b = 0.1$ nm, and $c = 3$ nm. A particle of electronic mass is trapped in this box. (a) What energy does belong to the lowest possible state? (b) List the different wave functions that belong to the first three levels? Given, $m_e = 9.11 \times 10^{-31}$ kg $= 0.511 MeV$

R.Q. 5.16 Show that the first derivative of the time-independent wave function is continuous even at points where V(x) as a finite discontinuity. (Answer: The condition of continuity of the 'logarithmic derivative' $(\partial/\partial x) Log\, u(x) = (\partial/\partial x) u'(x)/u(x)$).

R.Q. 5.17 A particle is enclosed in a rectangular box with impenetrable walls, inside which it can move freely. Find the eigen functions and possible values of the energy. What can be said about the degeneracy, if any, of the eigen functions?(Answers: $E_{n_x, n_y, n_z} = \frac{\hbar^2 \pi^2}{2m}\left(\frac{n_x^2}{a^2} + \frac{n_y^2}{b^2} + \frac{n_z^2}{c^2}\right)$.

If the ratio of any two sides is an irrational number, all the energy levels are non-degenerate; otherwise the energy spectrum is in general degenerate, i.e. if a cubical box is considered, $a = b = c$, $n_x^2 + n_y^2 + n_z^2 = 6$,has the level 3-fold degenerate; since the 3 linearly independent eigen functions have the same eigen value,. $E_{121} = E_{112} = E_{211} = 6\hbar^2\pi^2/2ma^2$. E_{111} is always non-degenerate).

R.Q. 5.18 Show that, in 3-D problems, the energy levels of a particle in a cubical box may be degenerate.

R.Q. 5.19 Consider a particle in a 1-D box of length "L" in a quantum state at time $t = 0$, given by $\psi(x,0) = \psi(0)$, $0 < x < L$, and $\psi(x,0) = 0$, otherwise., where $\psi(0)$ is a real constant. Find the expansion coefficients when $\psi(x,t)$ is expressed in terms of energy eigen states of the particle. (Answers: $c_i = \int \psi_i^* \Psi\, dx = \int \sqrt{\frac{2}{L}} Sin(n_i \pi x/L)$
$= \{\sqrt{2L}/(n_i\pi)^2 L\} [1 - Cos(n_i \pi x/L)]$).

R.Q. 5.20 The molecule $CH_2 = CH - CH = CH_2$, viz,.

But $-1,3-diene$ has the average length of the single and a double bond between C atoms is 0.14 nm. Assume that an electron belonging to one of the C atoms can move along the length of the molecule and could reach an average bond length beyond the terminal of the atom at each end, and assuming it to be thus trapped in a 1-D box. (a) Determine the energies of the first five levels; (b) Calculate the wavelength of the lines in the spectrum that would be found if the electron could move between these lengths.

(Answers: $L = (0.14x3) + 2x0.14/2 = 0.56$ nm, $m = 9.11 \times 10^{-31}$ kg .; ,

$E_2 = (4\pi^2 \hbar^2/8 m L^2) = 7.684 \times 10^{-19} J$, $E_3 = (9\pi^2 \hbar^2/8 m L^2) = 17.289 \times 10^{-19} J$,

$E_4 = (16\pi^2 \hbar^2/8 m L^2) = 30.736 \times 10^{-19} J$.: $\lambda_{E_1-E_2} \cong 576$ nm . $\lambda_{E_2-E_3} \cong 207$ nm

$\lambda_{E_3-E_4} \cong 148$ nm $\lambda_{E_4-E_5} \cong 114$ nm ; This *free electron model* gives the right order of magnitude for $\lambda_{obs} \cong 220$ nm; with $\lambda_{E_2-E_3} \cong 207$ nm .).

R.Q. 5.20 Determine the reflection coefficient and transmission coefficient of a particle from a rectangular potential wall. (a) Find the Wave function every where (unnormalized), (b) Normalize the wave function so that it corresponds to unit incident flux (1 particle / sec), (c) Solve the case for E <V_0, and discuss the significance of the result.

R.Q. 5.21 An electron with a kinetic energy of 10 eV at $x = -\infty$ is moving from left to right along the x-axis. The potential energy is V = 0, for x < 0, and V = 20 eV, for x > 0. Treat the electron as a 1-D plane wave. Write the Schrödinger equations for x < 0 and x > 0. Sketch the solutions in the two regions. What is the wavelength for x < 0? What are the boundary conditions at x = 0? What is the probability of finding the electron at x > 0? Given: $\hbar = 1.054 \times 10^{-34} J-s$, $c = 2.997 \times 10^8 ms^{-1}$, $m_e = 9.11 \times 10^{-31}$ kg . (Answer: $\lambda = 0.4$ nm).

R.Q. 5.22 A beam of electrons accelerated by a 5 V potential strikes on a square potential barrier of height 25 V and width $a = 5.2 \times 10^{-2}$ nm. What fraction of the incident beam gets through the barrier?

R.Q. 5.23 A beam of 10V electrons strikes on a rectangular potential barrier of height 5 V. Estimate the thickness for 100% transmission of the beam through the barrier.

R.Q. 5.24 An electron with a kinetic energy of 6 eV suddenly encounters an abrupt change in potential energy for which a 50% chance of reflection or transmission obtains. By how many eV has the potential changed?

R.Q. 5.25 A square well of width 2a = 5 fm is known to contain only one allowed energy level for electrons. What are the limits on the depth of this well?

R.Q. 5.26 Consider the potential shown. For particles with energy, $E > V_2$, incident from the left, calculate the transmission probability. Show that this probability is maximal if the region 0< x < a contains an integral number of half de Broglie wave lengths.

R.Q. 5.27 Calculate the transmission coefficient of a particle having total energy E, at the potential barrier given by V(x) = 0, for x< 0; V(x) = V_0, 0 < x < a; V(x) = 0, x > a, for the cases E >V_0, and 0 < E < V_0.

R.Q.5.28 Determine the reflection coefficient of a potential wall defined by $U(x) = U_0 / (1 + e^{-\beta x})$ such that the energy of the particle is E > U_0.

R.Q. 5.29 Determine the transmission coefficient for a potential barrier defined by the formula $U(x) = U_0 / \cosh^2 \beta x$, and the energy of the particle is restricted to E <U_0.

%%%%%%%%%%%

Chapter 6

EXACT SOLUTIONS: APPLICATIONS - 4
Linear Oscillator – Traditional Approach

Chapter 6

EXACT SOLUTIONS: APPLICATIONS - 4
Linear Oscillator – Traditional Approach

"Echoing the voice of Vedanta and all mystical thought that the fundamental search for reality takes man beyond the senses and the sensory world of phenomena", -
Dr. Frito Capra says (the author of <u>The Tao of Physics</u>')

6.1   INTRODUCTION

In this Chapter also, Quantum Mechanics is applied to simple but very general useful systems The objectives are to construct the Schrödinger Equation of the particle in a potential system under consideration; find the wave functions, the energy levels, probability distributions for the particle, and to apply the results of the various systems to practical problems in modern physics. At a first glance quantum mechanics seems a poor substitute for Newtonian Mechanics, but a closer analysis will show that Quantum Mechanics is a well-developed version of Classical Mechanics. It will be seen that Quantum Mechanics is the best effort to date to formulate a mechanics suitable for macroscopic and microscopic systems.

6.2   THE LINEAR HARMONIC OSCILLATOR, SHM:

6.2.1   Introduction

The simple harmonic oscillator is a physical system that is still a simplified abstraction but that has served as a theoretical model for the study of many real physical situations. Harmonic oscillation occurs when a system of some kind vibrates about an equilibrium configuration. Among the countless examples of such a system in both macroscopic and microscopic realms a few are:
(i) An Object supported by a spring,
(ii) An Object floating on a liquid,
(iii) Motion of atoms in a molecule (say, a diatomic molecule),
(iv) An atom in a crystal lattice, or,
(v) The quantization of fields.

This system provides one with one of the most fundamental problems in physics, enabling analysis of certain complicated systems in terms of normal modes of motion. The importance of simple harmonic motion lies very much in its role in classical as well as equally in quantum mechanics.

6.2.2   CLASSICAL Harmonic Oscillator:

By definition, a body is said to execute simple harmonic motion (shm), when (say, in an ideal spring) the restoring force, F(x) is proportional to the displacement of the body, x relative to the origin or its equilibrium position, x_0, *i.e.*, the relative distance apart, x is such that (Fig. 6.1).

6.2.2.1 Potential Energy - HOOKE'S LAW:
$$F(x) \propto x$$

6.2.2.2 Force
Any force of this feature can be expressed in a *Maclaurin's series* about x_o, as

$$\vec{F}(x) = \vec{F}(x)\big|_{x=x_o} + \frac{d}{dx}\vec{F}\big|_{x=x_o} x + \frac{1}{2}\frac{d^2}{dx^2}\vec{F}\big|_{x=x_o} x^2 + \cdots \quad (6.2.1)$$

At $x = x_o$, $\vec{F}(x) = $; i.e. the first term in equation (6.2.1) is zero.

$$\vec{F}(x) = \frac{d}{dx}\vec{F}\big|_{x=x_o} x$$

where the higher order terms are neglected.

If $\quad \frac{d}{dx}\vec{F}\big|_{x=x_o} x = -k =$ negative constant, $\quad (6.2.2)$

this Simple Harmonic Motion (SHM) is given by the <u>Hooke's Law</u>,

$$\boxed{\vec{F}(x) = -kx} \quad (6.2.3)$$

where k, called the **elastic constant**, has units of Nm^{-1}, represents force required to displace the body one unit of a distance, the *minus sign* implies the restoring force is always opposite in direction to the displacement.

6.2.2.3 The Elastic Potential Energy, V(x):
The classical Hamiltonian (H) for the body having mass, m and momentum, p_x,

$$H = T + V(x) = \frac{1}{2m} p_x^2 + V(x) \quad (6.2.4)$$

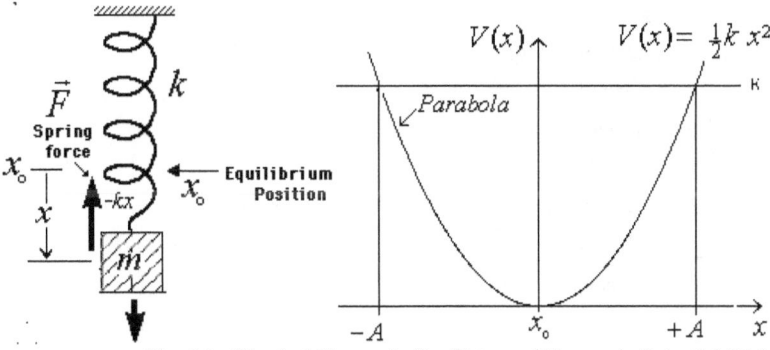

Fig. 6.1 Classical Harmonic Oscillator and Harmonic Potential V(x)

All oscillations are simple harmonic in character when their amplitudes are *sufficiently small*. For **Hooke's Law** of force, the potential energy, V(x) is given by

$$\vec{F}(x) = -\hat{\nabla} V(x) = -(d/dx)V(x)$$

∴ $\quad dV(x) = -\vec{F}(x)\, dx$.

Integrating (choosing the zero of the potential energy at the origin,

$$\int_0^x dV(x) = -\int_0^x \vec{F}(x)\, dx = -\int_0^x (-kx)\, dx$$

i.e., $\quad \boxed{V(x) = \frac{1}{2} k x^2} \quad (6.2.5)$

which is represented by **a parabola** (Fig 6.1).

6.2.2.2 Kinetic Energy
Thus according to equation (6.2.4), the energy of the oscillations
$$E = \tfrac{1}{2} m v^2 + \tfrac{1}{2} k x^2 \qquad (6.2..6)$$
At the end points of level R, which are the extrema for the zero kinetic energy,
i.e., at $x = \pm A$, $E = \tfrac{1}{2} k A^2$
The system executes SHM as long as there is no dissipative processes are present.

6.2.2.3 Kinematics of SHM
According to *Newton's II Law*, the **Equation of Motion** is $\vec{F}(x) = m \vec{a}$.

$\therefore \qquad \vec{F}(x) = m \left(\frac{d^2 x}{dt^2}\right) = -k x$

or, $\qquad \left(\frac{d^2 x}{dt^2}\right) + \frac{k}{m} x = 0$

Defining $\omega_c^2 = + \frac{k}{m}$ \qquad (6.2.7)

$$\left(\frac{d^2 x}{dt^2}\right) + \omega_c^2 x = 0 \qquad (6.2.8)$$

This is a standard form of the 2nd order differential equation, having solution,
$$x(t) = A \, Cos(\omega_c t + \varepsilon) \qquad (6.2.9)$$
$(\omega_c t + \varepsilon) =$ **Phase** of the oscillation,
$A =$ **Amplitude** of oscillation,
and $\omega_c =$ classical *angular frequency* of the oscillating body.

$$\boxed{\omega_c = \sqrt{k/m}} \qquad (6.2.10)$$

Using equation (6.2.9) in (6.2..6),
$$\frac{2E}{k A^2} = Sin^2 \omega_c t + (m \, \omega_c^2 / k) \, Cos^2 \omega_c t$$
which identity holds good only if
$$\frac{2E}{k A^2} = (m \, \omega_c^2 / k) = 1 \qquad (6.2.11)$$
The SHM motion in its full form become is given by the expresion
$$\boxed{x(t) = \sqrt{2E/k} \, Cos\{\sqrt{k/m} \, t\}} \qquad (6.2.12)$$
Energy $E = kA^2 = \tfrac{1}{2}(m \, \omega_c^2) A^2$

$$(6.2.11)$$

6.3 QUANTUM MECHANICAL Linear Harmonic Oscillator: TRADITIONAL APPROACH:

Even before one makes a detailed calculation, one may anticipate quantum mechanical modifications to the classical picture seen in Section 6.2.2:
(i) There will not be a continuous spectrum of allowed energies, but there will be a discrete spectrum,
(ii) The lowest energy of the system will be $E_0 \neq 0$,
(iii) There is certain probability of penetration of the potential well that the particle can go beyond the limits - A and +A.

6.3.1 Schrödinger Equation for SHM:
For 1–D oscillator,
$$\hat{H}\psi(x) = \hat{E}\psi(x) \tag{6.3.1}$$
The Oscillator potential is defined as
$$V(x) \begin{cases} = \frac{1}{2} k x^2, |x| > 0 \\ = 0 \quad\quad x = 0. \end{cases} \tag{6.3.2}$$

$$\nabla^2 \psi + a = \frac{2m}{\hbar^2}[E - V(x)]\,\psi(x) = 0 \tag{6.3.3}$$

Substituting equation (6.2.5) in (6.3.2)
$$\nabla^2 \psi(x) + [a - b^2 x^2]\psi(x) = 0 \tag{6.3.4}$$

Let $\quad a = \frac{2m}{\hbar^2} E \tag{6.3.5}$

and $\quad b^2 = \frac{2m}{\hbar^2}(\frac{1}{2})k = \frac{m}{\hbar^2}\omega_c^2 m$

or, $\quad b = \frac{m}{\hbar}\omega_c . \tag{6.3.6}$

Introduce a dimensionless coordinate,
$$q = b^{1/2} x \tag{6.3.7}$$
so that $(d\psi / dx) = (d\psi / dq)(dq / dx)$
$$= (d\psi / dq)(b^{1/2})(d^2\psi / dx^2)$$
$$= (d / dq)(d\psi / dx)(dq / dx)$$
$$= (d / dq)(b^{1/2} d\psi / dq)(b^{1/2})$$
$$= b\,(d^2\psi / dq^2) \tag{6.3.8}$$

Equation (6.3.4) simplifies to
$$(d^2\psi / dq^2) + [\tfrac{a}{b} - q^2]\,\psi(q) = 0. \tag{6.3.9}$$

Putting $\lambda = \frac{a}{b} = \frac{2E}{\hbar \omega_c} \tag{6.3.10}$

Substituting for λ in equation (6.3.9)
$$(d^2\psi / dq^2) + [\lambda - q^2]\,\psi(q) = 0 \tag{6.3.11}$$

All the quantities that appear in equation (6.3.11) are *dimensionless*. This <u>Schrödinger Equation for an oscillator</u> (6.3.11) is not easy to solve.

6.3.2 The Polynomial Solution (due to A. Sommerfeld, Wave Mechanics, Dutton, NY, 1929):
To solve equation (6.3.11) one has to examine $\psi(q)$.
Behavior of ψ(q) as $q \to$ very large, *i.e.* $q \to \infty$.

(b) As $q \to \infty$, $\quad \psi(q) = \psi_\infty(q)$,
$$\lambda\,\psi_\infty(q) \ll q^2 \psi_\infty(q),$$
i.e., $|q| \gg \lambda$ so that equation (6.3.11) in this limit becomes
$$(d^2\psi_\infty(q) / dq^2) = q^2 \psi_\infty(q). \tag{6.3.12}$$
Multiplying equation (6.3.12) by $2(d\psi / dq)$, one can write
$$(d\psi / dq)[d\psi(q) / dq]^2 - q^2(d / dq)\psi^2(q)] = 0.$$
Equivalently, $(d / dq)\{[d\psi(q) / dq]^2 - q^2\psi^2(q)]\} = -2q\,\psi^2(q)$

Neglecting the term $2q\,\psi^2$, a simplified equation is obtained as follows.
$$(d/dq)\{[d\psi/dq]^2 - q^2\psi^2]\} = 0.$$
i.e., $\quad (d/dq)[d\psi/dq]^2 = q^2\psi^2$.

Integrating, $\quad \int [d\psi/dq]^2 dq = \int q^2\psi^2 dq$, $[d\psi/dq]^2 - q^2\psi^2 = C$, constant.
$$\therefore d\psi/dq = [C + q^2\psi^2]^{1/2}.$$

6.3.3 Asymptotic Solution

Since both ψ(q) and ψ'(q) both must vanish at q = ∞, one must have C = 0. with
$$\underset{q \to \infty}{Lt}\ \psi(q) = 0.$$
Thus $\quad (d\psi/dq) = \pm q\,\psi$

whose solution acceptable at $q = \infty$ is
$$\psi_\infty(q) = e^{\pm \frac{1}{2}q^2}.$$

Now check that $2q\,\psi^2 = 2q\,e^{-q^2}$ is negligible compared with
$$d(q^2\psi^2)/dq = (d/dq)[q^2\,e^{-q^2}]$$
$$2q\,e^{-q^2} + [q^2\,e^{-q^2}](-2q) \cong (2q^3)\,e^{-q^2}$$
for large values of q. This means ψ <u>asymptotically</u> satisfies equation (6.3.12)
i.e., $\quad \underset{q \to \infty}{Lt}\ [d^2\psi_\infty/dq^2]/[q^2\psi_\infty] = 1.$

Since q = large, the choice of $e^{+\frac{1}{2}q^2}$ is unacceptable, and equation (6.3.12) has the solution,
$$\psi_\infty = e^{-\frac{1}{2}q^2} \tag{6.3.13}$$

$$(d\psi_\infty(q)/dq) = (d/dq)e^{-\frac{1}{2}q^2}$$
$$= (-q)e^{-\frac{1}{2}q^2}\ [d^2\psi_\infty(q)/dq^2] = (d/dq)(-q)e^{-\frac{1}{2}q^2}$$
$$= (q^2)e^{-\frac{1}{2}q^2} - e^{-\frac{1}{2}q^2} = (q^2 - 1)\,e^{-\frac{1}{2}q^2}.$$

$\underset{q \to \infty}{Lt}\ [d^2\psi_\infty/dq^2] = \underset{q \to \infty}{Lt}\ (q^2-1)\,e^{-q^2} \Rightarrow q^2\,e^{-q^2} \Rightarrow 0$, as q becomes large.

Equation (6.3.13) for $\psi_\infty(q)$ is called the APPROXIMATE or the ASYMPTOTIC solution of equation (6.3.11).

6.3.3 NEW SOLUTION

One may introduce a new function, f(q), i.e. split a factor,
$$\psi_\infty = e^{-q^2/2},$$
of the eigen function ψ(q) and substitute in
$$\psi(q) \cong f(q)\{e^{-q^2/2}\} \tag{6.3.14}$$
where f(q) is that part of ψ(q) that vanishes more rapidly than $\{e^{-q^2/2}\}$

Now equation (6.3.11) is $(d^2\psi/dq^2)+[\lambda-q^2]\psi(q)=0$

$$d\psi(q)/dq = (d/dq)[f(q)e^{-q^2/2}]$$
$$=[f(q)e^{-q^2/2}]+\frac{1}{2}(-2q)[f(q)e^{-q^2/2}]$$
$$=[e^{-q^2/2}][f'(q)-qf(q)](d^2\psi/dq^2)$$
$$=(d/dq)[e^{-q^2/2}][f'(q)-qf(q)]$$
$$=[e^{-q^2/2}][f''(q)-qf'(q)-f(q)]+[e^{-q^2/2}](f'(q)-qf(q)[-q]$$
$$=[e^{-q^2/2}][f''(q)-qf'(q)+q^2f(q)-f(q)]$$

Substituting this result in equation (6.3.11),
$$[f''(q)-qf'(q)+(q^2-1)f(q)+[\lambda-q^2]f(q)]=0,$$
giving the HERMITE'S DIFFERENTIAL EQUATION,

$$\boxed{[f''(q)-2qf'(q)+(\lambda-1)f(q)]=0} \qquad (6.3.14)$$

Though this equation may not appear to simple, in the discussions above one has accounted for the behavior of $\psi(q)$ as $q=\infty$. However, this equation can be solved by the **Method of Series**.

(a) Behavior of $\psi(q)$ as $q \to 0$, Series solution:
To solve the differential equation (6.3.12) subject to the condition that

$$\underset{q \to \infty}{Lt}[f(q)] = \psi_\infty = e^{-q^2/2} \qquad (6.3.15)$$

Assuming that $f(q)$ can be expanded in a POWER SERIES,

$$f(q) = \sum_{k=0}^{k=\infty} A_k q^k \qquad (6.3.16)$$

($A_0 \neq 0$) and determine the A_k -s.

6.3.4   RECURSION Formula:
Inserting this form of $f(q)$ in equation (6.3.14),

$$f(q) = \sum_{k=0}^{k=\infty} A_k q^k$$
$$f'(q) = \sum k A_k q^{k-1}$$
$$qf'(q) = q\sum k A_k q^{k-1} = \sum k A_k q^k f''(q)$$
$$= (d/dq)f'(q) = (d/dq)[\sum k A_k q^{k-1}]$$
$$= \sum k(k-1)A_k q^{k-2} = \sum (k+2)(k+1)A_{k+2} q^k$$

Substituting these in equation (6.3.14),
$$\sum(k+2)(k+1)A_{k+2}q^k + 2\sum k A_k q^k + (\lambda-1)\sum A_k q^k = 0.$$

It holds good for all values of q, or, in other words, this is an identity only if
$(k+2)(k+1)A_{k+2} + 2k A_k + (\lambda-1)A_k = 0$, for all values of k.

i.e., $(k+2)(k+1)A_{k+2} - (2k+1-\lambda)A_k = 0,$

$$\boxed{A_{k+2} = \frac{(2k+1-\lambda)}{(k+2)(k+1)} A_k} \qquad (6.3.17)$$

That means any series whose coefficients satisfy (6.3.17) will satisfy equation (6.3.14). Equation (6.3.17) is a <u>Recursion Formula</u>, *i.e.* '*one-after-another*' calculation of successfully the remaining coefficients, when the values of A_0 and A_1 are given. For given pair of A_0 and A_1, there are two possible Series, viz. EVEN and ODD Series, generated because of the Recursion Formula, which connects only terms that differ in the order of q by 2. The ratio of consecutive terms in either the two Series is

$$\frac{A_{k+2} \, q^{k+2}}{A_k \, q^k} = \frac{(2k+1-\lambda)}{(k+1)(k+2)}$$

These two Series are:

a)	Even Series	A_0 = finite	$A_1 = 0$
b)	Odd Series	$A_0 = 0$	A_1 = finite

(6.3.18)

That the two solutions *do not mix* is a consequence of the invariance of the Hamiltonian under reflection,

i.e., $\hat{H}(x) = \hat{H}(-x)$ or, $V(x) = V(-x)$.

6.3.4 Convergence of the Series $f(q)$:

Recall equation (6.3.16)

$$f(q) = \sum_{k=0}^{k=\infty} A_k q^k$$

If the Series given by (6.3.17) does not terminate after a finite number of terms, *i.e.*, by (6.3.14),

viz., $\psi(q) \equiv f(q)\{e^{-q^2/2}\}$

will diverge as $q \to$ very large, *i.e.* $q \to \infty$. The Recursion Formula (6.3.17), for large values of k, (say, $k \to \infty$),

$$\underset{k \to \infty}{Lt} \frac{A_{k+2}}{A_k} = \frac{2}{k}$$

i.e., $A_{k+2} \approx \frac{2}{k} A_k$.

6.3.5.1 To get further the correct form of $f(q)$

This Recurrence Relation holds good also for the coefficients of the Series $\{e^{-q^2}\}$. To examine consider the Series

$$\{e^{+q^2}\} = 1 + \frac{q^2}{1!} + \frac{q^4}{2!} + \cdot + \cdot + \cdot +$$

$$= \sum_{k=0}^{k=even} B_k q^k \qquad (6.3.19)$$

because the values of k are very large, the ratio of the coefficients of the successive terms,

$$B_{k+2} = \frac{(k/2)!}{;(k+2)/2]!} B_k$$

$$B_{k+2} = \frac{(k/2)!}{[(k/2)+1]!} B_k \approx \frac{2}{k} B_k$$

This means the coefficients of the Series $\{e^{+q^2}\}$ are similar to the Series $f(q)$. of equation (6.3.16) at large values of k. Thus the Series of (6.3.19) also satisfies the recurrence relation (6.3.17), which means $f(q)$ has a factor similar to (6.3.19). Thus for sufficiently large values of k,

$$\frac{A_{k+2}}{B_{k+2}} = \frac{A_k}{B_k} = C = \text{constant}$$

This means the solution $f(q)$ behaves like $\{e^{+q^2}\}$, where q = large, and

$$F(q) = (\text{a polynomial in } q) + C\{e^{+q^2}\}.$$

$$F(q) = g(q) + C\{e^{+q^2}\} \qquad (6.3.20)$$

where $g(q)$ = some polynomial of q that can be expanded in a Power Series that does not contain very high powers of q.

From equation (6.314) $\psi(q) \cong f(q)\{e^{-q^2/2}\}$

$$\psi(q) \cong \{g(q) + C\{e^{+q^2}\}\}\{e^{-q^2/2}\}.$$

$$\cong \{g(q)\{e^{-q^2/2}\}\} + C\{e^{+q^2/2}\} \qquad (6.3.21)$$

It is seen that diverges as $e^{+q^2/2}$ diverges when $q \to \infty$. In order to satisfy the boundary conditions as $q \to \pm\infty$, $f(q)$ behaves like $e^{+q^2/2}$. To avoid this one must have $f(q) = \sum_{k=0} A_k q^k$ = finite, i.e. the coefficients beyond A_k are zero, when the equation breaks off after a finite number of terms, leaving the first term only in equation (6.3.19), which is obviously bounded.

$$\psi(q) \cong g(q)\{e^{-q^2/2}\}. \qquad (6.3.22)$$

6.3.6 ENERGY Levels of the Oscillator:
Recalling (6.3.16) and (6.3.17)

$$f(q) = \sum_{k=0} A_k q^k$$

$$A_{k+2} = ((2k+1-\lambda_k)/(k+2)(k+1)) A_k.$$

an acceptable solution can thus be found, therefore, if the Recursion Formula terminates, that is, the Series $f(q)$ (6.3.16) terminates if the numerator of (6.3.17) is zero for $k > n$, whence

$$\boxed{\lambda_k = \lambda_n = (2n+1)} \qquad (6.3.23)$$

i.e., the series for $f(q)$ is broken off and becomes a polynomial in q if and only if equation (6.3.23) is satisfied. If n = even one gets an E*ven Series*, and if n = odd, the Series being terminated is odd and condition (6.3.18) applies. Therefore, the energy levels are given by equation (6.3.10) and (6.3.23), viz.,

$$\lambda_n = \frac{2 E_n}{\hbar \omega_c}$$

i.e., $\quad \lambda_n = 2 E_n / \hbar \omega_c = (2n+1)$.

$$\boxed{E_n = (n+\tfrac{1}{2}) \hbar \omega_c \; ; \; n = 0, 1, 2, 3, \ldots} \qquad (6.3.24)$$

n is called the QUANTUM NUMBER. The energy of a harmonic oscillator is thus quantized in steps of ($\hbar \omega_c$). There are thus DISCRETE, EQUALLY (*i.e.* EVENLY) SPACED eigen values (Fig. 6.2), unlike those of a particle in a box.
Equation (6.3.24) is the same as the one discovered by Planck for radiation field modes. This is because a decomposition of the EM Field into normal modes is essentially decomposition into harmonic oscillators that decoupled.

6.3.7. ZERO-POINT ENERGY, E_o

The lowest energy level of an oscillator is given by $n = 0$,

i.e., $\quad \boxed{E_o = \tfrac{1}{2} \hbar \omega_c} \qquad (6.3.25)$

The observations are:

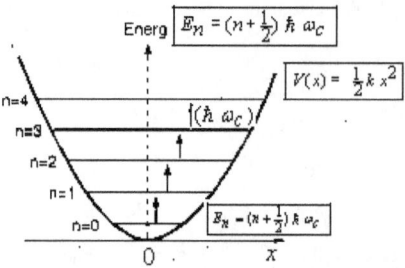

Fig. 6.2 Linear QM Oscillator Energy (Evenly spaced) levels.

Even the lowest state of a harmonic oscillator has some energy, the Zero-point energy, E_o. Its presence is a pure quantum mechanical effect and interpreted in terms of the *Uncertainty Principle*. It is the E_o that is responsible for the fact that

Helium remains liquid down to $T \approx 10^{-3} K$, at normal pressures. ω_c is larger for lighter atoms, which is why the effect is *not seen for Nitrogen*. E_o also depends on detailed features of the interatomic forces, which is why Liquid hydrogen does freeze. E_o will vanish in a world in which $h = 0$.

6.3.8 PROPERTIES of Function $f(q)$ in $\psi(q)$

Th functions (6.3.22), viz.

$$\psi(q) \cong g(q) \{e^{-q^2/2}\}$$

are called the HERMITE'S POLYNOMIAL.

6.3.8.1 Determination of $g(q)$:

The coefficients A_o and A_1 must be determined. Since for the harmonic oscillator, $V(x) = V(-x)$, the eigenfunction corresponding to bound states of energy, E_n will be either even or odd, which corresponds
either $A_1 = 0$, or $A_o \neq 0$, or to $A_1 \neq 0$ and $A_o = 0$; respectively, as in equation (6.3.18).
However, a convenient convention is

$\boxed{\text{Step \# 1}}$ to choose the coefficient of the term of highest order of q,

viz,. $A_n = 2^n$

which means in effect specifying A_o or A_1.

$\boxed{\text{Step \# 2}}$ The R*ecurrence Relation*is (6.3.17), viz.

$$A_{n+2} = ((2n + 1 - \lambda_n) / (n+2)(n+1)) A_n \qquad (6.3.17)$$

with $\lambda_n = (2n+1)$.   (6.3.23)

$\boxed{\text{Step \# 3}}$ gives the solution $g(q)$ the Hermite's polynomial, $H_n(q)$, as

$$H_n(q) = (2q)^n - \{\tfrac{n(n-1)}{1!}\}(2q)^{n-2} + \{\tfrac{n(n-1)(n-2)(n-3)}{2!}\}(2q)^{n-4} \qquad 6.3.26)$$

1) This polynomial for $g(q)$ are, except for normalization constants, the $H_n(q)$, the Hermite's Polynomial of *degree n in q*.
Eigen functions:

2) The nth Hermite's polynomial, $H_n(q)$ is a solution of the differential equation (6.3.14)
$$f''(q) - 2q f'(q) + (\lambda - 1) f(q) = 0$$
i.e., $\quad H_n''(q) - 2q H_n'(q) + (\lambda - 1) H_n(q) = 0 \quad (6.3.27)$

The Hermite's polynomials for a complete set in the range $-\infty < q < +\infty$, and may be defined either by the Recursion Formula (6.3.17), (6.3.18) and (6.3.22), or, more conveniently, as follows: $H_n(q)$ have been well investigated.

6.3.8.2 SOME convenient IDENTITIES (formulae), without proof:

To determine the coefficients $H_n(q)$ some identities are used. They are:

(1) Rodrigues's Formula
$$\boxed{H_n(q) = (-1)^n e^{+q^2} \frac{d^n}{dq^n}(e^{-q^2})} \quad (6.3.28)$$

2) Generating Function
$$S(q,s) = e^{-q^2 - (s-q)^2} = \sum \frac{H_n(q)}{n!} s^n \quad (6.3.29)a$$

Or, $\quad H_n(q) = \frac{d^n}{dq^n} S(q,s) \Big|_{s=0} \quad (6.3.29)b$

3) Differential Equation:
$H_n(q)$ satisfies equation (6.3.27) with $\lambda_n = (2n+1)$,
i.e., $\quad \boxed{H_n''(q) - 2 q H_n'(q) + 2n H_n(q) = 0} \quad (6.3.30)$

4) Recurrence Relation:
Differentiating with respect to s the Generating Function on gets
$$\boxed{H_{n+1}(q) - 2 q H_n(q) - 2n H_{n-1}(q) = 0}. \quad (6.3.31)$$

5) Differential Relation:
Differentiating with respect of q the Generating Function provides
$$\boxed{H_n'(q) - 2n H_{n-1}(q) = 0} \quad (6.3.32)$$

6) ORTHO-NORMALIZATION Relation:
$$\boxed{\int_{-\infty}^{+\infty} H_n(q) H_m(q) (e^{-q^2}) dq = \pi^{1/2} 2^n n! \, \delta_{nm}} \quad (6.3.33)$$

7) Weber-Hermite Function:
is the time independent general form of the nth eigen function: With
$$q = b^{1/2} x = (m \omega_C / \hbar)^{1/2} x \quad (6.3.8)$$
$$\boxed{\psi_n(q) = b^{1/4} [\pi^{1/2} 2^n n!]^{-1/2} H_n(q) (e^{-q^2/2})} \quad (6.3.34)$$
$$\boxed{\psi_n(q) = B_n H_n(q) (e^{-q^2/2})} \quad (6.3.35)$$
$$\boxed{B_n = \{(m \omega_C / \pi^2 \hbar)^{1/4} / [2^{n/2} \sqrt{n!}]\}}. \quad (6.3.36)$$

$$\psi_n(q) = \left\{(m\,\omega_C / \pi^2 \hbar)^{1/4} / [2^{n/2}\sqrt{n!}\,]\right\} H_n(q)\,(e^{-q^2/2}) \qquad (6.3.37)$$

The first 6 eigen functions of an oscillator are listed in Table 6.1.

TABLE 6.1   Eigen Functions of an Oscillator

n	$\lambda_n = (2n+1)$	E_n	$H_n(q)$	$\{\psi_n(q) = B_n\,H_n(q)\,(e^{-q^2/2})\}$
0	1	$\tfrac{1}{2}\hbar\omega_c$	$H_0(q) = 1$	$B_0\,(e^{-q^2/2})$
1	3	$3(\tfrac{1}{2}\hbar\omega_c)$	$(2q)$	$B_1\,(2q)\,(e^{-q^2/2})$
2	5	$5(\tfrac{1}{2}\hbar\omega_c)$	$(4q^2 - 2)$	$B_2\,(4q^2 - 2)(e^{-q^2/2})$
3	7	$7(\tfrac{1}{2}\hbar\omega_c)$	$(8q^3 - 12q)$	$B_3\,(8q^3 - 12q)\,(e^{-q^2/2})$
4	9	$9(\tfrac{1}{2}\hbar\omega_c)$	$(16q^4 - 48q^2 + 12)$	$B_4\,(16q^4 - 48q^2 + 12)(e^{-q^2/2})$
5	11	$11(\tfrac{1}{2}\hbar\omega_c)$	$(32q^5 - 150q^3 + 120q)$	$B_5\,(32q^5 - 150q^3 + 120q)\,(e^{-q^2/2})$

Fig 6.3 illustrates the form of the Eigen functions of a harmonic oscillator and probability densities for the first 4 quantum states, $n = 1$ to 3.

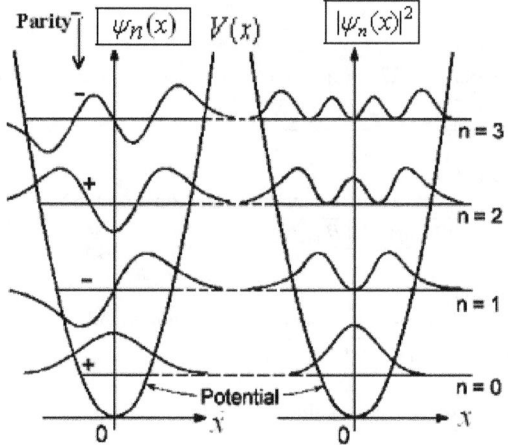

Fig 6.3 The form of the Eigen functions $\psi_n(x)$ and $|\psi_n(x)|^2$ of a harmonic oscillator $n = 0, 1, 2, 3$.

6.3.8.2 FEATURES of the Harmonic Oscillator Functions

Fig 6.3 shows the following *features*:

(i) The eigen functions are alternately symmetrical and anti-symmetrical about the origin, depending on $\psi_n(q)$ which has parity as that of $H_n(q)$ and since $\{e^{-q^2/2}\}$ is even.

(ii) The harmonic oscillator has '*zero-point energy*', in accord with the Uncertainty Principle. $E_0 = 0$ means $p_0 = 0$ and $x_0 =$ located exactly at the position of minimum potential energy, *i.e.* at $x = 0$. So $E_0 \neq 0$ means, $\Delta p_0 \Delta x_0 \geq \hbar/2$,

(iii) The even function (symmetric ones) has no Node whereas the odd ones have Nodes.

6.3.8.3. The Quantum Mechanical VIRIAL THEOREM:

The Virial Theorem states that for a system in which the particles interact according to the '*parabolic potential*', i.e. for a harmonic oscillator,

$$\boxed{<V> \ = \ <T> \ = \tfrac{1}{2} E_n = \tfrac{1}{2}(n+\tfrac{1}{2})\,\hbar\,\omega_c} \tag{6.3.38}$$

Worked out Example 6.1

Generate the Hermite's polynomial $H_2(q)$ where $q = b^{1/2}x$, $b = (m\,\omega_c/\hbar)$. Write $\psi_2(q)$ explicitly for a linear harmonic oscillator.

Solution:

Step # 1 The recurrence relation is $A_{k+2} = ((2k+1-\lambda_n)/(n+2)(n+1))\,A_k$; $\lambda_n = (2n+1)$.

Step # 2 For $H_2(q), n = 2$, and the highest coefficient is
$A_2 \cdot \lambda_n = (2n+1) = 2 \times 2 + 1 = 5$.
But $A_{k+2} = A_2$; $k = A_2/A_0 = [(2)(0)+1-5]/[(1)(2)] = -2$.

Step # 3 $\therefore H_2(q) = A_0 + A_2 q^2 = A_0[1+(A_2/A_0)q^2] = A_0[1+(-2)q^2]$

Step # 4 Choosing $A_n = 2^n$; $A_2 = 2^2 4$; $A_2 = 2^2 = 4$.; $\therefore A_0 = A_2/(-2) = -2$.

Step # 5 $\therefore H_2(q) = A_0 + A_2 q^2$; $H_2(q) = (4q^2 - 2)$

Step # 6 $\psi_2(q) = B_2 H_2(q)\,\{e^{-q^2/2}\} = = B_2(4q^2 - 2)\,\{e^{-q^2/2}\}$.

Worked out Example 6.2

Evaluate the expectation value of x for the linear harmonic oscillator in an arbitrary state. Given the recurrence formula, $H_{n+1}(q) - 2\,q\,H_n(q) + 2n\,H_{n-1}(q) = 0$; $q = b^{1/2}x$ and $b = (m\,\omega_c/\hbar)$.

Solution:

Step # 1 It is known that $<\hat{q}>_n = \int_{-\infty}^{+\infty} H_n(q)\,\hat{x}_n\,H_n(q)\,(e^{-q^2})\,dq$;

$\psi_n(q) = B_n H_n(q)\,\{e^{-q^2/2}\}$; $\int_{-\infty}^{+\infty} H_n(q)\,\hat{x}_n\,H_n(q)\,(e^{-q^2})\,dq = \pi^{1/2}\,2^n\,n!\,\delta_{nm}$ and $H_{n+1}(q) - 2\,q\,H_n(q) + 2n\,H_{n-1}(q) = 0$.

Step # 2 Pre-multiply by $H_n^*(q)\,(e^{-q^2})dq$ the recurrence relation and integrating over all space,

$\int_{-\infty}^{+\infty} H_n^*(q)\,(e^{-q^2})dq\,(H_{n+1}(q) - 2\,q\,H_n(q) + 2n\,H_{n-1}(q)) = 0$

i.e., $\int_{-\infty}^{+\infty} \{\psi_n^* \psi_{n+1}\,dq - \psi_n^*\,2q\,\psi_n + \psi_n^*\,2n\,\psi_{n-1}\}dq = 0$.

Step # 3 Applying the ortho-normal properties of $H_n(q)$, the above equation simplifies to $0 - 2<\hat{q}>_n + 0 = 0$, i.e. $<\hat{q}>_n = 0$

Step # 4 Because (6.3.8), $q = b^{1/2}x$, this means $<\hat{x}>_n = 0$ This is because $\psi_n(x)$ is a steady state.

6.4 DIPOLE MOMENT OF THE OSCILLATOR:

If the oscillating particle carries with it an electric charge, e,
Classical dipole moment,
$$\vec{p} = e\vec{q}. \tag{6.4.1}$$

Quantum Mechanically, $<\vec{p}> = \int_{-\infty}^{+\infty} \psi_n^*(q) \, [e\vec{q}] \, \psi_n(q) \, dq$ (6.4.2)

Worked out Example 6.3

Determine the lower limit of the possible values of the energy of an oscillator, using the Uncertainty Relation $\sqrt{(\Delta p_x)^2 (\Delta x)^2} \geq \hbar/2$.

Solution:

Step # 1 $\hat{H} = \hat{p}_x^2/2m + \frac{1}{2} m \omega_c^2 x^2$.

Step # 2 $E_{av} \geq <\hat{p}_x>^2/2m + \frac{1}{2} m \omega_c^2 <x>^2$.

Step # 3 Using $\sqrt{(\Delta p_x)^2 (\Delta x)^2} \geq \hbar/2$,

$<E> = <\Delta \hat{p}_x>^2/2m + \frac{1}{16} m \omega_c^2 \hbar^2 <\Delta \hat{p}_x>^{-2}$

Considering $<\Delta \hat{p}_x>^2$ as a variable, the minimum value of the above expression is

$E > \frac{1}{2} \hbar \omega_c$.

6.5 PARITY

The energy levels of 3-D oscillator can be written as

$E_n = E_n^x + E_n^y + E_n^z$

$= (n_x + \frac{1}{2}) \hbar \omega_c + (n_y + \frac{1}{2}) \hbar \omega_c + (n_z + \frac{1}{2}) \hbar \omega_c$

$= [n_x + n_y + n_z + 3(\frac{1}{2})] \hbar \omega_c$

$= [n + \frac{3}{2}] \hbar \omega_c$. (6.5.1)

Parity, $P = (-1)^{[n_x + n_y + n_z]}$ (6.5.2)

Worked out Example 6.4

Find the energy, parity, and degeneracy of the lowest 4 distinct groups of energy levels of a 3-D oscillator having quantum numbers n_x, n_y, n_z. Using the knowledge of parity and degeneracy of various states, deduce which values of angular momentum quantum number ℓ, are associated with $n = 2$ group of energy levels enumerated.

Solution:

Step # 1 It is known that $E_n = E_n^x + E_n^y + E_n^z = [n + \frac{3}{2}] \hbar \omega_c$; $P = (-1)^{[n_x + n_y + n_z]}$
Degeneracy., $n = 2$.

Step # 2 $E_n = [n + \frac{3}{2}] \hbar \omega_c = [2 + \frac{3}{2}] \hbar \omega_c = \frac{7}{2} \hbar \omega_c$

Step # 3 $P = (-1)^{[n_x + n_y + n_z]} = (-1)^2 = +1 =$ Even

Step # 4 Degeneracy of $n = 2$ state, $=$ # of states $= \ell(\ell+1) =, 6$

Step # 5 There is one distinct group of energy levels and are listed in Table 6.2.:

Table 6.2

$P=(-1)^{[n_x+n_y+n_z]}$	n_x,	n_y,	n_z	ℓ	$\ell(\ell+1)$
+1	1	1	0	2, 0	6
"	0	1	1		
"	1	0	1		
"	2	0	0		
"	0	2	0		
"	0	0	2		

6.6 CONCLUSIONS

For a harmonic oscillator the essential physical quantities are as shown in the diagram below.

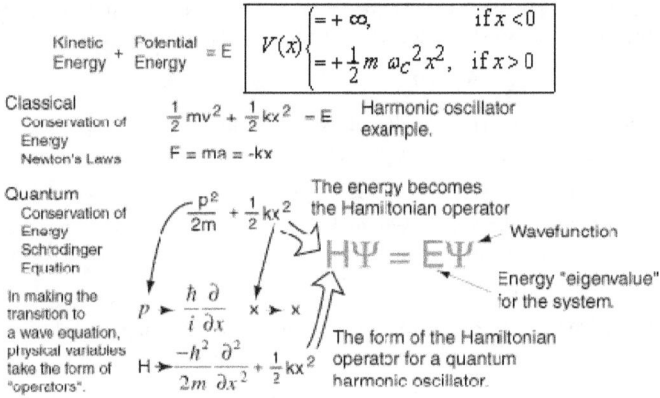

REVIEW QUESTIONS

R.Q. 6.1   Obtain $H_3(q)$, the Hermite's polynomial where $q = b^{\frac{1}{2}}x$, $b = (m \omega_c / h)$.
(Answer: $(8q^3 - 12q)$).

R.Q. 6.2   Calculate the value of $<\hat{p}_x>_n$ for the nth state of a harmonic oscillator. Given:
$H_n{}'(q) - 2n H_{n-1}(q) = 0$. $H_{n+1}(q) - 2q H_n(q) + 2n H_{n-1}(q) = 0$
(Answer: $<\hat{p}_x>_n = 0$).

R.Q. 6.3   Find $<T>_n$ for a linear harmonic oscillator. Given the identities,
$H_{n+1}(q) - 2q H_n(q) + 2n H_{n-1}(q) = 0$, and $H_n{}'(q) - 2n H_{n-1}(q) = 0$

(Answer: Find $<\hat{q}^2>_n = B_n^2[\pi^{1/2}2^n n!](2n+1)$; $<\hat{p}_x^2>_n = \hbar^2 b(2n+1) - \hbar^2 b<\hat{q}^2>_n$
$= \hbar m \omega_c (n+\tfrac{1}{2})$; $<T>_n = \tfrac{1}{2} \hbar m \omega_c (n+\tfrac{1}{2})$).

R.Q. 6.4 Find, $<V>_n$ for a linear harmonic oscillator. Given the identities, $H_{n+1}(q) - 2q H_n(q) + 2n H_{n-1}(q) = 0$, and $H_n'(q) - 2n H_{n-1}(q) = 0$ (Answer: Find $<\hat{q}^2>_n = B_n^2[\pi^{1/2}2^n n!](2n+1)$;; $<V>_n = \tfrac{1}{2} \hbar m \omega_c (n+\tfrac{1}{2})$).

R.Q. 6.5 Write down the Schrödinger Equation for an oscillator in the Momentum Representation. Find the wave function in momentum space. (Answer:
$(d^2/d\xi^2)\Phi(\xi) + (\lambda - \xi^2)\Phi(\xi) = 0$, where $\xi = \hat{p}_x/(m\omega_c \hbar)$;
$\Phi(\hat{p}_x) = \{1/\sqrt{m \omega_c \pi \hbar}\}[2^{n/2}\sqrt{n!}]H_n(\xi)(e^{-\xi^2/2})$).

R.Q. 6.6 Determine the probability distribution of the value of the momentum $|\Phi(\hat{p}_x)|^2$ for an oscillator.

R.Q. 6.7 The energy eigen values of a molecule indicate that the molecule is a 1-D harmonic oscillator. When the molecule de-excites from its first to the ground state a photon of $0.1\ eV$ energy is emitted. Assuming the oscillating part of the molecule as a proton, estimate the probability P that the proton is at a distance from the origin that not permissible to it by classical physics. Gven expression for error integral,
$\mathfrak{J} = \{1/\sqrt{2\pi}\}\int_{-\infty}^{+\infty} e^{-q^2/2} dq$ (Answer: $P = \left(1 - \{1/\sqrt{\pi}\}\int_{-\infty}^{+\infty} e^{-q^2} dq\right)$)

R.Q. 6.9 Evaluate the expectation value of the dipole moment of the oscillator for the following states: $\psi_n(q) = \psi_0(q)$; $\psi_n(q) = \psi_1(q)(e^{-i\omega t})$;
$\psi_n(q) = c_0 \psi_0(q) + c_1 \psi_1(q)(e^{-i\omega t})$; $|c_0|^2 + |c_1|^2 = 1$ and neither c_0 nor c_1 is 0.

R.Q. 6.10 Show by using the Virial theorem that in the quantum state n of the harmonic oscillator, the product of the position and the momentum uncertainties is given by $(\Delta p_x)_n (\Delta x)_n \geq (n+\tfrac{1}{2})\hbar$.

R.Q. 6.11 Calculate the expectation value $<x^2>_n$ in the nth state of a linear harmonic oscillator. Given;
$\int_{-\infty}^{+\infty} H_n(q) H_m(q) (e^{-q^2}) dq = \pi^{1/2} 2^n n!\ \delta_{nm}$ and $H_{n+1}(q) - 2q H_n(q) + 2n H_{n-1}(q) = 0$.
$x(t) = \sqrt{2E/k}\ Cos\{\sqrt{k/m}\ t\}$. Illustrate the Virial Theorem, and the Correspondence Principle.(Answers: $<\hat{q}^2>_n = B_n^2[\pi^{1/2}2^n n!](2n+1)$; $<\hat{p}_x^2>_n = \hbar^2 b(2n+1) - \hbar^2 b<\hat{q}^2>_n$
$= \hbar m \omega_c (n+\tfrac{1}{2})$; $<T>_n = \tfrac{1}{2}\hbar m \omega_c (n+\tfrac{1}{2})$; $<V>_n = \tfrac{1}{2}\hbar m \omega_c (n+\tfrac{1}{2})$;
$<x^2>_{Classical} = <x^2>_n = E_n/k$).

R.Q. 6.12 Calculate the possible energy levels of a particle of mass m moving in a 1-D potential, $V(x) = +\infty,\ if\ x < 0$; and $V(x) = +\tfrac{1}{2} m \omega_c^2 x^2,\ if\ x > 0$. with ω_c characteristic frequency
(Answer: $E_n = (n+\tfrac{1}{2})\hbar\omega_c$., $n = 0, 1, 2, 3..$).

R.Q. 6.13 In a 1-D oscillator, with characteristic frequency, ω_c, what is the energy and parity of the eigen state associated with quantum number n? What values may n have? (Answers: $E_n = (n+\tfrac{1}{2})\hbar\omega_c$; $P = (-1)^n$; $n = 0, 1, 2, 3..$).

R.Q. 6.14 The state of an oscillator having angular frequency, ω is represented by
$\psi_n(x) = e^{-i m \omega x^2/\hbar}$

What are the expectation values of the momentum and the position?

R.Q. 6.15 Given the expression for energy radiated per second by an accelerated charge, $dW/dt = \frac{2}{3}e^2 a^2 / c^3 \, erg\text{-}s^{-1}$; where a is acceleration of the charge. Using the Correspondence Principle, find the mean life of a highly excited quantum state of a simple harmonic oscillator in terms of the quantum number, n, of the state, the angular frequency, ω, the mass m and charge e of the oscillating particle. Transition Rule for the oscillator, $\Delta n = \pm 1$; $\psi_n(x) = e^{-im\omega x^2/\hbar}$; $\tau = (E_n - E_{n-1})/(dW/dt)$ (Answer:

$$\tau = \frac{3\,m\,c^3}{2\,(2n+1)\,e^2 \omega^2} \; s\,)$$

R.Q. 6.16 Determine the normalization constant, B_n of the harmonic oscillator, eigen function,

$\psi_n(q) = B_n \, H_n(q) \, (e^{-q^2/2})$; Rodrigue's Formula; $H_n(q) = (-1)^n e^{+q^2} \frac{d^n}{dq^n}(e^{-q^2})$;

$\int_{-\infty}^{+\infty} e^{-\lambda q^2} dq = \sqrt{\pi/\lambda}$ (Answer: $B_n = (m\omega_c / \pi^2 \hbar)^{1/4} / [2^{n/2}\sqrt{n!}]$).

R.Q. 6.17 Show that for a harmonic oscillator, $<x^4>_n = \frac{3}{4} x_o^4 (2n^2 + 2n + 1)$, where $x_o^2 = \hbar/\omega_c$.

&&&&&&&&&&&&

Chapter 7

EXACT SOLUTIONS - 4

QUANTUM MECHANICAL OSCILLATOR- II:

OPERATOR METHOD:

Chapter 7

EXACT SOLUTIONS - 4
QUANTUM MECHANICAL OSCILLATOR- II:
OPERATOR METHOD:

'Time, space, and causation are like the glass through which the Absolute is seen. ...
In the Absolute there is neither time, space, nor causation.' – Swami Vivekananda

"The Brahman of the Hindus, like the Dharmakaya of the Buddhists, and the Tao of the Taoists,can be seen, perhaps, as the ultimate unified field, from which spring not only the phenomena studied in physics, but all other phenomena as well" -
Dr. Fritjof Capra

"Brahman is life, Brahman is joy. Brahman is the void. ...Joy ,verily, that is the same as the void. The void, verily, that is the same as joy" -

Chandogya Upanishad, 4-10-4

7.1 PRELIMINARIES

Let us see how the great physicists used harmonic oscillators to establish beachheads to new physics.

1) Albert Einstein used harmonic oscillators to understand **specific heats of solids** and found that energy levels are quantized. This formed one of the key bridges between Classical and Quantum Mechanics.

2) Werner Heisenberg and Erwin Schrödinger formulated Quantum Mechanics. The role of harmonic oscillators in this process is well known.

3) Paul A. M. Dirac was quite fond of harmonic oscillators. He used oscillator states to construct **Fock space**. He was the first one to consider harmonic oscillator wave functions normalizable in the time variable. In 1963, Dirac used coupled harmonic

oscillators to construct a representation of the O(3,2) de Sitter group which is the basic scientific language for two-mode squeezed states.

4) Hediki Yukawa was the first one to consider a Lorentz-invariant differential equation, with momentum-dependent solutions which are Lorentz-covariant but not Lorentz-invariant. He proposed harmonic oscillators for **relativistic extended particles** five years before Hofstadter observed that protons are not point particles in 1955. Some people say he invented a string-model approach to particle physics.

5) Richard Feynman was also fond of harmonic oscillators. When he gave a talk in 1970 Washington Meeting of the American Physical Society, he stunned the audience by telling us not to use Feynman diagrams, but harmonic oscillators for quantum bound states. This figure illustrates what he said in 1970. Feynman diagrams applicable to running waves in Einstein's Lorentz-covariant world. Are Feynman's oscillators Lorentz-covariant? Yes in spirit, but there are many technical problems. Then can those problems be fixed. This is the question.

7.1.1   QUANTUM MECHANICAL OSCILLATOR

The harmonic oscillator plays a very important role in Quantum Mechanics, and it provides an essential element in the theory of radiation. The reason for this is that one may view the EM field as an infinite collection of oscillators. In Chapter 6, Section 6.2, the treatment was based on the traditional approach. In this Chapter to solve the oscillatory problem an alternative novel treatment, without involving any differential equation, is presented.

Consider $\hat{x}$, a single position observable and p_x, the canonical momentum of the particle executing harmonic oscillation. The Hamiltonian $\hat{H}$ has the form (6.2.4), viz.

$$\hat{H} = \hat{p}_x^2 / 2m + V(x) \tag{7.1.1}$$

$$V(x) = \tfrac{1}{2} k x^2 = \tfrac{1}{2} m \omega_c^2 x^2. \tag{7.1.2}$$

Quantum Mechanically (7.1.1) becomes

$$\hat{H} = \tfrac{1}{2m} \hat{p}_x^2 + \tfrac{1}{2} m \omega_c^2 \hat{x}^2 \tag{7.1.3}$$

where $\hat{x}$ and $\hat{p}_x$ are operators.

But let us not insist that $\hat{p}_x \rightarrow -i\hbar \hat{\nabla}_x$; but one knows the commutator relation between the dynamic variables, x and p_x as

$$[\hat{x}, \hat{p}_x] = (\hat{x}\hat{p}_x - \hat{p}_x\hat{x}) = i\hbar \tag{7.1.4}$$

7.1.2 New NON-HERMITIAN operators, $\hat{A}$ and $\hat{A}^\dagger$

From equation (7.1.3)

$$\hat{H} = \omega_c [\frac{1}{2m\omega_c} \hat{p}_x^2 + \frac{1}{2} m \omega_c \hat{x}^2]$$

$$\hat{H} = \omega_c [\sqrt{m\omega_c/2}\, x - i\hat{p}_x/\sqrt{2m\omega_c}\,][\sqrt{m\omega_c/2}\, x + i\hat{p}_x/\sqrt{2m\omega_c}\,].$$

$$\boxed{\hat{H} = \omega_c\, \hat{A}^\dagger \hat{A}}. \tag{7.1.5}$$

Since in Quantum Mechanics, equation (7.1.4) is valid, *i.e.* $\hat{x}$ and $\hat{p}_x$ do not commute each other, introducing the following notations.

$$\boxed{\hat{A} = \sqrt{m\omega_c/2}\, x - i\hat{p}_x/\sqrt{2m\omega_c}\,]} \tag{7.1.6}$$

$$\boxed{\hat{A}^\dagger = [\sqrt{m\omega_c/2}\, x + i\hat{p}_x/\sqrt{2m\omega_c}\,]} \tag{7.1.7}$$

The EQUATION OF MOTION of the harmonic oscillation (6.3.3) is

$$\nabla^2 \psi + a = \frac{2m}{\hbar^2}[E - \frac{1}{2}k x^2]\, \psi(x) = 0 \tag{7.1.8}$$

Let $\quad a = \frac{2m}{\hbar^2} E \tag{7.1.9}$

$$b = \frac{m}{\hbar} \omega_c \tag{7.1.10}$$

Equation (7.1.8) becomes

$$\nabla^2 \psi(x) + [a - b^2 x^2]\psi(x) = 0 \tag{7.1.11}$$

$$\boxed{\hat{q} = b^{1/2} \hat{x}} = \{\frac{m}{\hbar} \omega_c\}^{1/2} \hat{x}. \tag{7.1.12}$$

7.1.3 $\hat{q}, \hat{p},$ and $\hat{p}^2$ OPERATORS

However, for the EM field, it is not possible to give a simple meaning to q and $\psi(q)$. So let p is an operator such that

$$\hat{p} \to -i\hat{\nabla}_q = (-i/\hbar\, b^{1/2})(\hbar\, \partial/\partial x)$$

$$\boxed{\hat{p} \to -i\hat{\nabla}_q = (\hat{p}_x/\hbar\, b^{1/2})}. \tag{7.1.13}$$

$$\hat{p}^2 = (-i\hat{\nabla}_q)\cdot(-i\hat{\nabla}_q) = -\hat{\nabla}_q^2$$

$$= (-\partial/\partial q)\cdot(\partial/b^{1/2} \partial x) = -(\partial/b^{1/2}\partial x)\cdot(\partial/b^{1/2}\partial x)$$

$$= -(\partial^2/b\, \partial x^2) = (\hat{p}_x^2/\hbar^2\, b).$$

$$\boxed{\hat{p}^2 = -\hat{\nabla}_q^2 = (\hat{p}_x^2/\hbar^2 \, b)} \qquad (7.1.14)$$

Thus it is seen that $\hat{q}, \hat{p},$ and $\hat{p}^2$ are operators associated with $\hat{x}, \hat{p}_x$ and $\hat{p}_x^2$, respectively. Using equation (6.3.8)

The SCHRODINGER EQUATION of the quantum mechanical oscillator is

$$\boxed{(d^2\psi/dq^2)+[\tfrac{a}{b}-q^2]\,\psi(q)=0} \quad . \qquad (7.1.11)$$

Introducing the dimensionless parameter, λ

$$\lambda = \tfrac{a}{b} = \tfrac{2E}{\hbar \omega_c} = 2\,\varepsilon \qquad (7.1.15)$$

$$-\nabla^2 \psi(q)+[2\,\varepsilon-q^2]\,\psi(q)=0 \qquad (7.1.16)$$

In terms of $\hat{q}$ and $\hat{p}$, this equation becomes

$$(\hat{q}^2+\hat{p}^2)\,\psi(q) \equiv (\hat{p}^2+\hat{q}^2)\,\psi(q) = 2\,\varepsilon\,\psi(q) \qquad (7.1.17)$$

It can be shown that

$$[\hat{p},\hat{q}] = -1 \qquad (7.1.18)$$

7.1.4 NON-HERMITIAN OPERATORS, $\hat{a}$ and $\hat{a}^\dagger$

If $[\hat{p},\hat{q}] = 0$

Equation (7.1.17) becomes

$$\hat{H}\psi(q) = (\hat{q}^2+\hat{p}^2)\,\psi(q) \equiv (\hat{p}^2+\hat{q}^2)\,\psi(q) = 2\,\varepsilon\,\psi(q)$$

It could be factorized in the form,

$$(\hat{q}+i\hat{p})(\hat{q}-i\hat{p})\psi(q) = 2\,\varepsilon\,\psi(q)$$

On the other hand, $[\hat{p},\hat{q}] \neq 0$, the above equation becomes

$$(\hat{q}^2+\hat{p}^2)+i\,(\hat{p}\hat{q}-\hat{q}\hat{p}) = (\hat{q}^2+\hat{p}^2)+i(-i) \quad .$$

i.e., $\quad (\hat{q}+i\hat{p})(\hat{q}-i\hat{p}) = (\hat{q}^2+\hat{p}^2+1)$, or

Similarly $(\hat{q}-i\hat{p})(\hat{q}+i\hat{p}) = (\hat{q}^2+\hat{p}^2-1)$.

By adding these two equations,

$$(\hat{p}^2+\hat{q}^2) = \tfrac{1}{2}(\hat{q}+i\hat{p})(\hat{q}-i\hat{p}) + \tfrac{1}{2}(\hat{q}-i\hat{p})(\hat{q}+i\hat{p})$$

$$= \frac{1}{\sqrt{2}}(\hat{q}+i\hat{p})\frac{1}{\sqrt{2}}(\hat{q}-i\hat{p}) + \frac{1}{\sqrt{2}}(\hat{q}-i\hat{p})\frac{1}{\sqrt{2}}(\hat{q}+i\hat{p})$$

Define two new NON-HERMITIAN operators, *i.e.* dimensionless complex dynamic variables, $\hat{a}$ and $\hat{a}^\dagger$,

$$\boxed{\hat{a} = \frac{1}{\sqrt{2}}(\hat{q}+i\hat{p})}. \qquad (7.1.19)\ a$$

$$= \frac{1}{\sqrt{2}}(\hat{q}+i\,\partial/\partial q) = i\frac{1}{\sqrt{2}}(\hat{p}-i\hat{q}) \qquad (7.1.19)\ b$$

$$= \frac{1}{\sqrt{2m\hbar\omega_c}}(\hat{p}+i\,m\,\omega_c\hat{q})$$

$$= \frac{1}{\sqrt{2}}b^{1/2}(\hat{x}-i\hat{p}_x/m\,\omega_c)$$

$$\boxed{\hat{a}^\dagger = \frac{1}{\sqrt{2}}(\hat{q}-i\hat{p})}. \qquad (7.1.20)\ a$$

$$= \frac{1}{\sqrt{2}}(\hat{q}-i\,\partial/\partial q). \qquad (7.1.20)\ b$$

$$= i\frac{1}{\sqrt{2}}(\hat{p}+i\hat{q}). \qquad (7.1.20)\ c$$

$$= \frac{1}{\sqrt{2m\hbar\omega_c}}(\hat{p}-i\,m\,\omega_c\hat{q})$$

$$= \frac{1}{\sqrt{2}}b^{1/2}(\hat{x}+i\hat{p}_x/m\,\omega_c)$$

7.1.5: NEW FORM OF THE SCHRÖDINGER EQUATION for an oscillator:

Taking the product of the two operators (7.1.19) and (7.1.20)

$$\hat{a}\hat{a}^\dagger = \frac{1}{\sqrt{2}}(\hat{q}+i\hat{p})\frac{1}{\sqrt{2}}(\hat{q}-i\hat{p}) = \tfrac{1}{2}(\hat{q}^2+\hat{p}^2) = \hat{N} \qquad (7.1.21)$$

Then $\quad \hat{a}^\dagger\hat{a} = \tfrac{1}{2}(\hat{p}^2+\hat{q}^2) = \hat{N} \qquad (7.1.22)$

and $\quad \boxed{\hat{H} = (\hat{p}^2+\hat{q}^2) = (\hat{a}\hat{a}^\dagger + \hat{a}^\dagger\hat{a}) = 2\hat{N}} \qquad (7.1.23)$

<u>Schrödinger Equation for an oscillator</u> (7.1.17) becomes

$$\boxed{(\hat{a}\hat{a}^\dagger + \hat{a}^\dagger\hat{a})\,\psi(q) = 2\varepsilon\,\psi(q)} \qquad (7.1.24)$$

It is very easily shown that

$$\hat{A} = \hbar^{1/2}\hat{a}, \qquad (7.1.25)$$

$$\hat{A}^{\dagger} = \hbar^{1/2}\hat{a}^{\dagger} \tag{7.1.26}$$

and $\quad [\hat{A}, \hat{A}^{\dagger}] = \hbar \tag{7.1.27}$

7.2 The BASIC COMMUTATION RELATIONS:

The various commutation relations between the operators of a simple harmonic oscillator are given below, without proof.

$$[\hat{a}, \hat{a}^{\dagger}] = 1 \tag{7.2.1}$$

$$[\hat{a}, \hat{a}] = 0. \tag{7.2.2}$$

$$[\hat{a}^{\dagger}, \hat{a}^{\dagger}] = 0. \tag{7.2.3}$$

$$[\hat{H}, \hat{a}] = -\hbar\omega_c \hat{a} \tag{7.2.4}$$

$$[\hat{H}, \hat{a}^{\dagger}] = +\hbar\omega_c \hat{a}^{\dagger} \tag{7.2.5}$$

$$\hat{H} = \hbar\omega_c [\hat{a}\,\hat{a}^{\dagger} - \tfrac{1}{2}] \tag{7.2.6}$$

$$\hat{H} = \hbar\omega_c [\hat{a}^{\dagger}\,\hat{a} + \tfrac{1}{2}] \tag{7.2.7}$$

Worked out Example 7.1

Find $[\hat{H}, \hat{a}]$.

Solution:

$\boxed{\text{Step \# 1}}$ Equation (7.2.6) gives $\hat{H} = \hbar\omega_c[\hat{a}\,\hat{a}^{\dagger} - \tfrac{1}{2}]$.

$\boxed{\text{Step \# 2}}$ $[\hat{H}, \hat{a}] = \hbar\omega_c\,[(\hat{a}\,\hat{a}^{\dagger} - \tfrac{1}{2}), \hat{a}]$

$= [\hbar\omega_c(\hat{a}\,\hat{a}^{\dagger})\hat{a} - \tfrac{1}{2}\hbar\omega_c\hat{a}] - [\hat{a}\,\hbar\omega_c(\hat{a}\,\hat{a}^{\dagger} = \tfrac{1}{2}\hbar\omega_c\hat{a})]$

$= \hbar\omega_c(\hat{a}\,\hat{a}^{\dagger}\hat{a} - \hbar\omega_c\hat{a}\hat{a}\,\hat{a}^{\dagger})$

$= \hbar\omega_c[\hat{a}\,\hat{a}^{\dagger}\hat{a} - \hat{a}\hat{a}\,\hat{a}^{\dagger}]$

$= \hbar\omega_c\hat{a}\,[\hat{a}^{\dagger}\hat{a} - \hat{a}\,\hat{a}^{\dagger}]$

$= \hbar\omega_c\hat{a}\,[\hat{a}^{\dagger}, \hat{a}]$.

Step # 3 But $[\hat{a}, \hat{a}^\dagger] = 1$; So $[\hat{H}, \hat{a}] = \hbar\omega_c \hat{a}[-1]$

$$[\hat{H}, \hat{a}] = -\hbar\omega_c \hat{a}$$

7.3 ENERGY SPECTRUM OF THE OCILLATOR

7.3.1 To obtain the CONDITION $[\hat{a}^\dagger \psi_o(q) = 0]$

Let $\psi(q)$ is an eigen function, of $\hat{H}$, and that the form of $\psi_n(q)$ is unknown, and has eigen value, E_n.

The relations that are useful are:

Step # 1 **Classically** $\hat{H} = \hbar\omega_c (\hat{a}^\dagger \hat{a})$ (7.3.1)

Put $E / \hbar\omega_c = \varepsilon$. (7.3.2)

Step # 2 **Qunatum Mechanically** (7.2.6) and (7.2.7) state that

$$\hat{H} = \hbar\omega_c [\hat{a}\,\hat{a}^\dagger - \tfrac{1}{2}]$$ (7.3.3)

$$\hat{H} = \hbar\omega_c [\hat{a}^\dagger \hat{a} + \tfrac{1}{2}]$$ (7.3.4)

From (7.2.4) $[\hat{H}, \hat{a}] = -\hbar\omega_c \hat{a}$ (7.3.5)

∴ $[\hat{H}, \hat{a}^\dagger] = +\hbar\omega_c \hat{a}^\dagger$ (7.3.6)

Equations (7.3.1) and (7.3.2) differ by $\tfrac{1}{2}\hbar\omega_c$. So it indicates that

$$E_n' = E_n \pm \tfrac{1}{2}\hbar\omega_c.$$

Step # 3 $\hat{H}\psi_n(q) = E_n\psi_n(q) = \varepsilon_n \hbar\omega_c \psi_n(q)$

Step # 4 Using (7.3.6), $[\hat{H}, \hat{a}^\dagger]\psi_n(q) = \hat{H}\hat{a}^\dagger \psi_n(q) - \hat{a}^\dagger \hat{H}\psi_n(q)$.

∴ $\hat{H}\hat{a}^\dagger \psi_n(q) = \hat{a}^\dagger \hat{H}\psi_n(q) + [\hat{H}, \hat{a}^\dagger]\psi_n(q)$.

$$= \hat{a}^\dagger \hat{H}\psi_n(q) + [\hbar\omega_c \hat{a}^\dagger]\psi_n(q)$$

$$= \hat{a}^\dagger \{E_n + \hbar\omega_c\}\psi_n(q)$$

$$= \hat{a}^\dagger [(\varepsilon_n + 1)\hbar\omega_c]\psi_n(q).$$

$$= [(\varepsilon_n + 1)\hbar\omega_c]\{\hat{a}^\dagger \psi_n(q)\}.$$

Step # 5 Now (7.1.19) and (7.1.20) define, $\hat{a} = \frac{1}{\sqrt{2}}(\hat{q}+i\hat{p})$,

and $\quad \hat{a}^\dagger = \frac{1}{\sqrt{2}}(\hat{q}-i\hat{p})$.

$$\hat{a}\hat{a}^\dagger = \frac{1}{\sqrt{2}}(\hat{q}+i\hat{p})\frac{1}{\sqrt{2}}(\hat{q}-i\hat{p}) = \frac{1}{2}(\hat{q}^2+\hat{p}^2+1)$$

∴ $\quad 2\hat{a}\hat{a}^\dagger = (\hat{q}^2+\hat{p}^2+1)$ (7.3.7)

Step # 6 But (7.1.17)

$$(\hat{q}^2+\hat{p}^2)\,\psi_n(q) \equiv (\hat{p}^2+\hat{q}^2)\,\psi_n(q) = 2\,\varepsilon_n\,\psi_n(q) \quad (73.8)$$

i.e., $\quad (\hat{q}^2+\hat{p}^2)\,\psi_n(q) = 2\,\varepsilon_n\,\psi_n(q)$ (7.3.9)

Equation (7.3.7) may be written using (7.3.9)

As $\quad 2\hat{a}\hat{a}^\dagger \psi_n(q) = (\hat{q}^2+\hat{p}^2+1)\psi_n(q) = (2\,\varepsilon_n+1)\,\psi_n(q)$ (7.3.10)

Step # 7 Alternately, one may write

$$\hat{a}^\dagger \hat{a} = \frac{1}{\sqrt{2}}(\hat{q}+i\hat{p})\frac{1}{\sqrt{2}}(\hat{q}-i\hat{p}).$$

∴ $\quad 2\hat{a}^\dagger \hat{a}\,\psi_n(q) = (\hat{q}^2+\hat{p}^2-1)\psi_n(q)$ (7.3.11)

But $\quad (\hat{q}^2+\hat{p}^2)\psi_n(q) = (2\,\varepsilon_n)\,\psi_n(q)$

∴ $\quad 2\hat{a}^\dagger \hat{a}\,\psi_n(q) = (\hat{q}^2+\hat{p}^2-1)\psi_n(q) = (2\,\varepsilon_n-1)\,\psi_n(q)$ (7.3.12)

Step # 8 Pre-multiply equation (7.3.10) by $\hat{a}^\dagger$, then

$$2\hat{a}^\dagger \hat{a}\,\{\hat{a}^\dagger\psi_n(q)\} = (2\,\varepsilon_n +\!- 1)\,\{\hat{a}^\dagger\psi_n(q)\}$$

i.e., $\quad 2\hat{a}^\dagger \hat{a}\,\{\hat{a}^\dagger\psi_n(q)\} = [(2(\,\varepsilon_n+1) - 1]\,\{\hat{a}^\dagger\psi_n(q)\}$ (7.3.13)

this means $(\hat{a}^\dagger \hat{a})$ is Hermitian operator, $\hat{\mathbb{N}}$.

Step # 9 This equation can be satisfied only if either

$\quad \{\hat{a}^\dagger\psi_n(q)\} = 0$. (7.3.14)

or, $\quad \boxed{\{\hat{a}^\dagger\psi_n(q)\} = \psi_{n+1}(q)}$. (7.3.15)

Equation (7.3.13) becomes

$$2(\hat{a}^\dagger \hat{a}) \psi_{n+1}(q) = [(2(\varepsilon_n + 1) - 1] \psi_{n+1}(q)$$

which is identical for $\psi_{n-1}(q)$ in equation (7.2.5),

$$(\varepsilon_n + 1) = \varepsilon_{n+1}. \tag{7.3.16}$$

7.3.2 TO FIND THE SOLUTION OF THE CONDITION $\{\hat{a}^\dagger \psi_n(q)\} = 0$:

$$\{\hat{a}^\dagger \psi_n(q)\} = 0 \tag{7.3.17}$$

i.e., $\quad \{\frac{1}{\sqrt{2}}(\hat{q} - i\hat{p})\psi_n(q)\} = 0$

$$\{\frac{1}{\sqrt{2}}[(\hat{q} - (d/dq)]\psi_n(q)\} = 0$$

$$(d/dq)\,\psi_n(q)\} = \hat{q}\,\psi_n(q)$$

i.e., $\quad \psi_n(q) = N\, e^{+q^2/2}$

This solution is **unacceptable** as it diverges.

$$\therefore \quad \{\hat{a}^\dagger \psi_n(q)\} \propto \psi_{n+1}(q) \tag{7.3.18}$$

Thus given any solution $\psi_n(q)$, and the eigen value ε_n, it is always possible to generate a new state $\psi_{n+1}(q)$ by equation (7.3.18), with eigen values ε_{n+1}.

7.3.3 To obtain the CONDITION $\{\hat{a}\,\psi_n(q)\} = 0$:
Pre-multiplying equation (7.3.12), by $\hat{a}$

$$2\hat{a}\hat{a}^\dagger \{\hat{a}\,\psi_n(q)\} = (2\varepsilon_n - 1)\,\{\hat{a}\psi_n(q)\}$$

which means $(\hat{a}\hat{a}^\dagger) =$ Hermitian.

$$2(\hat{a}\hat{a}^\dagger)\{\hat{a}\,\psi_n(q)\} = [2(\varepsilon_n - 1) + 1]\,\{\hat{a}\psi_n(q)\} \tag{7.3.19}$$

Let either $\quad \{\hat{a}\,\psi_n(q)\} = 0,\tag{7.3.20}$

or, $\quad \{\hat{a}\,\psi_n(q)\} \propto \psi_{n-1}(q) \tag{7.3.21}$

Equation (7.3.16) is the same as equation (73.10) for $\psi_{n-1}(q)$ as it is

$$2(\hat{a}\hat{a}^\dagger)\psi_{n-1}(q) = [2(\varepsilon_n - 1) + 1]\,\psi_{n-1}(q)$$

provided that $(\varepsilon_n - 1) = \varepsilon_{n-1}$ (7.3.22)

7.3.4 The VACUUM or GROUND STATE, $\psi_o(q)$:

Thus given any solution of $\psi_n(q)$, with eigen value determined ε_n, it is possible to generate a new state of lower energy, $\psi_{n-1}(q)$, determined by equation (7.3.21)

$$\{\hat{a}\, \psi_n(q)\} \propto \psi_{n-1}(q)$$

with eigen value ($\varepsilon_n - 1$), unless $\psi_n(q)$ is the ground state; $\psi_o(q)$. In this case, it must satisfy equation (7.3.20)

$$\{\hat{a}\, \psi_n(q)\} = 0,$$

i.e., for $n = 0$, and $\{\hat{a}\, \psi_n(q)\} = 0$, becomes

$$\boxed{\{\hat{a}\, \psi_o(q)\} = 0} \quad (7.3.24)$$

Equation (7.3.24) is the equation, which defines the ground state of the oscillator system. Since in this state there is no phonon or photon ground state is known as the VACUUM STATE.

7.3.5 ZERO-POINT ENERGY, E_o:

Now as per (7.3.12) $\quad 2\hat{a}\hat{a}^\dagger\{\hat{a}\, \psi_n(q)\} = (2\varepsilon_n - 1)\{\hat{a}\psi_n(q)\}$

Putting $n = 0$, $\quad 2\hat{a}\hat{a}^\dagger\{\hat{a}\, \psi_o(q)\} = (2\varepsilon_o - 1)\{\hat{a}\psi_o(q)\}$ Because of equation (7.3.21)

$$2\hat{a}\hat{a}^\dagger\{\hat{a}\, \psi_o(q)\} = 0 = (2\varepsilon_o - 1)\{\hat{a}\psi_o(q)\}$$

i.e., $(2\varepsilon_o - 1) = 0$, or,

$$\varepsilon_o = \tfrac{1}{2}$$

By definition, (7.1.15) $E_n / \hbar\omega_c = \varepsilon_n$

$\therefore \quad \boxed{E_o = \tfrac{1}{2}\hbar\omega_c} \quad (7.3.25)$

7.3.6 EIGEN VALUE, E_n:

It is known (7.3.16) that

$$(\varepsilon_n + 1) = \varepsilon_{n+1}$$

$n = 0$, means $\varepsilon_1 = (\varepsilon_o + 1) = (\frac{1}{2} + 1)$

$n = 1$ $\varepsilon_2 = (\varepsilon_1 + 1) = (\frac{1}{2} + 2)$

$n = 2$ $\varepsilon_3 = (\varepsilon_2 + 1) = (\frac{1}{2} + 3)$

$n = n$ $\varepsilon_n = (\varepsilon_n + 1) = (\frac{1}{2} + n)$

∴ $\boxed{\varepsilon_n = (n + \frac{1}{2}), \; n = 0, 1, 2, 3, \ldots}$ (7.3.26)

By definition, (7.1.15) $E_n / \hbar\omega_c = \varepsilon_n$

∴ $\boxed{E_o = \varepsilon_o \, \hbar\omega_c = \frac{1}{2}\hbar\omega_c}$ (7.3.27)

This is in agreement with the set of DISCRETE spectrum of EIGEN VALUES found by the "TRADITIONAL METHOD". Fig. 7.1 displays the energy levels of the quantum oscillator. Thus we have succeeded in obtaining the discrete spectrum of a harmonic oscillator WITHOUT SOLVING *DIFFERENTIAL EQUATIONS.* of the oscillator.

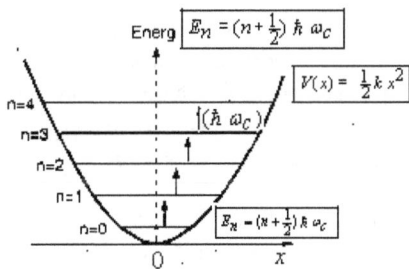

Fig. 7.1 Linear QM Oscillator Energy levels

7.4.1 NUMBER OPERATOR, $\hat{N}$, and OCCUPATION NUMBER, n:

Here n, the quantum number, may be regarded as the number of quanta of energy $\hbar\omega_c$ in the oscillator in state n, in addition to the **zero-point-energy**, $\frac{1}{2}\hbar\omega_c$.

For this reason n is known as the OCCUPATION NUMBER, of $\psi_n(q)$ and the eigen value of operator, $\hat{N}$, which is Hermitian in

$$\hat{N}\psi_n(q) = n\,\psi_n(q) \quad (7.4.1)$$

where $\hat{N} = \hat{a}^\dagger \hat{a}$, (7.4.2)

which is called the NUMBER OPERATOR, $\hat{N}$.

From (7.2.7)

$$\hat{H} = \hbar\omega_c[\hat{a}^\dagger \hat{a} + \tfrac{1}{2}]$$

$$\hat{H}\psi_n(q) = \hbar\omega_c[\hat{a}^\dagger \hat{a} + \tfrac{1}{2}]\psi_n(q)$$

$$\hat{H}\psi_n(q) = \hbar\omega_c[\hat{N}^\dagger + \tfrac{1}{2}]\psi_n(q)$$

$$\hat{H}\psi_n(q) = [n + \tfrac{1}{2}]\hbar\omega_c\,\psi_n(q).$$

Therefore, $\psi_n(q)$ is eigen function of both $\hat{H}$ and $\hat{N}$ operators.

The expression fort the energy levels is one of the most important features in quantum mechanics. It justifies Planck's explanation of the interaction of radiation (EM field) with matter, provided matter can be regarded as a collection of oscillators, each one emitting/absorbing radiation of its own frequency. The energy exchange is the restricted by the oscillator eigen values to take place in units of $\hbar\omega_c$, which is the Planck hypothesis. Q is now not related to the coordinate of the vibrating particle, but to the amplitude of, for example, a plane wave forming part of the decomposition of some field (like the phonon field or photon field). The function $\psi_n(q)$ of this partial wave amplitude is the wave function of a "state" with $n = n$, of the partial wave of energy $E_n = E_o + n\,\hbar\omega_c$, which requires the existence of n phonons/ photons of energy, $\hbar\omega_c$ in the partial wave example.

7.4.2   The EXPLICIT FORM of $\psi_o(q)$, the VACUUM STATE, of the oscillator
:(Derivation of wave functions explicitly)

The condition (7.3.24) is

$$\{\hat{a}\,\psi_o(q)\} = 0,$$

i.e. putting $n = 0$, $\{\hat{a}\,\psi_o(q)\} = 0$,

Insert the explicit expression for $\hat{a}$, from (7.1.19) is

$$\tfrac{1}{\sqrt{2}}(\hat{q} + i\,\partial/\partial q)\psi_o(q) = 0,$$

$$(\partial/\partial q)\psi_o(q) = -q\,\psi_o(q)$$

i.e., $[(d/\psi_o(q)/\psi_o(q)] = -q\, dq$

This is a first order differential equation, which on integration, reads

$$[\log \psi_o(q)] = (-q^2/2) + \text{Logarithamic Constant,N}..$$

$$[\log \psi_o(q)]/N_- = (-q^2/2)$$

$$\psi_o(q) = N_- \, e^{-q^2/2}. \tag{7.4.3}$$

$$\boxed{\psi_o(q) = N_- \, H_o(q) \, e^{-q^2/2}}.$$

where $H_0(q)$ is the zeroth Hermite's polynomial (which is unity)

and $N_- = \sqrt[4]{\pi}$ \hfill (7.4.4)

$$\boxed{\psi_o(q) = \sqrt[4]{\pi} \, H_o(q) \, e^{-q^2/2}} \tag{7.4.5}$$

7.4.3 The EXPLICIT FORM of $\psi_n(q)$ of the oscillator: (Derivation of wave functions explicitly)

Though the form of the wave function $\psi_o(q)$ is unknown, the successive eigen functions can be constructed, or generated, form $\psi_o(q)$ by repeated application of (7.3.18), viz.

$$\{\hat{a}^\dagger \psi_n(q)\} \propto \psi_{n+1}(q) \tag{7.4.6}$$

The ground state, $\psi_o(q)$ is an even function of q with no nodes; the first excited state is an odd function of q with one node. It is easy to verify that by repeated applications of (7.3.18), i.e., $\{\hat{a}^\dagger \psi_n(q)\} \propto \psi_{n+1}(q)\}$, that the other successive eigen functions have the general features derived in the previous procedure, i.e.

a) nth state has parity $(-1)^n$, with n- nodes,
b)   Particles in symmetric potentials have energy Eigen states which is *symmetric* (even parity), or *anti-symmetric* (odd parity), about the origin.

7.5.1 BOSE OPERATORS, $b^\dagger$ and b: (or, Ladder Operators/ Step-up and Step-down operators / Creation and Annihilation operators/ Raising and Lowering operators):

Defining the new operators, known as the **Bose operators**, $b^\dagger$ and b,

$$\hat{b} = \hat{a} = \frac{1}{\sqrt{2}}(\hat{q} + i\hat{p}) = \frac{1}{\sqrt{2}}(\hat{q} + \partial/\partial q) \qquad (7.5.1)$$

$$\hat{b}^\dagger \equiv \hat{a}^\dagger = \frac{1}{\sqrt{2}}(\hat{q} - i\hat{p}) = \frac{1}{\sqrt{2}}(\hat{q} - \partial/\partial q) \qquad (7.5.2)$$

are of particular interest, since they are of type which has not previously appeared. Sofar operators discussed represented some observable, typically for example, the energy operator, $\hat{H}$. The non-Hermitian operators, $\hat{a}$ and $\hat{a}^\dagger$, according to (7.3.18) and (7.3.21):

$$\{\hat{a}^\dagger \psi_n(q)\} \propto \psi_{n+1}(q) .$$

$$\{\hat{a} \psi_n(q)\} \propto \psi_{n-1}(q) .$$

represent the physical operation of shifting the oscillator, up or down, from one energy level to the next. Thus energy, in units of $\hbar\omega_c$, has to be either created to raise or annihilated (destroyed) to lower the state in the process. Consequently '$\hat{a}^\dagger$' is called a creation (raising) operator and '$\hat{a}$' is called a destruction (annihilation/ lowering) operator.

The two are together known as **Ladder or Bose operators**. They alternately play a very important role in the complete Quantum Mechanics of the interaction of radiation with electrons. The formalism of the Bose operators, $\hat{a}^\dagger$ and $\hat{a}$ will be of fundamental importance when quantization of the *photon field* (in QED) is made

$$\boxed{\psi_1(q) = \{\hat{a}^\dagger \psi_o(q)\}} \qquad (7.5.3)$$

By applying the operator $a^\dagger$ n times, one obtains the nth excited eigen function

$$\psi_n(q) = \{\hat{a}^\dagger\}^n \psi_o(q) \qquad (7.5.4)$$

Since equation (7.3.12), *viz.*,

$$2\hat{a}\hat{a}^\dagger \{\hat{a} \psi_n(q)\} = (2\varepsilon_n - 1) \{\hat{a}\psi_n(q)\} \qquad (7.5.5)$$

is homogeneous in $\psi_n(q)$, there remains, as usual with eigen functions, a free constant coefficient. One will choose its value to normalize $\psi_n(q)$, from now on normalization in the dimensionless co-ordinate, q, will be defined as

$$\int \hat{\psi}_n(q)^* \hat{\psi}_n(q) dq = 1 \qquad (7.5.6)$$

The normalization constant can be determined purely algebraically

$$\boxed{\psi_n(q) = \sqrt{\frac{1}{n!}} \{\hat{a}^\dagger\}^n \psi_o(q)} \qquad (7.5.7)$$

or $\quad \boxed{\{\hat{a}^\dagger\}^n \psi_0(q) = \sqrt{n!}\ \psi_n(q)}$ (7.5.8)

These are *wave packet* states of wave amplitude q in which mean values of q oscillates with frequency, ω_c.

It can further be shown that

$$\boxed{\hat{a}\ \psi_n(q) = \sqrt{n}\ \psi_{n-1}(q)} \quad (7.5.9)$$

$$\boxed{\hat{a}^\dagger \psi_n(q) = \sqrt{n+1}\ \psi_{n+1}(q)} \quad (7.5.10)$$

7.5 What is SECOND QUANTIZATION?

So far the discussions were confined to a single harmonic oscillator. The solution of the many body problems in Quantum Mechanics presents great difficulties. The solution of the Schrödinger Equation (TISE) in configuration space for several identical particles is practically hopeless. However, an alternative method is available which is known as "Second Quantization". In '*Ordinary Quantization*' (also called GARDEN QUANTIZATION) one replaces physical observables by operators that obey certain commutation relation. The properties of a system are calculated by allowing these operators to act on $\psi_n(q)$ obtained by solving the Schrödinger Equation.

First Quantization introduced the new constant, h; second quantization does not introduce a new constant; hence it is best regarded as a mathematical manipulation. The latter is based originally on the *analogy* between the wave packets and the EM Field. In **Second Quantization**, this $\psi_n(q)$ is interpreted as an operator acting on an abstract space. $\psi_n(q)$ becomes Annihilation Operator, $\hat{a}$, and it on acting annihilates the state $\psi_n(q)$. Likewise $\hat{\psi}_n(q)^*$ becomes the Creation Operator, $\hat{a}^\dagger$. The use of creation and annihilation operators to describe bosons abide by the commutation relations

$$[\hat{a}, \hat{a}^\dagger] = 1; \quad (7.5.11)$$

$$[\hat{a}, \hat{a}] = [\hat{a}^\dagger, \hat{a}^\dagger] = 0, \quad (7.5.12)$$

reflect to the possibility of having any number of particles in the same state. These are the properties of **bosons**. However, the use of creation and annihilation operators to describe **fermions** (particularly electrons), implies that they are to be treated as particles, yet the Quantum Mechanics treats them as waves. For bosons a classical wave theory on quantization becomes a particle theory. In the same way, the quantum mechanical wave function for a fermion may be treated as a matter field variable and quantization the gives a particle theory for which creation and annihilation operators are appropriate.

$$[\hat{b}_r, \hat{b}_s{}^\dagger]_+ = \delta_{rs} ; \tag{7.5.13}$$

$$[\hat{b}_r, \hat{b}_s]_+ = [\hat{b}_r{}^\dagger, \hat{b}_s{}^\dagger]_+ = 0, \tag{7.5.14}$$

where $[\hat{A}, \hat{B}]_+ = \hat{A}\hat{B} + \hat{B}\hat{A}$. This procedure is known as <u>Second Quantization</u>. For obvious reasons, its advantage is that particle number conservation disappears, and may be used to prove *Pauli's Exclusion Principle*. It is not the intention of the author to describe any further on Second Quantization than to introduce this procedure. (For more details in the interested readers are recommended to refer to the Chapter 26).

REVIEW QUESTIONS

R.Q. 7.1 Find the value of $[\hat{p}, \hat{q}]$. (Answer: $[\hat{p}, \hat{q}] = -1$)

R.Q. 7.2 The Hamiltonian of the harmonic oscillator can be written as $\hat{H} = \frac{1}{2}\hbar\omega_c[\hat{q}^2 - d^2/dq^2]$

$\hat{q} = \hat{x}(m\,\omega_c/\hbar)^{1/2}$. Show that $\hat{a} = \frac{1}{\sqrt{2}}(\hat{q} + i\hat{p}) = (\hat{q} + i\,\partial/\partial q)$; and $\hat{a}^\dagger = \frac{1}{\sqrt{2}}(\hat{q} - i\hat{p})$; $\hat{a}^\dagger = (\hat{q} - i\,\partial/\partial q)$, can be treated as annihilation and creation operators.

R.Q. 7.3 Show that $\hat{A} = \hbar^{1/2}\hat{a}$, $\hat{A}^\dagger = \hbar^{1/2}\hat{a}^\dagger$

R.Q. 7.4 Show that $[\hat{A}, \hat{A}^\dagger] = \hbar$

R.Q. 7.5 Find $[\hat{a}, \hat{a}^\dagger]$. (Answer: $[\hat{a}, \hat{a}^\dagger] = 1$}

R.Q. 7.6 Show that classically, the Hamiltonian of a harmonic oscillator can be written as $\hat{H} = \hat{a}^\dagger \hat{a}\,\hbar\omega_c$.

R.Q. 7.7 Show that the Hamiltonian of a harmonic oscillator can be written as $\hat{H} = (\hat{a}^\dagger\hat{a} + \frac{1}{2})\hbar\omega_c$ or

$$\hat{H} = (\tfrac{1}{2}\hbar\omega_c)[\hat{a}^\dagger\hat{a} + \hat{a}\,\hat{a}^\dagger]$$

R.Q. 7.8 Find $[\hat{H}, \hat{a}^\dagger]$ (Answer: $[\hat{H}, \hat{a}^\dagger] = +\hbar\omega_c\hat{a}^\dagger$ }.

R.Q. 7.9 Show that $\psi_1(q) = \sqrt{1/2}\ H_1(q)\ e^{-q^2/2}$, using the explicit method.

R.Q. 7.10 Arrive at an explicit expression for the harmonic oscillator function, $\psi_2(q)$}, by using the shift operators. Given: $\psi_o(q) = \sqrt[4]{\pi}\ H_o(q)\ e^{-q^2/2}$ (Answer: $\psi_2(q) = N_- \sqrt{1/2}\ (2q^2 - 1)\ e^{-q^2/2}$).

R.Q. 7.11 Show that $\hat{a}\ \psi_n(q) = \sqrt{n}\ \psi_{n-1}(q)$, for a harmonic oscillator. Given the identity, $H_n'(q) - 2n\ H_{n-1}(q) = 0$, and $\psi_n(q) = B_n\ H_n(q)\ (e^{-q^2/2})$.

R.Q. 7.12 Show that $\hat{a}^\dagger \psi_n(q) = \sqrt{n+1}\ \psi_{n+1}(q)$, for a harmonic oscillator. Given the identity $H_n'(q) - 2n\ H_{n-1}(q) = 0$, and $\psi_n(q) = B_n\ H_n(q)\ (e^{-q^2/2})$.

R.Q. 7.13 Show that $\psi_n(q) = \sqrt{\frac{1}{n!}}\ \{\hat{a}^\dagger\}^n \psi_o(q)$.

R.Q. 7.14 Define the non-Hermitian operators $\hat{a}$ and $\hat{a}^\dagger$. Obtain the Schrödinger's Equation in the operator form for a linear harmonic oscillator. (Answer: $(\hat{a}\hat{a}^\dagger + \hat{a}^\dagger \hat{a})\ \psi(q) = 2\varepsilon\ \psi(q)$).

R.Q. 7.15 Write down the essential commutation relations between $\hat{H}, \hat{a}$ and $\hat{a}^\dagger$. Using them arrive at the energy spectrum for a linear oscillator. ($[\hat{H}, \hat{a}] = -\hbar\omega_c \hat{a}$; $[\hat{H}, \hat{a}^\dagger] = +\hbar\omega_c \hat{a}^\dagger$; $\hat{H}\psi_n(q) = \varepsilon_n \hbar\omega_c \psi_n(q)$; $(\hat{q}^2 + \hat{p}^2)\ \psi_n(q) = 2\varepsilon_n\ \psi_n(q)$; $2\hat{a}\hat{a}^\dagger = (\hat{q}^2 + \hat{p}^2 + 1)$; $2\hat{a}^\dagger \hat{a}\ \psi_n(q) = (\hat{q}^2 + \hat{p}^2 - 1)\psi_n(q)$; $2(\hat{a}^\dagger \hat{a})\ \psi_{n+1}(q) = [(2(\varepsilon_n + 1) - 1]\ \psi_{n+1}(q)$; $(\varepsilon_n + 1) = \varepsilon_{n+1}$; $(\varepsilon_n - 1) = \varepsilon_{n-1}$; $\hat{H}\psi_n(q) = [n + \frac{1}{2}]\hbar\omega_c\ \psi_n(q)$ }.

%^%^%^%^%^^%^%

Chapter 8

EXACT SOLUTIONS - 5

SPHERICALLY SYMMETRIC SYSTEM-

PARTICLE ON A RING -RIGID ROTATOR

Chapter 8

EXACT SOLUTIONS - 5

SPHERICALLY SYMMETRIC SYSTEM-

PARTICLE ON A RING -RIGID ROTATOR

"We cannot solve the problem that we have created with same thinking that created them"

— Albert Einstein

8.1. INTRODUCTION:

This quantization of *angular momentum* is a fundamental property of quantum mechanical systems. It is known from Chapter 6 and 7 that quantization of angular momentum does not need to be imposed on a system but is a natural consequence of Schrödinger's model.

The simplest model of rotational motion that can be described quantum mechanically is a rigid rotor consisting of two masses held a fixed distance apart rotating about its Center of Mass. The rotation is assumed to be free of any outside potential energy. Angular momentum in quantum systems results from a direct correspondence of classical mechanical equations of rotational motion. In rotational motion the moment of inertia and angular velocity play the same role as the mass and velocity does in a straight moving body. In linear motion momentum and kinetic energy are given by $\vec{p} = m\vec{v}$ and $E = p^2/2m$, so a rigid rotor rotating with angular velocity ω and having a moment of inertia $I = \mu R^2$ exhibits angular momentum $|\vec{L}| = I\omega$; and an energy of $E = L^2/2I$. To convert these classical equations to quantum equations we simply need to replace the momentum with its quantum operator and solve Schrödinger's eigenvalue Equation.

Questions of uniqueness and continuity involve discussion of theoretical properties of *partial differential equations*, such as characteristics, domains of dependence and the maximum-minimum principle. The main purpose here is the construction of a solution of a partial differential equation subject to certain boundary and / or initial conditions. Construction techniques are:

(i) Direct Integration,
(ii) The method of Separation of Variables,
(iii) Fourier Series, and
(iv) Fourier Transforms.

The <u>method of Separation of Variables</u> was introduced and developed byJean le Rond d'Alembert, Daniel BernoulliandLeonhard Eulerduring the middle of the 18th Century. It is the oldest systematic technique (and still the most useful) for solving a partial differential equation.

8.2.1 The DUMB BELL MODEL of the DIATOMIC MOLECULE (Spherical top / Rotation of a ring):

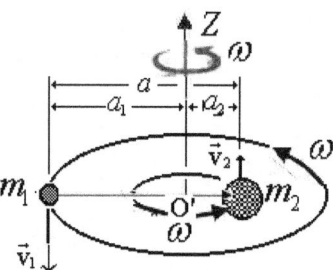

Fig. 8.1 'Dumb bell' shaped body of two masses rotating about O'

This quantum mechanical problem provides, for example, the basis of the Quantum Theory of the Rigid Rotator, which is of considerable importance in the study of the spectra of diatomic molecules, and many other fundamental physical problems. By a rigid rotor is understood a DUMB BELL-SHAPED body consisting of two spheres connected together by a 'slender rod' (Fig. 8.1).

Let m_1 and m_2 are the masses of the two spherical bodies,

a = separation between m_1 and m_2.

Consider that m_1 and m_2 rotate about a common Centre of gravity (or mass), O' with a common angular velocity, ω. The axis of rotation is perpendicular and on the line joining the two masses m_1 and m_2 and it divides this line in the inverse ratio of their masses (Fig 8.1). Let a_1 and a_2 are the distances of m_1 and m_2 from the axis of rotation. Here are two types of motion for the system: viz. (i) translational and (ii) rotational. Since we are not interested in the translational motion of the system in space, the Centre of gravity O' may be regarded as fixed (*i.e.* no external force is present) at the origin, O of the coordinate system.

8.2.1. REDUCTION OF THE TWO-BODY SYSTEM to the EQUIVALENT ONE-BODY problem:

Since O' is the Centre of mass (CM),

$m_1 a_1 = m_2 a_2$,

or $\quad a_1 / a_2 = m_2 / m_1$. \hfill (8.2.1)

$$a = a_1 + a_2 \quad (8.2.2)$$

$(a_1 + a_2)/a_2 = (m_1 + m_2)/m_1 = a/a_1$.

$a_1 = a [m_2 /(m_1 + m_2)]$.

Similarly, $a_2 = a [m_1 /(m_1 + m_2)]$.

Let $\vec{v}_1$ and $\vec{v}_2$ are the linear velocities of masses m_1 and m_2, respectively.

Then $\quad \vec{v} = a \, \omega$ \hfill (8.2.3)

Classical kinetic energy of the system, T

T = k.E. of the motion of the CM + k. E. of mass m_1 + k. E. of mass m_2

$$T = 0 + \tfrac{1}{2} m_1 (a_1 \omega^2) + \tfrac{1}{2} m_2 (a_2 \omega^2).$$

$$= \tfrac{1}{2} \omega^2 \left\{ [m_1 (a \, m_2 /(m_1 + m_2))]^2 + [m_2 (a \, m_1 /(m_1 + m_2))]^2 \right\}$$

$$= \tfrac{1}{2} \omega^2 \left\{ [a \, (m_1 + m_2)]^2 [m_1 m_2^2 + m_2 m_1^2] \right\}$$

$$= \tfrac{1}{2} \omega^2 \left\{ [a \, (m_1 + m_2)]^2 [m_1 m_2 (m_1 + m_2)] \right\}$$

$$= \tfrac{1}{2} a^2 \omega^2 \left\{ \frac{m_1 m_2}{(m_1 + m_2)} \right\}$$

$T = \tfrac{1}{2} \mu \, a^2 \omega^2 =$ Classical rotational k.E. of the system.

where $\quad \boxed{\mu = \left\{ \frac{m_1 m_2}{(m_1 + m_2)} \right\}}$ \hfill (8.2.4)

is the reduced mass of the system.

$$\mu = m_1 \left\{ 1 - \frac{m_1}{m_2} \right\}, \text{ as } (1 \mp x)^{-n} = [1 \pm n \, x + n \, (n+1) \, x^2 /2!) \pm .....]$$

8.2.2 MOMENT OF INERTIA OF THE RIGID ROTOR

Moment of inertia, $\Im_C$ about an axis Z perpendicular to distance a is

$$\Im = \mu\, a^2 \tag{8.2.5}$$

A '*spherical top*' is one for which the principal moments of inertia,

$$\Im_A = \Im_B = \Im_C = \Im. \tag{8.2.6}$$

8.2.3 LAPLACIAN ∇^2 in SPHERICAL COORDINATES:

The **Spherical Polar Coordinates** (r, θ, φ) of a point P, given as (x,y,z) in *Rectangular Coordinates*, (Fig. 8.2) have the following interpretation:
$r = |\vec{r}|$, length of the radius vector from the origin to the point

$$r = \sqrt{x^2 + y^2 + z^2}$$

θ = angle between the radius vector, $\vec{r}$ and $+Z$ - axis,
= Zenith angle (polar angle),
φ = angle between the projection of $\vec{r}$ in the xy plane and the $+X$- axis, measured in the direction shown in Fig 8.2(a)
= Azimuthal angle,

$$\varphi = \tan^{-1} y/x.$$

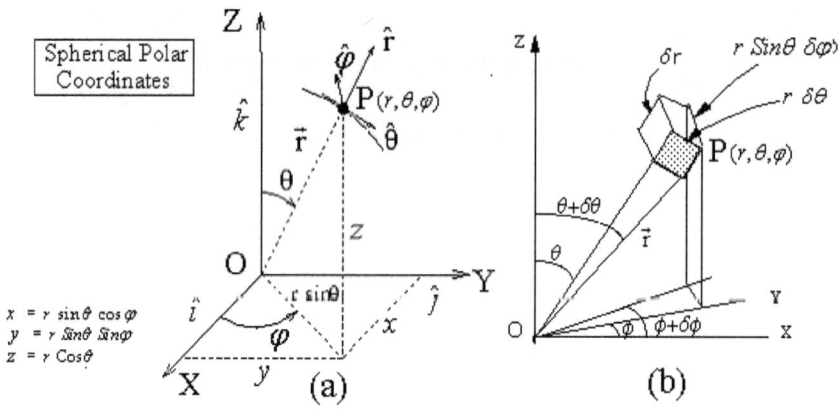

Fig. 8.2 The Spherical Polar Coordinates System.

Elemental volume
$$d\tau = dx\, dy\, dz = (r^2 dr)(\mathrm{Sin}\theta\, d\theta)\, d\varphi. \tag{8.2.7}$$
On the surface of a sphere whose centre is at O, lines of constant θ are like parallels of latitude on a globe (*i.e.*, $\theta = 90°$ at the equator, whereas the latitude of the equator is $0°$).

Lines of constant φ are like meridians of longitude taking the axis of the globe as +Z-axis, and +X-axis is at $\varphi = 0°$.

The Laplacian, ∇^2

$$\nabla^2 = \left[\frac{1}{r^2}\frac{\partial}{\partial r}\left(r^2 \frac{\partial}{\partial r}\right) + \frac{1}{r^2 \sin\theta}\frac{\partial}{\partial \theta}\left(\sin\theta \frac{\partial}{\partial \theta}\right) + \frac{1}{r^2 \sin\theta}\left(\frac{\partial^2}{\partial \varphi^2}\right)\right] \qquad (8.2.8)$$

8.2.4 SCHRODINGER EQUATION OF THE RIGID ROTOR

In Spherical Polar Coordinates, the particle in a central force field (field in which the potential depends upon the magnitude of the distance from a fixed point) is $V(r) = $ spherically symmetric, *i.e.* $V(r)$ has the symmetry of a sphere under any rotation of the system, is CENTRAL, and is independent of θ and φ,

i.e., $\qquad V(\vec{r}) = V(|\vec{r}|) = V(r)$

Expressing the Schrödinger Equation in terms of the spherical coordinates makes the problem considerably easier to solve. Choosing a <u>Right-Handed</u> Cartesian (x, y, z) and spherical polar (r, θ, φ) coordinate system the wave equation is

$$\nabla^2 \psi(r, \theta, \varphi) + \{2m/\hbar^2\}[E-V]\psi(r, \theta, \varphi) = 0. \qquad (8.2.9)$$

where for a rigid rotor,

$$r = \text{constant} \qquad (8.2.10)$$

Therefore the Schrödinger Equation becomes

$$\left\{(1/r^2 \sin\theta)[(\partial/\partial\theta)\sin\theta(\partial/\partial\theta)] + (1/r^2 \sin^2\theta)\partial^2/\partial\varphi^2\right\}\psi(r) + \left(\{2\mu/\hbar^2\}[E - V(r)]\right)\psi(r) = 0$$

The Hamiltonian $\hat{H}$

$$\hat{H} = T + V(x) = [1/2\mu][p_x^2] + 0$$

is an invariant under rotations.

8.3 To Construct the WAVEFUNCTION, $\psi(r)$ for the RIGID ROTOR:

In equation (8.2.10) only the angular part of $\hat{H}$ appears, because of condition (8.2.9). $\psi(r)$ can be constructed by the <u>Method of Separation of Variables</u>.
Now seek particular solutions of the form,

$$\boxed{\psi(r) = \psi(r, \theta, \varphi) = R(r) \cdot Y(\theta, \varphi)}. \qquad (8.3.1)$$

Equation (8.2.10) becomes
$$\{(1/\sin\theta)[(\partial/\partial\theta)\sin\theta(\partial/\partial\theta)] + (1/\sin^2\theta)\partial^2/\partial\varphi^2\}Y(\theta, \varphi)$$
$$= -\{2\mu r^2/\hbar^2\}[E - V(r)]Y(\theta, \varphi) \qquad (8.3.2)$$

$$= -\lambda, \text{ a constant} \qquad (8.3.3)$$

$$\lambda = \{2\mu r^2 / \hbar^2\} [E - V(r)] = \text{constant}. \tag{8.3.4}$$

8.3.1 SCHRODINGER EQUATION of the Azimuthal Part:

Defining the operator,

$$\hat{\Lambda} = \{(1/\sin\theta) [(\partial/\partial\theta) \sin\theta (\partial/\partial\theta)] + (1/\sin^2\theta) \partial^2/\partial\varphi^2\} \tag{8.3.5}$$

The Schrödinger Equation for the *rigid rotor* becomes

$$\boxed{[\hat{\Lambda} \; Y(\theta,\varphi)] / Y(\theta,\varphi) = -\lambda} \; . \tag{8.3.6}$$

This equation is independent of $V(r)$; and it plays an important role in potential theory. To solve this equation using again the method of separation of variables,

$$\boxed{Y(\theta,\varphi) = \Theta(\theta) \; \Phi(\varphi)} \tag{8.3.7}$$

where $\Theta(\theta)$ = polar part, and $\Phi(\varphi)$ = azimuthal part of Y.

Equation (8.3.6) becomes

$$\{[(1/\sin\theta) (\partial/\partial\theta) \sin\theta (\partial/\partial\theta)] + [(1/\sin^2\theta)(\partial^2/\partial\varphi^2)]\} \; \Theta(\theta).\Phi(\varphi) + \lambda \; \Theta(\theta).\Phi(\varphi) = 0 \tag{8.3.8}$$

After multiplication by $\sin^2\theta$, equation (8.3.8) leads to

$$\hat{\Lambda} = \{[\sin^2\theta / \Theta(\theta)] [(1/\sin\theta) (\partial/\partial\theta) \sin\theta (\partial/\partial\theta)] + \lambda\} \Theta(\theta)$$
$$= -[(\partial^2/\partial\varphi^2) \; \Phi(\varphi)] / \Phi(\varphi) \tag{8.3.9}$$

Since both the left and right hand sides of (8.3.9) are independent of each other, they become a pair of simultaneous ordinary differential equations, and are constants with respect to each other.

$$\{[\sin^2\theta / \Theta(\theta)] [(1/\sin\theta) (\partial/\partial\theta) \sin\theta (\partial/\partial\theta)] + \lambda \} \Theta(\theta) = [(\partial^2/\partial\varphi^2) \; \Phi(\varphi)] / \Phi(\varphi) = m^2 \tag{8.3.10}$$

say, where m^2 = a positive constant, chosen positive so as $\Phi(\varphi)$ to have periodic dependence. m^2 is chosen as the separation constant instead of m for later convenience.

8.3.2 The Azimuthal Factor of $\psi(r)$:

$$\boxed{[(d^2/d\varphi^2) \; \Phi(\varphi)] / \Phi(\varphi) = -m^2} , \tag{8.3.11}$$

This is familiar from the theory of ordinary differential equations. It has two linearly independent solutions:

$$\Phi(\varphi) = A_+ \; e^{+i\,m\,\varphi} + A_- \; e^{-i\,m\,\varphi},$$

i.e., $\qquad \Phi(\varphi) = A \; e^{+i\,|m|\,\varphi}$ \hfill (8.3.12)

(i) However, $\psi(r)$ has to be SINGLE-VALUED Function of the Cartesian Coordinates so as to be a physically reasonable solution. For this,

$$\Phi(\varphi) = \Phi(\varphi + 2\pi),$$

i.e., $\qquad A \; e^{+i\,|m|\,(\varphi + 2\pi)} = A \; e^{+i\,|m|\,(\varphi + 2\pi)}$

i.e., $\qquad e^{+i\,|m|\,2\pi} = 1$.

i.e., $\qquad m = 0, \pm 1, \pm 2, \pm 3, \ldots\ldots = $ an integer. \hfill (8.3.13)

(ii) In order that $\Phi(\varphi)$ is SQUARE INTEGRABLE, it must satisfy the **Born Condition**,

$$\int_0^{2\pi} \Phi(\varphi)^* \; \Phi(\varphi) \; d\varphi = 1$$

i.e., $\qquad \int_0^{2\pi} \{A \; e^{+i\,|m|\,\varphi}\Phi(\varphi)\}^* \{A \; e^{+i\,|m|\,\varphi}\} \; d\varphi = 1$

$$\int_0^{2\pi} \{A^2 d\varphi = 1$$

giving $A = 1/\sqrt{2\pi}$.

∴ $\boxed{\Phi(\varphi) = (1/\sqrt{2\pi})\, e^{+i|m|2\pi};\quad m = 0,\ \pm1,\ \pm2,\ \pm3,\ldots\ldots}$. (8.3.14)

= an integer

This is the familiar harmonic function.
The general normal azimuthal solution is the SUPERPOSITION (linear combination) of these functions (8.3.14)

$$\Phi(\varphi) = \sum_{m=-\infty}^{+\infty} c_m\, \hat{\Phi}_m(\varphi) = \sum_{m=-\infty}^{+\infty} c_m\, \frac{1}{\sqrt{2\pi}}\left(e^{+i m \varphi}\right) \quad (8.3.15)$$

8.4 THE POLAR FACTOR $\psi(r)$

8.4.1. Associated Legendre Equation

Consider the equation (8.3.8)

$$\{[(1/\sin\theta)\,(\partial/\partial\theta)\,\sin\theta\,(\partial/\partial\theta)] + [(1/\sin^2\theta)(\partial^2/\partial\varphi^2)]\}\,\Theta(\theta).\Phi(\varphi) + \lambda\,\Theta(\theta).\Phi(\varphi) = 0$$

$$\{[(1/\sin\theta)\,(\partial/\partial\theta)\,\sin\theta\,(\partial/\partial\theta)]\,\Theta(\theta).+ (\lambda - m^2/\sin^2\theta)\,\Theta(\theta). = 0. \quad (8.4.1)$$

This is a special type of ordinary differential equation, called the ASSOCIATED LEGENDRE Equation.

8.4.2 SOLUTION of the Associated Legendre Equation:

The solutions of which are the Associated Legendre Polynomials. To solve this it is necessary to carry out the following change of independent variable. Let

$$\rho = \cos\theta \quad (8.4.2)$$

$\sin\theta = (1 - \rho^2)^{1/2};\quad d\rho/d\theta = -\sin\theta$,

$(d\rho/d\theta)\,(d/d\rho) = -(1-\rho^2)^{1/2}(d/d\rho)$

$(1/\sin\theta)\,(d/d\theta) = -(d/d\rho)$.

Also, $\Theta(\theta) \to F(\rho)$

Equation (8.4.1) becomes,

$$(d/d\rho)\,[(1-\rho^2)\,(dF(\rho)/d\rho)\,] + [\lambda - m^2/(1-\rho^2)]\,F(\rho) = 0 \quad (8.4.3)$$

Let $\boxed{F(\rho) \propto (1-\rho^2)^{m/2} G(\rho)}$. (8.4.4)

were $G(\rho)$ is a polynomial in ρ. Substituting this and putting $(dF(\rho)/d\rho)$ and $F(\rho)$ in equation (8.4.3) one gets, after simplification,

$$G''\,(1-\rho^2) - G'\,(2m+2)\rho + G\,(\lambda - m - m^2) = 0 \quad (8.4.5)$$

Let its solution be a power series,

$$G(\rho) = \sum_{j=0}^{+\infty} a_j\, \rho^j;\quad a_j \neq 0 \quad (8.4.6)$$

8.4.3 RECURSION FORMULA for the Rigid Rotator:

Find $G'(\rho)$ and $G''(\rho)$.

$G'(\rho) = \sum a_j\, j\, \rho^{j-1}$,

$\rho\, G'(\rho) = \rho \sum a_j\, j\, \rho^{j-1} = \sum a_j\, j\, \rho^j$

$$G''(\rho) = (d/d\rho)\, G'(\rho) = (d/d\rho) \sum a_j j \, \rho^{j-1}$$
$$= \sum a_j j(j-1)\, \rho^{j-2} \sum (j+2)(j+1) a_{j+2}\, \rho^j$$

One gets on substitution,

$$\sum (m+h+1)(m+h+2)\, a_{h+2}\, \rho^{h+2} - \sum \lambda - (m+h+1)(m+h)\, a_h\, \rho^h = 0$$

where h = an integer, a value of j. Equating the coefficient of ρ to zero,

$$\boxed{\frac{a_{h+2}}{a_h} = \frac{\lambda - (m+h+1)(m+h)}{(m+h+1)(m+h+2)}} \tag{8.4.7}$$

This is the RECURSION FORMULA.
i.e,. any series which satisfies equation (8.4.7) will also satisfy equation (8.4.6).

8.4.4 ENERGY OF ROTATION of the Rigid Rotator:

Since G is to be finite, terms beyond a_h,

i.e., $\quad a_{h+1} = 0$

∴ $\quad a_{h+2} = 0$

In other words, $\lambda = (m+h+1)(m+h)$
Putting $\ell = (m+h) = 0, 1, 2, 3, ...$

$\quad \ell \geq m =$ an integer,

for well behaved solutions to exist.

$$\lambda_\ell = \ell(\ell + 1) \tag{8.4.8}$$

But $\quad \lambda_\ell = \{2\mu r^2 / \hbar^2\} [E_\ell - V(r)] =$ constant. $\tag{8.3.4}$

i.e., $\quad \lambda_\ell = \{2\Im / \hbar^2\} E_\ell$

ROTATIONAL ENERGY, E_ℓ

$$E_\ell = \{\ell(\ell+1)\, \hbar^2 / 2\Im\} \; ; \text{where } \ell = 0, 1, 2, 3,... \tag{8.4.9}$$

and $\ell \geq m =$ an integer.

8.4.4 QUANTIZATION OF ROTATIONAL ENERGY, E_ℓ :

Fig. 8.3 Rotational Energy levels, E_J of a Rigid Rotator

Replacing the spectroscopic notation by **Rotational Quantum Number** J

$$E_J = \{J(\ell+1)\hbar^2/2\Im\}\ ;\ \ell \to J = 0, 1, 2, 3, ...$$

and $\ell \geq m =$ an integer.
The restriction on ℓ requires that only those values of the energy given by this relation be allowed.
Therefore, the energy of a Rigid Rotator is QUANTIZED (Fig. 8.3).

8.4.5 ANGULAR MOMENTUM, $\vec{L}$ of the rotator:

Angular momentum, $\vec{L}$, by definition, is given by
$\vec{L}$ = moment of linear momentum.= (Radius of the arm) x (linear momentum).
$$\vec{L} = a\,(\mu\vec{v}) = (\mu\vec{v}a).$$
For the rigid rotator, total energy,
$$E = T + V(r) = \tfrac{1}{2}\mu\vec{v}^2 + 0 = \text{purely rotational energy.}$$
$$(E)(2\Im) = (\tfrac{1}{2}\mu\vec{v}^2)(2\mu a^2) = (\mu\vec{v}a)^2$$
$$= (\text{Angular momentum, }\vec{L})^2.$$
$$= \ell(\ell+1)\hbar^2$$
i.e., $\qquad \vec{L} = \sqrt{\ell(\ell+1)}\,\hbar$ \hfill (8.4.10)
Thus the angular momentum of the rigid rotor is QUANTIZED.

8.4.6 FERRER'S Associated LEGENDRE POLYNOMIAL of the 1st kind, $P_\ell^m(\rho)$

Physically acceptable solutions of equations (8.4.3) and (8.4.5), with $m \neq 0$, are obtained by differentiating m times equation (8.4.3), when $m = 0$, and one deduces readily that the solutions are the Associated Legendre Functions with $m \leq \ell$ when $\lambda_\ell = \ell(\ell+1)$ are the only ones acceptable solutions which are non-singular. Equation (8.4.4) is

$$\Theta(\theta) \to F(\rho) \propto (1-\rho^2)^{m/2}\,G(\rho). \tag{8.4.4}$$
$$\Theta_\ell^m(\theta) = (1-\rho^2)^{m/2}\,B_{\ell,m} P_\ell^m(\rho).$$
$$\boxed{\Theta_\ell^m(\theta) = B_{\ell,m}(1-\rho^2)^{m/2}\,(d^m/d\rho^m)P_\ell(\rho)}. \tag{8.4.11}$$

where $P_\ell^m(\rho)$ is called the Ferrer's <u>Associated Legendre Polynomial</u> in ρ of the First kind of degree ℓ and order m. It is well behaved for all finite values of ρ. $P_\ell(\rho)$ is called the <u>Legendre Polynomial</u> in ρ.

8.4.7 ASSOCIATED LEGENDRE POLYNOMIAL of the 2nd KIND, $Q_\ell^m(\rho)$:

The other solution is the *Associated Legendre Polynomial of the 2nd kind*, $Q_\ell^m(\rho)$, but it has singularity at $\rho = \pm 1$, i.e. it becomes infinite at these points, violating Postulate 1.

8.4.7.1 RODRIGUE'S FORMULA

$$G(\rho) = \frac{d^m P_\ell(\rho)}{d\rho^m} \tag{8.4.12}$$

For $m \geq 0$, and $m \leq \ell$

$P_\ell(\rho)$ is called the LEGENDRE POLYNOMIAL in ρ is the solution of equation (8.4.3), when $m = 0$, i.e. of the Legendre differential equation given by

$$\boxed{P_\ell(\rho) = [(-1)^\ell / 2^\ell \ell!] \frac{d^\ell (1-\rho^2)^\ell}{d\rho^2}} \tag{8.4.13}$$

This is known as RODRIGUE'S Formula.
$\rho = \pm 1$, means $\theta = 0$ or π.
The differential equation (8.4.3) in terms of ρ is then

$$\boxed{\frac{d}{d\rho}\left[(1-\rho^2)\frac{d}{d\rho}P_\ell^m(\rho)\right] + \left[\frac{\lambda - m^2}{(1-\rho^2)}\right]P_\ell^m(\rho) = 0} \tag{8.4.14}$$

The *Associated Legendre Polynomial of the 1st kind*,

$P_\ell^m(\rho)$ = finite, for all $-1 \leq \rho \leq +1$, and $\ell \geq |m| = 0, 1, 2, 3, ...$ (8.4.15)

$P_\ell^m(\rho)$ satisfies the ORTHO-NORMALITY CONDITION

$$\boxed{\int_{-1}^{+1} P_\ell^m(\rho) \cdot P_{\ell'}^m(\rho) d\rho = \left[\frac{(\ell+m)!}{(\ell-m)!}\frac{2}{2\ell+1}\right]\delta_{\ell\ell'}} \tag{8.4.16}$$

where the integration is carried out within the limits $\rho = \pm 1$.

8.5 ROTATIONAL SPECTRA:

8.5.1 SPHERICAL TOP DIATOMIC MOLECULE:

It was seen that for a spherical top,

$$I_A = I_B = I_C \equiv I \tag{8.2.6}$$

$$E_\ell = \{\ell(\ell+1)\hbar^2/2 I\}, \text{where } \ell = 0, 1, 2, 3, ... \tag{8.4.9}$$

and $\ell \geq |m| = 0, 1, 2, 3, ...; m \geq 0$ = an integer.

$(\hbar^2/2 I) = (\hbar 4\pi I)\hbar$

Defining $\boxed{B_0 = (\hbar/4\pi I)}$ (8.5.1)

Equation for rotational energy (4.4.9) corresponding to $Y(\theta,\varphi)$ can be expressed as

$$\boxed{E_{rotation} \equiv E_J = h B_0 J(J+1)} \tag{8.5.2}$$

For a molecule like Carbon Monoxide, CO, the ROTATIONAL QUANTUM NUMBER, J, replaces the quantum number ℓ.

$$J = \ell = (m + h) = 0, 1, 2, ... \tag{8.5.3}$$

m = magnitude of the projection of the $\vec{J}$ -vector on a given direction ($\theta = 0$).

$$m = \pm J, \pm(J-1), \pm(J-2), \pm......, 0$$

$$E_J = h B_0 J(J+1). \tag{8.5.4.}$$

$$I_A = I_B = I_C = I \tag{8.5.5}$$

and $\quad E_J = \frac{\hbar^2}{2I} J(J+1)$

8.5.2 ROTATIONAL ENERGY of a SYMMETRIC TOP:
There is also no difficulty in calculating the energy levels in the case where only two of the moments of inertia of the body are the same;

$$I_A = I_B \neq I_C. \qquad (8.5.6)$$

becomes $\quad E_J = \frac{\hbar^2}{2I_A} J(J+1) + \frac{\hbar^2}{2}\left[\frac{1}{I_C} - \frac{1}{I_A}\right] J_\zeta^2. \qquad (8.5.7)$

where J_ζ is the angular momentum component in a rotating system of coordinates.

8.5.3 ROTATIONAL ENERGY of an ASYMMETRICAL TOP:
For such a system,

$$I_A \neq I_B \neq I_C \qquad (8.5.8)$$

The calculation of energy levels in a general form is impossible.

8.6 PROPERTIES OF POLAR FUNCTIONS, $P_\ell^m(Cos\theta)$

8.6.1 ASSOCIATED LEGENDRE POLYNOMIALS, $P_\ell^m(Cos\theta)$:

TABLE 8.1 The First few Associated Legendre Polynomials $P_\ell^m(Cos\theta)$

ℓ	$P_\ell(Cos\theta)$	m	$P_\ell^m(Cos\theta)$
0	$P_0(\cos\theta) = 1$	0	$P_0^0(\cos\theta) = 1$
1	$P_1(\cos\theta) = \cos\theta$	0	$P_1^0(\cos\theta) = \cos\theta$
		1	$P_1^1(\cos\theta) = 1$
2	$P_2 = \frac{1}{2}(3\cos^2\theta - 1)$	0	$P_2^0(\cos\theta) = \frac{1}{2}(3\cos^2\theta - 1)$
		1	$P_2^1(\cos\theta) = 3\sin\theta\cos\theta$
		2	$P_2^2(\cos\theta) = 3\sin^2\theta$
3	$P_3 = \frac{1}{2}(5\cos^3\theta - 3\cos\theta)$	0	$P_3^0(\cos\theta) = \frac{1}{2}(5\cos^3\theta - 3\cos\theta)$
		1	$P_3^1(\cos\theta) = \frac{3}{2}(5\cos^2\theta - 1)\sin\theta$
		2	$P_3^2(\cos\theta) = 15\sin^2\theta\cos\theta$
		3	$P_3^3(\cos\theta) = 15\sin^3\theta$

For reasonably small values of ℓ and m, the Associated Legendre Polynomials, $P_\ell^m(Cos\theta)$ are rather simple functions: several of them are listed in Table 8.1.

8.6.2 DIFFERENTIAL FORMULA - Generation of $P_\ell^m(Cos\theta)$

The Legendre Polynomials $P_\ell(Cos\theta)$ may be used to generate the Associated Legendre Polynomials, $P_\ell^m(Cos\theta)$, by the <u>Differential Formula</u>

$$P_\ell^m(\rho) = (1-\rho^2)^{m/2} (d^m/d\rho^m) P_\ell(\rho); \ell = |m| = 0,1,2,3,4,....; m \geq 0 \qquad (8.4.11)$$

This permits one to derive any $P_\ell^m(\rho)$.

$$P_\ell(\rho) = [(-1)^\ell / 2^\ell \ell!] (d^\ell / d\rho^\ell) \{(1-\rho^2)^\ell\} \qquad (8.4.12)$$

which is <u>Rodrigue's Formula</u>.

Worked out Example 8.1

Use Rodrigue's Formula to show that $P_1(Cos\theta) = Cos\theta$. Given,

$$P_\ell(\rho) = [(-1)^\ell / 2^\ell \ell!] (d^\ell / d\rho^\ell) \{(1-\rho^2)^\ell\}.$$

Solution:

<u>Step # 1</u> Rodrigue's Formula is $P_\ell(\rho) = [(-1)^\ell / 2^\ell \ell!](d^\ell / d\rho^\ell)\{(1-\rho^2)^\ell\}$.

<u>Step # 2</u> $\ell = 1; P_{\ell=1}(\rho) = [(-1)^1 / 2^1 1!](d^1 / d\rho^1)\{(1-\rho^2)^1\} = (-1/2)(d/d\rho)\{(1-\rho^2)\}$.

$= (-1/2)(-2\rho) = \rho = Cos\theta$

8.7 RIGID ROTATOR WAVE FUNCTIONS:

8.7.1 SPHERICAL HARMONICS (TESSERAL HARMONICS):

The full eigen functions are simply products of the functions $\Theta_\ell^m(\theta)$ and $\Phi_m(\varphi)$,

i.e., $\qquad Y_\ell^m(\theta,\varphi) = \Theta_\ell^m(\theta) \cdot \Phi_m(\varphi) \qquad (8.7.1)$

$\Phi(\varphi) = (1/\sqrt{2\pi})\, e^{+i|m|2\pi}, \qquad (8.3.14)$

These functions, encountered also in the solution of boundary value problems in Electrostatics, are known as SPHERICAL HARMONICS of the ℓth degree and m^{th} order. They form a special factorizable type of surface harmonics.

8.7.2 NORMALIZED SPHERICAL HARMONICS $\hat{Y}_\ell^m(\theta,\varphi)$

The *normalized spherical harmonics* is given by the COMPLEX FUNCTION (not real function):

$$\hat{Y}_\ell^m(\theta,\varphi) = (-1)^m \left[\frac{\ell + \frac{1}{2}}{4\pi} \frac{(\ell-m)!}{(\ell+m)!} \right]^{1/2} P_\ell^m(\cos\theta)\, e^{i|m|\varphi} \qquad (8.7.2)$$

in the region Ω, defined by $\boxed{0 \le \varphi \le 2\pi,\ 0 \le \theta \le \pi}$

where the integral over the surface of the sphere (r = constant) is

$$\int_\Omega d\Omega = \int_0^{2\pi} d\varphi \int_0^\pi Sin\theta\, d\theta \qquad (8.7.3)$$

8.7.3 CONDON-SHORTLEY PHASE:

The factor $(-1)^m$ is a phase factor, often called the Condon-Shortley phase.

8.7.4 $(2\ell+1)$ DEGENERATE EIGENFUNCTIONS of the rigid rotor:

For each value of ℓ there are $2\ell+1$ spherical harmonics, $Y_\ell^m(\theta,\varphi)$, each with different m-value, $-\ell \le m \le +\ell$.

These are <u>degenerate</u> eigen functions of $\hat{L}^2$.

The spherical harmonics may look trivial, but they contain a wealth of physics of the problem: they completely describe the angular motion of any CENTRAL FORCE problem.

8.7.5 Few SPHERICAL HARMONICS, $Y_\ell^m(\theta,\varphi)$

Listed in Table 8.2 are the first few Spherical Harmonics, $Y_\ell^m(\theta,\varphi)$.

8.7.6 POLAR PLOTS

Polar plots are the probability distributions $\left| Y_\ell^m(\theta,\varphi) \right|^2$ versus distance x (y, z) in a coordinate plane. As an aid to visualizing the azimuthal part of the wave function, $Y_\ell^m(\theta,\varphi)$ are plotted in Fig. 8.2., for $\ell = 0$, 1, and 2, and all possible values of m.

TABLE 8.2   The First Few Spherical Harmonics, $Y_\ell^m(\theta,\varphi)$

$Y_\ell^m(\theta,\varphi)$ Symbol	Polar	Cartesian	Normalization Constant
Y_0^0	1	1	$\tfrac{1}{2}(1/\pi)^{1/2}$
Y_1^0	$\cos\theta$	z/r	$\tfrac{1}{2}(3/\pi)^{1/2}$
$Y_1^{\pm 1}$	$\mp(\sin\theta)e^{\pm i\phi}$	$\mp(x\pm iy)/r$	$\tfrac{1}{2}(3/2\pi)^{1/2}$
Y_2^0	$(3\cos^2\theta - 1)$	$(3z^2 - r^2)/r^2$	$\tfrac{1}{4}(5/\pi)^{1/2}$
$Y_2^{\pm 1}$	$\mp(\sin\theta)(\cos\theta)e^{\pm i\phi}$	$\mp z(x\pm iy)r^2$	$\tfrac{1}{2}(15/2\pi)^{1/2}$
$Y_2^{\pm 2}$	$(\sin^2\theta)e^{\pm 2i\phi}$	$(x\pm iy)^2/r^2$	$\tfrac{1}{4}(15/2\pi)^{1/2}$
Y_3^0	$(5\cos^3\theta - 3\cos\theta)$	$z(5z^2 - 3r^2)/r^3$	$\tfrac{1}{4}(7/\pi)^{1/2}$
$Y_3^{\pm 1}$	$\mp\sin\theta(5\cos^2\theta - 1)e^{\pm i\phi}$	$\mp(x\pm iy)(5z^2 - r^2)/r^3$	$\tfrac{1}{8}(21/\pi)^{1/2}$
$Y_3^{\pm 2}$	$(\sin^2\theta)(\cos\theta)e^{\pm 2i\phi}$	$z(x\pm iy)^2/r^3$	$\tfrac{1}{4}(105/2\pi)^{1/2}$
$Y_3^{\pm 3}$	$\mp(\sin^3\theta)e^{\pm 3i\phi}$	$\mp(x\pm iy)^3/r^3$	$\tfrac{1}{8}(35/\pi)^{1/2}$

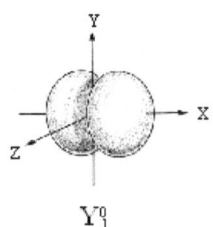

Fig 8.2 Polar plot of the spherical harmonic $Y_1^0(\theta,\varphi)$

8.7.5 PARITY of the $Y_\ell^m(\theta,\varphi)$

To check whether the Spherical Harmonics have parity even or odd, replace θ and φ by $(\pi - \theta)$ and $(\pi + \varphi)$, respectively, i.e.

Parity + means Even: $Y_\ell^m(\pi-\theta,\varphi) = +Y_\ell^m(\theta,\varphi)$.
Parity − means Odd: $Y_\ell^m(\pi-\theta,\varphi) = -Y_\ell^m(\theta,\varphi)$
Parity + means Even: $Y_\ell^m(\theta,\pi+\varphi) = +Y_\ell^m(\theta,\varphi)$.
Parity − means Odd: $Y_\ell^m(\theta,\pi+\varphi) = -Y_\ell^m(\theta,\varphi)$

In spherical harmonic functions, it turns out that the values of ℓ uniquely determines the parity of $Y_\ell^m(\theta,\varphi)$. Parity can otherwise be determined by means of the sign of:

$$\boxed{\text{Parity} = (-1)^\ell} \tag{8.7.4}$$

8.8 CONCLUSION

In this Chapter, the solution of the angular part of the Schrödinger Equation for purely radial force was obtained, and the results were used to discuss the rotation of systems, that have no radial motion.

REVIEW QUESTIONS

R.Q. 8.1 Calculate the reduced mass of the KCl molecule, for both of the naturally occurring isotopes of Cl in (a) u and (b) kg. (Answer: a) $^{39}K^{35}Cl; \mu = 18.4288\ u$; $^{39}K^{37}Cl; \mu = 18.9690\ u$

b) $^{39}K^{35}Cl; \mu = 3.06 \times 10^{-26}\ kg$; $^{39}K^{37}Cl; \mu = 3.1499 \times 10^{-26}\ kg$).

R.Q. 8.2 The bond length in KCl is 2.667 Å. Calculate the moment of inertia (in $kg-m^2$) and the rotational constant B (in cm^{-1}) for both $^{39}K^{35}Cl$ and $^{39}K^{37}Cl$. (Answers:
$^{39}K^{37}Cl; I = 2..24048 \times 10^{-45}\ kg-m^2$, $B = 0.124941\ cm^{-1}$;
$^{39}K^{35}Cl;, I = 2..17667 \times 10^{-45}\ kg-m^2$, $B = 0.128603\ cm^{-1}$).

R.Q. 8.3 Calculate the position of the center-of-mass in KCl for each of the two isotopic modifications, i.e. give the distances from the center-of-mass to each nucleus. The bond length in KCl is 2.667 Å. (Answers: $^{39}K^{35}Cl; r_K = 1.261$ Å, $r_{Cl} = 1.406$ Å;
$^{39}K^{37}Cl; r_K = 1.298$ Å, $r_{Cl} = 1.369$ Å).

R.Q. 8.4   Normalize the polar part, $\Theta_\ell^m(\theta)$ of the rigid rotator wave function. Given Rodrigue's formula, $P_\ell(\rho) = [(-1)^\ell / 2^\ell \ell!] (d^\ell / d\rho^\ell)[(1-\rho^2)^\ell]$.;
$\Theta_\ell^m(\theta) = B_{\ell,m}(1-\rho^2)^{m/2} (d^m / d\rho^m) P_\ell(\rho)$.

(Answer: Normalization consant is $\Theta_\ell^m(\theta) = (-1)^m [\ell + (1/2)] \left[\frac{(\ell-m)!}{(\ell+m)!}\right]^{1/2} P_\ell^m(Cos\theta)$).

R.Q. 8.5   Generate $P_2(Cos\theta)$ and $P_2^1(Cos\theta)$ Given Rodrigue's Formula
$P_\ell(\rho) = [(-1)^\ell / 2^\ell \ell!](d^\ell / d\rho^\ell)\{(1-\rho^2)^\ell\}$
(Answer: $P_2(Cos\theta) = (1/2)(3Cos^2\theta - 1)$; $P_2^1(Cos\theta) = -\sqrt{15/4}\ Sin\theta\ Cos\theta$)

R.Q. 8.6 Use Rodrigue's formula to show that $P_4(Cos\theta) = (1/8)(35\ Cos^4\theta - 30\ Cos^2\theta + 3)$ Given
$P_\ell(\rho) = [(-1)^\ell / 2^\ell \ell!](d^\ell / d\rho^\ell)\{(1-\rho^2)^\ell\}$.

R.Q. 8.7   Generate $P_2(Cos\theta)$ and $P_2^1(Cos\theta)$. Write explicitly the wave function of the Rigid Rotator in state $\ell = 2$ and $m = 1$. Given, $P_\ell(\rho) = [(-1)^\ell / 2^\ell \ell!](d^\ell / d\rho^\ell)\{(1-\rho^2)^\ell\}$;

$Y_\ell^m(\theta,\varphi) = (-1)^m \left[\frac{\ell+\frac{1}{2}}{4\pi} \frac{(\ell-m)!}{(\ell+m)!}\right]^{1/2} P_\ell^m(\cos\theta)\ e^{\ i |m| \varphi}$ (Answers:

$P_2(Cos\theta) = (1/2)(3Cos^2\theta - 1)$;
$P_2^1(Cos\theta) = -\sqrt{15/4}\ Sin\theta\ Cos\theta$; $Y_2^1(\theta,\varphi) = -\sqrt{15/8\pi}\ Sin\theta\ Cos\theta\ e^{+i\varphi}$)

&%&%&%&%&%

Chapter 9

EXACT SOLUTIONS 6

WAVE MECHANICS OF THE HYDROGEN ATOM

Chapter 9

EXACT SOLUTIONS 6

WAVE MECHANICS OF THE HYDROGEN ATOM

"The high destiny of the individual is to serve rather than to rule"   -Albert Einstein

"The great marvel is not the size of the enterprise, its secrecy or cost, but the achievement of scientific brains in putting together infinitely complex pieces of knowledge held by many men in different fields of science into a workable plan."   -President Truman

9.1   INTRODUCTION:

Schrödinger's first practical application of his Galilean Relativistic Wave Equation was not an investigation of square wells, barriers and such as the problems discussed earlier, in Chapters 5 – 8. On the other hand, he straightaway attacked the problem of the free hydrogen atom. In non-relativistic theory, this hydrogen atom is one of the few cases for which the Schrödinger Equation may be solved exactly.

In Chapter 8, the solution of the angular part of the Schrödinger Equation for purely radial force was obtained, and the results were used to discuss the rotation of systems, that have no radial motion. The hydrogen atom (H- atom, for brevity) is the simplest elemental atom of all in the Periodic Table of Elements, since it contains only one electron. So the simplest of all atomic systems to analyze mathematically is the H-atom. It is a system of two interacting particles (electron and proton), moving under their mutual interaction.

9.2   WAVE MECHANICS OF THE H - ATOM

The H - atom is a system of two interacting particles (electron and proton), moving under their mutual interaction. For simplicity assume that the nucleus and the electron both are spin less.

9.2.1 POTENTIAL ENERGY of the H-ATOM

Considering only the Coulomb potential between the electron and the nucleus, let

$+Ze$ = Nuclear charge, (Z = atomic number is unity for the H-atom),

$e = 1.6021 \times 10^{-19} C$ = electronic charge,

$M_p = 1.6726 \times 10^{-27} kg$ = nuclear mass = proton rest mass

$m_e = 9.109 \times 10^{-31} kg$ = electronic mass

$\varepsilon_0 = 8.85 \times 10^{-12} F\ m^{-1}$ = permittivity of free space

$k = (1/4\pi\varepsilon_0) = 8.99 \times 10^9\ Nm^2 C^{-2} = 8.99 \times 10^9\ F^{-1} m$

r = distance separating the two particles

Coulomb force, $\boxed{\vec{F}_{em} = \frac{1}{4\pi\varepsilon_0}\frac{-Ze^2}{r^2}}$ (9.2.1)

Hereafter, k will not be included for convenience (from below).

Coulomb potential, for H-Atom ($Z=1$),

$$\boxed{V(r) = \frac{-e^2}{r}}.$$ (9.2.2)

Equation (9.2.1.) has the symmetry of a sphere under any rotation of the system.

9.2.2 TOTAL ENERGY of the H-atom

The total energy,

$$E_T = V + T_T$$ (9.2.3)

Total Kinetic energy, T_T = K. E. of the nucleus, T_n + K. E. of the electron, T.

$$T_T = \frac{1}{2} M (\vec{v}_{nx}^2 + \vec{v}_{ny}^2 + \vec{v}_{nz}^2) + \frac{1}{2} m (\vec{v}_{ex}^2 + \vec{v}_{ey}^2 + \vec{v}_{ez}^2)$$ (9.2.4.)

where $\vec{v} \equiv \vec{v}_{ex}$ and $\vec{v}_{nx}$, respectively, stands for the velocity of the electron and the nucleus in the H-atom.

Since the Total Energy, E_T is onserved; it is convenient to use the expression for further calculation.

9.2.3 Separation of Variables

The fact that

(i) The attractive force (9.2.1) acts only along a straight line joining the two particles and

(ii) There are no external forces acting on the system, one may use very well separate the total energy into two parts: -

a) Energy of rotation of the atom as a whole, and
b) Energy due to rotation of the electron and the nucleus about a common Centre of Gravity (CG).

Representing the coordinates of the CG (called *Centre-of-Mass coordinates*), an *independent variable*, by

$$R_{CM} \equiv (X, Y, Z),$$

such that $X = (M_n x_n + m x_e)/M$; $Y = (M_n y_n + m y_e)/M$; $Z = (M_n z_n + m z_e)/M$. (9.2.5)

$$M = (M_n + m)$$

The corresponding linear momenta are:

$$\vec{P}_X = M(dX/dt) \; ; \; \vec{P}_Y = M(dY/dt) \; ; \; \vec{P}_Z = M(dZ/dt).$$

9.2.3.1 Reduced Mass, μ

Defining *Relative Coordinates*, r, another *independent variable* (position of the electron relative to the nucleus), by

$$r(x_e, y_e, z_e) = (r_n - r_e) \; ; \; x = (x_e - x_e), y = (y_e - y_e), z = (z_e - z_e) \quad (9.2.6)$$

It is known that the motion of two bodies about their CM can always be reduced to an equivalent ONE-BODY problem. Therefore, from expression (8.2.4)

$$\mu = M_n m / (M_n + m) \approx m \quad (9.2.7)$$

= the reduced mass of the H-atom.

The total energy, $E_T = V + T_e = \left(\{ (\vec{P}^2 / 2M) + (\vec{p}^2 / 2\mu) \} + \frac{-Z\,e^2}{r} \right) \quad (9.2.8)$

$$\vec{P}^2 = \vec{P}_X^2 + \vec{P}_Y^2 + \vec{P}_Z^2 \; ; \; \vec{p}^2 = \vec{p}_x^2 + \vec{p}_y^2 + \vec{p}_z^2.$$

9.2.3.2 Translational Energy, E_{CM}

In equation (9.2.8)

$$E_T = E_{CM} + E_{rel} \tag{9.2.9}$$

where $E_{CM} = (\vec{P}^2/2M)$ = due to translation of the CM, *i.e.* of M. (9.2.10)

9.2.3.3 Internal Energy, E_{rel}

In equation (9.2.8)

$$E_{rel} = \left(\vec{p}^2/2\mu + \frac{-Z\,e^2}{r}\right) \tag{9.2.11}$$

= a term depends on the distance between the two particles as well as the relative velocities.

This is also the *Internal Energy* that is due to the particle mass, μ.

9.2.3.4 The Role of V(r):

The motion of the electron relative to the nucleus is thus determined by the Coulomb interaction

$$V(r) = -Z\,e^2/r \tag{9.2.2}$$

which is the potential energy of the electron in the electric field of the nucleus. Can we now solve the problem of the motion of the electron by applying the Newtonian Equation of Motion

$$\vec{F} = d\vec{p}/dt\,? \quad \text{No!}$$

This is because we are analyzing this motion by means of Quantum Mechanics.

9.3 THE SCHRODINGER EQUATION

The Schrödinger Equation is set up for the motion, *viz.*

$$\nabla_T^2 \psi_T(\vec{r}) + \{(2m/\hbar^2)E_T\}\psi_T(\vec{r}) = 0. \tag{9.3.1}$$

i.e., $(\nabla_{CM}^2 + \nabla_{rel}^2)\psi_T(\vec{r}) + \{(2m/\hbar^2)E_T\}\psi_T(\vec{r}) = 0$

9.3.1 Factoring Out the CENTRE OF MASS MOTION For the ONE-ELECTRON ATOM

The hydrogen atom consists of two particles, the proton and the electron

Let $\psi_T(\vec{r}) \equiv \psi_T[(x,y,z),(X,Y,Z)] = \psi(x,y,z) \cdot \psi_{CM}(X,Y,Z)$ (9.3.2)

$$(\nabla_{CM}^2 \psi(R_{CM}) + \{(2M/\hbar^2)E_{CM}\}\psi(R_{CM}) = 0 \qquad (9.3.3)$$

This is the Equation of CM Motion for a one-electron atom, as a particle mass M with coordinates R_{CM} and energy E_{CM}. This equation is of course a *Free Particle Equation*.

9.3.2 The Relative Motion EQUATION (Factoring Out the Center of Mass Motion)

Since the potential depends only on the relative position, consider Equation (9.3.1) which contains

$$(\nabla_{rel}^2 \psi(x,y,z) + \{(2\mu/\hbar^2)E_{rel}\}\psi(x,y,z) = 0. \qquad (9.3.4)$$

Thus one could <u>decouple</u> the original *two-body* problem into two *one-body* problems, that of a free particle (CM) and that of a single particle of mass μ. Since $V(r)$ is a function of r rather than x, y, and z, it is not correct to substitute equation (9.2.2) in equation (9.3.4), if $(\nabla_{rel}^2) = (\nabla_x^2) = (\nabla_y^2) = (\nabla_z^2)$

The two alternatives are:

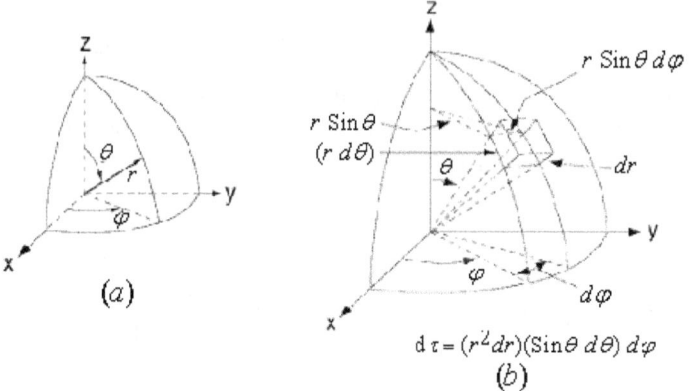

Spherical Polar Coordinates

Fig. 9.1 Right-handed Spherical Polar Coordinates System

(a) Either express $V(r) = V(x,y,z)$ by means of

$$(r^2) = (x^2 + y^2 + z^2) \qquad (9.3.5)$$

in the relative r coordinates,

b) Express the Schrödinger Equation in terms of the spherical polar coordinates, (r, θ, φ), defined earlier in Chapter 8, Section 8.2.3 (shown Fig. 9.1).

φ = Azimuth angle between the projection of r in the xy- plane and the +Z- axis, measured in the direction shown in Fig 9.1,

$$\varphi = \tan^{-1}(y/x). \tag{9.3.6}$$

$$d\tau = dx\, dy\, dz = (r^2 dr)(\text{Sin}\theta\, d\theta)\, d\varphi. \tag{9.3.7}$$

In Spherical Polar Coordinates, the particle in a central force field (field in which the potential depends upon the magnitude of the distance from a fixed point) is V(r) = spherically symmetric, i.e. V(r) has the symmetry of a sphere under any rotation of the system, is CENTRAL, and is independent of θ and φ,

i.e., $\quad V(\vec{r}) = V(|\vec{r}|) = V(r)$

Therefore, expressing the Schrödinger Equation in terms of the spherical coordinates makes the problem considerably easier to solve.

9.3.3   SCHRODINGER EQUATION (TISE) for the H-atom:

The algebra involved in $\nabla^2_{Cartesian} \rightarrow \nabla^2_{Spherical}$ is too lengthy for me to present here, and so reference is made to any good standard work on vector analysis (say, Arfken). Choosing a *Right-Handed Cartesian* (x, y, z) and Spherical polar (r, θ, φ) Coordinate System, the result is

$$(\nabla_{rel}^2) = (\nabla_x^2) = (\nabla_y^2) = (\nabla_z^2)$$

$$= \boxed{\nabla^2 = \left[\frac{1}{r^2}\frac{\partial}{\partial r}\left(r^2 \frac{\partial}{\partial r}\right) + \frac{1}{r^2 \text{Sin}\theta}\frac{\partial}{\partial \theta}\left(\text{Sin}\theta \frac{\partial}{\partial \theta}\right) + \frac{1}{r^2 \text{Sin}\theta}\left(\frac{\partial^2}{\partial \varphi^2}\right)\right]} \tag{9.3.8}$$

where $\nabla_r^2 = \frac{1}{r^2}\frac{\partial}{\partial r}\left(r^2 \frac{\partial}{\partial r}\right)$ \hfill (9.3.9)

The interest lies in the analysis of the Schrödinger Equation of a 'particle' of mass μ with coordinate r moving in a Coulomb (central) potential force with energy, $E_{rel} = E - V(r)$, given by equation (9.3.4).

Equation (9.3.4) can be written as

$$\left[\frac{1}{r^2}\frac{\partial}{\partial r}\left(r^2 \frac{\partial}{\partial r}\right) + \frac{1}{r^2 \text{Sin}\theta}\frac{\partial}{\partial \theta}\left(\text{Sin}\theta \frac{\partial}{\partial \theta}\right) + \frac{1}{r^2 \text{Sin}\theta}\left(\frac{\partial^2}{\partial \varphi^2}\right)\right]\psi(\vec{r})$$

$$+ \{(2\mu/\hbar^2)[E-V(r)]\,\psi(\vec{r}) = 0. \qquad (9.3.10)$$

$$\psi(\vec{r}) = \psi(\vec{r},\theta,\varphi)$$

9.4 Factoring Out the Angular Dependence: the Radial Wave Equation

9.4.1 SEPARATION OF VARIABLES:
Since V(r) is central; ψ(r) can be constructed by the method of 'separation of variables'.

Now seek particular solutions of the form,
$$\boxed{\psi(\vec{r}) = \psi(\vec{r},\theta,\varphi) = \mathfrak{R}(r) \cdot Y(\theta,\varphi)} \qquad (9.4.1)$$

where $\mathfrak{R}(r)$ = RADIAL function, depending on variable r only,

$Y(\theta,\varphi)$ = SPHERICAL HARMONICS (angular function) specifies dependence on θ and φ.

The angular functions, $Y(\theta,\varphi)$ are the same for all central force problems, and are independent of *V(r)*, as in the case of the Rigid Rotator, discussed in Chapter 8.

9.4.2 Legendian Operator, $\hat{\Lambda}$
It is known that *Legendrian Operator*, $\hat{\Lambda}$ (8.3.5) is
$$\hat{\Lambda} = \{(1/\sin\theta)\,[(\partial/\partial\theta)\,\sin\theta\,(\partial/\partial\theta)] + (1/\sin^2\theta)\,\partial^2/\partial\varphi^2\} \qquad (9.4.2)$$
It measures the **curvature** of a function relative to the surface of a sphere.

9.4.3 FACTORING OUT THE ANGULAR DEPENDENCE: RADIAL EQUATION

The angular part of the Schrödinger Equation
$$\boxed{[\hat{\Lambda}\,Y(\theta,\varphi)]\,/\,Y(\theta,\varphi) = -\lambda}. \qquad (9.4.3)$$
The partial differential equation (9.3.10) has, therefore, been reduced to a pair of simultaneous differential equations (9.4.3) and the one (9.4.4) given below.
$$\boxed{\frac{1}{\mathfrak{R}(r)}\left[\frac{d}{dr}\left(r^2\frac{d}{dr}\right)\right]\mathfrak{R}(r) + \{(2\mu r^2/\hbar^2)[E-V(r)]\,\mathfrak{R}(\vec{r}) = +\lambda} \qquad (9.4.4)$$

which is called the *Radial Equation*.

The angular equation (9.4.3.) has been solved already in the case of a Rigid Rotator, in Chapter 8.
The separation constant, given by equations (8.3.4) and (8.4.8), is
$$\lambda = \{2\mu r^2/\hbar^2\}\,[E - V(r)] = \lambda_\ell = \ell\,(\ell+1) \qquad (9.4.5)$$

9.4.4 EFFECTIVE POTENTIAL, V_{eff}, for the H-ATOM
The radial wave equation can be expressed in another form,
$$\boxed{-\frac{\hbar^2}{2\mu r^2}\left[\frac{d}{dr}\left(r^2\frac{d}{dr}\right)\right]\mathfrak{R}(r) + \left(\frac{\hbar^2}{2\mu r^2}+V(r)\right)\mathfrak{R}(\vec{r}) = E\,\mathfrak{R}(r)} \qquad (9.4.6)$$

It can be shown that for the H-atom from the following relations

$$\hat{H}\,\Re(\bar{r}) = \frac{\hat{p}^2}{2\mu}\Re(\bar{r}) + \left(\frac{\lambda\hbar^2}{2\mu r^2} + V(R)\right)\Re(\bar{r}) = E\,\Re(\bar{r}) \qquad (9.4.7)$$

and

$$\nabla_r^2\,\Re(\bar{r}) + \left(\frac{2\mu}{\hbar^2}[E - V(r)] + \frac{\ell(\ell+1)\hbar^2}{2\mu r^2}\right)\Re(\bar{r}) = 0 \qquad (9.4.8)$$

The electron is bound in a potential well that is given not by equation (9.2.2), *i.e.* purely Coulomb type alone but by the entire term within brackets in equation (9.4.8). This potential including the *centrifugal potential* is known in central force problems as the **Effective Potential**,

$$\boxed{V_{eff} = -\frac{Z e^2}{r} + \frac{\ell(\ell+1)\hbar^2}{2\mu r^2}} \qquad (9.4.9)$$
$$\text{Coulombic \& Centrifugal}$$

where the first term is the centrifugal potential. This varies as a function of both r and ℓ.(Fig. 9.2).

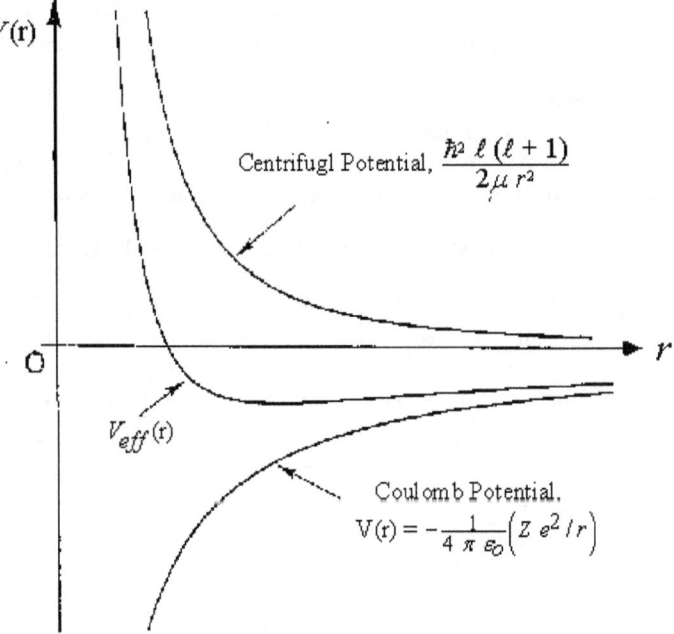

Fig. 9.2 The Effective Potential (9.4.9) for a Coulomb attractive potential and ℓ ≥ 1.

The Schrödinger Equation (9.4.8) can also be written as

$$\left(\frac{d^2}{dr^2} + \frac{2}{r}\frac{d}{dr}\right)\Re(\bar{r}) + \left(\frac{2\mu}{\hbar^2}[E - V(r)] + \frac{\ell(\ell+1)\hbar^2}{2\mu r^2}\right)\Re(\bar{r}) = 0 \qquad (9.4.10)$$

9.4.5 FURTHER SIMPLIFICATION of the Schrödinger Equation:

Equation (9.4.10) has to be simplified further by introducing a new unknown function,
$$\mathbb{R}(\vec{r}) = r\, \mathfrak{R}(\vec{r}) \tag{9.4.11}$$

Substituting, in (9.4.10) one gets

$$\mathbb{R}''(r) - (2\mu/\hbar^2)\left\{\frac{\ell(\ell+1)\hbar^2}{2\mu r^2} - \frac{Ze^2}{r}\right\}\mathbb{R}(r) = k^2\, \mathbb{R}(r). \tag{9.4.12}$$

$$\mathbb{R}''(r) - (2\mu/\hbar^2)[E - V_{eff}]\mathbb{R}(r) = 0 \tag{9.4.13}$$

since $\quad k^2 = -(2\mu/\hbar^2)\,E.$ \hfill (9.4.14)

This equation is the same form of the 1-D Schrödinger Equation, provided the equation (9.4.9) is regarded as $V(x)$. There arises two cases for the equation (9.4.10) with the total energy $E = T + V$, where $E > 0$, or $E < 0$.

$E > 0$, means $|E| > |V|$, since $E = T + V$, and one gets a FREE STATE for the electron in the H-atom.

When $E < 0$, $|E| < |V|$, since $E = T + V$, and one gets a BOUND STATE for the electron in the H-atom.

9.5 DISCRETE ENERGY SPECTRUM of Hydrogen; $E < 0$:

Solution to the Schrödinger Equation (9.4.12) is possible only by changing to dimensionless variables.

9.5.1. GOING TO A DIMENSIONLESS VARIABLE: NOTIONAL CHANGES:

To further simplify the equation, introduce the dimensionless variable
$$\sigma^2 = \left(\mu Z^2 e^4 / 2\,(-E)\,\hbar^2\right) \tag{9.5.1}$$

i.e.,
$$E = -(\mu Z^2 e^4 / 2\,\sigma^2\,\hbar^2) \tag{9.5.2}$$
$$\alpha^2 = (8\mu/\hbar^2)(-E) \tag{9.5.3}$$
$$\sigma = (2Z/\alpha)(1/a_o) \tag{9.5.4}$$
$$a_o = \hbar^2/\mu e^2 \tag{9.5.5}$$
$$\rho = \alpha r = (2Z/\sigma a_o)\, r = 2\,k\,r = (8\mu/\hbar^2)(-E)^{1/2} r \tag{9.5.6}$$
$$= \text{(a real parameter). } r.$$
$$r = \rho/\alpha = \rho(\sigma\hbar^2/2\mu Z e^2) = \rho(\sigma a_o/2Z) \tag{9.5.7}$$

with limits 0 and ∞, as that of r.

These notational changes have the effect of rendering the coefficients and independent variable of the equation (9.4.13) dimensionless. Thus one gets from (9.4.13)
$$\mathbb{R}''(r) - (2\mu/\hbar^2)[E - V_{eff}]\mathbb{R}(r) = 0$$

Further put $\mathbb{R}(\vec{r}) = \mathbb{R}(\rho) = \mathbb{S}(\rho)/\rho$ \hfill (9.5.8)

$$\mathbb{R}'(\rho) = \mathbb{S}(\rho)'/\rho - \mathbb{S}(\rho)'/\rho^2$$

$$\mathbb{R}''(\rho) = \mathbb{S}'''\rho/\rho - 2\,\mathbb{S}''(\rho)/\rho^2 + 2\,\mathbb{S}(\rho)/\rho^3$$

One then gets

$$\left[\frac{1}{\rho^2}\frac{d}{d\rho}\left(\rho^2 \frac{d}{d\rho}\right) + \left(\sigma/\rho - \frac{1}{4} - [\ell(\ell+1)/\rho^2]\right)\right]\mathbb{S}(\rho) = 0 \tag{9.5.9}$$

i.e., $\quad \dfrac{d^2}{d\rho^2}\mathbb{S}(\rho) + \dfrac{2}{\rho}\dfrac{d}{d\rho}\mathbb{S}(\rho) + \left(\sigma/\rho - \dfrac{1}{4} - [\ell(\ell+1)/\rho^2]\right)\mathbb{S}(\rho) = 0$ \hfill (9.5.10)

9.5.2 BEHAVIOUR OF $\mathbb{S}(\rho)$

9.5.2.1 Asymptotic Limits

The asymptotic solution to the equation (9.5.10) is to be examined.
In order to know something about the form of the solutions, examine first for
$\rho = \alpha r = 2 k r \to$ large. In this limit, the terms in $(1/\rho)$ and $(1/\rho^2)$ (in their dimensions)
will become negligible compared with $\frac{1}{4}$ or other terms and one finds

$$\frac{d^2}{d\rho^2}\mathbb{S}(\rho) + \left(-\frac{1}{4}\right)\mathbb{S}(\rho) = 0 \tag{9.5.11}$$

whose general solution is

$$\mathbb{S}(\rho) = A\, e^{+\rho/2} + B\, e^{-\rho/2}$$

with singular point as $\rho \to \infty$.
The only physically admissible finite solution of equation (9.5.11) is

$$\mathbb{S}(\rho) \cong e^{-\rho/2} \tag{9.5.12}$$

but only for $\rho \to \infty$.
One can write the solution of equation (9.5.10) as

$$\mathbb{S}(\rho) \cong \left(e^{-\rho/2}\right)\mathbb{F}(\rho) \tag{9.5.13}$$

where $\mathbb{F}(\rho)$ is that part of the solution that vanishes more rapidly than $\left(e^{-\rho/2}\right)$.

9.5.2.2 Limits at Which $\rho \to 0$

Consider next the second important limit, viz. $\rho \to 0$
In this limit the equation (9.5.10) becomes

$$\frac{d^2}{d\rho^2}\mathbb{S}(\rho)\bigl(-[\ell(\ell+1)/\rho^2]\bigr)\rho\mathbb{S}(\rho) = 0 \tag{9.5.14}$$

as $\rho \to 0$.
The solution of this equation, that is singular at $\rho \to 0$, has the limiting behaviour

$$\mathbb{S}(\rho) \sim \rho^\ell \tag{9.5.15}$$

9.5.2.3 Form of the Radial Function $\mathbb{S}(\rho)$

The results (9.5.13) and (9.5.15) indicate how all the radial functions, $\mathbb{S}(\rho)$ must behave
for very large and very small values of ρ; the Radial Functions must be of the form

$$\mathbb{S}(\rho) \cong \left(e^{-\rho/2}\right)\rho^\ell\, \mathbb{F}(\rho) \tag{9.5.16}$$

where $\mathbb{F}(\rho)$ is as yet unknown function of ρ that in the limits $\rho \to 0$ and $\rho \to \infty$ behaves
in such a way that $\mathbb{S}(\rho)$ is finite.

9.5.3 An Equation of $\mathbb{F}(\rho)$

By substituting equation (9.5.16) in equation (9.5.10), with a little algebra, leads to the
following equation for $\mathbb{F}(\rho)$

$$\left\{\rho\frac{d^2}{d\rho^2} + [2(\ell+1) - \rho]\frac{d}{d\rho} + [\sigma - (\ell+1)]\right\}\mathbb{S}'(\rho) = 0 \tag{9.5.17}$$

i.e., $\quad \rho\,\mathbb{F}''(\rho) + (j+1-\rho)\mathbb{F}'(\rho) + (q-j)\mathbb{F}(\rho) = 0 \tag{9.5.18}$

9.5.3.1 Power Series Solution of $F(\varrho)$:

Fortunately this results in a standard form

$$\mathbb{F}(\rho) = \sum_{j=0}^{\infty} a_j \, \rho^j, \quad a_o \neq 0 \tag{9.5.19}$$

The polynomials obtained in the solution of the radial equation of the H-atom can be seen as follows:

$$\boxed{\rho \, \mathbb{F}''(\rho) + (j+1-\rho)\mathbb{F}'(\rho) + (q-j)\mathbb{F}(\rho) = 0} \tag{9.5.18}$$

$$\boxed{j = 2(\ell+1), \quad q = \sigma + \ell} \tag{9.5.20}$$

Equation (9.5.18) is not having the self-adjoint form, *i.e.* not in the *Sturnmor- Liouville form*; but can be made so by multiplying it by $\left(e^{-\rho}\right) \rho^j$.

9.5.3.2 Associated LAGUERRE Differential Equation

Equation (9.5.18) is called the *Associated Laguerre Differential Equation*, whose solutions are the <u>Associated Laguerre Polynomials</u>, of degree *(q - j)* and order *j*, viz. $L_q^{\,j}(\rho)$, which is the jth derivative of the qth**Laguerre Polynomial**. They are defined only for integral values of q and j, satisfying the inequality,

$$j \leq q, \text{ an integer.} \tag{9.5.21}$$

$$\mathbb{F}(\rho) \equiv L_q^{\,j}(\rho) \tag{9.5.22}$$

9.5.4 RECURSION RELATION:

The Radial Function given by equation (9.5.16) can be written as

$$\mathbb{S}(\rho) \cong \left(e^{-\rho/2}\right) \rho^\ell \sum_{j=0}^{\infty} a_j \, \rho^j$$

$$= \left(e^{-\rho/2}\right) \rho^\ell L_q^{\,j}(\rho), \tag{9.5.23}$$

Let $\quad \mathbb{F}(\rho) \equiv L_q^{\,j}(\rho) = \sum_{j=0}^{\infty} a_j \, \rho^j \tag{9.5.19}$

$$\mathbb{F}'(\rho) = \sum j \, a_j \, \rho^{j-1} = \sum (j+1) \, a_{j+1} \, \rho^j$$

i.e., $\quad j\mathbb{F}'(\rho) = \sum j \, a_j \, \rho^j$

$$\mathbb{F}''(\rho) = \sum j(j-1) \, a_j \, \rho^{j-2} = \sum (j+1)j \, a_{j+1} \, \rho^{j-1},$$

$$\rho \mathbb{F}''(\rho) = \sum (j+1)j \, a_{j+1} \, \rho^j,$$

Substituting these terms in equation (9.5.17)
i.e.,

$$\sum (j+1)j \, a_{j+1} \, \rho^j + (j+1)\sum (j+1)j \, a_{j+1} \, \rho^j - \sum j \, a_j \, \rho^j + [\sigma - \ell(\ell+1)] \sum j \, a_j \, \rho^j = 0$$

i.e., $\quad \sum \left\{ a_{j+1} \, [j(j+1) + (j+1)2(\ell+1)] - a_j \, [j - \sigma - (\ell+1)] \right\} \rho^j = 0$

But $\rho^j \neq 0 \, [j(j+1) + (j+1)2(\ell+1)] \, a_{j+1} = [j - \sigma - (\ell+1)] \, a_j$.

i.e,. $\quad \boxed{\dfrac{a_{j+1}}{a_j} = \dfrac{[(j + \ell + 1 - \sigma)]}{[(j + \ell + 1)(j + \ell) + \ell(j+1)]}} \tag{9.5.24}$

This is known as the RECURSION FORMULA.

9.5.5 PRINCIPAL (Total) QUANTUM NUMBER, n of the H-atom:

If (9.5.19) does not terminate, one can see from the Recursion Formula that for *j* = large,

$$\dfrac{a_{j+1}}{a_j} \Rightarrow \dfrac{1}{j}$$

For acceptable solutions of equation (9.5.24) terminates only if the numerator of it is zero,

$[(j + \ell + 1 - \sigma)] = 0$, for $\sigma \geq n$.

i.e., $\boxed{\sigma \equiv n = (j + \ell + 1) = \text{integer}, 1, 2, 3, ...}$ (9.5.25)

as $j = 0, 1, 2, 3, ...;$

$\ell = 0, 1, 2, 3, ...$

n is called the PRINCIPAL QUANTUM NUMBER of the system.

9.5.6 DISCRETE ENERGY values of the H-atom:

From equation (9.5.2)

$$E_\sigma = -\mu Z^2 e^4 / 2\sigma^2 \hbar^2 \qquad (9.5.2)$$

Replacing the σ by n from (9.5.25) in (9.5.2)

$$E_n = \frac{-\mu Z^2 e^4}{2 n^2 \hbar^2} \qquad (9.5.26)$$

$$\boxed{E_n = -\frac{2\pi^2 \mu Z^2 e^4}{n^2 h^2}}$$

The energy levels of the H-atom are as shown in Fig 9.3.

9.5.6.1 RESTRICTIONS on n and ℓ, and radial quantum number, n':

Thus it is evident that

(a) The restriction on $\sigma \equiv n$, imposed by the boundary condition; has the effect of quantizing E_σ.

E_n is quantized as expected, and is independent of ℓ.

In Galilean Relativistic Theory, E_n does not explicitly depend on ℓ.

(b) Second condition is $j \leq q$, an integer (9.5.21).

$$q - j = n - \ell - 1 = n'$$

where n' is called the *radial quantum number*.

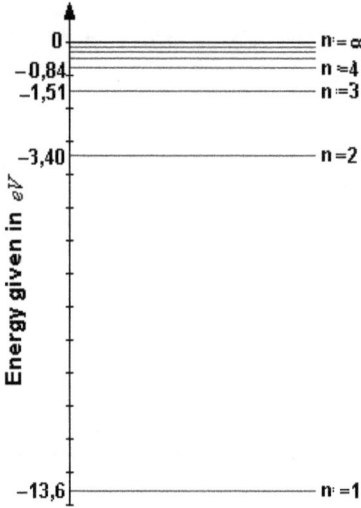

Fig. 9.3 Discrete Energy levels of electron on successive orbits in an H-atom

i.e., $(n+\ell)-(2\ell+1)=n-\ell-1=n'=0,1,2,3,....$
(9.5.27)
i.e., $q=(\sigma+\ell)=n+\ell$, i.e. $j=(2\ell+1)=$ always odd;
i.e., $(2\ell+1)\leq n+\ell$;
or, $n\geq \ell+1$
$\ell=0,1,2,...$ (9.5.28)
This is equivalent to the restriction,
$$\boxed{\ell \leq (n-1)}$$ (9.5.29)

9.6. PROPERTIES OF THE RADIAL FUNCTIONS:

The Radial wave Function is, from equations (9.4.11), (9.5.8) and (9.5.23):
$$\mathbb{R}(\bar{r}) = r\,\mathfrak{R}r)$$ (9.4.11)
$$\mathbb{R}(\bar{r}) = \mathfrak{R}(\rho) = \mathbb{S}(\rho)/\rho.$$ (9.5.8)
$$\mathbb{S}(\rho) \cong \left(e^{-\rho/2}\right)\rho^{\ell}\,\mathbb{F}(\rho)$$ (9.5.23)
$$\boxed{\mathfrak{R}(r) \equiv \mathfrak{R}_{n\ell}(r) = C_n\left(e^{-\rho/2}\right)\rho^{\ell}\,L^{2\ell+1}_{n+\ell}(\rho)}$$ (9.6.1)

with radial quantum number, n' such that
$$(n+\ell)-(2\ell+1)=n+\ell-1=n'=0,1,2,3,....$$ (9.5.27)
Putting $\sigma = n$, in the expression for ρ in equation (9.5.6)
$$\boxed{\rho = \alpha\,r = (2Z/n\,a_o)\,r}\quad .$$ (9.6.2)

The ASSOCIATED LAGUERRE Polynomials $L^{2\ell+1}_{n+\ell}(\rho)$ that form part of $\mathfrak{R}_{n\ell}(r)$ are rather simple polynomials for small n and ℓ, and they have
$$\boxed{n+\ell-1=n'\text{ radial modes (zeros)}...}.$$
These are listed in Table 9.1.

9.6.1 $L^j_q(\rho)$ related to LAGUERRE POLYNOMIALS:

$L^j_q(\rho)$ are related to simpler q^{th} Laguerre Polynomials, $L_q(\rho)$, by
$$\boxed{L^j_q(\rho) = \frac{d^j}{d\rho^j}\,L_q(\rho)}$$ (9.6.3)

9.6.2.1 Rodrigue's Generating Formula
$L_q(\rho)$ are generated *via*
$$\boxed{L_q(\rho) = \left(e^{+\rho}\right)\frac{d^q}{d\rho^q}\left(\rho^q e^{-\rho}\right)}$$ (9.6.4)

9.6.2.2 Rodrigues's Formula
$$L^j_q(\rho) = \left[\frac{(\rho^{-j})(e^{+\rho})}{q^j}\right]\frac{d^q}{d\rho^q}(\rho^{q+j})(e^{-\rho})$$ (9.6.5)

TABLE 9.1 The ASSOCIATED LAGUERRE Polynomials, $L_{n+\ell}^{2\ell+1}(\rho)$

$q=(n+\ell)$	$j=(2\ell+1)$	$L_{n+\ell}^{2\ell+1}(\rho)$
0	0	$L_0^0(\rho) = 1$ (not useful as q = 0)
1	0	$L_1^0(\rho) = (1-\rho)$
1	1	$L_1^1(\rho) = -1$
2	0	$L_2^0(\rho) = (2-4\rho+\rho^2)$
	1	$L_2^1(\rho) = (2\rho-4)$
	2	$L_2^2(\rho) = 2$
3	0	$L_3^0(\rho) = (6-18\rho+9\rho^2-\rho^3)$
	1	$L_3^1(\rho) = (-18+18\rho-3\rho^2)$
	2	$L_3^2(\rho) = (18-16\rho)$
	3	$L_3^3(\rho) = (-16)$
4	0	$L_4^0(\rho) = (24-96\rho+72\rho^2-16\rho^3+\rho^4)$

9.6.2.3 Recursion Formula

Since $L_q(\rho)$ and $L_q^j(\rho)$ are real functions of ρ,

$\mathfrak{S}(\rho) = \rho \, \mathfrak{R}(\rho) = $ real.

Since $L_q(\rho) = L_q^0(\rho)$

i.e., $j = 0$.

$L_q(\rho)$ satisfy $\boxed{L_{q+1}(\rho) = (2q+1-\rho) - q^2 L_{q-1}(\rho)}$ (9.6.6)

This is known as the **Recursion Formula**.

9.6.2.4 Differential Equation

Differential equation for the qth Laguerre Polynomial, $L_q(\rho)$ is

$\boxed{\rho L_q''(\rho) + (1-\rho)L_q'(\rho) + qL_q(\rho) = 0}$ (9.6.7)

9.6.2.5 NORMALIZATION of $\mathfrak{R}_{n\ell}(r)$

It is known from equation (9.4.1)

$\boxed{\psi(\vec{r}) = \psi(\vec{r},\theta,\varphi) = \mathfrak{R}_{n\ell}(r) \cdot Y_\ell^m(\theta,\varphi)}$

Normalization means

$\iiint_{0 \text{ to } \infty} |\hat{\psi}(\vec{r})|^2 \, d\tau = 1$

$\iiint_{0 \text{ to } \infty} |\hat{\psi}(\vec{r},\theta,\varphi)|^2 \, d\tau = 1$

Now from (9.3.7)

$d\tau = (r^2 dr)(\sin\theta \, d\theta) \, d\varphi$.   (8.2.7)

$\int_0^\infty r^2 dr \int_0^\pi \sin\theta \, d\theta \int_0^{2\pi} d\varphi |\hat{\psi}(\vec{r},\theta,\varphi)|^2 = 1$

Since, $\int_0^\infty |\hat{\mathfrak{R}}_{n\ell}(r)|^2 \, r^2 dr = 1$, and $\int_0^\infty |\mathfrak{S}(\rho)|^2 \, \rho^2 d\rho = 1$

$$\int_0^\infty |(e^{-\rho})| \rho^{2\ell} |L_{n+\ell}^{2\ell+1}(\rho)|^2 \rho^2 d\rho = \left\{ \frac{2n[(n+l)!]^3}{(n+l-1)!} \right\}$$

$$\int_0^\infty |\Re_{n\ell}(r)| r^2 dr = \left(\frac{na_0}{2Z}\right)^3 \int_0^\infty |S(\rho)|^2 \rho^2 d\rho$$

$$= \left(\frac{na_0}{2Z}\right)^3 \left\{ \frac{2n[(n+l)!]^3}{(n+l-1)!} \right\} \tag{9.6.8}$$

This means the normalization constant that normalizes $\Re_{n\ell}(r)$ is

$$C_n = \pm \sqrt{\left(\frac{2Z}{na_0}\right)^3 \left\{ \frac{[(n+l-1)!]}{[2n(n+l)!]^3} \right\}} \tag{9.6.9}$$

9.6.3.1 Why is the Negative Square Root Value?

The <u>Negative Square Root value</u> is taken for C_n by convention! This is so since

$(-1) = e^{i\pi}$ is the *Phase Factor*, and all $\psi(\bar{r})$-s are indeterminate to within a phase. Thus one may write the normalized Radial eigen Function as

$$\boxed{\Re(r) \equiv \Re_{n\ell}(r) = C_n \, e^{-\rho/2} \, \rho^\ell \, L_{n+\ell}^{2\ell+1}(\rho)}$$

$$= -\sqrt{\left(\frac{2Z}{na_0}\right)^3 \left\{\frac{[(n+l-1)!]}{[2n(n+l)!]^3}\right\}} \left(e^{-\rho/2}\right) \rho^\ell L_{n+\ell}^{2\ell+1}(\rho) \tag{9.6.10}$$

$$\boxed{\rho = \alpha \, r = (2Z/na_0) \, r}$$

$$\Re_{n\ell}(r) = C_n (\alpha r)^\ell \cdot L_{n+\ell}^{2\ell+1}(\alpha r) \cdot e^{-Zr/na_0} \tag{9.6.11 a}$$

$$\boxed{\hat{\Re}_{n\ell}(r) = -\sqrt{\left\{\frac{[(n+l-1)!]}{[2n(n+l)!]^3}\right\}} \cdot \left(\frac{2Z}{na_0}\right)^{\ell+3/2} \cdot L_{n+\ell}^{2\ell+1}\left(\frac{2Z}{na_0}r\right) \cdot r^\ell \left(e^{-Zr/na_0}\right)}$$

$$\tag{9.6.11 b}$$

9.6.4 EXPLICIT EXPRESSIONS for $\Re_{n\ell}(r)$ for $n = 1, 2,$ and 3

TABLE 9.2   Expressions for Radial Functions $\hat{\Re}_{n\ell}(r)$

n	ℓ	$\hat{\Re}_{n\ell}(r)$
1	0	$\hat{\Re}_{10}(r) = \{2\} (Z/a_0)^{3/2} e^{-Zr/a_0}$
2	0	$\hat{\Re}_{20}(r) = (Z/2a_0)^{3/2} [2-(Zr/a_0)] e^{-Zr/2a_0}$
	1	$\hat{\Re}_{21}(r) = \{\sqrt{1/3}\} (Z/2a_0)^{3/2} (Zr/a_0) e^{-Zr/2a_0}$
3	0	$\hat{\Re}_{30}(r) = \{2\} (Z/3a_0)^{3/2} [1-\frac{2}{3}(Zr/a_0) + \frac{2}{27}(Zr/a_0)^2] e^{-Zr/3a_0}$
	1	$\hat{\Re}_{31}(r) = \{4\sqrt{2/3}\} (Z/3a_0)^{3/2} [(Zr/a_0) - \frac{1}{6}(Zr/a_0)^2] e^{-Zr/3a_0}$
	2	$\hat{\Re}_{32}(r) = \{\frac{2}{27}\sqrt{2/5}\} (Z/3a_0)^{3/2} [(Zr/a_0)^2] e^{-Zr/3a_0}$

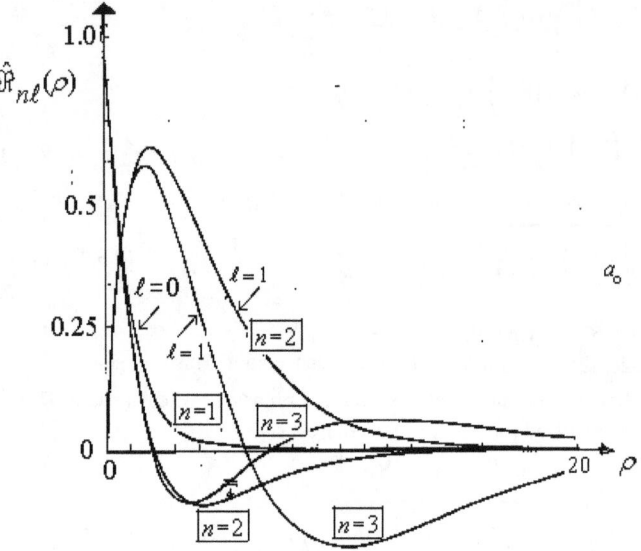

Fig. 9.4 Radial eigen Functions $\hat{\Re}_{n\ell}(\rho)$ for values of n = 1, 2 and 3.

Explicit expressions for $\hat{\Re}_{n\ell}(r)$ for n = 1, 2, and 3 are listed in Table 9.2 and illustrated in Fig. 9.4.

9.6.5 PROBABILITY DISTRIBUTIONS OF $\hat{\Re}_{n\ell}(r)$ NODES AND ZEROS

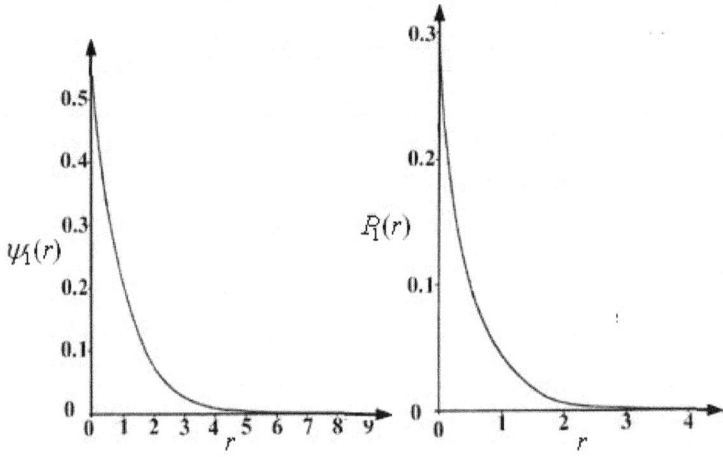

Fig. 9.5 The wave function $\psi_1(r)$ and probability $P_1(r) = |\psi_1(r)|^2$ versus r distribution for n = 1.

The probability distributions of $\hat{\mathfrak{R}}_{n\ell}(r)$, the Radial Functions, exhibit NODES, that is they vanish for certain values of r; in particular the function $\hat{\mathfrak{R}}_{n\ell}(r)$ has

$\boxed{n_r = (n-\ell-1)}$ RADIAL NODES,

and $\boxed{n_r = (n-\ell)}$ BUMPS (anti-nodes or zeros).

These are very clear from the illustrations in Fig. 9.5. and Fig 9.6

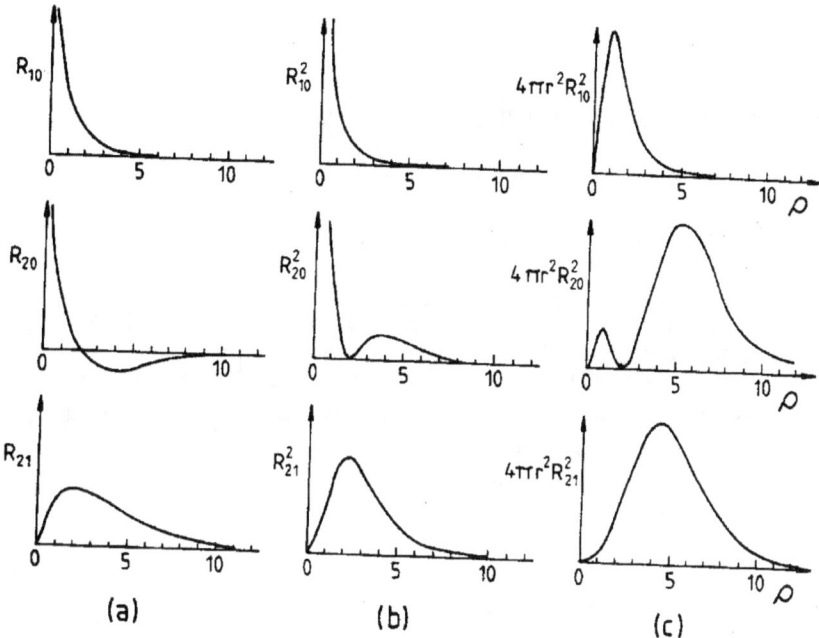

Fig. 9.6 (a) $\hat{\mathfrak{R}}_{n\ell}(r)$ (b) $|\hat{\mathfrak{R}}_{n\ell}(r)|^2$ and (c) Radial Charge density distribution $\{|\hat{\mathfrak{R}}_{n\ell}(r)|^2 \, 4\pi r^2\}$.

9.7. TOTAL WAVE FUNCTIONS OF THE H-ATOM

9.7.1 Angular Solutions (SPHERICAL HARMONICS), $Y_\ell^m(\theta,\varphi)$

The *angular equation* (9.4.3) is
$\hat{\Lambda}[Y(\theta,\varphi)]/Y(\theta,\varphi) = -\lambda$
whose solutions are the *Spherical Harmonics*, (8.7.2), viz.,

$$\boxed{\hat{Y}_\ell^m(\theta,\varphi) = (-1)^m \left[\frac{\ell+\frac{1}{2}}{4\pi} \frac{(\ell-m)!}{(\ell+m)!}\right]^{1/2} P_\ell^m(\cos\theta) \, e^{i|m|\varphi}} \qquad (8.7.2)$$

have been dealt with in detail in Chapter 8. These are the *normalized simultaneous* eigen functions of $\hat{\Lambda}$, viz., $\hat{L}^2$ and $\hat{L}_z$ with eigen values $\ell(\ell+1)\hbar^2$ and $m\hbar$, respectively.

9.7.2 The TOTAL WAVE FUNCTION of the H-atom

$$\psi(\vec{r}) = \psi(\vec{r}, \theta, \varphi) = \Re(r) \cdot Y(\theta, \varphi) \tag{9.4.1}$$

$$\psi(r) = \psi_{n,\ell,m}(r, \theta, \varphi) = \Re_{n,\ell}(r) \cdot Y_\ell^m(\theta, \varphi) \tag{9.7.1}$$

$$= \left\{ -\sqrt{\frac{[(n+l-1)!]}{[2n(n+l)!]^3}} \cdot \left(\frac{2Z}{n a_o}\right)^{\ell+3/2} \cdot L_{n+\ell}^{2\ell+1}\left(\frac{2Z}{n a_o} r\right) \cdot r^\ell \left(e^{-Zr/n a_o} \right) \right\}$$

$$\times \left[(-1)^m \left[\frac{\frac{\ell+1}{2} (\ell-m)!}{4\pi (\ell+m)!} \right]^{1/2} P_\ell^m(\cos\theta) \, e^{\,i\,|m|\,\varphi} \right] \tag{9.7.2}$$

9.8 INTERPRETATION OF THE QUANTUM NUMBERS

The total wave function (9.4.1) of the H-atom
$$\psi(\vec{r}) = \psi(\vec{r}, \theta, \varphi) = \Re(r) \cdot Y(\theta, \varphi)$$
Becomes $\psi(r) = \psi_{n,\ell,m}(r, \theta, \varphi) = \Re_{n,\ell}(r) \cdot Y_\ell^m(\theta, \varphi)$

9.8.1 One Quantum number per Degree of Freedom

It is seen from the earlier studies on 1-D simple quantum systems that <u>one quantum number</u> is required to describe the motion of the particle in 1-D problems. The H-atom is a 3-D problem, and it is seen that one gets 3 quantum numbers (n, ℓ, m) to describe the 3-D motion of the electron. The Hamiltonian, $\hat{H}$ commutes with the operators $\hat{L}^2$ and $\hat{L}_z$ so that observables with $\hat{H}$, $\hat{L}^2$ and $\hat{L}_z$ are separately *constants of motion*, and their expectation values are conserved, because

$$\frac{d}{dt}\langle\hat{\Omega}\rangle = 0$$

and if $\ [\hat{H}, \hat{\Omega}] = 0 \tag{9.8.1}$

9.8.2 PRINCIPAL QUANTUM NUMBER, n

It is interesting to consider the interpretation of the H-atom quantum numbers. The energy levels of hydrogen are completely specified by a single quantum number, n, and the *principal quantum number*.
$n = 1, 2, 3, \ldots$

9.8.2.1 Total Energy, E

The total energy E is positive, or has one of the negative values E_n, specified by (9.5.26)

$$E_n = -\frac{2\pi^2 \mu Z^2 e^4}{n^2 \hbar^2} \tag{9.8.2}$$

shows that this is precisely the same formula for the energy levels of the H-atom that Bohr obtained! Thus only negative values, the electron can have, indicating thee electron is bound to the nucleus, and E_n, is conserved. *i.e.* E_n = constant or discrete; n indicating quantization of the electron energy in the H-atom.

9.8.2.2 Shell Radius, $<r>_{n,\ell,m} \equiv <n>$

There is **one shell** for each value of the principal quantum number, with the

$\boxed{\textit{Shell Radius } <r>_{n,0,0} \equiv <n>}$. (9.8.3)

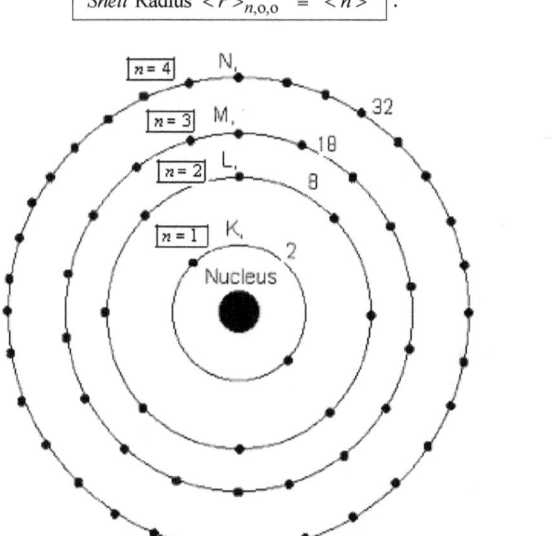

Fig 9.7 Shell symbol for Bohr Atom

defined by the value of n. It is conventional that the electron is said to occupy a specific GROUP, ENERGY LEVEL or ATOMIC SHELL. These shells are given **symbols** (Roman capital letters) as follows:

Principal Quantum number, $n =$	1	2	3	4	5.....
Shell Symbol	K	L	M	N	O.....

(9.8.4)

9.8.3 ORBITAL (Azimuthal) QUANTUM NUMBER, ℓ

9.8.3.1 Orbital Angular Momentum, $\hat{L}$

For a given energy, specified by n, the magnitude of the electron's orbital angular momentum, is given by the RIGHT HAND RULE (RH rule) in electricity and

$$\hat{L} = \sqrt{\ell(\ell+1)}\ \hbar \qquad (9.8.5)$$

where $\ell = 0, 1, 2, 3,.... (n-1)$; and $n > \ell$. The significance of ℓ is that it fixes the *magnitude* of the moment of momentum, which is a definite value with certainty. Like the total energy, E_n, electron's angular momentum is both CONSERVED (*i.e.* maintains constant value at all times- constant of motion) and QUANTIZED.

$$\tfrac{d}{dt}<\hat{L}> = 0$$

and if $[\hat{H}, \hat{L}] = 0$. (9.8.6)

It is customary to specify the electron's angular momentum states by a letter, according to the following scheme:

Orbital Quantum number, $\ell =$	0	1	2	3	.....
Sub-shell Designtion	s	p	d	f	.....

9.8.3.2 Radius of Sub-group, Sub-Level OR Sub-shell, $<r>_{n,\ell,o}$

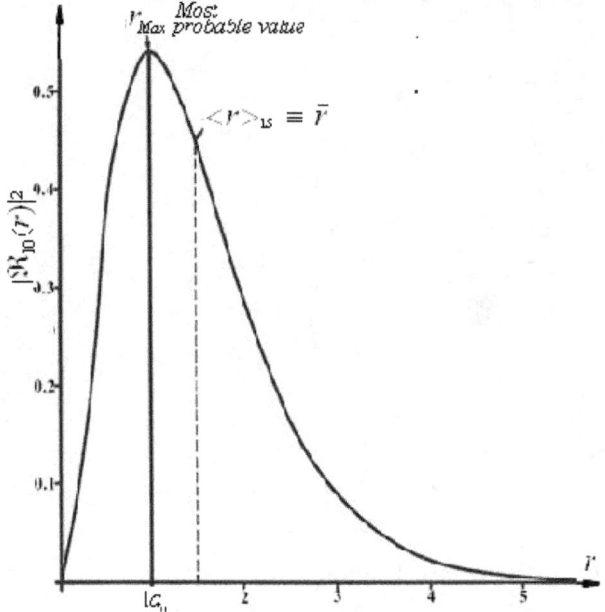

Fig. 9.8 Probability $|\Re_{10}(r)|^2$ distribution curve for $n = 1$.

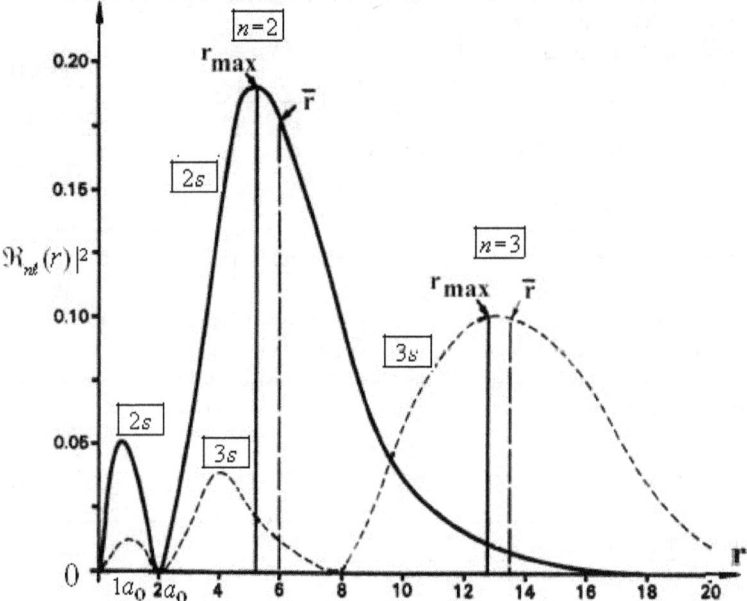

Fig. 9.9 Probability $|\Re_{no}(r)|^2$ distribution curves for $n = 2$ and $n = 3$.

Note: Notations for various radii:
Electrons that share a certain value of ℓ in a shell are said to occupy the same SUBGROUP, SUBLEVEL*l* or SUBSHELL. Thus there is one sub shell defined for each value of ℓ allowed by $\ell = 0, 1, 2, ..., (n-1)$.

$$\boxed{\text{Sub-}\textit{Shell}\text{ Radius } <r>_{n,\ell,o}} \qquad (9.8.7)$$

SHELL RADIUS is denoted by $<r>_{n,o,o}$

SUBSHELL RADIUS is denoted by $<r>_{n,\ell,o}$

In general, the radial distribution, $|\Re_{n\ell}(r)|^2$ does not have peak at $<r>_{n,\ell,o}$
Fig 9.6, Fig. 9.8 and Fig. 9.9 clearly show these.

9.8.3.2 Spectral Notation

It is customary to specify angular momentum states of electrons in an atom by a letter, with s corresponding to $\ell = 0$, p to $\ell = 1$, *etc.* according to scheme given below:

Orbital angular momentum : $\ell = 0\ 1\ 2\ 3\ 4\ 5\ 6.....,$

Angular momentum Electron State: s p d f g h i ....   (9.8.8)

Series   S P D F G H I.......

Here the capitals standing for **Sharp, Principal, Diffuse, Fundamental,** and the alphabetical sequence for $\ell > 3$.

In this notation an atomic state is conveniently denoted by the combination of n and ℓ, for example, $n = 2$, $\ell = 0$, is a 2s state, one in which $n = 4$, $\ell = 2$, is a 4d state.

9.8.3.3 Electron Configuration

The occupancy of the various sub-shells in a multi-electron atom is usually expressed with help of the notation as follows: Each sub-shell is identified by its principal quantum number n followed by the letter corresponding to its orbital quantum number, ℓ. A superscript after the letter indicates the number of electrons in that sub-shell ℓ. It is given by

$$\boxed{\text{\# of electrons in sub-shell, } \ell = 2\ell(\ell+1)} \qquad (9.8.9)$$

For example, the <u>electronic configuration</u> of sodium atom ($\ell = 11$), in the normal state, is written as

$$\boxed{\text{Na} \quad 1s^2\ 2s^2\ 2p^6\ 3s^1} \qquad (9.8.10)\text{a}$$

which means that there are two 1s ($n = 1$, $\ell = 0$) electrons, two 2s electrons ($n = 2$, $\ell = 0$), six 2p electrons, and one 3s electron.

The Electron Configuration of oxygen in the ground state is

$$\boxed{\text{O} \quad 1s^2\ 2s^2\ 2p^4} \qquad (9.8.10)\text{b}$$

9.8.4 MAGNETIC QUANTUM NUMBER, m_ℓ

9.8.4.1. SPACE (directional) QUANTIZATION:

For a given energy, specified by n, the magnitude of the electron's orbital angular momentum, is given by the vector,

$$\hat{L} = \sqrt{\ell(\ell+1)}\, \hbar \qquad (9.8.5)$$

ℓ determines the magnitude of the electron's angular momentum, $|\vec{L}|$. To describe it completely requires that the direction of $\vec{L}$ be specified as well as its magnitude. $\vec{L}$ is normal to the plane in which the rotational motion takes place. Its sense is given by the RH rule. An electron revolving around the nucleus is a minute current loop equivalent to a magnetic dipole of dipole moment, $\vec{\mu}_\ell$. This interacts with any magnetic field, $\vec{B}$ and causes the magnetic potential energy,

$$V_B = -\vec{\mu}_\ell \cdot \vec{B} = -\vec{\mu}_\ell\, \vec{B}\, \cos\theta = \mu_B\, \sqrt{\ell(\ell+1)}\, \vec{B}\, \cos\theta \qquad (9.8.11)$$

μ_B, being the Bohr Magneton. This means the direction of $\vec{L}$ is quantized with respect to the external field. This is referred to as Spatial Quantization. Since the magnetic axis of the atom could assume only positions described by the inclination θ, with the z-axis determined by the formula,

$$\cos\theta = m_\ell / \ell. \qquad (9.8.12)$$

Further it is known that $m_\ell = 0, \pm 1, \pm 2, \pm 3, ... \pm \ell$.

This means m_ℓ can have $(2\ell+1)$ values.

The Zeeman Effect is a vivid confirmation to *Spatial Quantization*.

A sub shell is filled or completed when the sum of the vectors, $\sum_{\text{o to }\ell} m_\ell = 0$

Also the sum of the vectors, $\sum_{\text{o to }\ell} m_\ell = 0$, for COMPLETED SUBGROUP. Hence, in order TO DETERMINE *the angular momentum of an atom, only those electrons, which are EXTERNAL TO THE CLOSED SHELLS, need be considered.*

9.8.4.2 MULTIPLICITY (Degeneracy):

Solving the problem of the H-atom by means of Quantum Mechanics has meant that one has to introduce three quantum numbers, *viz.* n, ℓ, m_ℓ, instead of just only n as in the Bohr Theory. Degeneracy is characteristic of the Coulomb potential (central potential), for a given n value there are ℓ values and for an ℓ value m_ℓ can have $(2\ell+1)$ values.

$$\boxed{\text{Degeneracy of Atomic level } n = \sum_{\ell=0}^{n-1}(2\ell+1) = n^2} \qquad (9.8.13)$$

Worked out Example 9.1

Confirm that the degeneracy of an atomic energy level of given principal quantum number, n is equal to n^2.

Solution:

Step # 1 For a given energy level, n, the orbital quantum number ℓ can have values from 0 to $(n-1)$. Also for a given ℓ value m_ℓ can have $(2\ell+1)$ values, viz. $0, \pm 1, \pm 2, \pm 3, ... \pm \ell$.

Step # 2 # of possible solutions for a given atomic *level* $n = \sum_{\ell=0}^{n-1}(2\ell+1)$.

$$= 1 + 3 + 5 + ... + (2n-1) = [2n(n-1)/2] + n = n^2$$

Since n determines the energy of the atom, a given (specified) energy state, n has n^2 eigen functions and n^2 corresponding eigen value solutions, but all having the same energy. The system is then said to be $\underline{n^2\text{-fold degenerate}}$, and the *Degree of Degeneracy* of the

n^{th} level known as <u>multiplicity</u> is n^2. (Note: if electron spin is also taken into account the degeneracy is n^2 multiplied by a factor of two).
The number of distinct degenerate states, for a given value of n, when all the four quantum numbers (viz. n, ℓ & m_ℓ, and s) are taken into account, is $2n^2$ (see Fig. 9.10).

| # of Distinct Degenerate States for a Shell, $n = 2n^2$ |

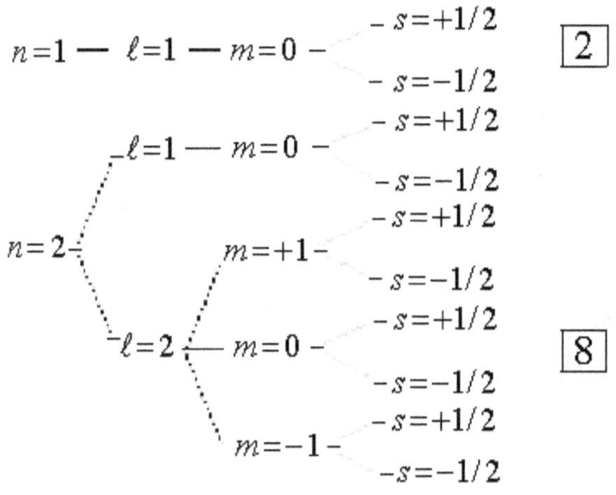

```
n=1 — ℓ=1 — m=0 — s=+1/2        [2]
                  — s=−1/2
                               − s=+1/2
              ℓ=1 — m=0 —
                               − s=−1/2
                               − s=+1/2
n=2 ·        m=+1 —
                               − s=−1/2
                               − s=+1/2
              ℓ=2 — m=0 —                [8]
                               − s=−1/2
                               − s=+1/2
              m=−1 —
                               − s=−1/2
```

Total number of states $2n^2$ per Shell

Fig. 9.10 Total number of stats per Shells $n = 1$ and $n = 2$.

9.8.4.3 Atomic Configuration

Capital letters are used to represent the **Total Angular Momentum of the Atom**, according to the following scheme:

Total Angular Momentun of ATOM	L = 0 1 2 3 4 5 .....
Spectral Notation	S P D F G H .....

The value of the total angular momentum of an atom, J, is written as a subscript at the lower right of the letter representing the particular L value of the atomic state. The number of possible values of J for a given value of L is written as a superscript at the upper left of the letter representing the L value. Thus

$^2P_{1/2}$, $^2P_{3/2}$., read "<u>doublet P one half</u>", etc.

or, $^3P_2$, $^3P_1$, $^3P_0$., read "<u>triplet P two</u>", and so on.

The Superscipt is an indication of the <u>multiplicity</u> of the terms of the **Atomic Configuration**.

Worked out Example 9.2

Evaluate the value of Bohr radius, a_o and the energy, E_o of the ground state of H-atom using the expression for the radial function.

Solution:

| Step # 1 | Given: $Z = 1$; $n = 1$., $\Re_{10} = C_1 e^{-Zr/a_o}$ $\Re_{10} = 2(Z/a_o)^{3/2} e^{-Zr/a_o}$ |

Step #2 The radial wave equation of the H-atom (9.4.10) is

$$\left(\frac{d^2}{dr^2}+\frac{2}{r}\frac{d}{dr}\right)\Re(\bar{r}) + \left(\frac{2\mu}{\hbar^2}[E-V(r)]+\frac{\ell(\ell+1)\hbar^2}{2\mu r^2}\right)\Re(\bar{r}) = 0$$

Step #3 It is required to evaluate a_o, which determines how fast the wave function goes to zero.

$$\frac{d}{dr}\Re_{1o} = \frac{d}{dr}[C_1 \, e^{-Z \, r/a_o}] = C_1(-1/a_o) \, e^{-Z \, r/a_o}$$

$$\frac{d^2}{dr^2} = C_1(1/a_o)^2 \, e^{-Z \, r/a_o} = C_1(1/a_o)^2 \Re_{1o}$$

$$\left(\frac{d^2}{dr^2}+\frac{2}{r}\frac{d}{dr}\right)\Re_{1o} + \left(\frac{2\mu}{\hbar^2}[E-V(r)]+\frac{\ell(\ell+1)\hbar^2}{2\mu r^2}\right)\Re_{1o} = 0$$

$$(1/a_o)^2 \Re_{1o} + (2/r)(-1/a_o)\Re_{1o} + \frac{2\mu}{\hbar^2}(E_1+e^2/r)\Re_{1o} = 0$$

i.e., $\left[(1/a_o)^2 - (2/ra_o) + \frac{2\mu}{\hbar^2}(E_1+e^2/r)\right]\Re_{1o} = 0$

Step #4 Since $\Re_{1o} \neq 0$, $\left[(1/a_o)^2 + \frac{2\mu}{\hbar^2}E_1 + [(\frac{\mu}{\hbar^2}e^2)-(1/a_o)\frac{2}{r}]\right] = 0$

Step #5 Because $r \neq 0$, for this equation to hold for all values of r, i.e. $0 < r < \infty$, both the terms must be zero, separately. $(1/a_o)^2 + \frac{2\mu}{\hbar^2}E_1 = 0$ and

$[(\frac{\mu}{\hbar^2}e^2)-(1/a_o)\frac{2}{r}] = 0$.

Step #6 $[(\frac{\mu}{\hbar^2}e^2)-(1/a_o)\frac{2}{r}] = 0$, whence $a_o = \hbar^2/(\mu e^2)$,

Step #7 $(1/a_o)^2 + \frac{2\mu}{\hbar^2}E_1 = 0$, gives $E_1 = -\hbar^2/(2\mu a_o^2)$ or

$$E_1 = (-\hbar^2/2\mu)(\mu e^2/\hbar^2)^2$$

$$E_1 = (-\mu e^4/2\hbar^2) = -13.6 eV.$$

9.9 ELECTRON PROBABILITY DENSITY

9.9.1 Bohr Model:

In Bohr's model of the H-atom, the electron is visualized as revolving around the nucleus in a circular path. In Spherical Polar Coordinate System, this means that the electron would always be found at a distance, r_n for the nth orbit (Fig. 9.10)

$$r_n = n^2 a_o \qquad (9.9.1)$$

from the nucleus and in the equatorial plane, $\theta = 90°$, while its azimuthal angle, $\varphi = \varphi(t)$ changes with time.

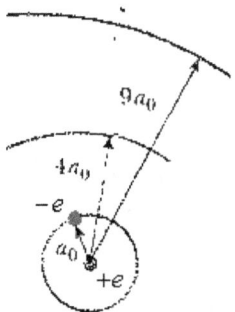

Fig. 9.10 Radii of various orbits $r_n = n^2 a_0$ for H-atom

9.9.2 QUANTUM (WAVEMECHANICAL) THEORY of the H-ATOM:

The picture of the H-atom according to Old Quantum Theory (Bohr model) is modified – no definite values of r, θ, and φ can be given. Only relative values of the probability of finding the electron at various locations can be found. It is known that (9.7.1)

$$\psi(r) = \psi_{n,\ell,m}(r,\theta,\varphi) = \mathfrak{R}_{n\,\ell}(r) \cdot Y_\ell^m(\theta,\varphi) \qquad (9.9.2)$$

The Probability electron density, $= |\psi(r)|^2 = |\mathfrak{R}_{n\,\ell}(r) \cdot Y_\ell^m(\theta,\varphi)|^2$

Probability of finding the electron at r from the nucleus and in an element of infinitesimal volume, $d\tau$ between r and r + dr,

$$P_{n\,\ell}(r) = |\mathfrak{R}_{n\,\ell}(r)|^2 d\tau = |\mathfrak{R}_{n\,\ell}(r)|^2 4\pi r^2 dr$$

$$P_{n\,\ell}(r \to 0) = 0,$$

$$P_{n\,\ell}(r \to \infty) = 0, \text{ as } d\tau \to \infty, \text{ and } \mathfrak{R}_{n\,\ell}(r \to \infty) = 0.$$

The probability distributions of $\mathfrak{R}_{n\,\ell}(r)$, the Radial Functions, exhibit *nodes*, that is they vanish for certain values of r; in particular the function $\mathfrak{R}_{n\,\ell}(r)$ has

$\boxed{n_r = (n-\ell-1)}$ RADIAL NODES, (9.9.3)

and $\boxed{n_r = (n-\ell)}$ BUMPS (anti-nodes or zeros). (9.9.4)

9.9.3 THE UNSOLD's THEOREM:

It states that, for any value of the orbital quantum number ℓ, the probability densities summed over all possible states, $m_\ell = 0, \pm 1, \pm 2, \pm 3, \ldots, \pm \ell$. Yield a constant independent of angles, θ, and φ; that is

$$\sum |Y_\ell^m(\theta,\varphi)|^2 = \text{Constant.} \qquad (9.9.5)$$

This theorem means that *every closed sub shell* atom or ion has a spherically symmetric distribution of electric charge.

Worked out Example 9.3
Find out the shell radius of the H-atom for $n = 1$.

Solution: Step # 1 $<r>_{n\ell} = \int_0^\infty \psi_{n\ell}(r)^* r \, \psi_{n\ell}(r) \, d\tau = \int_0^\infty |R_{n\ell}(r)|^2 \, r \, 4\pi r^2 dr$

For the ground state of H-atom, $n = 1$, $\ell = 0$, $Z = 1$,

Step # 2 On substitution, $<r>_{100} = 4\pi \frac{1}{4\pi} \int_0^\infty |2 (1/a_0)^{3/2} e^{-r/a_0}|^2 \, r^3 dr$

$<r>_{100} = \int_0^\infty |(4/a_0)^{3/2} e^{-r/a_0}|^2 \, r^3 dr = \int_0^\infty (4/a_0)^3 (-a_0/2) \, e^{-2r/a_0} \, r^3 dr$

$= (4/a_0)^3 [(a_0/2)^4 \, 3!] = (3/2)a_0$

Step # 3 Radius of the 1s sub shell $= <r>_{1s} = <r>_{100} = (3/2)a_0$ \hfill (9.9.2)

9.9.4 THE KRAMER'S RELATION (or Recurrence Relation):

The Recurrence Relation of the H-atom wave function and the expectation values are made use of to derive the Kramer's Relation.

$$\frac{s+1}{n^2} <r^s> - (2s+1) \, a_0 <r^{s-1}> + \tfrac{1}{4} s [(2\ell +1)^2 - s^2] \, a_0^2 <r^{s-2}> = 0 \quad (9.9.3)$$

where $s > (-2\ell - 3)$

or

$$\frac{s+2}{n^2} <r^{s+1}> - (2s+2) \, a_0 <r^s> + \tfrac{1}{4}(s+1)[(2\ell+1)^2 - (s+1)^2] \, a_0^2 <r^{s-1}> = 0 \quad (9.9.4)$$

with $s > -2 \, (\ell - 1)$.

It can be shown the following:

$<r^{-1}>_{n\ell} = \dfrac{1}{n^2 a_0}$ \hfill (9.9.5)

$<r>_{n\ell} = (a_0/2Z)[3n^2 - \ell(\ell+1)]$ \hfill (9.9.6)

$<r^2>_{n\ell} = (a_0^2/2Z^2)[5n^2 + 1 - 3\ell(\ell+1)] \, n^2$ \hfill (9.9.7)

$<r^2>_{n\ell} = (n^2 a_0^2/2Z^2)[(1+\tfrac{3}{2})[1 - \ell(\ell+1) - \tfrac{1}{3}]/n^2$ \hfill (9.9.8)

1s H-atom $r_{max} = a_0$. \hfill (9.9.9)

$(r^{-2})_{n\ell} = Z^2/[n^3 a_0^2 (\ell + \tfrac{1}{2})]$ \hfill (9.9.10)

$<(r^{-3})>_{n\ell} = Z^3/[n^3 a_0^3 \ell(\ell+\tfrac{1}{2})(\ell+1)]$ \hfill (9.9.11)

$\langle r \rangle_{1s} = \int_0^\infty R_{1s} \, r \, R_{1s} \, dr = 4\left(\dfrac{Z}{a_0}\right)^3 \int_0^\infty r^3 e^{-2Zr/a_0} dr = \dfrac{3a_0}{2Z}$

$\langle r \rangle_{2s} = \int_0^\infty R_{2s} \, r \, R_{2s} \, dr = \dfrac{1}{8}\left(\dfrac{Z}{a_0}\right)^3 \int_0^\infty r \, (2 - Zr/a_0)^2 \, e^{-Zr/a_0} dr = \dfrac{6a_0}{Z}$

$$\langle r \rangle_{2p} = \int_0^\infty R_{2p}\, r\, R_{2p}\, dr = \frac{1}{24}\left(\frac{Z}{a_0}\right)^5 \int_0^\infty r^5 e^{-Zr/a_0} dr = \frac{5a_0}{Z}$$

Worked out Example 9.4

Find the value of $\langle r^{-1} \rangle_{n\ell}$ for an atom, using Kramer's Relation.

Solution:

Step # 1 The Kramer Relation is

$$\frac{S+1}{n^2} \langle r^S \rangle - (2s+1) a_0 \langle r^{S-1} \rangle + \tfrac{1}{4} s [(2\ell+1)^2 - s^2] a_0^2 \langle r^{S-2} \rangle = 0$$

with the condition, $s > (-2\ell - 3)$.

Step # 2 For $s = 0$, $\frac{0+1}{n^2} \langle r^0 \rangle - (0+1) a_0 \langle r^{-1} \rangle + 0 = 0$ with the condition, $s > (-2\ell - 3)$.

Step # 3 $\frac{1}{n^2} - a_0 \langle r^{-1} \rangle = 0$, whence $\langle r^{-1} \rangle_{n\ell} = \frac{1}{n^2 a_0}$

Worked out Example 9.5

The wave function for hydrogen in 1s state, for which $n = 1$, $\ell = 0$, $m = 0$, is given by $\psi_{100} = \sqrt{1/4\pi}\, 2\, (1/a_0)^{3/2} e^{-Zr/a_0}$, where $a_0 = 0.053$ nm is the Bohr radius. Derive a formula for the probability density, and find where the electron is most likely to be in the H-atom.

Solution:

Step # 1 Given: $\psi_{100} = \sqrt{1/4\pi}\, 2\, (1/a_0)^{3/2} e^{-Zr/a_0}$.

Step # 2 The Probability electron density,

$$P_{n\ell}(r) = |\Re_{n\ell}(r)|^2 d\tau = |\Re_{n\ell}(r)|^2 4\pi r^2 dr,\quad P_{1s}(r) = |\Re_{1s}(r)|^2 4\pi r^2 dr.$$

$$= \frac{1}{4\pi} \int_0^\infty |2\,(1/a_0)^{3/2} e^{-r/a_0}|^2\, r^3 dr = 4\,(1/a_0)^3 e^{-2r/a_0}\, r^3$$

Step # 3 The most probable distance of the electron from the nucleus is $r = r_{max}$

Step # 4 r_{max} occurs when $\frac{d}{dr} P_{1s}(r)\big|_{r = r_{max}} = 0$.

$$\frac{d}{dr} P_{1s}(r) = \left\{(1/a_0) e^{-2r/a_0}\, r^2 + 2 r e^{-2r/a_0}\right\} = 0$$

$$\left\{(-r_{max}^2 / a_0) + r_{max}\right\} 2 e^{-2r/a_0} = 0.$$

i.e., for 1s H-atom $r_{max} = a_0$.

9.9.5 QUANTUM VIRIAL THEOREM for the H-atom

A theorem in classical mechanics which relates the kinetic energy of a system to the virial ($G = \langle p \cdot r \rangle$) of RJE Clausius, as defined below. The theorem can be generalized to quantum mechanics and has widespread application. It connects the average kinetic and potential energies for systems in which the potential is a power of the radius. In the common case that the forces are derivable from a power-law potential,

$$\boxed{V(r) \propto r^s},$$

where s is a constant..
To obtain the general Virial theorem, Assume,

$$(d/dt)<(r, p)> = 0$$

∴ $<T> = <r, (\partial/\partial r)V(r)> = s V(r)$

$$\boxed{2<T> = s V(r)} \quad (9.9.12)$$

Thus, in this case the virial theorem simply states that the kinetic energy is s/2 times the potential energy, the virial is just −s/2 times the potential energy.

If one calculates the *mean* kinetic energy and the *mean* potential energy of an electron in the ground state of the H-atom, one gets (see R.Q. 9.10),

$$<V>_{10} = 2 <E>_{10}.$$

$$<T>_{10} = - <E>_{10}.$$

The *Quantum mechanical Virial Theorem* states that for a system in which the particles interact according to **Coulomb's Law**,

$$\boxed{<T>_{n\ell} = -\tfrac{1}{2} <V>_{n\ell}}$$

Or, $\boxed{<V>_{n\ell} = 2 <E>_{n\ell} = -2 <T>_{n\ell}}$ (9.9.13)

Worked out Example 9.6

The radial wave function for an atom with one electron is given by
$\mathfrak{R}_{n\ell} = C_n 2 (Z/a_0)^{3/2} e^{-Z r/a_0}$. Find out the distance r where the distribution function attains maximum.

Solution:

$\boxed{\text{Step \# 1}}$ $\ell = n - 1$.

$\boxed{\text{Step \# 2}}$ # of radial nodes, (9.7.2), $n_r = (n - \ell - 1) = 0$,

$\boxed{\text{Step \# 3}}$ $P_{n\ell}(r) = |\mathfrak{R}_{n\ell}(r)|^2 r^2 dr$

$P_{n, n-1}(r) = |\mathfrak{R}_{n, n-1}(r)|^2 r^2$

$\boxed{\text{Step \# 4}}$ This is maximum when $(d/dr)P_{n, n-1}(r) = 0$

$\boxed{\text{Step \# 5}}$ i.e. $[2 n r^{2n-1} - (2Z/a_0)r^{2n}] e^{-Z r/a_0} = 0$

i.e., at $(r_n)_{Max} = n^2 a_0 / Z$ (9.9.14)

This is the same value that appears in the Bohr model.

9.10   ATOMIC ORBITALS (A O s):

9.10.1   STATES OF AN ATOM

9.10.1.1   Spherical Harmonics in Real Form (or Standing versus Rotating Waves)

It is customary to specify the angular momentum states of electron by a letter as described earlier in Section 9.8, viz. with s corresponding to $\ell = 0$, p to $\ell = 1$, etc. according to scheme (9.8.8), viz.,

Orbital angular momentum :	$\ell =$ 0 1 2 3 4 5 6. ....,
Angular momentum Electron State:	s p d f g h i ....
Series	S P D F G H I ......

(9.8.8)

with the capitals standing for Sharp, Principal, Diffuse, Fundamental, and the alphabetical sequence for $\ell > 3$.

9.10.1.2 Atomic Orbitals (A O s)

One electron wave function, for example, the H-atom wave function, $\psi_{n\ell m}(r)$ is referred to as an *Atomic Orbital*.

9.10.1.3 p-Orbitals

A state with n = 2, $\ell = 1$ is called a 2p state. In accordance with the spectroscopic notation, orbitals corresponding to $\ell = 1$ will be called *p-orbitals*. There are three 2p-states, which correspond to $m = +1, 0, -1$. Spectroscopically these may be designated as $2p_{+1}, 2p_o, 2p_{-1}$, states, respectively. The angular factors associated with these states are; aside from a numerical factor, (see, for example, on spherical harmonics).

i) $\quad Y_1^{+1} \equiv 2p_{+1} = \sqrt{3/8\pi}\ e^{+i\varphi}\ \text{Sin}\theta$, (9.10.1)

which depends only on θ. This is a rotating wave, i.e. a *non-standing wave* complex function.

ii) $\quad Y_1^0 \equiv 2p_o = \sqrt{3/4\pi}\ \text{Cos}\theta$, (9.10.2)

which a real function. This is a *standing wave* function.

iii) $\quad Y_1^{-1} \equiv 2p_{-1} = \sqrt{3/8\pi}\ e^{-i\varphi}\ \text{Sin}\theta$, (9.10.3)

which is a complex function. This is also a *non-standing wave* function.

9.10.2 LCAO (Linear Combination of Atomic Orbitals)

For many graphical purposes it is more convenient to replace these complex form for spherical functions by following <u>Linear Combination of degenerate standing waves / Atomic Orbitals</u> *(LCAO-s)*, because $2p_{+1}$ and $2p_{-1}$ are not standing wave functions, whereas $2p_o$ is. Standing waves are particularly useful in molecular physics.

Any combination such as

$$\psi_{n\ell m}(r) = [a\ \psi_{n\ell m=-1} + b\ \psi_{n\ell m=+1}] \quad (9.10.4)$$

is also a solution if one chooses a and b such that

$$\iiint \hat{\psi}_{n\ell}(r)^* \hat{\psi}_{n\ell}(r)\ d\tau = 1 \quad (9.10.5)$$

A convenient choice of the linear combination is illustrated for the 2p state as follows:

Worked out Example 9.7
Illustrate the choice of LCAOs in the case of 2p state.
Solution:

> **Step # 1** Given the AO-s (9.10.1), (9.10.2) and (9.10.4),;
> $z = r\ \text{Cos}\theta,\ x = r\ \text{Sin}\theta \text{Sin}\varphi,\ y = r\ \text{Sin}\theta \text{Cos}\varphi$
> **Step # 2** Obtain the LCAOs

$$\psi_{P_x} = p_x = -\sqrt{1/2}\ (p_{+1} - p_{-1}) \sim x \tag{9.10.6}$$

$$\psi_{P_z} = p_z = p_0 \sim z \tag{9.10.7}$$

$$\psi_{P_y} = p_y = +\sqrt{1/2}\ (p_{+1} + p_{-1}) \sim y \tag{9.10.8}$$

Step # 3 Thus the LCAO-s are just the Cartesian components of a unit vector in a Spherical Polar Coordinate System and are *standing waves*. These Spherical Harmonics in *real form* are **eigen functions** of $\hat{H}$ and $\hat{L}^2$, bot NOT of $\hat{L}_z$.

Step # 4 It can be verified that p_x, p_y, and p_z are *normalized and orthogonal* to one another. The designations; p_x, p_y, and p_z indicate that the angular part of these wave functions have their maximum values in the x, y, and z directions, respectively, *i.e.* the subscripts x, y, z indicate the behaviour of the real spherical harmonic in terms of Cartesian coordinates.

9.10.3 d – Orbitals

A state with $n = 3$, $\ell = 2$ is called a 3d state. In accordance with the spectroscopic notation, Orbitals corresponding to $\ell = 2$ will be called *d-Orbitals*. There are five 3d-states, which correspond to $m = 0$, ± 1, ± 2. *Spectroscopically*, these may be designated as $3d_{+2}$, $3d_{+1}$, $3d_0$, $3d_{-1}$, $3d_{-2}$ states, respectively.

As in the case of the p-Orbitals, as a convenient choice, one resorts to the LCAO approach to the *d-Orbitals*. All these have two angular nodes.

$$Y_2^{+2} \equiv 3d_{+2} = \sqrt{15/32\pi}\ e^{+i2\varphi}\ \text{Sin}^2\theta \equiv \sqrt{15/32\pi}\ (x+i\,y)^2/r^2 \tag{9.10.9}$$

$$Y_2^{+1} \equiv 3d_{+1} = -\sqrt{15/8\pi}\ e^{+i\varphi}\ \text{Sin}\theta\ \text{Cos}\theta$$

$$\equiv -\sqrt{15/8\pi}\ (x+i\,y)z/r^2 \tag{9.10.10}$$

$$Y_2^{0} \equiv 3d_0 = +\sqrt{5/16\pi}\ (3\,\text{Cos}^2\theta - 1) \equiv +\sqrt{15/32\pi}\ (x-i\,y)^2/r^2, \tag{9.10,11}$$

$$Y_2^{-1} \equiv 3d_{-1} = +\sqrt{15/8\pi}\ e^{-i\varphi}\ \text{Sin}\theta\ \text{Cos}\theta \equiv +\sqrt{15/8\pi}\ (x-i\,y)z/r^2, \tag{9.10.12}$$

$$Y_2^{-2} \equiv 3d_{-2} = = +\sqrt{15/32\pi}\ e^{-i2\varphi}\ \text{Sin}^2\theta \equiv +\sqrt{15/32\pi}\ (x-i\,y)^2/r^2 \tag{9.10.13}$$

As in the case of the *p-Orbitals*, as a convenient choice, one resorts to the LCAO approach to the d-Orbitals:

$$Y_2^{0} \equiv 3d_0 \quad \boxed{d_{z^2} \sim (3\,z^2 - r^2)}$$

$$\psi_{d_{xz}} = d_{xz} = +\sqrt{1/2}\ (d_{+1} + d_{-1})$$

$$\equiv +\sqrt{15/8\pi}\ \text{Cos}\theta\ \text{Sin}\theta\ \sqrt{1/2}\ [e^{+i\varphi} + e^{-i\varphi}]$$

$$\sim \text{Cos}\theta\ \text{Sin}\theta\ \text{Cos}\varphi \tag{9.10.14}$$

$$\therefore \boxed{d_{xz} = \sim xz}. \qquad (9.10.15)$$

$$\psi_{d_{yz}} = d_{yz} = -\sqrt{1/2}\,(d_{+1} - d_{-1})$$

$$\equiv -\sqrt{15/8\pi}\,\cos\theta\,\sin\theta\,(i)\sqrt{1/2}\,[e^{+i\varphi} - e^{-i\varphi}] \sim \cos\theta\,\sin\theta\,\sin\varphi$$

$$\therefore \boxed{d_{yz} = \sim yz} \qquad (9.10.16)$$

$$\psi_{d_{x^2-y^2}} = d_{x^2-y^2} = \sqrt{1/2}\,(d_{+2} + d_{-2}) \equiv +\sqrt{15/32\pi}\,\sin^2\theta\,\sqrt{1/2}\,[e^{+i2\varphi} + e^{-i2\varphi}]$$

$$\sim \sin^2\theta\,\cos 2\varphi \sim \sin^2\theta\,(\cos^2\varphi - \sin^2\varphi).$$

$$\therefore \boxed{d_{x^2-y^2} = \sim x^2 - y^2} \qquad (9.10.17)$$

$$\psi_{d_{xy}} = d_{xy} = (-i)\sqrt{1/2}\,(d_{+2} - d_{-2})$$

$$\equiv +\sqrt{15/32\pi}\,\sin^2\theta\,(-i)\sqrt{1/2}\,[e^{+i2\varphi} - e^{-i2\varphi}]$$

$$\sim \sin^2\theta\,\sin 2\varphi \sim \sin^2\theta\,\cos\varphi\,\sin\varphi$$

$$\therefore \boxed{d_{xy} = \sim xy} \qquad (9.10.18)$$

9.10.4 POLAR PLOTS

Polar plots are used to describe the angular distribution of the electron for non-s states ($\ell \neq 0$).

As an example, the plot representing an orbital $p_z = p_0$ characterized by the function, (9.10.7), viz.,

$$\psi_{p_z} = p_z = p_0 \sim z$$

will have its maximum value at $\theta = 0°$ and $\theta = 180°$, and will vanish at $90°$. This function is positive on one side of the XY plane and negative on the other side. Whenever an orbital has this property, the orbital is said to be ant-symmetric with respect to reflection in the XY plane

Such pictures of the angular dependence of the various Orbitals of the electron for particular states are very useful in the study of atoms and molecules (Fig. 9.10).

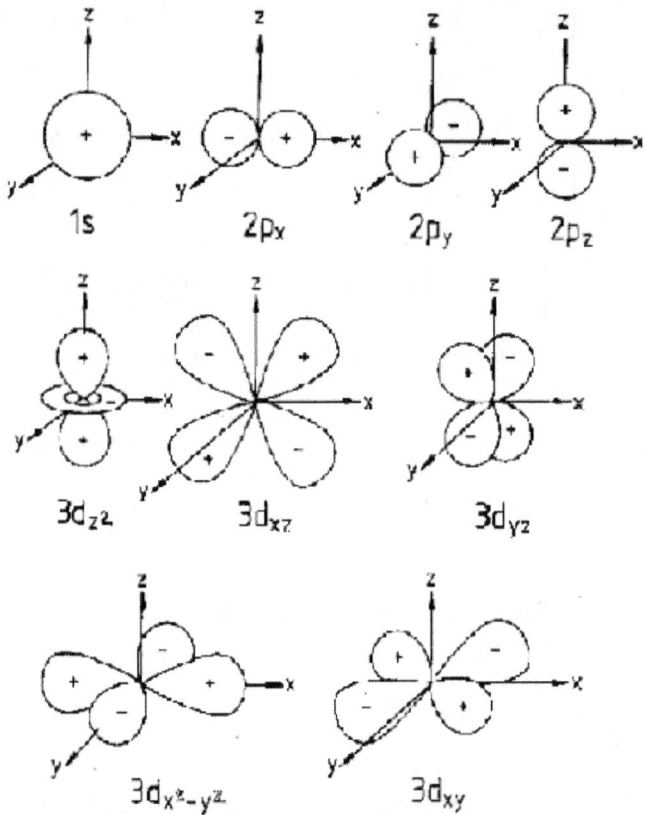

Polar diagrams for 1s, 2p, and 3d atomic orbitals showing the distributions

Fig. 9.10 Angular dependence of the various Orbitals with $n=1, n=2, n=3$.

9.10.5 MIXED STATE of an Atom

As seen earlier in the Bohr Theory of the H-atom, any atom in a specified state, $\psi_{n\ell m}$ will have its electron not in the oscillating condition. Therefore no radiation occurs when the atom is in specified quantum state. The wave function of an electron capable of existing in states n or m may be written as

$$\Psi = [a\,\psi_n + b\,\psi_m] \qquad (9.10.19)$$

where a^*a = probability that an electron is in a state, n,

b^*b = probability that an electron is in a state, m,

$$a^*a + b^*b = 1 \qquad (9.10.20)$$

if $\hat{\psi}_n$ = normalized

at time, $t = 0$, $a = 1$, $b = 0$. $\qquad (9.10.21)$

The state Ψ given by equation (9.10.19) is called a *mixed state* of the electron, and EM waves may be produced as a result of the electron making a <u>transition</u> from $n \longleftrightarrow m$

In a mixed state, both $a \neq 0$ and $b \neq 0$.

9.10.6 The SPECTRUM OF ATOMIC HYDROGEN: SELECTION RULES:

As seen earlier in the Bohr Theory of the H-atom, any atom in a specified state, $\psi_{n\ell m}$ will have its electron not in the oscillating condition. Therefore no radiation occurs when the atom is in specified quantum state. EM radiation may be produced as a result of the electron making a transition between two allowed states,

$$\psi_{n\ell m} \longleftrightarrow \psi_{n'\ell m} \qquad (9.10.22)$$

So far nothing was discussed about <u>selection rules</u>. It is shown below that these depend on the angular part of the functions, Ψ The selection rules derived below hold true for all atoms that have only one electron outside the closed shell - alkali metals, for example. The selections rules also hold good for more complex atoms provided they are interpreted in terms of eigen values with respect to the operator for the total electronic angular momentum of the atom.

The **instantaneous Dipole moment** of a H-atom is

$$\vec{\mu} = - <\vec{r}>, \qquad (9.10.23)$$

where $\vec{r}$ is the vectorial distance of the electron from the nucleus. The calculation of μ is illustrated in the

Worked out Example 9.7.

The <u>transition moment</u>, $\mathbb{R}$ is the

$$\mathbb{R} = \iiint (\psi_{n'\ell m})^* |r| \psi_{n\ell m} \, d\tau \qquad (9.10.24)$$

To evaluate this integral, each component must be evaluated separately. Thus the equation (9.10.24) becomes the three equations

$$\mathbb{R}_x = e \iiint (\psi_{n'\ell m})^* [r \, Sin\theta \, Cos\varphi] \, \psi_{n\ell m} \, r^2 \, Sin\theta \, dr \, d\theta \, d\varphi \qquad (9.10.25)$$

$$\mathbb{R}_y = e \iiint (\psi_{n'\ell m})^* \, \psi_{n\ell m} \, r^2 \, Sin\theta \, dr \, d\theta \, d\varphi \qquad (9.10.26)$$

$$\mathbb{R}_z = e \iiint (\psi_{n'\ell m})^* (rCos\theta) \, \psi_{n\ell m} \, r^2 \, Sin\theta \, dr \, d\theta \, d\varphi \qquad (9.10.27)$$

These integrals may be evaluated for general *Legendre polynomials* to obtain the Selection Rules

| $\Delta \ell = \pm 1$ |
| $\Delta m = 0, \pm 1$ |
| Δn = No restriction |

$$(9.10.28)$$

None of the mathematical procedure has gone through here.

9.10.6 PARITY of $Y_\ell^m (\theta, \varphi)$

To check whether the Spherical Harmonics have parity even or odd, replace θ and φ by $(\pi - \theta)$ and $(\pi + \varphi)$, respectively, It is seen from Chapter 8, Section 7 that

Parity + means Even: $Y_\ell^m(\pi-\theta,\varphi) = +Y_\ell^m(\theta,\varphi)$.
Parity − means Odd: $Y_\ell^m(\pi-\theta,\varphi) = -Y_\ell^m(\theta,\varphi)$
Parity + means Even: $Y_\ell^m(\theta,\pi+\varphi) = +Y_\ell^m(\theta,\varphi)$.
Parity − means Odd: $Y_\ell^m(\theta,\pi+\varphi) = -Y_\ell^m(\theta,\varphi)$

In spherical harmonic functions, it turns out that the values of ℓ uniquely determines the parity of $Y_\ell^m(\theta,\varphi)$. Parity can otherwise be determined by means of the sign of:

$$\text{Parity} = (-1)^\ell \tag{8.7.4}$$

9.11 CHEMICAL BONDING

9.11.1 Preliminaries

Atoms are perhaps the most fundamental entities (or species) in chemistry. Quantum Mechanics could account exactly for the energy levels of the H-atom and offered a theory that could be extended to more complicated atoms as ell. With enough mathematical effort exact agreement between theory and experiment could be obtained for helium and good results could be obtained for more complex atoms. Most important, the Quantum Mechanical picture of the atom led to a model that could rationalize many features of their structure; for example the ground state of a carbon atom was shown to be ^3P. Perhaps this Chapter might have induced some of the readers to get an additional insight into the interesting field of atomic bonding and molecular physics of special interest to chemists.

9.11.2 The physical significance of H-atom-like Orbitals

The two ways to discuss the Radial Functions, $\Re_{n\ell m}(r)$ are

(i) A plot of the function $|\Re_{n\ell}(r)|^2$ versus (r/a_o), the distance from the nucleus in atomic units, (r/a_o),

(ii) Integrate the over the angular variables and plot the resulting function, called the Radial Distribution Function,

$$\boxed{\iiint |\Re_{n\ell}(r)|^2 r^2 \sin\theta\, dr\, d\theta\, d\varphi = \iiint |\Re_{n\ell}(r)|^2 4\pi r^2\, dr}.$$

This gives the probability of finding the electron in a spherical shell of thickness dr at a distance r from the nucleus.

Important features of the H-atom wave functions to be noted from the above two plots are:

(1) It is known that the 1s state has only one bump (it is the largest probability)
(2) There are no nodes for the 2p function, It has ($\ell = n-1$, $\ell = 1$) also there is only one bump and the largest probability is next only to 1s.
(3) In the case of 2s ($\ell \neq n-1$, $\ell = 0$) there is one bump and a node, it has in addition penetration effect, i.e. there is a small peak at r $\sim a_0$, which is absent in 2p.
(4) On the other hand, 3s state ($\ell \neq n-1, \ell = 0$) has 2 nodes (zeros) in addition to one bump and two peaks.

9.11.3 HYBRIDIZATION:

A hybrid orbital is a LCAOs centred on a single atom.

9.11.3.1 *sp – hybrid*

This is an atomic wave function composed of equal properties of s- and p- Orbitals of the same atom, and an electron that occupies *sp*-orbital has 50% *s*-character and 50% *p*-character. The angle is $\varphi = \pi$ between the hybrids.
For two such equivalent hybrid Orbitals, say, h and h', $h = as + bp$, $h' = as + bp$;
$a^2 = Cos\varphi/(Cos\varphi - 1)$, $b^2 = 1 - a^2$
Generally, in any hybridized state the higher the proportion of p-character, the smaller is the angle, φ between the hybrids.

9.11.3.2 sp^3 *– hybrid*

Electron that occupies sp^3-hybrid has 25% s-character and 3 x 25% p-character. The angle φ between the hybrids is $\varphi = 109.5°$.

9.11.3.3 $sp^3 d^2$ *– hybrid*

An Orbital composed of s -, p -, and d - Orbitals in the ratio 1: 3:2 is known as the sp³d² hybrid.

9.11.3.4 π – **Orbital**:

It is formed from the overlap of p-Orbitals of adjacent atoms. Two electrons in a π – Orbital constitute the so called π – bond, due in-phase combination of the p-Orbitals.

9.11.3.5 π^+ – **Orbital**:

Anti-bonding π – Orbital is the out-of-phase combination of the p-Orbitals.

REVIEW QUESTIONS

R.Q. 9.1 Analyze the Schrödinger Equation for motion under central forces.

R.Q. 9.2 Show that for the H-atom $V_{eff} = \left(\ell(\ell+1)/(2\mu r^2)\right) - Z e^2/r$.

R.Q. 9.2 What is Energy and wavelength of a photon emitted when e^- drops from the $n = $ 4 to $n = 2$ energy level in a H atom? $c = 2.9979 \times 10^8\ ms^{-1}$, $h = 6.6256 \times 10^{-34}\ J-s$
(Answer: $1/\lambda = 2.0564 \times 10^6\ m^{-1}$, $\lambda = 486\ nm$; $E = 4.09 \times 10^{-19}\ J/photon$
$= 246 kJ/mole$ atoms)

R.Q. 9.3 How many number of nodes and bumps are there in the radial probability density curve for a given (n, ℓ), with ℓ the largest value? (Answers: # of radial nodes = 0. # of *bumps* = 1.

R.Q. 9.4 Given: $\psi_{100} = \sqrt{1/4\pi}\ 2\ (1/a_o)^{3/2} e^{-r/a_o}$, show that shell radius,
$<r>_{1s} = (3a_o/2)$

R.Q. 9.5 Derive the Recurrence Relation of the H-atom wave function expectation values, and derive Kramer's Relation. Given, $\left(\frac{d^2}{d\rho^2} + \frac{2}{\rho} \cdot \frac{1}{n^2} - \frac{\ell(\ell+1)}{\rho^2}\right) \Re_{n\ell}(\rho) = 0$, $\rho = r/a_0$, $a_0 = \hbar^2/(\mu e^2)$. (Answer:

$\frac{s+1}{n^2} <r^S> - (2s+1) a_0 <r^{S-1}> + \frac{1}{4}s[(2\ell+1)^2 - s^2] a_0^2 <r^{S-2}> = 0$

, with the condition, $s > (-2\ell - 3)$.)

R.Q. 9.6 Estimate the average value of the shell radius r of an atom with specification, (n, ℓ). The Kramer relation is

$\frac{s+1}{n^2} <r^S> - (2s+1) a_0 <r^{S-1}> + \frac{1}{4}s[(2\ell+1)^2 - s^2] a_0^2 <r^{S-2}> = 0$

;with the condition, $s > (-2\ell - 3)$. (Answer, $<r>_{n\ell} = (a_0/2Z)[3n^2 - \ell(\ell+1)]$)

R.Q. 9.7 An electron in the Coulomb field of a proton is in a state specified by (n, ℓ). Find $<r^2>_{n\ell}$, where r is the distance of the electron from the proton. Use the Kramer's relation (Answer:

$<r^2>_{n\ell} = (n^2 a_0^2 / 2Z^2)[(1+\frac{3}{2})[1-\ell(\ell+1) - \frac{1}{3}]/n^2)$.

R.Q. 9.8 Given: $\psi_{100} = \sqrt{1/4\pi} \; 2 \, (1/a_0)^{3/2} e^{-Zr/a_0}$.

Compute the molar diamagnetic susceptibility of atomic hydrogen given by $\chi = N_A e^2 <r^2>_{1s}/6mc^2$, N_A = Avogadro number.

R.Q. 9.9 An electron in the Coulomb field of a proton is in prepared in a state. Find out the most probable value of r of finding the electron is maximum. Do this explicitly for n =1. Use the result to differentiate the Quantum theory of H-atom from Bohr's theory of H-atom. Given, $\psi_{100} = \sqrt{1/4\pi} \; 2 \, (1/a_0)^{3/2} e^{-Zr/a_0}$, $a_0 = 0.053$ nm (Answer: 1s H-atom $r_{max} = a_0$., for Bohr theory, $r = a_0 = r_1$)

R.Q. 9.10 What is the probability of finding the electron in a H-atom anywhere in a sphere of radius a_0. Given, $\psi_{100} = \sqrt{1/4\pi} \; 2 \, (1/a_0)^{3/2} e^{-Zr/a_0}$ (Answer:

$P_{10}(r) = (1 - 5/e^2) = 0.32$)

R.Q. 9.11 Show that for the eigen states of the H-atom, $<r^{-1}>_{n\ell} = 1/n^2 a_0$. Then find $<V>_{n\ell}$, and compare this to the total energy. What is the relationship between $<V>_{n\ell}$ and $<T>_{n\ell}$? (Answer: $<E>_{n\ell} - \frac{1}{2}<T>_{n\ell}; <V>_{n\ell} = 2<E>_{n\ell} = -2<T>_{n\ell}$).(This relation confirms the Virial Theorem for H-atom).

R.Q. 9.12 Consider a H-atom of which the wave function at $t = 0$ is the following superposition of energy eigen functions,

$\psi_{n,\ell,m}(r,t); \; \psi_{n,\ell,m}(r,t=0) = \sqrt{1/14}\,[2\psi_{100}(r) - 3\,\psi_{200}(r) + \psi_{322}(r)]$.

(a) Is the wave function an eigen function of the parity operator? (b) What is the probability of finding the system in the ground state?, in the state (200)?, in the state (322)? (c) What are the $<\hat{H}>$?, $<\hat{L}^2>$?, and $<\hat{L}_z>$? (Answers: a) Yes, since $\psi_{100}(r)$, $\psi_{200}(r)$ & $\psi_{322}(r)$ all have even parity.($\ell = 0$ or even). (b) $P_{100}(r) = 2/7$;

$P_{200}(r) = 9/14$, $P_{322}(r) = 1/14$; c) $<\hat{H}> = 0.23 E_1$; $<\hat{L}^2> = 0.4$, $<\hat{L}_z> = 0.1$)

R.Q. 9.13 Generate the polynomial $L_1(\rho)$ and write down explicitly the radial wave function of the H-atom, for $n=1$. (Answer: $L_1(\rho) = 1$; AS
$\Re_{n\ell}(r) = C_n(\alpha r)^\ell \cdot L_{n+\ell}^{2\ell+1}(\alpha r) \cdot e^{-Zr/na_0}$; $\Re_{10} = 2(1/a_0)^{3/2} e^{-r/a_0}$).

R.Q. 9.14 Locate the radial mode of the 2s orbital in a H-atom.
(Answer: $(r_2)_{Max} = 4a_0$ This is where the radial mode of the 2s orbital occurs).

R.Q. 9.15 Show that for an H-atom, $<p^2> = -m^2 c^2 \alpha^2 [(Z^2/n^2) - 2Za_0/r]$. Use the Kramer's relation

R.Q. 9.16 Calculate the mean value of $(r^{-2})_{n\ell}$ in the quantum state (n, ℓ, m) of an electron in the Coulomb field of a proton. (Answer: $(r^{-2})_{n\ell} = Z^2/[n^3 a_0^2 (\ell + \frac{1}{2})]$)

R.Q. 9.17 A quantity of interest in spectroscopy is the average value of $(r^{-3})_{n\ell}$. Using the Kramer's Relation evaluate the $<(r^{-3})>_{n\ell}$ for an electron in a 2p orbital of a H-like atom. (Answer: $<(r^{-3})>_{n\ell} = Z^3/[n^3 a_0^3 \ell(\ell + \frac{1}{2})(\ell + 1)]$,
$<(r^{-3})>_{2p} = Z^3/[24 a_0^3]$)

R.Q. 9.18 Prove that the expectation value of $<(r^{-3})>_{n\ell}$ for the 2p state of hydrogen has the value $<(r^{-3})>_{2p} = Z^3/[24 a_0^3]$, where $a_0 = \hbar^2/e^2 \mu$. Use the Kramer's relation.

R.Q. 9.19 In spectroscopy an allowed electric dipole transition involves the electric dipole moment, $\bar{\mu}$. Calculate $\mu = -e <\bar{r}>$ using the state $\Psi(r,t) = [a\, \psi_{100}(r,t) + b\, \psi_{210}(r,t)]$, where $a = b = \sqrt{1/2}$, of the H-atom wave functions. Use Kramer's Relation. (Answers: $<\bar{r}> = (13/4) a_0$).

R.Q. 9.20 Show that the 3p eigen functions, $u_{31\pm1} = A_3\, e^{-ix} \sqrt{1/2}\, x\, (2-x)\, \sin\theta\, e^{\pm i\varphi}$, are orthogonal to the 3s eigen function, $u_{310} = A_3 \sqrt{2/3}\, e^{-ix}\, x\, (2-x)\, \cos\theta$;
$A_3 = \sqrt{1/\pi}(\mu e^2/n\hbar^2)^{3/2}$; $x = r(\mu e^2/n\hbar^2)$.

R.Q. 9.21 Show that for the H-atom, the transition $\psi_{1s} \longleftrightarrow \psi_{3s}$ is not allowed. (Answer: $\Delta\ell \neq ok$, $\Delta m = okayed$, $\Delta n = okayed$, The transition is disallowed)

R.Q. 9.22 Show that for the H-atom, the transition $\psi_{1s} \longleftrightarrow \psi_{3p}$ is allowed.

R.Q. 9.23 For a particle with angular momentum quantum number $\ell = 3$, sketch the allowed orientations of the angular momentum vector, relative to the z-axis. (Answer: An $\ell = 3$ state will have seven possible orientations of the orbital angular momentum vector, relative to the z-axis, corresponding to angles 30, 54.7, 73.2, 90, 106.7, 125.3, and 150 degrees.)

R.Q. 9.24 For a particle with angular momentum quantum numbers $\ell = 3$ and $m_\ell = 1$, what angle does the angular momentum vector make with the z-axis. (Answer: The angular momentum vector will be at $73.2°$ relative to the z-axis.).

R.Q. 9.25 Calculate the ionization energy of Be^{3+} (Answer: $Z = 4$ and $n = 1$ for the ground state. $R = 4^2 R_H = 217.6\ eV$.).

R.Q. 9.26 Calculate the frequency and wavelength of the radiation emitted when He^+ undergoes a transition from $n = 4 \rightarrow n = 3$. (Answer: $v = 6.3934 \times 10^{14}$ Hz, $\lambda = 468.9$ nm)

R.Q. 9.27 Determine whether Li^{2+} has an emission line in the visible region of the spectrum, i.e. $\lambda = 400$ nm to 900 nm. Give at least one possible wavelength along with the corresponding quantum numbers for the transition. (Answer: $Z = 3$, the Ritz formula shows that the transition, to fall in the visible region, must satisfy.

$(1/\lambda_{900}) \leq 9R\ [1/n_1^2 - 1/n_2^2] \leq (1/\lambda_{400})$. Many n_1 to n_2 pairs satisfy this relation, say for example $n_1 = 5$ to $n_2 = 4$).

R.Q. 9.28 At what distance from the nucleus is the electron most likely to be found in the following states of the H atom? (Answer: a) 1s b) 2s c) 2p. The radial probability distribution (after integrating over all angles) is $|R_{n\ell}(r)|^2\ 4\pi r^2$. The maximum in this function is a) 1s : $r = a_0/Z$, $r = a_0$ for hydrogen., b) 2s : $r = (3+\sqrt{5})a_0/Z$ and $r = (3 - \sqrt{5})a_0/Z$. there are two maxima, c) 2p: $r = 4a_0/Z = 4a_0$ for hydrogen).

R.Q. 9.29 Repeat problem R.Q. 9.27 for the He^+ atom. (Answer: a) 1s : $r = a_0/Z = a_0/2$, b) 2s : $r = (3 - \sqrt{5})a_0/2$ and $r = (3 + \sqrt{5})a_0/2$, there are two maxima. c) 2p : $r = 4a_0/Z = 2a_0$.)

R.Q. 9.30 Calculate the positions of the radial nodes in the a) 2s and b) 3s orbitals of the H atom. (Answer: a) 2s : $r = 2a_0/Z = 2a_0$ for hydrogen. b) 3s : $r = 3(3 - \sqrt{3})a_0/2Z$ and $r = 3(3 + \sqrt{3})a_0/2Z$. there are two nodes)

R.Q. 9.31 Calculate the average distance r from the nucleus in the following states of the H atom: a) 1s b) 2s c) 2p. (Answers: $<r>_{1s} = 3a_0/2Z$, $<r>_{2s} = 6a_0/Z$, $<r>_{2p} = 5a_0/Z$)

R.Q. 9.32 Give the term symbols for all the L-S states arising from the following configurations: a) 1s2s b) 2s2p c) 2p3d (Answers: All of these configurations may also be handled using the Clebsch-Gordon series method.

a) 1s2s Terms are ^1S and ^3S, b) 2s2p Terms are ^1P and ^3P, c) 2p3d Terms are ^1P, ^3P, ^1D, ^3D, ^1F and ^3F).

R.Q. 9.33 Give the spin-orbit coupling terms for each of the states in the previous problem (R.Q. 9.31). (Answers: a) $^1S_0$ and $^3S_1,...$ b) $^1P_1$, $^3P_0$, $^3P_1$ and $^3P_2$, c) $^1P_1$, $^3P_0$, $^3P_1$, $^3P_2$, $^1D_2$, $^3D_1$, $^3D_2$, $^3D_3$, $^1F_3$, $^3F_2$, $^3F_3$ and $^3F_4$).

R.Q. 9.34 Give all of the spectroscopically allowed transitions between the configurations $2s2p \rightarrow 2p3p$. (Answer: Allowed transitions: $2s2p\ ^1P \rightarrow 2p3p\ ^1S$ or ^1P or ^1D, $2s2p\ ^3P \rightarrow 2p3p\ ^3S$ or ^3P or ^3D).

&%&%&%&%&%&

Chapter 10

PARADOXES IN QUANTUM MECHANICS

Chapter 10

PARADOXES IN QUANTUM MECHANICS

"We cannot solve the problem that we have created with same thinking that created them"

(*From the American Museum of Natural History*)   - Albert Einstein

"To be confused about what is different and what is not, is to be confused about everything"

- David Bohm

10.1   INTRODUCTION

10.1.1   Paradox In Pysics - Distinguished

A statemnent or proposition seemingly self contradictory or absurd, but in reality a possible truth expressed, is termed a *paradox*. An example to this is '*more haste, less speed*'. Two ways are distinguished by which a paradox can arise in physics. A paradox is due to: (a) a fallacy <u>invoce</u>, when the line of reasoning is not satisfactory, and (2) a fallacy <u>inre</u>, when some underlying assumption is faulty.

10.1.2   The Pole and Barn Paradox

Consider a pole of certain length and a barn whose length is slightly shorter than that of the pole. If a pole-vaulter runs with such a pole through the barn at a relativistic speed, the pole and the vaulter will disappear inside the barn for an instance of time. But as the vaulter in concerned, it is the barn, which has undergone length contraction at such relativistic speeds. There are two events associated with the dynamics: (i) the disappearance of the back of the pole into the barn, and (ii) the emergence of the front of the pole from the barn. When one recognizes as to whether the precedence of the event over the other can depend on the frame of reference, the contradiction disappears. This is an example for **invoce** paradox.

10.1.3   Olber's Paradox

Assume that the Universe is infinite, static and uniform. Then one can come across an example for *inre* paradox. In this case, every line of sight from the surface of the Earth should intersect a Star; the night sky should appear as bight as the Sun's surface.

10.2 PARADOXES IN QUANTUM MECHANICS:

10.2.1 The Measurement Process – The Stern-Gerlach Experiment:

A famous experiment performed in 1922 by Otto Stern and Walther Gerlach, now called the *Stern-Gerlach experiment*, demonstrates the bare features of electron spin without any encumbrance from the orbital angular momentum, ℓ. Stern and Gerlach used silver atoms, however, Philipps and Taylor in 1927 repeated the experiment using hydrogen.

The experiment involves the force on a silver atom in a magnetic field, $\vec{B}_z$. This force due to the atom's magnetic dipole moment, along the z-axis, is

$$\mu_{\ell,z} = -(e/2m_e)\,\ell_z \propto \ell_z \tag{10.2.1}$$

The proportionality constant $= (g_e e / 2m_e)$

m_e being electronic mass and

the *electron-spin g-factor* $g_e = 2.0023193044$.

Similarly, for electron spin

$$\mu_{\ell,z} = -(e/2m_e)\,S_z \propto S_z. \tag{10.2.2}$$

Quantum mechanically,

$$\mu_{\ell,z} = -g_e(e\hbar/2m_e)\,m_e.$$

$$= -g_e \mu_B \, m_s \tag{10.2.3}$$

Substituting $g_e = 2$, and $m_s = \pm 1/2$, one gets

$$\mu_{\ell,z} = \pm\,(1.00)\,\mu_B \tag{10.2.4}$$

giving spin-up or spin-down cases. Because of the electron's spin magnetic moment, the hydrogen atom in the ground state ($n = 1$, $\ell = 0$, $\mu_{\ell,z} = 0$) behaves as a quantized dipole magnet.

If a magnetic dipole moment is placed in a *non-uniform magnetic field*, then there is a net force on the dipole. In the experiment the beam of atoms passes through an inhomogeneous magnetic field and then strikes a detector. The detector shows a pattern consisting of two distinct sub beams – due to the effect of quantization.

10.2.2 The Stern-Gerlach Experiment - An Idealized Measurement:

This experiment can be utilized as a model for the process of measurement in quantum mechanics. The object to be measured is the spin of the particles and the measuring apparatus is the position of the particle after traveling through the magnetic field, $\vec{B}_z$. The particle of *spin-up* and *spin-down* are deflected downwards and upwards, respectively; if z is the position of the pointer, for spin-up, $z < 0$, and for spin-down $z > 0$.

Let the initial spin wave function of such a particle can be put as

$$|\psi(0)> \ = \ |\psi(r,\theta,\varphi)>.|\chi(spin)>.|e^{-E_n t/\hbar}> \tag{10.2.5}$$

Symbolizing the object by letter O, the apparatus including the detector as A, and additional variable by Z to stand for all further macroscopic consequences that couple the state A of the apparatus.

$$|\psi(0)> \ = \ \sum_n C_n |O,n> \ |A> \ |Z> \tag{10.2.6}$$

After the beam has passed through the apparatus, (10.2.6) becomes

$$|\psi(0)> \ = \ \sum_n C_n |O,n> \ |A(n)> \ |Z(n)> \tag{10.2.7}$$

If one does not read off z, which always happens in practice, since one cannot keep track of all the macroscopic consequences, the density matrix of the mixture will be

$$\rho = \sum_n |C_n|^2 |O,n> \ |A(n)> \ |Z(n)><O(n)| \ <A(n)| \tag{10.2.8}$$

There occur two cases:

i) If one doesn't read off $A(n)$ either,

$$\rho = \sum_n |C_n|^2 |O,n><O,n|. \tag{10.2.9}$$

ii) If one reads off $A(n)$, then the probability of obtaining the particular reading $A(m)$ is

$$\rho = |A(m)> \ |O,m><O,m| \ <A(m)>|. \tag{10.2.10}$$

10.2.3 SCHRODINGER'S CAT and Quantum Reality

10.2.3.1 '*IN RE*' Nature

The key problem in the theory of measurement is the reduction of the wave function and, in particular, the question of when it takes place. This problem is illustrated quite drastically by means of the "Schrödinger's cat"

Fig 10.1 The Schrödinger Cat Paradox and Quantum Reality

Suppose that an ampoule of cyanide gas, a relay mechanism to break the ampoule and a Geiger Counter with radioactive atoms of $\tau_{1/2} = 1$ hr. such that within one hour there is a 50 – 50 chance of the gas being released from the ampoule. This device and the cat in a sealed box (let cat is alive in cyanide gas). Then after 1 hour there is equal chance of the cat being alive or dead (Fig 10.1).

Quantum mechanically, there is no way to achieve a $|\psi>$ for a living or dead cat. The cyanide gas, represented by $|\uparrow \& \downarrow>$ i.e. *spin-up* and *spin-down* particles, in a closed chamber with cat, is such that the $|\uparrow>$ gas particles would kill the cat, whereas the $|\downarrow>$ particles would not. Considering the effect of $|\uparrow>$ and $|\downarrow>$, which is analogous to the production of the $|\uparrow>$ and $|\downarrow>$ particles in the Stern-Gerlach experiment. Suppose the particle in the state $|\uparrow \& \downarrow>$ hits the cat, the state of spin and the cat makes a transition to $|\uparrow>|$ Dead cat $> + |\downarrow>|$ Living cat $>$ - pure state. When is it decided whether the cat is '*dead*' or '*alive*'? The answer is just when the observer opens the box! – An objective statement, independent of the conscious mind of the observer would be impossible. – What is the state of the observer himself in the quantum mechanical description?

According to the point of view just presented, the cat (together with the mechanism for killing it, *i.e.* cyanide particles) is linked to other macroscopic objects. These are influenced differently in the final states so that their respective wave functions do not overlap. For everything that follows, these macroscopic consequences are not recorded; the trace (like $Tr\ \rho = Tr\ A(n)\ \rho(t)$ is taken over them. The final state of the cat is described by a mixture of states corresponding to dead cat and living cat: the cat is either dead or living and not in a pure state, $|$ Dead cat $> + |$ Living cat $>$, which would include both possibilities. This is very counterintuitive but fundamental to quantum reality.

This clash between subjective experience and quantum theory has lead to much soul-searching. The Copenhagen Interpretation says Quantum Theory just describes our state of knowledge of the system and is essentially incomplete. Some people, following Hugh Everitt III, thought all the possibilities happen and there is a probability universe for each case. This is called the Many-worlds Interpretation. The cosmic wave function is thus in effect giving life to all the contingencies of contingencies as probability universes each with equal validity and there is no need for the wave function ever to collapse. It suffers from one difficulty which there are always elaborate explanations to circumvent. All the experience we have suggests just one possibility is chosen. The one we actually experience. Some scientists thus think collapse depends on a conscious observer. Others try to discover hidden laws which might provide the sub-quantum process which chooses one possibility rather than another, for example, David Bohm's idea of a particle piloted within a wave. This also has certain difficulties but we will examine a theory called the transactional interpretation which has features of all these ideas. In many considerations people try to pass the intrinsic problems of uncertainty away on the basis that in the large real processes we witness individual quantum uncertainties cancel in the law of averages of large numbers of particles. Chaotic processes are potentially able to inflate arbitrarily small fluctuations, so molecular chaos may inflate the fluctuations associated with quantum uncertainty.

10.2.3.2.1   '*IN VOCE*' NATURE: "The Unexpected Hanging":
A prisoner is sentenced on one Sunday to be hanged at noon on one of the following 7 days and to be kept ignorant as to which day it will be until that day arrives. Prisoner says this will never happen – since he is alive till Saturday. Reasoning backwards this way he concludes that he cannot be hanged on any day (as per the sentence). Nevertheless he was hanged on Friday!

10.3   The EPR (EINSTEIN, PODOLSKY & ROESN) PARADOX, EPR ARGUMENT, HIDDEN VARIABLE, THE BELL INEQUALITY

10.3.1   The EPR Argument, hidden variables:

Quantum Mechanics is characterized by indeterminacy. Einstein expressed his rejection of quantum theory by the remark: "*God does not play dice*". Einstein also rejected the fact that the value of non-diagonal observable is fixed only when an experiment is performed. This is reflected in the question: "*Is the man there when nobody looks*"? Hence there were repeated attempts to replace quantum theory by a statistical theory. Accordingly there exists hidden variables, whose values prescribe the values of all observables for any particular object, except that the hidden variables are unknown to the experimenter, thus yielding the *probabilistic* character of the theory. The probabilistic character of Quantum Mechanics would be quite analogous to that of *Classical Statistical Mechanics*, when the motion of all particles is in principle known.

Example: Consider a particle of Spin - $\frac{1}{2}$.

$$\hat{S}_z | \tfrac{1}{2} > = \tfrac{1}{2}\hbar | \tfrac{1}{2} > \qquad (10.3.1)$$

According to QM, z-component is not fixed. If one measures for N such particles, 50% of time the value $+\tfrac{1}{2}\hbar$, and 50% of time the value $-\tfrac{1}{2}\hbar$, are obtained.

According to the idea of hidden observables, for each particle, parameters unknown to one would determine when $+\tfrac{1}{2}\hbar$ or $-\tfrac{1}{2}\hbar$ results. Theses hidden observables would prescribe $\pm\tfrac{1}{2}\hbar$ each 50% of the time.

Einstein's thought experiments (to demonstrate the incompleteness of the QM description and to get around the indeterminism and the Uncertainty Relation) were each refuted by Niels Bohr.

An argument – sometimes referred to as a *paradox* – due to EP or simply 'EPR paradox', played a pivotal role in the discussion of variables; one considers this argument as reformulated by David Bohm.

Let two spin - $\tfrac{1}{2}$ particles in a singlet state

$$|0,0> = \sqrt{1/2} \, (|\uparrow > |\downarrow > - |\downarrow > |\uparrow >) \qquad (10.3.13)$$

be emitted from a source and move apart. Consider these two particles be separated by an arbitrarily large distance. One can find in the state (1) the following correlations in a measurement of the one-particle spin states.

i) if one measures the z-component of the spin and finds particles in $|\uparrow >$ the particle 2 has $|\downarrow >$ state,

ii) if one finds for particle 1 in $|\downarrow >$, then particle 2 in $|\uparrow >$,

iii) if one measures $\hat{S}_z$ then $+\tfrac{1}{2}\hbar$ for particle 1 implies that the value $-\tfrac{1}{2}\hbar$ for particle 2.

This expresses the non-locality nature of QM Theory. The experiment on particle 1 influences the result of the experiment on particle 2, although they are widely separated. The non-locality is a consequence of the existence of correlated many-particle states such as the direct product $|\uparrow > |\downarrow >$ and the fact that one can linearly superimpose such states.

EPR and the following argument in favour of hidden parameters in conjunction with the EPR thought experiment: By the measurement of S_z or S_x of particle 1, the values of S_z or S_x of particle 2 are known. Because of the separation of the particles, there was no influence on particle 2, and therefore, the values of S_z, S_x, etc. must have been fixed before the experiment. In the EPR argument, the consequences of the QM state (equation (10.3.13)) are used, but the inherent non-locality of quantum theory is denied.

The discussion of the EPR Paradox was useful for clarifying exactly what quantum theory implies, but neither Einstein nor Bohr was proposing that it could be used to make a direct test between the quantum and classical views f physical reality.

10.3.2 The Bell Inequality

In 1964 John Bell showed that the classical and quantum interpretations lead to different predictions in EPR type experiments. This opened the door for a series of increasingly rigorous tests of quantum theory, which could distinguish between *Einstein's local hidden variables* (naïve realism) and *conventional* quantum theory. Bell's theory was based on an idea for an EPR-type experiment proposed by David Bohm in 1951. Bohm simplified the ideas of EPR by applying them to a system involving correlated atomic spins. This had the advantage of giving a discrete set of spin axis directions rather than a correlation between continuous variables like position and momentum. Bell analyzed the correlation between **polarization** measurements using pairs of photons emitted in opposite directions from an excited atomic source.

Hidden variable theories assert that the photon polarizations states are objective properties of the individual photons determined at the moment of emission. Quantum Theory assumes that the two photons remain in a superposition of states until a measurement is carried out, at which point the wave function describing this superposition collapses and the polarization states are determined. Thus Quantum Theory leads to the conclusion that "**A measurement on one photon determines the instantaneous polarization states of both photons**". This generates a *correlation* between distant parts of the system, which is not present in the classical case.

Bell derived the inequality involving the frequencies of the different experimental results that would be *satisfied by* any classical hidden variables theory, but *violated by* Quantum Theory. This gave experimental physicist a simple prediction to test.

Out of the few EPR experiments, the experiment in 1976 by Alan Aspect can test between the predictions of separable hidden variables theories (for which Bell's inequality holds) and Quantum Mechanics (for which it is violated) . The Bell's inequality is *violated* in the Aspect experiment, and the results are In agreement with the predictions of Quantum Theory.

10.3.3 Against the COPENHAGEN Interpretation

Bohm began as a believer in the Copenhagen orthodoxy, probably because he studied under Bohr in Copenhagen. Thinking deeply on the status of quantum theory, and following the discussions he had with Einstein, he set about constructing an alternative *realistic hidden-variables* interpretation of quantum theory, based an idea initially proposed by Louis de Broglie in 1920s. Bohm developed this approach through a number of papers between 1950s and till his death in 1992. Accordingly, *hidden variables theory* can produce all the results of conventional theory and retain the reality of quantum objects.

10.4 QUANTUM INFORMATION: QUBITS (Quantum Bits):

By the 1990s advances in technology and experimental techniques led to a NEW PHYSICS (**Modern Physics** of the second half of the 20th Century) enable many of the thought experiments of the 1920s and 1930s to be carried out in the laboratory and opened the door to a variety of intriguing applications using the same counter intuitive ideas that challenged Einstein. Examples include *Quantum Cryptography*, **Quantum Computing**, *Quantum Teleportation* and Non-InteractiveMeasurements. As Steve Adams has put it, I quote; it is possible that we shall come to rely on quantum theory to secure our financial transactions and to allow rapid computations on **Super Quantum Computers**, or even to form images of objects we have never observed!

10.4.1 Quantum Computing

Quantum computing requires a 'quantum logic gate'. In conventional digital computing, data bits are zeros and ones, which must be stored. Any 'two-state device', such as on-off wall switch, could accomplish the task. Quantum Computing, however, uses **QBITS** in which the state is really a combination of two states. One of the ways being studied to produce a quantum logic gate uses a beryllium ion in an rf ion trap, the spin of the lone valence electron being the pair of states, either up or down (Phys. Rev. Lett., Dec. 18, 1995, p 4714) "*Quantum Chaos*"

The interface between chaos and Quantum Mechanics has become an area of challenge and interest because of the new ideas about the nature of chaos, introduced by quantum smoothing. It turns out that confined quantum systems from the nucleus to magnetically excited Atomic Orbitals (AOs) which should display chaos demonstrate a variety of subtle forms of repression of chaos which separates the energy levels, and converges in probability to the periodic repelling orbits called hidden in any chaotic system. Orbits with time thus tend to end on these periodic solutions. This phenomenon is called Quantum-'Scarring' of the wave function.

These constraints however begin to evaporate as soon as we leave confined systems and begin to enter the domain of unbound systems, such as electrons traversing a free molecular medium. This raises the distinct possibility that quantum chaos expressed in biological systems is right at the transition between the classical and quantum worlds. Chaotic systems may thus be able to amplify quantum effects into global fluctuations.

A key idea here is the "**square root of not**". Suppose we know an atom is excited by a certain amount of energy, but only shine a laser on it for half the time needed to provide this energy. Then the atom is in a superposition of the ground state and the excited state. If one then collapses the wave function, squaring it to its probability, as in

$P = \Psi^*\Psi$, it will be found to be in either the ground state or excited state with equal probability. One can use this uncertainty to perform a quantum calculation in the following way. Suppose one has a collection of such atoms which effectively form the 0s and 1s of a binary number 0 in the ground state and 1 in the excited state. If one then **partially** excites them all by giving them an energy only part way to the excited state, they each enter a superposition of excited and non-excited states and represent a superposition of all the binary numbers - e.g. for **two atom'bits'** - 00, 01, 10 and 11.

One can now devise a problem to solve - decrypting by factorizing a large number. One has a number register in two parts. The left part L is excited to a superposition. The right half R is designed to give the results of a *quantum factorizationremainder* of each of the possible numbers in L. These turn out to be periodic, so if we measure R we get one of the values.

To illustrate

a) Choose a random number x between 0 and n, then
b) Raise it to the power of the number in the L register.
c) Divide by n, and
d) Place the remainder in a second register.

It turns out that for increasing powers of x, the remainders form a repeating sequence. Because the number in the first register is different in each universe the result varies from universe to universe.

Worked out Example 10.1

Take the very simple case where the number to factorize, say 15, and $x = 2$.

Solution:

The powers of 2 give 2, 4. 8, 16, 32, 64, 128, 256 ...

Now divide by 15, and if the number won't go, keep the remainder.

That produces a repeating sequence 2, 4, 8, 1, 2, 4, 8, 1 ...

This in turn collapses L into a superposition of only those numbers with this particular value in R. Using a cunning trick with the L register which is consistent with wave functions; one can recombine them by interference to produce the Fourier Transform, replacing the wave by the *frequency spectrum*. When one now looks in L one finds a number which is the periodicity of the possible solutions in R. One can now use this quickly to find a factor of the number being sought.

The final observed value, the frequency f, has a good chance of revealing the factors of n from the expression $x^{f/2} - 1$ in the simple example above, the repeat sequence is the four values 2, 4, 8, 1, so the repeat frequency is 4. Thus **Shor's algorithm** produces the number: $(2^{4/2} - 1) = 3$ which is a factor of 15.

The essential principles of this calculation may pass over into a general problem solving paradigm –

(1) Make a superimposition of problem states,

(2) Find a representation of the solution which is periodic,

(3) Collapse this solution to one of its states,

(4) Transform the new superposition, and

(5) Measure the periodicity

to get the answer.

The idea that measurement of part of an entangled system may enable the whole system to collectively solve a problem connecting its entangled parts is the key.

^^*^*^*^*^*

Chapter 11

QUANTUM MECHANICS 3:

THE MATHEMATICAL FORMALISM

-MATRIX MECHANICS

Chapter 11

QUANTUM MECHANICS 3:
THE MATHEMATICAL FORMALISM
-MATRIX MECHANICS

"The plurality that we perceive is only an appearance, it is not real"

-Erwin Schrödinger

"Whatever you see as duality is unreal"   -Adi Shankara

11.1 INTRODUCTION

11.1.1 Schrödinger's Formalism

The "*Old Quantum Theory*" was superseded in 1925 by three different versions of quantum theory. One of these is the "*New Quantum Mechanics* or *Wave Mechanics*" (or Schrödinger Formalism). So far the discussions in this book were to formulate this more 'user-friendly' Wave Mechanics with the help of partial differential equations. The differential equations were to be solved for wave functions, $\Psi(r, t)$ which by themselves did not have any physical significance (i.e. represent any measurable quantities. But continuity of $\Psi(r)$ in position is an essential element in nature. Dynamical variables come out of this theory in the form of eigen values (discreteness). The fundamentals to this approach to Quantum Mechanics, first done for harmonic oscillator, are due to Erwin Schrödinger (1926), although it (developed from classical wave dynamics) is based heavily on the work of de Broglie's wave theory of matter. Degeneracy is a phenomenon that one encounters in systems described by partial differential equations. Schrödinger considered the physical entity (say, electron) as the de Broglie wave. This interpretation soon led to he difficulty since a wave may be partially reflected and partially transmitted at a boundary, but an electron cannot be split into two fractions for transmission and reflection. Max Born who proposed a statistical interpretation of the matter wave, which is now generally accepted, removed this difficulty. As a result of this it was rapidly developed into a general coherent system of mechanics, called Quantum Mechanics. It is an axiom in Quantum Mechanics that to every physical observable, there corresponds a hermitian operator and that all the set of eigen functions of a hermitian operator constitutes a *complete set*.

11.1.2 Heisenberg Formalism

Parallel to the advancement of W*ave Mechanics*, and even before, in 1925, in his historic paper, "On a quantum theoretical interpretation of kinematical and mechanical relations", Werner Heisenberg introduced a system of mechanics (developed from the Classical *Poisson brackets* of Hamiltonian Mechanics) in which the classical concepts of mechanics were drastically revised. He assumed that the atomic theory should emphasize observable quantities rather than the shapes of electronic orbits (Bohr's theory) in this formulation. Heisenberg adopted the *discrete quantum jumps* as the essential feature of the atomic phenomena. The Heisenberg approach spurns quantities that cannot be measured directly, such as probability amplitudes $\Psi(\mathbf{r}, t)$. Instead two discrete numbers, *i.e.* measurable quantities, are directly related with each other through the *calculus of matrices*, the rules of matrix algebra. So he used the algebra of matrices as the mathematical discipline. He found it necessary to introduce non-commutative algebra, say $[q, p] \neq 0$, in physics. He explained this successfully in 1925 in harmonic oscillator and explained the phenomenon of *zero-point energy*. It was M. Born and E. Jordan who extended this work to get $[q, p] = i\hbar$. The result was the development of the second version of quantum theory, known as <u>Matrix Mechanics</u> by Werner Heisenberg, Max Born, and Eccles Jordan at that time (1926).

11.1.3 Mathematical Equivalence of the Two Formalisms

Only later did Schrödinger recognize it how closely akin mathematically his wave mechanical and Heisenberg's matrix mechanical forms of the theory really are, *i.e.* the two are mathematically equivalent. Although both these formalisms are equivalent completely, it turns out that in some applications one is more convenient to use than the other.

11.1.4 Dirac's Formalism

Wolfgang Pauli, in January 1926, applied successfully the Heisenberg theory to the H-atom problem. In the same month PAM Dirac worked out a modified and more general formalism for the H-atom, based on these ideas, which is known as the <u>q-number Theory</u>. This is the third version of quantum theory. He also obtained the relation between classical Poisson Bracket $\{q, p\}$ and the quantum mechanical commutator $[q, p]$. Dirac has shown that the method of Schrödinger, seen in some earlier Chapters, becomes a special case.

11.1.5 Further Developments

In 1927, Dirac gave quantum mechanical description of the EM field. W. Pauli and C.G. Darwin introduced *spin* into Quantum Mechanics. Dirac in 1928 gave a *Relativistic Theory of Quantum Mechanics*, from which spin arose naturally. Measurement of **Lamb Shift** paved the way for the development of QED, and then to Quantized Fields (QFT).

11.2 MATHEMATICAL PRELIMINARIES

In the light of what have been seen in Section 11.1, the *concepts* of linear operators, linear spaces and vector spaces play a fundamental role.

The interpretations of the physics of quantum theory are based on the concept of the 'expectation value', which is essentially the *scalar product* of two vectors. The most thorough discussions of the foundations of Quantum Theory given by von Neumann and by Dirac are using the concepts of the theory of Hilbert space in which the scalar product plays a basic role. In the language of vector theory, the *operators* of the Schrödinger formalism appear as *matrices*, and the *wave functions* as representation of *vectors*

11.2.1 Linear Vector Spaces

Mathematical operations involving linear operators are often carried out by use of matrices; because *matrices transform in which linear relations among vectors are present*.

11.2.2 Elementary Ideas on Real (Ordinary) Vectors: *Ordinary Vector Algebra*.
Bold letter is used to indicate as the symbol of vector.

1) The representation of a vector $\vec{A}$ by an arrow means $\vec{A}$ starting from the origin O(0, 0, 0) terminates at the point P (x_1, y_1, z_1), in Cartesian Coordinates (*Right handed Cartesian Coordinate System*, O*xyz*). If an object located at O is displaced to point P, then the vector $\vec{A}$ is known as the **Displacement Vector**, $\vec{r}$ (Fig. 11.1).

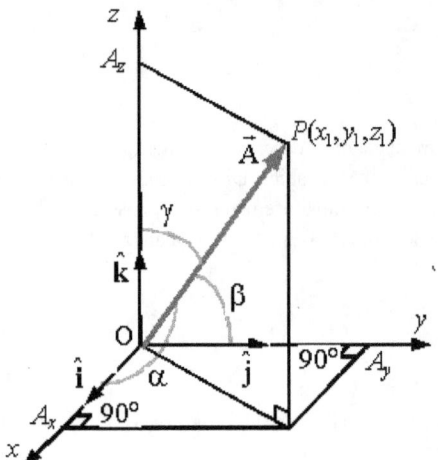

Fig. 11.1 Vector $\vec{A}$ depicted in a RH Coordinate system

2) $x_1 = \vec{r}\cos\alpha$; $y_1 = \vec{r}\cos\beta$; $z_1 = \vec{r}\cos\gamma$;   (11.2.1)

where $\cos\alpha$, $\cos\beta$, $\cos\gamma$ are called <u>direction cosines</u> and α, β, γ are the angles between $\vec{r}$ and the +x, +y, +z axes, respectively (Fig. 11.1); x_1, y_1, z_1 are called the projections (Cartesian components) of $\vec{r}$.

3) If $\hat{i}, \hat{j}, \hat{k}$ are the three unit vectors along the x, y, z directions,

$$\boxed{\vec{r} = (\hat{i}\, x_1 + \hat{j}\, y_1 + \hat{k}\, z_1)} = \text{Vector sum of the components of } \vec{r}. \quad (11.2.2)$$

4) <u>Scalar Product</u>: (*Inner product*, or *dot product*; Dot product symbol is "• "):

The scalar product of two vectors $\vec{A}$ and $\vec{B}$

$$\vec{A}\cdot\vec{B} = (\hat{i}\, A_x + \hat{j}\, A_y + \hat{k}\, A_z)\cdot(\hat{i}\, B_x + \hat{j}\, B_y + \hat{k}\, B_z) \quad (11.2.3)$$

for any set of axes x, y, z. Equivalently,

$$\boxed{\vec{A}\cdot\vec{B} = AB\,\cos\theta[= \angle(\vec{A},\vec{B})]} \quad (11.2.4)$$

The condition that $\vec{A}$ and $\vec{B}$ are perpendicular to each other is $\vec{A}\cdot\vec{B} = 0$.

5)   If $\hat{C} = \vec{A} + \vec{B}$,

$$\hat{C}^2 = (\vec{A} + \vec{B})^2 = A^2 + B^2 + 2\vec{A}\cdot\vec{B}$$

$$= A^2 + B^2 + 2 \cos\angle(\vec{A}, \vec{B})$$

Further one has $0 \cdot \vec{A} = 0$, \hfill (11.2.5)

$$\vec{A} \cdot \vec{B} = \vec{B} \cdot \vec{A},$$ \hfill (11.2.6)

$$(a\vec{A}) \cdot \vec{B} = a(\vec{A} \cdot \vec{B}) = \vec{A} \cdot (a\vec{B})$$ \hfill (11.2.7)

$$(\vec{A} + \vec{C}) \cdot \vec{B} = (\vec{A} \cdot \vec{B}) + (\vec{C} \cdot \vec{B}),$$ \hfill (11.2.8)

$$\vec{A} \cdot \vec{A} \geq 0,$$ \hfill (11.2.9)

$\vec{A} \cdot \vec{A} = 0$, implies $\vec{A} = 0$ \hfill (11.2.10)

Worked out Example 11.1

What is the angle between $\vec{A} = (1, 0, -5)$ and $\vec{B} = (3, 2, 4)$?

Solution: $\boxed{\text{STEP \# 1}}$ $\vec{A} \cdot \vec{B} = (\hat{i} A_x + \hat{j} A_y + \hat{k} A_z) \cdot (\hat{i} B_x + \hat{j} B_y + \hat{k} B_z)$

$\vec{A} \cdot \vec{B} = 3 + 0 - 20 = -17$

$\boxed{\text{STEP \# 2}}$ $\vec{A} \cdot \vec{B} = AB \cos[\theta = \angle(\vec{A}, \vec{B})]$

Angle, $\theta = \angle(\vec{A}, \vec{B})$; $-17 = 26 \times 29 \cos\theta$; which means $\theta - 128.3°$.

6) **Vector Product** or cross product (Vector product multiplication symbol is $\wedge$ or $\times$):

$$\boxed{\vec{a} \times \vec{b} \equiv \vec{a} \wedge \vec{b} \equiv a\, b\, \sin(\vec{a}, \vec{b})}.$$ \hfill (11.2.11)

where $\theta = \angle(\vec{a}, \vec{b})$ is the angle between the two vectors, and is consistent with the *right hand (RH) rule*.

Another way of defining the vector product is by taking an arbitrary set of axes with Unit vectors, $\hat{i}, \hat{j}, \hat{k}$

$$\vec{a} \wedge \vec{b} = (a_y b_z - b_z a_y)\hat{i} + (a_z b_x - b_x a_z)\hat{j} + (a_x b_y - b_y a_x)\hat{k}$$

It is used to define **torque**. For juggling with vector and scalar products, one gets the following three rules.

7) Triple Scalar Product

$$[\vec{A}\ \vec{B}\ \vec{C}] = \vec{A}\cdot(\vec{B}\wedge\vec{C}) = \vec{B}\cdot(\vec{C}\wedge\vec{A}) = \vec{C}\cdot(\vec{A}\wedge\vec{B})\,, \quad (11.2.12)$$

$$[\vec{A}\ \vec{B}\ \vec{C}] = (\hat{i}A_x + \hat{j}A_y + \hat{k}A_z)\begin{vmatrix} \hat{i} & \hat{j} & \hat{k} \\ B_x & B_y & B_z \\ C_x & C_y & C_z \end{vmatrix} \quad (11.2.13)$$

7) Triple Vector Product

$$\vec{A}\wedge\vec{B}\wedge\vec{C} = \vec{B}(\vec{A}\cdot\vec{C}) - \vec{C}(\vec{A}\cdot\vec{B}) \quad (11.2.14)$$

$$(\vec{B}\wedge\vec{C})\cdot(\vec{B}\wedge\vec{C}) = |\vec{B}\wedge\vec{C}|^2 = B^2 C^2 - (\vec{B}\cdot\vec{C})^2 .$$

8) If $\quad \vec{\nabla}\wedge(\vec{\nabla}\Phi) = (\vec{\nabla}\wedge\vec{\nabla}\Phi) = 0 \quad (11.2.15)$

this means the Curl of some vector expressed as a gradient of an appropriate scalar, the condition that its Curl is zero is customarily satisfied.

11.2.3 Linear Vector Space

The fundamental operations of ordinary vector algebra given in Section 11.2.2, also hold good, with the exception of the formation of vector products, in important operations which may be applied to functions of one or more independent variables which are defined and quadratically (twice) integrable (*i.e.* well behaved/ physically admissible quantum mechanically) over a common domain in the *configuration space*, whether the independent variables are familiar q or p of a dynamical system or not.

(a) Vector space defined

Thus a SET, V_n of elements $\{\vec{v}_i\}$ is called a VECTOR SPACE over a field F of non-zero elements (say *a, b, c, etc.*). The elements of a vector space are called **vectors**. The familiar 3-D space of positions vectors over the field of real numbers is an example of a vector space.

(b) Idea of LENGTH of a Vector

The idea of length of a vector finds expression in by forming the scalar product. The introduction of this operation converts the affine geometry of vectors into a metric geometry. Thus the Schrödinger Wave Equation *can be regarded as a representation of a vector*.

(c) **Scalar product** of Complex Vectors

It has been seen earlier that the scalar product has certain important properties. For a pair of ordinary vectors, $\vec{A}$ and $\vec{B}$,

$$\vec{A} \cdot \vec{B} = (\hat{i} A_1 + \hat{j} A_2 + \hat{k} A_3)(\hat{i} B_1 + \hat{j} B_2 + \hat{k} B_3)$$

$$\boxed{\vec{A} \cdot \vec{B} = (A_1 B_1 + A_2 B_2 + A_3 B_3) = \sum A_k B_k} \qquad (11.2.16)$$

With the ordinary vector algebra applied, the vector space, $V_n(F)$ over a field F is called an <u>inner scalar product space</u>. Similarly, in Euclidean space, E_n (a complex vector space of n-Dimensions is an ordered set of n complex numbers) a scalar product associates a complex number to any pair of vectors.

(d) Axioms

Defining the <u>hermitian scalar product</u> of two complex vectors, $\vec{A}$ and $\vec{B}$, by the equation

$$\boxed{(\vec{A}, \vec{B}) = \sum A_k^* B_k = \lambda}, \text{ a complex number.} \qquad (11.2.17)$$

Scalar product associates a complex number to any pair of vectors. Thus the introduction of the complex number insures that, when $\vec{B}$ is identified with $\vec{A}$, $(\vec{A} \cdot \vec{A}) =$ real and positive, or zero. This means relations (11.2.10) and (11.2.11) are invalid and instead

(9) $\boxed{(\vec{A}, \vec{B}) = (\vec{B}, \vec{A})^*}$ (11.2.18)

which means $(\vec{A}, \vec{B})$ is <u>anti-linear</u>(or *conjugate linear*) in its first argument, i.e.

(10) $(a \vec{A}, \vec{B}) = a^* (\vec{A}, \vec{B}) = (\vec{A}, a^* \vec{B})$ or

$$\boxed{(a \vec{A}, b \vec{B}) = a^* b (\vec{A}, \vec{B})} \qquad (11.2.19)$$

All the remaining laws for real vectors stated earlier hold good for complex vectors.

(11) $(\vec{C}, a \vec{A} + b \vec{B}) = a (\vec{C}, \vec{A}) + b (\vec{C}, \vec{B})$ (11.2.20)

11.2.4 LINEAR INDEPENDENCE

If λ, μ, ν are 3 *non-zero* scalars such that given the three vectors of the set $V_3(F)$, viz. $\vec{\alpha}, \vec{\beta}, \vec{\gamma}$ in the field F satisfy

$$\lambda \vec{\alpha} + \mu \vec{\beta} + \nu \vec{\gamma} = 0 \qquad (11.2.21)$$

then the set of vectors $\vec{\alpha}, \vec{\beta}, \vec{\gamma}$ are said to be <u>linearly independent set</u> in the field F, except for the trivial case,

$$\lambda = \mu = \nu = 0.$$

If λ, μ, ν are not of such a set, then

$$\lambda = \mu = \nu \neq 0$$

$$\boxed{\begin{array}{c} \lambda \vec{\alpha} + \mu \vec{\beta} + \nu \vec{\gamma} = 0 \\ \lambda = \mu = \nu \neq 0 \end{array}} \text{ – linearly independent set} \qquad (11.2.22)$$

This means $\vec{\alpha}, \vec{\beta}, \vec{\gamma}$ are said to be a set of <u>linearly independent vectors</u>, in F. In other words, none of these three vectors can be written as linear combinations of the other two. If a scalar a_{ii} is such that $a_{ii} = (\psi_i, \psi_i)$, then the vectors are linearly independent if the **gram determinant**, $|\Gamma| > 0$.

If $\qquad \lambda [\vec{\alpha} \vec{\beta} \vec{\gamma}] = \lambda \vec{\alpha} \cdot (\vec{\beta} \wedge \vec{\gamma})$

$$= \begin{vmatrix} \alpha_\lambda & \alpha_\mu & \alpha_\nu \\ \beta_\lambda & \beta_\mu & \beta_\nu \\ \gamma_\lambda & \gamma_\mu & \gamma_t \end{vmatrix}$$

then $\qquad |\Gamma| = [\vec{\alpha} \vec{\beta} \vec{\gamma}] > 0 \qquad (11.2.24)$

or, there is another way of dealing with linear independency of vectors.

Let $\vec{\delta} = (c_1 \hat{u} + c_2 \hat{v} + c_3 \hat{w}) = 0$, means $\hat{u}, \hat{v}$ and $\hat{w}$ are linearly dependent; *i.e.* the *Wronskian determinant*, W

$$|W| = \begin{vmatrix} \hat{u} & \hat{v} & \hat{w} \\ \partial u/\partial x & \partial v/\partial y & \partial w/\partial z \\ \partial^2 u/\partial x^2 & \partial^2 v/\partial y^2 & \partial^2 w/\partial z^2 \end{vmatrix} = 0 \qquad (11.2.25)$$

11.2.3.1 Theorem # 1

The set of vectors $\vec{\alpha}$, $\vec{\beta}$, and $\vec{\gamma}$ are said to be <u>linearly dependent set</u>, then

$$\boxed{|\Gamma| = [\vec{\alpha}\ \vec{\beta}\ \vec{\gamma}] = 0} \quad . \tag{11.2.26}$$

and $\vec{\alpha}$, $\vec{\beta}$, and $\vec{\gamma}$ are <u>coplanar</u> *with the origin*.

On the other hand, $\vec{\alpha}$, $\vec{\beta}$, and $\vec{\gamma}$ are **non-coplanar** *with the origin*, they are said to form a set of ***linearly independent*** vectors. $|\Gamma| > 0$, *i.e.*,

$$|W| = [\vec{\alpha}\ \vec{\beta}\ \vec{\gamma}] = \begin{vmatrix} \alpha_\lambda & \alpha_\mu & \alpha_\nu \\ \beta_\lambda & \beta_\mu & \beta_\nu \\ \gamma_\lambda & \gamma_\mu & \gamma_\iota \end{vmatrix} \neq 0$$

or $\boxed{\begin{array}{c} \vec{r} = \lambda\vec{\alpha} + \mu\vec{\beta} + \nu\vec{\gamma} \neq 0 \\ \lambda = \mu = \nu \neq 0 \end{array}}$ — linearly independent set & Non-coplanar (11.2.27)

11.2.3.2 DIMENSIONS OF A VECTOR SPACE: Theorem # 2

A vector space is said to be three-dimensions (3-D) if it contains precisely three linearly independent vectors. Then if $\vec{\alpha}$, $\vec{\beta}$, and $\vec{\gamma}$ are linearly independent triad (set V_3) then any other (4th) vector **r** may be written as

$$\vec{r} = \lambda\vec{\alpha} + \mu\vec{\beta} + \nu\vec{\gamma} \neq 0, \tag{11.2.27}$$

such that

$$\boxed{\begin{array}{l} \lambda = [\vec{r}\ \vec{\beta}\ \vec{\gamma}]/[\vec{\alpha}\ \vec{\beta}\ \vec{\gamma}] \\ \mu = [\vec{r}\ \vec{\gamma}\ \vec{\alpha}]/[\vec{\alpha}\ \vec{\beta}\ \vec{\gamma}] \\ \nu = [\vec{r}\ \vec{\alpha}\ \vec{\beta}]/[\vec{\alpha}\ \vec{\beta}\ \vec{\gamma}] \end{array}} \tag{11.2.28}$$

But the four vectors $\vec{r}$, $\vec{\alpha}$, $\vec{\beta}$, and $\vec{\gamma}$ are linearly dependent in 3- space. $\vec{\alpha}$, $\vec{\beta}$, and $\vec{\gamma}$ are known as <u>3-vectors</u> (*i.e.* 3-D vectors).

11.2.3.3 Theorem # 3 : Reciprocal Vectors

If $\vec{\alpha}$, $\vec{\beta}$, and $\vec{\gamma}$ are linearly independent triad (set $V_3(F)$) then there exists a <u>reciprocal triad</u>, $\vec{\alpha}'$, $\vec{\beta}'$, and $\vec{\gamma}'$ defined by

$$\boxed{\begin{aligned}\vec{\alpha}' &= [\vec{\beta} \wedge \vec{\gamma}] / [\vec{\alpha}\,\vec{\beta}\,\vec{\gamma}] \\ \vec{\beta}' &= [\vec{\gamma} \wedge \vec{\alpha}] / [\vec{\alpha}\,\vec{\beta}\,\vec{\gamma}] \\ \vec{\gamma}' &= [\vec{\alpha} \wedge \vec{\beta}] / [\vec{\alpha}\,\vec{\beta}\,\vec{\gamma}]\end{aligned}}$$
(11.2.29)

11.2.4 BASIS (Coordinate System) TRIAD OF VECTORS

11.2.4.1 PROPERTIES OF BASIS VECTORS: linear independency

A basis of a vector space is set of linearly independent vectors such that every vector in $V_3(F)$ is a linear combination of vectors in the basis.

i.e., $[\hat{i}\ \hat{j}\ \hat{k}] \neq 0$

In addition to the linear independency among basis vectors, there are three other properties. The introduction of the scalar product of vectors permits one to 'metrize' the vector space, i.e. to define length and distance. It is desirable to have the LENGTH (*i.e.* positive square root of norm or absolute magnitude) of a vector $\vec{X}$ is denoted as $\|\vec{X}\|$

$\|\vec{X}\| = +(\vec{X}, \vec{X})^{1/2}$ = a real number with positive sign. (11.2.30)

11.2.4.2 NORMALIZATION

$$\boxed{(\hat{i}, \hat{i}) = 1\,;\,(\hat{j}, \hat{j}) = 1;\,(\hat{k}, \hat{k}) = 1}$$
(11.2.31)

Equations (11.2.31) mean $\|\hat{i}\| = 1$, *etc.*

11.2.4.2 ORTHOGONALITY

$$\boxed{(\hat{i}, \hat{j}) = 0\,;\,(\hat{j}, \hat{k}) = 0;\,(\hat{k}, \hat{i}) = 0}$$
(11.2.32)

These mean that the three vectors are mutually orthogonal.

Thus the 3-vectors $(\hat{i}, \hat{j}, \hat{k})$ are said to have *unit modulus*, and they are both *orthogonal and normalized*. That is, they are ortho-normal.

Such a triad of 3-vectors $(\hat{i}, \hat{j}, \hat{k})$ which form an **ortho-normal set** is called a set of base vectors of the 3-space. $[\hat{i}\ \hat{j}\ \hat{k}] \neq 0$.

The Right-Handed Rectangular (Cartesian) Coordinate frame, Oxyz is shown in Fig. 11.2. The simplest property of 3-space is that an ortho-normal triad of vectors can be found which form a coordinate system (i.e. basis) to which all the

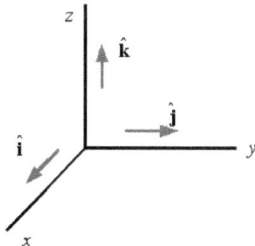

Fig. 11.2 Basis in RH Cartesian coordinate Frame

3-vectors. Such a set is also referred to act as an orthogonal basis of representation of a vector. The basis is clearly not unique. Thus x, y, z are Cartesian (Des Cartes, 1596 – 1650) coordinates; $(\hat{i}, \hat{j}, \hat{k})$ are <u>unit vectors,</u> then any **arbitrary vector**

$$\boxed{\vec{x} = (\hat{i}\,a + \hat{j}\,b + \hat{k}\,c)}. \tag{11.2.33}$$

11.2.4.3 n-SPACE

Many algebraic symbols can be conveniently sorted out into sets of n, say the numbers a_{1i} in the nth order determinant ($i = 1, 2, 3, .., n$). Such a set is called the *n*-components of 'vector' (the word is used in a *mathematical* and not physical sense) in space of n-dimensions, an *n*-space. Unlike the 3-D space, one should not try to visualize the geometry of an *n*-space. An *n*-space is algebraic in its conception. Just like the case of 3-space, an n-space is n dimensional because there exists *n* linearly independent <u>n-vectors,</u> but no $(n+1)$ *n*-vectors are linearly independent. It then implies that they form an n-dimensional coordinate system or basis. It should be emphasized that there is nothing unique about the choice of the basis, for any set of n linearly independent vectors are suitable for that purpose. But the most convenient choice, in general, is a set of n UNIT VECTORS, $\hat{e}_i$, defined by

$$\hat{e}_1 = \{1, 0, 0, \ldots 0, 0\}$$
$$\hat{e}_2 = \{0, 1, 0, \ldots 0, 0\}$$
$$\hat{e}_3 = \{0, 0, 1, \ldots 0, 0\}$$

$\cdot$

$\cdot$

$$\hat{e}_n = \{0, 0, 0, \ldots 0, 1\}$$

(11.2.34)

in $V_n(F)$. They are such that

$$\boxed{(\hat{e}_i, \hat{e}_j) = \delta_{ij}}.$$ (11.2.35)

Any arbitrary vector is expressed as

$$\boxed{\vec{X} = (x_1 \hat{e}_1 + x_2 \hat{e}_2 + \cdot + \cdot + x_n \hat{e}_n)}$$ (11.2.36)

Worked out Example 11.2

Find the scalar multipliers and write the vector [1,0,0] as a linear combination of the set of 3-vectors $\vec{X}^{(1)} = [1, 0, -1], \vec{X}^{(2)} = [0, 1, 0], and \vec{X}^{(3)} = [1, 0, 1]$.

Solution: $\boxed{\text{STEP \# 1}}$ Check if the Gram determinant $|\Gamma| > 0$.

$$|\Gamma| = \begin{vmatrix} 1 & 0 & -1 \\ 0 & 1 & 0 \\ 1 & 0 & 1 \end{vmatrix} = 2$$

$\boxed{\text{STEP \# 2}}$ Write the vector under question as a linear combination;

$$\vec{r} = [1, 0, 0] = \lambda \vec{X}^1 + \mu \vec{X}^2 + \nu \vec{X}^3.$$

$\boxed{\text{STEP \# 3}}$ check if $\vec{r} = \lambda \vec{X}^1 + \mu \vec{X}^2 + \nu \vec{X}^3 \neq 0$, as per equation (11.2.33).

For this determine the scalars using relations (11.2.28),

$$\lambda = [\vec{r} \vec{X}^{(2)} \vec{X}^{(3)}]/[\vec{X}^1 \vec{X}^2 \vec{X}^3] = [\vec{r} \vec{X}^{(2)} \vec{X}^{(3)}]/2 = 1/2$$
$$\mu = [\vec{r} \vec{X}^{(3)} \vec{X}^{(1)}]/[\vec{X}^1 \vec{X}^2 \vec{X}^3] = [\vec{r} \vec{X}^{(2)} \vec{X}^{(3)}]/2 = 0$$
$$\nu = [\vec{r} \vec{X}^{(3)} \vec{X}^{(1)}]/[\vec{X}^1 \vec{X}^2 \vec{X}^3] = [\vec{r} \vec{X}^{(2)} \vec{X}^{(3)}]/2 = 1/2$$

$\boxed{\text{STEP \# 4}}$: Write the vector.

$$\vec{r} = [1, 0, 0] = (1/2) \vec{X}^1 + (0) \vec{X}^2 + (1/2) \vec{X}^3 = (1/2)[1, 0, -1] + (1/2)[1, 0, 1].$$

11.2.4.5. *Theorem* Inner product of a vector with itself is positive definite.

Proof:

Consider two vectors $\vec{\psi}_a$ & $\vec{\psi}_b$

Let $\vec{\psi}_a = \sum a_i \hat{e}_i$ and $\vec{\psi}_b = \sum b_j \hat{e}_j$,

where $(\hat{e}_i, \hat{e}_j) = \delta_{ij}$. (11.2.35)

$\therefore \quad (\vec{\psi}_a, \vec{\psi}_b) = (\sum a_i \hat{e}_i, \sum b_j \hat{e}_j) = \sum\sum a_i^* \hat{e}_i (\hat{e}_i, \hat{e}_j)$

$= \sum\sum a_i^* b_j \delta_{ij} = \sum a_i^* b_i$

In particular, $(\vec{\psi}_a, \vec{\psi}_a) = \sum a_i^* a_i$

i.e., $\|\vec{\psi}_a\|^2 = \sum |a_i|^2$

$\therefore \quad (\vec{\psi}_a, \vec{\psi}_a) \geq 0$.

$(\vec{\Psi}, \vec{\Psi}) \geq 0$ (11.2.37)

11.2.4.6 CAUCHY-SCHWARZ's Inequality

The Cauchy-Schwarz inequality states that two vectors, $\vec{\psi}_a$ & $\vec{\psi}_b$, satisfy the relation

$$(\vec{\psi}_a, \vec{\psi}_b)(\vec{\psi}_b, \vec{\psi}_a) \geq |(\vec{\psi}_a, \vec{\psi}_b)|^2 \quad (11.2.38)$$

Proof: It is proved earlier that

$(\vec{\Psi}, \vec{\Psi}) \geq 0$. (11.2.37)

Let $\Psi = \vec{\psi}_a + b\,\vec{\psi}_b$.

$(\vec{\Psi}, \vec{\Psi}) = (\vec{\psi}_a + b\,\vec{\psi}_b, \vec{\psi}_a + b\,\vec{\psi}_b)$

$= (\vec{\psi}_a, \vec{\psi}_a) + b(\vec{\psi}_a, \vec{\psi}_b) + b^*(\vec{\psi}_b, \vec{\psi}_a) + b^* b(\vec{\psi}_b, \vec{\psi}_b)$

To minimize the left hand side for finding the corresponding *b*-value, one has to differentiate with respect to *b* (or *b**):

$$\{(\partial/\partial b)(\vec{\Psi}, \vec{\Psi})\}\big|_{min} = 0,$$

gives $\quad \{(\partial/\partial b)(\vec{\Psi}, \vec{\Psi})\}\big|_{min} = 0 + (\vec{\psi}_a, \vec{\psi}_b) + 0 + b^*(\vec{\psi}_b, \vec{\psi}_b)$

∴ $\quad (\vec{\psi}_a, \vec{\psi}_b) = -b^*_{min}(\vec{\psi}_b, \vec{\psi}_b).$

$$b^*_{min} = -(\vec{\psi}_a, \vec{\psi}_b)/(\vec{\psi}_b, \vec{\psi}_b)$$

Substitute this value of b^*_{min} into $(\vec{\Psi}, \vec{\Psi})$ to obtain

$$(\vec{\Psi}, \vec{\Psi}) = (\vec{\psi}_a, \vec{\psi}_a) + b(\vec{\psi}_a, \vec{\psi}_b)$$
$$+ [-(\vec{\psi}_a, \vec{\psi}_b)/(\vec{\psi}_b, \vec{\psi}_b)](\vec{\psi}_b, \vec{\psi}_a)$$
$$+ [-(\vec{\psi}_a, \vec{\psi}_b)/(\vec{\psi}_b, \vec{\psi}_b)]b(\vec{\psi}_b, \vec{\psi}_b) = 0$$

$$= (\vec{\psi}_a, \vec{\psi}_a) + b(\vec{\psi}_a, \vec{\psi}_b) + [(\vec{\psi}_a, \vec{\psi}_b)(\vec{\psi}_b, \vec{\psi}_a)/(\vec{\psi}_b, \vec{\psi}_b)] - b(\vec{\psi}_a, \vec{\psi}_b) \geq 0$$

$$(\vec{\Psi}, \vec{\Psi}) = (\vec{\psi}_a, \vec{\psi}_b)(\vec{\psi}_b, \vec{\psi}_b)] - |(\vec{\psi}_a, \vec{\psi}_b)|^2 \geq 0$$

∴ $\quad (\vec{\psi}_a, \vec{\psi}_b)(\vec{\psi}_b, \vec{\psi}_b)] \geq |(\vec{\psi}_a, \vec{\psi}_b)|^2$

This is <u>Cauchy-Schwarz inequality</u>. Only if $\vec{\psi}_b = \lambda\vec{\psi}_a$, λ = a scalar constant. when $\vec{\psi}_b \parallel \vec{\psi}_a$, or $\vec{\psi}_b$ and $\vec{\psi}_a$ are *Collinear*, the equality is satisfied.

11.2.5 HILBERT SPACE

In this Section the concept of a **Hilbert space** will be introduced. A Hilbert space is much the same type of object as n-Dimensional function space. A function space can be made an inner product space if one associates with any two functions f and g a scalar

$$(f, g) = \int f^*(x) g(x) dx,$$

and $\quad \| f \|^2 = (f, f) = \int | f(x) |^2 dx.$

11.3.1 COMPLETENESS RELATION

If a function space is not n-Dimensional for any finite number n, then it is said to be infinite-Dimensional, *i.e.* it contains arbitrarily large but countable number of functions so that the space is infinite-Dimensional. Here the *completeness relation* is

$$\underset{n\to\infty}{Lt}\ \|\vec{\Psi} - \sum_1^n a_i\ \hat{e}_i\ \| = 0,\qquad(11.2.38)$$

$a_i = (\vec{\Psi},\ \hat{e}_i)$ is satisfied. In physical applications this is equivalent to $\vec{\Psi} = \sum a_i\ \hat{e}_i$. i.e. every Cauchy sequence of elements belonging to an inner-product space has a limit which also belongs to the algebraic, Euclidean V_n, space of infinitely many dimensions known as the Hilbert space, $\mathfrak{A}$.

In Quantum Mechanics one is interested only in Hilbert space functions, defined in any interval $a < x < b$, i.e. $(\psi, \psi) = \|\psi\|^2 = \int \psi(x)^* \psi(x)\ dx = $ finite, always. Example is the complete set of Hermite polynomials of the harmonic oscillator.

11.3.2 PROPERTIES of a Hilbert space, $\mathfrak{A}$

(1) The space is linear; *i.e.* if $a = $ constant and φ is any element of the space, then $a\varphi$ is also an element of the space, addition of any two elements of the space is also an element of the space.
(2) There is an inner product, (φ, Ψ), for any two elements φ and Ψ in the space.
(3) For functions, defined in any interval $a \le x \le b$.

$$(\Psi,\Psi) = \|\Psi\|^2 = \int_a^b \Psi(x)^* \Psi(x)\ dx = \text{finite, always.}$$

(4) An element of $\mathfrak{A}$ of has a norm, that is related to the inner product,
$$(\text{norm of}\Psi)^2 = \|\Psi\|^2 = (\Psi,\Psi).$$
(5) $\mathfrak{A}$ is complete.

11.3.3 MATRICES

A *physical vector* such as, **velocity**, force or torque, is described by its components referred to a fixed set of axes. The idea of a set of physical variables as constituting a <u>vector</u> can be extended to non-symmetrical systems whose **dimensions** then refer to the # of degrees of freedom of the system, *i.e.* the variables needed to identify what state it is in. Two such vectors may be linearly related through an array of coefficients (mathematical quantities). There are many physical problems of interest that are most conveniently described by writing down an array of coefficients or matrix. These arrays are the dealt with according to a set of predetermined rules. Two types of arrays commonly used in Quantum Mechanics are determinants and matrices. Although knowledge of matrices is assumed, it will be useful to summarize the prominent properties of matrices.

11.3.3.1 Singular Matrix

The elements of a square matrix can be written and evaluated as a determinant. If det A = 0, the corresponding matrix A is called *singular matrix*. Since determinants do not exist for rectangular arrays, all rectangular matrices are, by definition, singular.

11.3.3.2 PROPERTIES of Matrices

i) **Equality**: Two matrices A and B are of the same order such that A = B, if and only if, $a_{ij} = b_{ij}$.

ii) **Addition**: A + B = C = B + A, means $(A + B)_{ij} = c_{ij} = a_{ij} + b_{ij}$.

Thus y = C x means

$$y_i = c_{ij} = (a_{ij} + b_{ij})x_i = a_{ij}x_i + b_{ij}x_i = Ax + Bx = Cx$$

(11.2.39)

Similarly, A + (B + C) = (A + B) + C.

iii) **Scalar Multiplication**: If a scalar, λ multiplies A, the λA means each element of A is multiplied by λ.

Example:

If $\lambda = 2$, and $A = \begin{pmatrix} 2 & 9 \\ 4 & 3 \end{pmatrix}$; $\lambda A = 2 \begin{pmatrix} 2 & 9 \\ 4 & 3 \end{pmatrix} = \begin{pmatrix} 4 & 18 \\ 8 & 6 \end{pmatrix}$

iv) **Matrix multiplication**: Product of two matrices A and B is given by the "**row-on-column**" Rule, viz. to find the i/j element of the product matrix C of A and B, (i.e. A B = C), multiply each element of row i in the left hand matrix (A) by the corresponding element in column j of the right hand matrix (B), and sum.

$$(AB)_{ij} = (\psi_i, AB\psi_i) = (\psi_i, A\sum \psi_k B_{kj}) = (\sum \psi_i, A\psi_k) B_{kj} = \sum (A_{ik} B_{kj})$$

i.e., $C_{ij} = \sum (A_{ik} B_{kj})$, or $c_{ij} = (a_{ik} b_{kj})$ (11.3.1)

v) Matrix multiplication is not commutative:

AB ≠ BA

$$A = \begin{pmatrix} 2 & 9 \\ 4 & 3 \end{pmatrix}, B = \begin{pmatrix} 1 & 5 \\ 7 & 2 \end{pmatrix}$$ (11.3.2)

$$AB = \begin{pmatrix} 2 & 9 \\ 4 & 3 \end{pmatrix}\begin{pmatrix} 1 & 5 \\ 7 & 2 \end{pmatrix} = \begin{pmatrix} 65 & 28 \\ 25 & 26 \end{pmatrix}; BA = \begin{pmatrix} 1 & 5 \\ 7 & 2 \end{pmatrix}\begin{pmatrix} 2 & 9 \\ 4 & 3 \end{pmatrix} = \begin{pmatrix} 22 & 24 \\ 22 & 69 \end{pmatrix}$$

i.e., $\boxed{AB \neq BA.}$ (11.3.2)

vi) Determinant of Product

$$AB| = |BA| = |A| \cdot |B|$$

$$|A| = -30, |B| = -33.$$

$$|AB| = (-30)(-33) = 990 = |BA|.$$

vii) NULL matrix, 0:

The identity operation for addition is called the null matrix, 0.

$$A + 0 = A, \text{ for all } A. \tag{11.3.3}$$

viii) UNIT (Idem) matrix, $\Im$: The identity operation for multiplication is called the unit matrix, $\Im$. The idem matrix is a diagonal matrix.

$$\Im = \begin{pmatrix} 1 & 0 & 0 \\ 0 & 1 & 0 \\ 0 & 0 & 1 \end{pmatrix} \tag{11.3.4}$$

ix) Symmetric Matrix

$$\boxed{a_{ij} = a_{ji}} \tag{11.3.5}$$

The matrix A is called symmetric matrix. A special case of symmetric matrix is the *diagonal matrix*.

x) Skew Symmetric Matrix

If $\boxed{a_{ij} = -a_{ji}}$ (11.3.6)

the matrix A is called Skew-symmetric matrix.

xi) Transpose of a matrix, A':

A' is a matrix A having rows and columns interchanged,

i.e., $\boxed{(a_{ij})' = a_{ji}}$

$$A = \begin{pmatrix} 1 & -1 & 2 \\ 3 & 0 & 1 \end{pmatrix}, \quad A' = \begin{pmatrix} 1 & 3 \\ -1 & 0 \\ 2 & 1 \end{pmatrix}$$

If A = A', then A is symmetric matrix.

$$(A + B)' = A' + B'. \tag{11.3.7}$$

$(A B)' = B' A'$. (11.3.8)

$(A x)' = x' A'$. (11.3.9)

xii) <u>Adjoint of matrix</u>, Adj A:*Adjoint* matrix is defined as the matrix, which is the transpose of the matrix with elements, which are *cofactors* of the original matrix A, i.e. it is the transpose of its cofactor matrix.

$$\text{Adj } A = \text{Cof}(A_{kj})' = \text{Cof}(A)' = \text{Cof}(A_{ji})$$ (11.3.10)

If $A = \begin{pmatrix} 2 & 5 & 3 \\ 3 & 1 & 2 \\ 1 & 2 & 1 \end{pmatrix}$; $\text{Cof } A = \begin{pmatrix} -3 & -1 & 5 \\ 1 & -1 & 1 \\ 7 & 5 & 13 \end{pmatrix}$; $\text{Adj } A = (\text{Cof } A)' = \begin{pmatrix} -3 & 1 & 7 \\ -1 & -1 & 5 \\ 5 & 1 & -13 \end{pmatrix}$

∴ xiii) <u>Inverse (Reciprocal) Matrix</u>, A^{-1}: A^{-1} is defined as the reciprocal of A, such that $\det A \neq 0$.

i.e., $$A^{-1}A = AA^{-1} = \Im$$ (11.3.11)

That is a matrix commutes with its inverse.

$$(A B C)^{-1} = C^{-1} B^{-1} A^{-1}.$$

∴ $$A^{-1} = (\text{Adj } A) / |A|.$$ (11.3.12)

If $A = \begin{pmatrix} 2 & 5 & 3 \\ 3 & 1 & 2 \\ 1 & 2 & 1 \end{pmatrix}$; $\text{Cof } A = \begin{pmatrix} -3 & -1 & 5 \\ 1 & -1 & 1 \\ 7 & 5 & 13 \end{pmatrix}$.

$\text{Adj } A = (\text{Cof } A)' = \begin{pmatrix} -3 & 1 & 7 \\ -1 & -1 & 5 \\ 5 & 1 & 13 \end{pmatrix}$ $\det A = 2(-3) - 5(1) + 3(5) = 4$.

∴ $A^{-1} = (\text{Adj } A) / |A| = (\frac{1}{4}) \begin{pmatrix} -3 & 1 & 7 \\ -1 & -1 & 5 \\ 5 & 1 & -13 \end{pmatrix}$

On the other hand, if one wants to find out the inverse of the matrix

$$\begin{pmatrix} 2 & 3i \\ 4 & 6i \end{pmatrix}$$

Calculation shows that the determinant of this matrix is zero, and so it is a singular matrix; under these circumstances none can write its inverse.

xiii) <u>Orthogonal Matrix</u>: Matrix A is orthogonal if

$$\boxed{A^{-1} = A'.} \qquad (11.3.13)$$

If $A = \begin{pmatrix} 1 & 2 & 3 \\ 1 & 3 & 5 \\ 1 & 5 & 12 \end{pmatrix}$; $Cof\ A = \begin{pmatrix} 11 & -7 & 2 \\ -9 & 9 & -3 \\ 1 & -2 & 1 \end{pmatrix}$

$Adj\ A = (Cof\ A)' = \begin{pmatrix} 11 & -9 & 1 \\ -7 & 9 & -3 \\ 1 & -2 & 1 \end{pmatrix}$

$\therefore \quad A^{-1} = Adj\ A/\ |A| = (\tfrac{1}{3}) \begin{pmatrix} 11 & -9 & 1 \\ -7 & 9 & -2 \\ 2 & -3 & 1 \end{pmatrix}$

$A^{-1}A = (\tfrac{1}{3}) \begin{pmatrix} 11 & -9 & 1 \\ -7 & 9 & -2 \\ 2 & -3 & 1 \end{pmatrix} \begin{pmatrix} 1 & 2 & 3 \\ 1 & 3 & 5 \\ 1 & 5 & 12 \end{pmatrix} = \begin{pmatrix} 1 & 0 & 0 \\ 0 & 1 & 0 \\ 0 & 0 & 1 \end{pmatrix} = \Im$, is satisfied;

So A^{-1} is inverse of A.

xv) <u>Hermitian Adjoint</u> (Hermitian conjugate), $A^\dagger$: Sofar the discussions were on matrices with real numbers. If the elements are in general complex numbers, corresponding to the symmetric matrix, $a_{ij} = a_{ji}$,

One may define a Hermitian adjoint of an arbitrary matrix A whose elements are such that

$$\boxed{(a_{ij})^* = a_{ji}} \qquad (11.3.14)$$

where a^* is the complex conjugate of a

i.e., $\quad A^\dagger = A^{*\prime}$ \qquad (11.3.15)

If $\quad A = \begin{pmatrix} 2+3i & 4-5i \\ 3 & 4i \end{pmatrix}$

$A^\dagger = A^{*\prime} = \begin{pmatrix} 2-3i & 4+5i \\ 3 & -4i \end{pmatrix}' = \begin{pmatrix} 2-3i & 3 \\ 4+5i & -4i \end{pmatrix}$

xvi) <u>Hermitian Self-Adjoint</u> (Hermitian)

This further implies that $\quad \boxed{(a_{ii})^* = a_{ii}}$

i.e., if $\quad \boxed{A^\dagger = A^{*\prime} = adj\ A = A} \qquad (11.3.16)$

then A is said to be <u>Hermitian</u>. Thus the diagonal elements of a Hermitian matrix are real.

$$A = \begin{pmatrix} a & b+ic \\ b-ic & d \end{pmatrix}; \quad A' = \begin{pmatrix} a & b-ic \\ b+ic & d \end{pmatrix}; \quad A'^* = \begin{pmatrix} a & b+ic \\ b-ic & d \end{pmatrix} = A$$

| Theorem | $(AB)^\dagger = B^\dagger A^\dagger$, | (11.3.17) |

xvii) <u>TRACE (Spur/ Character) of a matrix</u>, Tr A: Tr A is defined as the sum of the principal diagonal elements of matrix A.

$$\boxed{Tr\, A = \sum a_{ii} = A_{ii}} \tag{11.3.18}$$

Note: If a subscript occurs twice, i.e repeated, in an expression, then summation in all values of it is implied.

1. $\quad Tr(ABCD) = Tr(BCDA) = Tr(CDAB) = Tr(DABC)$.

2. $\quad Tr(AB - BA) = Tr(AB) - Tr(BA) = 0.$ (11.3.19)

3. $\quad Tr(A + B) = Tr(A) + Tr(B).$

xix) $\quad$ If $A = a, \quad B = \begin{pmatrix} b & c \\ d & e \end{pmatrix}; \quad C = \begin{pmatrix} f & g & h \\ i & j & k \\ l & m & n \end{pmatrix}$

the DIRECT SUM, $D = A \oplus B \oplus C = \begin{pmatrix} A & 0 & 0 \\ 0 & B & 0 \\ 0 & 0 & C \end{pmatrix}$ (11.3.20)

D is a block-diagonalized matrix, having two important properties: viz.

$\quad\quad\quad$ det D = (det A) (det B) (det C), and

$\quad\quad\quad$ Tr (D) = Tr (A) + Tr (B) + Tr (C).

xx) Direct Product

Tensor (or direct) Product Matrix, C of two matrices A and B is

$\quad\quad\quad C = A \otimes B .$ (11.3.21)

$$A = \begin{pmatrix} a & b \\ c & d \end{pmatrix} \text{ and } B = \begin{pmatrix} e & f & g \\ h & k & n \\ r & s & t \end{pmatrix}$$

$$D = A \oplus B = \begin{pmatrix} ae & af & ag & be & bf & bg \\ ah & ak & an & bh & bk & bn \\ ar & as & at & br & bs & bt \\ ce & cf & cg & de & ds & dt \\ ch & ck & cn & de & df & dg \\ cr & cs & ct & dr & ds & dt \end{pmatrix}$$

The element of D is $\boxed{D_{13,22} = b\,s = A_{12}\,B_{32}}$ (11.3.22)

xxi) Unitary Matrix, U:

If the matrix A has complex elements such that A^{-1} and adj A are identical, then A is called unitary, U matrix. ∴ $\boxed{U^{\dagger} = U^{-1} = (U^*)'}$ (11.3.23)

$$U^{\dagger} U^{-1} = \mathfrak{I} = U^{-1} U^{\dagger} \qquad (11.3.24)$$

If A is a (n x n) matrix, $\boxed{\det(-A) = (-1)^n \det A}$

11.3.3 LINEAR OPERATORS as Matrix Operators

The algebra of square matrices obeys the same fundamental laws as the operator algebra. In fact operators and matrices give two different realizations of a common abstract mathematical framework. It is known the *quadratically integrable (well behaved)* functions and complex vectors in Hilbert space. An operator is one, which transforms a function into another; so a matrix transforms a vector into another vector.

To every dynamical variable there corresponds a linear operator, and so there corresponds to every linear operator a matrix. In other words, <u>dynamical variables</u> *are matrix operators*.

11.4 DIRAC'S KET AND BRA NOTATION: (Dirac bracket notation)

11.4.1 Introduction

Integrals of wave function products over all space available to the particle,

viz., $\int \psi^* \psi \, d\tau$,

have already been discussed and used many times in all the earlier Chapters. The labour of writing them out can be greatly reduced by the introduction of a short hand notation, similar to that used by P. Dirac. This elegant notation is based of the observation that the order of the two factors in a (complex) scalar product is important, since in general

$$(\Psi_a, \Psi_b) \neq (\Psi_b, \Psi_a), \text{ but}$$

$$\boxed{(\Psi_a, \Psi_b) = (\Psi_b, \Psi_a)^*} \tag{11.2.32}$$

Although the absolute values of the two products are the same,

i.e., $\quad |(\Psi_a, \Psi_b)| = |(\Psi_b, \Psi_a)|$

It is known from the rules that $(\Psi_a, \Psi_b) \neq (\Psi_b, \Psi_a)$, and $(\Psi_a, \Psi_b) = (\Psi_b, \Psi_a)^*$

(a) $\quad \boxed{(\Psi_a, a\Psi_b) = a(\Psi_a, \Psi_b)} \tag{11.2.33}$

which show that the scalar product is *linear* with respect to the **POST-FACTOR**, Ψ_b, but

(b) $\quad (\Psi_a, \Psi_b) = (\Psi_b, \Psi_a)^*$

$$\boxed{(a\Psi_a, \Psi_b) = a^*(\Psi_a, \Psi_b)} \tag{11.2.33}$$

These two rules show that the scalar product is *not linear* with respect to the **PRE-FACTOR**, Ψ_a.

In fact, one has from rules (a) and (b) that

$$\boxed{(\Psi_c, a\Psi_a + b\Psi_b) = a(\Psi_c, \Psi_a) + b(\Psi_c, \Psi_b)} \tag{11.2.34}$$

and $\quad (a\Psi_a, \Psi_b) = a^*(\Psi_a, \Psi_b)$

that the scalar product is said to depend on the pre-factor in an anti-linear fashion.

11.4.2 Post-Factor and Pre-Factor Vector Spaces

This apparent asymmetry can be avoided if one thinks of the two factors as belonging to **two different Hilbert function spaces**. Each space is linear in itself, but the two are related to each other in an anti-linear manner. The two Hilbert Spaces are said to be *dual to each other*. One thus has a Hilbert space of post-factor vectors, and the other Hilbert Space of pre-factor vectors, but the two are not independent of each other. Clearly a new notation must be invented, because by merely writing ψ_a, one would not know whether this is to be the pre-factor or post-factor in a scalar product. In Quantum Mechanics to make the distinction between two function vectors in the two Hilbert spaces one might consider a notation like

$\Psi_a)$ for a <u>post-factor</u>, and

$(\Psi_a$ for a <u>pre-factor</u>.

11.4.3 DIRAC'S BOX NOTATION

Dirac has stylized the box notation and introduced two symbols in the form of a box into which all information can be put. So labeling by one single kernel letter as in the matrix calculus would not be necessary. Dirac's labeling of two kinds of function vectors are

$$|\,\rangle \text{ and } \langle\,|$$

for post-factor and pre-factors, respectively. This has the label within which consists of as many letters and ciphers as necessary to express all required information, i.e. one or more eigen values, or quantum numbers. Following Dirac one given vector-like set has its elements <u>kets</u>. The <u>KET</u> is written as a box of the form $|\,\rangle$. The function $\Psi_j(x)$, the post-factor in a scalar product, is represented by <u>ket symbol</u>

$$\Psi_j(x) \equiv |j\rangle, \qquad (11.4.1)$$

From the algebra of vector-like sets, a new vector-like set can be deduced from kets and complex numbers. This set is called <u>BRA</u> and symbolized by a box of the form $\langle\,|$ with a label within. Thus the pre-factor $(\Psi_j(x)$ in a scalar product, is represented by the <u>bra symbol</u>

$$(\Psi_j(x)) \equiv \langle j| \qquad (11.4.2)$$

One assumes that for every $|j\rangle$ in the <u>post-factor Hilbert space</u>, $\mathfrak{H}$ there corresponds a $\langle j|$ in the <u>pre-factor Hilbert space</u> $\mathfrak{H}$, and vice-versa, subject to the condition

$$|a\rangle + |b\rangle \longleftrightarrow \langle a| + \langle b|$$

and $\quad \lambda\,|a\rangle \longleftrightarrow \lambda^*\langle a| \qquad (11.4.3)$

where the double-headed arrow, $\longleftrightarrow$, indicates the *correspondence* between the two spaces. Taken by it, each one of the two spaces is a linear vector space satisfying all the vector rules of complex vectors.

Thus $\quad |a\rangle + |b\rangle = |b\rangle + |a\rangle \quad$ in one space

$$\langle a| + \langle b| = \langle b| + \langle a|| \qquad \text{in the other space.}$$

$$|a\rangle + |0\rangle = |a\rangle$$

$$|a\rangle + [\,|b\rangle + |c\rangle\,] = [|a\rangle + [|b\rangle] + |c\rangle = |a\rangle + |b\rangle + |c\rangle$$

$$\lambda (|a\rangle + |b\rangle) = \lambda |a\rangle + \lambda |b\rangle$$

$$a(b|a\rangle) = ab|a\rangle = (ab)|a\rangle \qquad (11.4.4)$$

n identical set of the rules above holds good for the bra, $\langle a |$.

14.4.4 Dual Vector Spaces

According to the algebra of vector-like sets, to every bra $\langle a |$ and ket $| b \rangle$ there belongs **one and only one complex number**, which will be called the *scalar product* of state vectors bra $\langle a |$ and ket $| b \rangle$, and be denoted by the

$$\textbf{Bra–c–ket symbol} \langle a || b \rangle$$

Here the bras comprise a space having the same dimensionality as the ket space and which is *dual* to it. These two spaces – the ket-space and the bra-space, *i.e.* the post-factor space and the pre-factor space are said to form a <u>dual vector *space*</u>. The connection between the dual spaces, as mentioned earlier, is given by the scalar product of the bra vector with the ket vector, such that

$$\langle a || b \rangle = (\Psi_a, \Psi_b) \equiv \langle a | b \rangle = \text{Scalar product, or bracket.} \qquad (11.4.5)$$

Here bracket means a bra vector closes a ket vector meaning bra-c-ket. Thus a state function, Ψ_a is called the state vector, $| a \rangle$, which may be represented by Ψ_a or f_a or g_a, and is called the ket vector $| a \rangle$. Its Hermitian adjoint state, $\Psi_a^\dagger$ or $f_a^\dagger$ or $g_a^\dagger$ is represented by the bra vector $\langle a |$; which is the complex conjugate, Ψ_a^* or Ψ_a^*).

i.e., $\boxed{\langle a | \equiv (| a \rangle)^*}$ \qquad (11.4.6)

In other words, <u>to each vector $| a \rangle$</u> there corresponds an adjoint vector <u>$\langle a | \equiv (| a \rangle)^*$</u>,

with component

$$\langle a || i \rangle \equiv (\langle i || a \rangle)^* \qquad (11.4.7)$$

The vector rules/equations, discussed earlier, can be transcribed in terms of kets without any major change.

(1) Thus linear operators, $\hat{L}$ and $\hat{M}$, associate with every arbitrary ket $| a \rangle$ another ket $| b \rangle$ such that

$$|b\rangle = \hat{L}|a\rangle \tag{11.4.8}$$

such that for arbitrary vectors $|a_1\rangle$ and $|a_2\rangle$ in the <u>ket vector space</u>,

$$\hat{L}(|a_1\rangle + |a_2\rangle) = \hat{L}|a_1\rangle + \hat{L}|a_2\rangle, \tag{11.4.9}$$

and $\quad \hat{L}(\lambda|a\rangle) = \lambda(\hat{L}|a\rangle)$ \hfill (11.4.10)

$$(\hat{L} + \hat{M})|a_1\rangle = \hat{L}|a_1\rangle + \hat{M}|a_1\rangle \tag{11.4.11}$$

$$(\hat{L}\,\hat{M})|a_1\rangle = \hat{L}(\hat{M}|a_1\rangle) \tag{11.4.12}$$

i.e,. generally, $\quad \hat{L}(\lambda|a_1\rangle + \mu|a_2\rangle) = \lambda\hat{L}|a_1\rangle + \mu\hat{L}|a_2\rangle$ \hfill (11.4.13)

(2) If $\hat{P}$ and $\hat{Q}$ are two linear operators in <u>bra vector space</u>

$$((\langle c_1| + \langle c_2|)\hat{Q} = \langle c_1|\hat{Q} + \langle c_2|\hat{Q} \tag{11.4.14}$$

and $\quad ((\langle c_1|\lambda)\hat{Q} = \lambda\langle c_1|\hat{Q}$ \hfill (11.4.15)

$$\langle c_1|(\hat{P} + \hat{Q}) = \langle c_1|\hat{P} + \langle c_1|\hat{Q} \tag{11.4.16}$$

$$\langle c_1|(\hat{P}\,\hat{Q}) = \hat{P}(\hat{Q}\langle c_1|) \tag{11.4.17}$$

(3) If $\hat{R}$ is a linear operator in *bra space* as well as in *ket space*, one may write

$$\langle c_1|\{|a_1\rangle\}\hat{R} = \{\langle c_1|\hat{R}\}|a_1\rangle = \langle c_1|\hat{R}|a_1\rangle. \tag{11.4.18}$$

This indicates the symmetry present in the dual space; and *the operator* $\hat{R}$ *acts either to the right or to the left*, whichever is convenient.

11.4.5 HERMITIAN AdjointOperator

The definition of the Hermitian adjoint operator is now expressed as

$$(\hat{A}\psi_a, \psi_c) = (\psi_a, \hat{A}^\dagger\psi_c)$$

becomes in the **bracket notation**

$$(\hat{A}\psi_a, \psi_c) = (\psi_a, \hat{A}^\dagger\psi_c) = \langle a|\hat{A}^\dagger|c\rangle = \langle c|\hat{A}|a\rangle \tag{11.4.19}$$

$$|b\rangle = \hat{A}|a\rangle \longleftrightarrow \langle b| = \langle a|\hat{A}^\dagger \tag{11.4.20}$$

is the general correspondence.

11.4.6 How to go over from Schrödinger's notation to Dirac's notation?

It is so simple that one has to replace

if $\quad \psi_n(x) \rightarrow |n\rangle$,

$\psi_m^*(x)\, dx \rightarrow \langle m|$

and operator $\quad \hat{\Omega} \rightarrow \hat{\Omega}$.

Thus a <u>Hermitian operator</u>, $\hat{\Omega}$ is characterized in the <u>Dirac's notation</u> by

$$\boxed{|b\rangle = \langle b|\hat{\Omega}|a\rangle^*} \qquad (11.4.21)$$

If $|j\rangle$ is an *eigen ket* (eigen vector) of the operator $\hat{\Omega}$ with *eigen value* Ω_j, the <u>eigen value equation</u> is expressed as $\boxed{\hat{\Omega}|j\rangle = \Omega_j|j\rangle}$ (11.4.22)

With these rules and conventions, the <u>old Schrödinger's notation and Dirac's bracket notation becomes equivalent</u>.

11.4.7. PHYSICAL INTERPRETATION of Kets and Bras:

According to Quantum Mechanics, a dynamical system can be in different 'states', and the result of the measurement of some *dynamical variable* (viz., a component of some physical quantity) depends on the state of the system it is in. But, contrary to Classical Mechanics, if the system is in arbitrary state the result of the measurement cannot be predicted. If a dynamic variable can be measured at all, it is only possible to predict the set of possible values and for each of them the probability that it may be the result of measurement. After the measurement it looks just as if the system had "jumped" into a state of special kind with respect to this variable, because in this new state the result of a second measurement of the same variable will always be the same as that of the first measurement. For example, in Classical Mechanics, the position of a point on the x-axis is given by one dynamic variable x and there are as many states as there are possible values of x. If the point is in state $x=3$, every measurement in this state gives with certainty the result $x=3$. But in Quantum Mechanics other states exist (and we know how to produce them) where there is only a certain probability for x, to be found by measurement for instance between 3 and $3\frac{1}{2}$.

Now in Matrix Calculus every linear operation is connected with a series of numerical values. Dynamical variables are always real, and we know that Hermitian operators have real eigen values only. Hence it seems opportune that every dynamical variable with a Hermitian operator (also called real operators) and to assume that the result of a measurement (if possible at all) can only be one of the eigen values. This leads to the assumption that every state corresponds to a ket (and its conjugate bra), determined to within a scalar factor. Every eigen value corresponds to a eigenket (or eigen vector) which corresponds to eigen function of the older terminology (or a set of eigenkets), we may expect the eigen states in which the result of the measurement is, with certainty, this eigen value. As to the other state, it will be necessary to assume that its corresponding ket can be expressed linearly in terms of those eigenkets whose eigen values can be possible results of a measurement in this state. But since this holds for every other state, this means that the operator must have so many eigenkets that every ket can be expressed linearly in them.

Thus if $|j\rangle$ is an eigen vector (eigenket) of the operator $\hat{\Omega}$ with eigen value Ω_j, the Eigenvalue Equation is $\hat{\Omega}|j\rangle = \Omega_j |j\rangle$

If $\{\hat{e}_j\}$ is an **orthonormal basis**, the general state function which is not an eigen function of such a Hermitian operator is, in older terminology,

$$\Phi_j = \sum c_{ij} \hat{e}_i \ . \tag{11.4.23}$$

A dynamical variable corresponding to such a Hermitian operator $\hat{\Omega}$ is called an "<u>observable</u>". Hence every "*measurable*" dynamic variable is an observable. The states of an observable and also the corresponding kets (and bras) are said to form a <u>complete set</u>.

i.e., $\quad |\Phi_j\rangle = \sum c_{ij} |\hat{e}_j\rangle = \sum c_{ij} |\hat{i}\rangle \tag{11.4.24}$

where, $\boxed{c_{ij} = \langle \hat{e}_i || \Phi_j \rangle = \langle \hat{i} || \Phi_j \rangle} \tag{11.4.25}$

In <u>Dirac's notation</u>, a closed bra-c-ket means a scalar quantity, while bra or ket (half bracket) means a vector.

Thus to every state of an ensemble specified by a complete set of quantum numbers, a', one associates an abstract vector, $|a'\rangle$. This set of vectors $\{|a'\rangle\}$ spans an abstract vector space (which is the Hilbert space when the number N of distinct values that the observable $\hat{A}$ can assume is infinite). By "<u>span</u>" we mean that if $|\beta\rangle$ is a vector associated with a state specified by the values of any other state of observables, there exists a unique expansion

$$|\beta\rangle = \sum c_{a'} |a'\rangle$$

where $c_{a'}$ is a complex number.

One also defines a dual Vector Space spanned by the dual vectors, $\langle a' |$. These dual vectors stand in one-to-one correspondence with the vectors $| a' \rangle$.

Worked out Example 11.3

Using the Dirac notation, show that the eigen values of a Hermitian operator are real. (Theorem 1: Property No.1 of a Hermitian operator).

Solution: STEP # 1   Given are: $\hat{Q}$ is a Hermitian operator; $| \psi_n(x) \rangle \equiv | \psi_n \rangle$ = an eigen ket; $\langle \psi_n || \psi_n \rangle \neq 0$; $\hat{Q} | \psi_n \rangle = q_n | \psi_n \rangle$; q_n = the eigen value of $\hat{Q}$, and

$$\langle \psi_m | \hat{Q} | \psi_n \rangle = \langle \psi_n | \hat{Q} | \psi_m \rangle^*. \qquad (11.3.21)$$

is the Hermiticity Condition.

STEP # 2 LHS $= \langle \psi_n | \hat{Q} | \psi_n \rangle = \langle \psi_n | (q_n | \psi_n \rangle) = q_n \langle \psi_n || \psi_n \rangle$

STEP # 3 RHS =
$\langle \psi_n | \hat{Q} | \psi_n \rangle^* = \{(\langle \psi_n | \hat{Q}) | \psi_n \rangle\}^* = \{(\langle \psi_n | q_n) | \psi_n \rangle\}^* = q_n^* \langle \psi_n || \psi_n \rangle$

STEP # 4   The Hermiticity Condition requires that LHS = RHS.

$\therefore \qquad q_n \langle \psi_n || \psi_n \rangle = q_n^* \langle \psi_n || \psi_n \rangle$.

i.e., $[q_n^* - q_n] \langle \psi_n || \psi_n \rangle = 0$.

STEP # 5 This is possible only if $[q_n^* - q_n] = 0$, Or, $q_n^* = q_n$

$$\boxed{q_n^* = q_n} \qquad (11.4.26)$$

which is possible only if q_n = a real quantity. Hence the theorem is proved.

Worked out Example 11.4

Using the Dirac notation, show that the eigen kets belonging to different eigen values of a Hermitian operator are orthogonal. (Theorem 2: Property No.2 of a Hermitian operator) (*i.e., zero overlap*):

Solution: STEP # 1   Given are: $\hat{Q}$ is a Hermitian operator; $|\psi_n(x)\rangle \equiv |\psi_n\rangle$ = an eigen ket; $\langle\psi_n||\psi_n\rangle \neq 0$; $[\langle\psi_m||\psi_m\rangle \neq 0$, $\hat{Q}|\psi_n\rangle = q_n|\psi_n\rangle$; $\hat{Q}|\psi_m\rangle = q_m|\psi_m\rangle$; q_n and q_m = the eigen values of $\hat{Q}$, i.e., $q^* = q$; and $q_n \neq q_m$

$$\langle\psi_m|\hat{Q}|\psi_n\rangle = \langle\psi_n|\hat{Q}|\psi_m\rangle^*. \tag{11.3.21}$$

is the Hermiticity Condition.

STEP # 2 LHS = $\langle\psi_m|\hat{Q}|\psi_n\rangle = \langle\psi_m|[\hat{Q}|\psi_n\rangle] = q_n\langle\psi_m||\psi_m\rangle]$.

STEP # 3 RHS

$\langle\psi_n|\hat{Q}|\psi_m\rangle^* = \{\langle\psi_n|[\hat{Q}|\psi_m\rangle]\}^* = \{\langle\psi_n|[q_m|\psi_m\rangle]\}^* = q_m\langle\psi_m||\psi_n\rangle$

STEP # 4 The Hermiticity Condition requires that LHS = RHS.

$\therefore \quad q_m\langle\psi_m||\psi_n\rangle = q_n\langle\psi_m||\psi_m\rangle$, i.e., $[q_m - q_n]\langle\psi_m||\psi_n\rangle = 0$

STEP # 5 Generally, $[q_m - q_n] \neq 0$, and so $[q_m - q_n]\langle\psi_m||\psi_n\rangle = 0$ is possible only if $\langle\psi_m||\psi_n\rangle = 0$.

$\therefore \quad \boxed{\langle\psi_m||\psi_n\rangle = 0} \tag{11.4.27}$

is known as the <u>Orthogonality Relation</u> between the two kets, $|\psi_n\rangle$ and $|\psi_m\rangle$ of the system.

11.5 REPRESENTATION OF KETS, BRAS AND OPERATORS IN MATRIX MECHANICS

11.5.1 Introduction

Let there is an eigenket equation in ket space with basis $\{|\hat{e}_j\rangle\}$

Any given vector $\quad |c\rangle = \sum c_i|\hat{e}_i\rangle \equiv \sum c_i|\hat{i}\rangle \tag{11.3.24}$

and $\quad |\hat{e}_i\rangle \equiv |\hat{i}\rangle$

forms an eigen vector of some observable A in the space,

$$\hat{A}|\hat{e}_i\rangle \equiv \hat{A}|\hat{i}\rangle = a_i|\hat{i}\rangle \qquad (11.5.1)$$

i.e., $\{|\hat{e}_j\rangle\}$ is a complete set of orthonormal basis vectors of some Hermitian operator $\hat{A}$ (*i.e.*, observable). Consider any arbitrary (general) vector, $|j\rangle \equiv |\Phi_j\rangle$, which is not an eigen vector of $\hat{A}$.

$$\hat{A}|j\rangle \ne a'|j\rangle$$

and $\qquad |j\rangle \equiv |\Phi_j\rangle = \sum c_{ij}|\hat{e}_i\rangle = \sum c_{ij}|\hat{i}\rangle \qquad (11.4.24)$

where, $c_{ij} = \langle \hat{e}_i \| \Phi_j \rangle = \langle \hat{i} \| j \rangle \qquad (11.4.25)$

are components of $|\Phi_j\rangle$ in the basis (*i.e.* representation) $\{\hat{e}_i\}$.

11.5.2 Matrix Notation of Ket $|j\rangle \equiv |\Phi_j\rangle$

It is now known that the arbitrary ket $|j\rangle$ is

$$|j\rangle \equiv |\Phi_j\rangle = \sum c_{ij}|\hat{e}_i\rangle = \sum c_{ij}|\hat{i}\rangle \qquad (11.4.24)$$

where, $c_{ij} = \langle \hat{e}_i \| \Phi_j \rangle = \langle \hat{i} \| j \rangle \qquad (11.4.25)$

So the arbitrary ket $|j\rangle$ is written as

$$|\Phi_j\rangle \equiv |j\rangle = \begin{pmatrix} c_{11} \\ c_{21} \\ c_{31} \\ \cdot \\ \cdot \\ \cdot \\ c_{n1} \end{pmatrix} = \begin{pmatrix} \langle 1 \| j \rangle \\ \langle 2 \| j \rangle \\ \langle 3 \| j \rangle \\ \cdot \\ \cdot \\ \cdot \\ \langle \tilde{n} \| j \rangle \end{pmatrix} \equiv \{\langle 1 \| j \rangle, \langle 2 \| j \rangle, \cdot, \cdot, \cdot, \cdot, \langle \hat{i} \| j \rangle\}$$

$$\boxed{|j\rangle \equiv |\Phi_j\rangle = \text{A Column Vector.}} \qquad (11.5.2)$$

A ket vector is a <u>column vector</u>.

11.5.3 Matrix Representation of Bra $\langle j | \equiv \langle \Phi_j |$

A bra, conjugate to (adjoint vector to) ket $|j\rangle$ is

$$\langle j | \equiv \langle \Phi_j | = (| j \rangle)^\dagger = [\sum c_{ij} | \hat{i} \rangle]^* = \sum c_{ij}^* \langle \hat{i} |$$

where $c_{ji} = c_{ij}^* = \langle \hat{e}_i \| \Phi_j \rangle^* = \langle \hat{i} \| j \rangle^*$

$$\langle j | \equiv \langle \Phi_j | = \left(c_{1j}^*, c_{2j}^*, c_{3j}^*, \cdot, \cdot, \quad , \cdot, c_{nj}^* \right)$$

$$= \left(\langle 1 \| j \rangle^*, \langle 2 \| j \rangle^*, \cdot, \cdot, \cdot, \quad , \langle \tilde{n} \| j \rangle^* \right).$$

$$\boxed{\langle j | \equiv \langle \Phi_j | = \text{A Row Vector.}} \qquad (11.5.3)$$

The whole column matrix and the row matrix are the seen to be 'representations' of the ket $| j \rangle$ and bra vector $\langle j |$, respectively, *i.e.* $| j \rangle$ is specified in the basis-representation $\{ \hat{e}_i \}$ of $\hat{A}$.

Thus once a given basis has been chosen, $| \Phi_j \rangle$ is completely specified by its components c_{ij}.

11.5.4 Scalar Product of Two Vectors, *viz.* Ket $| j \rangle$ and Bra $\langle j |$

To this ket $| j \rangle$ and bra vector $\langle j |$ there belongs the scalar product

$$\langle j \| j \rangle = \left(\sum c_{ij}^* \langle \hat{i} | \sum c_{ij} | \hat{i} \rangle \right)$$

$$= \left(\sum c_{ij}^* c_{ij} \langle \hat{i} \| \hat{i} \rangle \right) = \left(\sum |c_{ij}|^2 \langle \hat{i} \| \hat{i} \rangle \right)$$

$$= \left(\sum |c_{ij}|^2 \delta_{ii} \right) = \sum |c_{ij}|^2$$

$$\boxed{\langle j \| j \rangle = \left(\sum c_{ij}^* \langle \hat{i} | \sum c_{ij} | \hat{i} \rangle \right) = \sum |c_{ij}|^2} \qquad (11.5.4)$$

This is the well-known Closure Relation.

$$\boxed{\langle \hat{e}_i \| \hat{e}_i \rangle = \langle \hat{i} \| \hat{i} \rangle = \delta_{ii} = 1}. \qquad (11.5.5)$$

is the Normalization condition.

Further $\langle a \| b \rangle = \left(a_{1j}^*, a_{2j}^*, a_{3j}^*, \cdot, \cdot, \quad , \cdot, a_{nj}^* \right) \{ b_{1j}, b_{2j}, b_{3j}, \cdot, \cdot, \quad , \cdot, b_{nj} \}$

$$= \Big(\langle 1 \| j \rangle^*, \langle 2 \| j \rangle^*, \cdot, \cdot, \cdot, \cdot, \langle \hat{i} \| j \rangle^* \Big) \{ \langle 1 \| j \rangle, \langle 2 \| j \rangle, \cdot, \cdot, \cdot, \cdot, \langle \hat{i} \| j \rangle \}$$
(11.5.6)

Therefore in new notation, the <u>matrix representation</u> of the n-vectors in the basis $\{| \hat{e}_i \rangle\}$ is of the form (as described in Section 11.2.) of column vectors

$$| \hat{e}_1 \rangle = \{1, 0, 0, \ldots 0, 0\}$$
$$| \hat{e}_2 \rangle = \{0, 1, 0, \ldots 0, 0\}$$
$$| \hat{e}_3 \rangle = \{0, 0, 1, \ldots 0, 0\}$$
$$\cdot$$
$$\cdot$$
$$| \hat{e}_n \rangle = \{0, 0, 0, \ldots 0, 1\}$$
(11.2.45)

in $V_n(F)$.

They are such that $\langle \hat{e}_i \| \hat{e}_j \rangle = \delta_{ij}$. (11.2.46)

11.5.5 Matrix Representation of a Linear Operator, $\hat{\Omega}$

When a basis $\{\hat{e}_i\}$ is given in the N-vector space, which form eigen value equations for an operator $\hat{A}$, one has

$$\hat{A} | \hat{i} \rangle = a_i | \hat{i} \rangle$$
(11.5.1)

$$\hat{A} | j \rangle = a_j | j \rangle$$
(11.5.7)

A linear operator $\hat{\Omega}$ can be characterized by its effect on the basis vector in the N-space. To see this let (11.4.24)

$$| a \rangle \equiv | \Phi_a \rangle = \sum a_j | \hat{j} \rangle$$
(11.5.8)

$$| b \rangle \equiv | \Phi_b \rangle = \sum b_i | \hat{e}_i \rangle = \sum b_i | \hat{i} \rangle$$
(11.5.9)

From (11.4.20) and (11.5.9)

$$| b \rangle = \hat{\Omega} | a \rangle = \hat{\Omega} \sum a_j | j \rangle = \hat{\Omega} \sum a_j \hat{\Omega} | j \rangle$$
(11.5.10)

Multiplying by unity, viz. $\langle \hat{i} \| \hat{i} \rangle = \delta_{ii} = 1$,

$$| b \rangle = \sum a_j \delta_{ii} \hat{\Omega} | j \rangle = \sum\sum a_j | \hat{i} \rangle \langle \hat{i} | \hat{\Omega} | j \rangle$$

$$= \sum\sum a_j | \hat{i} \rangle \hat{\Omega}_{ij} = \sum | \hat{i} \rangle \Big(\sum \hat{\Omega}_{ij} a_j \Big) = \sum b_i | \hat{i} \rangle$$

where $b_i = \sum \hat{\Omega}_{ij} a_j$ (11.5.11)

This is the Matrix Equation which relates the scalar or expansion coefficients (components of a_j of the ket $|a\rangle$ and b_i of $|b\rangle = \hat{\Omega}|a\rangle$. This proves the contention that the effect of the operator, $\hat{\Omega}$ on any vector is known if all $\hat{\Omega}_{ij}$ are known. Equation (11.5.11) can be written conveniently in the matrix form as follows:

$$|b\rangle = \hat{\Omega}|a\rangle = \begin{pmatrix} b_1 \\ b_2 \\ \bullet \\ \vdots \\ b_N \end{pmatrix} = \begin{pmatrix} \Omega_{11} & \cdots & \Omega_{1N} \\ & & \\ & & \\ \vdots & & \vdots \\ \Omega_{N1} & \cdots & \Omega_{NN} \end{pmatrix} \begin{pmatrix} a_1 \\ a_2 \\ \bullet \\ \vdots \\ a_N \end{pmatrix}$$ (11.5.12)

one has $\therefore |b\rangle = \hat{\Omega}|a\rangle$ (11.5.10)

11.5.6 $\hat{A}$ -Representation defined

The choice of a particular basis determines the matrices in equation (11.4.10). In equations (11.4.2) and (11.4.3) the column/ row matrices represent the vectors $|a\rangle$ and $|b\rangle$ or $\langle a|$ and $\langle b|$, and the *square matrix*

$$\hat{\Omega} = (\hat{\Omega}_{ij}) = \begin{pmatrix} \Omega_{11} & \Omega_{12} & \bullet & \bullet & \Omega_{1N} \\ \hat{\Omega}_{21} & \Omega_{22} & \bullet & \bullet & \hat{\Omega}_{2Nj} \\ \hat{\Omega}_{31} & \hat{\Omega}_{32} & \bullet & \bullet & \hat{\Omega}_{3Nj} \\ \vdots & & \bullet & \Omega_{ij} & \bullet & \vdots \\ \Omega_{N1} & \Omega_{N2} & \bullet & \bullet & \Omega_{NN} \end{pmatrix}$$ (11.5.13)

is called the **matrix representation** of $\hat{\Omega}$. For this reason one says that all these matrices constitute a representation in the N-space. $\hat{\Omega}_{ij}$ in the matrix $(\hat{\Omega}_{ij})$ is called, therefore, a Matrix Element, given by

$$\boxed{\hat{\Omega}_{ij} = \langle \hat{i}|\hat{\Omega}|\hat{j}\rangle = \langle \hat{j}|\hat{\Omega}|\hat{i}\rangle^*}$$ (11.5.14)

i and j denoting the row # and column # in $\hat{\Omega}_{ij}$.

Since $\{\hat{e}_i\}$ form *Eigen ket Equations* with operator $\hat{A}$ (11.5.1) and (11.5.7), the whole matrix $(\hat{\Omega}_{ij})$ is said to be the $\hat{A}$ -REPRESENTATION of the linear operator $\hat{\Omega}$. Hence it is adopted here a given representation (*i.e.* a given set of orthonormal basis vectors

$\{|\hat{i}\rangle\}$ in which $|b\rangle$ and $|a\rangle$ are represented by vectors having components b_i and a_j, and the operator $\hat{\Omega}$ by a matrix $(\hat{\Omega}_{ij})$. This $\hat{A}$-representation is not unique. This is because the choice of a particular basis determines the matrices $(\hat{\Omega}_{ij})$. It is to be noted that in this representation, the operator $\hat{A}$ itself is represented by a diagonal matrix.

11.5.7 PROPERTIES of Matrix Representations

a) The Hermitian adjoints of operators are represented by the Hermitian adjoints of the matrix representations of the original operators.

b) Algebraic relations between vectors and operators lead to equivalent algebraic relations between their representative matrices (thus, in particular, the equations which define the eigenvalues λ_i and eigen vectors of operators become *Matrix Eigen value Equations*).

c) The trace (spur), $\text{Tr}(\hat{\Omega})$ of any Hermitian operator $\hat{\Omega}$, defined as $\sum \hat{\Omega}_{nn}$, is independent of the representation used to define it, i.e. $\boxed{\text{Tr}(\hat{\Omega}) = \sum \hat{\Omega}_{mn} = \text{invariant}}$

d) For any Hermitian operator $\hat{\Omega}$ its eigen values λ_i remain unchanged when the matrix undergoes a similarity transformation.

e) The determinant of $\hat{\Omega}$ $\boxed{\det(\hat{\Omega}) = |\hat{\Omega}| = \text{invariant}}$, of a matrix, independent of the representation used to define it when a similarity transformation is made.

f) The invariants, viz. $\text{Tr}(\hat{\Omega})$, and $\det(\hat{\Omega}) = |\hat{\Omega}|$, are not independent of the eigen values, λ_i. In fact, one may evaluate them in the system in which $\hat{\Omega}$ is diagonal, with the result

$$\boxed{\text{Tr}(\hat{\Omega}) = \sum \lambda_i} \tag{11.5.15}$$

and $\boxed{\det(\hat{\Omega}) = |\hat{\Omega}| = \prod \lambda_i}$ (11.5.16)

The eigen values λ_i are the only independent variables.

11.5.8 Projection Operator, $\hat{P}_i$

Suppose that $\hat{A}$ is an observable so that its eigenvectors $|\hat{e}_i\rangle$ form a complete set, $\{\hat{e}_i\}$. It is known that any arbitrary vector $|c\rangle \equiv |\Phi_c\rangle$ (in $\hat{A}$-representation) can be written as

$$|c\rangle \equiv |\Phi_c\rangle = \sum c_i |\hat{e}_i\rangle = \sum c_i |\hat{i}\rangle$$

and $\hat{A}|\hat{i}\rangle = a|\hat{i}\rangle$; where $c_i = \langle \hat{i} || c \rangle$.

Substituting this result

$$|c\rangle = \sum \langle \hat{i} || c \rangle |\hat{i}\rangle = \sum |\hat{i}\rangle \langle \hat{i} || c \rangle \qquad (11.5.17)$$

Now a typical term on the right hand side of (11.4.16) is a vector whose length is the scalar product of $|\hat{i}\rangle$ and $|c\rangle$ and whose direction is that of $|\hat{i}\rangle$; so the scalar product is the projection of $|c\rangle$ on $|\hat{i}\rangle$. The quantity

$$\boxed{\hat{P}_i = |\hat{i}\rangle\langle\hat{i}|} \qquad (11.5.18)$$

is the operator which performs this projection. This is a new linear operator called the <u>projection operator</u>, $\hat{P}_i$. It projects a vector on a subspace. Because of its structure it can operate on either a ket or a bra. In so doing it produces a new ket or a new bra, respectively.

Worked out Example 11.5

Obtain the result of a Projection Operator operating on a bra.

Solution: $\boxed{\text{STEP \# 1}}$ Let $\hat{P}_i = |\hat{i}\rangle\langle\hat{i}|$ is the Projection Operator; and $\langle b| =$ the given bra.

$\boxed{\text{STEP \# 2}} \therefore \langle b|\hat{P}_i = \langle b||\hat{i}\rangle\langle\hat{i}| = \beta\langle\hat{i}| = \langle b_i|$, say.

$\beta = \langle b || \hat{i} \rangle =$ a scalar.

$\boxed{\text{STEP \# 3}}$ $\langle b|$ is transformed to $\langle\hat{i}|$ and multiplied by β which is a scalar.

$\boxed{\text{STEP \# 4}}$ *i.e.* β is projection of $\langle b|$ along $\langle\hat{i}|$..

Worked out Example 11.6

Prove that the Projection Operator summed over any complete basis is the identity operator.

Solution: $\boxed{\text{STEP \# 1}}$ $\hat{P}_i = |\hat{i}\rangle\langle\hat{i}|$ is the projection operator, and $|\hat{i}\rangle =$ a basis vector.

$\boxed{\text{STEP \# 2}}$ $\hat{P}_i |\hat{i}\rangle = |\hat{i}\rangle$.

STEP # 3 For any vector, $|c\rangle = \sum c_i |\hat{i}\rangle$,

$$|c\rangle = \hat{P}_i |c\rangle = =|\hat{i}\rangle\langle\hat{i}|\sum c_i |\hat{i}\rangle = |\hat{i}\rangle\sum a_i = \sum a_i |\hat{i}\rangle.$$

STEP # 4 $\hat{P}_i \hat{P}_j |c\rangle = a_i \hat{P}_j |c\rangle = 0$, if $i \neq j$.

Consequently, Projection Operators for the basis vectors have the property
$\hat{P}_i \hat{P}_j = \hat{P}_j \hat{P}_i = 0$, if $i \neq j$

STEP # 5   For every $|c\rangle$; $\sum \hat{P}_i |c\rangle = \sum a_i |\hat{i}\rangle = |c\rangle$. Hence, $\sum \hat{P}_i = 3$

$\sum_{\text{complete basis}} \hat{P}_i = 3$ (11.5.22)

11.5.9 SPECTRUM of an Operator, $\hat{\Omega}$

Consider an operator corresponding $\hat{\Omega}$ to the observable Ω having eigen values λ_i (real), and having eigen vectors $|\hat{e}_i\rangle$. The corresponding Eigen value Equation is

$$\hat{\Omega}|\hat{i}\rangle = \lambda_i |\hat{i}\rangle.$$

The totality of the eigen values of $\hat{\Omega}$ is called the <u>Spectrum</u> of $\hat{\Omega}$. Since λ_i-s are all real one gets a '<u>discrete' spectrum</u>. If the eigen values are continuous range of values, a <u>Continuous Spectrum</u> results; and some cases a mixture of both discrete and continuous spectra results.

11.6. UNITARY AND SIMILARITY TRANSFORMATIONS IN QUANTUM MEHANICS:

11.6.1 Introduction

The significance of unitary operators in Quantum Mechanics is due to the following. One is aware of the fact that a given Hilbert Space has many bases. This is similar to the fact that 3-space is spanned by one of a continuum of triad basis vectors. One can obtain a new orthogonal triad basis in 3-space through a rotation of axes about the origin. In other words, these are effected through successive transformations, *i.e.* matrix multiplication. In order to understand this, a brief description of linear transformation is found necessary.

11.6.2 Linear Transformation

Let an *arbitrary vector* of an *n*-space $|\alpha^o\rangle$ is denoted by

$$|\alpha^o\rangle = \{x^o{}_1, x^o{}_2, x^o{}_3, \cdots, x^o{}_n\} \quad (11.6.1)$$

$$x_1 = \{a_{11}x^o{}_1 + a_{12}x^o{}_2 + a_{13}x^o{}_3 + \cdots \cdots + a_{1n} x^o{}_n\}$$
$$x_2 = \{a_{21}x^o{}_1 + a_{22}x^o{}_2 + a_{23}x^o{}_3 + \cdots \cdots + a_{2n} x^o{}_n\}$$
$$x_3 = \{a_{31}x^o{}_1 + a_{32}x^o{}_2 + a_{33}x^o{}_3 + \cdots \cdots + a_{3n} x^o{}_n\} \quad (11.6.2)$$
"
"
$$x_n = \{a_{n1}x^o{}_1 + a_{n2}x^o{}_2 + a_{n3}x^o{}_3 + \cdots \cdots + a_{nn} x^o{}_n\}$$

by which the scalar coordinates of $\mid \alpha^o \rangle$ are related to the coordinates of another vector, $\mid \alpha \rangle$

$$\mid \alpha \rangle = \{x_1, x_2, x_2, \cdots, \cdots, \cdots, x_n\} \quad (11.6.3)$$

by the scalar relations (11.6.2) (known as *simultaneous* or homogeneous *linear equations*

Equivalently, one may write the vector $\mid \alpha \rangle$, as related to the vector $\mid \alpha^o \rangle$, by the transformation in the unknown scalars $x_j (j = 1, 2, 3, \ldots, n)$.

This set of equations can be written in short hand form, in the matrix algebra, as

$$\begin{bmatrix} x_1 \\ x_2 \\ x_3 \\ " \\ " \\ x_n \end{bmatrix} = \begin{bmatrix} a_{11} & a_{12} & a_{13} & \cdots & \cdots & a_{1n} \\ a_{21} & a_{22} & a_{23} & \cdots & \cdots & a_{2n} \\ a_{31} & a_{32} & a_{33} & \cdots & \cdots & a_{3n} \\ " & & & & & \\ " & & & & & \\ a_{n1} & a_{n2} & a_{n3} & \cdots & \cdots & a_{nn} \end{bmatrix} \begin{bmatrix} x^o{}_1 \\ x^o{}_2 \\ x^o{}_3 \\ " \\ " \\ x^o{}_n \end{bmatrix} \quad (11.6.4)$$

or, $\boxed{\mid \alpha \rangle = \hat{A} \mid \alpha^o \rangle}$ (11.6.5)

One says that the transformation (11.6.5) is a Linear transformation in the n scalars, $x^o{}_1, x^o{}_2, x^o{}_3, \cdots, \cdots, \cdots, x^o{}_n$.

The Matrix Equation may be interpreted as defining the transformation $\hat{T}$ conveniently signified by

$$\boxed{\mid \alpha^o \rangle \longrightarrow \hat{T} \mid \alpha^o \rangle = \mid \alpha \rangle} \quad (11.6.6)$$

which carries the vector $\mid \alpha^o \rangle$ into the vector $\mid \alpha \rangle$.

11.6.3 Important Properties of a Linear Transformation

They are

(a) If k = a scalar, $\hat{A}\left(k \mid \alpha^\circ \right) = k\left(\hat{A} \mid \alpha^\circ \right)$ (11.6.7)

(b) $\hat{A}\left(\mid \alpha^\circ \rangle + \mid \beta^\circ \rangle\right) = \hat{A} \mid \alpha^\circ \rangle + \hat{A} \mid \beta^\circ \rangle$ (11.6.8)

Worked out Example 11.7

Rotation of the XY- axes about a fixed Z- axis through the angle φ changes the components (x, y) of the vector **r** to (x', y'). These new components are related to the original components through the rotation matrix $\hat{\Re}(\varphi)$. $\vec{r}' = \hat{\Re}(\varphi)\ \vec{r}$. Find the matrix $\hat{\Re}(\varphi)$.

Solution: Let the axes $O(x_1, x_2)$ and $O(x^\circ_1, x^\circ_2)$ be arranged so that P is any point in the plane with coordinates (x_1, x_2) or (x°_1, x°_2). Let the coordinates (x_1, x_2) are rotated counter-clockwise through an angle φ keeping $\vec{r}$ fixed. The relations between the components resolved in the old and those in the new are:

$$x^\circ_1 = x_1\ Cos\ \varphi + x_2\ Sin\ \varphi + x_3.0,$$

$$x^\circ_2 = x_1\ Sin\ \varphi + x_2\ Cos\ \varphi + x_3.0$$

$$x^\circ_3 = x_1(0) + x_2(0) + x_3.(1)$$

$a_{11} = a_{22} = Cos\ \varphi = Cos\ (x^\circ_1, x_1)$, $a_{12} = -a_{21} = Sin\ \varphi = Cos\ (x^\circ_1, x_2)$, $a_{33} = 1$, etc.

In matrix form,

$$\begin{pmatrix} Cos\ \varphi & Sin\ \varphi & 0 \\ -Sin\ \psi & Cos\ \psi & 0 \\ 0 & 0 & 1 \end{pmatrix} \begin{bmatrix} x_1 \\ x_2 \\ x_3 \end{bmatrix} = \begin{bmatrix} x^\circ_1 \\ x^\circ_2 \\ x^\circ_3 \end{bmatrix}$$

$$\mid X^\circ \rangle = \hat{\Re}(\varphi) \mid X \rangle$$

$$\hat{\Re}(\varphi) = \begin{pmatrix} Cos\ \varphi & Sin\ \varphi & 0 \\ -Sin\ \varphi & Cos\ \varphi & 0 \\ 0 & 0 & 1 \end{pmatrix} \quad (11.6.9)$$

11.6.3.1 Properties of ROTATIONS

The vector $\left| X^o \right\rangle$ has the magnitude $\left| X^o \right|$ given by

$$\left| X^o \right|^2 = (x_1^o)^2 + (x_2^o)^2 + (x_3^o)^2 \text{, in the } (x^o -) \text{ Coordinate System, and}$$

$$\left| X \right|^2 = (x_1)^2 + (x_2)^2 + (x_3)^2 \text{, in the } x- \text{ Coordinate System.}$$

(a)　It is seen that $\left| X^o \right|^2 = \left| X \right|^2$; i.e. the magnitude of a vector is scalar invariant under rotation of rectangular Cartesian coordinate axes. This means the operator, $\hat{\Re}(\varphi)$ preserves the length.

(b)　Taking the scalar product $\langle X^o || X \rangle = (x_1^2 + x_2^2) \, Cos\varphi$, i.e. the inner product of two vectors is an invariant under a rotation of rectangular coordinate axes.

11.6.4　TRANSFORMATION OF BASIS (or Representation)

So far the similarity between the geometry of the abstract Complex Vector Space (Unitary Space) and geometry in ordinary Euclidean (or real) Space was discussed. A Representation in the former Space corresponds to the introduction of a coordinate system in the latter. Now consider that a transformation form one Representation to another in the general space, just as one rotates a coordinate system in analytic geometry.

Let the coordinate equation, given by the operator equation,

$$\left| \Psi_b \right\rangle = \hat{S} \left| \Psi_a \right\rangle \qquad (11.6.10)$$

describes any linear operator $\hat{S}$ transforming a vector $\left| \Psi_a \right\rangle$ into another vector, $\left| \Psi_b \right\rangle$. This is followed by a second operator $\hat{\mathbb{R}}$ such that

$$\left| \Psi_c \right\rangle = \hat{\mathbb{R}} \left| \Psi_b \right\rangle \qquad (11.6.11)$$

One, therefore, gets

$$\left| \Psi_c \right\rangle = \hat{\mathbb{R}} \left| \Psi_b \right\rangle = \hat{\mathbb{R}} \left(\hat{S} \left| \Psi_a \right\rangle \right) = \hat{P} \left| \Psi_a \right\rangle \qquad (11.6.12)$$

Equation (11.6.12) can be interpreted as a linear transformation of the vector $\left| \Psi_a \right\rangle$ into $\left| \Psi_c \right\rangle$ by the linear operator P in n-space, defined by the basis $\{\left| \hat{e}_i \right\rangle\}$, where

$$|\Psi_a\rangle = \sum a_i |\hat{e}_i\rangle = \sum a_i |\hat{i}\rangle. \tag{11.6.13}$$

Now the question is whether it is possible to describe this transformation from a different coordinate system, *i.e.* using a different set of basis (since the above representation is not unique.). In component form, equation (11.6.12) is obtained as follows: Let $\{|\hat{e}_i^o\rangle\}$ and $\{|\hat{e}_i\rangle\}$ represent the OLD and NEW *sets of basis vectors*.

The new basis can be expressed in terms of the old ones: From

$$|b_i\rangle = \left|\sum \Omega_{ij} a_j\right\rangle = \hat{\Omega}|a_j\rangle. \tag{11.6.14}$$

Similarly, $|\hat{e}_k\rangle = \sum |\hat{e}_i^o\rangle (S_{ki})$ \hfill (11.6.15)

where S_{ik} are the matrix elements of the transformation matrix (S_{ik}). The square matrix (S_{ik}) defines the change of basis;

$$\hat{S} = (S_{ki}) = \begin{bmatrix} S_{11} & S_{12} & S_{13} & \cdots & S_{1n} \\ S_{21} & S_{22} & S_{23} & \cdots & S_{2n} \\ S_{31} & S_{32} & S_{33} & \cdots & S_{3n} \\ " & & & & \\ " & & & & \\ S_{n1} & S_{n2} & S_{n3} & \cdots & S_{nn} \end{bmatrix} \tag{11.6.16}$$

where the matrix element

$$\boxed{(S_{ki}) = \langle \hat{e}_i^o || \hat{e}_k \rangle} \tag{11.6.17}$$

A succession of two such changes in basis, $(\hat{S})$ and $(\hat{R})$, performed in this order is equivalent to a single transformation whose matrix $(\hat{P})$ is simply the product matrix

$$(\hat{P}) = (\hat{R})(\hat{S}).$$

116.5.1 UNITARY-SIMILARITY TRANSFORMATIONS and Operators transforming under a change of basis:

Now it is necessary to understand the manner in which operators transform under a change of basis. Let $\{|\hat{e}_i^o\rangle\}$ and $\{|\hat{e}_i\rangle\}$ represent the OLD and NEW sets of basis vectors in *n*-space.

$$\langle \hat{e}_i^o || \hat{e}_j^o \rangle = \delta_{ij}. \tag{11.6.18}$$

$$\langle \hat{e}_k || \hat{e}_\ell \rangle = \delta_{k\ell} \qquad (11.6.19)$$

Expanding theses two using the relation

$$| \hat{e}_k \rangle = \sum | \hat{e}_i^o \rangle \mathbb{S}_{k\,i} . \qquad (11.6.15)$$

$$\mathbb{S}_{k\,i} = \langle \hat{e}_i^o || \hat{e}_k \rangle \qquad (11.6.17)$$

$$(\mathbb{S}_{k\,i})^* = \langle \hat{e}_i^o || \hat{e}_k \rangle^* = (\mathbb{S}_{i\,k})$$

$$| \hat{e}_\ell \rangle = \sum | \hat{e}_j^o \rangle \mathbb{S}_{\ell\,j} \qquad (11.6.13)$$

Left hand side of (11.6.19),

$$\langle \hat{e}_k || \hat{e}_\ell \rangle = \left([\sum | \hat{e}_i^o \rangle \mathbb{S}_{k\,i}]^*, [\sum | \hat{e}_j^o \rangle \mathbb{S}_{\ell\,j}] \right)$$

$$= \sum \mathbb{S}_{k\,i} \mathbb{S}_{\ell\,j} \langle \hat{e}_i^o || \hat{e}_j^o \rangle$$

i.e.,. $\quad = \sum \mathbb{S}_{i\,k}^* \mathbb{S}_{\ell\,j} \langle \hat{e}_i^o || \hat{e}_j^o \rangle = \delta_{k\ell}$.

$$\sum \mathbb{S}_{i\,k}^* \mathbb{S}_{\ell\,j} \delta_{ij} = \sum \mathbb{S}_{i\,k}^* \mathbb{S}_{\ell\,j} \delta_{\ell\,i} = \delta_{k\ell}$$

Letting $\ell = k$, $\quad \sum \mathbb{S}_{i\,k}^* \mathbb{S}_{ki} = \delta_{kk} = 1$

Using the fact that $\mathbb{S}_{k\,i} = \mathbb{S}_{ik}^{*\,\prime} = \mathbb{S}_{i\,k}$

$$\sum \mathbb{S}_{k\,i}^{*\,\prime} \mathbb{S}_{ki} = 1 = \sum \hat{\mathbb{S}}^* \cdot \hat{\mathbb{S}} = \hat{\mathbb{S}}^{-1} \hat{\mathbb{S}} .$$

But $\quad \mathbb{S}_{k\,i} = \langle \hat{e}_i^o || \hat{e}_k \rangle$ and $\mathbb{S}_{k\,i}^{*\,\prime} = \mathbb{S}_{ik}{}'$, and $\hat{\mathbb{S}}^{-1} = \hat{\mathbb{S}}^\dagger$.

This means $\quad \hat{\mathbb{S}} = \hat{U}$, a Unitary Matrix. $\qquad (11.6.20)$

i.e., $\quad \hat{\mathbb{S}}^\dagger \hat{\mathbb{S}} = \hat{U}^\dagger \hat{U} = \hat{\mathfrak{I}} .$ $\qquad (11.6.21)$

The matrix transformation of one orthonormal set to another is Unitary. The real Unitary Matrix is called Orthogonal. Such changes of Representation are referred to as <u>Unitary Transformation</u> (analogous to the Orthogonal Transformations in ordinary (Euclidean) Space). Thus in a *Unitary Transformation*,

$$| \det \hat{U} |^2 = | \hat{U} |^2 = 1 .$$

In other words, if the transformation has to be unitary, the eigen vectors must be mutually orthogonal and the vice-versa. – a a Unitary Transformation means keeping the basis fixed $\hat{U}^\dagger \hat{U} = \hat{\Im}$ is the necessary and sufficient condition that the new basis is orthonormal. The same has the equivalent forms:

$$\hat{U} = (\hat{U}^{-1})^\dagger ; \quad \hat{U}^\dagger = \hat{U}^{-1} \text{ and } \hat{U}^{-1} (\hat{U}^\dagger)^{-1} = 1.$$

Instead of changing the Representation by means of a Unitary Transformation, one may also apply Unitary Transformation directly to the vectors and matrix operators, as described below.

$$1| \hat{e} \rangle = \hat{U} | \hat{e}^\circ \rangle.$$

Under a Unitary Transformation of vectors and operators, all scalar products remain unchanged.

11.6.6 UNITARY TRANSFORMATION of a STATE (an arbitrary) Vector:

To obtain the new components of an arbitrary state vector $| \Psi_a \rangle$ in the basis, one may write $| \Psi_a^\circ \rangle$ as

$$| \Psi_a^\circ \rangle = \sum a_i^\circ | \hat{e}_i^\circ \rangle = \sum a_i^\circ | \hat{i}^\circ \rangle$$

$$= \sum a_k | \hat{e}_k \rangle = \sum a_k | \hat{k} \rangle.$$

Using (11.6.15) $| \Psi_a^\circ \rangle = \sum a_k \{ \sum | \hat{e}_i^\circ \rangle (\mathbb{S}_{k\,i}) \} = \sum (\mathbb{S}_{k\,i}) a_k \{ \sum | \hat{e}_i^\circ \rangle \}$

The right hand side is identical to

$\sum a_i^\circ | \hat{e}_i^\circ \rangle$, i.e. $\sum (\mathbb{S}_{k\,i}) a_k \{ \sum | \hat{e}_i^\circ \rangle \} = \sum a_i^\circ | \hat{e}_i^\circ \rangle$, if

$$a_i^\circ = \sum (\mathbb{S}_{k\,i}) a_k. \tag{11.6.14}$$

i.e., $\quad | \Psi_a^\circ \rangle = \hat{\mathbb{S}} | \Psi_a \rangle$ \hfill (11.6.15)

or, $\{ a_1^\circ, a_2^\circ, a_3^\circ, \cdot, \cdot, \quad , \cdot, a_n^\circ \} = \hat{\mathbb{S}} \ (a_1, a_2, a_3, \cdot, \cdot, \quad , \cdot, a_n)$ (11.6.16)

Now $\quad \hat{\mathbb{S}}^{-1} \equiv \hat{\mathbb{S}}^\dagger$ \hfill (11.6.19)

This means $\hat{\mathbb{S}}$ = a Unitary Matrix, $\hat{U}$ \hfill (11.6.20)

$\therefore \quad \hat{S}^\dagger \hat{S} \quad \hat{U}^\dagger \hat{U} = \hat{\Im}$ (11.6.21)

Equation (11.5.15) becomes

$$| \Psi_a^\circ \rangle = \hat{U} | \Psi_a \rangle \qquad (11.6.22)$$

11.6.7 UNITARY TRANSFORMATION on an Operator:
Instead of changing the Representation by means of a Unitary Transformation, (11.6.22) viz.,

$$| \Psi_a^\circ \rangle = \hat{U} | \Psi_a \rangle,$$

or (11.6.15) $\quad | \Psi_a^\circ \rangle \therefore = \hat{S} | \Psi_a \rangle$

one may apply Unitary Transformation directly to vector $| \Psi \rangle$ as well as to the operator $\hat{X}$. To examine this let $| \alpha \rangle$ and $| \beta \rangle$ are two arbitrary ket vectors, and $\hat{X}$ is an observable, such that

$$\hat{X}^\circ | \alpha^\circ \rangle = | \beta^\circ \rangle \qquad (11.6.23)$$

Using Unitary Transformation $\hat{U}$ so that

$$\hat{U} | \alpha^\circ \rangle = | \alpha \rangle \qquad (11.6.24)$$

$$\hat{U} | \beta^\circ \rangle = | \beta \rangle \qquad (11.6.25)$$

Writing $\hat{X} | \alpha \rangle = | \beta \rangle \qquad (11.6.26)$

Using (11.6.24) in the left hand side of (11.6.26)

$$\hat{X} | \alpha \rangle = \hat{X} \{\hat{U} | \alpha^\circ \rangle\}$$

i.e., from equation (11.5.23), now right hand side of (11.5.26) is

$$| \beta \rangle = \hat{U} | \beta^\circ \rangle \quad \hat{U}\{\hat{X}^\circ | \alpha^\circ \rangle\} = \hat{X}\{\hat{U} | \alpha^\circ \rangle\} = \text{right hand side.}$$

So $\quad \hat{U}\{\hat{X}^\circ | \alpha^\circ \rangle\} = \hat{X}\{\hat{U} | \alpha^\circ \rangle\}$

Or, $\quad \hat{U}\{\hat{X}^\circ\} = \hat{X}\{\hat{U}\} \qquad (11.6.27)$

Using $\hat{U}^\dagger \hat{U} = \hat{\Im}$. (11.6.21)

In equation (11.6.26)

$$\hat{U}^\dagger\{\hat{X}|\ \alpha\rangle\} = \hat{U}^\dagger\{\hat{U}\hat{X}^\circ|\ \alpha^\circ\rangle\} = (\hat{U}^\dagger\hat{U})\{\hat{X}^\circ|\ \alpha^\circ\rangle\} = \hat{\Im}\{\hat{X}^\circ|\ \alpha^\circ\rangle\}.$$

Using equation (11.6.24) in the left hand side

i.e., $\quad \hat{U}^\dagger\{\hat{X}|\ \alpha\rangle\} = (\hat{U}\hat{X})|\ \alpha\rangle = \hat{U}^\dagger\ \hat{X}\ (\hat{U}|\ \alpha^\circ\rangle)$

$$= (\hat{U}^\dagger\ \hat{X}\ \hat{U})|\ \alpha^\circ\rangle = \hat{\Im}\hat{X}^\circ|\ \alpha^\circ\rangle$$

One, therefore, gets $\quad \boxed{\hat{X}^\circ = (\hat{U}^\dagger\ \hat{X}\ \hat{U})}$ (11.6.28)

Or, $\quad \boxed{\hat{X} = \hat{U}\ \hat{X}^\circ\ \hat{U}^\dagger}$ (11.6.29)

The two operators, $\hat{X}$ and $\hat{X}^\circ$, connected by $\hat{U}$ a Unitary Transformation are said to be UNITARY EQUIVALENTS.

Salient properties are:

(1) If $\hat{X}^\circ$ is Hermitian, then $\hat{X}$ is also Hermitian.

(2) $\quad \langle \alpha |\hat{X}| \alpha \rangle = \langle \alpha^\circ |\hat{X}^\circ| \alpha^\circ \rangle$

(3) The eigen values of $\hat{X}$ are also the eigen values of $\hat{X}^\circ$.

(4) If the $\hat{X}^\circ$ matrix rotates the vector $|\ \alpha\rangle$ into position $|\ \alpha^\circ\rangle$,

$$|\alpha\rangle = \hat{X}^\circ|\alpha^\circ\rangle;$$

$$|\beta\rangle = \hat{X}|\alpha\rangle;$$

$$\hat{S}|\beta\rangle = \hat{S}\hat{X}|\alpha\rangle,$$

Where $\hat{S}$ rotates the coordinates of $|\hat{e}_i^\circ\rangle$ to $|\hat{e}_i\rangle$.

Then $\quad \hat{X}^\circ = \hat{S}\ \hat{X}\ \hat{S}^{-1}$

and $\quad \hat{X} = \hat{S}^{-1}\hat{X}^\circ\ \hat{S}$

i.e., $\hat{X}$ is obtained from $\hat{X}^\circ$ by a Similarity Transformation.

In matrix equation, $\hat{S}^{-1} = \hat{U}$ and $\hat{S} = \hat{U}^\dagger$.

Thus the roles of the <u>Transformation Matrix</u>(operator) $\hat{S}$ and its Heremitian Conjugate $\hat{S}^\dagger$ have been interchanged in going from the

| Matrix Transformation equation →to→ Unitary Transformation equation |

The two ROTATIONS, one affecting only the basis, and the other keeping the basis fixed while rotating all vectors and operators, are equivalent only if they are performed in *opposite directions*, i.e. if one is in the inverse of the other.

A matrix $\hat{X}$ which commutes with its Hermitian conjugate is said to be NORMAL

$$\hat{X}^\dagger \hat{X} = \hat{X} \hat{X}^\dagger.$$

A matrix $\hat{X}$ can be <u>diagonalized</u> by a Unitary Transformation if, and only if, $\hat{X}$ is 'normal'. It is to be noted that Unitary Matrices as well as Hermitian matrices satisfy this condition.

11.6.8 | Theorem | A Hermitian matrix remains Hermitian under a Unitary Transformation.
Proof: Let the matrix $\hat{A}°$ is Hermitian.

$\hat{A}°$ is transformed into $\hat{A}$ by the Unitary Transformation, $\hat{U}$.

$$\hat{A} = \hat{U}^{-1} \hat{A}° \hat{U} \text{ (or } \hat{A} = \hat{U} \hat{A}° \hat{U}^\dagger).$$

Transposing, $\hat{A}' = (\hat{U}^{-1} \hat{A}° \hat{U})' = (\hat{U})' (\hat{A}°)' (\hat{U}^{-1})'$.

As $\hat{U}$ is unitary, and $\hat{A}°$ is Hermitian,

$$\hat{A}' = (\hat{U}^{-1} \hat{A}° \hat{U})^* = \hat{A}^*$$

So $\hat{A}$ is also Hermitian.

11.6.9 DIAGONALIZATION of Matrices: QuadraticFforms
(The fundamental problem in Quantum Mechanics).

The matrix notation very conveniently expresses the homogeneous quadratic function u (say) n variables:

$a_1, a_2, a_3, \cdot \: \cdot, \cdot \: \cdot, \cdot \: \cdot, a_n$

Let $\hat{X}$ is a symmetric operator in 3-space and

$$\vec{r} = \{ a_1, a_2, a_3 \} = \text{a vector}$$

with respect to an orthonormal basis, $\{ \hat{e}_i, \hat{e}_i, \hat{e}_i \}$, in 3-space. The quadratic surface is given by

$$u(a_1, a_2, a_3) = \vec{r} \, \hat{X} \, \vec{r}$$

$$= (a_1, a_2, a_3) \begin{pmatrix} x_{11} & x_{12} & x_{13} \\ x_{21} & x_{22} & x_{23} \\ x_{31} & x_{32} & x_{33} \end{pmatrix} \begin{pmatrix} a_1 \\ a_2 \\ a_3 \end{pmatrix}.$$

$$= x_{11}a_1^2 + x_{22}a_2^2 + x_{33}a_3^2 + 2x_{12}a_1a_2 + 2x_{32}a_3a_2 + 2x_{13}a_1a_3 \quad (11.630)$$

which is a homogeneous equation of the 2nd degree in the variables, a_1, a_2, a_3. When

$$u(a_1, a_2, a_3) = \vec{r} \, \hat{X} \, \vec{r} > 0, \text{ for } \vec{r} \neq 0,$$

then the quadratic form is POSITIVE DEFINITE

Now consider the effect on $u(a_1, a_2, a_3)$ of a transformation via a non-singular matrix.

Let $\hat{S}$ = non-singular matrix, which transforms $| \vec{r} \rangle \longrightarrow | \vec{r}° \rangle$,

where $\quad | \vec{r} \rangle = \hat{S} | \vec{r}° \rangle,$ \hfill (11.6.31)

or $\quad \langle \vec{r} | = \langle \vec{r}° | \hat{S}'.$ \hfill (11.6.32)

Then one obtains $u(a_1, a_2, a_3)$

$$= u(a_1°, a_2°, a_3°) = [\langle \vec{r}° | \hat{S}'] \hat{X} [\hat{S} | \vec{r}° \rangle]$$

$$= \langle \vec{r}° | [\hat{S}' \, \hat{X} \, \hat{S}] | \vec{r}° \rangle = \langle \vec{r}° | \hat{X}° | \vec{r}° \rangle \qquad (11.6.33)$$

i.e., $\quad \hat{X}° = [\hat{S}' \, \hat{X} \, \hat{S}]$ = a new matrix in quadratic form. \hfill (11.6.34)

If the transformation matrix is also orthogonal,

$$\hat{X}° = [\hat{S}' \, \hat{X} \, \hat{S}] = [\hat{S}^{-1} \, \hat{X} \, \hat{S}]$$

$\hat{S}$ is orthogonal means one has to find a new orthonormal basis $\{ \hat{e}_i°, \hat{e}_i°, \hat{e}_i° \}$, in the space in which $\vec{r}$ exists with respect to which

$$u(a_1°, a_2°, a_3°) = x_{11}°(a_1°)^2 + x_{22}°(a_2°)^2 + x_{33}°(a_3°)^2.$$

$$= (a_1°, a_2°, a_3°) \begin{pmatrix} x_{11}° & 0 & 0 \\ 0 & x_{22}° & 0 \\ 0 & 0 & x_{33}° \end{pmatrix} \begin{pmatrix} a_1° \\ a_2° \\ a_3° \end{pmatrix} \quad (11.6.35)$$

$u(a_1°, a_2°, a_3°)$ is said to be the CANONICAL FORM of $u(a_1, a_2, a_3)$. The problem is thus regarded as finding an orthonormal basis with respect to which $\hat{X}$ is diagonal. This is the fundamental problem in Quantum Mechanics.

$$\hat{X}° = \hat{X}°_d = \begin{pmatrix} x_{11}° & 0 & 0 \\ 0 & x_{22}° & 0 \\ 0 & 0 & x_{33}° \end{pmatrix} \quad (11.6.36)$$

is said to be the classical canonical form of a matrix $\hat{X}$ for the case when ALL THE CHARACTERISTIC ROOTS OF $\hat{X}$ ARE DISTINCT. Thus the process of subjecting the symmetric matrix $\hat{X}$ to Similarity orthogonal Transformation so that the transform of $\hat{X}$ is DIAGONAL will effect a transformation of the surface to its PRINCIPAL AXES.

For matrices of dimensions up to (4 x 4); and for higher order matrices, Jacobi's method of diagonalization is used with iterative techniques for feeding in Computers.

11.6.10 Properties of a Unitary Transformation

(i) The Secular Equation of the given matrix $\hat{\Omega}$ is invariant,

i.e., $\boxed{|\hat{\Omega} - \lambda \hat{O}| = 0 = \text{Invariant}}$.

(ii) It follows that the EIGEN VALUES $\boxed{\lambda_i = \text{Invariant}}$

(iii) The determinant, which is the product of eigen values, $\boxed{det(\hat{\Omega}) = \Pi \lambda_i = \text{Invariant}}$

(iv) The trace (spur), which is the sum of the eigen values, i.e.
$\boxed{\text{Tr}(\hat{\Omega}) = \sum \lambda_i = \text{Invariant}}$.

(v) The product of two Unitary Transformations is a Unitary Transformation.

(vi) Further one has the following useful identities:

$\boxed{det(\hat{U}) = 1}$.

$$\boxed{\mathrm{Tr}(\hat{U}\,\hat{X}^\circ_d\,\hat{U}^\dagger) = \mathrm{Tr}(\hat{X})}$$

$$\boxed{\det(\hat{U}\,\hat{X}^\circ_d\,\hat{U}^\dagger) = \det(\hat{X})}$$

$$\boxed{(\hat{U}\,\hat{X}^\circ_d\,\hat{U}^\dagger) = (\hat{U}\,\hat{X}\,\hat{U}^\dagger)}.$$

(vii) $\hat{U}$ is the unitary operator which represents a rotation of a given physical system about a given axis; i.e. $\hat{U}$ is such that the state vectors and observables of the rotated system are related to those of the original system by the relations

$$|\alpha\rangle = \hat{S}|\alpha^\circ\rangle$$

and $\boxed{\hat{X} = \hat{S}^{-1}\hat{X}^\circ_d\hat{S} = \hat{U}\,\hat{X}^\circ_d\,\hat{U}^\dagger}$;

or $\boxed{\hat{X}^\circ_d = \hat{S}\,\hat{X}\,\hat{S}^{-1} = \hat{U}^\dagger\,\hat{X}\,\hat{U}}$.

Worked out Example 11.8

Find an orthonormal set of basis for a given symmetric matrix,

$$\hat{T} = \begin{pmatrix} 5 & 10 & 8 \\ 10 & 2 & -2 \\ 8 & -2 & 11 \end{pmatrix}, \text{ with respect to a set of orthonormal basis; } \{\hat{i},\hat{j},\hat{k}\}$$

Solution: STEP # 1 Given: Matrix $\hat{T}$ and orthonormal set of basis vectors: $\{\hat{i},\hat{j},\hat{k}\}$.

i.e., $(1, 0, 0), (0, 1, 0), (0, 0, 1)$.

STEP # 2 Find the eigen values λ_i of $\hat{T}$:

$\hat{T}$ is a matrix of rank $r=3$. Since $r=$ finite, the eigen values are the roots of the *Secular Equation* (**Cayley – Hamilton Theorem**), $(\hat{T} - \lambda\,\hat{\Im}) = 0$

$$(\hat{T} - \lambda\,\hat{\Im}) = \begin{pmatrix} 5-\lambda & 10 & 8 \\ 10 & 2-\lambda & -2 \\ 8 & -2 & 11-\lambda \end{pmatrix} = 0$$

STEP # 3 i.e. $(5-\lambda)\begin{vmatrix} 2-\lambda & -2 \\ -2 & 11-\lambda \end{vmatrix} - 10\begin{vmatrix} 10 & -2 \\ 8 & 11-\lambda \end{vmatrix} + 8\begin{vmatrix} 10 & 2-\lambda \\ 8 & -2 \end{vmatrix} = 0$

$(5-\lambda)\,[(2-\lambda)(11-\lambda)-4] - 10\,[110 - \lambda + 16] + 8\,[-20 - 16 + 8\lambda] = 0$.

i.e., $\quad (5-\lambda)\,[(28\lambda^2 - 13\lambda) - 10\,[126 - 10\lambda] + 8\,[8\lambda - 36] = 0$.

$-\lambda^3 + 18\lambda^2 + 81\lambda - 1458 = 0$.

STEP # 4 Factorizing yields, $(\lambda + 9)(\lambda - 9)(\lambda - 18) = 0$.

The roots are: $\lambda = -9;\ +9;\ +18$.

These are the eigen values of the given matrix operator $\hat{T}$

$\lambda_1 = -9;\ \lambda_2 = +9;\ \lambda_3 = +18$

Since these three values are different they belong to <u>non-degenerate</u> case.

STEP # 5 To find the eigen vector, $|\,\bar{\psi}_1\,\rangle$ corresponding to $\lambda = \lambda_1$:

When $\lambda = \lambda_1 = -9$:

Any arbitrary vector $|\,\bar{\psi}_1\,\rangle = \bar{r}_1 = \{r_1 \ell_1 \hat{i} + r_1 m_1 \hat{j} + r_1 n_1 \hat{k}\}$

where ℓ_1, m_1, n_1 are '*direction cosines*', such that $\ell_1^2 + m_1^2 + n_1^2 = 1$, and $\bar{r}_1 = \{\ell_1, m_1, n_1\}$ is a ket $|\,\bar{\psi}_1\,\rangle$.

STEP # 6 To find ℓ_1, m_1, n_1, construct a set of homogeneous equations for $\lambda_1 = -9$,

viz., $\quad (\hat{T} - \lambda_1 \hat{\mathfrak{I}})\,|\,\bar{\psi}_1\,\rangle = \begin{pmatrix} 5+9 & 10 & 8 \\ 10 & 2+9 & -2 \\ 8 & -2 & 11+9 \end{pmatrix} \begin{pmatrix} \ell_1 \\ m_1 \\ n_1 \end{pmatrix} = 0$

i.e., $\quad \ell_1 : m_1 : n_1 = \begin{vmatrix} 11 & -2 \\ -2 & 20 \end{vmatrix} : \begin{vmatrix} -10 & -2 \\ 8 & 20 \end{vmatrix} : \begin{vmatrix} 10 & 11 \\ 8 & -2 \end{vmatrix} = 2 : -2 : -1$

STEP # 7 $|\,\bar{\psi}_1\,\rangle$ = Column vector, $\{\,2, -2, -1\,\}$.

STEP # 8 On normalization, $|\,\hat{\psi}_1\,\rangle = (1/9)^{1/2}\,(2\hat{i} - 2\hat{j} - 1\hat{k})$.

STEP # 9   To find the eigen vector, $|\,\bar{\psi}_2\,\rangle$ corresponding to $\lambda = \lambda_2 = +9$

Let $\quad |\,\bar{\psi}_2\,\rangle = \bar{r}_2 = \{r_2 \ell_2 \hat{i} + r_2 m_2 \hat{j} + r_2 n_2 \hat{k}\}$,

where ℓ_2, m_2, n_2 are '*direction cosines*',

and $|\bar{\psi}_2\rangle = \bar{r}_2 = \{\ell_2, m_2, n_2\}$ is a ket. As in STEP # 4, for $\lambda_2 = +9$,

$$(\hat{T} - \lambda_2 \hat{\mathfrak{I}})|\bar{\psi}_2\rangle = \begin{pmatrix} -4 & 10 & 8 \\ 10 & -7 & -2 \\ 8 & -2 & 2 \end{pmatrix} \begin{pmatrix} \ell_2 \\ m_2 \\ n_2 \end{pmatrix} = 0,$$

i.e., $\ell_2 : m_2 : n_2 = \begin{vmatrix} -7 & -2 \\ -2 & 2 \end{vmatrix} : - \begin{vmatrix} -10 & -2 \\ 8 & 2 \end{vmatrix} : \begin{vmatrix} 10 & -7 \\ 8 & -2 \end{vmatrix} = -1 : -2 : +2$

$|\bar{\psi}_2\rangle = $ Column vector, $\{-1, -2, +2\}$.

On normalization, $|\hat{\psi}_2\rangle = (1/9)^{1/2} \left(-\hat{i} - 2\hat{j} + 2\hat{k}\right)$.

STEP # 10 To find the eigen vector, $|\bar{\psi}_3\rangle$ corresponding to $\lambda = \lambda_3 = +18$:

When $\lambda = \lambda_3 = +18$:

Let $|\bar{\psi}_3\rangle = \bar{r}_3 = \{r_3\ell_3\hat{i} + r_3 m_3 \hat{j} + r_3 n_3 \hat{k}\}$,

where ℓ_3, m_3, n_3 are 'direction cosines', As in STEP # 4, for $\lambda_3 = +18$,

$$(\hat{T} - \lambda_3 \hat{\mathfrak{I}})|\bar{\psi}_3\rangle = \begin{pmatrix} -13 & 10 & 8 \\ 10 & -16 & -2 \\ 8 & -2 & -7 \end{pmatrix} \begin{pmatrix} \ell_3 \\ m_3 \\ n_3 \end{pmatrix} = 0$$

$\ell_3 : m_3 : n_3 = \begin{vmatrix} -16 & -2 \\ -2 & -7 \end{vmatrix} : - \begin{vmatrix} -10 & -2 \\ 8 & -7 \end{vmatrix} : \begin{vmatrix} 10 & -6 \\ 8 & -2 \end{vmatrix} = +2 : +1 : +2$

$|\bar{\psi}_3\rangle = $ Column vector, $\{+2 : +1 : +2\}$.

On normalization, $|\hat{\psi}_3\rangle = (1/9)^{1/2} \left(2\hat{i} + 1\hat{j} + 2\hat{k}\right)$.

STEP # 11 The Orthogonal Vectors required are:

$|\hat{\psi}_1\rangle = (1/9)^{1/2} \left(2\hat{i} - 2\hat{j} - 1\hat{k}\right)$

$|\hat{\psi}_2\rangle = (1/9)^{1/2} \left(-\hat{i} - 2\hat{j} + 2\hat{k}\right)$,

$|\hat{\psi}_3\rangle = (1/9)^{1/2} \left(2\hat{i} + 1\hat{j} + 2\hat{k}\right)$.

STEP # 12 To verify if the three vectors, $||\bar{\psi}_1\rangle, |\bar{\psi}_2\rangle \& |\bar{\psi}_3\rangle$ are normalized as can be seen from

$\langle \tilde{\psi}_1 || \tilde{\psi}_1 \rangle = 1$, $\langle \tilde{\psi}_2 || \tilde{\psi}_2 \rangle = 1$ & $\langle \tilde{\psi}_3 || \tilde{\psi}_3 \rangle = 1$.

To check orthogonality:

$$\langle \tilde{\psi}_2 || \tilde{\psi}_1 \rangle = (1/9)^{1/2} \left(-\hat{i} - 2\hat{j} + 2\hat{k} \right) (1/9)^{1/2} \left(2\hat{i} - 2\hat{j} - 1\hat{k} \right)$$

$$= (1/9)(2 - 4 + 2) = 0,$$

$$\langle \tilde{\psi}_3 || \tilde{\psi}_1 \rangle = (1/9)^{1/2} \left(2\hat{i} + 1\hat{j} + 2\hat{k} \right) (1/9)^{1/2} \left(2\hat{i} - 2\hat{j} - 1\hat{k} \right).$$

$$= (1/9)(4 - 2 - 2) = 0$$

$$\langle \tilde{\psi}_3 || \tilde{\psi}_2 \rangle = (1/9)^{1/2} \left(2\hat{i} + 1\hat{j} + 2\hat{k} \right) (1/9)^{1/2} \left(-\hat{i} - 2\hat{j} + 2\hat{k} \right)$$

$$= (1/9)(-2 - 2 + 4) = 0.$$

STEP # 13 Because the eigen values of a matrix are the diagonal elements of its diagonal matrix one can guess that the diagonalized form of $\hat{T}$ is

$$\hat{T}_d^\circ = \begin{pmatrix} -9 & 0 & 0 \\ 0 & +9 & 0 \\ 0 & 0 & +18 \end{pmatrix}$$

11.6.11 The SCHMIDT ORTHOGONALIZATION Procedure

Till the previous section the discussions were based on non-degenerate eigen states.

Such eigen vectors of a Hermitian operator are automatically orthogonal.

However, it is frequently true that two or more linearly independent eigen vectors $|\hat{\psi}_n\rangle$ correspond to the same eigen value, λ. The eigen value is the said to be degenerate. Here an orthogonal set has to be constructed from a set of degenerate vectors.

Let $|\hat{\psi}_1\rangle, |\hat{\psi}_2\rangle, |\hat{\psi}_3\rangle, ..., |\hat{\psi}_n\rangle$ are the arbitrary degenerate set of eigen vectors. For this let the normalized eigen vectors be unit vectors in a vector space. To construct an arbitrary vector $|\Psi\rangle$ another vector orthogonal to a given unit vector $|\bar{u}\rangle$, all one has to do is to remove from $|\Psi\rangle$ its component parallel to $|\bar{u}\rangle$, viz..
$|\bar{u}\rangle\langle\bar{u}||\Psi\rangle$.

The result is $|\Psi'\rangle = |\Psi\rangle - |\bar{u}\rangle\langle\bar{u}||\Psi\rangle$.

$|\Psi'\rangle$ = linearly dependent, not normalized, not orthogonal

In analogy with this, let $|\Psi_1\rangle$ is normalized, and denoted by $|\Psi_{1n}\rangle$.

Let $|\bar{\psi}_1\rangle \equiv |\hat{\psi}_1\rangle$ = the first vector of the new set.

Since $\langle \hat{\psi}_1 || \hat{\psi}_1 \rangle = 1$ (11.6.37)

$\langle \bar{\psi}_1 || \bar{\psi}_1 \rangle = 1$.

Defining the scalar product,

$\langle \hat{\psi}_1 || \bar{\psi}_2 \rangle = a_{21}$, (11.6.38)

choose $|\hat{\psi}_2\rangle \equiv |\bar{\psi}_2\rangle - a_{21} |\hat{\psi}_1\rangle$ (11.6.39)

One can demand $\langle \hat{\psi}_1 || \hat{\psi}_2 \rangle = 0$.

This means to examine if $|\hat{\psi}_2\rangle$ satisfies this condition.

Therefore, $\langle \bar{\psi}_1 || \bar{\psi}_1 \rangle = \langle \hat{\psi}_1 || \bar{\psi}_2 \rangle - a_{21} \langle \bar{\psi}_1 || \bar{\psi}_1 \rangle = a_{21} - a_{21} = 0$. (11.6.40)

So $|\hat{\psi}_1\rangle$, $|\hat{\psi}_2\rangle$ are orthogonal.

Now normalize $|\bar{\psi}_2\rangle$ by setting $\langle \hat{\psi}_2 || \hat{\psi}_2 \rangle = 1$. (11.6.41)

Defining the scalar products, $\langle \hat{\psi}_1 || \bar{\psi}_3 \rangle = a_{31}$, (11.6.42)

$\langle \hat{\psi}_2 || \bar{\psi}_3 \rangle = a_{32}$, (11.6.43)

choose $|\hat{\psi}_3\rangle = |\bar{\psi}_3\rangle - a_{31} |\hat{\psi}_1\rangle - a_{32} |\hat{\psi}_2\rangle = 1$ (11.6.44)

One can demand $\langle \hat{\psi}_1 || \hat{\psi}_3 \rangle = 0$;

$\langle \hat{\psi}_2 || \hat{\psi}_3 \rangle = 0$;

and $\langle \hat{\psi}_3 || \hat{\psi}_3 \rangle = 1$.

From equation (11.6.42), taking the scalar product this means to examine if $|\hat{\psi}_2\rangle$ satisfies this condition.

Therefore,

$$\langle \hat{\psi}_1 \| \hat{\psi}_3 \rangle = \langle \hat{\psi}_1 \| \hat{\psi}_3 \rangle - a_{31} \langle \hat{\psi}_1 \| \hat{\psi}_1 \rangle - a_{32} \langle \hat{\psi}_1 \| \hat{\psi}_2 \rangle = a_{31} - a_{31} - 0 = 0.$$

Similarly,

$$\langle \hat{\psi}_2 \| \hat{\psi}_3 \rangle = \langle \hat{\psi}_2 \| \hat{\psi}_3 \rangle - a_{31} \langle \hat{\psi}_2 \| \hat{\psi}_1 \rangle - a_{32} \langle \hat{\psi}_2 \| \hat{\psi}_2 \rangle = a_{32} - 0 - a_{32} = 0.$$

∴ $| \hat{\psi}_1 \rangle, | \hat{\psi}_2 \rangle$, and $| \hat{\psi}_3 \rangle$ form an orthogonal set of vectors, and so on.

It will be noticed that although this is one possible way of constructing an orthonormal set the vectors, $\{ | \hat{\psi}_i \rangle \}$ are not unique.

11.7   TYPES OF PICTURES (EQUATIONS OF MOTION)

11.7.1   INTRODUCTION

In Quantum Mechanics, the state of a physical system at any given time is described by a unit vector in a Hilbert Space, in which sets of axes can be defined by the eigen vectors of complete sets of observables of the system.

Any change with time in the state of the system can be investigated

i)   By keeping the basis (axes) fixed and allowing the state vector to rotate, or
ii) By allowing the state vector fixed and permitting the axes to rotate, or
iii) By allowing simultaneous rotation of the state vector and the axes,

using in each case the appropriate *Equations of Motion* of the vectors concerned. Either one of these modes of description of quantum phenomena is called the Representation, or this is not to be confused with the Representation of vectors and operators of vector spaces by matrices.

The three possibilities described in the previous paragraph are called, respectively, the **Schrödinger**, the **Heisenberg** and the **Interaction**Pictures. In practice one adopts the 'Representation', which lends itself best to the solution of each particular problem.

The discussions so far in the previous Chapters gave a general scheme of relations between states and dynamical variables for a dynamical system at one instant of time. To get a complete theory of dynamics one must also consider the connections between different instants of time. When one makes an observation on the dynamical system, the state of the system gets changed in an unpredictable way, but in between observations causality applies, in Quantum Mechanics as in classical mechanics, and the system is governed by equations of motion which make the state at one time determine the state at a later time.

11.7.2 The EVOLUTION OPERATOR, $\hat{T}(t,t_0) \Rightarrow \hat{U}(t,t_0)$

11.7.2.1. Consider a dynamical system.

(1) $|t_0\rangle$ = state of the system at time $t = t_0$.   (11.7.1)

$|t\rangle$ = state of the system at time t,

$\hat{A}$ = an observable operator.

Then $\hat{A}|a', t_0\rangle = a'|a', t_0\rangle$ (11.7.2)

(2) $|t\rangle$ at $t > t_0$ arises for $|t_0\rangle$ by the action of a time development operator, $\hat{T}(t,t_0)$ in Hilbert Space.

$$\hat{T}(t,t_0) \equiv \hat{T}(t = t_0) \qquad (11.7.3)$$

is the unitary operator in question.

Because the system is considered to be ISOLATED, $\hat{T}$ can only depend on $(t - t_0)$.

$$\boxed{|a't_0, t\rangle = \hat{T}(t - t_0)|a', t_0\rangle} \qquad (11.7.4)$$

(3)   Consider a time t' such that $t > t' > t_0$. Let the system evolves from $|a', t_0\rangle$ until $t + t'$, into the state represented by

$$|a't_0, t'\rangle = \hat{T}(t' - t_0)|a', t_0\rangle \qquad (11.7.5)$$

This state could then be allowed to evolve until t, at which time it is

(4) $|a't_0, t\rangle = \hat{T}(t - t')|a't_0, t'\rangle$ (11.7.6)

The state given by equation (11.6.6) is clearly identical with that given by equation (11.7.4), because equation (11.7.6) is $|a't_0, t\rangle = \hat{T}(t-t')[\hat{T}(t'-t_0)|a', t_0\rangle]$

i.e., $|a't_0, t\rangle = \hat{T}(t - t_0)|a', t_0\rangle$

(5) $\hat{T}(t, t_0) \equiv \hat{T}(t - t') \hat{T}(t' - t_0)$ (11.7.7)

which is the *group property* of the time development unitary operator,

since $(t - t') + (t' - t_0) = (t - t_0)$.

(6) Setting $t = t_o$.

$$\hat{T}(t_o - t_o) \equiv \hat{T}(t_o, t_o) = \hat{T}(0) = \hat{\Im} . \qquad (11.7.8)$$

It means that $\hat{T}(t_0 - t') \hat{T}(t' - t_0) = \hat{\Im}$

∴ $\hat{T}(t-t')^{-1} = \hat{T}(t'-t) = \hat{T}(t-t')^{\dagger}$

i.e., $\hat{T}(t, t_0) \Rightarrow \hat{U}(t, t_0) =$ unitary. $\qquad (11.7.9)$

(7) Consider $(t - t_o) \Rightarrow$ infinitesimally small.

∴ $(t - t_o) \Rightarrow \delta t$, and $t_o = 0$.

$\hat{U}(t + \delta t, t_0) = \hat{U}(\delta t)$,

Now $\hat{U}(0) = 1$, $\hat{U}(\delta t) = \hat{U}(0) + \delta \hat{U}$.

$$\hat{U}(\delta t) = 1 + \delta \hat{U} . \qquad (11.7.10)$$

Unitary then implies $\hat{\Im} \approx 1 + (\delta \hat{U} + \delta \hat{U}^{\dagger}) + \hat{O}[(\delta \hat{U})^2]$.

Hence one requires $(\delta \hat{U} + \delta \hat{U}^{\dagger}) = 0$ $\qquad (11.7.11)$

This means $\delta \hat{U} =$ anti-Hermitian. $= i$ x Hermitian operator. $\qquad (11.7.12)$

(8) Further, let $t' = t - \delta t$,

If $\delta t =$ small, $\hat{U}(t, t - \delta t) = \delta \hat{U}(\delta t)$

$\hat{U}(\delta t_1) \hat{U}(\delta t_2) = \hat{U}(\delta t_1 + \delta t_2)$

∴ $\delta \hat{U}(\delta t) \approx \delta t$ $\qquad (11.7.13)$

Because of (11.7.12) and (11.7.13)

$\boxed{\delta \hat{U} = (-i / \hbar) \, \delta t \text{ [a time independent operator, } \hat{H}]}$

One can write $\hat{U}(t, t - \delta t) = \delta \hat{U}(\delta t) = 1 - (-i / \hbar) \, \delta t \, \hat{H}$.

∴ $\delta \hat{U} = \{(-i \hat{H} \, \delta t / \hbar)\}$ $\qquad (11.7.14)$

where $\hat{H}$ is the time independent Hamiltonian operator, of the system. IT IS INTRODUCED FOR COVENIENCE *to ensure that* $\hat{H}$ *has the dimension of energy.*

11.7.2.2 FUNDAMENTAL LAW of EVOLUTION in Quantum Mechanics
Using the Group Property,

$$\hat{U}(t+\delta t) - \hat{U}(t) = [\hat{U}(\delta t) - 1]\,\hat{U}(t)$$

$$= d\hat{U}(t) = \{(-i\,\hat{H}\,dt/\hbar)\}\,\hat{U}(t)$$

or, $\quad (\partial/\partial t)\hat{U}(t) = \{(-i\,\hat{H}/\hbar)\}\,\hat{U}(t) \quad\quad$ (11.7.15)

since energy is conserved for a <u>conservative</u> system,

$[\hat{H},\hat{H}] = 0$,

$$\hat{U}(t) = e^{\{-i\,\hat{H}\,t/\hbar\}}. \quad\quad (11.7.16)$$

i.e., $\quad \boxed{\hat{U}(t-t_o) = e^{\{-i\,\hat{H}\,(t-t_o)/\hbar\}}}$.

because of the initial condition, $\hat{U}(0) = 1$.

If the system were <u>non-conservative</u>, then $\boxed{\hat{H} = \hat{H}(t)}$.

$$\boxed{\hat{U}(t,t_o) \ne e^{\{-i\,\hat{H}\,(t-t_o)/\hbar\}}}$$

$\hat{U}$ is also defined by $\quad \boxed{\hat{U}(t,t_o) = 1 - \{(-i/\hbar)\}\int \hat{H}.\,\hat{U}(t',\,t_o)\,dt'} \quad$ (11.7.17)

Either of the equation (11.7.15) or (11.7.16) or (11.7.17) is known as the expression for the FUNDAMENTAL LAW OF EVOLUTION in Quantum Mechanics. An equivalent expression for this law is the SCHRODINGER EQUATION.

REVIEW QUESTIONS

R.Q. 11.1 Given two vectors: $\vec{x} = [2\,\hat{e}_1 + i\,\hat{e}_2 + (4 - 3\,i)\,\hat{e}_3]$ and $\vec{y} = [(1+i)\,\hat{e}_1 + 2i\,\hat{e}_2 + (1 - i)\,\hat{e}_3]$

Find the scalar products $(\vec{x},\vec{y})$ and $(\vec{y},\vec{x})$. (Answers: $(\vec{x},\vec{y}) = (11 + i)$;

$(\vec{y},\vec{x}) = (11 - i)$).

R.Q.11.2   Show that the three 3-vectors $\mathbf{X}^1 = [1, 0, -1]$, $\mathbf{X}^2 = [0, 1, 0]$ and $\mathbf{X}^3 = [1. 0. 1]$ are linearly independent. (Answer: $|\Gamma| = 2 (> 0)$, i.e., $[\vec{X}^1\ \vec{X}^2\ \vec{X}^3] \neq 0$ Thus the three vectors are linearly independent).

R.Q. 11.3 Determine which of the following sets of functions are linearly dependent, and find the relations between the functions in these cases. (a) $u = x$; $v = x^2$; and $w = x^3$; (b) $u = e^x$, $v = e^{-x}$, and $w = 2\ Coshx$. (Answer: (a) not linearly dependent, (b) $\Psi = \sum c_i f_i(x)$, with $c_1 = 1, c_2 = 1, c_3 = -2$).

R.Q. 11.4 Write the matrix representation of the given operator, $\hat{P} = [(d/dx) + k]$ and eigen ket $\left| \sqrt{1/2a}\ e^{-i(n\pi/a)x} \right\rangle$. (Answer: $\hat{P}_{mn} = \langle m | \hat{P} | n \rangle = \{i\ (n\pi/a) + k\}\ \delta_{mn}$, so $\hat{P}$ is diagonal).

R.Q. 11.5   Obtain the result of a Projection Operator operating on a ket vector. (Answer: $\hat{P}_i = | i \rangle\langle i | | c \rangle = | c_1 \rangle$).

R.Q. 11.6   Show that the Projection Operator $\hat{P}_i = \begin{pmatrix} 1 & 0 & 0 \\ 0 & 1 & 0 \\ 0 & 0 & 1 \end{pmatrix}$ operating on a vector projects it on a next lowerdimensional subspace. (Answer: Assume $\langle c | = \{a, b, c\}$, $\langle c_1 | = \langle c | \hat{P}_i$ is a 2-D vector).

R.Q. 11.7 Prove that the multiple operation of Projection Operators, viz. $\hat{P}_i^n = \hat{P}_i$.

R.Q. 11.8 If $\hat{P}_i$ is a Projection Operator, show that $\hat{P}_i^n - \hat{P}_i = 0$.

R.Q. 11.9 If $\hat{P}_i$ is a Projection Operator, show that. $(1 - \hat{P}_i^2) = 1 - \hat{P}_i$.

R.Q. 11.10 Prove that the Projection Operator has no inverse.(Answer: Since $| c \rangle = | \hat{e}_1 \rangle + | \hat{e}_2 \rangle$,and $| c \rangle = \hat{P}_i | c \rangle + (1 - \hat{P}_i) | c \rangle$).

R.Q. 11.11 Find the eigen values of the Projection Operator, $\hat{P}_i$.(Answer: 0 or 1).

R.Q. 11.12   For arbitrary vectors $\langle n |$ and $| m \rangle$, demonstrate the hermiticity of the momentum operator, $\hat{p}_z$.

R.Q. 11.13   For arbitrary vectors $\langle n |$ and $| m \rangle$, show that $\langle n | (-\hbar^2 \nabla^2) | m \rangle = \langle m | (-\hbar^2 \nabla^2) | n \rangle^*$.

R.Q. 11.14 For arbitrary vectors $\langle n |$ and $| m \rangle$, show that i.e. $\hat{p}^2 = (-\hbar^2 \nabla^2)$, the momentum operator is Hermitian..

R.Q. 11.15 Given the quadric surface: $5x^2 + 2y^2 + 11z^2 + 20xy + 16xz - 4yz = 0$.

Express this in the matrix form $\hat{A}$. Find the quadric surface in its principal axes.

(Answer: $\hat{A} = \begin{pmatrix} 5 & 10 & 8 \\ 10 & 2 & -2 \\ 8 & -2 & 11 \end{pmatrix}$, $(x^2/9)+(y^2/9)-(z^2/18) = 1$).

R.Q. 11.16 A system contains an observable which has the corresponding operator, A given by $\hat{A} = \begin{pmatrix} 1+i & 1 \\ -i & 0 & 0 \\ 1 & 0 & 0 \end{pmatrix}$.. (a) Is the operator $\hat{A}$ Hermitian? (b) Obtain the eigen values of $\hat{A}$, (c) Find the unitary matrix that will diagonalize $\hat{A}$, (d) Establish that the eigen vectors comprising $\hat{U}$ are orthonormal, (e) Show that the unitary nature of $\hat{U}$, (f) Show that $\hat{A}° = \hat{U}^{-1}\hat{A}\hat{U}$ is diagonal. (Answers: a) $\hat{A}$ is hermitian; b) $\lambda_1 = 0, \lambda_2 = -1, \lambda_3 = +2$;

c) $\hat{S} = \hat{U}^\dagger \begin{pmatrix} 0 & \sqrt{1/3} & -2\sqrt{1/6} \\ \sqrt{1/2} & \sqrt{1/3}\,i & -\sqrt{1/6}\,i \\ -\sqrt{1/2}\,i & -\sqrt{1/3} & \sqrt{1/6} \end{pmatrix}$ d) eigen vectors are orthonormal; e) $\hat{U}^\dagger \hat{U} = \hat{\Im}$;

f) $\hat{A}_d = \hat{U}^{-1}\hat{A}\hat{U}\lambda_1 = \begin{pmatrix} 0 & 0 & 0 \\ 0 & -1 & 0 \\ 0 & 0 & +2 \end{pmatrix}$).

R.Q. 11.17 Find the characteristic vectors of the operator for the central conic surface $5x^2 + 2z^2 - 4zx = 6$.(Answer: Conic surface is $\hat{A} = \begin{pmatrix} 5 & -2 \\ -2 & 2 \end{pmatrix}$; $\lambda_1 = 1, \lambda_2 = 6$; $\hat{\psi}_1 = \sqrt{1/5}\{\,1,\,2\,\}, \hat{\psi}_2 = \sqrt{1/5}\{\,-2,\,1\,\}$).

R.Q. 11.18 Show that for any (2x2) matrix has two eigen vectors and corresponding eigen values, and the eigen values are not orthogonal (the eigen values are not necessarily real)
(Answer: $\hat{A} = \begin{pmatrix} 2 & 4 \\ 1 & 2 \end{pmatrix}$; $\lambda_1 = 0, \lambda_2 = 4$; $\hat{\psi}_1 = \sqrt{1/5}\{\,2,\,-1\,\}\hat{\psi}_2 = \sqrt{1/5}\{\,2,\,1\,\}$).

R.Q. 11.19 For Nitrogen the 2s orbital does not have the correct form at r = 0; it vanishes there. Nor is it orthogonal to the 1s orbital. Correct it by modifying it.
$|\hat{\psi}_{1s}\rangle = |\,1s\,\rangle = \sqrt{1/4\pi}\,2(Z/a_o)^{3/2}\,e^{-Zr/a_o}$,
$|\hat{\psi}_{2s}\rangle = |\,2s\,\rangle = \sqrt{(1/32\,\pi\,a_o)}\,\{2-(Z/a_o)\}^{3/2}\,e^{-Zr/a_o}$.

R.Q. 11.20 An energy level E_1 of a physical system is found to be 3-fold degenerate, with vectors $\psi_1 = \{\,1,0,2,2\,\}, \psi_2 = \{\,1,1,0,1\,\}, \psi_3 = \{\,1,1,0,0\,\}$. Construct a set of orthonormal vectors, which are linear combinations of the vector whose components are: $\psi_1 = \{\,1,0,2,2\,\}, \psi_2 = \{\,1,1,0,1\,\}, \psi_3 = \{\,1,1,0,0\,\}$, respectively.

(Answers: $a_{21} = \pm 1, a_{31} = 1/3, a_{32} = \pm\sqrt{2}\,(5/6)$ $\hat{\psi}_1 = \sqrt{1/9}\{\,1,0,2,2\,\}$,
$\hat{\psi}_2 = \pm\sqrt{1/2}\{\,2/3,\,1,\,-2/3,\,1/3\,\}, \hat{\psi}_3 = \pm\sqrt{2/6}\{\,2,\,1,\,2,\,3\,\}$).

R.Q. 11.21 Examine if an observable Q of a physical system can be described by $\hat{Q} = \begin{pmatrix} 0 & 1 & 0 \\ 1 & 0 & 0 \\ 0 & 0 & 1 \end{pmatrix}$.

R.Q. 11.22 Show that the vectors $\vec{v}_1 = \frac{1}{\sqrt{2}}(\vec{u}_1 + \vec{u}_2)$, and $\vec{v}_2 = \frac{1}{\sqrt{2}}(\vec{u}_1 - \vec{u}_2)$ are ortho-normal if $\vec{u}_1$ and $\vec{u}_2$ are linearly independent.

R.Q. 12.23 Show that a Projection operator $\hat{P}$ satisfied the inequality $0 \le \hat{P} \le 1$.

and it is seen that the system is degenerate. (a) Find the eigen values and eigen vectors. (b) Construct the transformation matrix to show explicitly that it will diagonalize $\hat{Q}$.

(Answers: $\lambda_1 = -1, \lambda_2 = +1, \lambda_3 = +1$, $\hat{\psi}_1 = \sqrt{1/2}\{1, -1, 0\}$, $\hat{\psi}_2 = \sqrt{1/2}\{1, 1, 0\}$,

$a_{21} = 0, a_{31} = 0, a_{32} = \sqrt{2}$, $\frac{1}{2}\begin{pmatrix} 1 & 1 & 0 \\ -1 & 1 & 0 \\ 0 & 0 & 1 \end{pmatrix}$).

R.Q. 12.24 Give an example to show that the Hilbert space of quantum is infinite dimensional.

R.Q. 12.25 Discuss the Bra and Ket notations for Vectors.

&&*&*&*&*&*

Chapter 12

TYPES OF PICTURES (EQUATIONS OF MOTION):

SCHRODINGER, HEISENBERG AND

INTERACTION PICTURES

Chapter 12

TYPES OF PICTURES (EQUATIONS OF MOTION): SCHRODINGER, HEISENBERG AND INTERACTION PICTURES

An expert is someone who knows some of the worst mistakes

that can be made in his subject, and how to avoid them

Werner Heisenberg

12.1 INTRODUCTION

In Quantum Mechanics, the state of a physical system at any given time is described by a unit vector in a Hilbert Space, in which sets of axes can be defined by the eigen vectors of complete sets of observables of the system.

Any change with time in the state of the system can be investigated

(i) By keeping the basis (axes) fixed and allowing the state vector to rotate,

(ii) Or, by allowing the state vector fixed and permitting the axes to rotate, or

(iii) By allowing simultaneous rotation of the state vector and the axes,

using in each case the appropriate equations of motion of the vectors concerned. Either one of these modes of description of quantum phenomena is called the REPRESENTATION; or this is not to be confused with the representation of vectors and operators of vector spaces by matrices.

The three possibilities described in the previous paragraph are called, respectively, the SCHRODINGER, the HEISENBERG and the INTERACTION "Pictures". In practice one adopts the 'Representation', which lends itself best to the solution of each particular problem.

The discussions so far in the previous Chapters gave a general scheme of relations between states and dynamical variables for a dynamical system at one instant of time. To get a complete theory of dynamics one must also consider the connections between different instants of time. When one makes an observation on the dynamical system, the state of the system gets changed in an unpredictable way, but in between observations causality applies, in Quantum Mechanics as in Classical Mechanics, and the system is governed by equations of motion which make the state at one time determine the state at a later time.

12.2 The EVOLUTION OPERATOR, $\hat{T}(t,t_o) \Rightarrow \hat{U}(t,t_o)$:

12.2.1 Time Independent $\hat{H}$

Consider a dynamical system in which a particular state of the system in motion through out the time during which the system is left undisturbed.

$$|t\rangle = \text{state of the system at time } t. \tag{12.2.1}$$

Assume that in addition to the specification of the physical values actually obtained in a measurement, one would specify the time at which each observation is made. Let the successful measurements can be carried out with arbitrarily small time delays, i.e. that there is a continuous, real variable t, $-\infty < t < +\infty$, which labels the sequence of measurements. It is to be noted that the parameter t, appearing in theory, is not an eigen value of any Hermitian operator.

(1) $\quad |t_o\rangle = \text{state of the system at time } t = t_o.$ (12.2.2)

$\quad |t\rangle = \text{state of the system at time } t,$

$\hat{A}$ = an observable operator.

Then $\quad \hat{A}|a', t_o\rangle = a'|a', t_o\rangle$ (12.2.3)

(2) $|t\rangle$ at $t > t_o$ arises for $|t_o\rangle$ by the action of a time development operator, $\hat{T}(t,t_o)$ in Hilbert Space.

$$\hat{T}(t,t_o) \equiv \hat{T}(t = t_o) \tag{12.2.4}$$

is the unitary operator in question.

Because the system is considered to be ISOLATED, $\hat{T}$ can only depend on $(t = t_o)$.

$$\boxed{|a't_o, t\rangle = \hat{T}(t - t_0)|a', t_0\rangle} \tag{12.2.5}$$

(3) Consider a time t' such that $t > t' > t_0$. Let the system evolves from $|a', t_0\rangle$ until $t + t'$, into the state represented by

$$|a't_0, t'\rangle = \hat{T}(t' - t_0)|a', t_0\rangle \qquad (12.2.6)$$

This state could then be allowed to evolve until t, at which time it is

(4) $\quad |a't_0, t\rangle = \hat{T}(t - t')|a't_0, t'\rangle.$ \hfill (12.2.7)

The state given by equation (12.2.7) is clearly identical with that given by equation (12.2.5), because equation (12.2.7) is

$$|a't_0, t\rangle = \hat{T}(t-t')[\hat{T}(t'-t_0)|a', t_0\rangle]$$

i.e., $\quad |a't_0, t\rangle = \hat{T}(t-t_0)|a', t_0\rangle$

(5) $\quad \hat{T}(t, t_0) \equiv \hat{T}(t-t')\,\hat{T}(t'-t_0)$ \hfill (12.2.8)

which is the group property of the time development Unitary Operator, since

$$(t-t') + (t'-t_0) = (t-t_0)$$

(6) Setting $t = t_0$. $\quad \hat{T}(t_0 - t_0) \equiv \hat{T}(t_0, t_0) = \hat{T}(0) = \hat{\Im}$. \hfill (12.2.9)

It means that $\quad \hat{T}(t_0 - t')\,\hat{T}(t' - t_0) = \hat{\Im}$

$\therefore \quad \hat{T}(t-t')^{-1} = \hat{T}(t'-t) = \hat{T}(t-t')^{\dagger}$.

i.e., $\quad \hat{T}(t, t_0) \Rightarrow \hat{U}(t, t_0) =$ unitary. \hfill (12.2.10)

(7) Consider $(t - t_0) \Rightarrow$ infinitesimally small.

$\therefore (t - t_0) \Rightarrow \delta t$, and $t_0 = 0$.

$\hat{U}(t + \delta t, t_0) = \hat{U}(\delta t)$,

Now $\quad \hat{U}(0) = 1$, $\hat{U}(\delta t) = \hat{U}(0) + \delta \hat{U}$

$\therefore \quad \hat{U}(\delta t) = 1 + \delta \hat{U}$ \hfill (12.2.11)

Unitary then implies $\hat{\Im} = 1 + (\delta\hat{U} + \delta\hat{U}^{\dagger}) + \hat{O}[(\delta\hat{U})^2]$.

Hence one requires $(\delta\hat{U} + \delta\hat{U}^{\dagger}) = 0$. \hfill (12.2.12)

This means $\quad \delta\hat{U} =$ anti-Hermitian. $= i \times$ Hermitian operator. \hfill (12.2.13)

(8)   Further, let $t' = t - \delta t$,

If δt = small, $\hat{U}(t, t-\delta t) = \delta\hat{U}(\delta t)$

$$\hat{U}(\delta t_1)\hat{U}(\delta t_2) = \hat{U}(\delta t_1 + \delta t_2).$$

∴ $\delta\hat{U}(\delta t) \approx \delta t$.   (12.2.14)

Because of (12.2.13) and (12.2.14)

$$\boxed{\delta\hat{U} = (-i/\hbar)\, \delta t \text{ [a time independent operator, } \hat{H}]}.$$

One can write $\hat{U}(t, t-\delta t) = \delta\hat{U}(\delta t) = 1 - (-i/\hbar)\, \delta t\, \hat{H}$

∴ $\delta\hat{U} = \{(-i\hat{H}\,\delta t/\hbar)\}$ (12.2.14)

where $\hat{H}$ is the TIME INDEPENDENT Hamiltonian Operator of the system. *It is introduced for* CONVENIENCE *to ensure that* $\hat{H}$ *has the dimension of energy*.

12.2.2 FUNDAMENTAL LAW OF EVOLUTION in Quantum Mechanics

Using the group property,

$$\hat{U}(t+\delta t) - \hat{U}(t) = [\hat{U}(\delta t) - 1]\, \hat{U}(t)$$

$$= d\hat{U}(t) = \{(-i\hat{H}\, dt/\hbar)\}\, \hat{U}(t).$$

or, $(\partial/\partial t)\hat{U}(t) = \{(-i\hat{H}/\hbar)\}\, \hat{U}(t)$ (12.2.15)

since energy is conserved for a <u>conservative system</u>, $[\hat{H}, \hat{H}] = 0$,

$$\hat{U}(t) = e^{\{-i\hat{H}t/\hbar\}}.$$ (12.2.16)

i.e., EXPLICITLY, $\boxed{\hat{U}(t-t_0) = e^{\{-i\hat{H}(t-t_0)/\hbar\}}}$
(12.2.17)

Because of the initial condition, $\hat{U}(0) = 1$.

If the system were <u>non-conservative</u>, then $\boxed{\hat{H} = \hat{H}(t)}$.

$$\boxed{\hat{U}(t,t_o) \neq e^{\{-i\hat{H}(t-t_o)/\hbar\}}}$$

$\hat{U}$ is also defined by $\boxed{\hat{U}(t,t_o) = 1 - \{(-i/\hbar)\}\int \hat{H}\cdot\hat{U}(t', t_o)\, dt'}$ (12.2.18)

Either of the equation (12.2.15) or (12.2.16) or (12.2.18) is known as the expression for the FUNDAMENTAL LAW of Evolution in Quantum Mechanics. *Equivalently* an expression for this law is the Schrödinger Equation.

12.3 The SCHRODINGER PICTURE (The S-picture)

12.3.1 Initial wave function, the state function, and Schrödinger Equation

Consider a particular state of the system in motion through out the time during which the system is left undisturbed (For convenience ket notation is used for the state function).

$|\,t\,\rangle$ = state of the system at time t.

Assume that in addition to the specification of the physical values actually obtained in a measurement, one would specify the time at which each observation is made. Let the successful measurements can be carried out with arbitrarily small time delays, *i.e.*, that there is a continuous, real variable t, $-\infty < t < +\infty$, which labels the sequence of measurements. It is to be noted that the parameter t, appearing in theory, is not an eigen value of any Hermitian operator.

Consider an isolated system.

a) at time $t = t_o$:

An observer prepares a variety of states $|\,a',\,t_0\,\rangle, |\,a'',\,t_0\,\rangle$, *etc.* for an observable operator $\hat{A}$.

There will be a complete set of such states, for $\hat{A}$.

b) at time $t > t_o$:

In general the pure state will not correspond to the original vector $|\,a',\,t_0\,\rangle$, but will correspond to some other vector.

If $|\,a'\,t_0,\,t\,\rangle$ = the moving ket;

One has
$$[\hat{A}-a']| a', t_0 \rangle = 0 \tag{12.3.1}$$

When $t > t_0$;
$$[\hat{A}-a']| a't_0, t \rangle \neq 0 \quad . \tag{12.3.2}$$

The expectation value of $\hat{A}$ at time t, i.e. $t = t$,
$$\langle \hat{A} \rangle_t = \langle a't_0, t | \hat{A} | a't_0, t \rangle \neq a' \tag{12.3.3}$$

Nevertheless, the actual result of any $\hat{A}$-*measurement* (i.e. measurement of the observable $\hat{A}$) at time $t > t_0$ must still be one of the values: $a', a'', ..$; though the probability of obtaining this particular result will be neither 0 nor 1, as it was at time $t = t_0$. In other words, if one writes the operator $\hat{A}$ in terms of its matrix in the $| a' t_0, t \rangle$-**representation**, and carries out he digonalization of this matrix, one must again find the eigen values $a', a'', ...$. It is, therefore, natural to assert that $| a' t_0, t \rangle$ is obtained from $| a' t_0 \rangle$ by a Unitary Transformation, because the unitary equivalent observables have the same spectra, *i.e.* such a transformation does not alter the spectrum $\{ a' \}$.

$$| a't_0, t \rangle = \hat{T}(t'-t_0) | a' t_0 \rangle . \tag{12.2.5}$$

$$\hat{T}(t_0 - t_0) \equiv \hat{T}(t_0, t_0) = \hat{T}(0) = \hat{\Im} . \tag{12.2.9}$$

$$\therefore \quad \hat{T}(t-t_0) \Rightarrow \hat{U}(t-t_0) , \tag{12.2.10}$$

$$| a't_0, t \rangle = \hat{U}(t'-t_0) | a' t_0 \rangle \tag{12.3.4}$$

$| a't_0, t \rangle$ satisfies the Schrödinger Equation (S - equation),

$$[\hat{H}_S - i\hbar (\partial / \partial t)] | a't_0, t \rangle = 0. \tag{12.3.5}$$

with the initial condition $| a't_0, t_0 \rangle = | a't_0 \rangle$, because $\hat{U}(0) = 1$,

$$\therefore \quad \hat{H} = \hat{H}_S$$

where the subscript S in $\hat{H}$ denotes the Hamiltonian in the Schrödinger Picture.

Using the explicit expression for $\hat{U}(t-t_0)$ from (12.2.17), viz.

$$\hat{U}(t-t_0) = e^{\{-i\hat{H}(t-t_0)/\hbar\}} . \tag{12.2.17}$$

In equation (12.3.4) $| a't - t_0 \rangle = e^{\{-i\hat{H}(t-t_0)/\hbar\}} | a't_0 \rangle$ \hfill (12.3.6)

The equation assumes a very simple form when the Hamiltonian is itself a number of the complete set of observables that specifies the initial state, $|a't_o\rangle$. When this is the case one shall write

$$|E\ a't_o\rangle \text{ instead of } |a't_o\rangle,$$

where E is the eigen values of the constants of the motion. Since the system is conservative one, *i.e.* its energy as represented by the $\hat{H}$, does not explicitly depend upon the time,

$$|E a't - t_o\rangle = e^{\{-i E (t-t_o)/\hbar\}} |E\ a't_o\rangle. \qquad (12.3.7)$$

These eigen kets of $\hat{H}$, therefore, belong to the same ray for all values of t. They represent the same physical states at different times and are, therefore, stationary. Thus for time independent observable, say, $\hat{B}$, and the results of observation

$$\langle\hat{B}\rangle_t = \langle E'\ a't_0, t\ |\hat{B}| E''\ a''t_o, t\rangle = e^{\{-i (E''-E') (t-t_o)/\hbar\}} \langle E'\ a't_0, t\ |\hat{B}| E''\ a''t_o, t\rangle \qquad (12.3.8)$$

In the earlier Chapters, so far, all the considerations were with time independent operators (like $\hat{p}, \hat{q}, etc.$), which can act on wave functions (which in the general case is time dependent). However, the wave functions had to obey the Time Dependent Schrödinger Equation (TDSE),

$$[\hat{H}_S - i\hbar\ (\partial/\partial t)]\ |a't_o, t\rangle_S = 0 \qquad (12.3.5)$$

This Representation is called the <u>Schrödinger Picture</u>, with $\hat{H}_S(\hat{p}_k, \hat{q}_k)$, the Hamiltonian of the system.

The Schrödinger Equation of motion is extremely valuable in developing approximate methods for evaluating scattering cross sections. It is left unchanged in form upon transforming to a Matrix Representation. A Representation in which the base functions (base vectors) are time independent is known as the *Schrödinger Picture*. Thus keeping the axes fixed and allowing the state vector to rotate one can investigate any change with time in the state of the system.

12.3.2   EQUATION OF MOTION of an operator $\hat{Q}_S$ in the S-picture

It is known that for any time dependent state $|a't_o, t\rangle_S$, the corresponding Equation of Motion is

$$i\hbar\ (\partial/\partial t)\ |a't_o, t\rangle_S = \hat{H}_S\ |a't_o, t\rangle_S. \qquad (12.3.5)$$

$$-i\hbar (\partial/\partial t) \langle a't_o, t |_S = \hat{H}_S \langle a't_o, t |_S \qquad (12.3.6)$$

For any operator, $\hat{Q}_S$ not <u>explicitly</u> dependent on time t,

$$\frac{d}{dt}\langle \hat{Q}_S \rangle = \frac{d}{dt}\left[\langle a't_o, t |_S \hat{Q}_S | a't_o, t \rangle_S \right].$$

$$= \left\{ (\partial/\partial t)\langle a't_o, t |_S \right\} \hat{Q}_S | a't_o, t \rangle_S + \left[\langle a't_o, t |_S \{(\partial/\partial t)\hat{Q}_S\}| a't_o, t \rangle_S \right]$$

$$+\langle a't_o, t |_S \hat{Q}_S \{ (\partial/\partial t)| a't_o, t \rangle_S \}.$$

Using equation (12.3.6)

$$\frac{d}{dt}\langle \hat{Q}_S \rangle = \left\{ (1/-i\hbar) \hat{H}_S \langle a't_o, t |_S \right\} \hat{Q}_S | a't_o, t \rangle_S$$

$$+ \left[\langle a't_o, t |_S \{(\partial/\partial t)\hat{Q}_S\}| a't_o, t \rangle_S \right]$$

$$+\langle a't_o, t |_S \hat{Q}_S \{ (1/i\hbar)\hat{H}_S | a't_o, t \rangle_S \}.$$

$$= \left[\langle a't_o, t |_S \{(\partial/\partial t)\hat{Q}_S\}| a't_o, t \rangle_S \right]$$

$$+(1/i\hbar)\langle a't_o, t |_S \left(\hat{Q}_S\hat{H}_S - \hat{H}_S \hat{Q}_S \right)| a't_o, t \rangle_S$$

$$= \left[\langle a't_o, t |_S \{(\partial/\partial t)\hat{Q}_S\}| a't_o, t \rangle_S \right]$$

$$+(1/i\hbar)\langle a't_o, t |_S \left[\hat{Q}_S, \hat{H}_S \right] | a't_o, t \rangle_S$$

$$\therefore \quad \boxed{\frac{d}{dt}\langle \hat{Q}_S \rangle = \langle (\partial/\partial t)\hat{Q}_S \rangle + (1/i\hbar)\langle \left[\hat{Q}_S, \hat{H}_S \right] \rangle} \qquad (12.3.7)$$

This is the <u>EQUATION OF MOTION OF OPERATOR</u>, $\hat{Q}_S$.

If $\hat{Q}_S$ is explicitly not time dependent, then

$$(\partial/\partial t)\hat{Q}_S = 0.$$

and $\quad \left[\hat{Q}_S, \hat{H}_S \right] = 0$

$$\therefore \quad \frac{d}{dt}\langle \hat{Q}_S \rangle = 0,$$

In the S-Picture, $\hat{Q}_S$ = a Constant of the Motion.

Thus (12.3.5) is $\hat{H}_S | a'_{t_o}, t \rangle_S = (i\hbar \dfrac{d}{dt}) | a'_{t_o}, t \rangle_S$

$(d/dt) | a'_{t_o}, t \rangle_S$ is a matrix whose each element is the time derivative of the corresponding element of the matrix $| a'_{t_o}, t \rangle_S$.

Thus in the S-Picture the basis vectors are time independent, *i.e.*, the operators are time independent too. The state vectors are generally time dependent, and they satisfy the Schrödinger Equation., and determine the time evolution of the system.

Worked out Example 12.1

Show that, in the S-picture, the total energy is conserved in a conservative system.

Solution: STEP # 1 Given: The S-Picture, $(\partial/\partial t)\langle \hat{H}_S \rangle = 0$, and (12.3.7) is

$$(d/dt)\langle \hat{Q}_S \rangle = \langle (\partial/\partial t)\hat{Q}_S \rangle + (1/i\hbar)\langle [\hat{Q}_S, \hat{H}_S] \rangle$$

STEP # 2 So using $\hat{Q}_S = \hat{H}_S$, the equation of motion (12.3.7) becomes

$$(d/dt)\langle \hat{H}_S \rangle = + (1/i\hbar)\langle [\hat{H}_S, \hat{H}_S] \rangle$$

i.e., $[\hat{H}_S, \hat{H}_S] = 0$.

i.e., the TOTAL ENERGY IS A CONSTANT OF THE MOTION.

This is analogue to conservation of energy in a conservative system.

12.4 The HEISENBERG PICTURE (H-Picture)

Another very important way of discussing time development is with the Heisenberg Picture, known also as the H-Picture. In Quantum Optics and laser theory as well as in other field of Quantum Mechanics this Representation is used. Consider a set of vectors $\{ | \hat{\psi}_j (r, t) \rangle \}$ which is an orthonormal set at time $t = 0$.

$\{ | \hat{\psi}_j (r, t) \rangle \}$ satisfies the Schrödinger Equation.

$$\left| \hat{\psi}_j\ (r,\ t) \right\rangle_H = \hat{U}^{-1}(t - t_o) \left| \hat{\psi}_j\ (r,\ t_o) \right\rangle_S . \tag{12.4.1}$$

12.4.1. To prove that the state vector, $\left| \hat{\psi}_j\ (r,\ t) \right\rangle_H$ = fixed:

From S-picture, let

$$\left| \hat{\psi}_j\ (t) \right\rangle_S = \hat{U}(t - t_o) \left| \hat{\psi}_j\ (t_o) \right\rangle_S \tag{12.3.4}$$

One has $\hat{U}(t,\ t_o) = \hat{U}^{-1}(t_o, t)$.

$$\hat{U}(t,\ t_1)\ U(t_1, t_o) = \hat{U}(t,\ t_o) . \tag{12.4.2}$$

$$\hat{U}(t_o, t_o) = 1 . \tag{12.4.3}$$

From $\hat{\Omega}_H(t) = \hat{U}^{-1}(t, t_o)\ \hat{\Omega}_S \hat{U}(t, t_o)$ \tag{12.4.4}

One gets on substitution of equation (12.3.4) in (12.4.1)

$$\left| \hat{\psi}_j(t) \right\rangle_H = \hat{U}^{-1}(t,\ t_o)[\hat{U}(t,\ t_o) \left| \hat{\psi}_j(t_o) \right\rangle_S = \left\{ \hat{U}^{-1}(t,\ t_o)\ \hat{U}(t,\ t_o) \right\} \left| \hat{\psi}_j(t_o) \right\rangle_S$$

$$= \hat{U}(t_o,\ t_o) \left| \hat{\psi}_j(t_o) \right\rangle_S = \left| \hat{\psi}_j(t_o) \right\rangle_S .$$

= Constant with respect to time.

$\therefore$ $\left| \hat{\psi}_j(t) \right\rangle_H = \hat{U}^{-1}(t,\ t_o) \left| \hat{\psi}_j(t) \right\rangle_S = \left| \hat{\psi}_j(t_o) \right\rangle_S .$ \tag{12.4.5}

12.4.2 TIME INDEPENDENCY of the Orthonormality of a Set of Vectors, $\left\{ \left| \hat{\psi}_j(t) \right\rangle_H \right\}$

For the state vector,

$$\hat{H}_H \left| \hat{\psi}_j(t) \right\rangle_H = i\hbar\ (\partial/\partial t) \left| \hat{\psi}_j(t) \right\rangle_H \tag{12.4.6}$$

Taking the scalar product with $\left\langle \hat{\psi}_j(t) \right|_H$

$$\left\langle \hat{\psi}_j(t) \right|_H \hat{H}_H \left| \hat{\psi}_j(t) \right\rangle_H = i\hbar \left\langle \hat{\psi}_j(t) \right|_H (\partial/\partial t) \left| \hat{\psi}_j(t) \right\rangle_H . \tag{12.4.7}$$

Alternatively (12.4.7), taking the complex conjugate

$$\left\{ \left\langle \hat{\psi}_j(t) \right|_H \hat{H}_H \left| \hat{\psi}_j(t) \right\rangle_H \right\}^* = (i\hbar)^* \left\{ \left\langle \hat{\psi}_j(t) \right|_H (\partial/\partial t) \left| \hat{\psi}_j(t) \right\rangle_H \right\}^* .$$

$$\left\langle \hat{H}_H\ \hat{\psi}_j(t) \right|_H \left| \hat{\psi}_j(t) \right\rangle_H = (-i\hbar) \left\langle (\partial/\partial t)\ \hat{\psi}_j(t) \right|_H \left| \hat{\psi}_j(t) \right\rangle_H .$$

$\hat{H}_H$ being Hermitian,

$$\langle \hat{\psi}_j(t) |_H | \hat{\psi}_j(t) \rangle_H = (i\hbar) \langle (\hat{\psi}_j(t) |_H (\partial/\partial t) | \hat{\psi}_j(t) \rangle_H \quad (12.4.9)$$

Subtracting (12.4.8) from (12.4.9),

$$0 = (i\hbar)\left[\langle (\hat{\psi}_j(t) |_H (\partial/\partial t) | \hat{\psi}_j(t) \rangle_H - \langle (\hat{\psi}_j(t) |_H (\partial/\partial t) | \hat{\psi}_j(t) \rangle_H \right] 0$$

$$= (i\hbar)(d/dt) \langle (\hat{\psi}_j(t) |_H | \hat{\psi}_j(t) \rangle_H \quad (12.4.10)$$

This means $\quad \langle (\hat{\psi}_j(t) |_H | \hat{\psi}_j(t) \rangle_H = \delta_{jj}. \quad (12.4.11)$

i.e., the orthonormality of the set of vectors does not change with time.

This holds good for all time. Therefore, the vector $| \hat{\psi}_j(t) \rangle_H$ can be used to obtain a Matrix Representation.

Let $\quad |\Psi(t)\rangle_H = \sum c_j | \hat{\psi}_j(r,t) \rangle_H = \hat{U}^\dagger(t) | \hat{\psi}_j(r, t) \rangle_S \quad (12.4.12)$

because of equation (12.4.1), *viz.,* $| \hat{\psi}_j(r,t) \rangle_H = \hat{U}^\dagger(t) | \hat{\psi}_j(r, t) \rangle_S$.

Each term in the sum satisfies the Schrödinger Equation, the sum with constant coefficients, c_j will, therefore, satisfy the Schrödinger Equation, and the Representation of the wave function; $\hat{\psi}$ (solution of the Schrödinger Equation) is 'TISE'. The coefficients of $\hat{\psi}$, which are the Representation, are time independent.

$$| \hat{\psi}_j(r,t) \rangle_H = \hat{U}^\dagger(t) | \hat{\psi}_j(r, t_o) \rangle_S.$$

$\therefore \quad (d/dt) | \hat{\psi} \rangle_H = 0 \quad (12.4.13)$

i.e., the <u>STATE vector is FIXED</u> and the <u>BASIS VECTORS forming axes are allowed to ROTATE</u>.

12.4.3 The HEISENBERG EQUATION OF MOTION (Time Variation) of Operator, $\hat{Q}_H$

In the H-Picture, consider equation (12.4.1)

$$| \hat{\psi}_j(r,t) \rangle_H = \hat{U}^{-1}(t, t_o) | \hat{\psi}_j(r, t) \rangle_S$$

Consider an operator, $\hat{Q}_H$, is time dependent, *i.e.* $\hat{Q}_H = \hat{Q}_H(t)$, and $\hat{Q}_S = \hat{Q}_S(0)$.

$$\hat{Q}_H(t) = \hat{U}^{-1}(t, t_o)\hat{Q}_S(0)\hat{U}(t, t_o) . \qquad (12.4.14)$$

$$\langle \hat{Q}_H(t) \rangle = \langle \hat{\psi}_j(t) |_H \hat{Q}_H(t) | \hat{\psi}_j(t) \rangle_H .$$

with matrix elements

$$\hat{Q}_{ij}|_H = \langle \hat{\psi}_j \| \hat{Q}_H \hat{\psi}_i \rangle_H \qquad (12.4.15)$$

Since from equation (12.2.16)

$$\hat{U}(t, t_0) = e^{\{-i\hat{H}(t-t_0)/\hbar\}}$$

and $\quad \hat{Q}_H(t) = \hat{U}^{-1}(t)\,\hat{Q}_S(0)\,\hat{U}(t)$

becomes $\quad \hat{Q}_H(t) = e^{\{+i\hat{H}t/\hbar\}}\,\hat{Q}_S(0)\,e^{\{-i\hat{H}t/\hbar\}} . \qquad (12.4.17)$

$$(d/dt)\hat{Q}_{ij}|_H = (d/dt)\langle \hat{\psi}_j \| \hat{Q}_H \hat{\psi}_i \rangle_H .$$

$$|\hat{Q}_H(t)\,\hat{\psi}_i\rangle_H + \langle \hat{\psi}_j \,|(\partial/\partial t)\hat{Q}_H(t)|\,\hat{\psi}_i\rangle_H$$

$$+ \langle \hat{\psi}_j \,|\hat{Q}_H(t)|(\partial/\partial t)\,\hat{\psi}_i\rangle_H . \qquad (12.4.18)$$

Defining the <u>Heisenberg operator</u>

$$\hat{H}_H = \hat{U}(t_o, t)\,\hat{H}_S(0)\,\hat{U}^\dagger(t_0, t) \qquad (12.4.19)$$

and $\quad \hat{Q}_S(t) = \hat{U}(t_o, t)\left((\partial/\partial t)\,\hat{Q}_S(t)\right)\hat{U}^\dagger(t_0, t) ,$

But $\quad |\hat{H}_H\,\hat{\psi}_i\rangle_H = \hat{H}_H|\,\hat{\psi}_i\rangle_H\,(i\hbar)|(\partial/\partial t)\,\hat{\psi}_i(t)\rangle_H$

$$\langle \hat{\psi}_i \ddot{H}_H | = \langle \hat{\psi}_i (i\hbar)(\partial/\partial t)\,|_H .$$

∴ $\quad (d/dt)\hat{Q}_{ij}|_H = \langle (-i/\hbar)\,\hat{H}_H\,\hat{\psi}_j \| \hat{Q}_H\,\hat{\psi}_i\rangle + \langle \hat{\psi}_j \|(-i/\hbar)\,\hat{Q}_H\,\hat{H}_H\,\hat{\psi}_i\rangle$

$$+ \langle \hat{\psi}_j \,|[(\partial/\partial t)\,\hat{Q}_H]|\,\hat{\psi}_i\rangle$$

$$. = (i/\hbar)\,\langle \hat{\psi}_j |\hat{H}_H\hat{Q}_H|\,\hat{\psi}_i\rangle + (-i/\hbar)\langle \hat{\psi}_j \|\hat{Q}_H\,\hat{H}_H\,\hat{\psi}_i\rangle$$

$$+ \langle \hat{\psi}_j \,|[(\partial/\partial t)\,\hat{Q}_H]|\,\hat{\psi}_i\rangle$$

$$= \langle \hat{\psi}_j | (i/\hbar) [\hat{H}_H, \hat{Q}_H] | \hat{\psi}_i \rangle + \langle \hat{\psi}_j | [(\partial/\partial t) \hat{Q}_H] | \hat{\psi}_i \rangle$$

∴ $(d/dt) \hat{Q}_{ij}\big|_H = \langle \hat{\psi}_j | (i/\hbar) [\hat{H}_H, \hat{Q}_H] + [(\partial/\partial t) \hat{Q}_H] | \hat{\psi}_i \rangle$. (12.4.20)

Since $\hat{H}_H$ is Hermitian, and the basic commutation relations are unchanged by a Unitary Transformation. Thus the matrix relation

∴ $(d/dt) \hat{Q}_H(t) = (i/\hbar) [\hat{H}_H, \hat{Q}_H] + [(\partial/\partial t) \hat{Q}_H]$. (12.4.21)

is valid. From this relation emerges

$$\boxed{(i/\hbar\, d/dt)\, \hat{Q}_H(t) = [\hat{H}_H, \hat{Q}_H] + (i\hbar)(\partial/\partial t)\hat{Q}_H}$$ (12.4.22)

This is the <u>HEISENBERG EQUATION OF MOTION for an operator</u>, $\hat{Q}_H$

The quantum mechanical <u>commutator</u> and the corresponding classical <u>Poisson Brackets</u> are related through $\boxed{[\hat{Q}_H, \hat{H}_H] = (i\hbar)\{q, h\}}$. (12.4.23)

The CLASSICA ANALOGUE of the equation (12.4.22) is

∴ $(d/dt)\hat{Q}_H(t) = \{\hat{H}_H, \hat{Q}_H\} + (\partial/\partial t)\hat{Q}_H$ (12.4.24)

If $\hat{Q}_H \neq \hat{Q}_H(t)$, explicitly, then $(\partial/\partial t)\hat{Q}_H = 0$.

Equation (12.4.22) becomes

$$\boxed{(i\hbar\, d/dt)\, \hat{Q}_H(t) = [\hat{H}_H, \hat{Q}_H]}$$ (12.4.25)

This equation replaces the <u>S -Equation of Motion in the H-Picture</u>.

This relation is valid for any operator provided it is *time-independent* in the S-Picture.

In the H-Picture, the eigen vectors of the observable $\hat{Q}_H$, change in time according to

$$\boxed{|q', t\rangle_H = \hat{U}^{-1}(t, t_o) |q', t_o\rangle_S}$$ (12.4.26)

Taking transition from S-Picture to H-Picture, initially many different choices of $\hat{U}$ are possible leading to different Pictures of quantum dynamics besides the S-Picture. The H-Picture is obtained if
$\hat{U}(t, t_o) \equiv \hat{T}(t, t_o) = \hat{T}^\dagger(t, t_o) = \hat{T}^{-1}(t_o, t)$

$$\boxed{\mid q', t_o \rangle_S = \hat{U}(t, t_o) \mid q', t_o \rangle_H} \quad (12.4.27)$$

Since $\hat{U}^\dagger(t, t_0) = e^{\{+i\hat{H}(t-t_0)/\hbar\}}$ (12.2.16)

Differentiating with respect to t, equation (12.4.27) and since $(d/dt)\mid q', t_o \rangle_S = 0$

One gets the EQUATION OF MOTION as

$$\boxed{(i\hbar\, \partial/\partial t)\mid q', t \rangle_H = \hat{H}_H \mid q', t \rangle_H}$$
(12.4.28)

or, $\boxed{\hat{E} \mid q', t \rangle_H = -\hat{H}_H \mid q', t \rangle_H}$. (12.4.29)

This is very similar to the equation of motion in the S-Picture, but for the **minus sign**. The two Pictures are related to each other in much the same way as the two descriptions of the rotation of a rigid body with respect to a coordinate system, one with body moving and the coordinate system fixed, and the other with the body kept at rest and the coordinates moving in the opposite direction.

12.4.4 CONSERVATIVE SYSTEM

If the Hamiltonian, $\hat{H}_H$ of the system is not explicitly time dependent, $\hat{H}_H = $ Constant of the Motion, *i.e.* the expectation value of total energy of the system is conserved and the system is said to be conservative.

Higher order derivatives can be computed by a repetition of the calculation; thus

$$i\hbar\,(d/dt)^2 \langle \hat{Q}_H \rangle = \langle [\,[\hat{Q}_H, \hat{H}_H]\,, \hat{H}_H\,] \rangle \quad (12.4.30)$$

and so forth.

These equations give one the very important theorem that *"the Expectation value of an observable, $\hat{\Omega}$ which commutes with the Hamiltonian, $\hat{H}$ is a CONSTANT OF THE MOTION"*.

$\therefore \quad (d/dt)\langle \hat{\Omega} \rangle = 0$, if $[\hat{\Omega}, \hat{H}] = 0$. (12.4.31)

12.4.5 RELATION BETWEEN QUANTUM MECHANICS AND CLASSICAL MECHANICS

The differences are:

(1) Classical Mechanics deals with average values; while Quantum Mechanics deals with expectation values,
(2) There is no harm in thinking expectation values classically. But one cannot bring classical pictures when thinking about details of motion.

12.5 DIRAC REPRESENTATION (Interaction picture) or I – Picture

12.5.1 Introduction

A third way to describe time development was first used by P. Dirac, and later by Sin-Itiro Tomonaga and Julian Schwinger. This is the Interaction Picture (I - Picture), which combines the advantages of both the S- and H- Pictures. Any problem in Quantum Mechanics essentially consists of more or less complete and more or less precise determination of the properties of the Unitary Operator $\hat{U}(t, t_o)$. All the predictions of the theory are given by the matrix elements, $|\langle \psi_n(t_o) | \hat{U}(t, t_o) | \psi_n(t_o) \rangle|^2$. The I – Picture is convenient as a point of departure for *Time Dependent Perturbation Theory*.

12.5.2 Ket vectors and operators in the Interaction Picture:

The time dependent problems can be generally separated into two classes:

i) The Hamiltonian $\hat{H}$ is explicitly time dependent, *eg.* an atom exposed to a time dependent external field.

ii) The Hamiltonian $\hat{H}$ is time independent, but the state initially prepared by the experimenter is not a stationary state. *eg.* collision problems.

The theoretical treatment of cases (i) and (ii) is usually very similar. One can always transform a $\hat{H}$ describing (ii) into a problem involving a time dependent $\hat{H}$. So the total interaction Hamiltonian, $\hat{H}^I(t)$ can be divided into two parts, *viz.* free part, $\hat{H}_o^I$ and the perturbation part, $\hat{H}_1^I$.

Denote, $\hat{H}_o^I = \hat{H}_o(0)$, and $\hat{H}_1^I = \hat{H}_1(t)$.

Let $\hat{H}_o$ for the corresponding $\hat{H}$ - Equation of Motion is easily solvable.

$$\hat{H}^I(t) = \hat{H}_o(0) + \hat{H}_1(t) \tag{12.5.1}$$

Choose an orthonormal set of base functions which satisfies the S-Equation (12.3.5) with $\hat{H}^I(t)$ as the Hamiltonian.

$$\left[i\hbar \, (\partial/\partial t) - \hat{H}^I(t)\right] | \, a', t \, \rangle_S = 0 \ . \tag{12.5.2}$$

Introducing the ket, $\| \, a', t \, \rangle_S$ when $t < t_o$, as obtained from equation (12.3.5).

Further in analogy with the H-Picture let in the I-Picture the state vectors are obtained by

$$| \, a', t \, \rangle_I = \hat{U}^\dagger(t) \, | \, a', t \, \rangle_S , \tag{12.5.3}$$

$$\hat{U}^\dagger(t, t_0) = e^{\{+i\hat{H}(t-t_0)/\hbar\}} . \tag{12.2.16}$$

$$\boxed{| \, a', t \, \rangle_I = e^{\{+i\hat{H}t/\hbar\}} | \, a', t \, \rangle_S} \tag{12.5.4}$$

It is to be noted that $| \, a', t \, \rangle_I$ = time independent, if $\hat{H}_1(t) = 0$. To substitute into equation (12.5.2):

Let $\quad \Phi = \hat{U}\varphi ,$

$$(d/dt)\Phi = (d/dt)\hat{U}\varphi + \hat{U}(d/dt)\varphi .$$

But $\quad \left[\hat{H}_o(0) + \hat{H}_1(t)\right]\Phi = i\hbar \, (\partial/\partial t)\Phi \approx \hat{H}_o(0)\Phi$

$$(d/dt)\hat{U} = (i/\hbar) \, \hat{H}_o(0)\Phi .$$

So $\quad \hat{U} = e^{\{-i\hat{H}_o t/\hbar\}}$

So equation (12.5.4) can be written as

$$\boxed{| \, a', t \, \rangle_S = e^{\{-i\hat{H}_o t/\hbar\}} | \, a', t \, \rangle_I} \tag{12.5.5}$$

Now $\quad \left[i\hbar \, (\partial/\partial t) - \hat{H}^I(t)\right] | \, a', t \, \rangle_S = 0 \tag{12.5.2}$

Pre-multiplying this by $e^{\{+i\hat{H}_o t/\hbar\}}$, the equation of evolution of state vectors in the I-picture, as result of the $\hat{H}_1(t)$ of $\hat{H}^I(t)$, becomes

$$\left[i\hbar \, (\partial/\partial t) - \hat{H}^I(t)\right] | \, a', t \, \rangle_I = 0 \ . \tag{12.5.6}$$

where $\quad \hat{H}^I(t) = e^{\{+i\hat{H}_o t/\hbar\}} \hat{H}_1(t) \, e^{\{-i\hat{H}_o t/\hbar\}} \tag{12.5.7}$

i.e., all the operators in the I-Picture evolve due to the free part, $\hat{H}_o(0)$.

Thus equation (12.5.7) is the S-Equation of a system having an explicitly t-dependent Hamiltonian, which is the case (i), mentioned earlier. Finally when one uses equation (12.5.4) to (12.5.7) all operators $\hat{Q}$ in the I-picture have the time-dependence.

$$\boxed{\hat{Q}^I(t) = e^{\{+i\hat{H}_o t/\hbar\}} \hat{Q}_1(t) \, e^{\{-i\hat{H}_o t/\hbar\}}} \quad (12.5.8)$$

$$= \hat{U}^\dagger(t)\,\hat{Q}_1(t)\,\hat{U}(t)$$

Equations (12.5.4) to (12.5.8) define the Interaction Picture (so named because $\hat{H}_1^I(t)$ is generally taken as the term in the Hamiltonian representing the interaction between distinct systems). The I-Picture is distinguished from the S- and H- Pictures in that both the states and the observables move in time. The states $|a', t\rangle_I$, for different values of t are, as always, related by a unitary parameter, $\hat{U}(t)$,

$$|a', t_2\rangle_I = \hat{U}_I(t_2, t_1) \,|a', t_1\rangle_I \quad (12.5.9)$$

The relation between $\hat{U}_I(t_2, t_1)$ and $\hat{U}_S(t_2, t_1)$ can be found from

$$\hat{U}_S(t_2, t_1) = e^{\{-i\hat{H}_o t_2/\hbar\}} \, \hat{U}_I(t_2, t_1) \, e^{\{+i\hat{H}_o t_1/\hbar\}} \quad (12.5.10)$$

$\hat{U}_I(t_2, t_1)$ should satisfy the differential equation

$$(i\hbar\,\partial/\partial t)\,\hat{U}_I(t_2, t_1) + \hat{U}_I(t_2, t_1)\,\hat{H}_o = 0 \quad (12.5.11)$$

with $\quad \hat{U}_I(t_1, t_1) = \hat{\Im}$

12.5.3 TOMONAGA-SCHWINGER Equation of Motion

The equations of motion of any time-dependent matrix operator, $\hat{Q}^I(t)$ are:

$$\boxed{(i\hbar)(d/dt)\,\hat{Q}_I(t) = \left[\,\hat{Q}^I(t),\,\hat{H}_o^I\,\right] + (i\hbar)(\partial/\partial t)\,\hat{Q}^I(t)} \quad (12.5.12)$$

This is the <u>Tomonaga - Schwinger Equation of Motion</u>, *i.e.* in the I-picture the state vectors evolve due to the perturbative part of the Hamiltonian operator, $\hat{H}_1(t)$ and the operator due to the free part, $\hat{H}_o^I$. Thus the operators $\hat{Q}^I(t)$ change their form in time, but this change is determined only by $\hat{H}_o^I$.

It should be noted that equations (12.5.5) and (12.5.12) reduce to those of the H-Picture, i.e. time independent $|\psi\rangle$ and $(d/dt)\hat{Q}$, given by equation (12.4.24), when $\hat{H}_1 = 0$. The interaction representation is particularly useful when $\hat{H}_1(t) = small$, i.e. when $\hat{H}_1(t)$ affects the eigen values only slightly. Under these conditions approximation methods, like perturbation techniques, can be employed. The significant advantage of the I-Picture is that it facilitates greatly practical calculations when complicated interactions are present.

12.6 COMPARISION of the S-, H-, and Interaction Pictures

The Equations of Motion of the state vector $|a', t\rangle$ and of any observable $\hat{Q}_S(0)$ of a physical system in each of the three Pictures are listed in the Table 12.1. The subscripts S, H, and I denote the Schrödinger, Heisenberg and Interaction Pictures, respectively. $\hat{U}(t,t_o)$ and $\hat{U}^{(o)}(t,t_o)$ are the Unitary Operators satisfying the differential equations,

$$(i\hbar\,\partial/\partial t)\hat{U}(t,t_o) = \hat{H}\,\hat{U}(t,t_o)$$

$$(i\hbar\,\partial/\partial t)\hat{U}^{(o)}(t,t_o) = \hat{H}_o\,\hat{U}^{(o)}(t,t_o).$$

With the initial conditions

$$\hat{U}(t_o,t_o) = \hat{U}(0) = 0.$$

$$\hat{U}^{(o)}(t_o,t_o) = \hat{U}^{(o)}(0) = 0.$$

If $\quad \hat{H}^I_o(0) \equiv \hat{H}_o(0)$,

and $\quad \hat{H}^I_1(t) \equiv \hat{H}_1(t)$

$$\hat{H}(t) = \hat{H}_o(0) + \hat{H}_1(t)$$

one gets as solutions

$$\hat{U}(t,t_o) = e^{\left\{-i\hat{H}(t-t_o)/\hbar\right\}}.$$

$$\hat{U}^{(o)}(t,t_o) = e^{\left\{-i\hat{H}_o(t-t_o)/\hbar\right\}}.$$

TABLE 12.1   Comparison of S-, H-, and I- Pictures

S-picture   Basis vectors = fixed.

State vectors $\mid a't_o, t \rangle_S$ = move,

$\mid a't_o, t \rangle_S = \hat{U}(t, t_o) \mid a't_o \rangle_S$ (12.3.4)

$\mid a't_o, t \rangle_S = (i\hbar \, d/dt) \mid a't_o, t \rangle_S$ (12.3.5)

(S- Equation).

Operators $\hat{Q}_S$ are fixed, i.e. $(\partial/\partial t) \hat{Q}_S = 0$,

$[\hat{Q}_S, \hat{H}_S] = 0$, $(\partial/\partial t) \langle \hat{Q}_S \rangle = 0$

$(d/dt) \langle \hat{Q}_S \rangle = (1/i\hbar) \langle [\hat{Q}_S, \hat{H}_S] \rangle$

When the initial conditions are chosen such that for $t = t_o$, the S-, H-, and I- Pictures coincide.

$\boxed{\mid a', t_o \rangle_S = \mid a' \rangle_H = \mid a', t_o \rangle_I}$

$\boxed{\hat{Q}_S(0) = \hat{Q}_H(t_o) = \hat{Q}_I(t_o)}$.

H-picture   Basis vectors = move.

State vectors $\mid a't_o, t \rangle_H$ = fixed, $(\partial/\partial t) \mid a't_o, t \rangle_H = 0$

$\mid \psi_i(r, t) \rangle_H = \hat{U}^\dagger(t) \mid \psi_i(r, t_o) \rangle_S$ (12.4.13)

$(i\hbar \, \partial/\partial t) \mid q', t \rangle_H = -\hat{H}_H \mid q', t \rangle_H$ (12.4.28)

(H-Equation).

$\hat{E} \mid q', t \rangle_H = -\hat{H}_H \mid q', t \rangle_H$ (12.4.29)

Operators $\hat{Q}_H$ are time dependent

$\hat{Q}_H(t) = \hat{U}^\dagger(t, t_o) \hat{Q}_S(0) \hat{U}(t, t_o)$ (12.4.14)

$(d/dt) \hat{Q}_H(t) = [\hat{Q}_H, \hat{H}_H] + (\partial/\partial t) \hat{Q}_H$ (12.4.24)

If $\hat{Q}_H \neq \hat{Q}_H(t)$, explicitly, then $(\partial/\partial t) \hat{Q}_H = 0$

$(i\hbar)(d/dt) \hat{Q}_H(t) = [\hat{Q}_H, \hat{H}_H]$ (12.4.25)

I - picture   Basis vectors = move.

State vectors $| a't_o, t \rangle_I$ = move, $(\partial/\partial t)| a't_o, t \rangle_I \neq 0$

$$\hat{H}^I(t) = \hat{H}_o(0) + \hat{H}_1(t) \tag{12.5.1}$$

$$| a', t \rangle_I = \hat{U}^\dagger(t) | a', t \rangle_S \tag{12.5.3}$$

$$\hat{Q}^I(t) = \hat{U}^\dagger(t, t_o) \hat{Q}_I(0) \hat{U}(t, t_o) \tag{12.5.8}$$

$$(i\hbar \partial/\partial t)| a', t \rangle_I = \hat{H}^I(t) | a', t \rangle_I \tag{12.5.14}$$

$$(i\hbar)(d/dt) \hat{Q}_I(t) = \left[\hat{Q}^I(t), \hat{H}^I_o \right] + (i\hbar)(\partial/\partial t) \hat{Q}^I(t) \tag{12.5.12}$$

(Tomonaga - Schwinger Equation)

Operators $\hat{Q}^I(t)$ change their form in time, as determined only by $\hat{H}^I_o$.

REVIEW QUESTIONS

R.Q. 12.1 Find out if the linear momentum in 1-D system is a constant of the motion.

(Answer: $(d/dt)\langle \hat{p} \rangle = (i/\hbar) \langle [\hat{H}, \hat{p}] \rangle = \langle -(dV/dx) \rangle = \langle \vec{F} \rangle$)

R.Q. 12.2   Heisenberg Equation of Motion leads to the Hamilton's canonical equations in Classical Mechanics. Prove it.

R.Q. 12.3 If the Hamiltonian of a physical system is given in terms of the variables, $\hat{q}_k$ and $\hat{p}_k$, prove that the canonical equations: $(\partial/\partial t)\hat{p}_k = -(\partial\hat{H}/\partial\hat{q}_k)$; and $(\partial/\partial t)\hat{q}_k = +(\partial\hat{H}/\partial\hat{p}_k)$, are obtained from the Heisenberg Equation of Motion.

R.Q. 12.4 Show that in the I-Picture the time development of the ket, $| a', t_1 \rangle_I$, is specified by a dynamical equation of the Schrödinger type. (Answer:

$$(i\hbar \partial/\partial t)| a', t \rangle_I = \hat{H}^I_1(t) \left\{ e^{\{+i\hat{H}_o t/\hbar\}} | a', t \rangle_S \right\} = \hat{H}^I(t) | a', t \rangle_S)$$

R.Q. 12.5 Show that, if the observables $\hat{B}, \hat{Q}, \hat{R}$, satisfy the commutation relation $[\hat{B}, \hat{Q}] = i\hat{R}$, in the S-Picture, this relation is valid also in the other Pictures. (Answers: $[\hat{B}_S, \hat{Q}_S] = i\hat{R}_S, [\hat{B}_H, \hat{Q}_H] = i\hat{R}_H, [\hat{B}_I, \hat{Q}_I] = i\hat{R}_I$).

R.Q. 12.6 Distinguish between Schrödinger and Heisenberg pictures of time development. Show that in both the pictures, expectation values of observables obey identical equations of motion.

R.Q. 12.7 Show that in a Heisenberg picture the operator $\hat{Q} = \hat{P} \sin\omega t - m\omega \hat{x} \cos\omega t$ is time independent for a simple harmonic oscillator? Whether is it a constant of motion?

&%&%&%&%&%&

Chapter 13

THE HARMONIC OSCILLATOR – REVISITED:

MATRIX FORMULATION

Chapter 13

THE HARMONIC OSCILLATOR – REVISITED:

MATRIX FORMULATION

"We cannot solve the problem that we have created with same thinking that created them"

- Albert Einstein

13.1 INTRODUCTION

The **harmonic oscillator** plays a very important role in Quantum Mechanics, and it provides an essential element in the theory of radiation. The reason for this is that one may view the EM field as an infinite collection of oscillators. In Section 6.2, the treatment was based on the *traditional approach*. In Chapter 7 to solve the oscillatory problem an alternative novel treatment, *viz. operator method*, without involving any differential equation, is presented. In the present Chapter the *matrix method* will be used to study the quantum mechanical oscillator.

It was seen in Chapter 11 that in a linear vector space,

An arbitrary vector

$$|j\rangle \equiv |\Psi_j(x)\rangle = \text{a column matrix,}$$

$$\langle j| \equiv \langle \Psi_j(x)| = \text{a row matrix,}$$

giving its components along a complete set of orthonormal basis vectors, $\{|\hat{e}_j\rangle\}$ which span the space.

$$\boxed{\langle \hat{e}_i \| \hat{e}_j \rangle = \delta_{ij}}.$$

They also form eigen kets of an operator $\hat{A}$, *i.e.*

$$\hat{A}|\hat{e}_i\rangle \equiv \hat{A}|\hat{i}\rangle = a_i|\hat{i}\rangle$$

i.e. one can write $|j\rangle \equiv |\Phi_j\rangle = \sum c_{ij} |\hat{e}_j\rangle = \sum c_{ij} |\hat{i}\rangle$,

where, $\boxed{c_{ij} = \langle \hat{e}_i || \Phi_j \rangle = \langle \hat{i} || j \rangle}$.

(b) Furthermore, each linear operator, $\hat{Q}$, to which these vectors $|\Phi_j\rangle$ are subject, can be represented by a matrix defined on the same basis vectors, i.e. $\hat{A}$- representation.

(c) The effect of applying a given operator, $\hat{Q}$, to a vector $|\Phi_j\rangle$, then, can be determined by multiplying the Matrix Representation of the operator, $\hat{Q}$, times the column matrix $|\Phi_j\rangle$, of the vector components;

$$|\Phi_k\rangle = \hat{Q} |\Phi_j\rangle.$$

(d) It is known from Chapter 7 that the energy eigen functions of the harmonics oscillator form a complete orthonormal set, the only complication being that the set is denumerably infinite. One can regard these eigen functions as eigen vectors and as basis vectors in Hilbert space.

(e) The important quantities and relations in a linear oscillator, studied in Chapter 6 and Chapter 7 are listed below:

The Hamiltonian of a linear oscillator is

$$\boxed{\hat{H} = \tfrac{1}{2m}\hat{p}_x^2 + \tfrac{1}{2} m \omega_c^2 \hat{x}^2} \tag{7.1.3}$$

$$\boxed{\hat{q} = b^{1/2}\hat{x} = \{\tfrac{m}{\hbar}\omega_c\}^{1/2}\hat{x}} \tag{7.1.12}$$

$$\boxed{\hat{p} \rightarrow -i\hat{\nabla}_q = (\hat{p}_x / \hbar\, b^{1/2})} \tag{7.1.13}$$

$$\hat{p}_x \rightarrow -i\hbar\hat{\nabla}_x \tag{3.6.2}$$

In the H-picture, the equations of motion are:

$$(d/dt)\hat{p}_k(t) = (i/\hbar)\left[\hat{H}_H, \hat{p}_k(t)\right] = m\,\omega_c\,\hat{x}(t),$$

and $\quad (d/dt)\hat{x}_k(t) = (i/\hbar)\left[\hat{H}_H, \hat{x}_k(t)\right] = (1/m)\hat{p}_k(t) \tag{12.4.21}$

These actually reduce to one equation if one introduces the non-Hermitian operator

$$\hat{a} = \tfrac{1}{\sqrt{2}}(\hat{q} + i\hat{p}) = \tfrac{1}{\sqrt{2}}(\hat{q} + i\,\partial/\partial q) = i\tfrac{1}{\sqrt{2}}(\hat{p} - i\hat{q}) \tag{7.1.19}$$

$$= \frac{1}{\sqrt{(2m\hbar\omega_c)}} [\hat{p} + im\omega_c\hat{q}] - i\frac{1}{\sqrt{2}} b^{1/2} [\hat{x} - i\hat{p}_x/m\omega_c] .$$

$$\hat{a}^\dagger = \frac{1}{\sqrt{2}} (\hat{q} - i\hat{p}) = \frac{1}{\sqrt{2}} (\hat{q} - i\partial/\partial q) = \sqrt{(1/2)} (q - \partial/\partial q) \tag{7.1.20}$$

$$= \frac{1}{\sqrt{(2m\hbar\omega_c)}} [\hat{p} - im\omega_c\hat{q}] - i\frac{1}{\sqrt{2}} b^{1/2} [\hat{x} + i\hat{p}_x/m\omega_c]$$

because they satisfy, individually,

$$[(d/dt) \pm i\omega_c][\hat{a}(t) \text{ or } \hat{a}^\dagger(t)] = 0 ..$$

This integrates to $\hat{a}^\dagger(t) = \hat{a}(0) e^{i\omega_c t}$

where $\hat{a} = \hat{a}(0)$, is evaluated at $t = 0$, at which time the S- and H- pictures are identical.

$$\boxed{\hat{H} = \tfrac{1}{2}\hbar\omega_c\,(\hat{a}^\dagger\hat{a} + \hat{a}\hat{a}^\dagger) = (\hat{q}^2 + \hat{p}^2)} .$$

$$[\hat{a},\hat{a}^\dagger] = 1. \tag{7.1.28}$$

$$\boxed{\hat{H} = \hbar\omega_c\,(\hat{a}^\dagger\hat{a} + \tfrac{1}{2})} \tag{7.1.30}$$

Defining a new operator, $\hat{N}$,

$$\hat{a}^\dagger\hat{a} = \tfrac{1}{2}(\hat{p}^2 + \hat{q}^2) = \hat{N}. \tag{7.1.22}$$

$$[\hat{H},\hat{a}] = -\hbar\omega_c\hat{a}. \tag{7.1.32}$$

$$[\hat{H},\hat{a}^\dagger] = +\hbar\omega_c\hat{a}^\dagger. \tag{7.1.33}$$

$$\boxed{E_n = \varepsilon_n\hbar\omega_c = (n+\tfrac{1}{2})\hbar\omega_c ,\ n = 0, 1, 2, 3,\ldots} \tag{7.2.14}$$

$$\hat{N}|n\rangle = n|n\rangle . \tag{7.3.1}$$

$$\boxed{|0\rangle = \sqrt[4]{\pi}\, H_0(q)\, e^{-q^2/2}} \tag{7.4.3}$$

$$|1\rangle = \hat{a}^\dagger|0\rangle = \sqrt{\tfrac{1}{2}}\, H_1(q)\, e^{-q^2/2} \tag{7.4.4}$$

$$|n\rangle = (\hat{a}^\dagger)^n|0\rangle = B_n\, H_n(q)\, e^{-q^2/2} .$$

$$\hat{a}|n\rangle = \sqrt{n}\,|n-1\rangle \tag{7.4.8}$$

$$\hat{a}^\dagger \mid n \rangle = \sqrt{n+1} \mid n+1 \rangle. \tag{7.4.9}$$

$$\boxed{\mid n \rangle = (n!)^{-1/2} [\hat{a}^\dagger]^n \mid 0 \rangle}. \tag{7.4.10}$$

$$H_{n+1}(q) - 2\, q\, H_n(q) - 2n\, H_{n-1}(q) = 0 \tag{6.3.31}$$

$$H_n'(q) - 2n\, H_{n-1}(q) = 0 \tag{6.3.32}$$

$$\int_{-\infty}^{+\infty} H_n(q)\, H_m(q)\, (e^{-q^2})\, dq = \pi^{1/2}\, 2^n\, n!\; \delta_{nm} \tag{6.3.33}$$

Virial Theorem is

$$\boxed{<V>_n \;=\; <T>_n \;=\; \tfrac{1}{2} E_n = \tfrac{1}{2}(n+\tfrac{1}{2})\, \hbar\, \omega_c} \tag{6.3.38}.$$

13.2 OSCILLATOR PURE STATES

13.2.1 ENERGY ($\hat{H}$-) REPRESENTATION of $\hat{H}$

Consider the energy matrix $\hat{E}$ of the harmonic oscillator. This ($\hat{E}$), of course, the matrix of the $\hat{H}$, using the oscillator wave functions, as the basis vector $\mid i \rangle$.

i.e. $\hat{H} \mid i \rangle = E_i \mid i \rangle$, and $\hat{H} \mid j \rangle = E_j \mid j \rangle$;

Each <u>matrix element</u> of $\hat{H}$ is given by

$$\boxed{\hat{H}_{ij} = E_{ij} = \langle i \mid E_j \mid j \rangle = E_j \langle i \mid\mid j \rangle = E_j\, \delta_{ij}} \tag{13.2.1}$$

Since both $\langle i \mid$ and $\mid j \rangle$ are eigen vectors of $\hat{H}$, the effect of the operator $\hat{H}$, is merely to generate the eigen value E_{ij}, which, being a number, can be moved outside the scalar product symbol. This *Kronecker delta* will be zero for all off-diagonal matrix elements, E_{ij} ($i \neq j$), because of the orthogonality of the oscillator wave functions.

$$\boxed{E_{nn} = E_n = (n + \tfrac{1}{2})\, \hbar \omega_c\,,\; n = 0,\, 1,\, 2,\, 3, \ldots} \tag{7.3.26}$$

If one labels the rows and columns in the order

$\mid \phi_o \rangle,\, \mid \phi_1 \rangle,\, \mid \phi_2 \rangle,\, \mid \phi_3 \rangle$, *etc.* the energy matrix appears as follows:

$$E_{nn} = E_n = (n + \tfrac{1}{2})\, \hbar \omega_c = \tfrac{1}{2}\, \hbar \omega_c\, (2n+1) \tag{13.2.2}$$

$$\hat{H} = \tfrac{1}{2}\hbar\omega_c \begin{pmatrix} 1 & 0 & 0 & 0 & . & . & . & 0 \\ 0 & 3 & 0 & 0 & . & . & . & 0 \\ 0 & 0 & 5 & 0 & . & . & . & 0 \\ & & 0 & 7 & 0 & & & 0 \\ & & & & 9 & & & 0 \\ & & & & & & & \\ 0 & & & & & & & 0 \\ 0 & & & & & & & (2n+1) \end{pmatrix} \quad (13.2.3)$$

This matrix illustrates the important fact that an operator has a diagonal representation when its eigen vectors constitute the basis. That is why $\hat{H}$ is diagonal in its own ($\hat{H}-$) Representation (or Coordinate Representation).

13.2.1 ENERGY ($\hat{H}-$) REPRESENTATION of $\hat{x}$

Consider the matrix of the operator $\hat{x}$. When solving the oscillator problem, one has taken

$$\hat{q} = b^{1/2}\hat{x} = \{\tfrac{m}{\hbar}\omega_c\}^{1/2}\hat{x} \qquad (7.1.12)$$

as independent variable, and the state of the oscillator is expressed by a function of q, viz. $|\psi_i\rangle$

$$|\Psi\rangle = \sum c_i |\hat{\psi}_i\rangle; \quad \text{and } c_i = \langle \hat{\psi}_i || \Psi \rangle,$$

Now consider $\hat{x}$. From (7.1.19) and (7.1.20),

$$(\hat{a}+\hat{a}^\dagger) = 2\sqrt{\tfrac{1}{2}}\,\hat{q}; \text{ and } \hat{q} = b^{1/2}\hat{x},$$

$$(\hat{a}-\hat{a}^\dagger) = 2\sqrt{\tfrac{1}{2}}\,\hat{p}; \text{ and } \hat{p} = (\hat{p}_x/\hbar\, b^{1/2}). \text{ One gets } \therefore$$

$$\sqrt{n}\,|n-1\rangle = (2\sqrt{\tfrac{1}{2}}\,\hat{q})|n\rangle + \sqrt{n+1}\,|n+1\rangle = 0.$$

$$\therefore \quad \hat{q}\,|n\rangle = \sqrt{\tfrac{n}{2}}\,|n-1\rangle + \sqrt{\tfrac{n+1}{2}}\,|n+1\rangle. \qquad (13.2.4)$$

but $\quad |\Psi\rangle = \sum c_n |\hat{\psi}_n\rangle$ = a vector with coefficients, c_n.

$\hat{q}\,|\Psi\rangle = \sum c'_n |\hat{\psi}_n\rangle$ = a vector with coefficients, c'_n.

i.e., operator $\hat{q}$ transforms a vector with expansion coefficient c_n into a vector with expansion coefficient c'_n.

$$\hat{q} \, c_n = c'_n = c'_{n-1}\sqrt{\tfrac{n}{2}} + c_{n+1}\sqrt{\tfrac{n+1}{2}} \,. \tag{13.2.5}$$

or, since $\hat{q} \mid \Psi \rangle = \sum c'_n \mid \hat{\psi}_n \rangle$

$$\therefore c'_n = \sum \{\langle n \mid \hat{q} \mid k \rangle c_k\} = \sum \hat{q}_{nk} \, c_k \,. \tag{13.2.6}$$

where $\hat{q}_{nk} = \langle n \mid \hat{q} \mid k \rangle = \sqrt{\tfrac{n}{2}} \, \delta_{n-1,\,k} + \sqrt{\tfrac{n+1}{2}} \, \delta_{n+1,\,k}$ (13.2.7)

This implies the non-vanishing matrix elements *viz.* the TRANSITION PROBABILITIES, of $\hat{q}$:

$$\hat{q}_{n,n-1} = \langle n \mid \hat{q} \mid n-1 \rangle = \sqrt{\tfrac{n}{2}} \,. \tag{13.2.8}$$

$$\hat{q}_{n,n+1} = \langle n \mid \hat{q} \mid n+1 \rangle = \sqrt{\tfrac{n+1}{2}} \,. \tag{13.2.9}$$

While all the other elements vanish, including the expectation values,

$$\hat{q}_{n,n} = \langle n \mid \hat{q} \mid n \rangle = 0 \,. \tag{13.2.11}$$

$$\hat{q} = \begin{pmatrix}
0 & \sqrt{\tfrac{1}{2}} & 0 & 0 & \cdot & \cdot & \cdot & 0 \\
\sqrt{\tfrac{1}{2}} & 0 & 1 & 0 & \cdot & 0 & \cdot & 0 \\
0 & 1 & 0 & \sqrt{\tfrac{3}{2}} & 0 & \cdot & \cdot & 0 \\
0 & 0 & \sqrt{\tfrac{3}{2}} & 0 & \sqrt{2} & 0 & & 0 \\
0 & 0 & 0 & \sqrt{2} & 0 & & & 0 \\
0 & & & & & & & 0 \\
0 & & & & & & \sqrt{n-\tfrac{1}{2}} & \\
0 & & & & & \sqrt{n-\tfrac{1}{2}} & 0 &
\end{pmatrix} \tag{13.2.12}$$

Since $\hat{q} = b^{1/2}\hat{x} = \{\tfrac{m}{\hbar}\omega_c\}^{1/2}\hat{x}$, one can write the matrix representation of the operator $\hat{x}$, by dividing the $\hat{q}$ – matrix by $b^{1/2} = \{\tfrac{m}{\hbar}\omega_c\}^{1/2}$.

$$\hat{x}_{n,n-1} = \langle n \mid \hat{x} \mid n-1 \rangle = b^{-1/2}\sqrt{\tfrac{n}{2}} \,. \tag{13.2.13}$$

$$\hat{x}_{n,n+1} = \langle n \mid \hat{x} \mid n+1 \rangle = b^{-1/2}\sqrt{\tfrac{n+1}{2}} \,. \tag{13.2.14}$$

and $\quad \hat{x}_{n,n} = \langle n | \hat{x} | n \rangle = 0$. (13.2.15)

$$\hat{x} = \left(\frac{\hbar}{m\omega_c}\right)^{1/2} \begin{pmatrix} 0 & \sqrt{1/2} & 0 & 0 & . & . & . & 0 \\ \sqrt{1/2} & 0 & 1 & 0 & . & 0 & . & 0 \\ 0 & 1 & 0 & \sqrt{3/2} & 0 & . & . & 0 \\ 0 & 0 & \sqrt{3/2} & 0 & \sqrt{2} & 0 & & 0 \\ 0 & 0 & 0 & \sqrt{2} & 0 & & & 0 \\ 0 & & & & & & & 0 \\ 0 & & & & & & & \sqrt{n-1/2} \\ 0 & & & & & \sqrt{n-1/2} & 0 \end{pmatrix}$$ (13.2.16)

Notice that $\hat{x}$ is not diagonal. This was to be expected because the basis vectors do not form eigen value equations with $\hat{x}$. This is the $\hat{H}$- Representation of the coordinate operator of a harmonic oscillator.

13.2.2 ENERGY REPRESENTATION of the Momentum operator, $\hat{p}_x$

Consider the operator, $\hat{p}_x$.

$| \Psi \rangle = \sum c'_n | \hat{\psi}_n \rangle$.

$\hat{p}_q | \Psi \rangle = -i \sum c'_n (\partial/\partial q) | \hat{\psi}_n \rangle$ (13.2.17)

from *mathematical jugglery* using $\hat{a}$ and $\hat{a}^\dagger$,

$(d/dq) | n \rangle = \sqrt{\frac{n}{2}} | n-1 \rangle - \sqrt{\frac{n+1}{2}} | n+1 \rangle$. (13.2.18)

This leads to $\quad \hat{p}_q | \Psi \rangle = \sum i\, c_{n-1} \sqrt{\frac{n}{2}} - i\, c_{n+1} \sqrt{\frac{n+1}{2}} | n \rangle$. (13.2.19)

$(\hat{p}_q)_{n,k} = \langle n | \hat{p}_q | k \rangle = i \sqrt{\frac{n}{2}} \delta_{n-1,k} - \sqrt{\frac{n+1}{2}} \delta_{n+1,k}$. (13.2.20)

One, therefore, gets the non-vanishing matrix elements of $\hat{p}_q$,

$(\hat{p}_q)_{n,n-1} = \langle n | \hat{p}_q | n-1 \rangle = i \sqrt{\frac{n}{2}}$. (13.2.21)

$$(\hat{p}_q)_{n,n+1} = \langle n | \hat{p}_q | n+1 \rangle = -i\sqrt{\tfrac{n+1}{2}}.\tag{13.2.22}$$

and $\quad (\hat{p}_q)_{n,n} = \langle n | \hat{p}_q | n \rangle = 0.\tag{13.2.23}$

$$\hat{p}_q = \begin{pmatrix} 0 & -i\sqrt{\tfrac{1}{2}} & 0 & 0 & \cdot & \cdot & \cdot & 0 \\ +i\sqrt{\tfrac{1}{2}} & 0 & -i & 0 & \cdot & 0 & \cdot & 0 \\ 0 & +i & 0 & -i\sqrt{\tfrac{3}{2}} & 0 & \cdot & \cdot & 0 \\ 0 & 0 & +i\sqrt{\tfrac{3}{2}} & 0 & -i\sqrt{2} & 0 & & 0 \\ 0 & 0 & 0 & +i\sqrt{2} & 0 & 0 & & 0 \\ 0 & & & & & & & 0 \\ 0 & & & & & & & -i\sqrt{n-\tfrac{1}{2}} \\ 0 & & & & & & +i\sqrt{n-\tfrac{1}{2}} & 0 \end{pmatrix}\tag{13.2.24}$$

$\hat{p}_x = \hat{p}_q\, \hbar\, b^{1/2}$

$\langle n | \hat{p}_x | k \rangle = (i\hbar b^{-2})(n-k)\langle n | \hat{x} | k \rangle$

$$= \hbar b^{1/2}\left\{ i\sqrt{\tfrac{n}{2}}\,\delta_{n-1,k} - i\sqrt{\tfrac{n+1}{2}}\,\delta_{n+1,k} \right\}.\tag{13.2.25}$$

13.2.3 ENERGY REPRESENTATION of the operators $\hat{a}$ and $\hat{a}^\dagger$

The Ladder Operators of a harmonic oscillator were described in detail in Chapter 7. They are non-Hermitian ones.

As in the case of operator $\hat{q}$ (section 13.2) one can show that

$$\hat{a}^\dagger_{k,n} = \langle k | \hat{a}^\dagger | n \rangle = \sqrt{\tfrac{n+1}{2}}\,\delta_{k,n+1}.\tag{13.2.26}$$

$$\hat{a}_{k,n} = \langle k | \hat{a} | n \rangle = \sqrt{n}\,\delta_{k,n-1}\tag{13.2.27}$$

and $\quad \hat{a}^\dagger_{n,n} = \langle n | \hat{a}^\dagger | n \rangle = 0.\tag{13.2.28}$

$$\hat{a}_{n,n} = \langle n | \hat{a} | n \rangle = 0.\tag{13.2.29}$$

$$\hat{a} = \begin{pmatrix} 0 & \sqrt{1} & 0 & 0 & . & . & . & 0 \\ 0 & 0 & \sqrt{2} & 0 & .0. & . & & 0 \\ 0 & 0 & 0 & \sqrt{3} & 0. & . & . & 0 \\ 0 & 0 & 0 & 0 & \sqrt{4} & & & 0 \\ 0 & 0 & 0 & 0 & 0 & & & 0 \\ 0 & & & & & & & 0 \\ 0 & & & & & & \sqrt{n-1} & 0 \\ 0 & & & & & 0 & & 0 \end{pmatrix} \quad (13.2.30)$$

$$\hat{a}^\dagger = \begin{pmatrix} 0 & 0 & 0 & 0 & . & 0. & . & 0 \\ \sqrt{1} & 0 & 0 & 0 & . & 0. & . & 0 \\ 0 & \sqrt{2} & 0 & 0 & & 0. & . & . & 0 \\ 0 & 0 & \sqrt{3} & 0 & & 0 & & 0 \\ 0 & 0 & 0 & \sqrt{4} & & 0 & & 0 \\ 0 & & & & \sqrt{5} & & & 0 \\ 0 & & & & & & 0 & 0 \\ 0 & 0 & 0 & 0 & & 0 & \sqrt{n-1} & 0 \end{pmatrix} \quad (13.2.31)$$

$\hat{a}$ and $\hat{a}^\dagger$ are seen to be real matrices.

Worked out Example 13.1

Write the Matrix Representation of the operators, $\hat{a}(t), \hat{a}^\dagger(t)$ and $\hat{q}(t)$ for a harmonic oscillator, in the H-picture.

Solution: ⎯Step # 1⎯ Arrange for getting the Energy Representations of $\hat{a}$ and $\hat{a}^\dagger$, as given by (13.2.30) and (13.2.31).

⎯Step # 2⎯ Multiply by the temporal exponential factor as given below.

$$\hat{a}(t) = \hat{a}(0) \, e^{i\,\omega_c\,t}. \qquad (13.2.32)$$

$$\hat{a}^\dagger(t) = \hat{a}^\dagger(0) \, e^{-i\,\omega_c\,t} \qquad (13.2.33)$$

$$\hat{q}(t) = \{\hbar/m\,\omega_c\}^{1/2}[\hat{a}(t) + a^\dagger(t)] \qquad (13.2.34)$$

13.3   OCSILLATOR IN A MIXED STATE

13.3.1 Energy Representations of other operators in an oscillator in a general state:

It can be shown that

$$\langle n | \hat{x}^2 | n \rangle = \left(\hbar/2m\omega_c\right)\langle [\hat{a}-\hat{a}^\dagger][\hat{a}^\dagger-\hat{a}]\rangle = \left(\hbar/2m\omega_c\right)\langle [\hat{a}\hat{a}^\dagger + \hat{a}^\dagger\hat{a}]\rangle_n \quad (13.3.3)$$

$$\langle n | \hat{p}^2 | n \rangle = \left(\hbar m \omega_c/2\right)\langle [\hat{a}+\hat{a}^\dagger]^2\rangle_n = \left(\hbar m \omega_c/2\right)\langle [\hat{a}\hat{a}^\dagger + \hat{a}^\dagger\hat{a}]\rangle_n. \quad (13.3.2)$$

$$\langle \hat{x}^2 \rangle_n = E_n / \left(m\omega_c^2\right). \quad (13.3.3)$$

$$\langle \hat{p}^2 \rangle_n = m E_n. \quad (13.3.4)$$

$$\Delta \hat{p}_x = \sqrt{\langle \hat{p}_x^2 \rangle - \langle \hat{p}_x \rangle^2} \quad (13.3.5)$$

$$\Delta \hat{p}_x \cdot \Delta \hat{x} = (n+\tfrac{1}{2})\hbar$$

Worked out Example 13.2

Find the expectation value of the energy $<E>$ of a harmonics oscillator prepared in the state $|\Psi(z,t)\rangle = \sqrt{\tfrac{1}{2}}\{|\hat{\psi}_0(z,t)\rangle + |\hat{\psi}_1(z,t)\rangle\}$, where $|\hat{\psi}_0(z,t)\rangle$ and $|\hat{\psi}_1(z,t)\rangle$ are the ground and first excited eigen kets and both of them are normalized.

Solution: Step # 1 Given: The time dependent state of the system,

$$|\Psi(z,t)\rangle = \sqrt{\tfrac{1}{2}}\{|\hat{\psi}_0(z,t)\rangle + |\hat{\psi}_1(z,t)\rangle\}, \quad |\Psi(z,t)\rangle = |\Psi(z)\rangle e^{-iEt/\hbar}.$$

$$\langle \hat{\psi}_0(z,t) || \hat{\psi}_0(z,t)\rangle = 1 \,;\, \langle \hat{\psi}_1(z,t) || \hat{\psi}_1(z,t)\rangle = 1.$$

Step # 2 $\hat{H}|\hat{\psi}_0(z,t)\rangle = E_0 |\hat{\psi}_0(z,t)\rangle;\; \hat{H}|\hat{\psi}_1(z,t)\rangle = E_1 |\hat{\psi}_1(z,t)\rangle$

$$\hat{H} = \tfrac{1}{2}\hbar\omega_c \begin{pmatrix} 1 & 0 \\ 0 & 3 \end{pmatrix}$$

Step # 3 $|\Psi(z,t)\rangle$ can be written as a matrix representation,

$$\left\{\sqrt{\tfrac{1}{2}} e^{-iE_0 t/\hbar}, \sqrt{\tfrac{1}{2}} e^{-iE_1 t/\hbar}\right\}$$

Step # 4 To find the expectation value of the energy $<E>=<\hat{H}>$,

$$<\hat{H}> = \langle \Psi(z,t) | \hat{H} | \Psi(z,t) \rangle$$

$$\hat{H} = \sqrt{\tfrac{1}{2}} \left\{ e^{i E_0 t/\hbar}, e^{i E_1 t/\hbar} \right\} \tfrac{1}{2} \hbar \omega_c \begin{pmatrix} 1 & 0 \\ 0 & 3 \end{pmatrix} \sqrt{\tfrac{1}{2}} \begin{pmatrix} e^{-i E_0 t/\hbar} \\ e^{-i E_1 t/\hbar} \end{pmatrix}$$

$$= \tfrac{1}{2} \left(\tfrac{1}{2} \hbar \omega_c \right)(1+3) = \hbar \omega_c \quad .$$

13.4. q-DEFORMED OSCILLATOR

The so called q-deformed algebras have been object of interest in the physics and mathematical physics and a great effort has been devoted to its understanding and development (for example, Biederharn, 1995). Basically they are deformed versions of the standard Lie algebras, which h are recovered as the deformation parameter $q \to 1$. The deformed algebras encompass a set of symmetries that is richer than that of the standard Lie algebras.

Boson Harmonic Oscillator (BHO) is the normal harmonic oscillator whose energy $\hbar \omega$ is called a phonon, because phonons obey Bose-Einstein statistics. In Quantum group (Biederharn, 1995) treatment q-deformation of the motion and $\hbar$-quantization are two independent concepts. In BHO it is seen that

$$[a, a^\dagger] \text{ and } \hat{H} = \tfrac{\hbar \omega}{2} (\hat{a}\hat{a}^\dagger + \hat{a}^\dagger \hat{a}) = \hbar \omega (\hat{N} + \tfrac{1}{2})$$

satisfactorily explain experimental results. Biedenharn defined new generators, $\hat{a}_q^\dagger$ as q-creation and $\hat{a}_q$ as q-destruction operators.

In q-BHO,

$$E_q(n) = \tfrac{1}{2} \hbar \omega \, [(n+1)_q + n_q],$$

which are not equi-distant energy levels but are q-dependent unlike the BHO.(Ref.: Annamma John & Devanarayanan, 1997). It has been found that q-deformed oscillator is a useful tool in quantum field theory, since it constitutes a structure more compatible with interactionms. The number q is viewed as convergence parameter and canbe used to regulate divergence in field theory calculations.

REVIEW QUESTIONS

R.Q. 13.1 A harmonic oscillator is in the mixed state given by

$|\Psi(z,t)\rangle = \left\{ \sqrt{\frac{1}{10}}|\hat{\psi}_0(z,t)\rangle + \frac{2}{\sqrt{10}}|\hat{\psi}_1(z,t)\rangle + \frac{2}{\sqrt{10}}|\hat{\psi}_2(z,t)\rangle + \frac{1}{\sqrt{10}}|\hat{\psi}_3(z,t)\rangle \right\}$ Using the matrix method calculate $<\hat{x}>, <\hat{p}_x>$ and $<E>$. (Answers:

$<\hat{x}> = \sqrt{\hbar/m\omega_c}\left(\frac{2}{10}\right)\left\{\sqrt{\frac{1}{2}} + \sqrt{\frac{1}{2}} + 1 + 1 + \sqrt{\frac{3}{2}} + \sqrt{\frac{3}{2}}\right\}$;

$<\hat{p}_x> = \sqrt{\hbar b}\left\{i\sqrt{\frac{1}{2}}\frac{2}{\sqrt{10}}\frac{1}{\sqrt{10}} + \left[-i\frac{1}{\sqrt{2}}\frac{1}{\sqrt{10}} + i\frac{2}{\sqrt{10}}\right]\frac{2}{\sqrt{10}} + \left[-i\frac{2}{\sqrt{10}} + i\sqrt{\frac{3}{2}}\frac{2}{\sqrt{10}}\right]\frac{2}{\sqrt{10}} - i\sqrt{\frac{3}{2}}\frac{2}{\sqrt{10}}\frac{1}{\sqrt{10}}\right\}$;

$<E> = 2\hbar\omega_c$)

R.Q. 13.2 A quantum oscillator is in the vacuum state

$|0\rangle = \{1, 0, 0, 0, 0, \ldots, 0\} B_0 H_0(q)e^{-q^2/2}e^{-iE_0 t/\hbar}$,

Use $|n\rangle = \sqrt{n!}\,(a^\dagger)^n|0\rangle$, to find $|1\rangle$ and $|2\rangle$, where

$$\hat{a}^\dagger = \begin{pmatrix} 0 & 0 & 0 & 0 & 0. & & 0 \\ \sqrt{1} & 0 & 0 & 0 & .0. & & 0 \\ 0 & \sqrt{2} & 0 & 0 & 0. & \cdots & 0 \\ 0 & 0 & \sqrt{3} & 0 & 0 & & 0 \\ 0 & 0 & 0 & \sqrt{4} & 0 & & 0 \\ 0 & 0 & 0 & 0 & \sqrt{5} & & 0 \\ 0 & & & & & & 0 \\ 0 & 0 & 0 & 0 & 0 & \sqrt{n-1} & 0 \end{pmatrix}$$

Examine the relation .

(Answers: $|1\rangle = \{0, \sqrt{2}, 0, 0, 0, 0, \ldots, 0\} B_1 H_1(q)e^{-q^2/2}e^{-iF_1 t/\hbar}$;

$|2\rangle = (1/\sqrt{2!})\{0, 0, \sqrt{3}, 0, 0, 0, 0, \ldots, 0\} B_2 H_2(q)e^{-q^2/2}e^{-iE_2 t/\hbar}$; $<0|1> = 0$;

$<0|2> = 0$; $<2|1> = 0$).

R.Q. 13.3 Find the expectation value $<\hat{H}>, <\hat{p}_x^2>, <\hat{x}^2>, <\hat{x}>$, and $<\hat{p}_x>$ and hence calculate $\Delta\hat{p}_x \cdot \Delta\hat{x}$ of a quantum oscillator whose state vector is given by

$|\Psi(z,t)\rangle = \sqrt{\frac{1}{6}}\{1, 2, 1, 0, 0, 0, \ldots, 0\}$. Use the matrix method, to find $\hat{H}$, $\hat{a}$, and $a^\dagger$. (Answers:

$<\hat{H}> = \frac{3}{2}\hbar\omega_c$; $<\hat{p}_x^2> = \frac{3}{2}m\hbar\omega_c$; using Virial Theorem, $<\hat{x}^2> = \frac{3}{2}\hbar/m\omega_c$;

$<\hat{x}> = \frac{2}{3}\sqrt{\hbar/m\,\omega_c}(\sqrt{\frac{1}{2}}+1)$; $<\hat{p}_x> = \frac{1}{3}\sqrt{\hbar\,m\,\omega_c}\,(1-i)$;

R.Q. 13.4 Show that $[\hat{x},\hat{H}] = i\hbar\,(\hat{p}/m)$, if $\hat{H}$ is the Hamiltonian for a one-dimensional oscillator.

&&*&*&*&*&*&*

Chapter 14

INVARIANCE PRINCIPLE AND CONSERVED QUANTITIES IN QUANTUM MECHANICS

Chapter 14

INVARIANCE PRINCIPLE AND CONSERVED QUANTITIES IN QUANTUM MECHANICS

"To be confused about what is different and what is not, is to be confused about everything"   - David Bohm

"The concept of substances has disappeared from fundamental physics"

- Sir Arthur Eddington

14.1 INTRODUCTION

Many of the regularities in physics may be expressed as CONSERVATION LAWS, each of which states that the MAGNITUDE of some physical quantity is CONSTANT. Universal Conservation Laws are those of energy and momentum. For Quantum Mechanics, conservation laws are in addition to the Universal energy conservation, momentum conservation and that of electric charge have to be valid.

14.1.1 SYMMETRY and Invariance

What is the connection between Symmetry Principles (or Invariance Principle) and Conservation Laws in Quantum Mechanics? The symmetry operations with which one is concerned are transformations of the dynamical variables that leave the time-independent Hamiltonian, $\hat{H}$ of an 'isolated system' invariant. It will be shown in this Chapter that for each operation that leaves the $\hat{H}$ invariant there is a corresponding dynamical variable that is a constant of the motion and is conserved.

The concepts are, for a physical system there exists

a) (i) A set of symmetries satisfied by it, (ii) the laws of physics for the system is invariant with respect to these symmetries,
b) There exists an explicit transformation law, and
c) The dynamical law is covariant under the symmetry transformation.

14.1.2 Symmetry of the Hamiltonian, $\hat{H}$ GENERATOR explained

To each space-time coordinate used to describe a system, there corresponds a UNITARY TRANSFORMATION, U of the wave function of the system.

If the coordinate transformation depends continuously on a parameter as in the important cases of translations and rotations in space, the associated unitary operator can be expressed in terms of a HERMITIAN OBSERVABLE quantity called a GENERATOR. The important physical observables such as linear momentum and angular momentum can be interpreted in this way as generators. Further, if the coordinate transformation leaves the system invariant then the corresponding generator commutes with the $\hat{H}$ of the system. That is the generator is a CONSTANT OF THE MOTION. This is stated as

$$[\hat{H}, \text{Generator}] = 0. \tag{14.1.1}$$

If the generator commutes with $\hat{H}$, then it commutes with any function of $\hat{H}$.

The fact that a quantized system permits symmetry means that there will be different ways to formally describe the system, ways that, however, are equivalent. If the NEW system is equivalent in all respects to the OLD one, then they are connected, if and only if, by either a UNITARY or an ANTI-UNITARY TRANSFORMATION, $\hat{U}$. Thus if there exists a symmetry $\hat{S}$, it will be described by a unitary or anti-unitary operator, $\hat{U}_s$, such that

$$|\Psi\rangle = \hat{U}_S |\Psi\rangle^\circ, \tag{14.1.2}$$

and $\quad \hat{Q}(t) = \hat{U}(t, t_o) \hat{Q}^\circ(t) \hat{U}^\dagger(t, t_o). \tag{14.1.3}$

where $|\Psi\rangle^\circ$ and $\hat{Q}^\circ(t)$ represent the state and the observable in the original (OLD) description of the system. Either of the equations (14.1.2) or (14.1.3) is a unitary transformation, and is called QUANTUM MECHANICAL CANONICAL TRANSFORMATION. Conservation laws of this type, viz. angular momentum and electric charge, are to be distinguished from which apply to idealized system to which real situations may or may not approximate, and these arise in quantum mechanical description of the interaction between elementary particles.

It is known that the Invariance Principles of elementary particle physics fall into two broad classes – (a) SPACE-TIME SYMMETRIES such as *uniform translations*, with *Lorentz Transformations*, which arise from the existence of equivalent space-time frames of reference, (b) INTERNAL SYMMETRIES such as Isospin, SU (3), Charge conjugation. In the first class one has transformations of two types: those, which involve the time (t) coordinate and those, which do not.

14.2. COORDINATE TRANSFORMATIONS and STATE VECTOR TRANSFORMATIONS:

In setting up quantum mechanical (QM) formulism it was implicitly chosen a frame of reference with respect to which the coordinates are measured. The form of the state vector and operator will in general change if one uses a different frame of reference. How does this happen is to be described below. The equations of transformation, which define the change of one FRAME OF REFERENCE Σ to the second Σ', give the coordinate of point $P(x',y',z')$ referred to Σ' in terms of $P(x, y, z)$ referred to Σ. This is symbolized by

$$\begin{aligned} x &\longrightarrow x' = f(x,y,z), \\ y &\longrightarrow y' = g(x,y,z), \\ z &\longrightarrow z' = h(x,y,z). \end{aligned} \qquad (14.2.1)$$

If the transformation is a UNIFORM TRANSLATION by (a_x, a_y, a_z), then

$$\begin{aligned} x &\longrightarrow x' = x + a_x \\ y &\longrightarrow y' = y + a_y \\ z &\longrightarrow z' = z + a_z. \end{aligned} \qquad (14.2.2)$$

A second example is a ROTATION IN A PLANE (say, x-plane).

The inversion of the coordinate system, or parity transformation, is a third example.

If the system is in a state described by the wave function $|\,a,x,y,z\,\rangle$ relative to the frame of reference Σ, one defines the transformed state $|\,a,x',y',z'\,\rangle'$ as that of the new coordinates referred to Σ', which has the same value as the old state function at the same point of space. This is expressed by

$$|\,a,x',y',z'\,\rangle' = |\,a,x,y,z\,\rangle, \qquad (14.2.3)$$

which simply states that the probability amplitude for finding the particle at a specified point of space is the same whichever coordinate system is used. If one can solve equation (14.2.1), for x, y, z in terms of x', y', z' which shall be expressed by

$$\begin{aligned} x &= f(x',y',z'), \\ y &= g(x',y',z'), \\ z &= h(x',y',z'), \end{aligned} \qquad (14.2.4)$$

Then one gets $|\,a,x',y',z'\,\rangle' = |\,a, f(x',y',z'), g(x',y',z')y, h(x',y',z')\,\rangle$

which serves to define $| a, x', y', z' \rangle'$. This for the case of translation (14.2.2),

$$| a, x', y', z' \rangle' = | a, (x'-a_x), (y'=a_y), (z'-a_z) \rangle .$$

Dropping the **prime** throughout,

$$| a, x, y, z \rangle' = | a, f(x, y, z), g(x, y, z)y, h(x, y, z) \rangle$$

where $| a, x, y, z \rangle'$ is the new state function referred to the frame Σ.

The relation between a coordinate transformation (uniform translation by a_x and a state function transformation $| a, x', y', z' \rangle \longrightarrow | a, x', y', z' \rangle'$ is shown in Fig. 14.1.

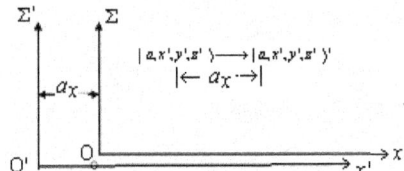

Fig 14.1 Relation between a Coordinate Transformation and a Wavefunction Transformation

$| a, x' \rangle$ looks the same relative to Σ as $| a, x' \rangle'$ does relative to Σ'.

To denote that $| a, x, y, z \rangle'$ arises from $| a, x, y, z \rangle$, one writes it as

$$| \Psi \rangle' = \hat{U}_S | \Psi \rangle, \qquad (14.1.2)$$

where $\hat{U}_S$ is an operator which depends on the transformation called SYMMETRY TRANSFORMATION of the system, which is the basis for the law of behavior of a system

$$\Sigma \longrightarrow \Sigma', \textit{viz.}, \text{the equation (14.2.1)}.$$

14.3. SPATIAL TRANSLATION and CONSERVATION OF LINEAR MOMENTUM:

What is the connection between the translation properties of a system and the conservation of linear momentum? Symmetry property gives rise to the Conservation Law.

14.3.1. Unitary Translation Operator, $\hat{U}_T(\vec{a})$

Space r and time t are homogeneous; without which no law of nature can be described. The expression for homogeneous space-time is "*translation symmetry*". Let me assume that the physical properties of an isolated system can not be altered by an arbitrary translation vector, $\hat{T}(\vec{a})$, where $\vec{a}$ is a vector. In other words, a translation vector, $\hat{T}(\vec{a})$, causes the system to be undisturbed, but the origin of position coordinates, $\vec{r}^{\,o}$ at t_o (initial time) is displaced by an amount $(-\vec{a})$; $\vec{a}$ is independent of $\vec{r}$.

$$\vec{r} \longrightarrow \vec{r}^{\,o} + \vec{a}^{\,o}, \tag{14.3.1}$$

i.e. $\quad \vec{r} = \hat{T}(\vec{a})\,\vec{r}^{\,o} \tag{14.3.2}$

$$\vec{r} \equiv \vec{r}^{\,o} + \vec{a} \tag{14.3.3}$$

Such a transformation is called <u>Active Transformation</u>; if the coordinates were not changed then it is a <u>Passive Transformation</u>. Inversely,

$$\vec{r}^{\,o} \longrightarrow \vec{r} - \vec{a}^{\,o} \tag{14.3.4}$$

i.e., $\quad \vec{r}^{\,o} = \hat{T}^{-1}(\vec{a})\,\vec{r}, \tag{14.3.5}$

$$\vec{r}^{\,o} \equiv \vec{r} - \vec{a} \tag{14.3.6}$$

Let the system of a single particle be represented at time $t = t_o$ by the one particle spatial state $\left|\Psi^o(\vec{r}^{\,o})\right\rangle \equiv \left|\Psi^o(\vec{r}^{\,o}, t_o)\right\rangle$. After effecting a translation on $\left|\Psi^o(\vec{r}^{\,o})\right\rangle$ by an operator $\hat{U}_T(\vec{a})$ corresponding to $\hat{T}(\vec{a})$ one gets

$$\left|\Psi(\vec{r}, t_o)\right\rangle = \hat{U}_T(\vec{a})\left|\Psi^o(\vec{r}, t_o)\right\rangle \tag{14.3.7}$$

as in equation (14.1.2). This is equivalent to shifting the origin of the coordinates by $(-\vec{a})$. From equation (14.3.2), and

$$\left|\Psi(\vec{r}, t_o)\right\rangle = \hat{U}_T(\vec{a})\left|\Psi^o(\vec{r}^{\,o}, t_o)\right\rangle$$

$$\left|\Psi(\vec{r}, t_o)\right\rangle = \left|\Psi^o\left(\hat{T}^{-1}(\vec{a})\,\vec{r}, t_o\right)\right\rangle = \left|\Psi^o(\vec{r} - \vec{a}, t_o)\right\rangle \tag{14.3.8}$$

This means $\left|\Psi(\vec{r}, t_o)\right\rangle$ and $\left|\Psi^o(\vec{r} - \vec{a}, t_o)\right\rangle$ are equivalent.

The physical properties of the system cannot be altered by such a transformation. So $\hat{U}_T(\vec{a})$ is a <u>Unitary operator</u> indeed !

14.3.2 To find an EXPLICIT FORM of $\hat{U}_T(\vec{a})$ for One Particle System:

Let $|\Psi(\vec{r}, t_o)\rangle$ and $|\Psi^o(\vec{r}^o, t_o)\rangle$ are the two descriptions of the system.

Consider first the effect of an infinitesimal translation $\delta\vec{a}$, on $|\Psi^o(\vec{r}^o, t_o)\rangle$.

From equation (14.3.8), viz., $|\Psi(\vec{r}, t_o)\rangle = |\Psi^o(\vec{r} - \vec{a}, t_o)\rangle$

an infinitesimal translation, $\delta\vec{a}$, means

$$|\Psi(\vec{r}, t_o)\rangle = |\Psi^o(\vec{r} - \delta\vec{a}, t_o)\rangle.$$

It is known that the <u>Taylor Series</u> expansion of the state function can be written as

$$|\Psi(\vec{r}, t_o)\rangle = |\Psi^o(\vec{r} - \delta\vec{a}, t_o)\rangle = |\Psi^o(\vec{r}, t_o)\rangle$$

$$- \delta\vec{a}_x(\partial/\partial x)|\Psi^o(\vec{r}, t_o)\rangle - \delta\vec{a}_y(\partial/\partial y)|\Psi^o(\vec{r}, t_o)\rangle - \delta\vec{a}_z(\partial/\partial z)|\Psi^o(\vec{r}, t_o)\rangle$$

$$+ \frac{1}{2!}(\delta\vec{a}_x)^2(\partial^2/\partial x^2)|\Psi^o(\vec{r}, t_o)\rangle + \cdots \qquad (14.3.9)$$

Keeping only up to Ist order term, equation (14.3.9) becomes

$$|\Psi(\vec{r}, t_o)\rangle = |\Psi^o(\vec{r}, t_o)\rangle$$

$$- \delta\vec{a}_x(\partial/\partial x)|\Psi^o(\vec{r}, t_o)\rangle - \delta\vec{a}_y(\partial/\partial y)|\Psi^o(\vec{r}, t_o)\rangle - \delta\vec{a}_z(\partial/\partial z)|\Psi^o(\vec{r}, t_o)\rangle$$

$$= (I - \delta\vec{a}\cdot\hat{\nabla})|\Psi^o(\vec{r}, t_o)\rangle. \qquad (14.3.10)$$

On account of equation (14.3.7) $|\Psi(\vec{r}, t_o)\rangle = \hat{U}_T(\vec{a})|\Psi^o(\vec{r}, t_o)\rangle$.

$$\therefore \qquad \hat{U}_T(\delta\vec{a})|\Psi^o(\vec{r}, t_o)\rangle = (I - \delta\vec{a}\cdot\hat{\nabla})|\Psi^o(\vec{r}, t_o)\rangle.$$

i.e., $\qquad \hat{U}_T(\delta\vec{a}) = (I - \delta\vec{a}\cdot\hat{\nabla}) \qquad (14.3.11)$

But Linear Momentum, $\hat{p} = -i\hbar\hat{\nabla} \qquad (14.3.12)$

$$\hat{U}_T(\delta\vec{a}) = \hat{U}_T(\vec{r} - \delta\vec{a}, \vec{r}) = [I - (i/\hbar)\delta\vec{a}\cdot\hat{p}] \qquad (14.3.13)$$

This shows that the *Linear Momentum Operator*, $\hat{p}$, is the <u>Generator</u> of the Infinitesimal Translations, $\delta\vec{a}$.

A finite translation = a successive infinitesimal translations in steps of $\delta\vec{a}$.

Let $\quad \delta\vec{a} = \vec{a}/n$. $\hfill (14.3.14)$

n = integer, $\to \infty$.

From equation (14.3.13) and to get the finite transformation, $U_T(\vec{a})$ one has to iterate the infinitesimal transformation, $\hat{U}_T(\delta\vec{a})$.

i.e., $\quad \hat{U}_T(\vec{a}) = \underset{n\to\infty}{Lt} \left[I - (i/\hbar) \frac{\vec{a}}{n} \cdot \hat{p} \right]^n$ $\hfill (14.3.15)$

Since $e^{-\beta} = \underset{n\to\infty}{Lt} \left[I - \frac{\beta}{n} \right]^n$ $\hfill (14.3.16)$

Equations (14.3.15) yields

$$\hat{U}_T(\vec{a}) = e^{-(i/\hbar)\frac{\vec{a}}{n}\cdot\hat{p}} = e^{-\vec{a}\cdot\hat{\nabla}} \hfill (14.3.17)$$

$$= [I - \vec{a} \cdot \text{grad}]$$

$\therefore \quad \hat{U}_T(\vec{a}) = e^{-\vec{a}\cdot\hat{\nabla}} = \left[I - \vec{a}\cdot\hat{\nabla} - \frac{1}{2}\vec{a}\cdot\hat{\nabla}^2 - \cdots \right]$ $\hfill (14.3.18)$

If $\quad |\hat{p}\,\Psi(\vec{r},\,t_o)\rangle = \hat{p}_o\,|\Psi^o(\vec{r},\,t_o)\rangle$. $\hfill (14.3.19)$

$\hat{U}_T(\vec{a})\,|\Psi^o(\vec{r},\,t_o)\rangle = e^{-(i/\hbar)\vec{a}\cdot\hat{p}}\,|\Psi^o(\vec{r},\,t_o)\rangle$

$= \left[I - \vec{a}\cdot\hat{\nabla} - \frac{1}{2}\vec{a}\cdot\hat{\nabla}^2 - \cdots \right] |\Psi^o(\vec{r},\,t_o)\rangle$.

$= |\Psi^o(\vec{r}-\vec{a},\,t_o)\rangle$. $\hfill (14.3.20)$

$= |\Psi^o(\vec{r},\,t_o)\rangle$, with a change in phase.

The transformation law of the fundamental observables, $\hat{p}_o$ of the system determines the operator $\hat{U}_T(\vec{a})$ up to a phase factor. The phase of $\hat{U}_T(\vec{a})$ has no physical significance. This means $\hat{U}_T(\vec{a})$ does not change the state of the system.

Conversely, <u>if the state of the system is unaltered by a spatial translation it is an eigenstate of the linear momentum operator.</u>

14.3.3 For a N-particle system

As in the case for a single particle with momentum $\hat{p}$

$$\hat{U}_T(\delta\vec{a}) = \hat{U}_T(\vec{r} - \delta\vec{a}, \vec{r}) = \left[I - (i/\hbar)\,\delta\vec{a}\cdot\hat{p}\right] \tag{14.3.13}$$

For N particles, the Total Momentum Operator $\hat{P}$ is

$$\hat{P} = \hat{p}_1 + \hat{p}_2 + \hat{p}_3 + \cdots + \hat{p}_N. \tag{14.3.21}$$

Correspondingly, $\quad \hat{U}_T(\delta\vec{a}) = \left[I - (i/\hbar)\,\delta\vec{a}\cdot\hat{P}\right]$

$$\therefore \quad \hat{U}_T(\vec{a}) = e^{-(i/\hbar)\frac{\vec{a}}{n}\cdot\hat{P}}. \tag{14.3.22}$$

Since the Hamiltonian $\mathcal{H}$ of an isolated system is invariant under any symmetry transformation, and so Spatial Translation, $\hat{T}(\vec{a})$,

$$\hat{H} = \hat{H}^\circ$$

Further, $\left\langle \Psi^\circ(\vec{r}^\circ, t_o) \mid \hat{H}^\circ \mid \Psi^\circ(\vec{r}^\circ, t_o) \right\rangle = \left\langle \Psi(\vec{r}^\circ, t_o) \mid \hat{H} \mid \Psi(\vec{r}^\circ, t_o) \right\rangle$

and $\quad \hat{U}_T(\vec{a}) \mid \Psi^\circ(\vec{r}^\circ, t_o) \rangle = \mid \Psi(\vec{r}^\circ, t_o) \rangle.$

$$\therefore \quad \hat{H} = \hat{U}_T(\vec{a})\,\hat{H}^\circ\,\hat{U}_T^\dagger(\vec{a}) = \hat{H}^\circ. \tag{14.3.23}$$

using $\hat{U}_T(\delta\vec{a}) = \left[I - (i/\hbar)\,\delta\vec{a}\cdot\hat{P}\right],$

$$\hat{U}_T(\vec{a})\,\hat{H}^\circ\,\hat{U}_T^\dagger(\vec{a}) = \left[I - (i/\hbar)\,\delta\vec{a}\cdot\hat{P}\right]\hat{H}^\circ\left[I + (i/\hbar)\,\delta\vec{a}\cdot\hat{P}\right]$$

$$= \left\{ \hat{H}^\circ - (i/\hbar)\,\delta\vec{a}\cdot\left[\hat{P}, \hat{H}^\circ\right] \right\}.$$

From equation (14.3.23), one gets, therefore,

$$\hat{H}^\circ = \left\{ \hat{H}^\circ - (i/\hbar)\,\delta\vec{a}\cdot\left[\hat{P}, \hat{H}^\circ\right] \right\}.$$

i.e., $\quad (i/\hbar)\,\delta\vec{a}\cdot\left[\hat{P}, \hat{H}^\circ\right] = 0.$

In other words, $\left[\hat{P}, \hat{H}^\circ\right] = 0.$ \hfill (14.3.24)

This means the $\boxed{\text{Total Momentum } \hat{P} = \text{A Constant of the Motion}}$.

This constant of the motion is a physical observable.

But $\hat{U}_T(\vec{a})$, is only unitary; and physical observables must correspond to **Hermitian operator**. Since the general equation giving time dependence of $\hat{P}$ is

$$\tfrac{d}{dt} <\hat{P}> = -(i/\hbar)\left\langle [\hat{P}, \hat{H}^\circ] \right\rangle + \left\langle \tfrac{\partial \hat{P}}{\partial t} \right\rangle = 0 \qquad (14.3.25)$$

Is the definition of an observable $\hat{P}$ to be a constant of motion of an *isolated system* with $\hat{H}^\circ$, if $\left\langle \tfrac{\partial \hat{P}}{\partial t} \right\rangle = 0$ (no explicit time dependence of $\hat{P}$).

Thus the Conservation of the Total Momentum, of an isolated sytem, results from the Invariance of its $\hat{H}^\circ$ under Spatial Translation, $\hat{T}(\vec{a})$.

14.4. TEMPORAL TRANSLATIONS AND ENERGY CONSRVATION: EVOLUTION OPERATOR: $\hat{U}_T(t, t_o)$.

14.4.1. Expression for $\delta\hat{U}(\delta t)$

Consider a dynamical system in which a particular state of the system in motion through out the time during which the system is left undisturbed. It is known from equation (12.2.1)

$$|\psi(t)\rangle = \text{state of the system at time } t.$$

t is a continuous, real variable, $-\infty < t < +\infty$, which labels the sequence of measurements. It is to be noted that the parameter t, appearing in theory, is not an eigen value of any Hermitian operator.

$|\psi(t)\rangle$ at $t > t_o$ arises for $|\psi(t_o)\rangle$ by the action of a time development operator, $\hat{T}(t, t_o)$ in **Hilbert Space**.

$$\hat{T}(t, t_o) \equiv \hat{T}(t - t_o)$$

is the Unitary Operator in question.

Because the system is considered to be ISOLATED, $\hat{T}$ can only depend on $(t - t_o)$.

$$|\psi(t_o, t)\rangle = \hat{T}(t - t_o)|\psi(t_o)\rangle \qquad (12.2.5)$$

Consider a time t' such that $t > t' > t_0$. Let the system evolves from $|a', t_0\rangle$ until $t+t'$, into the state represented by $|\psi(t_0,t)\rangle = \hat{T}(t-t')|\psi(t_0,t')\rangle$

This state could then be allowed to evolve until t, at which time it is

$$|\psi(t_0,t)\rangle = \hat{T}(t-t')|\psi(t_0,t')\rangle \qquad (12.2.7)$$

The state given by equation (12.2.7) is clearly identical with that given by equation (12.2.5), because equation (12.2.7) is

$$|\psi(t_0,t)\rangle = \hat{T}(t-t')\left[\hat{T}(t'-t_0)|\psi(t_0)\rangle\right]$$

i.e., $\quad |\psi(t_0,t)\rangle = \hat{T}(t-t_0)|\psi(t_0)\rangle$. and

$$\hat{T}(t, t_0) = \hat{T}(t-t')\,\hat{T}(t'-t_0). \qquad (12.2.8)$$

which is the group property of the time development Unitary Operator,

since $\quad (t-t') = (t'-t_0) = (t-t_0)$.

Setting $t = t_0$. $\hat{T}(t_0 - t_0) = \hat{T}(t_0, t_0) = \hat{T}(0) = \hat{\Im}$.

It means that $\quad \hat{T}(t_0 - t')\,\hat{T}(t'-t_0) = \hat{\Im}$

$\therefore \qquad \hat{T}(t-t')^{-1} = \hat{T}(t', t) = \hat{T}^\dagger(t-t')$.

i.e., $\quad \hat{T} \equiv \hat{U} =$ unitary, i,e, $\hat{T}(t'-t_0) \Rightarrow \hat{U}(t'-t_0)$,

Consider $(t-t_0) \Rightarrow$ infinitesimally small, i.e., $(t-t_0) \Rightarrow \delta t$,

and at $t_0 = 0 \qquad \hat{U}(t_0 + \delta t, t_0) = \hat{U}(\delta t)$,

Now $\quad \hat{U}(0) = 1$,

$\hat{U}(\delta t) \approx \hat{U}(0) + \delta\hat{U}$.

$\hat{U}(\delta t) \approx 1 + \delta\hat{U}$.

Unitary then implies $\hat{\Im} \approx 1 + \delta\hat{U}$; but $\hat{\Im} \approx 1 + (\delta\hat{U} + \delta\hat{U}^\dagger) + \hat{O}[(\delta\hat{U})^2]$.

Hence one requires $(\delta\hat{U} + \delta\hat{U}^\dagger) = 0$. $\qquad (12.2.12)$

This means $\quad \boxed{\delta\hat{U} = \text{anti-Hermitian} = i \times \text{Hermitian operator}} \qquad (12.2.13)$

Further, let $t' = t - \delta t$, and if δt = small,

$$\hat{U}(t, t-\delta t) = \delta \hat{U}(\delta t),$$

$$\hat{U}(\delta t_1)\hat{U}(\delta t_2) = \hat{U}(\delta t_1 + \delta t_2)$$

∴ $\quad \delta \hat{U}(\delta t) \propto \delta t \qquad (12.2.14)$

Because of (12.2.13) and (12.2.14)

$$\delta \hat{U}(\delta t) = (-i/\hbar)\delta t \cdot \text{[a time dependent operator, } \hat{H}].$$

One can write $\quad \hat{U}(t, t-\delta t) = \delta \hat{U}(\delta t) = \{1 - (-i/\hbar)\delta t \cdot \hat{H}(t)\}.$

$$\delta \hat{U}(\delta t) = (-i\hat{H}\delta t/\hbar).$$

where is $\hat{H}$ the time independent Hamiltonian operator, of the system. It is introduced for convenience to ensure that $\hat{H}$ has the DIMENSIOIN OF ENERGY.

14.4.2. FUNDAMENTAL LAW of Evolution in Quantum Mechanics:

Using the Group Property,

$$\hat{U}(t+\delta t) - \hat{U}(t) = [\hat{U}(\delta t) - 1]\hat{U}(t) = d\hat{U}(t) = \{(-i\hat{H}\,dt/\hbar)\}\hat{U}(t)$$

or, $\quad (\partial/\partial t)\hat{U}(t) = \{(-i\hat{H}/\hbar)\}\hat{U}(t) \qquad (12.2.15)$

since energy is conserved for a conservative system,

$$[\hat{H}, \hat{H}] = 0,$$

$$\hat{U}(t) = e^{\{-i\hat{H}t/\hbar\}}. \qquad (12.2.16)$$

i.e., $\quad \boxed{\hat{U}(t-t_0) = e^{\{-i\hat{H}(t-t_0)/\hbar\}}}. \qquad (12.2.17)$

because of the initial condition, $\hat{U}(0) = 1$.

If the system were non-conservative, then $\boxed{\hat{H} = \hat{H}(t)}$;

$$\boxed{\hat{U}(t, t_0) \neq e^{\{-i\hat{H}(t-t_0)/\hbar\}}}$$

$\hat{U}$ is also defined by

$$\boxed{\hat{U}(t,t_o) = 1 - \{(-i/\hbar)\}\int \hat{H}\cdot \hat{U}(t', t_o)\, dt'} \qquad (12.2.18)$$

Either of the equation (12.2.15) or (12.2.16) or (12.2.18) is known as the EXPRESSION FOR THE FUNDAMENTAL LAW OF EVOLUTION in Quantum Mechanics. An equivalent expression for this law is the SCHRODINGER EQUATION.

Thus an EVOLUTION OPERATOR, $\hat{U}(t,t_o)$ is DEFINED such that for state function $|\psi(t_o)\rangle$ at time t_o and $|\psi(t_1)\rangle$ at time t_1, then

$$|\psi(t_1)\rangle = \hat{U}(t_1,t_o)\cdot |\psi(t_o)\rangle \qquad (14.4.1)$$

$$|\psi(t)\rangle = \hat{U}(t,t_1)\cdot |\psi(t_1)\rangle$$

$$|\psi(t)\rangle = \hat{U}(t_1,t_o)\, \hat{U}(t,t_1)\cdot |\psi(t_o)\rangle = \hat{U}(t,t_o)\cdot |\psi(t_o)\rangle \qquad (14.4.2)$$

because of the initial condition, $\boxed{\hat{U}(t_o,t_o) = \hat{3}}$ \qquad (14.4.3)

or $\qquad \hat{U}(t_o,t_o) = \hat{U}(t,t')\,\hat{U}(t',t_o)$ \qquad (14.4.4)

$$\hat{U}(t,t_o) \ne e^{\{-i\hat{H}(t-t_o)/\hbar\}}. \qquad (14.4.5)$$

U is also defined by $\hat{U}(t,t_o) = 1 - \{(i/\hbar)\,\hat{H}(t_o)\cdot(t-t_o)\}$. \qquad (14.4.6)

where $\hat{H}(t_o)$ is, therefore, the '<u>Generator</u>' of a *finite* Unitary Transformation

$|\psi(t_o)\rangle$ to $|\psi(t_o,t)\rangle$, i.e. a *finite* time translation, described by $\hat{U}(t,t_o)$.

using $\quad \hat{U}(\delta t) = \left[1-(i/\hbar)\,\delta t\cdot\hat{H}\right]$,

$$\hat{U}(\delta t)\,\hat{H}^o\,\hat{U}^\dagger(\delta t) = \left[1-(i/\hbar)\,\delta t\cdot\hat{H}\right]\hat{H}^o\left[1+(i/\hbar)\,\delta t\cdot\hat{H}\right] = \{\hat{H}^o - (i/\hbar)\,\delta t\cdot[\hat{H},\hat{H}^o]\}.$$

From equation (14.3.23), one gets, therefore,

$$\hat{H}^o = \{\hat{H}^o - (i/\hbar)\,\delta t\cdot[\hat{H},\hat{H}^o]\}.$$

i.e., $\qquad (i/\hbar)\,\delta t\cdot[\hat{H},\hat{H}^o] = 0$.

In other words, $[\hat{H},\hat{H}^o] = 0$. \qquad (14.3.24)

This means the $\underline{\hat{H} = \text{a Constant of the Motion}}$.

This Constant of the Motion is a *physical observable*.

But $\hat{U}(\delta t)$ is only unitary; and physical observables must correspond to Hermitian operator. Since the general equation giving time dependence of $\hat{H}$ is

$$\frac{d}{dt}\langle\hat{H}\rangle = -(i/\hbar)\langle[\hat{H},\hat{H}^\circ]\rangle + \left\langle\frac{\partial\hat{H}}{\partial t}\right\rangle = 0 \tag{14.3.25}$$

is the definition of an observable $\hat{H}$ to be a constant of motion of an isolated system with $\hat{H}^\circ$, and $\langle(\partial[\hat{H}/\partial t)\rangle = 0$ (no explicit time dependence of $\hat{H}$).

Thus the CONSERVATION OF THE TOTAL ENERGY of an isolated system IS A CONSEQUENCE OF THE UINVARIANCE of its $\hat{H}^\circ$ with respect to TEMPORAL TRANSLATION.

$$|\psi(t)\rangle = \hat{U}(t,t_0)|\psi(t_0)\rangle = e^{\{-i\hat{H}(t-t_0)/\hbar\}}|\psi(t_0)\rangle \tag{14.4.7}$$

$$= \left[I - (i/\hbar)\hat{H}(t-t_0) + \tfrac{1}{2}\hat{H}(i/\hbar)\hat{H}(t-t_0)^2 + \cdots\right]|\psi(t_0)\rangle.$$

$$= \left[I - (i/\hbar)E(t-t_0) + \cdots\right]|\psi(t_0)\rangle.$$

$$|\psi(t)\rangle = e^{\{-iE(t-t_0)/\hbar\}}|\psi(t_0)\rangle. \tag{14.4.8}$$

14.5 SPACE INVERSION / SPATIAL REFLECTION AND PARITY CONSERVATION, $\hat{P}$

14.5.1. Reflection through the origin of the basis is also known as Parity operation. There corresponds to this a Unitary Operator called the Parity Operator, $\hat{P}$. For a single particle state function, $|\psi(\vec{r})\rangle$,

$$s\vec{r} \longrightarrow (-\vec{r}),$$

$$\hat{P}|\psi(\vec{r})\rangle = |\psi(-\vec{r})\rangle, \tag{14.5.1}$$

Similarly for a many particle system,

$$\hat{P}|\psi(\vec{r}_1,\vec{r}_2,\vec{r}_3,\ldots,\vec{r}_N)\rangle = |\psi(-\vec{r}_1,-\vec{r}_2,-\vec{r}_3,\ldots,-\vec{r}_N)\rangle,$$

The Parity Operator $\hat{P}$ is hermitian since, for any two states $|\psi_m(\vec{r})\rangle$ and $|\psi_n(\vec{r})\rangle$,

$$\langle\psi_n(\vec{r})|\hat{P}|\psi_m(\vec{r})\rangle = \langle\psi_n(\vec{r})||\psi_n(-\vec{r})\rangle = \langle\psi_n(-\vec{r})||\psi_n(\vec{r})\rangle$$

$$= \hat{P} \langle \psi_n(\vec{r}) \| \psi_m(\vec{r}) \rangle \quad (14.5.2)$$

Further, $\hat{P}^2 = \hat{\Im}$ \hfill (14.5.3)

The eigen values of $\hat{P}$ are ± 1, corresponding to the eigen states even or odd, respectively.

Consider $| \psi_+(\vec{r}) \rangle$ = even,

and $\quad | \psi_-(\vec{r}) \rangle$ = odd.

$$\hat{P} | \psi_+(\vec{r}) \rangle = | \psi_+(-\vec{r}) \rangle = +1 | \psi_+(\vec{r}) \rangle$$

$$\hat{P} | \psi_-(\vec{r}) \rangle = | \psi_-(-\vec{r}) \rangle = -1 | \psi_-(\vec{r}) \rangle \quad (14.5.4)$$

Further $\langle \psi_+(\vec{r}) \| \psi_-(\vec{r}) \rangle$ should equal to

$$\langle \psi_+(\vec{r}) \| \psi_-(\vec{r}) \rangle = \langle \psi_+(-\vec{r}) \| \psi_-(-\vec{r}) \rangle = -\langle \psi_+(\vec{r}) \| \psi_-(\vec{r}) \rangle.$$

Therefore $\langle \psi_+(\vec{r}) \| \psi_-(\vec{r}) \rangle = 0$. \hfill (14.5.5)

which means *orthogonality*, according to which $\| \psi_+(\vec{r}) \rangle$ and $| \psi_-(\vec{r}) \rangle$ belong to different eigen values (+1 and –1) of $\hat{P}$. They also form **a complete set**,

$$| \psi(\vec{r}) \rangle = | \psi_+(\vec{r}) \rangle + | \psi_-(\vec{r}) \rangle \quad (14.5.6)$$

$$| \psi_+(\vec{r}) \rangle = \sqrt{\tfrac{1}{2}} \{ | \psi(\vec{r}) \rangle + | \psi(-\vec{r}) \rangle \} = even, \quad (14.5.7)$$

$$| \psi_-(\vec{r}) \rangle = \sqrt{\tfrac{1}{2}} \{ | \psi(\vec{r}) \rangle - | \psi(-\vec{r}) \rangle \} = odd. \quad (14.5.8)$$

14.5.2 EFFECT OF $\hat{P}$ on $\vec{r}$ and $\vec{p}$

Consider the two new vectors $\vec{r}$ and $\vec{p}$.

$$\hat{P} \vec{r}^\circ \hat{P}^\dagger = \vec{r} = (-\vec{r}^\circ) \quad (14.5.9)$$

and $\quad \hat{P} \vec{p}^\circ \hat{P}^\dagger = \vec{p} = (-\vec{p}^\circ). \quad (14.5.10)$

But $\quad \hat{P} = \hat{P}^\dagger$ means Unitary property of $\hat{P}$.

$\hat{P}$ operator operating on is equivalent to a right hand coordinate system into a left hand side coordinate system. This shows that

$$[\hat{P}, \hat{H}] = 0. \quad (14.5.11)$$

The operator $\hat{P}$ is a significant one *in weak interactions of elementary particles, where it is violated.* On the other hand, *in the atomic and other nuclear systems* $\hat{P}$ *is conserved.*

14.6 SPATIAL ROTATION AND CONSERVATION OF ANGULAR MOMENTUM:

14.6.1 Spatial rotation operator, $\mathcal{R}_{\delta\varphi}$:

Consider the effect of an ANTI-CLOCKWISE rotation through an angle $\delta\varphi$ about the z-axis (Fig. 14.2) on the basis (Cartesian Coordinates) (x, y, z). Under the operation $\hat{\mathbb{R}}_{\delta\varphi}$ the basis ($\vec{r}$ $Sin\theta$ $Cos\varphi$, $\vec{r}$ $Sin\theta$ $Sin\varphi$, $\vec{r}$ $Cos\theta$) changes to, $\vec{r} = \hat{\mathbb{R}}_{\delta\varphi}$ $\vec{r}^o$

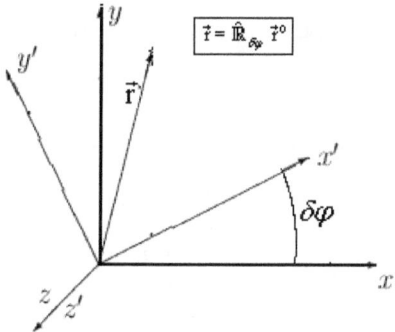

Fig. 14.2. Anti-clockwise rotation through an angle, $\delta\varphi$

$$\vec{r} = \hat{\mathbb{R}}_{\delta\varphi} \ \vec{r}^o (x,y,z) = (\vec{r} \ Sin\theta \ Cos(\varphi-\delta\varphi), \vec{r} \ Sin\theta \ Sin(\varphi-\delta\varphi), \vec{r} \ Cos\theta) , \quad (14.6.2)$$

$$= \begin{cases} \vec{r} \ Sin\theta \ Cos\varphi \ Cos\delta\varphi + \vec{r} \ Sin\theta \ Sin\varphi \ Sin\delta\varphi), \\ \vec{r} \ Sin\theta \ Sin\varphi \ Cos\delta\varphi - \vec{r} \ Sin\theta \ Cos\varphi \ Sin\delta\varphi, \ \vec{r} \ Cos\theta) \end{cases}$$

$$= \begin{cases} \vec{r} \ Sin\theta \ Cos\varphi + \vec{r} \ \delta\varphi \ Sin\theta \ Sin\varphi, \\ \vec{r} \ Sin\theta \ Sin\varphi - \vec{r} \ \delta\varphi \ Sin\theta \ Cos\varphi, \ \vec{r} \ Cos\theta) \end{cases}$$

$$\hat{\mathbb{R}}_{\delta\varphi} \ \vec{r}^o (x,y,z) = (x + y \ \delta\varphi, \ y - x \ \delta\varphi, \ z) = (x, y, z) - (-y, x, 0) \ \delta\varphi \quad (14.6.3)$$

where use of the following is made:

$$Sin \ \delta\varphi = \delta\varphi + \cdots, \ Cos \ \delta\varphi = 1 - \tfrac{1}{2}(\delta\varphi)^2 + \cdots \quad (14.6.4)$$

Thus under the infinitesimal rotation $\delta\varphi$ the basis differs from the initial condition only by an infinitesimal amount. That, of course, is hardly unexpected. Many of the results can be derived by the algebraic methods from the commutation rules.

14.6.2 The EXPLICIT FORM of $\hat{U}(\hat{\mathbb{R}}_{\delta\varphi})$

Assume the space is isotropic. An arbitrary clockwise rotation can be given by specifying the direction of a single axis and the magnitude of rotation about the axis. Consider the motion of the coordinate system (basis) through an Euler angle of rotation, $\vec{\alpha} = (\alpha_x, \alpha_y, \alpha_z)$, such that the direction of the 'vector' is that of the arbitrary axis of rotation with unit vector **n** and its magnitude $|\vec{\alpha}| = |\alpha, \hat{n}|$.

$$|\vec{\alpha}| = \sqrt{(\alpha_x^2 + \alpha_y^2 + \alpha_z^2)}, \tag{14.6.5}$$

is the angle of rotation in the right hand frame (or clockwise) about the z-axis.

$$|\alpha, \hat{n}| \le \pi.$$

So a rotation about the z-axis by angle α_z is symbolized by

$$|\alpha, \hat{n}| = (0, 0, \alpha_z) \tag{14.6.6}$$

For simplicity let the particle system is one of single particle. A vector $\vec{r}^o$ is moved to vector **r** under rotation, which is represented by

$$\vec{r}^o \Rightarrow \hat{\mathbb{R}}_\alpha \Rightarrow \vec{r}, \tag{14.6.7}$$

or
$$\left| X^o \right\rangle \Rightarrow \hat{\mathbb{R}}_\varphi \Rightarrow \left| X \right\rangle$$

$$\hat{\mathbb{R}}_\alpha \equiv \hat{\mathbb{R}}(\alpha_z) = \begin{pmatrix} \cos\alpha_z & \sin\alpha_z & 0 \\ -\sin\alpha_z & \cos\alpha_z & 0 \\ 0 & 0 & 1 \end{pmatrix} \tag{14.6.8}$$

Let the infinitesimal value in α_z is represented by $\delta\alpha_z$,

Since $\vec{r}^o \Rightarrow \hat{\mathbb{R}}_{\delta\alpha} \Rightarrow \vec{r}$, \tag{14.6.9}

i.e., $\vec{r}^o = \hat{\mathbb{R}}_{\delta\alpha} \vec{r}$, \tag{14.6.10}

i.e. , $\vec{r} = \vec{r}^o + \delta\alpha_z \, \hat{n} \wedge \vec{r}^o$, \tag{14.6.11}

But the state $\left| \psi^o(\vec{r}) \right\rangle = \left| \psi^o(\vec{r}^o) \right\rangle$ \tag{14.6.12}

The effect of the rotation of the basis of the observer on the state $\left| \psi^o(\vec{r}^o) \right\rangle$ is, the transformed new state,

$$\left| \psi^o(\vec{r}) \right\rangle = \left| \psi^o(x, y, z) \right\rangle,$$

where $x = x^o \cos \delta\alpha_z + y^o \sin \delta\alpha_z = x^o - y^o \delta\alpha_z$

$y = -x^o \sin \delta\alpha_z + y^o \cos \delta\alpha_z = x^o \delta\alpha_z + y^o.$

$z = z^o$ \hfill (14.6.13)

Equation (14.6.10) means that

$$\vec{r} = (\hat{R}_{\delta\alpha})^{-1} \vec{r}^o = (\hat{R}_{\delta\alpha})^{-1} \begin{pmatrix} x^o \\ y^o \\ z^o \end{pmatrix} = \begin{pmatrix} x^o - y^o \delta\alpha_z \\ y^o + x^o \delta\alpha_z \\ z^o \end{pmatrix} \quad (14.6.14)$$

where $(\hat{R}_{\delta\alpha})^{-1}$ is obtained by changing $\delta\alpha_z$ to $-\delta\alpha_z$.

If one writes $\hat{U}(\delta\alpha_z)$ for the Unitary Operator that effects a change from $\left| \psi^o(\vec{r}^o) \right\rangle$ to $\left| \psi^o(\vec{r}) \right\rangle$,

i.e., $\quad \left| \psi^o(\vec{r}) \right\rangle = \hat{U}(\delta\alpha_z) \left| \psi^o(\vec{r}^o) \right\rangle$ \hfill (14.6.15)

But $\quad \left| \psi^o(\vec{r}) \right\rangle = \hat{U}(\delta\alpha_z) \left| \psi^o(\hat{R}_{\delta\alpha}^{-1} \vec{r}^o) \right\rangle = \left| \psi^o(\vec{r}^o - \delta\alpha_z \hat{n} \wedge \vec{r}^o) \right\rangle$ \hfill (14.6.16)

because of equation (14..6.11).

Using the Taylor's expansion, as in equation (14.3.9) one gets

$$\left| \psi^o(\vec{r}^o - \delta\alpha_z \hat{n} \wedge \vec{r}^o) \right\rangle = \left| \psi^o(\vec{r}^o) \right\rangle - (\delta\alpha_z \hat{n} \wedge \vec{r}^o)[\hat{\nabla} \left| \Psi^o(\vec{r}^o) \right\rangle + \cdots$$

$$= [(1 - \delta\alpha_z \hat{n} \cdot \vec{r}^o \wedge \hat{\nabla})] \left| \Psi^o(\vec{r}) \right\rangle \quad (14.6.17)$$

Now, $\quad \vec{r}^o \wedge \hat{\nabla} = [\vec{r}^o \wedge (h/i)\hat{\nabla}] (i/h) = (i/h)[\vec{r}^o \wedge \hat{p}]$ \hfill (14.6.18)

$$[\vec{r}^o \wedge \hat{p}] = \begin{vmatrix} i & j & k \\ x^o & y^o & z^o \\ \hat{p}_x & \hat{p}_y & \hat{p}_z \end{vmatrix} = \begin{vmatrix} \hat{\ell}_x \\ \hat{\ell}_y \\ \hat{\ell}_z \end{vmatrix} = \vec{L} \quad (14.6.19)$$

Defining the orbital angular momentum, $\vec{L}$ by

$$\vec{r}^o \wedge \hat{V} = (i/\hbar)[\vec{r}^o \wedge \hat{p}] = (i/\hbar)\vec{L} \qquad (14.6.20)$$

equation (14.6.15) becomes

$$\left| \psi^o(\vec{r}^o) \right\rangle = [(I - \delta\alpha_z \ \hat{n} \cdot \vec{r}^o \wedge \hat{V})] \left| \Psi^o(\vec{r}) \right\rangle = [(I - \delta\alpha_z \ \hat{n} \cdot (i/\hbar)\vec{L}] \left| \Psi^o(\vec{r}) \right\rangle$$

$$= [(I - \delta\alpha_z \ (i/\hbar) \ \hat{n} \cdot \vec{L}] \left| \Psi^o(\vec{r}) \right\rangle$$

$$= [(I - \delta\alpha_z \ (i/\hbar) \ \vec{L}_z] \left| \Psi^o(\vec{r}) \right\rangle \qquad (14.6.21)$$

Comparing (14.6.16) and (14.6.21),

$$\hat{U}(\delta\alpha_z) = [(I - \delta\alpha_z \ (i/\hbar) \ \vec{L}_z]$$

Putting $\delta\alpha_z = \alpha_z / m$, $\qquad (14.6.22)$

For finite rotation one has to iterate the infinitesimal transformation m times,

$$\left| \Psi^o(\vec{r}) \right\rangle \cong \underset{m \to \infty}{Lt} \left[I - (i/\hbar) \frac{\alpha_z}{m} \cdot \vec{L}_z \right]^m \left| \Psi^o(\vec{r}) \right\rangle \qquad (14.6.23)$$

Since $e^{-\beta} = \underset{m \to \infty}{Lt} \left[I - \frac{\beta}{m} \right]^m \qquad (14.3.16)$

Equations (14.6.23) yields

$$\therefore \quad \left| \Psi^o(\vec{r}) \right\rangle \cong e^{-(i/\hbar) \ \alpha \cdot \vec{L}} \left| \Psi^o(\vec{r}) \right\rangle \qquad (14.6.24)$$

This means $\hat{U}_n(\alpha_z) = e^{-(i/\hbar) \ \alpha_z \cdot \vec{L}_n}$. $\qquad (14.6.25)$

Arbitrary rotation will be represented in the Hilbert space of states by

$$\hat{U}_n(\alpha_z) = e^{-(i/\hbar) \ \alpha_z \cdot \vec{L}_n}.$$

This means the orbital angular momentum operator $\vec{L}_z$ may be called the generator of infinitesimal rotation. Except when they are performed about the same axis rotations do not in general commute.

14.6.3 To find a CONSERVATION LAW for ANGULAR MOMENTUM

It is stated earlier that

$$\left| \psi^o(\vec{r}) \right\rangle = \hat{U}(\delta\alpha_z) \left| \psi^o(\vec{r}^o) \right\rangle \tag{14.6.15}$$

$$\langle \psi || \psi \rangle = \langle \hat{U}_n \psi^o || \hat{U}_n \psi^o \rangle = \langle \psi^o | \hat{U}_n^\dagger \hat{U}_n | \psi^o \rangle$$

$$= \langle \psi^o | \hat{\mathfrak{I}} | \psi^o \rangle = \langle \psi^o || \psi^o \rangle.$$

Now, $\hat{H}$ for the system must be an invariant for spatial rotation, *i.e.*,

$$\hat{H} = \hat{H}^o,$$

∴ $\qquad \hat{H} = \hat{U}_n \hat{H}^o \hat{U}_n^\dagger = \hat{H}^o$

∴ $\qquad \hat{U}_n \hat{H}^o = \hat{H}^o \hat{U}_n,$

and $\qquad [\hat{U}_n, \hat{H}^o] = 0.$ $\hfill(14.6.26)$

Because any angle of rotation,

$$\alpha_z = m\, \delta\alpha_z, \tag{14.6.22}$$

And $\hat{L}_n$ is the Generator of Infinitesimal Rotation about the **n**-axis, it follows that invariance of $\hat{H}$ under rotations implies that, for an isolated system of STRUCTURE LESS particle,

$$[\hat{L}_n, \hat{H}^o] = 0. \tag{14.6.27}$$

so that the total angular momentum of the system is conserved. Conversely, if the $\hat{H}$ of a system of *structure less* particles commutes with the total orbital angular momentum, $\hat{L}_n$ then $\hat{H}$ is invariant under spatial rotations.

If the particle has both intrinsic and orbital angular momenta, $\hat{S}$ and $\hat{L}$ operators, the total angular momentum operator, $\hat{J}$ is

$$\hat{J} = \hat{L} + \hat{S} \tag{14.6.28}$$

Then the Unitary Operator corresponding to equation (14.6.25) will be

$$\hat{U}_n(\alpha_z) = e^{-(i/\hbar)\, \alpha_z \cdot \hat{J}_n}. \tag{14.6.29}$$

$$[\hat{J}_n, \hat{H}^o] = 0. \tag{14.6.30}$$

Thus *the total angular momentum, $\hat{J}$ of a particle having structure is conserved*.

14.7. TIME REVERSAL (TEMPORAL REFLECTION) INVARIANCE AND COMPLEX CONJUGATION:

14.7.1 Time Reversal, $\hat{\mathfrak{I}}$

Consider a structure less particle of mass m moving in a potential, V(r). Let a time reversal, which is a discrete transformation, τ is effected on the system as represented by

$$t \longrightarrow (\tau) \longrightarrow -t \tag{14.7.1}$$

The *Equation of Motion* of the particle is

$$(i\hbar)(\partial/\partial t)| \psi(\vec{r}, t) \rangle = \hat{H}| \psi(\vec{r}, t) \rangle. \tag{14.7.2}$$

$$(i\hbar \tfrac{\partial}{\partial t})| \psi(\vec{r}, t) \rangle = \left[-\tfrac{\hbar^2}{2m}\nabla^2 + V(r)\right]| \psi(\vec{r}, t) \rangle. \tag{14.7.3}$$

and $\quad (i\hbar \tfrac{\partial}{\partial t})| \psi(\vec{r}, -t) \rangle = \left[-\tfrac{\hbar^2}{2m}\nabla^2 + V(r)\right]| \psi(\vec{r}, -t) \rangle \tag{14.7.4}$

It follows that for coordinate space,

$$| \psi'(\vec{r}, t) \rangle = | \psi^*(\vec{r}, -t) \rangle = \hat{K}| \psi(\vec{r}, -t) \rangle. \tag{14.7.5}$$

Similarly, for momentum state (in k-space),

$$| \varphi'(p, t) \rangle = | \varphi^*(-p, -t) \rangle. \tag{14.7.6}$$

Here $\hat{K}$ denotes the operator for **complex conjugation**. $\hat{K}$ is anti-unitary and such that

$$\hat{K}^2 = \hat{\mathfrak{I}}; \text{ or } \hat{K} = \hat{K}^\dagger. \tag{14.7.7}$$

In general, however, Hamiltonian $\hat{H}$ is not real, and one supposes that there exists a unitary operator $\hat{U}_\tau$ such that

$$\hat{U}_\tau \hat{H}^* \hat{U}_\tau^\dagger = \hat{H} \tag{14.7.8}$$

$$\hat{U}_\tau \longrightarrow \hat{\mathfrak{I}};, \text{ when } \hat{H} \longrightarrow \text{Real}.$$

when $\quad (i\hbar \tfrac{\partial}{\partial t})| \psi^*(\vec{r}, -t) \rangle = \hat{H}^*| \psi^*(\vec{r}, -t) \rangle$

is operated on by $\hat{U}_\tau$, the solution is

$$|\psi'(\vec{r},t)\rangle = \hat{U}_\tau | \psi^*(\vec{r},-t)\rangle$$
$$= \hat{U}_\tau \hat{K} | \psi(\vec{r},-t)\rangle = \hat{\Im} | \psi(\vec{r},-t)\rangle \qquad (14.7.9)$$

Thus one gets $\hat{\Im} = \hat{U}_\tau \hat{K}$ = the **Time Reversal Operator**, which is anti-unitary.

If the intrinsic property of the particle is defined as spin, $\hat{s}$

$$\hat{s} = \tfrac{1}{2}\hbar\,\hat{\sigma} \qquad (14.7.10)$$

$\hat{\sigma}$ are **Pauli matrices**, then it can be shown that

$$\hat{\Im} = \left\{ e^{-(i/\hbar)\,\hat{\pi}\cdot\hat{s}} \right\} \hat{K} \qquad (14.7.11)$$

But $\qquad \vec{r}' = \hat{\Im}\,\vec{r}\,\hat{\Im}^\dagger,$ $\qquad\qquad\qquad\qquad\qquad (14.7.12)$

has to be obeyed, and

$$\vec{p}' = \hat{\Im}\,\vec{p}\,\hat{\Im}^\dagger = -\vec{p}, \qquad (14.7.13)$$

$$[\hat{U}_\tau, \vec{r}\,] = 0, \qquad (14.7.14)$$

$$[\hat{U}_\tau, \vec{p}\,] = 0, \qquad (14.7.15)$$

Further, $(i\hbar\tfrac{\partial}{\partial t})|\psi'(\vec{r},t)\rangle = \hat{\Im}\,\hat{H}\,\hat{\Im}^\dagger | \psi'(\vec{r},t)\rangle \qquad (14.7.16)$

This means $[\hat{\Im}, \hat{H}] = 0, \qquad\qquad\qquad\qquad (14.7.17)$

Thus <u>the conservation of Time Reversal $\hat{\Im}$ has to be obeyed. This law is violated only in weak interactions.</u>

14.7.2 TIME REVERSAL, $\hat{\Im}$ and COMPLEX CONJUGATION, $\hat{K}$

The time reversal operation $\hat{\Im}$ presents many analogues with complex conjugation. Two linear operators that are time reversal transforms one of the other is will henceforth be called COMPLEX CONJUGATES. The classical **Lagrangian**, $L(q',q)$ is a second-degree polynomial with respect to the velocities q'.

$$L(q',q) = L(-q',q)$$

Systems of isolated particle have always this symmetry property. In an external electric field (static) the same may be preserved. But a magnetic field introduces a linear coupling with respect to the velocities and, therefore, destroys it.

$$\therefore \quad \hat{H}(p,r) = \left[\frac{p^2}{2m} + V(r)\right] = \hat{H}(-p,r) \ .$$

Therefore, all solutions $\vec{r}(t)$ of the *Equations of Motion* are reversible with respect to time; defining

$$\vec{r}_{rev}(t) = \vec{r}(-t)$$

is also a solution of the Equation of Motion.

Consider the two cases,

$$\vec{r}_a(t) = \vec{r}_b(-t) \ ,$$

$$\vec{r}_a(t) = -\vec{r}_b(-t) \ ,$$

$$\vec{p}_{rev}(t) = -\vec{p}(-t) \ .$$

If $\hat{\mathfrak{J}}$ is the time-reversal operator, for angular momentum components

$$\hat{\mathfrak{J}} \ (\vec{r} \wedge \vec{p}) \ \hat{\mathfrak{J}}^\dagger = -(\vec{r} \wedge \vec{p}) \ . \tag{14.7.18}$$

For spin, $\hat{\mathfrak{J}} \ (s) \ \hat{\mathfrak{J}}^\dagger = -s$

$$\hat{\mathfrak{J}} \ (J) \ \hat{\mathfrak{J}}^\dagger = -J \ .$$

$\hat{\mathfrak{J}}$ is anti-linear. $\left| \ \psi(t) \ \right\rangle_{rev} (t) = \hat{\mathfrak{J}} \left| \ \psi(-t) \ \right\rangle$

14.8.1 CONSERVATION LAWS AND CONSTANTS OF THE MOTION

For many transformations, independent of time, the action of $\hat{T}$ (an associated operator) is defined independently of the law of motion of the state vector to which it is applied.

Example: If L is a spatial transformation, $\hat{T}$ is a certain function of the operators of reflection, infinitesimal translation and infinitesimal rotation, *i.e.* a certain function of $\Pi, J,$ and p; $\hat{T}$ therefore commutes with $\hat{\gamma}^o$.

Since $\hat{\gamma}^o\left((i\frac{\partial}{\partial t})-\hat{H}\right) \equiv \hat{B}(A) - m$

$[\hat{T}, \hat{B}(A)] = 0$

is equivalent to $[\hat{T}, \hat{H}] = 0$. (14.8.1)

Thus if potential $A_\mu(x)$ invariant under translation,

$$[\hat{p}, \hat{H}] = 0, \qquad (14.8.2)$$

and there is conservation of linear momentum, p.

If $A_\mu(x)$ is spherically symmetric,

$$[\hat{J}, \hat{H}] = 0, \qquad (14.8.3)$$

and the TOTAL ANGULAR MOMENTUM, $\bar{J}$ IS CONSERVED.

If $A_\mu(x)$ is invariant under reflection in the origin,

$$[\hat{\Pi}, \hat{H}] = 0, \qquad (14.8.4)$$

and the PARITY, $\hat{\Pi}$ is conserved.

These are listed in Table 14.1.

Table 14.1.

	Symmetry Transformation	Concerned Generator
(1)	Spatial Translation	Linear Momentum, p.
(2)	Rotation	Angular Momentum, J_n.
(3)	Space Inversion	Parity, $\hat{\Pi}$.
(4)	Time Translations	Hamiltonian, $\hat{H}$ (total energy).

REVIEW QUESTIONS

R.Q.14.1   A system is given to be symmetrical abo ut the X–axix. What is the observable that will be a constant of motion for the system?

R.Q. 14.2   Discuss the relationship between infinitesimal rotations and Angular momentum. Hence deduce the commutation rules for the components of the angular momentum.

&*&*&*&*&*&*&

Chapter 15

ANGULAR MOMENTUM – PROPERTIES

Chapter 15

ANGULAR MOMENTUM – PROPERTIES

"Whatever you see as duality is unreal" -   Adi Shankara
"The plurality tha we perceive is only an appearance., it is not real"
                                           - Erwin Schrodinger

15.1 INTRODUCTION

The physical significance of the quantum numbers ℓ and m_ℓ is yet to be explained. The fact that these were obtained in solving the angular part of the Schrödinger Equation suggests that they might have something to do with the Angular Momentum of an atom. To pursue this idea one translates the classical expressions of the angular momentum into the language of Quantum Mechanics. Angular Momentum is an important integral of motion in the Classical Mechanics of a particle in a central field. It also plays a role of great importance in Quantum Mechanics because of the invariance of the Hamiltonian, $\hat{H}$ for free atom (Chapter 14) to the group of rotations in the 3-D Space.

15.1.1 POLAR VECTORS: Classical Cartesian Components

In Classical Mechanics the linear momentum vector, $\vec{p}$ (Fig. 15.1) and radial vector, $\vec{r}$, relative to the origin O of a point mass (structure less) particle, are defined as below:
If $(\hat{i}, \hat{j}, \hat{k})$ are the three Unit Vectors along the x, y, z directions,

$$\vec{r} = (\hat{i}\, x_1 + \hat{j}\, y_1 + \hat{k}\, z_1), \quad \boxed{\text{Vector form}}. \quad (15.1.1)$$

$$\equiv \begin{pmatrix} x_1 \\ y_1 \\ z_1 \end{pmatrix} \quad \boxed{\text{Matrix form}} \quad (15.1.2)$$

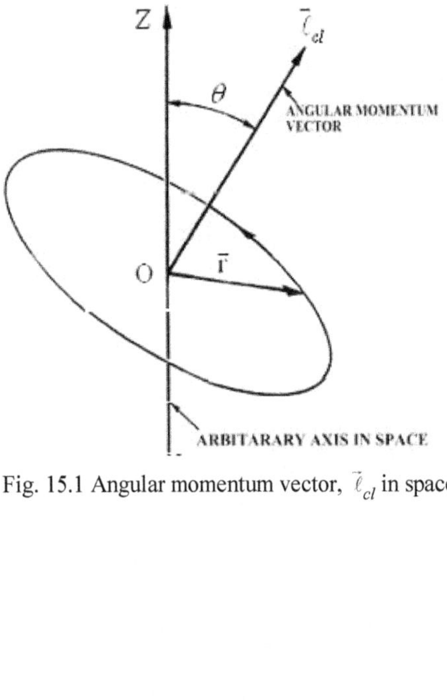

Fig. 15.1 Angular momentum vector, $\vec{\ell}_{cl}$ in space

$$\vec{p} = (\hat{i}\, p_x + \hat{j}\, p_y + \hat{k}\, p_z) \quad \boxed{\text{Vector form}} \qquad (15.1.3)$$

$$\equiv \begin{pmatrix} p_x \\ p_y \\ p_z \end{pmatrix} \quad \boxed{\text{Matrix form}} \qquad (15.1.4)$$

$\vec{r}$ and $\vec{p}$ = both are polar (ordinary vectors). They do change their sign under inversion. They are odd, *i.e.* classical operators.

15.1.2 The ANGULAR MOMENTUM OPERATOR: Axial vector: Classical Cartesian Components
In Classical Mechanics the Angular Momentum vector, $\vec{\ell}_{cl}$ around origin O of a *point mass* (structure less) particle, with Linear Momentum vector, $\vec{p}$ and radial vector, $\vec{r}$, relative to the origin, is defined as

$$\vec{\ell}_{cl} = \vec{r} \wedge \vec{p}, \qquad (15.1.5)$$
$$= (\hat{i}\, x_1 + \hat{j}\, y_1 + \hat{k}\, z_1) \wedge (\hat{i}\, p_x + \hat{j}\, p_y + \hat{k}\, p_z),$$

$$[\vec{r} \wedge \hat{p}] = \begin{vmatrix} i & j & k \\ x & y & z \\ \hat{p}_x & \hat{p}_y & \hat{p}_z \end{vmatrix} \quad \boxed{\text{determinant form}} \qquad (15.1.6)$$

$$\vec{\ell}_{cl} = \vec{r} \wedge \vec{p} \equiv [\hat{i}\,(y\,p_z - z\,p_y) + \hat{j}\,(z\,p_x - x\,p_z) + \hat{k}\,(x\,p_y - y\,p_x)] \quad \boxed{\text{Vector form}} \quad (15.1.7)$$

$$\equiv \begin{pmatrix} (y\,p_z - z\,p_y) \\ (z\,p_x - x\,p_z) \\ (x\,p_y - y\,p_x) \end{pmatrix} \quad \boxed{\text{Matrix form}} \qquad (15.1.8)$$

Here it should be understood that
$\vec{\ell}_{cl}$ = a PSEUDO-VECTOR (or AXIAL VECTOR).
It does not change sign under inversion and is an even, *ie.* vector operator.
$\vec{r}$ and $\vec{p}$ both are POLAR (ordinary vectors), that is why
$$\vec{\ell}_{cl} = \vec{r} \wedge \vec{p} = -\vec{p} \wedge \vec{r}.$$

Worked out Example 15.1

If $\vec{r} = (2\hat{i} + \hat{j})\, m$, and $\vec{p} = (5\hat{i} + 2\hat{j})$ kg m s^{-1}, Find $\vec{\ell}_{cl}$.

Solution: Step # 1 Given $\vec{\ell}_{cl} = \vec{r} \wedge \vec{p}$

Step # 2 $\quad [\vec{r} \wedge \hat{p}] = \begin{vmatrix} i & j & k \\ 2 & 1 & 0 \\ 5 & 2 & 0 \end{vmatrix}$

$\vec{\ell}_{cl} = [\hat{i}\,(1 \times 0 - 0 \times 2) + \hat{j}\,(0 \times 5 - 2 \times 0) + \hat{k}\,(2 \times 2 - 1 \times 5)] = -\hat{k}$ units ($kg\, m^2\, s^{-1}$).

Or, alternatively Step # 3 It is known that $[\vec{r} \wedge \hat{p}] = \begin{pmatrix} (y\,p_z - z\,p_y) \\ (z\,p_x - x\,p_z) \\ (x\,p_y - y\,p_x) \end{pmatrix}$

Step # 4 $= \begin{pmatrix} (5 \times 0 - 0 \times 2) \\ (0 \times 5 - 2 \times 0) \\ (2 \times 2 - 1 \times 5) \end{pmatrix} = \begin{pmatrix} 0 \\ 0 \\ -1 \end{pmatrix};$

Step # 5 i.e, $\vec{\ell}_{cl} = -\hat{k}$ units ($kg\ m^2\ s^{-1}$), meaning the $\vec{\ell}_{cl}$-vector is directed perpendicular to both the X-component of $\vec{r}$ and Y-component of $\vec{p}$ and in the negative Z-axis, having magnitude 1 unit of angular momentum.

15.1.3 Angular Momentum – QUANTUM MECHANICAL

By choosing a specific representation (Chapter 11) of the operators the equivalent equation of (15.1.8) may be written in Quantum Mechanics. By substituting in equation (15.1.8) the vector operators of $\hat{q}$ and $\hat{p}_q$, in <u>Position Representation</u> –

i.e. $\quad \hat{q} \longrightarrow \hat{q} = \{x, y, z\}$

and $\quad \hat{p}_q \longrightarrow (-i\hbar)(\partial/\partial \hat{q}) = (-i\hbar)\hat{\nabla}_q$. $\hfill (15.1.9)$

Now $\quad \vec{\ell}_{cl} = \vec{r} \wedge \vec{p}$,

To get the quantum mechanical equivalent of the classical $\vec{\ell}_{cl}$,

$$\vec{\ell} = \vec{r} \wedge (-i\hbar)\hat{\nabla}_q$$

$$\vec{\ell} = (\hat{i}\, x_1 + \hat{j}\, y_1 + \hat{k}\, z_1) \wedge (\hat{i}\, \hat{\nabla}_x + \hat{j}\, \hat{\nabla}_y + \hat{k}\, \hat{\nabla}_z)$$

$$\vec{\ell} = (-i\hbar) \begin{vmatrix} i & j & k \\ x & y & z \\ \hat{\nabla}_x & \hat{\nabla}_y & \hat{\nabla}_z \end{vmatrix} \quad \boxed{\text{determinant form}} \quad (15.1.10)$$

$$\vec{\ell} = (-i\hbar)[\hat{i}\,(y\,\hat{\nabla}_z - z\,\hat{\nabla}_y) + \hat{j}\,(z\,\hat{\nabla}_x - x\,\hat{\nabla}_z) + \hat{k}\,(x\,\hat{\nabla}_y - y\,\hat{\nabla}_x)]] \quad \boxed{\text{Vector form}} \quad (15.1.11)$$

$$\equiv (-i\hbar) \begin{pmatrix} (y\,\hat{\nabla}_z - z\,\hat{\nabla}_y) \\ (z\,\hat{\nabla}_x - x\,\hat{\nabla}_z) \\ (x\,\hat{\nabla}_y - y\,\hat{\nabla}_x) \end{pmatrix} \quad \boxed{\text{Matrix form}} \quad (15.1.12)$$

$$\equiv \begin{pmatrix} \hat{\ell}_x \\ \hat{\ell}_y \\ \hat{\ell}_z \end{pmatrix} \hfill (15.1.13)$$

$$\hat{\ell} = (\hat{i}\,\hat{\ell}_x + \hat{j}\,\hat{\ell}_y + \hat{k}\,\hat{\ell}_z) \hfill (15.1.14)$$

$$\boxed{\vec{\ell} = \vec{r} \wedge \vec{p} \equiv [\hat{i}\,(\hat{y}\,\hat{p}_z - \hat{z}\,\hat{p}_y) + \hat{j}\,(\hat{z}\,\hat{p}_x - \hat{x}\,\hat{p}_z) + \hat{k}\,(\hat{x}\,\hat{p}_y - \hat{y}\,\hat{p}_x)]}$$

where $\hat{\ell}_x$, $\hat{\ell}_y$, $\hat{\ell}_z$ are the Cartesian components of Angular Momentum along x, y, and z directions. These expressions show that they are moments of Linear Momentum about some point Further, it is known that (Chapter 4)

$$[\hat{z}, \hat{p}_z] = i\hbar \hfill (15.1.15)$$

15.1.4 CARTESIAN COMPONENTS of $\hat{\ell}$

In Position Representation, the Cartesian components, viz. $\hat{\ell}_x$, $\hat{\ell}_y$, $\hat{\ell}_z$ of Angular Momentum, $\hat{\ell}$ along x, y, and z directions are:

$$\hat{\ell}_x = (-i\hbar)\,(y\,\hat{\nabla}_z - z\,\hat{\nabla}_y),$$

$$\hat{\ell}_y = (-i\hbar)\,(z\,\hat{\nabla}_x - x\,\hat{\nabla}_z),$$

$$\hat{\ell}_z = (-i\hbar)\,(x\,\hat{\nabla}_y - y\,\hat{\nabla}_x). \hfill (15.1.16)$$

15.2 PROPERTIES OF ℓ-VECTORS

Many of the results can be derived by means of algebraic methods from Commutation Rules. A central aspect of operators is their commutation properties, whether observables can be simultaneously specified. They control the time evolution of the system.

15.2.1 COMMUTATION RELATIONS of the components of $\hat{\ell}$

Some of the mathematical properties of Angular Momentum operators, $\hat{\ell}$, will be examined first. A vector, ℓ, is called an Angular Momentum Operator if its components are observables which satisfy the commutations rules to be obtained as follows:

15.2.1.1 COMMUTATOR of the Operators $\hat{\ell}_x$, $\hat{\ell}_y$, and $\hat{\ell}_z$ in the Position Representation: $[\hat{\ell}_x, \hat{\ell}_y]$

The commutator can be obtained as in the worked out example that follows:

Worked out Example 15.2

Show that the value of $[\hat{\ell}_x, \hat{\ell}_y]$ is $+ i\hbar \hat{\ell}_z$

Solution: Step # 1 It is known that $[\hat{z}, \hat{p}_z] = i\hbar$, and

$$[\hat{\ell}_x, \hat{\ell}_y] = (\hat{\ell}_x\hat{\ell}_y - \hat{\ell}_y\hat{\ell}_x) = [(y p_z - z p_y), (z p_x - x p_z)]$$

Step # 2 Expanding the commutator,

$$[\hat{\ell}_x, \hat{\ell}_y] = [\hat{y}\hat{p}_z, \hat{z}\hat{p}_x] - [\hat{y}\hat{p}_z, \hat{x}\hat{p}_z] - [\hat{z}\hat{p}_y, \hat{z}\hat{p}_x] + [\hat{z}\hat{p}_y, \hat{x}\hat{p}_z]$$

$$= \hat{y}[\hat{p}_z, z]\hat{p}_x - [\hat{y}, \hat{x}]\hat{p}_z^2 - \hat{z}^2[\hat{p}_y, \hat{p}_x] + \hat{x}[\hat{z}, \hat{p}_z]\hat{p}_y.$$

Step # 3 Using $[\hat{z}, \hat{p}_z] = i\hbar$

$$[\hat{\ell}_x, \hat{\ell}_y] = \hat{y}\hat{p}_x(-i\hbar) - (0)\hat{p}_z^2 - \hat{z}^2(0) + \hat{x}\hat{p}_y(+i\hbar) = (+i\hbar)(\hat{x}\hat{p}_y - \hat{y}\hat{p}_x)$$

$$= + i\hbar \hat{\ell}_z = \text{another operator}.$$

∴ $\boxed{[\hat{\ell}_x, \hat{\ell}_y] = + i\hbar \hat{\ell}_z}$. (15.2.1)

i.e. $\boxed{[\hat{\ell}_y, \hat{\ell}_x] = - i\hbar \hat{\ell}_z}$ (15.2.1)

The commutators other than (15.2.1) can be derived in the same way, but one can realize easily that the operators are defined by cyclically permuting the subscripts successively. So the cyclic permutations of $\hat{\ell}_x$, $\hat{\ell}_y$, and $\hat{\ell}_z$ (by keeping in mind the cyclic orders of the indices of axes) will give the commutators:

$\boxed{[\hat{\ell}_y, \hat{\ell}_z] = + i\hbar \hat{\ell}_x}$. (15.2.2)

$\boxed{[\hat{\ell}_z, \hat{\ell}_x] = + i\hbar \hat{\ell}_y}$. (15.2.3)

Thus three operators, $\hat{\ell}_x$, $\hat{\ell}_y$, and $\hat{\ell}_z$ do not commute to one another (mutually), and these can not be measured with certainty (or specified) simultaneously
These three Commutation Rules may be combined in a single Law as

$\boxed{\hat{\ell} \wedge \hat{\ell} = + i\hbar \hat{\ell}}$. (15.2.4)

or, $\boxed{[\hat{\ell}_i, \hat{\ell}_j] = + i\hbar \hat{\ell}_k = + i\hbar \varepsilon_{ijk} \hat{\ell}_k}$. (15.2.5)

$$\boxed{\hat{\ell} \wedge \hat{\ell} \neq 0}$$

given by equation (15.2.4) shows that $\hat{\ell}$ is a VECTOR OPERATOR and NOT the classical vector.

15.2.1.2 RELATION between COMPONENTS of $\hat{\ell}$-vector and COORDINATES of particle: $[\hat{\ell}_x, \hat{y}]$, etc.

Unlike in Classical Mechanics where there are no restrictions on the magnitudes of $\hat{\ell}$, ω or $\hat{\ell}_x$, $\hat{\ell}_y$, and $\hat{\ell}_z$, in Quantum Mechanics the properties of observables are extensively modified.

eg. $[\hat{z}, \hat{p}_z] = +i\hbar$, etc.

Consider the commutator $[\hat{\ell}_x, \hat{y}]$

The commutator can be obtained as in the worked out example for $[\hat{\ell}_x, \hat{y}]$ that follows:

Worked out Example 15.3

Find out the value of $[\hat{\ell}_x, \hat{y}]$

Solution: $\boxed{\text{Step \# 1}}$ It is known that $[\hat{z}, \hat{p}_z] = i\hbar$, and

$$\vec{\ell} = [\hat{i}(\hat{y}\hat{p}_z - \hat{z}\hat{p}_y) + \hat{j}(\hat{z}\hat{p}_x - \hat{x}\hat{p}_z) + \hat{k}(\hat{x}\hat{p}_y - \hat{y}\hat{p}_x)]$$

$\boxed{\text{Step \# 2}}$ $[\hat{\ell}_x, \hat{y}] = [(\hat{y}\hat{p}_z - \hat{z}\hat{p}_y), \hat{y}] = [\hat{y}\hat{p}_z, \hat{y}] - [\hat{z}\hat{p}_y, \hat{y}]$

$= \hat{y}[\hat{p}_z, \hat{y}] - \hat{z}[\hat{p}_y, \hat{y}] = \hat{y}(0) - \hat{z}(-i\hbar) = +i\hbar \hat{z}$.

$$\boxed{[\hat{\ell}_x, \hat{y}] = +i\hbar \hat{z}} \tag{15.2.6}$$

Similarly one gets $\boxed{[\hat{\ell}_y, \hat{z}] = +i\hbar \hat{x}}$. \quad (15.2.7)

$\boxed{[\hat{\ell}_z, \hat{x}] = +i\hbar \hat{y}}$. \quad (15.2.8)

$\boxed{[\hat{\ell}_x, \hat{x}] = 0}$. \quad (15.2.9)

$\boxed{[\hat{\ell}_y, \hat{y}] = 0}$. \quad (15.2.10)

$\boxed{[\hat{\ell}_z, \hat{z}] = 0}$ \quad (15.2.11)

12.2.1.2 COMMUTATORS between the $\vec{\ell}$– and $\vec{p}$- vectors

Relations similar to (15.2.1) to (15.2.11) can be written between $\vec{\ell}$– and $\vec{p}$- vectors, by keeping in mind the cyclic orders of the indices of axes, viz. $x \to y$, $y \to z$, $z \to x$.

Thus $\boxed{[\hat{\ell}_x, \hat{p}_x] = 0}$. \quad (15.2.12)

$\boxed{[\hat{\ell}_x, \hat{p}_y] = +i\hbar \hat{p}_z}$. \quad (15.2.13)

$\boxed{[\hat{\ell}_x, \hat{p}_z] = -i\hbar \hat{p}_y}$ \quad (15.2.14)

The corresponding relations for $\hat{\ell}_y$ are

$\boxed{[\hat{\ell}_y, \hat{p}_x] = -i\hbar \hat{p}_z}$. \quad (15.2.15)

$$\boxed{[\hat{\ell}_y, \hat{p}_y] = 0}. \tag{15.2.16}$$

$$\boxed{[\hat{\ell}_y, \hat{p}_z] = +i\hbar \hat{p}_x}. \tag{15.2.17}$$

The commutators corresponding to $\hat{\ell}_z$ are

$$\boxed{[\hat{\ell}_z, \hat{p}_x] = +i\hbar \hat{p}_y}. \tag{15.2.18}$$

$$\boxed{[\hat{\ell}_z, \hat{p}_y] = -i\hbar \hat{p}_x}. \tag{15.2.19}$$

$$\boxed{[\hat{\ell}_z, \hat{p}_z] = 0}. \tag{15.2.20}$$

15.3 SHIFT OPERATORS (LADDER / STEP-UP AND STEP-DOWN / RAISING AND LOWERING / CONSTRUCTION AND DESCTRUCTION OPERATORS): $\hat{\ell}_+$ and $\hat{\ell}_-$

15.3.1 DEFINITIONS of $\hat{\ell}_+$ and $\hat{\ell}_-$

Instead of $\hat{\ell}_x$, $\hat{\ell}_y$, and $\hat{\ell}_z$ it is often more convenient to use two further operators $\hat{\ell}_+$ and $\hat{\ell}_-$, which are complex combinations of $\hat{\ell}_x$ and $\hat{\ell}_y$.

Step-up operator, $\boxed{\hat{\ell}_+ = \hat{\ell}_x + i\hat{\ell}_y}$, (15.3.1)

Step-down operator, $\boxed{\hat{\ell}_- = \hat{\ell}_x - i\hat{\ell}_y}$, (15.3.2)

such that $\boxed{\hat{\ell}_+ = (\hat{\ell}_-)^\dagger}$ (15.3.3)

$\hat{\ell}_+$ is the adjoint of $\hat{\ell}_-$ operator.

The inverse relations are useful and these are

$$\boxed{\hat{\ell}_x = \tfrac{1}{2}(\hat{\ell}_+ + \hat{\ell}_-)} \tag{15.3.4}$$

$$\boxed{\hat{\ell}_y = \tfrac{1}{2i}(\hat{\ell}_+ - \hat{\ell}_-)} \tag{15.3.5}$$

15.3.2 COMMUTATORS BETWEEN the $\hat{\ell}_z, \hat{\ell}_+$ and $\hat{\ell}_-$

Consider $[\hat{\ell}_z, \hat{\ell}_+]$ through the worked out example that follows:.

Worked out Example 15.4

What is the commutator between $\hat{\ell}_z$ and $\hat{\ell}_+$?

Solution: Step # 1 Given that $\hat{\ell}_\pm = \hat{\ell}_x \pm i\hat{\ell}_y$

Step # 2 $[\hat{\ell}_z, \hat{\ell}_+] = \hat{\ell}_z \hat{\ell}_+ - \hat{\ell}_+ \hat{\ell}_z = \hat{\ell}_z(\hat{\ell}_x + i\hat{\ell}_y) - (\hat{\ell}_x + i\hat{\ell}_y)\hat{\ell}_z$

$= (\hat{\ell}_z \hat{\ell}_x - \hat{\ell}_x \hat{\ell}_z) + i(\hat{\ell}_z \hat{\ell}_y - \hat{\ell}_y \hat{\ell}_z)$

$= [\hat{\ell}_z, \hat{\ell}_x] + i[\hat{\ell}_z, \hat{\ell}_y] = i\hbar\hat{\ell}_y + i(-i\hbar \hat{\ell}_x)$

Step # 3 $= i\hbar(\hat{\ell}_y - i\hat{\ell}_x) = \hbar(\hat{\ell}_x + i\hat{\ell}_y)$

$= +\hbar \hat{\ell}_+$

$$[\hat{\ell}_z,\hat{\ell}_+] = +\hbar\,\hat{\ell}_+ \qquad (15.3.6)$$

In the same way one can show that

$$[\hat{\ell}_z,\hat{\ell}_-] = -\hbar\,\hat{\ell}_- . \qquad (15.3.7)$$

$$[\hat{\ell}_+,\hat{\ell}_-] = +2\hbar\,\hat{\ell}_z . \qquad (15.3.8)$$

15.3.2 The SQUARE OF THE TOTAL ANGULAR MOMENTUM OPERATOR, $\hat{\ell}^2$ and the magnitude of $\hat{\ell}$.

Now, if one knows $\hat{\ell}_z$ (say) can one know the magnitude of Angular Momentum? In analogy with Classical Mechanics, from the square of the three components of $\hat{\ell}$, one can form the operator, $\hat{\ell}^2$:

$$\hat{\ell}^2 = \hat{\ell}_x^2 + \hat{\ell}_y^2 + \hat{\ell}_z^2 , \qquad (15.3.9)$$

where $\hat{\ell}^2$ is a *hermitian* operator. $\hat{\ell}^2$ is an additional property of Angular Momentum, *i.e.* the magnitude of $\hat{\ell}$.

$$|\hat{\ell}| = \sqrt{\hat{\ell}^2} = \sqrt{(\hat{\ell}\cdot\hat{\ell})} . \qquad (15.3.10)$$

$\hat{\ell}^2$ commutes with $\hat{\ell}_x, \hat{\ell}_y$ and $\hat{\ell}_z$ separately.

i.e.,
$$[\hat{\ell}^2,\hat{\ell}_x] = 0 \qquad (15.3.11)$$

$$[\hat{\ell}^2,\hat{\ell}_y] = 0 \qquad (15.3.12)$$

$$[\hat{\ell}^2,\hat{\ell}_z] = 0 . \qquad (15.3.13)$$

Further, $[\hat{\ell}^2,\hat{\ell}_\pm] = 0$. $\qquad (15.3.14)$

The three commutation rules (15.3.11) to (15.3.14) can be expressed in generalized form as

$$[\hat{\ell}^2,\hat{\ell}_i] = 0 . \qquad (15.3.14)$$

Or as $[\hat{\ell}_I,\hat{\ell}^2] = 0$, where $i = x, y, z$ and $\pm$. $\qquad (15.3.14)$

It is easily shown that

$$\hat{\ell}_x[\hat{\ell}_x,\hat{\ell}_z] = \hat{\ell}_x^2 \hat{\ell}_z - \hat{\ell}_x \hat{\ell}_z \hat{\ell}_x \qquad (15.3.15)$$

$$[\hat{\ell}_x,\hat{\ell}_z]\hat{\ell}_x = \hat{\ell}_x \hat{\ell}_z \hat{\ell}_x - \hat{\ell}_z \hat{\ell}_x^2 . \qquad (15.3.16)$$

$$[\hat{\ell}_x^2,\hat{\ell}_z] = \hat{\ell}_x[\hat{\ell}_x,\hat{\ell}_z] + [\hat{\ell}_x,\hat{\ell}_z]\hat{\ell}_x \qquad (15.3.17)$$

One gets the identities

$$\hat{\ell}_+\hat{\ell}_- = \hat{\ell}^2 - \hat{\ell}_z^2 + \hbar\,\hat{\ell}_z \qquad (15.3.18)$$

$$\hat{\ell}_-\hat{\ell}_+ = \hat{\ell}^2 - \hat{\ell}_z^2 - \hbar\,\hat{\ell}_z . \qquad (15.3.19)$$

i.e.,
$$\hat{\ell}^2 = \hat{\ell}_+\hat{\ell}_- + \hat{\ell}_z^2 - \hbar\,\hat{\ell}_z \qquad (15.3.20)$$

$$\hat{\ell}^2 = \hat{\ell}_-\hat{\ell}_+ + \hat{\ell}_z^2 + \hbar\,\hat{\ell}_z . \qquad (15.3.21)$$

It can be shown that $[\hat{\ell}^2, \hat{\ell}_z] = 0$

i.e., there are states of a system in which $\hat{\ell}^2$ and $\hat{\ell}_z$ are simultaneously specified.

15.4. ANGULAR MOMENTA IN SPHERICAL COORDINATES

To express the operators $\hat{\ell}_x, \hat{\ell}_y$ and $\hat{\ell}_z$ in terms of **r**, θ, φ, the following have to be found. There are various methods to arrive at the expressions. The one that appeals to the author is used below:

$$|\bar{r}| = r = \sqrt{(x^2 + y^2 + z^2)}, \tag{15.4.1}$$

$$\boxed{\begin{array}{l} x = r\, Sin\theta\, Cos\varphi \\ y = r\, Sin\theta\, Sin\varphi \\ z = r\, Cos\theta \end{array}} \tag{9.3.7}$$

$$\varphi = \tan^{-1}\frac{y}{x}. \tag{9.3.6}$$

$$d\tau = dx\, dy\, dz = (r^2 dr)(Sin\theta\, d\theta)(d\varphi)$$

$$\frac{\partial r}{\partial z} = \frac{\partial}{\partial z}\sqrt{(x^2+y^2+z^2)} = \frac{1}{2}2z(x^2+y^2+z^2)^{-1/2} = \frac{z}{r}$$

$$= \frac{r}{r} Cos\theta$$

$$\boxed{\frac{\partial r}{\partial z} = Cos\theta}. \tag{15.4.2}$$

$$\frac{\partial r}{\partial y} = \frac{\partial}{\partial y}\sqrt{(x^2+y^2+z^2)} = \frac{1}{2}2y(x^2+y^2+z^2)^{-1/2} = \frac{y}{r}$$

$$= \frac{r}{r} Sin\theta\, Sin\varphi$$

$$\boxed{\frac{\partial r}{\partial y} = Sin\theta\, Sin\varphi} \tag{15.4.3}$$

$$\frac{\partial r}{\partial x} = \frac{\partial}{\partial x}\sqrt{(x^2+y^2+z^2)} = \frac{1}{2}2x(x^2+y^2+z^2)^{-1/2} = \frac{x}{r}$$

$$= \frac{r}{r} Sin\theta\, Cos\varphi$$

$$\boxed{\frac{\partial r}{\partial x} = Sin\theta\, Cos\varphi} \tag{15.4.4}$$

$$\boxed{\frac{\partial \theta}{\partial x} = \frac{1}{r} Cos\theta\, Cos\varphi} \tag{15.4.5}$$

$$\boxed{\frac{\partial \theta}{\partial y} = \frac{1}{r} Cos\theta\, Sin\varphi} \tag{15.4.6}$$

$$\boxed{\frac{\partial \theta}{\partial z} = \frac{1}{r}(-Sin\theta)} \tag{15.4.7}$$

$$\boxed{\frac{\partial \varphi}{\partial x} = -\frac{1}{r}\frac{Sin\varphi}{Sin\theta}} \tag{15.4.8}$$

$$\hat{\ell}_x = (-i\hbar)(y\hat{\nabla}_z - z\hat{\nabla}_y),$$
$$\hat{\ell}_y = (-i\hbar)(z\hat{\nabla}_x - x\hat{\nabla}_z),$$
$$\hat{\ell}_z = (-i\hbar)(x\hat{\nabla}_y - y\hat{\nabla}_x). \tag{15.1.16}$$

$$\hat{\ell}_x = (-i\hbar)(y\hat{\nabla}_z - z\hat{\nabla}_y)$$

$$= (-i\hbar)\left\{ \begin{array}{l} (r\, Sin\theta\, Sin\varphi)\left[\frac{\partial r}{\partial z}\frac{\partial}{\partial r} + \frac{\partial \theta}{\partial z}\frac{\partial}{\partial \theta} + \frac{\partial \varphi}{\partial z}\frac{\partial}{\partial \varphi}\right] \\ -r\, Cos\theta\left[\frac{\partial r}{\partial y}\frac{\partial}{\partial r} + \frac{\partial \theta}{\partial y}\frac{\partial}{\partial \theta} + \frac{\partial \varphi}{\partial y}\frac{\partial}{\partial \varphi}\right] \end{array} \right\}$$

$$\hat{\ell}_x = (-i\hbar)\left\{(-\sin\theta)\frac{\partial}{\partial\theta} - \cot\theta\,\cos\varphi\frac{\partial}{\partial\varphi}\right\} \qquad (15.4.9)$$

$$\hat{\ell}_y = (-i\hbar)\left\{(-\cos\varphi)\frac{\partial}{\partial\theta} - \cot\theta\,\sin\varphi\frac{\partial}{\partial\varphi}\right\} \qquad (15.4.10)$$

$$\hat{\ell}_z = (-i\hbar)\left\{\frac{\partial}{\partial\varphi}\right\} \qquad (15.4.11)$$

$$\hat{\ell}^2 = \hat{\ell}_x{}^2 + \hat{\ell}_y{}^2 + \hat{\ell}_z{}^2 = \hat{\ell}_x\hat{\ell}_x + \hat{\ell}_y\hat{\ell}_y + \hat{\ell}_z\hat{\ell}_z,$$

Substituting and simplifying one gets

$$\boxed{\hat{\ell}^2 = (-\hbar^2)\left[\frac{1}{\sin\theta}\frac{\partial}{\partial\theta}\left\{\sin\theta\frac{\partial}{\partial\theta}\right\} + \frac{1}{\sin^2\theta}\frac{\partial^2}{\partial\theta^2}\right]} \qquad (15.4.12)$$

$$\boxed{\hat{\ell}_\pm = \hat{\ell}_x \pm i\,\hat{\ell}_y} \qquad (15.3.1)$$

on substitution for $\hat{\ell}_x$ and $\hat{\ell}_y$,

$$\boxed{\hat{\ell}_\pm = \hbar\,e^{\pm i\varphi}\left\{\pm\frac{\partial}{\partial\theta} + i\cot\theta\frac{\partial}{\partial\varphi}\right\}} \qquad (15.4.13)$$

15.5 SPECTRA OF OPERATORS $\hat{\ell}_z$ and $\hat{\ell}^2$

15.5.1 EIGEN VALUES OF ANGULAR MOMENTUM OPERATOR $\hat{\ell}_z$: (To show that m_ℓ is the eigen value of $\hat{\ell}_z$).

In order to determine the eigen values of the component, in some direction, of the angular momentum of a particle, it is convenient to use the expression for its operator in spherical coordinates, in Position Representation, taking the direction in question as the polar (z-) axis. Let the state function is

$$|\psi(\vec{r})\rangle \equiv |\psi(r,\theta,\varphi)\rangle = |f(r,\theta)\cdot\Phi(\varphi)\rangle \qquad (15.5.1)$$

where $f(r,\theta)$ is the arbitrary function of r and θ.

$$\hat{\ell}_z = (-i\hbar)\left\{\frac{\partial}{\partial\varphi}\right\} \qquad (15.4.11)$$

$$\hat{\ell}_z|\psi(r,\theta,\varphi)\rangle = (-i\hbar)\left\{\frac{\partial}{\partial\varphi}\right\}|\psi(r,\theta,\varphi)\rangle = (-i\hbar)\left\{\frac{\partial}{\partial\varphi}\right\}|f(r,\theta)\cdot\Phi(\varphi)\rangle$$

$$(-i\hbar)\left\{\frac{\partial}{\partial\varphi}\right\}|f(r,\theta)\cdot\Phi(\varphi)\rangle = \hat{\ell}_z|f(r,\theta)\cdot\Phi(\varphi)\rangle$$

$$-i\left\{\frac{d}{d\varphi}\right\}|\Phi(\varphi)\rangle = m|\Phi(\varphi)\rangle \qquad (15.5.2)$$

This satisfies requirements of an eigen value equation; and where m (more correctly, m_ℓ) is the eigen value of the operator $\hat{\ell}_z/\hbar$.

Solving (15.5.2) one gets

$$|\Phi(\varphi)\rangle = A\,e^{\pm im\varphi} \qquad (15.5.3)$$

On normalization,

$$|\Phi(\varphi)\rangle \equiv |m_\ell\rangle = \sqrt{\frac{1}{2\pi}}\,e^{\pm i|m|\varphi}, \qquad (8.3.14)$$

If this function satisfies the $\boxed{\text{POSTULATE \# 1}}$ (Chapter 4), then

$$|\Phi(\varphi)\rangle \equiv |\Phi(\varphi+2\pi)\rangle.$$

i.e,. $\quad A\,e^{\pm im\varphi} \equiv A\,e^{\pm im(\varphi+2\pi)}.$

This is possible only if $e^{\pm i m (\varphi+2\pi)} = 1$.

i.e., $\boxed{m \equiv m_\ell = 0, \pm 1, \pm 2, etc.}$

$\boxed{\hat{\ell}_z | m_\ell \rangle = m_\ell \hbar | m_\ell \rangle; \quad m \equiv m_\ell = 0, \pm 1, \pm 2, etc.}$. (15.5.4)

Thus the eigen values of $\hat{\ell}_z$ are integers, +ve or –ve. Since the direction of the z-axis is not wave distinct, it is clear that the same result is obtained for $\hat{\ell}_x$ and $\hat{\ell}_y$. In fact, however, the only common eigen function of the operators $\hat{\ell}_x$, $\hat{\ell}_y$ and $\hat{\ell}_z$ corresponds to the simultaneous values $\hat{\ell}_x = \hat{\ell}_y = \hat{\ell}_z = 0$, when $\hat{\ell} = 0$. If one of the operators $\hat{\ell}_x, \hat{\ell}_y$ and $\hat{\ell}_z \neq 0$, then $\hat{\ell}_x, \hat{\ell}_y$ and $\hat{\ell}_z$ have no common eigen function.

15.5.2 EIGEN VALUES of $\hat{\ell}_\pm$

$$\hat{\ell}_z = (-i\hbar)\left\{\frac{\partial}{\partial\varphi}\right\} \quad (15.4.11)$$

$$\hat{\ell}_\pm = \hbar e^{\pm i\varphi}\left\{\pm\frac{\partial}{\partial\theta} + i\cot\theta \frac{\partial}{\partial\varphi}\right\} \quad (15.4.13)$$

$$\hat{\ell}_z | m_\ell \rangle = m_\ell \hbar | m_\ell \rangle \quad (15.5.4)$$

$$Y_\ell^m(\theta,\varphi) = \Theta_\ell^m(\theta)\cdot\Phi(\varphi) \quad (8.7.1)$$

It is known that by operating on $Y_\ell^m(\theta,\varphi)$ by $\frac{\partial}{\partial\varphi}$,

$$\frac{\partial}{\partial\varphi}\left| Y_\ell^m(\theta,\varphi) \right\rangle = i m \left| Y_\ell^m(\theta,\varphi) \right\rangle .$$

The action of $\frac{\partial}{\partial\theta}$ on $\left| Y_\ell^m(\theta,\varphi) \right\rangle$ may be studied by using

$$\left| \ell m_\ell \right\rangle = \left| Y_\ell^m(\theta,\varphi) \right\rangle = (-1)^m \sqrt{\frac{(\ell+\frac{1}{2})}{4\pi}}\sqrt{\frac{(\ell-m)!}{(\ell+m)!}} P_\ell^m(\cos\theta) \cdot e^{\pm i |m|\varphi} \quad (8.7.2)$$

and the **Recurrence Relations** (satisfied by $P_\ell^m(\rho = \cos\theta)$)

$$(1-\rho^2)\frac{d}{d\rho}P_\ell^m(\rho) = (1-\rho^2)P_\ell^{m+1}(\rho) - m\rho P_\ell^m(\rho)$$

$$= \sqrt{(1-\rho^2)}(\ell+m)(\ell-m+1)P_\ell^m(\rho) + m\rho P_\ell^m(\rho) \quad (15.5.5)$$

satisfied by the *Associated Legendre Function* $P_\ell^m(\rho)$ s.

In this way it is found that

$$\boxed{\hat{\ell}_\pm | \ell m \rangle = \hbar\sqrt{[\ell(\ell+1) - m(m\pm 1)]} \, | \ell (m\pm 1) \rangle} \quad (15.5.6)a$$

$$\boxed{\hat{\ell}_\pm | \ell m \rangle = \hbar\sqrt{[(\ell\mp m) \cdot (\ell\pm m+1)]} \, | \ell (m\pm 1) \rangle} \quad (15.5.6)b$$

15.5.3 THE RESULT OF OPERATORS $\hat{\ell}_x$ and $\hat{\ell}_y$ on $| \ell m_\ell \rangle$

The result of the action of the operators $\hat{\ell}_x$ and $\hat{\ell}_y$ on $| \ell m_\ell \rangle$ is thus immediately obtained, as

$$\hat{\ell}_x = \frac{1}{2}(\hat{\ell}_+ + \hat{\ell}_-) \quad (15.3.4)$$

$$\hat{\ell}_y = \frac{1}{2i}(\hat{\ell}_+ - \hat{\ell}_-) \quad (15.3.5)$$

15.5.4 EIGEN VALUES $\ell (\ell+1) \hbar^2$ of the Angular Momentum Operator $\hat{\ell}^2$

15.5.4.1 EFFECT OF THE SHIFT OPERATORS and $\hat{\ell}^2$ on the state of the system:

Define first the quantum number m_ℓ through the statement

$$\hat{\ell}_z | \; m_\ell \; \rangle = m_\ell \hbar \; | \; m_\ell \; \rangle \qquad (15.5.4)$$

where $| \; m_\ell \; \rangle$ is the stationary states belonging to degenerate level and distinguished by *eigen value* $m_\ell \hbar$. Now let me operate on the equation (15.5.4) by operator $\hat{\ell}_+$,

$$\hat{\ell}_+ \left(\hat{\ell}_z | \; m_\ell \; \rangle \right) = \hat{\ell}_+ \left(m_\ell \hbar \; | \; m_\ell \; \rangle \right) = m_\ell \hbar \; \hat{\ell}_+ | \; m_\ell \; \rangle \qquad (15.5.6)$$

$$[\hat{\ell}_z, \hat{\ell}_+] = +\hbar \; \hat{\ell}_+. \qquad (15.3.6)$$

$$[\hat{\ell}_z, \hat{\ell}_+] | \; m_\ell \; \rangle = +\hbar \; \hat{\ell}_+ | \; m_\ell \; \rangle$$

i.e., $\quad \hat{\ell}_+ \hat{\ell}_z | \; m_\ell \; \rangle = \left(\hat{\ell}_z \hat{\ell}_+ - \hbar \; \hat{\ell}_+ \right) | \; m_\ell \; \rangle$

using equation (15.5.6) $\left(\hat{\ell}_z \hat{\ell}_+ - \hbar \; \hat{\ell}_+ \right) | \; m_\ell \; \rangle = m_\ell \hbar \left(\hat{\ell}_+ | \; m_\ell \; \rangle \right)$

$$\therefore \quad \hat{\ell}_z \left(\hat{\ell}_+ | \; m_\ell \; \rangle \right) = (m_\ell + 1)\hbar \left(\hat{\ell}_+ | \; m_\ell \; \rangle \right) \qquad (15.5.7)$$

This means that $\left(\hat{\ell}_+ | \; m_\ell \; \rangle \right)$ is, apart from a normalization constant, a new eigen state of $\hat{\ell}_z$ having the new eigen value $(m_\ell + 1)\hbar$.

Similarly, using $\quad [\hat{\ell}_z, \hat{\ell}_-] | \; m_\ell \; \rangle = -\hbar \; \hat{\ell}_- | \; m_\ell \; \rangle \qquad (15.3.7)$

one gets $\quad \hat{\ell}_z \left(\hat{\ell}_- | \; m_\ell \; \rangle \right) = (m_\ell - 1)\hbar \left(\hat{\ell}_- | \; m_\ell \; \rangle \right) \qquad (15.5.8)$

Since $[\hat{\ell}^2, \hat{\ell}_z] = 0, \qquad (15.3.13)$

one can write from (15.5.6)

$$| \; m_\ell \pm 1 \; \rangle = \text{a constant} \times \left(\hat{\ell}_\pm | \; m_\ell \; \rangle \right) \qquad (15.5.9)$$

$| \; \ell \; m_\ell \; \rangle$ is also an eigen state of $\hat{\ell}^2$ operator.

Let $\quad \hat{\ell}^2 | \; \ell \; m_\ell \; \rangle = \Lambda \hbar^2 | \; \ell \; m_\ell \; \rangle \qquad (15.5.10)$

where Λ is a dimensionless function of ℓ and m_ℓ, say. One has to examine if the new eigen state of $\hat{\ell}_z$ viz. $\left(\hat{\ell}_\pm | \; m_\ell \; \rangle \right)$, satisfies this equation as $[\hat{\ell}^2, \hat{\ell}_z] = 0$, has to be satisfied.

i.e. $\hat{\ell}^2$ and $\hat{\ell}_z$ can be taken as a compatible set, and $| \; \ell \; m_\ell \; \rangle$ as the eigen kets of $\hat{\ell}_z$ and $\hat{\ell}^2$ *simultaneously*.

15.5.4.2 To FIND THE LIMITATIONS ON THE NUMBER of states of the Angular Momentum:

Now $\quad (\hat{\ell}_x^2 + \hat{\ell}_y^2) | \; m_\ell \; \rangle = (\hat{\ell}^2 - \hat{\ell}_z^2) | \; m_\ell \; \rangle = (\Lambda - m_\ell^2) \hbar^2 | \; m_\ell \; \rangle$

Since $\hat{\ell}_x, \hat{\ell}_y, \hat{\ell}_z$ and $\hat{\ell}^2$ are hermitian operators, with real expectation values,

$\langle \hat{\ell}_x^2 \rangle, \langle \hat{\ell}_y^2 \rangle$ and $\langle \hat{\ell}_z^2 \rangle$ must be positive numbers.

Thus $\quad \Lambda > m_\ell^2 \qquad (15.5.11)$

This sets the limits on the number of times that either $\hat{\ell}_+$ or $\hat{\ell}_-$ may be applied to a given $| \; m_\ell \; \rangle$, and thus the set of eigen states must terminate with a minimum and a maximum eigen value of $\hat{\ell}_z$ for each ℓ -value. For convenience, let

$\ell_1 = m_{\max}$, and $\ell_2 = m_{\min}$, $\qquad (15.5.12)$

$\hat{\ell}_+ | \; \ell_1 \; \rangle = 0, \qquad (15.5.13)$

$\hat{\ell}_- | \; \ell_2 \; \rangle = 0 \qquad (15.5.14)$

Let n be the integer representing the number of steps between $| \; \ell_1 \; \rangle$ and $| \; \ell_2 \; \rangle$, *i.e.*,   # of steps above $|\ 0\rangle$ state $= \ell_1$,

        # of steps below $|\ 0\rangle$ state $= \ell_2$,

such that $n = m_{max} + m_{min} = \ell_1 + (-\ell_2) = \ell_1 - \ell_2 = 0$ or an integer.   (15.5.15)

Now, $\hat{\ell}_-\hat{\ell}_+ = (\hat{\ell}^2 - \hat{\ell}_x^2 - \hbar\hat{\ell}_z)$ (15.3.19)

$\hat{\ell}_-\hat{\ell}_+ |\ \ell_1\rangle = (\hat{\ell}^2 - \hat{\ell}_x^2 - \hbar\hat{\ell}_z) |\ \ell_1\rangle$

i.e., $(\hat{\ell}^2 - \hat{\ell}_z^2 - \hbar\hat{\ell}_z) |\ \ell_1\rangle = 0$

$(\Lambda - \hat{\ell}_1^2 - \hbar\hat{\ell}_1) |\ \ell_1\rangle = 0$.

Solving this,

$\Lambda = (\hat{\ell}_1^2 + \hat{\ell}_1) = \hat{\ell}_1(\hat{\ell}_1 + 1)$, and

$\Lambda = (\hat{\ell}_2^2 + \hat{\ell}_2) = \hat{\ell}_2(\hat{\ell}_2 + 1)$.

$\hat{\ell}^2 |\ \ell_1\rangle = \Lambda\hbar^2 |\ \ell_1\rangle = \hat{\ell}_1(\hat{\ell}_1 + 1)\hbar^2 |\ \ell_1\rangle$

$\hat{\ell}^2 |\ \ell_2\rangle = \Lambda\hbar^2 |\ \ell_2\rangle = \hat{\ell}_2(\hat{\ell}_2 + 1)\hbar^2 |\ \ell_2\rangle$

or, $\Lambda = \hat{\ell}_1(\hat{\ell}_1 + 1) = \hat{\ell}_2(\hat{\ell}_2 + 1)$ (15.5.16)

$[\hat{\ell}^2, \hat{\ell}_\pm] = 0$.   (15.3.14)

$\hat{\ell}^2 \hat{\ell}_+ |\ \ell\ m_\ell\rangle = \hat{\ell}_+ \hat{\ell}^2 |\ \ell\ m_\ell\rangle = \hat{\ell}_+ \Lambda\hbar^2 |\ \ell\ m_\ell\rangle$

Thus one can see that $\hat{\ell}_+$ is a <u>Step Up Operator</u>, if $|\ \ell\ m_\ell\rangle$ is an eigen ket of $\hat{\ell}_z$ and $\hat{\ell}^2$, operation on $|\ \ell\ m_\ell\rangle$ with $\hat{\ell}_+$ will generate a new eigen ket associated with the same eigen value of $\hat{\ell}^2$, but an eigen value of $\hat{\ell}_z$ plus one $\hbar$, *i.e.*, $(m_\ell + 1)\hbar$.

Writing again $\hat{\ell}_z (\hat{\ell}_+ |\ m_\ell\rangle) = (m_\ell + 1)\hbar\ (\hat{\ell}_+ |\ m_\ell\rangle)$ (15.5.7)

Operating again this equation by $\hat{\ell}_+$, one gets

$\hat{\ell}_+ [\hat{\ell}_z (\hat{\ell}_+ |\ m_\ell\rangle)] = \hat{\ell}_+ [(m_\ell + 1)\hbar\ (\hat{\ell}_+ |\ m_\ell\rangle)]$

$[\hat{\ell}_z, \hat{\ell}_+] = +\hbar\ \hat{\ell}_+$.   (15.3.6)

$\hat{\ell}_+ [\hat{\ell}_z (\hat{\ell}_+ |\ m_\ell\rangle)] = \hat{\ell}_+ [\hat{\ell}_+ \hat{\ell}_z |\ m_\ell\rangle] - \hbar\ (\hat{\ell}_+ |\ m_\ell\rangle) = \hat{\ell}_+^2 (\hat{\ell}_z |\ m_\ell\rangle) - \hbar(\hat{\ell}_+^2 |\ m_\ell\rangle)$

$= \hat{\ell}_+^2 (\hat{\ell}_z |\ m_\ell\rangle) - \hbar(\hat{\ell}_+^2 |\ m_\ell\rangle)$

$\hat{\ell}_+ \{(m_\ell + 1)\hbar(\hat{\ell}_+ |\ m_\ell\rangle)\} = (m_\ell + 1)\hbar(\hat{\ell}_+^2 |\ m_\ell\rangle)$

i.e., $\hat{\ell}_+^2 \hat{\ell}_z |\ m_\ell\rangle = (m_\ell + 2)\hbar(\hat{\ell}_+^2 |\ m_\ell\rangle)$ (15.5.17)

Also as $[\hat{\ell}^2, \hat{\ell}_+^2] = 0$,

$\hat{\ell}^2 (\hat{\ell}_+^2 |\ m_\ell\rangle) = \hat{\ell}_+^2 \hat{\ell}^2 |\ m_\ell\rangle = \Lambda\hbar^2 \hat{\ell}_+^2 |\ m_\ell\rangle$

Once again one gets a new simultaneous eigen ket of $\hat{\ell}_z$ and $\hat{\ell}^2$. Generalizing these results, one may get

$\hat{\ell}^2 (\hat{\ell}_\pm^2 |\ m_\ell\rangle) = \Lambda\hbar^2 \{\hat{\ell}_\pm^2 |\ m_\ell\rangle\}$ (15.5.18)

and (15.5.7) becomes

$$\boxed{\hat{\ell}_z (\hat{\ell}_\pm^S |\ m_\ell\rangle) = (m_\ell \pm s)\hbar\ (\hat{\ell}_+^S |\ m_\ell\rangle)}$$ (15.5.19)

$\Lambda = \hat{\ell}_1(\hat{\ell}_1 + 1) = \hat{\ell}_2(\hat{\ell}_2 + 1)$ (15.5.16)

This implies $|\hat{\ell}_1| = |\hat{\ell}_2| \equiv \ell$.
Or that $\hat{\ell}_1 = -(\hat{\ell}_2 + 1)$
This contradicts the positive value of $\hat{\ell}_1$ and $\hat{\ell}_2$ and so is rejected.
One has $n = \ell_1 - \ell_2 = 0$, or an integer.
So $\hat{\ell}_1 = \hat{\ell}_2$.
$$\Lambda = \ell(\ell+1) \tag{15.5.20}$$
and (15.5.10) becomes
$$\boxed{\hat{\ell}^2 | \ell, m_\ell \rangle = \ell(\ell+1) \hbar^2 | \ell, m_\ell \rangle} \tag{15.5.21}$$

15.5.4.3 PREDICTION OF HALF-INTEGRAL VALUES for ℓ in quantum systems:

Consider $n = m_{max} + m_{min} = \ell_1 + (-\ell_2) = \ell_1 - \ell_2 = 0$ or an integer. (15.5.15)
$n = \ell_1 - \ell_2 = \ell - \ell = 0$ or an integer.
$$\ell = \frac{n}{2} = 0, \frac{1}{2}, 1, \frac{3}{2}, 2, \frac{5}{2}, 3, \text{ etc }. \tag{15.5.22}$$

This means $\ell = \frac{n}{2}$ is an integer or a 'half-integer', i.e. ladder is symmetrical.
In classical angular momentum, ℓ = integer only is meaningful physically; so n = even integer. ℓ = maximum value of the observable component of total $\hat{\ell}^2$. Thus one can never measure a component of ℓ equal in magnitude to the total ℓ. $\hat{\ell}^2$ has eigen values ranging from $\ell_1 = \ell$, to $\ell_2 = -\ell$, i.e. $(2\ell+1)$ values.

$$\hat{\ell}_+ | \ell\, \ell \rangle = 0. \tag{15.5.23}$$
$$\hat{\ell}_- | \ell\, -\ell \rangle = 0. \tag{15.3.24}$$

15.5.4.4 EFFECT OF SHIFT OPERATORS- a Cosmic Game?:

Consider equation (15.5.19), viz.,
$$\hat{\ell}_z \left(\hat{\ell}_\pm^S | m_\ell \rangle \right) = (m_\ell \pm s) \hbar \left(\hat{\ell}_+^S | m_\ell \rangle \right)$$

Thus one can see that $\hat{\ell}_+$ is a Step Up Operator, if $| m_\ell \rangle$ is an eigen ket of both $\hat{\ell}_z$ and $\hat{\ell}^2$, operation on $| \ell\, m_\ell \rangle$ with $\hat{\ell}_+$ will generate a new eigen ket associated with the same eigen value of $\hat{\ell}^2$, but an eigen value of $\hat{\ell}_z$ plus one $\hbar$, i.e., $(m_\ell +1) \hbar$. That is, one unit of angular momentum is added to system when a Step Up operator operates on the system, or the effect of the Step Down Operator on the state is to create a state whose angular momentum is destroyed by one unit.
This is a very useful and interesting process of the shift operators on the state of the system. The original idea may be an excellent parallel to this aspect of the shift operators of Angular Momentum with the Hindu Mythological Game, very commonly played among children in India, on the "*Vaikunda Ekadesi Day*". The cosmic system has the '*Paramapada*', i.e., highest state $| \ell_1 \rangle = | \ell, +\ell \rangle = |$ Swarga (Lord's Feet $\rangle$, whereas its '*lowest state*'

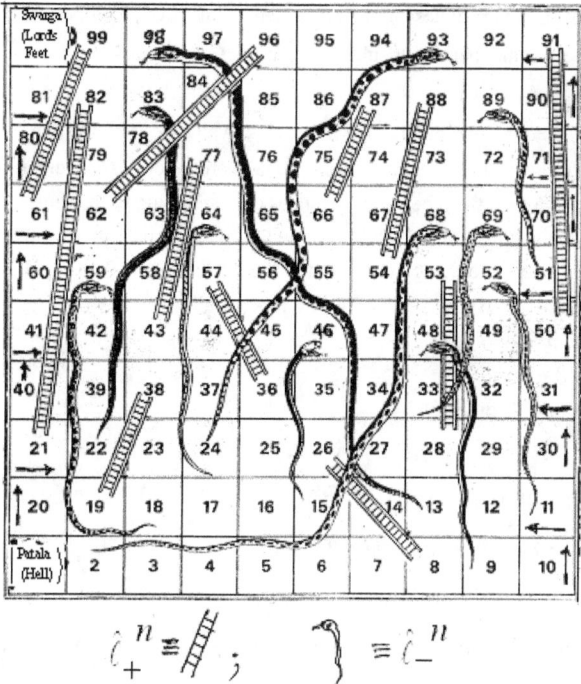

Fig. 15.3 Cosmic Game. 'Ladder' and 'Serpent' symbolize the two *Cosmic Operators*, $\hat{\ell}_+^n$ and $\hat{\ell}_-^n$; Uppermost state $|\ell_1\rangle \equiv |\ell, +\ell\rangle = |$ Swarga (Lord's Feet) $\rangle$; lowest state
$|\ell_2\rangle \equiv |\ell, -\ell\rangle = |$ Patala (Hell) $\rangle$.

$|\ell_2\rangle \equiv |\ell, -\ell\rangle = |$ Patala (Hell) $\rangle$; corresponding to the top or bottom of the ladder, respectively. 'Ladder' and 'serpent' symbolize the two COSMIC OPERATORS, $\hat{\ell}_+$ and $\hat{\ell}_-$. Here ladders of different lengths, varying in the number 's' of steps, represent the step up operators, $\hat{\ell}_+^n$; whereas the Step Down Operators $\hat{\ell}_-^n$ are symbolized by the serpents of different lengths. Starting with the player at the state $|$ Patala (Hell) $\rangle$ he climbs up determined by the dice that he plays, and he climbs down whenever he reaches the mouth of a serpent. The game is so common that there was a time when the cosmic chart appeared as advertisements of corn products in India by its manufacturer. The cosmic chart is reproduced in Fig. 15.3 for those interested.

15.6. SIMULTANEOUS EIGEN VALUES OF OPERATORS $\hat{\ell}^2$ and $\hat{\ell}_z$

15.6.1 To approach the problem

In the S-Picture the wave functions of a H-atom were found as solutions of the 2nd order partial differential equation (*viz.* the Schrödinger Equation). In this section solving a 1st order equation will discover them. This is very much simpler and corresponds to a technique used in the harmonic oscillator by operator method (Chapter 7).
Define first the quantum number m through the statement

$$\hat{\ell}_z | m_\ell \rangle = m_\ell \hbar | m_\ell \rangle \qquad (15.5.4)$$

where $|\,\ell\, m_\ell\,\rangle$ is the stationary states belonging to degenerate level and distinguished by eigen value $m_\ell \,\hbar$, of the z-component of angular momentum $\hat{\ell}_z$. Then assume

$$\hat{\ell}^2 |\,\ell\, m_\ell\,\rangle = \ell\,(\ell+1)\,\hbar^2 |\,\ell\, m_\ell\,\rangle \tag{15.5.21}$$

where ℓ = orbital quantum number, related to the eigen value of $\hat{\ell}^2$ by the expression $\ell\,(\ell+1)\,\hbar^2$. Thus $|\,\ell\, m_\ell\,\rangle$ is the eigen state of both $\hat{\ell}^2$ and $\hat{\ell}_z$. There are in total $(2\ell+1)$ eigen states; and if any one of these is known then the other 2ℓ eigen states corresponding to the same ℓ value can be obtained by repeated applications of $\hat{\ell}_+$ and $\hat{\ell}_-$. In particular, if the known state is either $|\,\ell,+\ell\,\rangle$ or $|\,\ell,-\ell\,\rangle$, corresponding to the top or bottom of the ladder, the repeated application of the same operator will generate the whole set of states. So it is sufficient to find $|\,\ell,+\ell\,\rangle$. It is evident, therefore, that one has to solve

$$\hat{\ell}_+ \,|\,\ell,+\ell\,\rangle = 0. \tag{15.5.23}$$

The explicit form of $\hat{\ell}_+$ is given by

$$\hat{\ell}_\pm = \hbar\, e^{\pm i\varphi} \left\{ \pm \frac{\partial}{\partial \theta} + i\,\mathrm{Cot}\theta\, \frac{\partial}{\partial \varphi} \right\} \tag{15.4.13}$$

15.6.2 To find out the UPPERMOST STATE $|\,\ell,+\ell\,\rangle$

$$\hat{\ell}_+ \,|\,\ell,+\ell\,\rangle = \hbar\, e^{+i\varphi} \left\{ +\frac{\partial}{\partial \theta} + i\,\mathrm{Cot}\theta\, \frac{\partial}{\partial \varphi} \right\} |\,\ell,+\ell\,\rangle = 0.$$

Or
$$\left\{ +\frac{\partial}{\partial \theta} + i\,\mathrm{Cot}\theta\, \frac{\partial}{\partial \varphi} \right\} |\,\ell,+\ell\,\rangle = 0 \tag{15.6.1}$$

To solve this partial differential equation the SEPARATION OF VARIABLES Technique is used.

Let $\quad |\,\ell,+\ell\,\rangle = |\,P(\theta)\,\rangle \cdot |\,\Phi(\varphi)\,\rangle \tag{15.6.2}$

$$\left\{ +\frac{\partial}{\partial \theta} + i\,\mathrm{Cot}\theta\, \frac{\partial}{\partial \varphi} \right\} |\,P(\theta)\,\rangle \cdot |\,\Phi(\varphi)\,\rangle = 0$$

$$\left\{ \frac{\partial}{\partial \theta} \right\} |\,P(\theta)\,\rangle \cdot |\,\Phi(\varphi)\,\rangle + \left\{ i\,\mathrm{Cot}\theta\, \frac{\partial}{\partial \varphi} \right\} |\,P(\theta)\,\rangle \cdot |\,\Phi(\varphi)\,\rangle = 0$$

i.e., $\quad \frac{1}{|\,P(\theta)\,\rangle}\left\{ \frac{\partial}{\partial \theta} \right\} |\,P(\theta)\,\rangle + \frac{1}{|\,\Phi(\varphi)\,\rangle}\left\{ i\,\mathrm{Cot}\theta\, \frac{\partial}{\partial \varphi} \right\} |\,\Phi(\varphi)\,\rangle = 0$

$$\frac{1}{\mathrm{Cot}\theta\,|\,P(\theta)\,\rangle}\left\{ \frac{d}{d\theta} \right\} |\,P(\theta)\,\rangle = -\frac{i}{|\,\Phi(\varphi)\,\rangle}\left\{ \frac{d}{d\varphi} \right\} |\,\Phi(\varphi)\,\rangle = k,\text{ a Separation Constant}.$$

Thus one gets two equations, viz.

$$\left\{ \frac{d|\,\Phi(\varphi)\,\rangle}{|\,\Phi(\varphi)\,\rangle} \right\} = ik\, d\varphi \tag{15.6.3}a$$

$$\left\{ \frac{d|\,P(\theta)\,\rangle}{|\,P(\theta)\,\rangle} \right\} = k\,\mathrm{Cot}\theta\, d\theta \tag{15.6.3}b$$

Integrating the two equations,

$$|\,\Phi(\varphi)\,\rangle = A\, e^{ik\varphi} \tag{15.6.4}$$

$$d|\,P(\theta)\,\rangle = |\,P(\theta)\,\rangle\, k\,\mathrm{Cot}\theta\, d\theta$$

$$d|\,P(\theta)\,\rangle = |\,P(\theta)\,\rangle\, k\,\frac{\mathrm{Cos}\theta}{\mathrm{Sin}\theta}\, d\theta$$

$$d|\,P(\theta)\,\rangle = |\,P(\theta)\,\rangle\, k\,\frac{\mathrm{Sin}^{k-1}\theta\, d\mathrm{Sin}\theta}{\mathrm{Sin}^k\theta}$$

$$\frac{d|\,P(\theta)\,\rangle}{|\,P(\theta)\,\rangle} = d[\log \mathrm{Sin}^k\theta],$$

$\log|P(\theta)\rangle = -\log Sin^k\theta$, and

$$|P(\theta)\rangle \propto Sin^k\theta, \qquad (15.6.5)$$

The complete solution is obtained by combining (15.6.4) and (15.6.5)

$$\boxed{|\ell,+\ell\rangle = |P(\theta)\rangle \cdot |\Phi(\varphi)\rangle = C_\ell^{+\ell} Sin^k\theta \, e^{ik\varphi}}. \qquad (15.6.6)$$

where $C_\ell^{+\ell}$ is normalization constant.

15.6.3 Find $C_\ell^{+\ell}$ is normalization constant

Since $\hat{\ell}_z |\ell,+\ell\rangle = \ell\hbar |\ell,+\ell\rangle$

and $\hat{\ell}^2 |\ell,+\ell\rangle = \ell(\ell+1)\hbar^2 |\ell,+\ell\rangle$,

It can be shown that $k = \ell$,

The complete form of $|\ell,+\ell\rangle$ is

$$\boxed{|\ell,\ell\rangle = |P(\theta)\rangle \cdot |\Phi(\varphi)\rangle = C_\ell^\ell Sin^\ell\theta \, e^{i\ell\varphi}}. \qquad (15.6.7)$$

15.6.4 To find out the 2ℓ eigen states corresponding to $m_\ell = (2\ell+1)$ values:

It is required to get use the step down operator on the state $|\ell,\ell\rangle$.

15.6.4.1 To find out the LOWERSTATE $|\ell, \ell$ -s$\rangle$:

It is known that

$$\hat{\ell}_- = \hbar e^{-i\varphi} \left\{ -\frac{\partial}{\partial\theta} + i\cot\theta \frac{\partial}{\partial\varphi} \right\} \qquad (15.4.13)$$

When $\hat{\ell}_-$ operates on a state it leaves the eigen value of $\hat{\ell}^2$ unchanged, and only steps down the state, as shown by the relation,

$$\hat{\ell}_- |\ell, m_\ell\rangle = \hbar\sqrt{[\ell(\ell+1) - m_\ell(m_\ell - 1)]} \, |\ell, (m_\ell-1)\rangle \qquad (15.5.6)$$

i.e., $\hat{\ell}_- |\ell, m_\ell\rangle = B_{\ell\ell} \hbar |\ell, (m_\ell-1)\rangle \qquad (15.6.8)$

where $B_{\ell\ell} = \sqrt{[\ell(\ell+1) - m_\ell(m_\ell-1)]}. \qquad (15.6.9)$

$|\ell,\ell\rangle = |C_\ell^\ell Sin^\ell\theta \, e^{i\ell\varphi}\rangle \qquad (15.6.7)$

$\therefore \quad \hat{\ell}_- |\ell,\ell\rangle = B_{\ell\ell} \hbar |\ell, (\ell-1)\rangle, \qquad (15.6.10)$

$B_{\ell\ell} = \sqrt{[\ell(\ell+1) - \ell(\ell-1)]} = \sqrt{2\ell}. \qquad (15.6.11)$

$B_{\ell\ell} \hbar |\ell, (\ell-1)\rangle = \hat{\ell}_- |C_\ell^\ell Sin^\ell\theta \, e^{i\ell\varphi}\rangle$.

$= \hbar e^{-i\varphi} \left\{ -\frac{\partial}{\partial\theta} + i\cot\theta \frac{\partial}{\partial\varphi} \right\} |C_\ell^\ell Sin^\ell\theta \, e^{i\ell\varphi}\rangle$

$= -C_\ell^\ell \hbar |e^{-i\varphi} \{\ell Sin^{\ell-1}\theta \cos\theta - i(i\ell)\cot\theta Sin^\ell\theta\} e^{i\ell\varphi}\rangle$.

$= -2C_\ell^\ell \hbar\ell |Sin^{\ell-1}\theta \cos\theta \, e^{i(\ell-1)\varphi}\rangle. \qquad (15.6.12)$

Using (15.6.10) and (15.6.11),

$|\ell, \ell-1\rangle = \frac{1}{\hbar\sqrt{2\ell}} \hat{\ell}_- |\ell,\ell\rangle$

$= -\sqrt{2\ell} \, C_\ell^\ell |Sin^{\ell-1}\theta \cos\theta \, e^{i(\ell-1)\varphi}\rangle$.

This leads to $\boxed{C_\ell^\ell = \frac{1}{\sqrt{2\ell}\,\ell!}\sqrt{\frac{2\ell+1}{4\pi}}}. \qquad (15.6.13)$

15.6.4.2 To find $|\ell, m_\ell\rangle$ (alternate approach)

$$|\ell, \ell\rangle = |C_\ell^\ell \operatorname{Sin}^\ell\theta\, e^{i\ell\varphi}\rangle. \tag{15.6.6}$$

$$\hat{\ell}_-|\ell, \ell\rangle = B_{\ell\ell}\,\hbar\,|\ell,(\ell-1)\rangle, \tag{15.6.10}$$

$$|\ell,(\ell-1)\rangle = \{C_\ell^{\ell-1}/C_\ell^\ell \hbar\}\hat{\ell}_-|\ell, \ell\rangle$$

$$= -C_\ell^{\ell-1} e^{-i\varphi}\left\{-\frac{\partial}{\partial\theta} + i\operatorname{Cot}\theta\,\frac{\partial}{\partial\varphi}\right\}|\operatorname{Sin}^\ell\theta\, e^{i\ell\varphi}\rangle$$

$$= (-1)\,C_\ell^{\ell-1}\left|\,e^{i(\ell-1)\varphi}\left\{\frac{d}{d\theta} + \ell\operatorname{Cot}\theta\right\}\operatorname{Sin}^\ell\theta\,\right\rangle$$

But $\quad\left\{\dfrac{d}{d\theta} + \ell\operatorname{Cot}\theta\right\}f(\theta) = \dfrac{d}{d\theta}[\operatorname{Sin}^\ell\theta\,f(\theta)]/\operatorname{Sin}^\ell\theta$

i.e., $\quad|\ell,(\ell-1)\rangle = (-1)\,C_\ell^{\ell-1}\left|\,\dfrac{1}{\operatorname{Sin}^\ell\theta}\,e^{-i(\ell-1)\varphi}\left\{\dfrac{d}{d\theta}\operatorname{Sin}^{2\ell}\theta\right\}\right\rangle$ (15.6.14)

For the sth application of $\hat{\ell}_-$,

$$|\ell,(\ell-s)\rangle = \left(\tfrac{1}{\hbar}\right)^s\left[C_\ell^{\ell-s}/C_\ell^\ell\right]\hat{\ell}_-^s|\ell,\ell\rangle$$

$$(-1)^s\,C_\ell^{\ell-s}\left|\,\dfrac{1}{\operatorname{Sin}^\ell\theta}\,e^{-i(\ell-s)\varphi}\left\{\dfrac{d^s}{d\theta^s}\operatorname{Sin}^{2\ell}\theta\right\}\right\rangle$$

$$= (-1)^s\,C_\ell^{\ell-s}\,\dfrac{1}{\operatorname{Sin}^{\ell-s}\theta}\,e^{-i(\ell-s)\varphi}\left\{\dfrac{1}{\operatorname{Sin}\theta}\,\dfrac{d^s}{d\theta^s}\operatorname{Sin}^{2\ell}\theta\right\}$$

Substituting $\quad\rho = \cos\theta,\ \dfrac{d}{d\theta} = -\operatorname{Sin}\theta\,\dfrac{d}{d\rho}$

Then $\quad|\ell,(\ell-s)\rangle = C_\ell^{\ell-s}\left[\dfrac{1}{1-\rho^2}\right]^{(\ell-s)/2} e^{-i(\ell-s)\varphi}\,\dfrac{d^s}{d\rho^s}(1-\rho^2)^\ell$

Putting $\quad\ell - s = m_\ell$

$$|\ell,(\ell-s)\rangle \equiv |\ell, m_\ell\rangle$$

$$= C_\ell^{m_\ell}\left[\dfrac{1}{1-\rho^2}\right]^{m_\ell/2} e^{-i m_\ell\varphi}\,\dfrac{d^{\ell-m_\ell}}{d\rho^{\ell-m_\ell}}(1-\rho^2)^\ell. \tag{15.6.15}$$

Defining the Associated Legendre Polynomials, $P_\ell^{m_\ell}(\rho)$

$$P_\ell^{m_\ell}(\rho) = (1-\rho^2)^{m_\ell/2}\,\dfrac{d^{m_\ell}}{d\rho^{m_\ell}}P_\ell(\rho);\qquad \ell \geq |m_\ell| = 0, 1, 2, 3, \ldots; . \tag{8.4.11}$$

$$P_\ell(\rho) = \dfrac{(-1)^\ell}{2^\ell\,\ell!}\,\dfrac{d^\ell}{d\rho^\ell}(1-\rho^2)^\ell. \tag{8.4.13}$$

This is known as Rodrigues's Formula.
$\rho = \pm 1$, means $\theta = 0$ or π.

$$P_\ell^{m_\ell}(\rho) = (-1)^{m_\ell}\left[\dfrac{1}{1-\rho^2}\right]^{m_\ell/2}\,\dfrac{d^{\ell-m_\ell}}{d\rho^{\ell-m_\ell}}(\rho^2 - 1)^\ell. \tag{15.6.16}$$

$$\boxed{\,|\ell, m_\ell\rangle = (-1)^{m_\ell}\,C_\ell^{m_\ell}\,P_\ell^{m_\ell}(\rho)\,e^{-i m_\ell\varphi}\,}. \tag{15.6.17}$$

REVIEW QUESTIONS

R.Q. 15.1   What does the quantity $(\hat{z}\hat{p}_x - \hat{x}\hat{p}_z)$ stands for?(Answer: $\hat{\ell}_y$)

R.Q. 15.2   Show that the value of $[\hat{\ell}_z, \hat{\ell}_x] = +i\hbar\hat{\ell}_y$

R.Q. 15.3   Find the value of $[\hat{\ell}_y, \hat{\ell}_z]$. (Answer: $+i\hbar\hat{\ell}_x$)

R.Q. 15.4   Evaluate the commutator $[\hat{\ell}_y, \hat{z}], [\hat{\ell}_x, \hat{x}]$ and $[\hat{x}, \hat{\ell}_z]$. (Answer: $+i\hbar\hat{x}$; 0 and $-i\hbar\hat{y}$)

R.Q. 15.5   Show that $\hat{\ell}_z$ and $\hat{z}$ commute mutually.

R.Q. 15.6   Do $\hat{x}$ and $\hat{\ell}_z$ commute each other? What about $\hat{\ell}_z$ and $\hat{\ell}_y$? (Answer: No, $[\hat{x}, \hat{\ell}_z] = -i\hbar\hat{y}$; $[\hat{\ell}_z, \hat{\ell}_y] = -i\hbar\hat{\ell}_x$).

R.Q. 15.7   Evaluate the commutator bracket $[\hat{\ell}_+, \hat{\ell}_-]$. (Answer: $+2\hbar\hat{\ell}_z$)

R.Q. 15.8   Find $[\hat{\ell}_x^2, \hat{\ell}_z]$. (Answer: $\hat{\ell}_x[\hat{\ell}_x, \hat{\ell}_z] + [\hat{\ell}_x, \hat{\ell}_z]\hat{\ell}_x$)

R.Q. 15.9   Prove that $\hat{\ell}^2$ commutes with $\hat{\ell}_z$. What does it signify? (Answer: there are states of a system in which $\hat{\ell}^2$ and $\hat{\ell}_z$ are simultaneously specified).

R.Q. 15.10   Show the identity $\hat{\ell}^2 = \hat{\ell}_+\hat{\ell}_- + \hat{\ell}_z^2 - \hbar\hat{\ell}_z$.

R.Q. 15.11   Given the eigen state $|\ell\, m_\ell\rangle$ of operators $\hat{\ell}^2$ and $\hat{\ell}_z$ having eigen values $\ell(\ell+1)\hbar^2$ and $m_\ell\hbar$, respectively. Show that $|\Theta\rangle = (\hat{\ell}_x + i\hat{\ell}_y)|\ell\, m_\ell\rangle$ is likewise an eigen ket of $\hat{\ell}^2$ and $\hat{\ell}_z$ and determine the eigen values. Show that if $\ell = 0$, the state $|\ell\, m_\ell\rangle$ is also an eigen state of $\hat{\ell}_x$ and $\hat{\ell}_y$

R.Q. 15.12   Show that the trace of the matrix that represents a component of angular momentum is zero.

&&*&*&*&*&*&*

Chapter 16

ANGULAR MOMENTUM – MATRICES
Explicit dealing of Orbital Angular Momentum

Chapter 16

ANGULAR MOMENTUM – MATRICES
Explicit dealing of Orbital Angular Momentum

We cannot predict what comes out of a singularity ... It is a disaster for science.
                                                                - Stephen Hawking

16.1   INTRODUCTION:

Any eigen state of a system is expressed in general form by $|\gamma \ell m_\ell \rangle$ where γ is some invisible quantum number of the system, and ℓ and m_ℓ are the usual orbital angular momentum and orbital magnetic quantum numbers. The eigen value of $\hat{\ell}^2$ is $\ell(\ell+1)\hbar^2$ according to equation (15.5.21) and that of $\hat{\ell}_z$ and $\hat{\ell}^2$ is $m_\ell \hbar$ (15.5.4), as usual; *i.e.* the spherical harmonics are simultaneous eigen states of $\hat{\ell}_z$ and $\hat{\ell}^2$.

It is now desired to derive the Matrix Representations of the angular momentum operators, discussed in Chapter 15. It will turn out that one can do this without ever specifying a particular set of states. One shall start from the algebraic (Commutation) Relations between the operators that were derived in Chapter 15.

Since the algebraic relations between operators remain valid for their Matrix Representations, it is known that the commutators must also be valid for the Angular Momentum Matrices.

$$[\hat{\ell}_x, \hat{\ell}_y] = +i\hbar \hat{\ell}_z \tag{15.2.1}$$

$$[\hat{\ell}_y, \hat{\ell}_z] = +i\hbar \hat{\ell}_x. \tag{15.2.2}$$

$$[\hat{\ell}_z, \hat{\ell}_x] = +i\hbar \hat{\ell}_y. \tag{15.2.3}$$

$$[\hat{\ell}^2, \hat{\ell}_x] = 0 \tag{15.3.11}$$

$$[\hat{\ell}^2, \hat{\ell}_y] = 0 \tag{15.3.12}$$

$$[\hat{\ell}^2, \hat{\ell}_z] = 0. \tag{15.3.13}$$

Now for all of the discussions in this Chapter, it will be convenient if *a Representation is chosen such that both* $\hat{\ell}_z$ *and* $\hat{\ell}^2$ *are* DIAGONAL *matrices*.

$$\hat{\ell}_z | m_\ell \rangle = m_\ell \hbar | m_\ell \rangle; \quad m \equiv m_\ell = 0, \pm 1, \pm 2, \text{etc.} \tag{15.5.4}$$

$$[\hat{\ell}^2, \hat{\ell}_\pm] = 0. \tag{15.3.14}$$

Also $[\hat{\ell}^2, \hat{\ell}_+^2] = 0,$

$$\hat{\ell}^2 | \ell, m_\ell \rangle = \ell(\ell+1)\hbar^2 | \ell, m_\ell \rangle \tag{15.5.21}$$

16.2   UNAMBIGUOUS DETERMINATIONS OF ANGULAR MOMENTUM MATRICES

It is very essential to assume that both $\hat{\ell}_z$ and $\hat{\ell}^2$ are diagonal and that the Commutation Relations (15.2.1) to (15.2.3) are valid.

16.2.1 To find the matrices for $\hat{\ell}_z$ and $\hat{\ell}^2$

Let $\{\,|\,\ell\, m_\ell\,\rangle\,\}$ form a complete set and form a basis.

$\hat{\ell}_z$ is hermitian and forms eigen value equation,

$$\hat{\ell}_z|\,\ell\, m_\ell\,\rangle = m_\ell\, h\,|\,\ell\, m_\ell\,\rangle \tag{15.5.4}$$

Matrix element of $\hat{\ell}_z$ is

$$\langle\,\ell',\, m_{\ell'}\,|\,\hat{\ell}_z\,|\,\ell,\, m_\ell\,\rangle = \langle\,\ell',\, m_{\ell'}\,|\,m_\ell\, h\,|\,\ell\, m_\ell\,\rangle$$
$$= m_\ell\, h\,\langle\,\ell',\, m_{\ell'}\,\|\,\ell\, m_\ell\,\rangle = m_\ell\, h\,\delta_{\ell'\ell}\cdot\delta_{m'm} \tag{16.2.1}$$

This matrix element shows that only diagonal elements, viz. those for which the Kronecker is non-vanishing, exist. The off-diagonal elements are characterized by either $\ell = \ell'$ or $m_\ell = m_{\ell'}$ (or both), which are, therefore, zero. Using the matrix element one can develop the Matrix Representation as shown below:

In the treatment given above, thus the diagonal elements are the eigen values of the operator. The index increases by one unit downward every $(2\ell + 1)$ rows. The index m_ℓ decreases by one unit on going from row to row, starting with $m_\ell = \ell$, in the upper row of each ℓ – row matrix.

$$\hat{\ell}_z = \begin{pmatrix} \text{matrix shown in image} \end{pmatrix} \hbar \tag{16.2.2}$$

(matrix structure: block-diagonal with subspace of $(2\ell+1) = 3$ Dimensions for $\ell=1$ with diagonal entries $+1, 0, -1$; subspace of $(2\ell+1) = 5$ Dimensions for $\ell=2$ with diagonal entries $+2, +1, 0, -1, -2$; subspace of $(2\ell+1) = 7$ Dimensions for $\ell=3$ with diagonal entries $+3, +2, +1, 0, -1, -2, -3$; plus the $\ell=0$ entry of 0)

Worked out Example 16.1

What is the matrix element $\langle\,0,\,0\,|\,\hat{\ell}_z\,|\,0,\,0\,\rangle$?

Solution: Step # 1: Given $\langle\,\ell',\, m_{\ell'}\,|\,\hat{\ell}_z\,|\,\ell,\, m_\ell\,\rangle = m_\ell\, h\,\delta_{\ell'\ell}\cdot\delta_{m'm}$; $\ell = 0$, $m_\ell = 0$;

Step # 2: $\langle\,0,\,0\,|\,\hat{\ell}_z\,|\,0,\,0\,\rangle = m_\ell\, h\,\delta_{\ell'\ell}\cdot\delta_{m'm} = 0$.

16.2.2 To find the MATRIX for $\hat{\ell}^2$

Let $\{\,|\,\ell,\, m_\ell\,\rangle\,\}$ is a complete set and form a basis.

$\hat{\ell}^2$ is hermitian and forms eigen value equation,

$$\hat{\ell}^2 | \ell, m_\ell \rangle = \ell(\ell+1) \hbar^2 | \ell, m_\ell \rangle \qquad (15.5.4)$$

Matrix element of $\hat{\ell}^2$ is

$$\langle \ell', m_\ell' | \hat{\ell}^2 | \ell, m_\ell \rangle = \langle \ell', m_\ell' | \ell(\ell+1) \hbar^2 | \ell, m_\ell \rangle$$
$$= \ell(\ell+1) \hbar^2 \delta_{\ell'\ell} \cdot \delta_{m'm} \qquad (16.2.3)$$

The matrix element shows that only diagonal elements, *viz.* those for which the Kronecker is non-vanishing, exist. The off-diagonal elements are characterized by either $\ell = \ell'$ or $m_\ell = m_\ell'$ (or both), which are, therefore, zero. Using the matrix elements one can develop the Matrix Representation as shown below:

$$\hat{\ell}^2 = \begin{pmatrix}
0 & 0 & 0 & 0 & 0 & 0 & 0 & 0 & 0 & 0 & 0 & 0 & 0 & 0 & 0 & 0 \\
0 & 2 & 0 & 0 & 0 & 0 & 0 & 0 & 0 & 0 & 0 & 0 & 0 & 0 & 0 & 0 \\
0 & 0 & 2 & 0 & 0 & 0 & 0 & 0 & 0 & 0 & 0 & 0 & 0 & 0 & 0 & 0 \\
0 & 0 & 0 & 2 & 0 & 0 & 0 & 0 & 0 & 0 & 0 & 0 & 0 & 0 & 0 & 0 \\
0 & 0 & 0 & 0 & 6 & 0 & 0 & 0 & 0 & 0 & 0 & 0 & 0 & 0 & 0 & 0 \\
0 & 0 & 0 & 0 & 0 & 6 & 0 & 0 & 0 & 0 & 0 & 0 & 0 & 0 & 0 & 0 \\
0 & 0 & 0 & 0 & 0 & 0 & 6 & 0 & 0 & 0 & 0 & 0 & 0 & 0 & 0 & 0 \\
0 & 0 & 0 & 0 & 0 & 0 & 0 & 6 & 0 & 0 & 0 & 0 & 0 & 0 & 0 & 0 \\
0 & 0 & 0 & 0 & 0 & 0 & 0 & 0 & 6 & 0 & 0 & 0 & 0 & 0 & 0 & 0 \\
0 & 0 & 0 & 0 & 0 & 0 & 0 & 0 & 0 & 12 & 0 & 0 & 0 & 0 & 0 & 0 \\
0 & 0 & 0 & 0 & 0 & 0 & 0 & 0 & 0 & 0 & 12 & 0 & 0 & 0 & 0 & 0 \\
0 & 0 & 0 & 0 & 0 & 0 & 0 & 0 & 0 & 0 & 0 & 12 & 0 & 0 & 0 & 0 \\
0 & 0 & 0 & 0 & 0 & 0 & 0 & 0 & 0 & 0 & 0 & 0 & 12 & 0 & 0 & 0 \\
0 & 0 & 0 & 0 & 0 & 0 & 0 & 0 & 0 & 0 & 0 & 0 & 0 & 12 & 0 & 0 \\
0 & 0 & 0 & 0 & 0 & 0 & 0 & 0 & 0 & 0 & 0 & 0 & 0 & 0 & 12 & 0 \\
0 & 0 & 0 & 0 & 0 & 0 & 0 & 0 & 0 & 0 & 0 & 0 & 0 & 0 & 0 & 12
\end{pmatrix} \hbar^2$$

(rows labeled by ℓ, m_ℓ: $\ell=0, m_\ell=0$; $\ell=1, m_\ell=+1,0,-1$; $\ell=2, m_\ell=+2,+1,0,-1,-2$; $\ell=3, m_\ell=+3,+2,+1,0,-1,-2,-3$. Columns labeled similarly by ℓ', m_ℓ'. The $\ell=1$ block is a Subspace of $(2\ell+1)=3$-Dimensions; the $\ell=3$ block is a Subspace of $(2\ell+1)=7$-Dimensions.)

(16.2.4)

As in the previous case, the diagonal elements are the eigen values of the operator. The index ℓ increases by one unit downward every $(2\ell+1)$ rows. The index m_ℓ decreases by one unit on going from row to row, starting with $m_\ell = \ell$, in the upper row of each ℓ-row matrix.

The matrices $\hat{\ell}_z$ and $\hat{\ell}^2$ have been evaluated in the Representation in which they have <u>diagonal form</u>.

16.2.3 MATRIX REPRESENTATIONS of $\hat{\ell}_+$ and $\hat{\ell}_-$

One still does not know the physical significance of $\hat{\ell}_+$ and $\hat{\ell}_-$ operators. So it is required that one investigate these operators. These two enable one to know the $\hat{\ell}_x$ and $\hat{\ell}_y$ operators. To find out the Representations of $\hat{\ell}_+$ and $\hat{\ell}_-$ is difficult. It is known that

$$\boxed{\hat{\ell}_\pm | \ell, m_\ell \rangle = \hbar \sqrt{[\ell(\ell+1) - m_\ell(m_\ell \pm 1)]} \; | \ell, (m_\ell \pm 1) \rangle} \qquad (15.5.6)\text{a}$$

$$\boxed{\hat{\ell}_\pm | \ell, m_\ell \rangle = \hbar \sqrt{[(\ell \mp m_\ell)\cdot(\ell \pm m_\ell + 1)]} \; | \ell, (m_\ell \pm 1) \rangle} \qquad (15.5.6)\text{b}$$

$$\hat{\ell}_- = \begin{pmatrix} m_\ell': & 0 & +1 & 0 & -1 & +2 & +1 & 0 & -1 & -2 & +3 & +2 & +1 & 0 & -1 & -2 & -3 \\ \ell': & 0 & & 1 & & & & 2 & & & & & & 3 & & & \end{pmatrix}$$

(Matrix 16.2.5 for $\hat{\ell}_-$ showing block-diagonal structure with subspaces of $(2\ell+1)=3$-Dimensions and $(2\ell+1)=7$-Dimensions, with nonzero entries $\sqrt{2}, \sqrt{2}$ in the $\ell=1$ block; $2, \sqrt{6}, \sqrt{6}, 2$ in the $\ell=2$ block; $\sqrt{6}, \sqrt{10}, \sqrt{12}, \sqrt{12}, \sqrt{10}, \sqrt{6}$ in the $\ell=3$ block, all multiplied by $\hbar$.)

(16.2.5)

Matrix elements of $\hat{\ell}_+$ and $\hat{\ell}_-$ are

$$\langle \ell', m_\ell' | \hat{\ell}_\pm | \ell, m_\ell \rangle = \langle \ell', m_\ell' | \hbar\sqrt{[(\ell \mp m_\ell)\cdot(\ell \pm m_\ell + 1)]} | \ell, (m_\ell \pm 1) \rangle$$
$$= \hbar\sqrt{[(\ell \mp m_\ell)\cdot(\ell \pm m_\ell + 1)]}\, \delta_{\ell'\ell}\cdot\delta_{m_\ell', m_\ell \pm 1}. \qquad (16.2.6)$$

$$\hat{\ell}_+ = \begin{pmatrix} m_\ell': & 0 & +1 & 0 & -1 & +2 & +1 & 0 & -1 & -2 & +3 & +2 & +1 & 0 & -1 & -2 & -3 \\ \ell': & 0 & & 1 & & & & 2 & & & & & & 3 & & & \end{pmatrix}$$

(Matrix 16.2.7 for $\hat{\ell}_+$ with analogous structure; upper-diagonal entries $\sqrt{2}, \sqrt{2}$ in $\ell=1$; $2, \sqrt{6}, \sqrt{6}, 2$ in $\ell=2$; $\sqrt{6}, \sqrt{10}, \sqrt{12}, \sqrt{12}, \sqrt{10}, \sqrt{6}$ in $\ell=3$; all multiplied by $\hbar$.)

(16.2.7)

The Kronecker, $\delta_{m_\ell' m_\ell \pm 1}$, indicates that all the diagonal elements adjacent to the leading (principal) diagonal of $\hat{\ell}_+$ and $\hat{\ell}_-$ matrices are non-zero. These are shown below:
$\hat{\ell}_+$ is, being the complex conjugate of $\hat{\ell}_-$, it is obtained by simply reflecting the elements of $\hat{\ell}_-$ across the diagonal. It is given below:

Worked out Example 16.2

Calculate $\langle \ell', m_\ell' |[\hat{\ell}_+, \hat{\ell}_-]| \ell, m_\ell \rangle$.

Solution: $\boxed{Step\ \#\ 1}$ It is known (15.3.8) that $[\hat{\ell}_+, \hat{\ell}_-] = +2\hbar \hat{\ell}_z$, and (15.5.4) that

$$\hat{\ell}_z | \ell\, m_\ell \rangle = m_\ell \hbar | \ell\, m_\ell \rangle$$

$\boxed{Step\ \#\ 2}$ $\langle \ell', m_\ell' |[\hat{\ell}_+, \hat{\ell}_-]| \ell, m_\ell \rangle$

$= \langle \ell', m_\ell' |(+2\hbar \hat{\ell}_z)| \ell, m_\ell \rangle = +2\hbar \langle \ell', m_\ell' | \hat{\ell}_z | \ell, m_\ell \rangle$

$\boxed{Step\ \#\ 3}$ $\langle \ell', m_\ell' |[\hat{\ell}_+, \hat{\ell}_-]| \ell, m_\ell \rangle = +2\hbar \langle \ell', m_\ell' | m_\ell \hbar | \ell\, m_\ell \rangle$

$= +2 m_\ell \hbar^2 \langle \ell', m_\ell' || \ell, m_\ell \rangle = +2 m_\ell \hbar^2 \delta_{\ell' \ell} \cdot \delta_{m_\ell' m_\ell}$

16.2.4 Formulation of matrices of the components of angular momentum, $\hat{\ell}_x$ and $\hat{\ell}_y$

It is known (15.3.4) and (15.3.5) that

$$\hat{\ell}_x = \tfrac{1}{2}(\hat{\ell}_+ + \hat{\ell}_-)$$

$$\hat{\ell}_y = \tfrac{1}{2i}(\hat{\ell}_+ - \hat{\ell}_-)$$

Thus using matrix additions of (16.2.5) and (16.2.6) according to the scheme above, one gets

$$\hat{\ell}_x = \tfrac{1}{2}\hbar \begin{pmatrix}
0 & 0 & 0 & 0 & 0 & 0 & 0 & 0 & 0 & 0 & 0 & 0 & 0 & 0 & 0 & 0 \\
0 & 0 & \sqrt{2} & 0 & 0 & 0 & 0 & 0 & 0 & 0 & 0 & 0 & 0 & 0 & 0 & 0 \\
0 & \sqrt{2} & 0 & \sqrt{2} & 0 & 0 & 0 & 0 & 0 & 0 & 0 & 0 & 0 & 0 & 0 & 0 \\
0 & 0 & \sqrt{2} & 0 & 0 & 0 & 0 & 0 & 0 & 0 & 0 & 0 & 0 & 0 & 0 & 0 \\
0 & 0 & 0 & 0 & 0 & 2 & 0 & 0 & 0 & 0 & 0 & 0 & 0 & 0 & 0 & 0 \\
0 & 0 & 0 & 0 & 2 & 0 & \sqrt{6} & 0 & 0 & 0 & 0 & 0 & 0 & 0 & 0 & 0 \\
0 & 0 & 0 & 0 & 0 & \sqrt{6} & 0 & \sqrt{6} & 0 & 0 & 0 & 0 & 0 & 0 & 0 & 0 \\
0 & 0 & 0 & 0 & 0 & 0 & \sqrt{6} & 0 & 2 & 0 & 0 & 0 & 0 & 0 & 0 & 0 \\
0 & 0 & 0 & 0 & 0 & 0 & 0 & 2 & 0 & 0 & 0 & 0 & 0 & 0 & 0 & 0 \\
0 & 0 & 0 & 0 & 0 & 0 & 0 & 0 & 0 & \sqrt{6} & 0 & 0 & 0 & 0 & 0 & 0 \\
0 & 0 & 0 & 0 & 0 & 0 & 0 & 0 & \sqrt{6} & 0 & \sqrt{10} & 0 & 0 & 0 & 0 & 0 \\
0 & 0 & 0 & 0 & 0 & 0 & 0 & 0 & 0 & \sqrt{10} & 0 & \sqrt{12} & 0 & 0 & 0 & 0 \\
0 & 0 & 0 & 0 & 0 & 0 & 0 & 0 & 0 & 0 & \sqrt{12} & 0 & \sqrt{12} & 0 & 0 & 0 \\
0 & 0 & 0 & 0 & 0 & 0 & 0 & 0 & 0 & 0 & 0 & \sqrt{12} & 0 & \sqrt{10} & 0 & 0 \\
0 & 0 & 0 & 0 & 0 & 0 & 0 & 0 & 0 & 0 & 0 & 0 & \sqrt{10} & 0 & \sqrt{6} & 0 \\
0 & 0 & 0 & 0 & 0 & 0 & 0 & 0 & 0 & 0 & 0 & 0 & 0 & \sqrt{6} & 0 & 0 \\
\end{pmatrix}$$

with rows/columns labeled by $\ell = 0, 1, 2, 3$ and $m_\ell' = 0, +1, 0, -1, +2, +1, 0, -1, -2, +3, +2, +1, 0, -1, -2, -3$.

Subspace of $(2\ell+1) = 3$-Dimensions

Subspace of $(2\ell+1) = 7$-Dimensions

(16.2.8)

$$\hat{\ell}_y = \frac{1}{2}\hbar i \begin{pmatrix} \ell'/\ell & m_\ell'/m_\ell & 0 & +1 & 0 & -1 & 2 & 1 & 0 & -1 & -2 & 3 & 2 & 1 & 0 & -1 & -2 & -3 \\ & & 0 & 1 & & & & 2 & & & & & & & 3 & & & \\ 0 & 0 & 0 & 0 & 0 & 0 & 0 & 0 & 0 & 0 & 0 & 0 & 0 & 0 & 0 & 0 & 0 & 0 \\ & +1 & 0 & 0 & -\sqrt{2} & 0 & 0 & 0 & 0 & 0 & 0 & 0 & 0 & 0 & 0 & 0 & 0 & 0 \\ 1 & 0 & 0 & \sqrt{2} & 0 & -\sqrt{2} & 0 & 0 & 0 & 0 & 0 & 0 & 0 & 0 & 0 & 0 & 0 & 0 \\ & -1 & 0 & 0 & \sqrt{2} & 0 & 0 & 0 & 0 & 0 & 0 & 0 & 0 & 0 & 0 & 0 & 0 & 0 \\ & +2 & 0 & 0 & 0 & 0 & 0 & -2 & 0 & 0 & 0 & 0 & 0 & 0 & 0 & 0 & 0 & 0 \\ & +1 & 0 & 0 & 0 & 0 & 2 & 0 & -\sqrt{6} & 0 & 0 & 0 & 0 & 0 & 0 & 0 & 0 & 0 \\ 2 & 0 & 0 & 0 & 0 & 0 & 0 & \sqrt{6} & 0 & -\sqrt{6} & 0 & 0 & 0 & 0 & 0 & 0 & 0 & 0 \\ & -1 & 0 & 0 & 0 & 0 & 0 & 0 & \sqrt{6} & 0 & -2 & 0 & 0 & 0 & 0 & 0 & 0 & 0 \\ & -2 & 0 & 0 & 0 & 0 & 0 & 0 & 0 & 2 & 0 & 0 & 0 & 0 & 0 & 0 & 0 & 0 \\ & +3 & 0 & 0 & 0 & 0 & 0 & 0 & 0 & 0 & 0 & -\sqrt{6} & 0 & 0 & 0 & 0 & 0 \\ & +2 & 0 & 0 & 0 & 0 & 0 & 0 & 0 & 0 & \sqrt{6} & 0 & -\sqrt{10} & 0 & 0 & 0 & 0 \\ & +1 & 0 & 0 & 0 & 0 & 0 & 0 & 0 & 0 & 0 & \sqrt{10} & 0 & -\sqrt{12} & 0 & 0 & 0 \\ 3 & 0 & 0 & 0 & 0 & 0 & 0 & 0 & 0 & 0 & 0 & 0 & \sqrt{12} & 0 & -\sqrt{12} & 0 & 0 \\ & -1 & 0 & 0 & 0 & 0 & 0 & 0 & 0 & 0 & 0 & 0 & 0 & \sqrt{12} & 0 & -\sqrt{10} & 0 \\ & -2 & 0 & 0 & 0 & 0 & 0 & 0 & 0 & 0 & 0 & 0 & 0 & 0 & \sqrt{10} & 0 & -\sqrt{6} \\ & -3 & 0 & 0 & 0 & 0 & 0 & 0 & 0 & 0 & 0 & 0 & 0 & 0 & 0 & \sqrt{6} & 0 \end{pmatrix}$$

Subspace of $(2\ell+1) = 7$-Dimensions

(16.2.9)

Worked out Example 16.3

Write the matrix representation of $\hat{\ell}_x$ for $\ell = 3$.

Solution: Step # 1 | $\ell = 3$; The matrix will have $(2\ell + 1) = 7$-Dimensions

Step # 2 | $\hat{\ell}_x = \frac{1}{2}\hbar \begin{bmatrix} 0 & \sqrt{6} & 0 & 0 & 0 & 0 & 0 \\ \sqrt{6} & 0 & \sqrt{10} & 0 & 0 & 0 & 0 \\ 0 & \sqrt{10} & 0 & \sqrt{12} & 0 & 0 & 0 \\ 0 & 0 & \sqrt{12} & 0 & \sqrt{12} & 0 & 0 \\ 0 & 0 & 0 & \sqrt{12} & 0 & \sqrt{10} & 0 \\ 0 & 0 & 0 & 0 & \sqrt{10} & 0 & \sqrt{6} \\ 0 & 0 & 0 & 0 & 0 & \sqrt{6} & 0 \end{bmatrix}$

Worked out Example 16.4

Write the matrix representation of $\hat{\ell}_x, \hat{\ell}_y$ and $\hat{\ell}_z$ for $\ell = 2$. Using matrix algebra, verify that the these matrices satisfy the equation $\hat{\ell}^2 = \hat{\ell}_x^2 + \hat{\ell}_y^2 + \hat{\ell}_z^2$

Solution: Step # 1 | $\ell = 2$

Step # 2

$\hat{\ell}_x = \frac{1}{2}\hbar \begin{bmatrix} 0 & 2 & 0 & 0 & 0 \\ 2 & 0 & \sqrt{6} & 0 & 0 \\ 0 & \sqrt{6} & 0 & \sqrt{6} & 0 \\ 0 & 0 & \sqrt{6} & 0 & 2 \\ 0 & 0 & 0 & 2 & 0 \end{bmatrix}$; $\hat{\ell}_y = \frac{1}{2}\hbar \begin{bmatrix} 0 & -2 & 0 & 0 & 0 \\ +2 & 0 & -\sqrt{6} & 0 & 0 \\ 0 & +\sqrt{6} & 0 & -\sqrt{6} & 0 \\ 0 & 0 & +\sqrt{6} & 0 & -2 \\ 0 & 0 & 0 & +2 & 0 \end{bmatrix}$; $\hat{\ell}_z = \hbar \begin{bmatrix} +2 & 0 & 0 & 0 & 0 \\ 0 & +1 & 0 & 0 & 0 \\ 0 & 0 & 0 & 0 & 0 \\ 0 & 0 & 0 & -1 & 0 \\ 0 & 0 & 0 & 0 & -2 \end{bmatrix}$

Step # 3 Squaring each $\hat{\ell}_x$, $\hat{\ell}_y$ and $\hat{\ell}_z$ and adding them together,

$$\hat{\ell}_x^2 = \tfrac{1}{2}\hbar \begin{bmatrix} 0 & 2 & 0 & 0 & 0 \\ 2 & 0 & \sqrt{6} & 0 & 0 \\ 0 & \sqrt{6} & 0 & \sqrt{6} & 0 \\ 0 & 0 & \sqrt{6} & 0 & 2 \\ 0 & 0 & 0 & 2 & 0 \end{bmatrix} \times \tfrac{1}{2}\hbar \begin{bmatrix} 0 & 2 & 0 & 0 & 0 \\ 2 & 0 & \sqrt{6} & 0 & 0 \\ 0 & \sqrt{6} & 0 & \sqrt{6} & 0 \\ 0 & 0 & \sqrt{6} & 0 & 2 \\ 0 & 0 & 0 & 2 & 0 \end{bmatrix} = \tfrac{1}{4}\hbar^2 \begin{bmatrix} 4 & 0 & 2\sqrt{6} & 0 & 0 \\ 0 & 10 & 0 & 6 & 0 \\ 2\sqrt{6} & 0 & 12 & 0 & 0 \\ 0 & 6 & 0 & 10 & 0 \\ 0 & 0 & 0 & 0 & 4 \end{bmatrix}$$

Step # 4 Similarly for $\hat{\ell}_y^2$ and $\hat{\ell}_z^2$

Step # 5 Adding the three squared matrices

$$\hat{\ell}^2 = \hat{\ell}_x^2 + \hat{\ell}_y^2 + \hat{\ell}_z^2 = \hbar^2 \begin{bmatrix} +6 & 0 & 0 & 0 & 0 \\ 0 & +6 & 0 & 0 & 0 \\ 0 & 0 & +6 & 0 & 0 \\ 0 & 0 & 0 & +6 & 0 \\ 0 & 0 & 0 & 0 & +6 \end{bmatrix}$$

REVIEW QUESTIONS

R.Q. 16.1 What is the matrix element $\langle 2, 0 | \hat{\ell}_z | 2, 2 \rangle$? (Answer: $2\hbar$)

R.Q. 16.2 Calculate $\langle \ell', m_\ell' | [\hat{\ell}_z, \hat{\ell}_-] | \ell, m_\ell \rangle$. (Answer:

$-\hbar^2 \sqrt{[(\ell \mp m_\ell)\cdot(\ell \pm m_\ell + 1)]}\ \delta_{\ell'\ell}\cdot\delta_{m_\ell' m_\ell \pm 1}$, or

$-\hbar^2 \sqrt{[\ell(\ell+1) - m_\ell(m_\ell \pm 1)]}\ \delta_{\ell'\ell}\cdot\delta_{m_\ell' m_\ell \pm 1}$).

R.Q. 16.3 Calculate $\langle 2, 1 | \hat{\ell}_+ | 2, 0 \rangle$. (Answer: $\hbar\sqrt{6}$)

R.Q. 16.4 Calculate $\langle 2, 2 | \hat{\ell}_+^2 | 2, 0 \rangle$. (Answer: $\hbar\sqrt{24}$)

R.Q. 16.5 Calculate $\langle 2, 0 | \hat{\ell}_+\hat{\ell}_- | 2, 0 \rangle$. (Answer: $6\hbar^2$)

R.Q. 16.6 Calculate $\langle 2, 0 | \hat{\ell}_-^2 \hat{\ell}_z \hat{\ell}_+^2 | 2, 0 \rangle$. (Answer: $48\hbar^5$)

R.Q. 16.7 Operate on $| 2, 2 \rangle$ with $\hat{\ell}_-$ successively to obtain $| 2, 0 \rangle$. (Answer: $\hat{\ell}_-^2 | 2, 2 \rangle = \hbar^2 \sqrt{24} | 2, 0 \rangle$).

R.Q. 16.8 The Hamiltonian of a system is given by $\hat{H} = \frac{1}{2I_1}\hat{\ell}_x^2 + \frac{1}{2I_2}\hat{\ell}_y^2 + \frac{1}{2I_3}\hat{\ell}_z^2$,

Find the eigen values of $\hat{H}$, when $\ell = 2$. I-s are the moments of inertia of the system along specified axes. Take $I_1 = I_2 = I_3$ (Answer:

$$\hat{H} = \tfrac{3}{4}\hbar^2 \begin{bmatrix} +6 & 0 & 0 & 0 & 0 \\ 0 & +6 & 0 & 0 & 0 \\ 0 & 0 & +6 & 0 & 0 \\ 0 & 0 & 0 & +6 & 0 \\ 0 & 0 & 0 & 0 & +6 \end{bmatrix}).$$

R.Q. 16.9 The Hamiltonian of a system is given by $\hat{H} = \frac{1}{2I_1}\hat{\ell}_x^2 + \frac{1}{2I_2}\hat{\ell}_y^2 + \frac{1}{2I_3}\hat{\ell}_z^2$,

Find the eigen values of $\hat{H}$, when $\ell = 2$. I-s are the moments of inertia of the system along specified axes. Take $I_1 = I_2 = I_3$ (Answer:

$$\hat{H} = \tfrac{3}{4}\hbar^2 \begin{bmatrix} +6 & 0 & 0 & 0 & 0 \\ 0 & +6 & 0 & 0 & 0 \\ 0 & 0 & +6 & 0 & 0 \\ 0 & 0 & 0 & +6 & 0 \\ 0 & 0 & 0 & 0 & +6 \end{bmatrix}).$$

R.Q. 16.10   If $|w\rangle = \sqrt{\tfrac{1}{26}}\begin{pmatrix} 1 \\ 4 \\ -3 \end{pmatrix}$ is the state vector of a system with Angular Momentum 1 unit, what is the probability that a measurement $\hat{\ell}_x$ yields the value 0?

(Answer; $\ell = 1$; $\hat{\ell}_x = \tfrac{1}{2}\hbar\begin{pmatrix} 0 & \sqrt{2} & 0 \\ \sqrt{2} & 0 & \sqrt{2} \\ 0 & \sqrt{2} & 0 \end{pmatrix}$; $\langle \hat{\ell}_x \rangle = \langle w | \hat{\ell}_x | w \rangle$)

R.Q. 16.11   Determine the matrix representation of $(\hat{\ell}_x \hat{\ell}_y + \hat{\ell}_y \hat{\ell}_x)$ for a system with $\ell = 1$ (Answer: .

$$(\hat{\ell}_x \hat{\ell}_y + \hat{\ell}_y \hat{\ell}_x) = \tfrac{1}{4i}(\hat{\ell}_- \hat{\ell}_+ + \hat{\ell}_+ \hat{\ell}_-) ; \quad = \tfrac{1}{2i}\hbar^2 \begin{pmatrix} 1 & 0 & 0 \\ 0 & 2 & 0 \\ 0 & 0 & 1 \end{pmatrix}).$$

R.Q. 16.12   Determine the matrices of the operators $\hat{\ell}_x, \hat{\ell}_y$ and $\hat{\ell}_z$ in the basis formed by the set of states given below: $\tfrac{3}{8\pi} P_1^1(\rho) \cdot e^{i\varphi}$; $\tfrac{3}{8\pi} P_1(\rho)$; $\tfrac{3}{8\pi} P_1^1(\rho) \cdot e^{-i\varphi}$.

R.Q. 16.13   Evaluate the matrix element of $\langle \ell, m_\ell +2 | [\hat{\ell}_x] | \ell, m_\ell \rangle$, $\hat{\ell}_x$ for $\ell = 1$, $m_\ell = -1$. (Answer:

$\langle \ell, m_\ell +2 | [\hat{\ell}_x] | \ell, m_\ell \rangle = \tfrac{1}{2}\langle \ell, m_\ell +2 | \hbar \sqrt{[\ell(\ell+1) - m_\ell(m_\ell -1)]} | \ell, m_\ell -1 \rangle; \sqrt{\tfrac{1}{2}}\hbar$).

R.Q. 16.14   Find the Unitary Matrix $\hat{U}$ that will diagonalize $\hat{\ell}_x$. Perform this for a 3x3 matrix. (Answer: $\ell = 1$;

$\hat{\ell}_x = \tfrac{1}{2}\hbar\begin{pmatrix} 0 & \sqrt{2} & 0 \\ \sqrt{2} & 0 & \sqrt{2} \\ 0 & \sqrt{2} & 0 \end{pmatrix}$; $\lambda_1 = -9, \lambda_2 = +9, \lambda_3 = +18$

R.Q. 16.15   Show that $[\hat{\ell}_x, \hat{r}^2] = [\hat{\ell}_z, \hat{r}^2]$, where $\hat{r}^2 = x^2 + y^2 + z^2$. (Answer:

$[\hat{\ell}_x, \hat{y}^2] = +i\hbar\,\hat{z}\,\hat{y}; [\hat{\ell}_x, \hat{z}^2] = -i\hbar\,\hat{y}\,\hat{z}$; $[\hat{\ell}_z, \hat{x}^2] = +i\hbar\,\hat{y}\,\hat{x}; [\hat{\ell}_z, \hat{y}^2] = -i\hbar\,\hat{x}\,\hat{y}; [\hat{\ell}_z, \hat{z}^2] = 0$; as $[\hat{\ell}_z, \hat{\ell}_x] = +i\hbar\,\hat{\ell}_y$ and $[\hat{\ell}_x, \hat{y}] = +i\hbar\,\hat{z}$ thereby proving the result).

R.Q. 16.16   Obtain the matrices for the operators $\hat{J}^2$, $\hat{J}_z$, $\hat{J}_+$, and $\hat{J}_-$.

&&*&*&*&*&*

Chapter 17

INTRINSIC ANGULAR MOMENTUM PARTICLE WITH SPIN - $\frac{1}{2}$

Chapter 17

INTRINSIC ANGULAR MOMENTUM
PARTICLE WITH SPIN - $\frac{1}{2}$

"In science one tries to tell people, in such a way as to be understood by everyone, something that no one ever knew before. But in poetry, it's the exact opposite"
— PAM Dirac

17.1.1 INTRODUCTION

Orbital angular momentum, $\vec{\ell}$ has been only dealt with explicitly (*i.e.* using the position and momentum operators) so far. These operators can represent the moment of inertia of a mass point, without internal structure. In all low energy interactions, the wave mechanical description is complete with reference to the simple model of a point particle in a given external field, and such gross picture could solve many problems of atomic and nuclear physics. But the simple model was unable to account for many of the finer details.

Classically, ℓ is defined as

$$\vec{\ell}_{cl} = \vec{r} \wedge \vec{p}, \tag{15.1.5}$$

with linear momentum vector, $\vec{p}$ and radial vector, $\vec{r}$, connecting the Centre of Mass of the particle to the point to which the Angular Momentum is referred. Classically, ℓ can take on any value. But Quantum mechanically, $\hat{\ell}_z$ can assume only certain orientations with respect to a given direction (*Spatial Quantization*).

However, it has been seen that the formalism based on commutation rules between components of conventional orbital, say,

$$[\hat{\ell}_x, \hat{\ell}_y] = i\hbar \hat{\ell}_z. \tag{15.2.1}$$

and hermiticity of $\hat{\ell}$, permitted (implied) either integral or half-integral values of ℓ; the restriction to INTEGRAL value resulted from

i) An EXPLICIT FORM OF THE OPERATOR, *i.e.* say for $\hat{\ell}_z$, viz.

$$\boxed{\hat{\ell}_z = (\hat{x}\hat{p}_y - \hat{y}\hat{p}_x) = -i\hbar\,(\partial/\partial\varphi)}. \tag{15.1.16}$$

ii) The requirement of a SINGLE-VALUED wave function.

The result of HALF-INTEGRAL value for ℓ followed directly from the commutation relations for the components of angular momentum, *i.e.*

$$\boxed{[\hat{\ell}_x, \hat{\ell}_y] = i\hbar \hat{\ell}_z}, \text{ etc.}$$

If the corresponding elements can be found in nature, this would suggest that these COMMUTATORS (15.2.1) REPRESENT A MORE FUNDAMENTAL ASPECT OF ANGULAR MOMEMENTUM than do the two results (15.1.16) and the postulate mentioned above (ii).

17.1.1 EXPERIMENTAL CONFIRMATION that the result of HALF-INTEGRAL value for ℓ followed directly from the commutation relations

Experiment has confirmed that this is indeed the case and that there exists elements of reality corresponding to the previously excluded half-integral values for ℓ. These are relating to the spin angular momentum of the particle. It is clear that if this generalization is to be admitted, new coordinates must be admitted. Or, the requirement of single-valuedness cannot be met. The new coordinates represent INTERNAL DEGREES OF FREEDOM of the particle.

17.1.2 QM OPERATOR may contain terms which vanish in the classical limit, $h \longrightarrow 0$

For spin angular momentum, it is found empirically that the quantum numbers may take either integral or half-integral values. The most natural quantum mechanical description of spin is accomplished with the help of relativistic generalizations of Schrödinger Equation. Spin-0 particles are described by the Klein-Gordon Equation, $Spin-\frac{1}{2}$ particles by the Dirac Equation where Dirac welded Quantum Mechanics to Relativity. In the deduction of the Schrödinger Equation (Chapter 3) the spin was missed, for it was assumed that the classical limit is valid, and things that disappear in that limit might be missed.

The existence of spin indicates that Quantum Mechanics corresponding to classical dynamical quantities cannot always be obtained by the elementary rule: replace the classical quantities $\vec{r}$ and $\vec{p}$ by QM operators, $\vec{r}$ and $(-i\hbar\hat{\nabla}_q)$, and symmetrize. Evidently, in addition to the above prescription one must stipulate that the QM operator may contain terms that vanish in the classical limit, $h \longrightarrow 0$.

17.1.3 The Stern-Gerlach Experiment and Interpretation by Goudsmit and Uhlenbeck

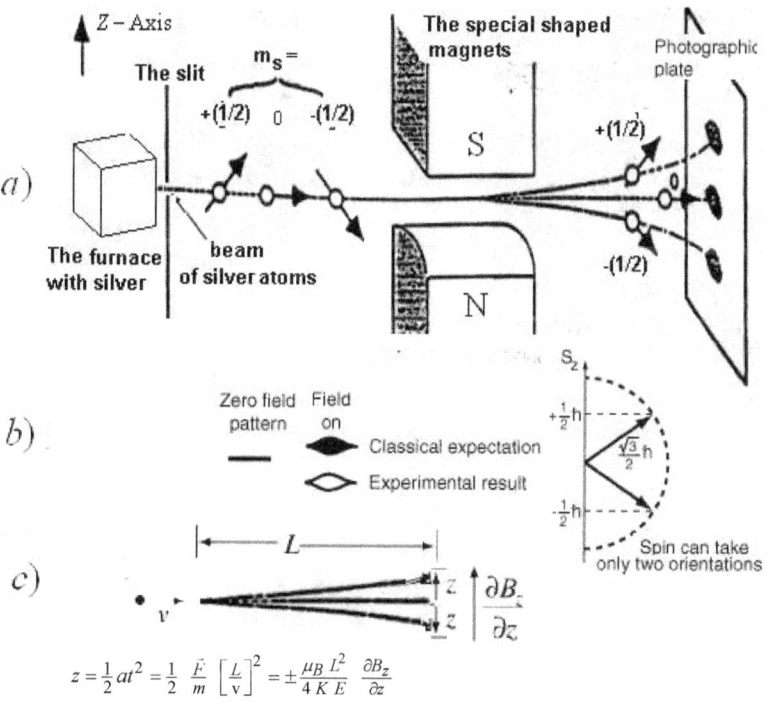

Fig 17.1   The Stern-Gerlach experiment and self-explanatory features

In 1922, Otto Stern and Walter Gerlach gave experimental evidence for the intrinsic magnetic moment (Spatial Quantization), whereas S.A. Goudsmit & G.E. Uhlenbeck's examination (1925) of the emission spectra, the concept of the spin of electron was put forward. They used it in 1928 to explain the fine structure splitting of hydrogen. Therefore, the correct interpretation was given to the Stern-Gerlach observations (on neutral silver atoms in an inhomogeneous magnetic field) only after Goudsmit and Uhlenbeck were led by a wealth of spectroscopic evidence to hypothesize the existence of an electron spin and intrinsic magnetic moment. Fig 17.1 is self-explanatory of the experimental set up as well as the interpretation of results. Whether one can understand the occurrence of intrinsic angular momentum and magnetic moment in terms of the structure of the elementary particles? The Dirac's Relativistic Theory of the Electron (gave internal ½ spin value for electron) will provide one with a deeper understanding of these properties, for some particles. This is because spin is a relativistic phenomenon, having no classical counterpart; and the Schrödinger Formalism of QM is built in the classical Hamiltonian function.

17.1.4 The POLARIZATION OF WAVES

As the magnitude of the magnetic moment, $\vec{\mu}$ shows, the presence of an intrinsic spin and magnetic moment of electron is directly related to the finite value of h;

$$\vec{\mu} = m\, \vec{\mu}_B$$

where $\vec{\mu}_B$ is the unit called <u>Bohr Magneton</u>.

Hence this is a *quantum effect*, and Classical Physics cannot be expected to be of much help in guiding one to the proper description of the spin. The '*Wave Mechanics*' was developed on the basis of the correspondence between the momentum p of a particle and its wavelength λ suggests that one may determine WHAT WAVE FEATURE CORRESPONDS TO THE ELECTRON SPIN. Polarization experiments suggest that the wave must be represented by a wave function, which under spatial rotations transforms neither as scalar nor as a vector, but in a more complicated way. In addition to the *x, y, z* coordinates, the wave function must depend on at least one other dynamic variable to permit the description of a intrinsic $\vec{\mu}_B$ and Intrinsic Angular Momentum which the electron possesses. Since both the polarization of the waves and the lining up of the particle spins are aspects of an orientation in space of the electron let it a wave or a particle, Wave Mechanics may account for both observations. So also the vector properties of EM waves are closely related to the spin of photons. The spin is a dynamic variable, σ, having two distinct values $\sigma = \pm 1$.

17.1.5 SPIN OPERATORS

W. Pauli introduced the Pauli matrices and his Formalism gives the essential features of the proper Dirac theory of the electron. It may be repeated that the quantization of $\vec{\ell}_{cl} = \vec{r} \wedge \vec{p}$, leads to integral values of ℓ and hence to odd values of the number of its possible orientations. It was, therefore, a surprise when the **alkali spectra** showed unmistakably doublets. Two orientations demand $2\ell + 1 = 2$, or $\ell = \tfrac{1}{2}$. Many attempts were made before 1924 to explain this half-integer. Pauli found the first half of the correct solution in 1924; he suggested that the electron possesses a classically non-describable two-valued ness, but he did not associate a physical picture with this property. The second half of the solution was provided by Uhlenbeck and Goudsmit (1925), who postulated a spinning electron. The two-valued ness then arises from the two different directions of rotation.

Of course, one way has to be found to incorporate the value ½ into QM. It is easy to see that QM operators that correspond to $\vec{\ell}_{cl} = \vec{r} \wedge \vec{p}$, satisfy the commutation relations. $[\ell_x, \ell_y] = i\hbar \ell_z$, etc. It is postulated that the Commutation Relations above are more fundamental than the classical definition, $\ell_{cl} (= \mathbf{r} \wedge \mathbf{p})$,. To express this fact, the symbol ℓ is reserved for orbital angular momentum. J is assumed to satisfy the Commutation Relations

$$[\hat{\ell}_x, \hat{\ell}_y] = i\hbar \hat{\ell}_z, etc.$$

The consequences of these (Commutation Relations of J) can be explained by using algebraic techniques. The result is the vindication of Pauli's and of Goudsmit and Uhlenbeck's proposals. J satisfies the eigen value equations analogous to the ones for $\hat{\ell}$.

$$\hat{J}_z | m_J \rangle = m_z \hbar | m_J \rangle \quad (15.5.4)$$

$$\hat{J}^2 | j, m_J \rangle = j(j+1) \hbar^2 | j, m_J \rangle \quad (15.5.21)$$

However, $J = 0, \frac{1}{2}, 1, \frac{3}{2}, 2, \frac{5}{2},\ldots$. $\quad (15.5.22)$

For each J, m can assume $(2j+1)$ values from $-j$ to $+j$. The particular value of J depends not only on the system, but also on the reference point to which J is referred.

17.2. DESCRIPTION OF PARTICLE WITH $Spin = \frac{1}{2}$ (i.e., $Spin-\frac{1}{2}$ particle)

All stable particles (eg. electrons, protons, positrons, neutrons) have spin $= \frac{1}{2}$, and, therefore, it is of special interest.

As seen earlier in the Chapter, there are two possible orientations for spin $= \frac{1}{2}$, the spin wave function of such a particle can be put as

$$| \psi \rangle = | \psi(r,t,\varphi_m) \rangle = \begin{pmatrix} a_{+\frac{1}{2}}(r,t) \\ a_{-\frac{1}{2}}(r,t) \end{pmatrix}. \quad (17.2.1)$$

$$= | \psi(r,\theta,\varphi) \rangle = | \chi(spin) \rangle \cdot | e^{-iE_n t/\hbar} \rangle$$

$$| \psi(r,\theta,\varphi) \rangle = | \Re_{n\ell}(r) \cdot e^{-iE_n t/\hbar} \rangle \cdot | \ell, m_\ell \rangle \cdot | s, m_s \rangle \quad (17.2.2)$$

For spin, $s = \frac{1}{2}$, $m_s = \pm\frac{1}{2}$.

$$\boxed{| \psi(r,\theta,\varphi) \rangle \propto | \tfrac{1}{2}, \pm\tfrac{1}{2} \rangle}. \quad (17.2.3)$$

17.2.1. SPIN OPERATOR $\hat{S}_z$ DEFINED

The $Spin-\frac{1}{2}$ offers an excellent illustration of the extra-ordinary convenience for some purposes in QM the use of matrix algebra. Consider an electron in a magnetic field, B. z- be the direction along which the magnetic field is chosen. Then the electron has a magnetic moment, parallel to its spin axis.

$$\mu \approx -\left(\frac{e}{2mc}\right) S.$$

The z-component of the Intrinsic Magnetic Moment, μ_z is evidently $\bar{\mu}_S = \pm \bar{\mu}_B$, and this can be described by a hermitian matrix,

$$\hat{\mu}_S = -\bar{\mu}_B \begin{pmatrix} 1 & 0 \\ 0 & -1 \end{pmatrix}$$

$$\hat{\mu}_S = -\left(\frac{e}{mc}\right)\tfrac{1}{2}\hbar \begin{pmatrix} 1 & 0 \\ 0 & -1 \end{pmatrix} = -\left(\frac{e}{mc}\right) \hat{S}_z \quad (17.2.4)$$

$$\hat{S}_z = \tfrac{1}{2}\hbar \begin{pmatrix} 1 & 0 \\ 0 & -1 \end{pmatrix} \qquad (17.2.5)$$

$\hat{S}_z$ is called <u>spin operator</u>.

i.e. $\vec{\mu} = -(\tfrac{e}{2mc})\vec{S}$

As in the case of orbital angular momentum (15.5.4) and (15.5.21), one may write

$$\boxed{\hat{S}_z | m_s \rangle = m_s \hbar | m_s \rangle} \qquad (17.2.6)$$

$$\boxed{\hat{S}^2 | s, m_s \rangle = s(s+1)\hbar^2 | s, m_s \rangle} \qquad (17.2.7)$$

Therefore, one may introduce the three components of the spin operator, $\hat{S}_x, \hat{S}_y, \hat{S}_z$, and defined by their Commutation Relations,

$$\boxed{[\hat{S}_x, \hat{S}_y] = i\hbar \hat{S}_z}, \qquad (17.2.8)$$

$$\boxed{[\hat{S}_y, \hat{S}_z] = i\hbar \hat{S}_x}, \qquad (17.2.9)$$

$$\boxed{[\hat{S}_z, \hat{S}_x] = i\hbar \hat{S}_y}, \qquad (17.2.10)$$

$$\boxed{[\hat{S}_z, \hat{S}_\pm] = \pm\hbar \hat{S}_\pm}, \qquad (17.2.11)$$

$$\boxed{[\hat{S}_+, \hat{S}_-] = 2\hbar \hat{S}_z}, \qquad (17.2.12)$$

$$\boxed{[\hat{S}^2, \hat{S}] = 0}, \qquad (17.2.13)$$

$$\boxed{\hat{S}^2 = \hat{S}_x^2 + \hat{S}_y^2 + \hat{S}_z^2} \qquad (17.2.14)$$

For the shift operators, $\hat{S}_\pm$,

$$\langle s', m_s' | \hat{S}_\pm | s, m_s \rangle = \hbar\sqrt{[(s\mp m_s)(s\pm m_s +1)]}\,\delta_{s's}\cdot\delta_{m_s' m_s\pm 1} \qquad (17.2.15)$$

17.2.2 SPIN MATRICES

The simplest mathematical representations of these are by means of (2x2) matrices.

$$\langle s', m_s' | \hat{S}_z | s, m_s \rangle = \langle s', m_s' | m_s \hbar | s, m_s \rangle$$
$$= m_s \hbar \cdot \delta_{s's}\cdot\delta_{m_s' m_s} \qquad (17.2.16)$$

For spin, $s = \tfrac{1}{2}$, $m_s = \pm\tfrac{1}{2}$.

So $\hat{S}_z = \tfrac{1}{2}\hbar \begin{pmatrix} 1 & 0 \\ 0 & -1 \end{pmatrix} \qquad (17.2.17)$

$$\langle \tfrac{1}{2}, m_s' | \hat{S}_z | \tfrac{1}{2}, m_s \rangle = \langle \tfrac{1}{2}, m_s' | (\tfrac{1}{2}\hbar) | \tfrac{1}{2}, m_s \rangle = \hbar\cdot 1\cdot \delta_{m_s' m_s\pm 1} \qquad (17.2.18)$$

Raising Operator,

$$\hat{S}_+ = \hbar \begin{pmatrix} 0 & 1 \\ 0 & 0 \end{pmatrix} \qquad (17.2.19)$$

So Lowering Operator,

$$\hat{S}_- = \hbar \begin{pmatrix} 0 & 0 \\ 1 & 0 \end{pmatrix} \qquad (17.2.20)$$

$\hat{S}_x$ and $\hat{S}_y$ can be obtained from the shift operators

$$\hat{S}_x = \tfrac{1}{2}(\hat{S}_+ + \hat{S}_-) \tag{17.2.21}$$

$$\hat{S}_y = \tfrac{1}{2i}(\hat{S}_+ - \hat{S}_-) \tag{17.2.22}$$

$$\hat{S}_x = \tfrac{1}{2}\hbar \begin{pmatrix} 0 & 1 \\ 1 & 0 \end{pmatrix} \tag{17.2.23}$$

$$\hat{S}_y = \tfrac{1}{2}\hbar \begin{pmatrix} 0 & -i \\ +i & 0 \end{pmatrix} \tag{17.2.24}$$

by using *Pauli's Spin Matrices*, $\hat{\sigma}_x, \hat{\sigma}_y, \hat{\sigma}_z$.

This Representation is written as

$$\boxed{\hat{S} = \tfrac{1}{2}\hbar\,\hat{\sigma}},$$

Or $\quad\boxed{\hat{\sigma} = \tfrac{2}{\hbar}\hat{S}}.$ (17.2.25)

This is the definition of the Pauli Matrices, $\hat{\sigma}$ which are more conventional than the spin matrices.

$$\hat{\sigma}_x = \begin{pmatrix} 0 & 1 \\ 1 & 0 \end{pmatrix} \tag{17.2.26}$$

$$\hat{\sigma}_y = \begin{pmatrix} 0 & -i \\ +i & 0 \end{pmatrix} \tag{17.2.27}$$

$$\hat{\sigma}_z = \begin{pmatrix} 1 & 0 \\ 0 & -1 \end{pmatrix} \tag{17.2.28}$$

These are completely determined by the commutations relations, given below, and the selection of <u>eigen spinors</u> of $\hat{\sigma}_z$, *i.e.* $\hat{\alpha}$ and $\hat{\beta}$.

17.2.3 SIMPLE PROPERTIES OF PAULI'S MATRICES

1) Pauli's matrices satisfy the commutation relations:

$$[\hat{\sigma}_x, \hat{\sigma}_y] = 2i\,\hat{\sigma}_z, \tag{17.2.29}$$

$$[\hat{\sigma}_y, \hat{\sigma}_z] = 2i\,\hat{\sigma}_x, \tag{17.2.30}$$

$$[\hat{\sigma}_z, \hat{\sigma}_x] = 2i\,\hat{\sigma}_y, \tag{17.2.31}$$

$$\hat{\sigma} \wedge \hat{\sigma} = 2i\,\hat{\sigma}, \tag{17.2.32}$$

2) The auxiliary matrices constructed are:

$$\hat{\sigma}_x = \tfrac{1}{2}(\hat{\sigma}_+ + \hat{\sigma}_-) \tag{17.2.33}$$

$$\hat{\sigma}_y = \tfrac{1}{2i}(\hat{\sigma}_+ - \hat{\sigma}_-) \tag{17.2.34}$$

$$[\hat{\sigma}_z, \hat{\sigma}_\pm] = \pm 2\,\hat{\sigma}_\pm, \tag{17.2.35}$$

$$[\hat{\sigma}_+, \hat{\sigma}_-] = 4\,\hat{\sigma}_z \tag{17.2.36}$$

They must satisfy

$$[\hat{S}_x, \hat{S}_y] = i\hbar\,\hat{S}_z \tag{17.2.8}$$

3) $\quad\boxed{\hat{\sigma}^2 = \hat{\sigma}_x^2 + \hat{\sigma}_y^2 + \hat{\sigma}_z^2}.$

$$\hat{\sigma}_x^2 = \hat{\sigma}_y^2 = \hat{\sigma}_z^2 = \begin{pmatrix} 1 & 0 \\ 0 & 1 \end{pmatrix} = \hat{\Im} \tag{17.2.37}$$

Because $s = \tfrac{1}{2}$, $m_s = \pm\tfrac{1}{2}$ are the only eigen values, the *Cayley-Hamilton Theorem* implies

$$\boxed{\hat{S}_x{}^2 = \hat{S}_y{}^2 = \hat{S}_z{}^2 = \tfrac{1}{4}\hbar^2 \hat{\Im}}. \tag{17.2.38}$$

$$\boxed{\hat{S}^2 = \hat{S}_x{}^2 + \hat{S}_y{}^2 + \hat{S}_z{}^2 = \tfrac{3}{4}\hbar^2 \hat{\Im}} \tag{17.2.39}$$

$$\boxed{\hat{S}^2 = s(s+1)\hbar^2 \,\hat{\Im}}. \tag{17.2.40}$$

$$\boxed{[\hat{S}^2, \hat{S}\,] = 0} \tag{17.2.13}$$

Pauli's Matrices, besides being hermitian, are also Unitary Matrices. The eigen values of $\hat{S}^2$, viz. $s(s+1)\hbar^2$ are characteristic of a particle and does not change by definition.

4) It can be seen that any two Pauli's Matrices anti-commute:
$$(\hat{\sigma}_x \pm i\,\hat{\sigma}_y)^2 = (\hat{\sigma}_x{}^2 - \hat{\sigma}_y{}^2) \pm i\,(\hat{\sigma}_x\hat{\sigma}_y + \hat{\sigma}_y\hat{\sigma}_x) = 0.$$

i.e.
$$\hat{\sigma}_x\hat{\sigma}_y = -\hat{\sigma}_y\hat{\sigma}_x = i\hat{\sigma}_z. \tag{17.2.41}$$
$$\hat{\sigma}_y\hat{\sigma}_z = -\hat{\sigma}_z\hat{\sigma}_y = i\hat{\sigma}_x$$
$$\hat{\sigma}_z\hat{\sigma}_x = -\hat{\sigma}_x\hat{\sigma}_z = i\hat{\sigma}_y.$$

i.e. $(\hat{\sigma}_x\hat{\sigma}_y + \hat{\sigma}_y\hat{\sigma}_x) = 0$, etc. the cyclical permutations.

5) The traces of all Pauli's matrices vanish.
$$Tr(\hat{\sigma}_x) = Tr(\hat{\sigma}_y) = Tr(\hat{\sigma}_z) = 0. \tag{17.2.42}$$
and $\quad Det(\hat{\sigma}_x) = Det(\hat{\sigma}_y) = Det(\hat{\sigma}_z) = 1. \tag{17.2.43}$

6) The 4 matrices $\hat{\Im}, \hat{\sigma}_x, \hat{\sigma}_y,$ and $\hat{\sigma}_z$ are linearly independent, and any (2x2) arbitrary matrix $\hat{A}$ can be represented as
$$\hat{A} = \lambda_0\,\hat{\Im} + \lambda_1\hat{\sigma}_x + \lambda_2\hat{\sigma}_y + \lambda_3\hat{\sigma}_z. \tag{17.2.44}$$

These are relations peculiar to the $Spin-\tfrac{1}{2}$ representations only, and do not hold good for $Spin-1$ matrices.

7) For any two vectors $\vec{B}$ and $\vec{C}$
$$(\hat{\sigma}\cdot\vec{B})(\hat{\sigma}\cdot\vec{C}) = \vec{B}\cdot\vec{C} + i\,\hat{\sigma}\cdot(\vec{B}\wedge\vec{C}). \tag{17.2.45}$$

17.2 SPINORS: SPIN-UP AND SPIN-DOWNSTATES

The eigen states of $\hat{S}_z$ will be represented by two component column vector (i.e. a 2-vector), called EIGEN SPIN WAVE FUNCTION, or EIGEN SPINOR. Eigen spinor is thus an eigen vector. There are only two possible eigen states or orientations of the spin for a $Spin-\tfrac{1}{2}$ particle, with respect to a fixed axis, say the z-axis. These are commonly referred to as the parallel (spin-up) and anti parallel (spin-down) to z.

If the eigen states are vectors then the operators must be matrices.

17.3.1 Determination of the Eigen Spinors

$$\hat{S}_z = \tfrac{1}{2}\hbar \begin{pmatrix} 1 & 0 \\ 0 & -1 \end{pmatrix} \tag{17.2.5}$$

$$\hat{S}_z |\,m_s\,\rangle = m_s\,\hbar |\,m_s\,\rangle \tag{17.2.6}$$

$$m_s = \pm\tfrac{1}{2}\hbar$$

Assume that the eigen states of $\hat{S}_z$ are represented by a 2-component column vector. Consider a 2-D complex vector space, called the Spin space, spanned by the *2-vectors* of the type

$$\begin{pmatrix} u \\ v \end{pmatrix}$$

where u and v are complex numbers.

$$\hat{S}_z \begin{pmatrix} u \\ v \end{pmatrix} = m_s \hbar \begin{pmatrix} u \\ v \end{pmatrix} \tag{17.3.1}$$

L.H.S. $\hat{S}_z \begin{pmatrix} u \\ v \end{pmatrix} = \tfrac{1}{2}\hbar \begin{pmatrix} 1 & 0 \\ 0 & -1 \end{pmatrix}\begin{pmatrix} u \\ v \end{pmatrix} = \tfrac{1}{2}\hbar \begin{pmatrix} u \\ -v \end{pmatrix}$ (17.3.2)

R.H.S. $= m_s \hbar \begin{pmatrix} u \\ v \end{pmatrix} = \pm\tfrac{1}{2}\hbar \begin{pmatrix} u \\ v \end{pmatrix}$ (17.3.3)

Because of equation (17.3.1), (17.3.2) and (17.3.3) are equal. This means

$$\begin{pmatrix} u \\ -v \end{pmatrix} = \pm \begin{pmatrix} u \\ v \end{pmatrix}$$

i.e. $\begin{pmatrix} u \\ 0 \end{pmatrix}$ or $\begin{pmatrix} 0 \\ v \end{pmatrix}$ (17.3.4)

i.e. $\begin{pmatrix} 1 \\ 0 \end{pmatrix}$ or $\begin{pmatrix} 0 \\ 1 \end{pmatrix}$ (17.3.5)

Thus the + eigen solution ($m_s = +\tfrac{1}{2}\hbar$) has v = 0, whereas the −ve solution ($m_s = -\tfrac{1}{2}\hbar$) has u = 0.

The case $s = \tfrac{1}{2}$ is so important that it has proved useful to introduce special state vectors called spinors on the *2-D spin space*.

These two eigen spinors are referred to as the spin-up and spin-down states of the particle. These are symbolized by one of the following:

$$|\chi(\text{up})\rangle \equiv |\chi(\uparrow)\rangle \equiv |\chi_\uparrow\rangle \equiv |\chi_+\rangle \equiv |\alpha\rangle \equiv \begin{pmatrix} 1 \\ 0 \end{pmatrix} \equiv \left|\tfrac{1}{2}, +\tfrac{1}{2}\right\rangle \equiv \left|+\tfrac{1}{2}\right\rangle \tag{17.3.6}$$

$$|\chi(\text{down})\rangle \equiv |\chi(\downarrow)\rangle \equiv |\chi_\downarrow\rangle \equiv |\chi_-\rangle \equiv |\beta\rangle \equiv \begin{pmatrix} 0 \\ 1 \end{pmatrix} \equiv \left|\tfrac{1}{2}, -\tfrac{1}{2}\right\rangle \equiv \left|-\tfrac{1}{2}\right\rangle \tag{17.3.7}$$

17.2.3 SIMULTANEOUS EIGEN SPINORS of $\hat{S}^2$ and $\hat{S}_z$

$|\alpha\rangle$ and $|\beta\rangle$ are simultaneous eigen spinors of the spin operators $\hat{S}^2$ and $\hat{S}_z$. Accordingly one needs two basis vectors, $|\alpha\rangle$ and $|\beta\rangle$ in a spin space to describe the state of an electron at each point in space. A general (arbitrary) spinor (vector), χ_k, in spin space can be expanded in this complete set, by a linear superposition of $|\alpha\rangle$ and $|\beta\rangle$ is expressed as

$$|\chi_1\rangle = \begin{pmatrix} \alpha_1 \\ \alpha_2 \end{pmatrix} = (\alpha_{1+}|\alpha\rangle + \alpha_{1-}|\beta\rangle) \tag{17.3.8}$$

where α_{1+} and α_{1-} are the expansion coefficients and may be complex scalars.

The probability of finding the electron in the state, $|\chi_1\rangle$ is given by $|\alpha_{1+}|^2$ and $|\alpha_{1-}|^2$, respectively, in the spin-up ($|\alpha\rangle$) and spin-down ($|\beta\rangle$) states,

$|\chi_1\rangle$ has to be normalized for acceptance.

i.e., $\langle \chi_1||\chi_1\rangle = (|\alpha_{1+}|^2 + |\alpha_{1-}|^2) = 1$ (17.3.9)

Consider two spinors $|\chi_k\rangle$, with $k = 1$, and $k = 2$.

The two spinors $|\chi_1\rangle$ and $|\chi_2\rangle$ are said to be orthogonal if the scalar product

$\langle \chi_2||\chi_1\rangle = 0$. (17.3.10)

$$\langle \chi_1 | \chi_k \rangle = \delta_{k1}. \qquad (17.3.11)$$

Further, it can be shown that

$$\hat{S}_z | \alpha \rangle = \tfrac{1}{2} \hbar | \alpha \rangle \qquad (17.3.12)$$

$$\hat{S}_z | \beta \rangle = -\tfrac{1}{2} \hbar | \beta \rangle \qquad (17.3.13)$$

$$\hat{S}_+ | \alpha \rangle = 0, \qquad (17.3.13)$$

$$\hat{S}_- | \alpha \rangle = \hbar | \beta \rangle \qquad (17.3.14)$$

$$\hat{S}_- | \beta \rangle = 0.$$

$$\hat{S}_+ | \beta \rangle = \hbar | \alpha \rangle.$$

In general, for an operator,

$$\langle \alpha | \hat{S}_+ | \beta \rangle = \hbar.$$

$$\langle \beta | \hat{S}_- | \alpha \rangle = \hbar.$$

$$\langle \hat{S}_x \rangle = 0, \qquad (17.3.17)$$

$$\langle \hat{S}_x \rangle^2 + \langle \hat{S}_y \rangle^2 = \tfrac{1}{4} \hbar^2 \qquad (17.3.18)$$

$$\hat{S}^2 = \tfrac{1}{4} \hbar^2 \hat{\sigma} \cdot \hat{\sigma} = \tfrac{3}{4} \hbar^2 \mathfrak{J}.$$

Worked out Example 17.1

Using the *explicit* form of $\hat{S}^2$, for electrons, show that $\hat{S}^2 | \tfrac{1}{2}, +\tfrac{1}{2} \rangle = \tfrac{3}{4} \hbar^2 | \tfrac{1}{2}, +\tfrac{1}{2} \rangle$.

Solution: Step #1 It is known that for electrons, $s = \tfrac{1}{2}$, $m_s = \pm \tfrac{1}{2}$; $| s, m_s \rangle = | \tfrac{1}{2}, +\tfrac{1}{2} \rangle = | \alpha \rangle$;

$$\hat{S}^2 = \tfrac{3}{4} \hbar^2 \mathfrak{J} = \tfrac{3}{4} \hbar^2 \begin{pmatrix} 1 & 0 \\ 0 & 1 \end{pmatrix}; \; | \alpha \rangle = \begin{pmatrix} 1 \\ 0 \end{pmatrix}; \; \mathfrak{J} = \begin{pmatrix} 1 & 0 \\ 0 & 1 \end{pmatrix}$$

Step #2 $\therefore \quad \hat{S}^2 | \tfrac{1}{2}, +\tfrac{1}{2} \rangle = \tfrac{3}{4} \hbar^2 \begin{pmatrix} 1 & 0 \\ 0 & 1 \end{pmatrix} | \tfrac{1}{2}, +\tfrac{1}{2} \rangle$

$$= \tfrac{3}{4} \hbar^2 \mathfrak{J} | \tfrac{1}{2}, +\tfrac{1}{2} \rangle.$$

Since $\mathfrak{J}$ is a unit matrix, $\mathfrak{J} | s, m_s \rangle = | s, m_s \rangle$.

$$\therefore \quad \hat{S}^2 | \tfrac{1}{2}, +\tfrac{1}{2} \rangle = \tfrac{3}{4} \hbar^2 | \tfrac{1}{2}, +\tfrac{1}{2} \rangle.$$

Or, $\quad \mathfrak{J} | s, m_s \rangle = \begin{pmatrix} 1 & 0 \\ 0 & 1 \end{pmatrix} \begin{pmatrix} 1 \\ 0 \end{pmatrix} = \begin{pmatrix} 1 \\ 0 \end{pmatrix}$

Step #3 $\hat{S}^2 | \tfrac{1}{2}, +\tfrac{1}{2} \rangle = \tfrac{3}{4} \hbar^2 \begin{pmatrix} 1 & 0 \\ 0 & 1 \end{pmatrix} \begin{pmatrix} 1 \\ 0 \end{pmatrix} = \tfrac{3}{4} \hbar^2 | \tfrac{1}{2}, +\tfrac{1}{2} \rangle$

Now one has to examine if these two vectors satisfy Postulate 1. It is interesting to observe that if $\varphi \to \varphi + 2\pi$, the solutions change sign. This is characteristic of S = Odd $-\tfrac{1}{2}$ spin eigen spin wave functions (Fermions states); although this does not violate QM, since -1 is just a phase factor. It does mean that no classical macroscopic wave packet can be constructed that has odd half-integral angular momentum.

REVIEW QUESTIONS

R.Q. 17.1 Find the eigen values and eigen vectors of $(\hat{S}_x + \hat{S}_y)$, for a system with $Spin-\frac{1}{2}$.

R.Q. 17.2 Find the effect of $\hat{S}_x$ and $\hat{S}_y$ on $|\alpha\rangle$ and $|\beta\rangle$.

R.Q. 17.3 Calculate $<\hat{S}_x> = \langle\psi|\hat{S}_x|\psi\rangle$, choosing $|\psi\rangle = c_1|\alpha\rangle e^{-i E_n t_1/\hbar} + c_2|\beta\rangle e^{-i E_n t_2/\hbar}$.

where $\left(|c_1|^2 + |c_2|^2\right) = 1$.

R.Q. 17.4 Choosing $|\psi\rangle = c_1|\alpha\rangle e^{-i E_n t_1/\hbar} + c_2|\beta\rangle e^{-i E_n t_2/\hbar}$, where $\left(|c_1|^2 + |c_2|^2\right) = 1$. show that $<\hat{S}> = <\hat{S}_x> + <\hat{S}_y> + <\hat{S}_z>$ fulfills the equation $(d/dt)<\hat{S}_x> = \mu <\hat{S}_y> B_z$. Given: $B = (0, 0, B_z)$.

R.Q. 17.5 Prove that $e^{-i \sigma_y \theta/2} = Cos(\theta/2) + i \sigma_y(\theta/2)$.

R.Q. 17.6 State the 4 spin operators for an electron. What are their eigen values? (Answers: The 4 spin operators are $\hat{S}_x, \hat{S}_y, \hat{S}_z$ and $\hat{S}^2$; $\pm\frac{1}{2}\hbar, \pm\frac{1}{2}\hbar, \pm\frac{1}{2}\hbar$ and $\frac{3}{4}\hbar^2$)..

R.Q. 17.7 Determine the eigen states of the operator, $\left(Cos\,\varphi + \hat{S}_y\,Sin\varphi\right)$ for an electron. (Answers: Eigen values are $\lambda = \pm 1$; respectively, $\sqrt{\frac{1}{2}}\begin{pmatrix}e^{-i\varphi}\\e^{+i\varphi}\end{pmatrix}$ and $\sqrt{\frac{1}{2}}\begin{pmatrix}e^{-i\varphi}\\e^{-i\varphi}\end{pmatrix}$)

R.Q. 17.8 Comment on the statement: "The spin is non-relativistic, but the Spin-Orbit Coupling is".

R.Q. 17.9 Write down the expression for the Z-component of the spin of the Dirac particle in terms of the Dirac matrices. Hence show that the spin of the Dirac particle is *half*.

&^&*&*&*&*&*

Chapter 18

ADDITION OF ANGULAR MOMENTA
TOTAL ANGULAR MOMENTUM, C.G. COEFFICIENTS

Chapter 18

ADDITION OF ANGULAR MOMENTA
TOTAL ANGULAR MOMENTUM, C.G. COEFFICIENTS

"Even though the realms of religion and science in themselves are clearly marked off from each other, nevertheless there exists between the strong reciprocal relationships and dependencies. The situation may be expressed by an image. Science without religion is lame, religion without science is blind"
Albert Einstein

18.1 INTRODUCTION

Addition of Angular Momenta has found importance in all areas of modern physics, *viz.* atomic spectroscopy, nuclear physics and particle scattering. It illustrates the concept of *change of basis* seen in the H-Picture. From the previous Chapters one has known that there are two types of angular momenta – orbital, ℓ and spin, s ($=\frac{1}{2}$) of an electron. Thus the concept of Total Angular Momentum, J for the physical system occurs, and this J has to be evaluated by means of an appropriate combination of ℓ and s to explain the various phenomena seen in the experimental observations in atomic/ molecular/ nuclear/ particle physics and other areas of physics. It is now time for a more systematic to learn the formal theory of Angular Momentum addition.

From the Total Angular Momentum operator J, one can construct the operator for the square of the total angular momentum, J^2 all the relations given below for the $\hat{\ell}$ – operator:

$$[\hat{\ell}_x, \hat{\ell}_y] = +i\hbar\,\hat{\ell}_z. \tag{15.2.1}$$

$$\hat{\ell} \wedge \hat{\ell} = i\hbar\,\hat{\ell}. \tag{15.2.4}$$

$$\hat{\ell}_\pm = \hat{\ell}_x \pm i\,\hat{\ell}_y, \tag{15.3.1}$$

$$[\hat{\ell}_z, \hat{\ell}_+] = +\hbar\,\hat{\ell}_+ \tag{15.3.6}$$

In the same way one can show that

$$[\hat{\ell}_z, \hat{\ell}_-] = -\hbar\,\hat{\ell}_- \tag{15.3.7}$$

$$[\hat{\ell}_+, \hat{\ell}_-] = +2\hbar\,\hat{\ell}_z. \tag{15.3.8}$$

$$\hat{\ell}^2 = \hat{\ell}_x^2 + \hat{\ell}_y^2 + \hat{\ell}_z^2, \tag{15.3.9}$$

$$\hat{\ell}_+\hat{\ell}_- = \hat{\ell}^2 - \hat{\ell}_z^2 + \hbar\,\hat{\ell}_z. \tag{15.3.18}$$

$$\hat{\ell}_-\hat{\ell}_+ = \hat{\ell}^2 - \hat{\ell}_z^2 - \hbar\,\hat{\ell}_z. \tag{15.3.19}$$

$$[\hat{\ell}^2, \hat{\ell}_z] = 0. \tag{15.3.13}$$

Further, $[\hat{\ell}^2, \hat{\ell}_\pm] = 0$. \hfill (15.3.14)

$$\hat{\ell}_z \mid m_\ell \rangle = m_\ell\,\hbar \mid m_\ell \rangle \tag{15.5.4}$$

$$\hat{\ell}_\pm \mid \ell, m_\ell \rangle = \hbar\sqrt{[\ell(\ell+1) - m_\ell(m_\ell \pm 1)]} \mid \ell\,(m_\ell \pm 1)\rangle \tag{15.5.6a}$$

$$\hat{\ell}_\pm \mid \ell, m_\ell \rangle = \hbar\sqrt{[(\ell\mp m_\ell)\cdot(\ell\pm m_\ell +1)]} \mid \ell, (m_\ell \pm 1)\rangle \tag{15.5.6b}$$

$$\hat{\ell}^2 \mid \ell, m_\ell \rangle = \ell\,(\ell+1)\,\hbar^2 \mid \ell, m_\ell \rangle \tag{15.5.21}$$

All these relations listed above hold good when S or J operator replaces ℓ.

Eg. $\hat{j}^2 = \hat{j}_x^2 + \hat{j}_y^2 + \hat{j}_z^2$,

$\hat{j} \wedge \hat{j} = i\hbar \hat{j}$.

18.2 VECTOR MODEL FOR COMBINING ANGULAR MOMENTA

Angular Momentum of all types and magnetic moments are vectors; before the introduction of QM, a vector representation of atomic models showing the various orientated positions of the angular momentum vectors and magnetic moment vectors with respect to a vanishing magnetic field previously set up was used. Vector diagrams for the composition of orbit and spin for different possible states of the electron in a state are used to display.

One may consider a situation in which one deals with a system involving a multi-electron atom, with operators, $\hat{\ell}_1, \hat{\ell}_2, \hat{\ell}_3, etc.$ and $\hat{s}_1, \hat{s}_2, \hat{s}_3, etc.$

The interaction energy between $\vec{\mu}_S$ in a magnetic field $\vec{B}$ is

$$E_{\vec{B}} = -\tfrac{1}{2}\vec{\mu}_S \cdot \vec{B} = -\tfrac{1}{2} K \bar{\mu}_B \vec{S} \cdot \vec{B} \propto \vec{S} \cdot \vec{L} \qquad (18.2.1)$$

Since $\vec{L} \propto \vec{B}$

where $\vec{L}$ acts only on x, y, z, but $\vec{S}$ couples the two spinor components. A term of the form $\vec{S} \cdot \vec{L}$ is referred to a SPIN-ORBIT COUPLING, and this arises in atoms as a magnetic and relativistic correction in the electro-static central potential. In the presence of $\vec{S} \cdot \vec{L}$ interaction, $\vec{L}$ is no longer a *Constant of the Motion*.

18.2.1 NO SPIN-ORBIT INTERACTION – $\vec{S} \cdot \vec{L} = 0$

The total wave function $|\psi\rangle$ given in

$$|\psi\rangle \equiv |\psi(r,\theta,\varphi)\rangle \equiv |n, \ell, m_\ell, s, m_s\rangle = |\chi(spin)\rangle \cdot \left| e^{-iE_n t/\hbar} \right\rangle$$
$$= \left| \mathfrak{R}_{n\ell}(r) \cdot e^{-iE_n t/\hbar} \right\rangle \cdot |\ell, m_\ell\rangle \cdot |s, m_s\rangle \qquad (17.2.2)$$

For an electron possessing both orbital and spin angular momenta assume that there is no interaction between them.

Then, $\vec{L} = \sum \vec{\ell}_i$, and $\vec{S} = \sum \vec{s}_i$

Further, $[\hat{L}, \hat{S}] = 0$. $\qquad (18.2.2)$

Hence the above wave function, $|\psi\rangle$ is an eigen function of both $\hat{L}_z$ and $\hat{S}_z$, meaning m_ℓ and m_s, are good quantum numbers, *i.e.* projections of $\vec{L}$ and $\vec{S}$ are *constants of the motion*.
So the Total Angular Momentum is given by $\vec{j} - \vec{j}$ coupling, viz.

$$\vec{j} = \vec{\ell}_i + \vec{s}_i$$
$$\vec{J} = \sum \vec{j}_i (= \vec{\ell}_i + \vec{s}_i),$$

There is the so-called <u>spin degeneracy</u>.

18.2.2 SPIN-ORBIT INTERACTION is present, $\vec{S} \cdot \vec{L} \neq 0$, and <u>Russell-Saunders coupling</u> ($\vec{L} - \vec{S}$ coupling)

In reality, there is an interaction between $\vec{L}$ and $\vec{S}$ – called the Spin-Orbit Interaction. This removes the spin degeneracy.

$$[\vec{S} \cdot \vec{L}, \hat{L}] \neq 0, \qquad (18.2.3)$$
$$[\vec{S} \cdot \vec{L}, \hat{S}] \neq 0. \qquad (18.2.4)$$

i.e., $\vec{L} \cdot \vec{S}$ does neither commute with $\hat{L}$ nor with $\hat{S}$.

So $|\psi\rangle$ given above is no longer correct wave function, and m_ℓ and m_s, are no longer independently quantized.: they cease to be good quantum numbers. This means both $\vec{L}$ and $\vec{S}$ cannot be treated independently as a Constant of the Motion. However, if there is no external torque acting on the isolated system (like an atom) the Total Angular Momentum, $\vec{J}$, is defined to be just the vector sum

$$\vec{J} = \vec{L} + \vec{S} \tag{18.2.5}$$

where $\vec{L}$ and $\vec{S}$ each satisfy the Commutation Relations Since they operate on different variables, $[\hat{L}, \hat{S}] = 0$. Hence $\vec{J}$ satisfies the Commutation Relations. Vectors, $\vec{L}$ and $\vec{S}$, both precess about the vector $\vec{J}$ at the same angular velocity.

where $\quad \vec{J}_x = \vec{L}_x + \vec{S}_x;$ \hfill (18.2.6)

$\vec{J}_y = \vec{L}_y + \vec{S}_y;$

$\vec{J}_z = \vec{L}_z + \vec{S}_z.$

Now, $\quad \vec{J} = \vec{L} + \vec{S}$ \hfill (18.2.7)

is a Constant of the Motion. The quantum condition on angular momentum now applies to $\hat{J}^2$ and $\hat{J}_z$ instead of to $\hat{L}^2, \hat{S}^2, \hat{L}_z$ and $\hat{S}_z$, respectively. That is one can define the eigen vectors $|j, m_j\rangle$ such that

$$\boxed{\hat{J}^2 |j, m_j\rangle = j(j+1)\hbar^2 |j, m_j\rangle} \tag{18.2.8}$$

$$\boxed{\hat{J}_z |m_j\rangle = m_j \hbar |m_j\rangle} \tag{18.2.9}$$

The general and conserved quantum numbers are j and m_j; where $j = (\ell+s), \ldots\ldots, (\ell-s)$. For each j-value, m_j has $(2j+1)$ values. For a <u>single-electron atom</u>, the 2 values are $j = (\ell \pm \frac{1}{2})$.

The new angular momentum eigen vectors replace the old ones. The new eigen vectors are the linear combinations of the old ones which diagonalize the matrices $\hat{J}^2$ and $\hat{J}_z$. They are listed in Table 18.1.

TABLE 18.1

Orbital State		Eigenkets
ℓ – value	j - value	$|j, m_j\rangle$
s	$\frac{1}{2}$	$\left|\frac{1}{2}, \frac{1}{2}\right\rangle, \left|\frac{1}{2}, -\frac{1}{2}\right\rangle$
p	$\frac{3}{2}$	$\left|\frac{3}{2}, \frac{3}{2}\right\rangle, \left|\frac{3}{2}, +\frac{1}{2}\right\rangle, \left|\frac{3}{2}, -\frac{3}{2}\right\rangle, \left|\frac{3}{2}, -\frac{1}{2}\right\rangle :$
	$\frac{1}{2}$	$\left|\frac{1}{2}, \frac{1}{2}\right\rangle, \left|\frac{1}{2}, -\frac{1}{2}\right\rangle$
d	$\frac{5}{2}$	$\left|\frac{5}{2}, \frac{5}{2}\right\rangle, \left|\frac{5}{2}, +\frac{3}{2}\right\rangle, \left|\frac{5}{2}, +\frac{1}{2}\right\rangle,$
		$\left|\frac{5}{2}, -\frac{5}{2}\right\rangle, \left|\frac{5}{2}, -\frac{3}{2}\right\rangle, \left|\frac{5}{2}, -\frac{1}{2}\right\rangle$
	$\frac{3}{2}$	$\left|\frac{3}{2}, \frac{3}{2}\right\rangle, \left|\frac{3}{2}, -\frac{1}{2}\right\rangle, \left|\frac{3}{2}, -\frac{3}{2}\right\rangle, \left|\frac{3}{2}, -\frac{1}{2}\right\rangle$

18.2.3 MATRIX REPRESENTATION of $\hat{J}^2$

Let $|j, m_j\rangle$ be a complete set and form a basis.

$\hat{J}^2$ is hermitian and forms eigen value equation,

$$\hat{J}^2 |j, m_j\rangle = j(j+1)\hbar^2 |j, m_j\rangle \tag{18.2.8}$$

Matrix element of $\hat{J}^2$ is

$$\langle j', m_j' | \hat{J}^2 | j, m_j \rangle = \langle j', m_j' | [j(j+1)\hbar^2] | j, m_j \rangle$$
$$= [j(j+1)\hbar^2] \delta_{j'j} \delta_{m_j'm_j}. \tag{18.2.10}$$

This matrix element shows that only diagonal elements, *viz.* those for which the Kronecker is non-vanishing, exist. The off-diagonal elements are characterized by either $j = j'$, or by $m_j = m_j'$ (or both), which are, therefore, zero. Using the matrix element one can develop the matrix representation as shown below:

$$\hat{J}^2 = \begin{pmatrix} \frac{3}{4} & 0 & & & & & & & & & & & & \\ 0 & \frac{3}{4} & & & & & & & & & & & & \\ & & 2 & 0 & 0 & & & & & & & & & \\ & & 0 & 2 & 0 & & & & & & & & & \\ & & 0 & 0 & 2 & & & & & & & & & \\ & & & & & \frac{15}{4} & 0 & 0 & 0 & & & & & \\ & & & & & 0 & \frac{15}{4} & 0 & 0 & & & & & \\ & & & & & 0 & 0 & \frac{15}{4} & 0 & & & & & \\ & & & & & 0 & 0 & 0 & \frac{15}{4} & & & & & \\ & & & & & & & & & 6 & 0 & 0 & 0 & 0 \\ & & & & & & & & & 0 & 6 & 0 & 0 & 0 \\ & & & & & & & & & 0 & 0 & 6 & 0 & 0 \\ & & & & & & & & & 0 & 0 & 0 & 6 & 0 \\ & & & & & & & & & 0 & 0 & 0 & 0 & 6 \end{pmatrix} \hbar^2 \tag{18.2.11}$$

with $j = \frac{1}{2}, 1, \frac{3}{2}, 2, \ldots$ and Subspace of $(2J+1) = 5$ Dimensions.

In the treatment given above, thus the diagonal elements are the eigen values of the operator. The index increases by one unit downward every *(2j +1)* rows. The index m_j decreases by one unit on going from row to row, starting with $m_j = j$, in the upper row of each *j*-row matrix.

Further, $\hat{J}^2 = (\hat{L} + \hat{S})^2 = \hat{L}^2 + \hat{S}^2 + 2\hat{L} \cdot \hat{S}$ \hfill (18.2.12)

18.2.4 TO FIND THE MATRIX for $\hat{J}_z$

Let $\{|j, m_j\rangle\}$ is a complete set and it forms a basis.

$\hat{J}_z$ is hermitian and forms eigen value equation,

$$\hat{J}_z |j, m_j\rangle = m_j \hbar |j, m_j\rangle \tag{18.2.13}$$

Matrix element of $\hat{J}_z$ is

$$\langle j', m_j' | \hat{J}_z | j, m_j \rangle = \langle j', m_j' | (m_j \hbar) | j, m_j \rangle = (m_j \hbar) \, \delta_{j'j} \, \delta_{m_j'm_j} \quad (18.2.14)$$

This matrix element shows that only diagonal elements, *viz.* those for which the Kronecker is non-vanishing, exist. The off-diagonal elements are characterized by either $j = j'$, or by $m_j' = m_j$ (or both), which are, therefore, zero. Using the matrix element one can develop the Matrix Representation as shown below:

In the treatment given above, thus the diagonal elements are the eigen values of the operator. The index increases by one unit downward every $2j + 1$ rows. The index m_j decreases by one unit on going from row to row, starting with $m_j = j$, in the upper row of each *j*-row matrix.

$$\hat{J}_z = \begin{pmatrix} \frac{1}{2} & 0 & & & & & & & & & & & & & \\ 0 & -\frac{1}{2} & & & & & & & & & & & & & \\ & & +1 & 0 & 0 & & & & & & & & & & \\ & & 0 & 0 & 0 & & & & & & & & & & \\ & & 0 & 0 & -1 & & & & & & & & & & \\ & & & & & +\frac{3}{2} & 0 & 0 & 0 & & & & & & \\ & & & & & 0 & +\frac{1}{2} & 0 & 0 & & & & & & \\ & & & & & 0 & 0 & -\frac{1}{2} & 0 & & & & & & \\ & & & & & 0 & 0 & 0 & -\frac{3}{2} & & & & & & \\ & & & & & & & & & +2 & 0 & 0 & 0 & 0 \\ & & & & & & & & & 0 & +1 & 0 & 0 & 0 \\ & & & & & & & & & 0 & 0 & 0 & 0 & 0 \\ & & & & & & & & & 0 & 0 & 0 & -1 & 0 \\ & & & & & & & & & 0 & 0 & 0 & 0 & -2 \end{pmatrix} \hbar \quad (18.2.15)$$

(Subspace of $(2J+1) = 5$ Dimensions)

Worked out Example 18.1

What is the matrix element $\langle \frac{3}{2}, 0 | \hat{J}_z | \frac{3}{2}, -\frac{1}{2} \rangle$?

Solution: Step # 1   It is known that $\langle j', m_j' | \hat{J}_z | j, m_j \rangle = \langle j', m_j' | (m_j \hbar) | j, m_j \rangle$
$= (m_j \hbar) \, \delta_{j'j} \, \delta_{m_j'm_j}$.

Step # 2 $\langle \frac{3}{2}, 0 | \hat{J}_z | \frac{3}{2}, -\frac{1}{2} \rangle = \langle \frac{3}{2}, 0 | (-\frac{1}{2}) \hbar | \frac{3}{2}, -\frac{1}{2} \rangle = (-\frac{1}{2}) \hbar \, \delta_{j'j} \, \delta_{m_j'm_j} = 0$.

18.2.4 STEP-UP MATRIX, $\hat{J}_+$

$$\hat{J}_z = \begin{pmatrix} \text{matrix as shown} \end{pmatrix} \hbar \qquad (18.2.16)$$

$$\hat{J}_- = (\hat{J}_+)^\dagger. \qquad (18.2.17)$$

18.2.5 MATRIX of $\hat{J}_y$

It is known that $\hat{J}_\pm = \hat{J}_x \pm i\hat{J}_y$

$$\hat{J}_y = \frac{1}{2i}(\hat{J}_+ - \hat{J}_-) \qquad (18.2.18)$$

$$\hat{J}_x = \frac{1}{2}(\hat{J}_+ + \hat{J}_-) \qquad (18.2.19)$$

$$\hat{J}_y = \begin{pmatrix} \text{matrix as shown} \end{pmatrix} \cdot \tfrac{1}{2}i\hbar \qquad (18.2.20)$$

18.3 ADDITION OF TWO ANGULAR MOMENTA – GENERAL CASE

For more than one electron,

'*Russell-Saunders*' Coupling (L.S Coupling) would give

$$\hat{L} = \sum \hat{\ell}_i, \tag{18.3.1}$$

and $\quad \hat{S} = \sum \hat{s}_i, \tag{18.3.2}$

and $\quad \hat{J} = (\hat{L} + \hat{S}). \tag{18.3.3}$

j-j coupling would give

$$\hat{j}_i = \hat{\ell}_i + \hat{s}_i, \tag{18.3.4}$$

$$\hat{J} = \sum \hat{j}_i = \sum (\hat{\ell}_i + \hat{s}_i), \tag{18.3.5}$$

There is the so-called **spin degeneracy**.

18.3.1. TO FIND THE COMMUTATION RELATIONS and EIGEN VALUE EQUATIONS

Consider an isolated system involving two angular momentum operators.

Classically, if one operator of angular momentum is $\vec{J}_1$ and the second is $\vec{J}_2$ acting in two different spaces, the total angular momentum $\vec{J}$ of the combined system is simply

$$\vec{J} = \vec{J}_1 + \vec{J}_2, \tag{18.3.6}$$

Quantum mechanically, $\hat{J}_1$ & $\hat{J}_2$, each of which may be either a spin or an orbital momentum or a combination of these. $\hat{J}_1$ & $\hat{J}_2$ satisfy either equation (18.3.3) or (18.3.5). Further, $\hat{J}_1$ is defined in a space of $(2j_1+1)$ (2-dimensions; when j_1 = maximum projection of $\vec{J}_1$. Like wise, $\vec{J}_2$ is defined in a space of $(2j_2+1)$)-dimensions; when j_2 = maximum projection of $\vec{J}_2$. This happens if one handles two different particles or if one treats the $\hat{L}$ and $\hat{S}$ of a single particle. $\hat{J}_1$ & $\hat{J}_2$ are two operators that satisfy the Commutation Relations, *viz.* the vector commutation relations,

$$\hat{J}_1 \wedge \hat{J}_1 = i\hbar \hat{J}_1. \tag{18.3.7}$$

$$\hat{J}_2 \wedge \hat{J}_2 = i\hbar \hat{J}_2. \tag{18.3.8}$$

Let $\quad [\hat{J}_1, \hat{J}_2] = 0 \tag{18.3.9}$

The two systems being isolated,

$$\hat{J} \wedge \hat{J} = i\hbar \hat{J}. \tag{18.3.10}$$

Therefore, $\vec{J} = \vec{J}_1 + \vec{J}_2 \tag{18.3.6}$

is also an Angular Momentum. There arise two cases:

(i) For the case of two isolated systems, assume that

$$\hat{J}^2 |j, m_j\rangle = j(j+1)\hbar^2 |j, m_j\rangle \tag{18.2.8}$$

$$\hat{J}_z |j, m_j\rangle = m_j \hbar |j, m_j\rangle \tag{18.2.13}$$

i.e.,
$$\hat{J}_1^2 |j_1, m_{j_1}\rangle = j_1(j_1+1)\hbar^2 |j_1, m_{j_1}\rangle \tag{18.3.11}$$

$$\hat{J}_{1z} |j_1, m_{j_1}\rangle = m_{j_1} \hbar |j_1, m_{j_1}\rangle \tag{18.2.14}$$

$$\hat{J}_2^2 |j_2, m_{j_2}\rangle = j_2(j_2+1)\hbar^2 |j_2, m_{j_2}\rangle \tag{18.3.15}$$

$$\hat{J}_{2z} |j_2, m_{j_2}\rangle = m_{j_2} \hbar |j_2, m_{j_2}\rangle \tag{18.2.16}$$

For the combine system, the UNCOUPLED VECTOR is

$$|j_1, j_2, m_{j_1}, m_{j_2}\rangle = |j_1, m_{j_1}\rangle \cdot |j_2, m_{j_2}\rangle = |m_{j_1}, m_{j_2}\rangle \tag{18.3.17}$$

Consider the case for which

$$\hat{J}^2 = (\hat{J}_1 + \hat{J}_2)^2 = \hat{J}_1^2 + \hat{J}_2^2 + 2\hat{J}_1 \cdot \hat{J}_2. \tag{18.3.18}$$

Since $\quad [\hat{J}_1^2, \hat{J}_1] = 0, \tag{18.3.19}$

$$[\hat{J}_2{}^2, \hat{J}_2] = 0 \quad (18.3.20)$$

and $\quad [\hat{J}_1, \hat{J}_2] = 0 \quad (18.3.9)$

i.e., $\quad [\hat{J}_1{}^2, \hat{J}\ ] = 0, \quad (18.3.21)$

$$[\hat{J}_2{}^2, \hat{J}\ ] = 0 \quad (18.3.21)$$

$$[\hat{J}_1{}^2, \hat{J}_z] = 0, \quad (18.3.22)$$

$$[\hat{J}_2{}^2, \hat{J}_z] = 0 \quad (18.3.23)$$

$$\boxed{[\hat{J}_i, \hat{J}_j] = i\hbar\, \varepsilon_{ijk}\, \hat{J}_k} \quad (18.3.24)$$

where $\hat{J}_z = \hat{J}_{1z} + \hat{J}_{2z}$, etc.

Therefore, $\hat{J}_1{}^2, \hat{J}_2{}^2, \hat{J}_{1z}$ and $\hat{J}_{2z}$ constitute a set of commuting operators. Their common eigen vectors may be denoted as the UNCOUPLED VECTOR,

$$|j_1, j_2, j, m\rangle \equiv |j, m\rangle. \quad (18.3.25)$$

$$\hat{J}_1{}^2 |j_1, j_2, j, m\rangle \equiv j_1(j_1+1)\hbar^2 |j_1, j_2, j, m\rangle \quad (18.3.26)$$

Similarly for $\hat{J}_2{}^2, \hat{J}^2$, and $\hat{J}_z$

viz., $\quad \hat{J}_2{}^2 |j_1, j_2, j, m\rangle \equiv j_2(j_2+1)\hbar^2 |j_1, j_2, j, m\rangle \quad (18.3.27)$

$$\hat{J}^2 |j_1, j_2, j, m\rangle \equiv j(j+1)\hbar^2 |j_1, j_2, j, m\rangle \quad (18.3.28)$$

$$\hat{J}_z |j_1, j_2, j, m\rangle \equiv m\hbar |j_1, j_2, j, m\rangle \quad (18.3.29)$$

For fixed j_1 and j_2, there are $(2j+1)$ vectors $|j_1, j_2, j, m\rangle$.

Also there are $(2j_1+1)(2j_2+1)$ of the vector $|j_1, j_2, j, m\rangle$.

18.3.2 TO FIND THE RELATION BETWEEN TWO SETS OF VECTORS

Uncoupled vector (using old bases):

$$|j_1, j_2, m_1, m_2\rangle = |j_1, m_1\rangle |j_2, m_2\rangle \equiv |m_1, m_2\rangle \quad (18.3.17)$$

Coupled vector (using new base):

$$|j_1, j_2, j, m\rangle \equiv |j, m\rangle. \quad (18.3.25)$$

$|j, m\rangle$ and $|m_1, m_2\rangle$ are different bases in the same Hilbert space.

Classically, $|\bar{J}|$ has magnitude between

$$(|\bar{J}_1| + |\bar{J}_2|) \text{ and } (|\bar{J}_1| - |\bar{J}_2|) \quad (18.3.26)$$

Let $\quad |\bar{J}_1| > |\bar{J}_2|$.

So $j = (j_1+j_2), \ldots\ldots\ldots, (j_1-j_2)$.

For fixed values of j_1 and j_2,

# of vectors $|j_1, j_2, j, m\rangle$, $N = \sum (2j+1)$

$$= [2(j_1+j_2)+1] + [2(j_1+j_2)-1] + \ldots\ldots + [2(j_1-j_2)+1]$$

$$= N_1 \cdot N_2 \cdot N_3 \cdots\cdots N_{(2j+1)}.$$

$$= (2j+1) \times (\text{\# of brackets}) + [2(j_1+j_2-1)+\ldots-j_2\ ]$$

$$= (2j_1+1)(2j_2+1) \quad (18.3.27)$$

= the same as the # of vectors $|j_1, j_2, m_1, m_2\rangle$ also.

Since it is known that

$$|\varphi_j\rangle = \sum c_{ij} |\hat{e}_i\rangle = \sum c_{ij} |\hat{i}\rangle \quad (11.3.24)$$

where, $c_{ij} = \langle \hat{e}_i \| \varphi_j \rangle \equiv \langle \hat{i} \| \varphi_j \rangle$. (11.3.25)

$$| \varphi_j \rangle = \sum | \hat{i} \rangle \langle \hat{i} \| \varphi_j \rangle$$ (18.3.28)

$$| j_1, j_2, j, m \rangle = \sum\sum | j_1, j_2, m_1, m_2 \rangle \langle j_1, j_2, m_1, m_2 \| j_1, j_2, j, m \rangle$$ (18.3.29)

Expansion coefficient $= \langle j_1, j_2, m_1, m_2 \| j_1, j_2, j, m \rangle$ (18.3.30)

i.e., $| j, m \rangle = \sum\sum | m_1, m_2 \rangle \langle m_1, m_2 \| j, m \rangle c$ (18.3.31)

Since the quantum numbers j_1 and j_2 are the same in every term in equation (18.3.31). Thus the addition problem consists of determining
i) The eigen vectors, and
ii) The eigen values of $\hat{J}^2$ and $\hat{J}_z$ in terms of those separate, independent angular momentum operators.

18.4 CLEBSCH-GORDAN COEFFICIENTS :(or C.G. coefficients or VECTOR ADDITION coefficients) / WIGNER / the 3j symbols)

18.4.1 ADDITION OF TWO ANGULAR MOMENTA, $\hat{J}_1$ and $\hat{J}_2$:
Clebsch-Gordan coefficients / 3j symbols

It was seen in the discussions earlier in this Chapter that how several angular momenta can be combined using one of the <u>Vector Coupling models</u>. This approach, augmented by <u>Hund's Rules</u>, enables one to account for many details of the spectra of multi-electron atoms without knowing the exact eigen functions of the total angular momentum. However, it is frequently necessary to know these eigen functions in order to use a representation in which the square of the angular momentum is diagonal. Quantum mechanically, $\hat{J}_1$ and $\hat{J}_2$, each of which may be either a spin or an orbital momentum or a combination of these. $\hat{J}_1$ and $\hat{J}_2$ satisfy either equation (18.3.3) or (18.3.5). Further, $\hat{J}_1$ is defined in a space of $(2j_1+1)$-dimensions; when j_1 = maximum projection of $\hat{J}_1$, i. e. maximum value of m_1. The eigen kets $| j_1, m_1 \rangle$ constitute the basis on which $\hat{J}_1^2$ and $\hat{J}_{1z}$ are diagonal.

Likewise $\hat{J}_2$ is defined in a space of $(2j_2+1)$-dimensions; when j_2 = maximum projection of $\hat{J}_2$, i.e. maximum value of m_2. The eigen kets $| j_2, m_2 \rangle$ constitute the basis permitting a diagonal representation for $\hat{J}_2^2$ and $\hat{J}_{2z}$. $\hat{J}_1$ and $\hat{J}_2$ are two operators that satisfy the commutation relations, viz. the vector commutation relations, listed in the previous Section.
One must find out what happens when a 'new space' is defined by combining $\hat{J}_1$ and $\hat{J}_2$, to write $\hat{J} = \hat{J}_1 + \hat{J}_2$ (equation, 18.3.6). The 'new space' is called a PRODUCT SPACE and it has dimensions
$N = (2j_1+1)(2j_2+1)$ (equation: 18.3.27). This product space of dimensions N, contains $(2n+1)$ subspaces of dimensions $N_1 \cdot N_2 \cdot N_3 \cdot \cdots \cdot N_{(2j+1)}$, where

$N_1 = (2j+1) = [2(j_1+j_2)+1]$ (18.4.1)
$N_2 = (2j-1) = [2(j_1+j_2)-1]$ (18.4.2)
$N_{(2j+1)} = [2(j_1+j_2)+1]$. (18.4.3)

The matrix representation of $\hat{J}^2$, say, an (N x N) matrix is given below:
$\hat{J}_2^2 | j_1, j_2, j, m \rangle \equiv j(j+1) \hbar^2 | j_1, j_2, j, m \rangle$ (18.3.28)
$j = j_1 + j_2$

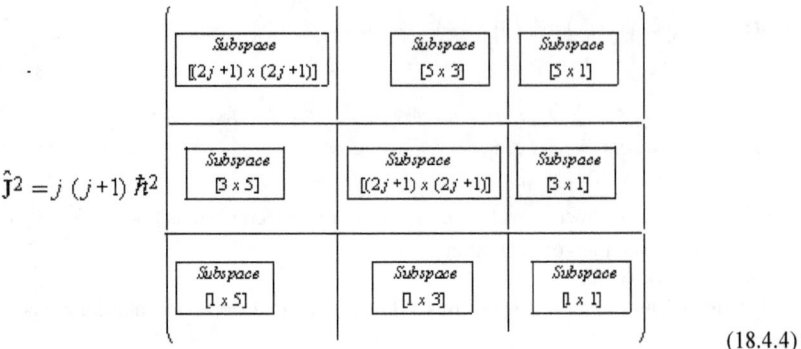

(18.4.4)

It is assumed here that $j_1 > j_2$.

In order to *diagonalize* the matrices $\hat{J}^2$ and $\hat{J}_z$,

| Step # 1 | One must obtain the simultaneous eigen kets of $\hat{J}^2 = (\hat{J}_1 + \hat{J}_2)^2$ and $\hat{J}_z = \hat{J}_{1z} + \hat{J}_{2z}$

| Step # 2 | Express the new eigen kets in terms of kets which are known already, *viz.*

(a) $|j_1, m_1\rangle$, of $\hat{J}_1^2$ and $\hat{J}_{1z}$, and

(b) $|j_2, m_2\rangle$, of $\hat{J}_2^2$ and $\hat{J}_{2z}$.

| Step # 3 | Find a ket of some linear combination of the basis $|j_1, m_1\rangle$ and $|j_2, m_2\rangle$. (Because the coordinates of $\hat{J}_1$ and $\hat{J}_2$ are independent in the old Representation) such that $\hat{J}^2$ and $\hat{J}_z$, operating on this new ket give eigen values $j(j+1)\hbar^2$ and $m\hbar$, respectively. In other words the eigen kets of the product space are written as

$$|j_1, j_2, j, m\rangle \equiv |j, m\rangle. \tag{18.3.25}$$

$$\hat{J}^2 |j_1, j_2, j, m\rangle \equiv j(j+1)\hbar^2 |j_1, j_2, j, m\rangle \tag{18.3.28}$$

$$\hat{J}_z |j_1, j_2, j, m\rangle \equiv m\hbar |j_1, j_2, j, m\rangle \tag{18.3.29}$$

It was seen that

$$\hat{J}_1^2 |j_1, m_1\rangle \equiv j_1(j_1+1)\hbar^2 |j_1, m_1\rangle \tag{18.3.11}$$

$$\hat{J}_{1z} |j_1, m_1\rangle \equiv m_1\hbar |j_1, m_1\rangle \tag{18.2.14}$$

$$\hat{J}_2^2 |j_2, m_2\rangle \equiv j_2(j_2+1)\hbar^2 |j_2, m_2\rangle \tag{18.3.15}$$

$$\hat{J}_{2z} |j_2, m_2\rangle \equiv m_2\hbar |j_2, m_2\rangle \tag{18.2.16}$$

| Step # 4 | Define the new eigen kets

$$|j_1, j_2, m_1, m_2\rangle = |j_1, m_1\rangle \cdot |j_2, m_2\rangle \equiv |m_1, m_2\rangle \tag{18.3.17}$$

| Step # 5 |: Now $|j_1, j_2, j, m\rangle$ will be a linear combination of the $|j_1, j_2, m_1, m_2\rangle$.

| Step # 6 | $|j_1, j_2, m_1, m_2\rangle$ kets form a complete ortho-normal basis so that one may define a projection operator

$$\hat{P} = \sum |j_1, j_2, m_1, m_2\rangle\langle j_1, j_2, m_1, m_2|$$
$$= \{\sum |j_1, m_1\rangle\langle j_1, m_1|\}\{\sum |j_2, m_2\rangle\langle j_2, m_2|\} = 1 \tag{18.4.5}$$

In terms of the original eigen kets

$$|j_1, j_2, j, m\rangle = \sum |j_1, j_2, m_1, m_2\rangle\langle j_1, j_2, m_1, m_2 | j_1, j_2, j, m\rangle \tag{18.4.6}$$

Since j_1 and j_2, are fixed numbers for a specific calculation, the notation may be simplified by omitting them from the kets. Maximum m_1 value = j_1, (18.4.7)
Maximum m_2 value = j_2, (18.4.8)
Maximum m value = $j = j_1 + j_2$, (18.4.9)
Assume that the condition (18.4.8), viz. $m = m_1 + m_2$, is always satisfied.
Then $|j_1, j_2, j, m\rangle \equiv |j, m\rangle = \sum |m_1, m_2\rangle \langle m_1, m_2 || j, m\rangle$ (18.4.10)
where the scalars are
$$\langle m_1, m_2 || j, m\rangle = \langle j_1, j_2, m_1, m_2 || j_1, j_2, j, m\rangle$$ (18.4.11)
These are merely coefficients that tell how much each old eigen ket $|j_1, j_2, j, m\rangle$ contributes to the new eigen ket $|j, m\rangle$. These form the matrix element s of the Unitary Transformation carrying one from the uncoupled to the coupled Representation, written as
$$|j_1, j_2, j, m\rangle = \sum\sum |j_1, j_2, m_1, m_2\rangle \langle j_1, j_2, m_1, m_2 || j_1, j_2, j, m\rangle$$ (18.4.6)
These coefficients, $\langle j_1, j_2, m_1, m_2 || j_1, j_2, j, m\rangle$, are called the ANGULAR MOMENTUM COUPLING COEFFICIENTS, or CLEBSCH-GORDAN or 'vector addition' or simply the 'C-coefficients'. Many alternative notations exist for them (Rose, M. E., Brink & Satchler, Condon & Shortley).

The properties of the C.G. Coefficients are derived directly from their definition as a Unitary Transformation carrying one from the $\{|j_1, j_2, m_1, m_2\rangle$, or $\hat{J}_1^2, \hat{J}_{1z}, \hat{J}_2^2$ and $\hat{J}_{2z}\}$ to the $\{|j_1, j_2, j, m\rangle$, or $\hat{J}_1^2, \hat{J}_2^2, \hat{J}^2$ and $\hat{J}_z\}$ Representation. For instance, by applying $\hat{J}_z = \hat{J}_{1z} + \hat{J}_{2z}$, to both sides of equation (18.4.6) one can establish
$$\langle m_1, m_2 || j, m\rangle = 0, \text{ for } m \neq m_1 + m_2.$$ (18.4.12)a
$$|j, m\rangle = \sum |m_1, m_2\rangle \langle m_1, m_2 || j, m\rangle,$$
when $j = j_1 + j_2$, $m = j$, there is only one term in this sum. This one vector is all that one needs to construct the rest of the vectors.
∴ $m = m_1 + m_2$, (18.4.12)b
The equation (18.4.12)a may then be put in the form
$$\langle m - m_2, m_2 || j, m\rangle = 0, \text{ unless } |j_1 - j_2| \leq j \leq (j_1 + j_2).$$ (18.4.12)c

The explicit calculation of the C.G. coefficients proceeds through the study of the action of the raising and lowering operators, $\hat{J}_\pm = (\hat{J}_1 + \hat{J}_2)_\pm$, on equation (18.4.6). General formulae as well as other properties are given in books on Angular Momentum, and the author does not intend to present them here.

TABLE 18.2 Values of $\langle m_1, m_2 || j, m\rangle = \langle j_1, \frac{1}{2}, m - m_2, m_2 || j_1, \frac{1}{2}, j, m\rangle$

j	$m_2 = \frac{1}{2}$	$m_2 = -\frac{1}{2}$
$j_1 + \frac{1}{2}$	$\sqrt{\frac{(j_1 + m + \frac{1}{2})}{(2j_1 + 1)}}$	$\sqrt{\frac{(j_1 - m + \frac{1}{2})}{(2j_1 + 1)}}$
$j_1 - \frac{1}{2}$	$-\sqrt{\frac{(j_1 - m + \frac{1}{2})}{(2j_1 + 1)}}$	$\sqrt{\frac{(j_1 + m + \frac{1}{2})}{(2j_1 + 1)}}$

Many of these C-coefficients are available in tabular form. A short Table listing the explicit values of the C.G coefficients for $\hat{J}_2 = \frac{1}{2}$ and 1 are reproduced in Table 18.2 and Table 18.3. These are frequently used.
The C- coefficients are such that
$\langle m_1, m_2 \mid\mid j, m \rangle = \langle j_1, j_2, m_1, m_2 \mid\mid j_1, j_2, j, m \rangle = 0$,
unless $j = (j_1 + j_2); m = j$.

TABLE 18.3   Values of $\langle m_1, m_2 \mid\mid j, m \rangle = \langle j_1, 1, m - m_2, m_2 \mid\mid j, m \rangle$

j	$m_2 = 1$	$m_2 = 0$	$m_2 = -1$
$j_1 + 1$	$\sqrt{\dfrac{(j_1 + m)(j_1 + m + 1)}{(2j_1 + 1)(2j_1 + 2)}}$	$\sqrt{\dfrac{(j_1 - m)(j_1 + m + 1)}{(2j_1 + 1)(2j_1 + 2)}}$	$\sqrt{\dfrac{(j_1 - m)(j_1 - m + 1)}{(2j_1 + 1)(2j_1 + 2)}}$
j_1	$-\sqrt{\dfrac{(j_1 + m)(j_1 - m + 1)}{(2j_1)(2j_1 + 1)}}$	$\sqrt{\dfrac{m}{j_1(j_1 + 1)}}$	$\sqrt{\dfrac{(j_1 - m)(j_1 + m + 1)}{(2j_1)(2j_1 + 1)}}$
$j_1 - 1$	$\sqrt{\dfrac{(j_1 - m)(j_1 - m + 1)}{(2j_1)(2j_1 + 1)}}$	$-\sqrt{\dfrac{(j_1 - m)(j_1 + m)}{(j_1)(2j_1 + 1)}}$	$\sqrt{\dfrac{(j_1 - m + 1)(j_1 - m)}{(2j_1)(2j_1 + 1)}}$

3j symbols:
The C.G. coefficient $\langle j_1, j_2, m - m_2, m_2 \mid\mid j_1, j_2, j, m \rangle$ is slightly modified and the resulting quantity is known as a '3j-symbol', defined by

$$\begin{pmatrix} j_1 & j_2 & j \\ m_1 & m_2 & m \end{pmatrix}$$

where, by comparison of equations (18.4.12)a and (18.4.12)c, j_1, j_2 and j form a triangle and $m + m_1 + m_2 = 0$ for non-vanishing 3 j-symbol.

18.4.2   ADDITION OF THREE ANGULAR MOMENTA, $\hat{J}_1$, $\hat{J}_2$ and $\hat{J}_3$:
RECAH coefficients and 6 j symbols
   Recah Coefficients are used in the case of the combination of three Angular Momenta. However, this will not be discussed here as this outside the scope of this book.

18.4.3   Addition of Four Angular Momenta, $\hat{J}_1$, $\hat{J}_2$, $\hat{J}_3$ and $\hat{J}_4$: 9 j symbols.
9 j-symbols are used in the case of the combination of four Angular Momenta. However, this will also not be discussed here as this outside the scope of this book.

Worked out Example 18.2
   The Angular Momentum of two isolated systems are $j_1 = 3$, and $j_2 = 1$, If these two were to combine find out the dimensions of various subspaces of angular momentum matrices of the product space.

Solution: Step # 1   Given that $j_1 = 3$, $j_2 = 1$, $n = 2$; i.e., $j_1 > j_2$. $N_1 = (2j + 1) = [2(j_2 + j_2) + 1)$;
$N_2 = (2j - 1) = [2(j_2 + j_2) - 1)$; $N_{2j+1} = [2(j_1 - j_2) + 1]$ Product space dimensions is $N = [2j + 1]$ which is formed of subspaces of $N_1, N_2, \cdot, \cdot, \cdot, N_{2j+1}$ -Ds

Step # 2   The dimension of the product space $N = (2j_1 + 1)(2j_2 + 1) = (2 \times 3 + 1)(2 \times 1 + 1) = 7 \times 3 = 21$.

| Step # 3 | # of subspaces = (n + 1) = (2 + 1) = 3
| Step # 4 | Dimension of subspace1 = $N_1 = (2j+1) = [2(j_2+j_2)+1)] = [2(3+1)+1] = 9$
| Step # 5 | Dimension of subspace2 = $N_2 = (2j-1) = [2(j_2+j_2)-1)] = [2(3+1)-1] = 7$
| Step # 6 | Dimension of subspace3 = $N_{2j+1} = [2(j_1-j_2)+1] = [2(3-1)+1] = 5$.
| Step # 7 | The dimension of the product space $N = 9 + 7 + 5 = 21$

Worked out Example 18.3

Evaluate the new eigen kets $|j, m\rangle$ and the Wigner Coefficients for $j_1 = 1$, and $j_2 = 1$

Solution | Step # 1 | Given that $j_1 = 3, j_2 = 1, n = 2$; i.e., $j_1 > j_2$. $N_1 = (2j+1) = [2(j_2+j_2)+1)]$;
$N_2 = (2j-1) = [2(j_2+j_2)-1)]$; $N_{2j+1} = [2(j_1-j_2)+1]$ Product space dimensions is $N = [2j+1]$
which is formed of subspaces of $N_1, N_2, \cdot, \cdot, \cdot, N_{2j+1}$-Ds

| Step # 2 |. $N_1 = 5, N_2 = 3$, and $N_3 = 1$; $N = 5+3+1 = 9$; i.e., 9 eigen kets in both the old and new bases;

| Step # 3 |

$$\hat{J}^2 = 6\hbar^2 \begin{pmatrix} \text{Subspace \#1} \\ (5 \times 5) \text{ Matrix} \\ \\ \text{Subspace \#2} \\ (3 \times 3) \text{ Matrix} \\ \\ \text{Subspace \#3} \\ (1 \times 1) \text{ Matrix} \end{pmatrix}$$

| Step # 4 | The 9 old kets are:

$|1, 1\rangle, |1, 0\rangle, |1, -1\rangle, |0, 1\rangle, |0, 0\rangle, |-1, -1\rangle, |-1, 0\rangle, |-1, 1\rangle, |0, -1\rangle$

| Step # 5 | New kets: which are PRIMED for convenience

$j=2 \to |2, 2\rangle', |2, 1\rangle', |2, 0\rangle', |2, -1\rangle', |2, -2\rangle$;
$j=1 \to |1, 1\rangle', |1, 0\rangle', |1, -1\rangle'$; $j=0 \to |0, 0\rangle'$;

These are 9 new kets;

| Step # 6 | But $|2, 2\rangle' = |1, 1\rangle'$ **since** $m_1 = m_2 = 1$, $m = m_1 + m_2 = 2$, and $m_1 \| m_2$;
The state $|2, -2\rangle' = |-1, -1\rangle'$, since $m_1 = m_2 = -1, m = m_1 + m_2 = -2$.

| Step # 7 | Using $\hat{J}_\pm$, and starting with $|2, 2\rangle'$,

i.e., $\hat{J}_- |2, 2\rangle'$ or $\hat{J}_+ |2, -2\rangle'$ one gets

1) $|2, 1\rangle' = \sqrt{\frac{1}{2}}\{|0, 1\rangle + |1, 0\rangle\}$;

2) $|2, 0\rangle' = \sqrt{\frac{1}{2}}\{|-1, 1\rangle + 2|0, 0\rangle + |1, -1\rangle\}$.

| Step # 8 | In like manner

3) $|2, -1\rangle' = \sqrt{\frac{1}{6}}\{|0, -1\rangle + |-1, 0\rangle\}$

4) $|2, -2\rangle' = |-1, -1\rangle$

TABLE 18.4   Values of $\langle 1, 1, m - m_2, m_2 \| j, m \rangle$

Diagram Representation	New Ket $\| j, m \rangle'$	Old ket $\| j, m \rangle$	Wigner Coefficient $\langle 1, 1, m - m_2, m_2 \| j, m \rangle$
↑ ↓	$\| 2, 2 \rangle'$	$\| 1, 1 \rangle$	1
	$\| 2, 1 \rangle'$	$\| 0, 1 \rangle$	$\sqrt{\frac{1}{2}}$
		$\| 1, 0 \rangle$	$\sqrt{\frac{1}{2}}$
	$\| 2, 0 \rangle'$	$\| 1, -1 \rangle$	$\sqrt{\frac{1}{6}}$
		$\| 0, 0 \rangle$	$2\sqrt{\frac{1}{6}}$
		$\| -1, 1 \rangle$	$\sqrt{\frac{1}{6}}$
	$\| 2, -1 \rangle'$	$\| 0, -1 \rangle$	$\sqrt{\frac{1}{2}}$
		$\| -1, 0 \rangle$	$\sqrt{\frac{1}{2}}$
	$\| 2, -2 \rangle'$	$\| -1, -1 \rangle$	1
	$\| 1, 1 \rangle'$	$\| 0, 1 \rangle$	$\sqrt{\frac{1}{2}}$
		$\| 1, 0 \rangle$	$-\sqrt{\frac{1}{2}}$
	$\| 1, 0 \rangle'$	$\| 1, -1 \rangle$	$\sqrt{\frac{1}{2}}$
		$\| 0, 0 \rangle$	0
		$\| -1, -1 \rangle$	$-\sqrt{\frac{1}{2}}$
	$\| 1, -1 \rangle'$	$\| 0, -1 \rangle$	$\sqrt{\frac{1}{2}}$
		$\| -1, 0 \rangle$	$-\sqrt{\frac{1}{2}}$
	$\| 0, 0 \rangle'$	$\| 1, -1 \rangle$	$\sqrt{\frac{1}{3}}$
		$\| 0, 0 \rangle$	$-\sqrt{\frac{1}{3}}$
		$\| -1, 1 \rangle$	$\sqrt{\frac{1}{3}}$

Step # 9   It may be noted that all the nine original kets have been used in forming the first five primed kets. It turns out that all new kets keeping the same m value will be linear combinations of the same group of old kets, each of which satisfies the condition, $m = m_1 + m_2$.

5) $\| 1, 1 \rangle' = \sqrt{\frac{1}{2}} \{ \| 0, 1 \rangle - \| 1, 0 \rangle \}$.

Step # 10 6) find $\hat{J}_- \| 1, 1 \rangle'$ giving $\| 1, 0 \rangle' = \sqrt{\frac{1}{2}} \{ \| 1, -1 \rangle - \| -1, 1 \rangle \}$.

Step # 11 7) $\hat{J}_- \| 1, 0 \rangle'$ gives $\| 1, -1 \rangle' = \sqrt{\frac{1}{2}} \{ \| 0, -1 \rangle - \| -1, 0 \rangle \}$

Step # 12 8) To find $|0, 0\rangle'$, it must be linear combination of $|1, -1\rangle, |-1, 1\rangle$ and $|0, 0\rangle$, and it must be orthogonal to both $|2, 0\rangle'$ and $|1, 0\rangle'$. The result is

$$|0, 0\rangle' = \sqrt{\tfrac{1}{3}}\{|1, -1\rangle - |0, 0\rangle + |-1, 1\rangle\}$$

Step # 13 The Wigner coefficients, for $j_1 = 1$, and $j_2 = 1$, are listed in Table 18.4.

18.4.4 TENSOR OPERATORS AND THE WIGNER-ECKART THEOREM:

In Chapter 14, spatial rotation of vector quantity was discussed. In equation (14.6.15), viz.

$$|\psi^o(\bar{r})\rangle = \hat{U}_n(\alpha_z) |\psi^o(\bar{r}^o)\rangle \qquad (14.6.15)$$

an operator (or matrix) $\hat{U}_n(\delta\alpha_z)$ which acts in function space (say, *Hilbert space*), and characterize a rotation $\hat{\Re}_\alpha \equiv \hat{\Re}(\alpha_z)$ in *configuration space*. For convenience introduce an operator $\hat{T}(\Re_\alpha) \equiv \hat{U}_n(\alpha_z)$. Thus the operation

$\hat{T}(\Re_\alpha) |\psi^o(\bar{r}^o)\rangle$ yields a new ket which from equation (14.6.5), must satisfy

$$\hat{T}(\Re_\alpha) |\psi^o(\bar{r}^o)\rangle = |\psi^o(\bar{r})\rangle \qquad (18.5.1)$$

18.5.1 Spherical Tensors

One can construct from orbital angular momentum eigen kets, $|\ell, m_\ell\rangle s$, and from spin eigen kets (spinors), $|\ell, m_\ell\rangle s$, new eigen kets of the total angular momentum $\hat{J} = \hat{L} + \hat{S}$. The procedures for doing this are well known from the theory of angular momentum, which were seen in the earlier Chapters. If one considers the addition of $\hat{J}_1$ and $\hat{J}_2$ corresponding to angular momenta of two different particles or alternatively to orbital and internal angular momenta for one particle, the one has to take into account two possible basis sets. These are the 'uncoupled representation' and 'coupled representation'.

The matrix transformation from one to the other involves C-G.coefficients, $\langle j_1, j_2, m - m_2, m_2 || j_1, j_2, j, m\rangle$,

i.e. $\langle j_1, j_2, \hat{J} || m_1, m_2, m\rangle$, which is covered in Section 18.5.

For the problem at hand, the uncoupled representation is simply given by the product of spherical harmonics and the spherical basis vectors. The form eigen kets of the four relevant operators for the uncoupled representation:

$$\hat{\ell}_z |\ell, m_\ell\rangle = m_\ell \hbar |\ell, m_\ell\rangle \qquad (15.5.4)$$

$$\hat{\ell}^2 |\ell, m_\ell\rangle = \ell(\ell+1) \hbar^2 |\ell, m_\ell\rangle \qquad (15.5.21)$$

$$\hat{S}_z |s, m_\ell\rangle = m_s \hbar |s, m_s\rangle \qquad (17.2.6)$$

$$\hat{S}^2 |s, m_s\rangle = s(s+1) \hbar^2 |s, m_s\rangle \qquad (17.2.7)$$

Using the C.G. coefficients carry the out a transformation to the COUPLED REPRESENTATION, the new eigen kets the total angular momentum, $\hat{J}$, are defined as

$$\hat{T}_{J\ell M}(\bar{r}) = \langle \ell, s, J, M || m, \mu, M\rangle |\ell, m_\ell\rangle |\mu, m_s\rangle \qquad (18.5.2)$$

These three component quantities are known as VECTOR SPHERICAL HARMONICS. Then

$$\hat{J}^2 \hat{T}_{J\ell M}(\bar{r}) = J(J+1) \hbar^2 \hat{T}_{J\ell M}(\bar{r}) \qquad (18.5.3)$$

$$\hat{J}_z \hat{T}_{J\ell M}(\bar{r}) = M \hbar \hat{T}_{J\ell M}(\bar{r}) \qquad (18.5.4)$$

and also $\quad \hat{\ell}^2 \hat{T}_{J\ell M}(\bar{r}) = \ell(\ell+1) \hbar^2 \hat{T}_{J\ell M}(\bar{r}) \qquad (18.5.5)$

$$\hat{S}^2 \hat{T}_{J\ell M}(\vec{r}) = s(s+1)\hbar^2 \hat{T}_{J\ell M}(\vec{r}) \tag{18.5.6}$$

The symmetric properties of the spherical harmonics and of the spinors together with that of the C.G. coefficients, one sees that

$$\hat{T}_{J\ell M}(\vec{r})^* = (-1)^{M+\ell+s+J} \hat{T}_{J\ell M}(\vec{r}) \tag{18.5.7}$$

18.5.2 THE WIGNER-ECKART THEOREM

It is common to take the matrix elements of an irreducible tensor operator $\hat{T}_{J\ell M}(\vec{r})$ of rank L between the initial and final angular momentum states $|j, m\rangle$ and $|j', m'\rangle$. The WIGNER-ECKART THEOREM enables one to separate the magnetic quantum number dependence of the matrix element in the form of the C.G. coefficient.

18.5.2.1 STATEMENT OF THE WIGNER-ECKART THEOREM

In a representation $\{\hat{J}^2, \hat{J}_z\}$ in which the base vectors are $|\tau, \hat{J}, M\rangle$, the matrix element $\langle \tau, \hat{J}, M | \hat{T}_{J\ell M}(\vec{r})^k | \tau', \hat{J}', M' \rangle$ is equal to the product of the Clebsch-Gordon coefficient, $\langle \tau', k, M, q \| \hat{J}, M \rangle$ and a quantity $\langle \tau, \hat{J} \| \tau', \hat{J}' \rangle$, independent of M, M' and q,

i.e. $\langle \tau, \hat{J}, M | \hat{T}_{J\ell M}(\vec{r})^k | \tau', \hat{J}', M' \rangle$

$$= \sqrt{\tfrac{1}{2}j+1} \langle \tau, \hat{J} | \hat{T}_{J\ell M}(\vec{r})^k | \tau', \hat{J}' \rangle \langle \tau', k, M', q \| J, M \rangle \tag{18.5.8}$$

$\langle \tau, \hat{J} | \hat{T}_{J\ell M}(\vec{r})^k | \tau', \hat{J}' \rangle$ is called the *reduced matrix element*.
The same is represented by

$$\langle j', m' | \hat{T}_{LM}(\vec{r}) | j, m \rangle = \langle j L m M \| j', m' \rangle \{ | j' \rangle \hat{T}_{LM}(\vec{r}) | j \rangle \} \tag{18.5.8a}$$

From the properties of the C.G. Coefficient is obvious that conservation of Angular Momentum $j + L = j'$ and $m' = M + m$ follows.

18.5.2.2 DERIVATION of the Theorem

An irreducible tensor $\hat{T}_{JL;M}(\vec{r})^k$ of rank L is defined as set of (2L + 1) functions (M = -L to +L), which under the (2L+1)-dimensional representation of the rotation group, transforms as follows:

$$\hat{\mathfrak{R}}_\alpha \hat{T}_{JL;M}(\vec{r})^{\eta-1} = \sum D^L_{M'M}(\theta\,\varphi\,\psi)\,\hat{T}_{JL;M}(\vec{r}) \tag{18.5.9}$$

where the rotation operator, $\hat{\mathfrak{R}}_\alpha$, rotates the coordinate system by an angle α about the direction $\hat{n}$, $\vec{J}$ being the Total Angular Momentum.

$$\hat{\mathfrak{R}}_\alpha = e^{(-i\alpha\hat{n}\cdot\vec{J})} \tag{18.5.10}$$

An equivalent definition of an irreducible tensor is that the set $\hat{T}_{JL;M}(\vec{r})$ satisfy the Commutation Relation

$$[\hat{J}_\pm, \hat{T}_{JL;M}(\vec{r})]_- = \hbar\sqrt{(j\mp m)(j\pm m+1)}\,|\hat{T}_{JL;M}(\vec{r})\rangle \tag{18.5.11}$$

$$[\hat{J}_\pm, \hat{T}_{JL;M}(\vec{r})]_- = M\hbar\,|\hat{T}_{JL;M}(\vec{r})\rangle \tag{18.5.12}$$

In considering the matrix elements of (18.5.11) and (18.5.12), from equation (18.5.12) one has

$$\langle j', m' | [\hat{J}_\pm, \hat{T}_{JL;M}(\vec{r})] | j, m \rangle = M\hbar \langle j', m' | \hat{T}_{JL;M}(\vec{r}) | j, m \rangle$$

$$= M\hbar \langle j', m' | \hat{J}_\pm \hat{T}_{JL;M}(\vec{r}) | j, m \rangle - M\hbar \langle j', m' | \hat{T}_{JL;M}(\vec{r}) \hat{J}_\pm | j, m \rangle$$

$$= m'\hbar \langle j', m' | \hat{T}_{JL;M}(\vec{r}) | j, m \rangle - m\hbar \langle j', m' | \hat{T}_{JL;M}(\vec{r}) | j, m \rangle.$$

which gives $\quad (m'-m-M)\hbar \langle j', m' | \hat{T}_{JL;M}(\vec{r}) | j, m \rangle = 0 \tag{18.5.13}$

Therefore, for $\langle j', m' | \hat{T}_{JL;M}(\vec{r}) | j, m \rangle \neq 0$, one must have
$$m' = m + M.$$
From equation (18.5.11), one has
$$\langle j', m' |[\hat{J}_\pm , \hat{T}_{JL;M}(\vec{r})]| j, m \rangle$$
$$= \langle j', m' | \hat{J}_\pm \hat{T}_{JL;M}(\vec{r}) | j, m \rangle - \langle j', m' | \hat{T}_{JL;M}(\vec{r}) \hat{J}_\pm | j, m \rangle$$
$$= \hbar \sqrt{(L \mp M)(L \pm M + 1)} \langle j', m' | \hat{T}_{JL;M}(\vec{r}) | j, m \rangle$$

$$= \hbar \sqrt{\{j' \pm m'\}(j' \mp m' + 1)} \langle j', m' \mp 1| \hat{T}_{JL;M}(\vec{r}) | j, m \rangle$$
$$- \hbar \sqrt{\{j \mp m\}(j \pm m + 1)} \langle j', m'| \hat{T}_{JL;M}(\vec{r}) | j, m \pm 1 \rangle \qquad (18.5.14)$$
where in the last expression the use of the properties of $J_\pm$ was made.
From equations (18.5.13) and (18.5.14) are satisfied only if
$$m' = m + M \pm 1.$$
Consider next the coupling of angular momentum states $|j, m\rangle$ and $|L, M\rangle$ to form the composite state $|j', m'\rangle$, as implied in equation (18.5.8).
$$|j', m'\rangle = \sum \langle J, m, M || j', m' \rangle \langle j, m || L, M \rangle \qquad (18.5.15)$$
Operating with on $\hat{J}_\mp' = (\hat{J}_\mp + \hat{J})$ both sides of this equation, one gets
$$\hbar \sqrt{\{j' \pm m'\}(j' \mp m' + 1)} |j', m' \mp 1\rangle$$
$$= \sum \hbar \sqrt{\{j \pm m\}(j \mp m + 1)} \langle J, L, m, M || j', m' \rangle |j, m \mp 1\rangle |L, M\rangle$$
$$+ \sum \hbar \sqrt{(L \pm M)(L \mp M + 1)} \langle J, L, m, M || j', m' \rangle |j, m\rangle |L, M \mp 1\rangle$$
$$= \sum \hbar \sqrt{\{j' \pm m'\}(j' \mp m' + 1)} \langle J, L, m'', M' || j', m' \mp 1\rangle |j, m''\rangle |L, M'\rangle .(18.5.16)$$
where in the last line equation (18.5.15) is used to express the first line. Multiplying equation (18.5.16) through by $\langle j, m''| \langle L, M'|$, using the orthonormality condition of the states, and relabeling the z-components, one obtains
$$\hbar \sqrt{\{j \mp m\}(j \pm m + 1)} \langle J, L, m \pm 1, M || j', m' \rangle$$
$$+ \hbar \sqrt{\{L \mp M\}(L \mp M + 1)} \langle J, L, m, M \pm 1 || j', m' \rangle$$
$$= \hbar \sqrt{\{j' \pm m'\}(j' \mp m' + 1)} \langle J, L, m, M ||(j', m' \mp 1\rangle \qquad (18.5.17)$$
By comparing equations (18.5.14) and (18.5.17), it is clear that the matrix elements $\langle \tau, \hat{J}, M | \hat{T}_{J\ell M}(\vec{r})^k | \tau', \hat{J}', M'\rangle$, i.e. $\langle j', m' | \hat{T}_{J\ell M}(\vec{r})^k | j, m \rangle$ are proportional to the C.G. coefficient $\langle J, L, m, M ||(j', m')\rangle$, which completely takes into account the m, m', M dependence of the matrix element. Hence the result equation (18.5.8) follows. The factor $\langle J, L, m, M ||(j', m')\rangle$ depends only on the rank of the tensor $\hat{T}_{JLM}(\vec{r})$, and the physical nature of the tensor is contained in the reduced matrix element
$$\langle \tau, \hat{J} | \hat{T}_{J\ell M}(\vec{r})^k | j', \hat{J}' \rangle \equiv \langle j' | \hat{T}_{J\ell M}(\vec{r})^k | j \rangle.$$

REVIEW QUESTIONS

R.Q. 18.1 Given the angular momentum of two isolated systems as $j_1 = 1$, and $j_2 = 1$. If these two were to combine find out the dimensions of various the subspaces of angular momentum matrices of the product space. Give the Matrix Representation of $\hat{J}_z$. Answers: $N_1 = 5, N_2 = 3,$ and $N_3 = 1$; $N = 5+3+1 = 9$; $\hat{J}_z |j_1, j_2, j, m\rangle \equiv m\hbar |j_1, j_2, j, m\rangle$; $m = m_1 + m_2 = 2$

$\hat{J}_z = 2\hbar$

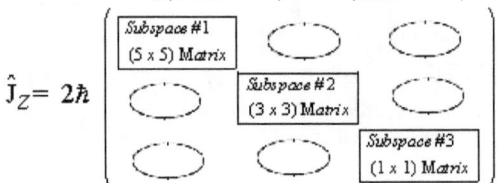

R.Q. 18.2 Combine two unit Angular Momenta.
R.Q. 18.3 Illustrate how the Wigner coefficients may be used to determine the eigen kets of the Spin-Orbit Coulomb repulsion Hamiltonian, $\hat{H} = \sum -\frac{\hbar^2}{2m}\nabla_i^2 - \frac{Ze^2}{r_i} + \varsigma\, \vec{\ell}_i \cdot \vec{s}_i + \sum \frac{e^2}{r_{ij}}$, of a many-electron atom, for the np^2 electron configuration. Do it for L-S coupling.
R.Q. 18.4 Draw up tables of the quantum numbers m_1, m_2 and ℓ, corresponding to different values of m for the case where $\ell_1 = 2$ and $\ell_2 = 1$. (Answers: Table

TABLE 18.5

m_1, m_2	m	ℓ
(2, 1)	1	3
(2, 0), (1, 1)	2	3, 2
(2, 1), (1, 0), (0, 1)	1	3, 2, 1
(1, -1), (1, 0), (0, 1)	1	3, 2, 1
(0, -1), (-1, 0), (-2, 1)	-1	3, 2, 1
(-2, 0), (-1, -1)	-2	3, 2
(-2, 1)	-3	3

).

&&*&*&*&*&*&*&*&*

Chapter 19

PERTURBATION THEORY OF THE STATIONARYSTATE

Chapter 19

PERTURBATION THEORY OF THE STATIONARY STATE

"To be confused about what is different and what is not, is to be confused about everything"
- David Bohm
"For the wise all 'things are wiped away"
-Gautama Buddha

19.1 INTRODUCTION

In practice it is possible to find exact solutions to the Schrödinger Equation, *i.e.* the eigen value equation, for only a very few potentials, V(r), or systems. These are the *Harmonic Oscillator*, *Rigid Rotator*, bound states of a *Particle in a Box* and a *Single Electron Atom* (H-atom). Even here discrepancies occur due to

(i) Additional interactions, such as the Spin-Orbit and Spin-Spin Interactions,

(ii) Complications may also arise when an EM Field is present,

(3) Further difficulty appears when additional particles are added to the system (many body systems).

In such cases an exact solution is possible in the unrealistic case of completely non-interacting particles. Probably this is approximately exact if there are interactions, which were only the weak intermolecular forces. Thus often an exact mathematical solution of the Schrödinger Equation cannot be found for other problems. Usually one has to seek approximate solutions, to obtain the eigen values and eigen functions for interacting potentials that do not lead to exactly solvable equations. In fact no physical system can be studied without approximation. Two main methods are used for this purpose: the linear VARIATION METHOD, to be discussed in Chapter 20, and PERTURBATION THEORY, to be dealt with in this Chapter. In the approximate treatment of many body Schrödinger Equations the WKB Approximation is also made use of. Each of these techniques is useful for specific problems. Perturbation Methods apply when the problem is to be solved differs but little from a situation with a known solution. Perturbation Theories are of two kinds:

(a) TIME-INDEPENDENT (Stationary state or bound state) and

(b) TIME-DEPENDENT. The time independent perturbation theory will be covered up in this Chapter.

19.2 TIME INDEPENDENT PERTURBATION (OR STATIONARY STATE/ OR THE RAYLEIGH-SCHRODINGER EXPANSION) THEORY

The state of a system is mainly determined by some strong interaction.
The real system is to be regarded as a modification of one of the model systems whose solutions are known already. The Hamiltonian, $\hat{H}$ of the real system

$$\hat{H} = \hat{H}^{(o)} + \hat{H}^{(1)} . \tag{19.2.1}$$

where $\hat{H}^{(o)}$ = the IDEALIZED Hamiltonian whose solutions are known.
$\hat{H}^{(1)}$ = the TIME INDEPENEDENT Perturbation Term in the Hamiltonian due to additional interactions (external influence) and the effects of which cannot be ignored. $\hat{H}^{(1)}$ is a hermitian operator like ideal $\hat{H}^{(o)}$

If $\boxed{\hat{H}^{(1)} < \hat{H}^{(o)}}$, (19.2.2)

i.e. $\hat{H}^{(o)}$ is dominant in $\hat{H}$, the corrections to both the eigen values and eigen functions resulting from $\hat{H}^{(1)}$, will also be small, and Perturbation Theory can be used.
One can write for the real system,

$$\boxed{\hat{H} = \hat{H}^{(o)} + \alpha \hat{H}^{(1)}}.$$ (19.2.3)

The real parameter α is not fundamental, but is added for convenience to 'turn on' and 'turn off' the perturbation, $\hat{H}^{(1)}$.
$\alpha = 0 \sim 1$.

Consider the choice of a set of ortho-normal basis functions, $\left\{ \left| \hat{\psi}_n^o \right\rangle \right\}$, such that

$$\boxed{\left\langle \hat{\psi}_m^o \middle| \hat{\psi}_n^o \right\rangle = \delta_{mn}}.$$ (19.2.4)

where $\left| \hat{\psi}_n^o \right\rangle$ are the unperturbed eigen functions having unperturbed energies $E^o{}_n$. In mathematical terms, this means that they form the solutions of the IDEAL (zeroth order) problem

$$\hat{H}^{(o)} \left| \hat{\psi}_n^o \right\rangle = E^o{}_n \left| \hat{\psi}_n^o \right\rangle$$ (19.2.5)

i.e. the spectrum of $\hat{H}$, varies continuously with α, coinciding with the spectrum of $\hat{H}^{(o)}$ when $\alpha = 0$.
The new problem to be solved is

$$\boxed{\hat{H} \left| \hat{\psi}_n \right\rangle = E_n \left| \hat{\psi}_n \right\rangle}$$ (19.2.6)

where $\left| \hat{\psi}_n \right\rangle$ and E_n are perturbed stationary state functions and energies, respectively. Using (19.2.3), this becomes $(\hat{H}^{(o)} + \alpha \hat{H}^{(1)}) \left| \hat{\psi}_n \right\rangle = E_n \left| \hat{\psi}_n \right\rangle$ (19.2.7)

19.3 NON-DEGENERATE CASE

19.3.1 SETTING UP the solution

Assume that both $\left| \hat{\psi}_n^o \right\rangle$ and $\left| \hat{\psi}_n \right\rangle$ are

i) Discrete, and
ii) Non-degenerate sets of time-independent functions, and
iii) There is one-to-one correspondence between the members of each set as below; thus

$$\left| \hat{\psi}_1 \right\rangle \xrightarrow{\alpha \to 0} \left| \hat{\psi}_1^o \right\rangle,$$
$$\left| \hat{\psi}_2 \right\rangle \xrightarrow{\alpha \to 0} \left| \hat{\psi}_2^o \right\rangle, \text{ etc.}$$ (19.3.1)

This is stated better mathematically, as

$$\boxed{\underset{\alpha \to 0}{Lt} \left| \hat{\psi}_n \right\rangle = \left| \hat{\psi}_n^o \right\rangle}$$ (19.3.2)

$$\underset{\alpha \to 0}{Lt} E_n = E_n^{(o)}$$ (19.3.3)

In the context of (19.3.1), and $\alpha \hat{H}^{(1)}$ sufficiently small, it is reasonable to expand $|\hat{\psi}_n\rangle$ and E_n (which are functions of α) as rapidly converging Power Series in α.

$$|\hat{\psi}_n\rangle = |\hat{\varphi}_n^{(o)}\rangle + \alpha|\hat{\varphi}_n^{(1)}\rangle + \alpha^2|\hat{\varphi}_n^{(2)}\rangle + \alpha^3|\hat{\varphi}_n^{(3)}\rangle + \cdots + \alpha^n|\hat{\varphi}_n^{(n)}\rangle. \quad (19.3.4)$$

$$E_n = A_n^{(o)} + \alpha A_n^{(1)} + \alpha^2 A_n^{(2)} + \alpha^3 A_n^{(3)} + \cdots + \alpha^n A_n^{(n)} \quad (19.3.5)$$

where because of equations (19.3.2) and (19.3.3)

$$\boxed{|\hat{\varphi}_n^{(o)}\rangle \equiv |\psi_n^o\rangle; \quad A_n^{(o)} = E_n^{(o)}} \quad (19.3.6)$$

$$|\hat{\psi}_n\rangle = |\psi_n^o\rangle + \alpha|\hat{\varphi}_n^{(1)}\rangle + \alpha^2|\hat{\varphi}_n^{(2)}\rangle + \alpha^3|\hat{\varphi}_n^{(3)}\rangle + \cdots + \alpha^n|\hat{\varphi}_n^{(n)}\rangle. \quad (19.3.7)$$

$$E_n = E_n^{(o)} + \alpha A_n^{(1)} + \alpha^2 A_n^{(2)} + \alpha^3 A_n^{(3)} + \cdots + \alpha^n A_n^{(n)}. \quad (19.3.8)$$

It is assumed those successive terms of these series get smaller so that the Series converges.

19.3.2 To find the CORRECTIONS TO ENERGY and WAVE FUNCTION:
Substituting equation (19.3.8) in (19.2.7),

$$(\hat{H}^{(o)} + \alpha \hat{H}^{(1)})|\hat{\psi}_n\rangle = E_n|\hat{\psi}_n\rangle \quad (19.2.7)$$

$$(\hat{H}^{(o)} + \alpha \hat{H}^{(1)})|\hat{\psi}_n\rangle$$

$$= (\hat{H}^{(o)} + \alpha \hat{H}^{(1)})\left[|\psi_n^o\rangle + \alpha|\hat{\varphi}_n^{(1)}\rangle + \alpha^2|\hat{\varphi}_n^{(2)}\rangle + \alpha^3|\hat{\varphi}_n^{(3)}\rangle + \cdots + \alpha^n\right]$$

$$= E_n|\hat{\psi}_n\rangle$$

$$= \left[E^o_n + \alpha A_n^{(1)} + \alpha^2 A_n^{(2)} + \alpha^3 A_n^{(3)} + \cdots + \alpha^n A_n^{(n)}\right]$$

$$(\hat{H}^{(o)} + \alpha \hat{H}^{(1)})\left[|\psi_n^o\rangle + \alpha|\hat{\varphi}_n^{(1)}\rangle + \alpha^2|\hat{\varphi}_n^{(2)}\rangle\right]$$

$$= \left[E_n^{(o)} + \alpha A_n^{(1)} + \alpha^2 A_n^{(2)}\right]\left[|\psi_n^o\rangle + \alpha|\hat{\varphi}_n^{(1)}\rangle + \alpha^2|\hat{\varphi}_n^{(2)}\rangle\right]$$

i.e., $\left[\hat{H}^{(o)} - E_n^{(o)}\right]|\psi_n^o\rangle + \alpha\left\{\hat{H}^{(1)}|\psi_n^o\rangle + \hat{H}^{(o)}|\hat{\varphi}_n^{(1)}\rangle - A_n^{(1)}|\psi_n^o\rangle - E_n^{(o)}|\hat{\varphi}_n^{(1)}\rangle\right\}$

$$+ \alpha^2\left\{\hat{H}^{(o)}|\hat{\varphi}_n^{(2)}\rangle + \hat{H}^{(1)}|\hat{\varphi}_n^{(1)}\rangle - E_n^{(o)}|\hat{\varphi}_n^{(2)}\rangle - A_n^{(1)}|\hat{\varphi}_n^{(1)}\rangle - A_n^{(2)}|\psi_n^o\rangle\right\}$$

$$+ \alpha^3\{\cdots\cdots\} + \cdots\cdots = 0 \quad (19.3.9)$$

Since α = an arbitrary parameter, the coefficient of each power of α can be set equal to zero. i.e.

$$\boxed{\left[\hat{H}^{(o)} - E^o_n\right]|\psi_n^o\rangle = 0} \quad (19.3.10)$$

$$\boxed{\hat{H}^{(1)}|\psi_n^o\rangle + \hat{H}^{(o)}|\hat{\varphi}_n^{(1)}\rangle - A_n^{(1)}|\psi_n^o\rangle - E^o_n|\hat{\varphi}_n^{(1)}\rangle = 0}, \quad (19.3.11)$$

$$\boxed{\hat{H}^{(o)}|\hat{\varphi}_n^{(2)}\rangle + \hat{H}^{(1)}|\hat{\varphi}_n^{(1)}\rangle - E_n^{(o)}|\hat{\varphi}_n^{(2)}\rangle - A_n^{(1)}|\hat{\varphi}_n^{(1)}\rangle - A_n^{(2)}|\psi_n^o\rangle = 0} \quad (19.3.12)$$

19.3.3 ZEROTH ORDER PERTURBATION

$$\left[\hat{H}^{(o)} - E^o_n\right]|\psi_n^o\rangle = 0 \quad (19.3.10)$$

This equation is identical to that of (19.2.5). The solutions are exact ones, viz. $E_n^{(o)}$ and $|\psi_n^o\rangle$.

19.3.2.1 FIRST ORDER Perturbation Theory
If one assumes α = small, one can satisfy the equation (19.3.11),

$$\hat{H}^{(1)}\left|\hat{\psi}_n^o\right\rangle + \hat{H}^{(o)}\left|\hat{\varphi}_n^{(1)}\right\rangle - A_n^{(1)}\left|\hat{\psi}_n^o\right\rangle - E_n^{(o)}\left|\hat{\varphi}_n^{(1)}\right\rangle = 0, \tag{19.3.11}$$

This may be solved by expanding the fist order correction to the wave function in terms of unperturbed eigen functions, $|\psi^o{}_n\rangle$, i.e. $\left|\hat{\varphi}_n^{(1)}\right\rangle$ values in equation (19.3.7) may be expressed as a linear combination of the eigen functions of $\hat{H}^{(o)}$, i.e. the ortho-normal set $\left\{\left|\hat{\psi}_n^o\right\rangle\right\}$.

$$\boxed{\left|\hat{\varphi}_n^{(1)}\right\rangle = \sum_i a_{ni}^{(1)}\left|\hat{\psi}_i^o\right\rangle} \tag{19.3.13}$$

which is exact. $\left|\hat{\varphi}_n^{(1)}\right\rangle$ are transformed from $\left|\hat{\psi}_i^o\right\rangle$. $a_{ni}^{(1)}$ = complex constants.

On substitution (19.3.11) becomes

$$\hat{H}^{(1)}\left|\hat{\psi}_n^o\right\rangle + \hat{H}^{(o)}\sum_i a_{ni}^{(1)}\left|\hat{\psi}_i^o\right\rangle - A_n^{(1)}\left|\hat{\psi}_n^o\right\rangle - E_n^{(o)}\sum_i a_{ni}^{(1)}\left|\hat{\psi}_i^o\right\rangle = 0$$

i.e. $\quad \hat{H}^{(1)}\left|\hat{n}\right\rangle_o + \hat{H}^{(o)}\sum a_{ni}^{(1)}\left|\hat{i}\right\rangle_o - A_n^{(1)}\left|\hat{n}\right\rangle_o - E_n^{(o)}\sum a_{ni}^{(1)}\left|\hat{i}\right\rangle_o = 0$

Multiplying throughout by ${}_o\langle k |$ (instead of pre-multiplying by $\hat{\psi}_k^o$, post-multiplying by $d\tau$ and integrating),

$${}_o\langle k|\hat{H}^{(1)}|\hat{n}\rangle_o + {}_o\langle k|\hat{H}^{(o)}\sum a_{ni}^{(1)}|\hat{i}\rangle_o$$
$$- {}_o\langle k|A_n^{(1)}|\hat{n}\rangle_o - {}_o\langle k|E_n^{(o)}\sum a_{ni}^{(1)}|\hat{i}\rangle_o = 0.$$

$${}_o\langle k|\hat{H}^{(1)}|\hat{n}\rangle_o + \sum a_{ni}^{(1)}(E_n^{(o)}){}_o\langle k\|\hat{i}\rangle_o$$
$$- {}_o\langle k|A_n^{(1)}|\hat{n}\rangle_o - \sum a_{ni}^{(1)}{}_o\langle k\|\hat{i}\rangle_o E_n^{(o)} = 0$$

$${}_o\langle k|\hat{H}^{(1)}|\hat{n}\rangle_o + \sum a_{ni}^{(1)}(E^o{}_n)\delta_{ki} = A_n^{(1)}\delta_{ki} + \sum a_{ni}^{(1)} E_n^{(o)}\delta_{ki}.$$

$${}_o\langle k|\hat{H}^{(1)}|\hat{n}\rangle_o + a_{nk}^{(1)}(E_k^{(o)}) = A_n^{(1)}\delta_{kn} + a_{nk}^{(1)} E_n^{(o)},$$

as $\quad a_{ni}^{(1)}\cdot\delta_{ki} = a_{nk}^{(1)}$

$$\hat{H}_{kn}^{(1)} + a_{nk}^{(1)}(E_k^{(o)}) = A_n^{(1)}\delta_{kn} + a_{nk}^{(1)}(E_n^{(o)}), \tag{19.3.14}$$

$\hat{H}_{kn}^{(1)} = {}_o\langle k|\hat{H}^{(1)}|\hat{n}\rangle_o$ is the matrix element of the operator $\hat{H}^{(1)}$ with the unperturbed eigen functions as the basis. Then

$$\hat{H}_{kn}^{(1)} + a_{nk}^{(1)}\left(E_k^{(o)} - E_n^{(o)}\right) = A_n^{(1)}\delta_{kn} \tag{19.3.15}$$

Two cases can be considered.

19.3.4.1 **Case # 1** When $k = n$, from equation (19.3.15)

$$A_n^{(1)}\delta_{kn} = \hat{H}_{kn}^{(1)} + a_{nk}^{(1)}\left(E_k^{(o)} - E_n^{(o)}\right)$$

$$\boxed{A_n^{(1)} = \hat{H}_{kn}^{(1)}}. \tag{19.3.16}$$

This important result shows that the first order perturbation correction in the energy $E^o{}_n$ of the nth eigen state, $\left|\hat{\psi}_n^o\right\rangle$, is the diagonal matrix element corresponding to the nth row and n^{th} column of the matrix of the perturbation, $\hat{H}^{(1)}$. Now one can write the approximate eigen value, corrected to first order, $E_n^{(1)}$ is obtained from (19.3.8) and (19.3.16) as

$$\boxed{E_n \approx E_n^{(1)} \cong E_n^{(1)} + \hat{H}_{nn}^{(1)}}. \qquad (19.3.17)$$

This is the first order-perturbed energy.

19.3.4.2 **Case # 2** When $k \neq n$, equation (19.3.15) becomes

$$\hat{H}_{kn}^{(1)} + a_{nk}^{(1)}\left(E_k^{(o)} - E_n^{(o)}\right) = 0,$$

$$\boxed{a_{nk}^{(1)} = \frac{\hat{H}_{nn}^{(1)}}{\left(E_n^{(o)} - E_k^{(o)}\right)}}. \qquad (19.3.18)$$

where the CONDITION $\left|\hat{H}_{kn}^{(1)}\right| \ll \left(E_n^{(o)} - E_k^{(o)}\right)$ be satisfied.

This is how the first-order corrections to the wave function $\left|\hat{\psi}_n\right\rangle$ is made.

The first-order perturbed wave function, $\left|\psi_n^{(1)}\right\rangle$ is from (19.3.7) and (19.313)

$$\boxed{\left|\hat{\psi}_n\right\rangle \approx \left|\hat{\psi}_n^{(1)}\right\rangle = \left|\hat{\psi}_n^{(o)}\right\rangle + \left|\hat{\varphi}_n^{(1)}\right\rangle}$$

$$\cong \left|\hat{\psi}_n^{(o)}\right\rangle + \sum a_{ni}^{(1)} \left|\hat{\psi}_i^{(o)}\right\rangle$$

$$\boxed{\left|\hat{\psi}_n\right\rangle \cong \left|\hat{\psi}_n^{(o)}\right\rangle + \sum \frac{\hat{H}_{nn}^{(1)}}{\left(E_n^{(o)} - E_k^{(o)}\right)}\left|\hat{\psi}_i^{(o)}\right\rangle} \qquad (19.3.19)$$

19.3.5 FEATURES

a) Each wave function that has a non-zero matrix element with $\left|\hat{\psi}_n^{(o)}\right\rangle$ contributes to the new wave function of the nth state. Hence, when the matrix of the perturbation has non-vanishing, off-diagonal elements, this means that the perturbation has produced some interference or mixing of the original wave functions so that they are no more orthogonal.

The AMOUNT OF THIS MIXING = function of $\left|\hat{H}_{kn}^{(1)}\right|$ and $\left|\left(E_n^{(o)} - E_k^{(o)}\right)\right|$.

a) This reveals that states that are widely separated in energy are not expected to interfere to any large degree.

b) The denominator of (19.3.19) $\neq 0$, since degeneracy is excluded.

c) The first order correction to energy can either raise or lower the energy from $E_n^{(o)}$.

d) $\left|\hat{\varphi}_n^{(1)}\right\rangle = \sum a_{ni}^{(1)} \left|\hat{\psi}_i^o\right\rangle$ (19.3.13) means that $\left|\hat{\psi}_i^o\right\rangle$ the unperturbed eigen function are transformed to $\left|\hat{\varphi}_n^{(1)}\right\rangle$. The same matrix $\hat{A} = [a_{nk}^{(1)}]$, which transforms $\left|\hat{\psi}_i^o\right\rangle$ to $\left|\hat{\varphi}_n^{(1)}\right\rangle$, will transform the matrix $\hat{H}$ into its diagonal form through $\hat{A}\hat{H}\hat{A}^\dagger = $ diagonal $\hat{H}$ (i.e. $\hat{H}^{(o)}$).

19.4 SECOND ORDER Perturbation Theory

19.4.1 Equation relating quantities for the second-order corrections
The second order perturbation equation is

$$\hat{H}^{(o)}\left|\hat{\varphi}_n^{(2)}\right\rangle + \hat{H}^{(1)}\left|\hat{\varphi}_n^{(1)}\right\rangle - E_n^{(o)}\left|\hat{\varphi}_n^{(2)}\right\rangle - A_n^{(1)}\left|\hat{\varphi}_n^{(1)}\right\rangle - A_n^{(2)}\left|\hat{\psi}_n^o\right\rangle = 0. \qquad (19.3.12)$$

Further, $\left|\hat{\psi}_n\right\rangle = \left|\hat{\psi}_n^o\right\rangle + \alpha\left|\hat{\varphi}_n^{(1)}\right\rangle + \alpha^2\left|\hat{\varphi}_n^{(2)}\right\rangle + \alpha^3\left|\hat{\varphi}_n^{(3)}\right\rangle + \cdots + \alpha^n\left|\hat{\varphi}_n^{(n)}\right\rangle \qquad (19.3.7)$

$$E_n = E_n^{(o)} + \alpha A_n^{(1)} + \alpha^2 A_n^{(2)} + \alpha^3 A_n^{(3)} + \cdots + \alpha^n A_n^{(n)} \qquad (19.3.8)$$

Assume that $\left|\hat{\varphi}_n^{(2)}\right\rangle$ values of equation (19.3.7) may be expressed as a linear combination of the ortho-normal set of eigen functions $\left\{\left|\hat{\psi}_n^o\right\rangle\right\}$,

$$\boxed{\left|\hat{\varphi}_n^{(2)}\right\rangle = \sum a_{nj}^{(2)} \left|\hat{\psi}_j^o\right\rangle} \qquad (19.4.1)$$

and already $\left|\hat{\varphi}_n^{(1)}\right\rangle = \sum a_{ni}^{(1)} \left|\hat{\psi}_i^o\right\rangle$ \qquad (19.3.13)

$a_{nj}^{(2)}$ are expansion coefficients, in general, complex.
Substituting these in equation (19.3.12)

$$\hat{H}^{(o)} \sum a_{nj}^{(2)} \left|\hat{\psi}_j^o\right\rangle + \hat{H}^{(1)} \sum a_{ni}^{(1)} \left|\hat{\psi}_i^o\right\rangle - E_n^{(o)} \sum a_{nj}^{(2)} \left|\hat{\psi}_j^o\right\rangle$$
$$- A_n^{(1)} \sum a_{ni}^{(1)} \left|\hat{\psi}_i^o\right\rangle - A_n^{(2)} \left|\hat{\psi}_n^o\right\rangle = 0$$

Multiplying by $\left\langle\hat{\psi}_k^o\right|$ through out the equation,

$$\left\langle\hat{\psi}_k^o\left|\hat{H}^{(o)}\sum a_{nj}^{(2)}\right|\hat{\psi}_j^o\right\rangle + \left\langle\hat{\psi}_k^o\left|\hat{H}^{(1)}\sum a_{ni}^{(1)}\right|\hat{\psi}_i^o\right\rangle - \left\langle\hat{\psi}_k^o\left|E_n^{(o)}\sum a_{nj}^{(2)}\right|\hat{\psi}_j^o\right\rangle$$
$$-\left\langle\hat{\psi}_k^o\left|A_n^{(1)}\sum a_{ni}^{(1)}\right|\hat{\psi}_i^o\right\rangle - \left\langle\hat{\psi}_k^o\left|A_n^{(2)}\right|\hat{\psi}_n^o\right\rangle = 0$$

$$\left\langle\hat{\psi}_k^o\left|\hat{H}^{(o)}\sum a_{nj}^{(2)}\right|\hat{\psi}_j^o\right\rangle + \left\langle\hat{\psi}_k^o\left|\hat{H}^{(1)}\sum a_{ni}^{(1)}\right|\hat{\psi}_i^o\right\rangle$$
$$= \left\langle\hat{\psi}_k^o\left|E_n^{(o)}\sum a_{nj}^{(2)}\right|\hat{\psi}_j^o\right\rangle + \left\langle\hat{\psi}_k^o\left|A_n^{(1)}\sum a_{ni}^{(1)}\right|\hat{\psi}_i^o\right\rangle + \left\langle\hat{\psi}_k^o\left|A_n^{(2)}\right|\hat{\psi}_n^o\right\rangle.$$

i.e., $\sum a_{nj}^{(2)} \left\langle\hat{\psi}_k^o\left|\hat{H}^{(o)}\right|\hat{\psi}_j^o\right\rangle + \sum a_{ni}^{(1)} \left\langle\hat{\psi}_k^o\left|\hat{H}^{(1)}\right|\hat{\psi}_i^o\right\rangle$

$= \sum a_{nj}^{(2)} \left\langle\hat{\psi}_k^o\left|E_n^{(o)}\right|\hat{\psi}_j^o\right\rangle + \left\langle\hat{\psi}_k^o\left|A_n^{(1)}\sum a_{ni}^{(1)}\right|\hat{\psi}_i^o\right\rangle + \left\langle\hat{\psi}_k^o\left|A_n^{(2)}\right|\hat{\psi}_n^o\right\rangle.$

$\sum a_{nj}^{(2)} {}_o\langle k|\hat{H}^{(o)}|\hat{j}\rangle_o + \sum a_{ni}^{(1)} {}_o\langle k|\hat{H}^{(1)}|\hat{j}\rangle_o$

$= \sum a_{nj}^{(2)} {}_o\langle k|E_n^{(o)}|\hat{j}\rangle_o + {}_o\langle k|A_n^{(1)}\sum a_{ni}^{(1)}|\hat{j}\rangle_o + {}_o\langle k|A_n^{(2)}|\hat{n}\rangle_o$

$\sum a_{nj}^{(2)} E_j^{(o)} \delta_{kj} + \sum a_{ni}^{(1)} \hat{H}_{kj}^{(1)} = \sum a_{nj}^{(2)} E_n^{(o)} \delta_{kj} + \sum a_{ni}^{(1)} A_n^{(1)} \delta_{kj} + A_n^{(2)} \delta_{kn}.$

$a_{nk}^{(2)} E_k^{(o)} + \sum a_{ni}^{(1)} \hat{H}_{kj}^{(1)} = a_{nk}^{(2)} E_n^{(o)} + \sum a_{ni}^{(1)} A_n^{(1)} \delta_{kj} + A_n^{(2)} \delta_{kn}.$

$a_{nk}^{(2)} \left(E_k^{(o)} - E_n^{(o)}\right) + \sum a_{ni}^{(1)} \hat{H}_{kj}^{(1)} = \sum a_{nk}^{(1)} A_n^{(1)} + A_n^{(2)} \delta_{kn}.$ \qquad (19.4.2)

19.3.3 SECOND-ORDER ENERGY CORRECTION
For $k = n$, in equation (19.4.2)

$a_{nn}^{(2)}(0) + \sum a_{ni}^{(1)} \hat{H}_{ni}^{(1)} = \sum a_{nn}^{(1)} A_n^{(1)} + A_n^{(2)} \cdot 1.$

$A_n^{(2)} = \sum a_{ni}^{(1)} \hat{H}_{ni}^{(1)} - \sum a_{nn}^{(1)} A_n^{(1)} = \sum a_{ni}^{(1)} \hat{H}_{ni}^{(1)} - a_{nn}^{(1)} \hat{H}_{nn}^{(1)}$

$= \sum_{i \neq n} a_{ni}^{(1)} \hat{H}_{ni}^{(1)}$ (where $i \neq n$) because of (19.3.16)

$A_n^{(2)} = \sum_{i \neq n} a_{ni}^{(1)} \hat{H}_{ni}^{(1)} = \sum_{i \neq n} \frac{\hat{H}_{in}^{(1)}}{\left(E_n^{(o)} - E_i^{(o)}\right)} \hat{H}_{ni}^{(1)},$

because of (19.3.18)

$$\boxed{A_n^{(2)} = \sum_{i \neq n} \frac{\hat{H}_{in}^{(1)} \hat{H}_{ni}^{(1)}}{\left(E_n^{(0)} - E_i^{(0)}\right)} = \sum_{i \neq n} \frac{\left|\hat{H}_{ni}^{(1)}\right|^2}{\left(E_n^{(0)} - E_i^{(0)}\right)}}$$ (19.4.3)

using (19.3.8),

$$\boxed{E_n \approx E_n^{(2)} \cong E_n^{(0)} + A_n^{(1)} + A_n^{(2)}}$$

$$\boxed{E_n \approx E_n^{(2)} \cong E_n^{(0)} + \hat{H}_{nn}^{(1)} + \sum_{i \neq n} \frac{\left|\hat{H}_{ni}^{(1)}\right|^2}{\left(E_n^{(0)} - E_i^{(0)}\right)}}.$$ (19.4.4)

i.e., $E_n^{(2)}$ can ONLY LOWER THE ENERGY of the ground state.

19.3.4 SECOND-ORDER CORRECTION TO WAVE FUNCTION

When $k \neq n$, equation (19.4.2) becomes

$$a_{nk}^{(2)} \left(E_k^{(0)} - E_n^{(0)}\right) + \sum a_{ni}^{(1)} \hat{H}_{kj}^{(1)} = \sum a_{nk}^{(1)} A_n^{(1)} + A_n^{(2)} \cdot (0).$$

$$a_{nk}^{(2)} \left(E_k^{(0)} - E_n^{(0)}\right) + \sum a_{ni}^{(1)} \hat{H}_{ki}^{(1)} - \sum a_{nk}^{(1)} A_n^{(1)} = 0$$

$$a_{nk}^{(2)} \left(E_n^{(0)} - E_k^{(0)}\right) = \sum a_{ni}^{(1)} \hat{H}_{ki}^{(1)} - \sum a_{nk}^{(1)} A_n^{(1)}$$

$$a_{nk}^{(2)} = \frac{\left\{\sum a_{ni}^{(1)} \hat{H}_{ki}^{(1)} - \sum a_{nk}^{(1)} A_n^{(1)}\right\}}{\left(E_n^{(0)} - E_k^{(0)}\right)}$$

$$\boxed{a_{nk}^{(2)} = \sum_{(i \neq n);(k \neq n)} \frac{\left\{\sum a_{in}^{(1)} \hat{H}_{ki}^{(1)}\right\}}{\left(E_n^{(0)} - E_i^{(0)}\right)\left(E_n^{(0)} - E_k^{(0)}\right)} - \sum \frac{\hat{H}_{kn}^{(1)} \cdot \hat{H}_{nn}^{(1)}}{\left(E_n^{(0)} - E_k^{(0)}\right)^2}}$$ (19.4.5)

$$\left|\hat{\psi}_n\right\rangle \approx \left|\hat{\psi}_n^2\right\rangle \cong \left|\hat{\psi}_n^o\right\rangle + \left|\hat{\varphi}_n^{(1)}\right\rangle + \left|\hat{\varphi}_n^{(2)}\right\rangle.$$

$$\boxed{\left|\hat{\psi}_n\right\rangle \approx \left|\hat{\psi}_n^o\right\rangle + \left|\hat{\varphi}_n^{(1)}\right\rangle + \sum_{k \neq n} a_{nk}^{(1)} \left|\hat{\psi}_k^o\right\rangle + \sum_{k \neq n} a_{nk}^{(2)} \left|\hat{\psi}_k^o\right\rangle}$$ (19.4.6)

with the expansion coefficients from (19.3.18) & (19.4.5).
The complexity of the expression (19.4.6) indicates that, for wave function corrections, only First-Order Perturbation Theory is used, and for energy correction, beyond Second Order Perturbation is not used.

19.3.5 APPLICATIONS of the Perturbation Theory
Let us test out this scheme using a particularly simple example

19.3.5.1 Particle in a Box

Worked out Example 19.1
Illustrate the essential features of the First Order Perturbation Energy correction on the one-dimensional box bounded by an infinite potential barrier with a linear potential gradient added as perturbation.

Solution: $\boxed{\text{STEP \# 1}}$ It is known that $\hat{\psi}_n(x,t) = \sqrt{\frac{2}{a}} \, Sin(\frac{n\pi x}{a}) \cdot e^{-i\omega t}$. (5.3.18)

$$E_n = \hbar^2 (\tfrac{n\pi}{a})^2 / 2m, \text{ where } n = 0, \pm 1, \pm 2, \ldots \quad (5.3.11);$$

$$E_n \approx E_n^{(1)} \cong E_n^{(1)} + \hat{H}_{nn}^{(1)} \quad (19.3.17)$$

$$V(x) = \tfrac{1}{2} m \omega^2 x^2 + \tfrac{x}{a} V_1; \; V(x) = 0, \text{ at } x = 0; V(x) = V_1, \text{ at } x = a.$$

The perturbation Hamiltonian $\hat{H}^{(1)} = \tfrac{x}{a} V_1; \; A_n^{(1)} = \hat{H}_{kn}^{(1)}$

STEP # 2 $\hat{H}_{nn}^{(1)} = {}_o\langle \hat{n} | \hat{H}^{(1)} | \hat{n} \rangle_o$

$$\hat{H}_{nn}^{(1)} = \langle \hat{\psi}_n^o | \hat{H}^{(1)} | \hat{\psi}_n^o \rangle = = \langle \hat{\psi}_n^o | [\tfrac{x}{a} V_1] | \hat{\psi}_n^o \rangle$$

$$= \left(\tfrac{V_1}{a}\right) \int_0^a \sqrt{\tfrac{2}{a}} \, Sin(\tfrac{n\pi x}{a}) \cdot x \sqrt{\tfrac{2}{a}} \, Sin(\tfrac{n\pi x}{a}) \, dx$$

STEP # 3 $= \left(\tfrac{V_1}{a}\right) \tfrac{2}{a} \int_0^a x \, Sin^2(\tfrac{n\pi x}{a}) \, dx \quad (19.3.16)$

$$= \left(\tfrac{V_1}{a}\right) \tfrac{2}{a} \left(\tfrac{a}{n\pi}\right) \int_0^a x \, Sin^2(\tfrac{n\pi x}{a}) \, d(\tfrac{n\pi x}{a}). \text{ Put } \varsigma = (\tfrac{n\pi x}{a}),$$

$$= \left(\tfrac{V_1}{a}\right) \left(\tfrac{2}{a}\right) \left(\varsigma^2\right) \left\{\tfrac{\varsigma^2}{4} - \tfrac{\varsigma \, Sin 2\varsigma}{4} - \tfrac{Cos 2\varsigma}{8}\right\}_0^a = \left(\tfrac{V_1}{2}\right) \quad (19.4.7)$$

STEP # 4 $E_n \approx E_n^{(1)} \cong E_n^{(1)} + \hat{H}_{nn}^{(1)} = \{\hbar^2 (\tfrac{n\pi}{a})^2 / 2m\} + \left(\tfrac{V_1}{2}\right) \quad (19.4.7)$

19.3.5.2 Anharmonic Oscillator

Worked out Example 19.2

Given a system whose Hamiltonian is given by $\hat{H} = \hat{H}^{(0)} + \hat{H}^{(1)} = [\tfrac{\hat{p}_x^2}{2m} + \tfrac{1}{2} kx^2] + \alpha x^3$ ..Find the second order correction to the energy spectrum of ground state, using the perturbation theory.

Solution: STEP # 1 It is known that from equations (5.3.18), (5.3.11), (19.3.17)

$$\hat{\psi}_n(x,t) = \sqrt{\tfrac{2}{a}} \, Sin(\tfrac{n\pi x}{a}) \cdot e^{-i\omega t} \; ; \; E_n = \hbar^2 (\tfrac{n\pi}{a})^2 / 2m, \text{ where } n = 0, \pm 1, \pm 2, \ldots \quad ;$$

$$E_n \approx E_n^{(1)} \cong E_n^{(1)} + \hat{H}_{nn}^{(1)} \; ; \; V(x) = \tfrac{1}{2} m \omega^2 x^2 + \tfrac{x}{a} V_1; \; V(x) = 0, \text{ at } x = 0; V(x) = V_1, \text{ at } x = a. \text{ The perturbation}$$

Hamiltonian $\hat{H}^{(1)} = \tfrac{x}{a} V_1; \; A_n^{(1)} = \hat{H}_{kn}^{(1)}$

The unperturbed Hamiltonian, $\hat{H}^{(0)} = \tfrac{1}{2} kx^2$ with eigen values $E_n = (n + \tfrac{1}{2}) \hbar \omega_c$

$$| \hat{\psi}_n(q) \rangle = \sqrt{\tfrac{m \omega_c}{\pi^2 \hbar}} / \tfrac{1}{2^{n/2} \sqrt{n!}} \, H_n(q) \cdot e^{-\tfrac{1}{2} q^2} \quad (6.2.40);$$

$$\hat{a} = \sqrt{\tfrac{1}{2} (\hat{q} + i\hat{p})} \; (7.1.19) \; ; \; \hat{a}^\dagger = \sqrt{\tfrac{1}{2} (\hat{q} - i\hat{p})} \quad (7.1.20)$$

$$| n \rangle = B_n \, H_n(q) \cdot e^{-\tfrac{1}{2} q^2} \; ; \; \hat{a} | n \rangle = \sqrt{n} \, | n-1 \rangle \; (7.4.8); \; \hat{a}^\dagger | n+1 \rangle = \sqrt{n+1} \, | n \rangle \quad (7.4.9);$$

Given that

$$\hat{H} = \hat{H}^{(0)} + \hat{H}^{(1)} = [\tfrac{\hat{p}_x^2}{2m} + \tfrac{1}{2} kx^2] + \alpha x^3$$

STEP # 2 Perturbation term, $\hat{H}^{(1)} = \alpha x^3$

$$E_n \approx E_n^{(2)} \cong E_n^{(0)} + A_n^{(1)} + A_n^{(2)} \cong E_n^{(0)} + \hat{H}_{nn}^{(1)} + A_n^{(2)}. \quad (19.4.4)$$

$$A_n^{(2)} = \sum_{i \neq n} \frac{|\hat{H}_{ni}^{(1)}|^2}{\left(E_n^{(0)} - E_i^{(0)}\right)}$$

$\hat{H}_{ni}^{(1)} = \langle \hat{\psi}_n^{(0)} | \hat{H}^{(1)} | \hat{\psi}_i^{(0)} \rangle = {}_0\langle n | (\alpha x^3) | i \rangle_0 = {}_0\langle n | (\alpha q / b^{1/2}) | i \rangle_0$.

Keeping only the terms i =1 and i = 3, for the unperturbed ground state, shows that there are only two non-vanishing matrix elements $\hat{H}_{01}^{(1)}$ and $\hat{H}_{03}^{(1)}$.

STEP # 3 **To evaluate** $\hat{H}_{01}^{(1)}$ **and** $\hat{H}_{03}^{(1)}$

$\hat{q}^3 = \{\sqrt{\frac{1}{2}}(\hat{a}+\hat{a}^\dagger)\}^3 = \frac{1}{2}\sqrt{\frac{1}{2}}\{\hat{a}^3 + (\hat{a}^\dagger)^3 + \hat{a}^2\hat{a}^\dagger + \hat{a}^\dagger\hat{a}^2 + \hat{a}(\hat{a}^\dagger)^2 + (\hat{a}^\dagger)^2\hat{a} + \hat{a}(\hat{a}^\dagger)\hat{a} + (\hat{a}^\dagger)\hat{a}(\hat{a}^\dagger)\}$

STEP # 4 Applying (7.4.8), (7.4.9) and (7.4.10), $\hat{a}^3 | n \rangle = \sqrt{(n+1)!} | 0 \rangle$

$\hat{q}^3 | m \rangle = \frac{1}{2}\sqrt{\frac{1}{2}}\{\hat{a}^3 + (\hat{a}^\dagger)^3 + \hat{a}^2\hat{a}^\dagger + \hat{a}^\dagger\hat{a}^2 + \hat{a}(\hat{a}^\dagger)^2 + (\hat{a}^\dagger)^2\hat{a} + \hat{a}(\hat{a}^\dagger)\hat{a} + (\hat{a}^\dagger)\hat{a}(\hat{a}^\dagger)\} | m \rangle$

$\langle n | \hat{q}^3 | m \rangle = \frac{1}{2}\sqrt{\frac{1}{2}}\langle n | \{\hat{a}^3 + (\hat{a}^\dagger)^3 + \hat{a}^2\hat{a}^\dagger + \hat{a}^\dagger\hat{a}^2 + \hat{a}(\hat{a}^\dagger)^2 + (\hat{a}^\dagger)^2\hat{a} + \hat{a}(\hat{a}^\dagger)\hat{a} + (\hat{a}^\dagger)\hat{a}(\hat{a}^\dagger)\} | m \rangle$

$\sqrt{8}\langle n | \hat{q}^3 | m \rangle = \sqrt{(m+1)(m+2)(m+3)!}\delta_{n,m+3} + \sqrt{m(m-1)(m-2)!}\delta_{n,m-3}$
$\qquad + 3m(m+1)\sqrt{(m+1)}\,\delta_{n,m+1} + 3m\sqrt{m}\,\delta_{n,m-1}$ \hfill (19.4.12)

STEP # 5 The matrix $\hat{H}_{nm}^{(1)}$ from the above equation is thus obtained:

$\hat{H}_{nm}^{(1)} = {}_0\langle n | (\alpha x^3) | m \rangle_0 = {}_0\langle n | (\alpha q^3 / b^{3/2}) | m \rangle_0 = (\alpha^3 / b^{3/2}) {}_0\langle n | (q^3) | m \rangle_0$ (19.4.13)

From equation (19.4.12) it is clear that

$\langle n | (q^3) | m \rangle \neq 0$, only for $n - m = \pm 1$. i.e. $\langle n-1 | (q^3) | n \rangle = 3 [\frac{1}{2}\frac{\alpha}{b}]^{3/2}$

$\langle n | (q^3) | n-3 \rangle = [\frac{1}{2}\frac{\alpha}{b}]^{3/2}[n(n-1)(n-2)]^{1/2}$

STEP # 6

$$\hat{H}_{nm}^{(1)} = [\frac{1}{2}\frac{\alpha}{b}]^{1/2} \begin{bmatrix} 0 & \sqrt{3/2} & 0 & \sqrt{3} & 0 & 0 & 0 & \text{etc} \\ \sqrt{3/2} & 0 & 6 & 0 & \sqrt{12} & 0 & 0 & \text{etc} \\ 0 & 6 & 0 & 9\sqrt{3/2} & 0 & \sqrt{30} & 0 & \text{etc} \\ \sqrt{3} & 0 & 9\sqrt{3/2} & 0 & 12\sqrt{12} & 0 & 0 & \text{etc} \\ 0 & \sqrt{12} & 0 & 12\sqrt{2} & 0 & 15\sqrt{5/2} & 0 & \text{etc} \\ 0 & 0 & \sqrt{30} & 0 & 15\sqrt{5/2} & 0 & 0 & \text{etc} \\ 0 & 0 & 0 & 0 & 0 & 0 & 0 & \text{etc} \\ 0 & 0 & 0 & 0 & 0 & 0 & 0 & \text{etc} \end{bmatrix}$$ (19.4.14)

STEP # 7 $\hat{H}_{01}^{(1)} = 3\,\alpha\,[\frac{1}{2b}]^{3/2}$ \hfill (19.4.15)

$\hat{H}_{03}^{(1)} = \sqrt{6}\,\alpha\,[\frac{1}{2b}]^{3/2} = \frac{1}{2}\alpha\,[\frac{3}{b^3}]^{1/2}$. \hfill (19.4.16)

STEP # 8 Then $A_0^{(2)} = \sum_{i \neq 0} \frac{|\hat{H}_{0i}^{(1)}|^2}{[E_0^0 - E_i^0]} = [\frac{\alpha^2}{8b^3}]\left(3^2 / [\frac{1}{2}\hbar\omega_c - \frac{3}{2}\hbar\omega_c] + 6/[\frac{1}{2}\hbar\omega_c - \frac{7}{2}\hbar\omega_c]\right)$

$\qquad = -[\frac{\alpha^2}{8b^3}]\left([\frac{9}{\hbar\omega_c}] + [\frac{6}{3\hbar\omega_c}]\right) = -\frac{11\,\alpha^2}{8b^3(\hbar\omega_c)}$ \hfill (19.4.17)

STEP # 10 $E_n \approx E_n^{(2)} \cong E_n^{(0)} + A_n^{(1)} + A_n^{(2)}$

$$E_0^{(2)} \cong E_0^{(o)} + A_0^{(1)} + A_0^{(2)} \qquad (19.4.4)$$

$$\cong \frac{1}{2}\hbar\omega_c + 0 - \frac{11\,\alpha^2}{8\,b^3(\hbar\omega_c)}$$

19.5 THE NORMAL STATE OF HELIUM ATOM:- Application of First Order Perturbation Theory to a system of Three Interacting Particles:

It is now time for one to study the energy levels of a two-electron atom, principally Helium. The helium atom consists of a nucleus of charge + 2e, *i.e.* its atomic number $Z = 2$, and two electrons. Thus helium atom is an example of a three-body system – a system of three interacting particles. So it is very difficult to analyze it within Classical Mechanics.
It is necessary to label the two electrons as 1 and 2.
Assume that no forces other than the EM forces (Coulomb force to a very good approximation) are necessary to describe the dynamics of a helium atom with the help of Quantum Mechanics. That is, the contributions, *viz.*
(i)   The spin-orbit interaction term,
(ii)   Spin-spin interaction term,
(iii)   Small effects connected with the motion of the nucleus and
(iv)   Relativistic effects
are neglected.

19.5.1 Schrödinger Equation for HELIUM ATOM
Let the nucleus is placed at the Origin of Spherical Polar Coordinate System.
$\mathbf{r}_1$ and $\mathbf{r}_2$ are the position vectors of electrons 1 and 2, respectively (Fig. 19.1).

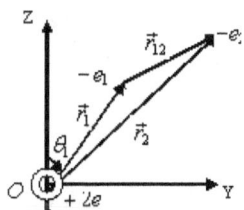

Fig. 19.1. Helium atom

Let the electrons do not spin.
The Hamiltonian of the atom is

$$\hat{H} = \frac{\hat{p}_1^2}{2m} + \frac{\hat{p}_2^2}{2m} + Ze^2\left(\frac{1}{r_1} + \frac{1}{r_2}\right) + \frac{e^2}{|r_{12}|} \qquad (19.5.1)$$

$$= \left(-\frac{\hbar^2}{2m}\right)(\nabla_1^2 + \nabla_2^2) - \left(\frac{Ze^2}{r_1} + \frac{Ze^2}{r_2}\right) + \frac{e^2}{|\vec{r}_1 - \vec{r}_2|} \qquad (19.5.2)$$

where *m* is the mass of the electron.
For the purpose of simplifying the calculation of the integrals involved in problems of this type, it is convenient to use atomic units, *i.e.* in terms of the Bohr Radius, a_0,

$$a_0 = \hbar^2/me^2$$
$$x_1 \to a_0 X_1, \quad r_1 \to a_0 R_1,$$

$$x_2 \to a_0 X_2, \quad r_2 \to a_0 R_2,$$
$$x_{12} \to a_0 X_{12}, \quad r_{12} \to a_0 R_{12},$$
$$\therefore \quad \nabla_1^2 = (\nabla_{X1}^2 + \nabla_{X2}^2 + \nabla_{X3}^2) = (1/a_0^2)(\nabla_{x1}^2 + \nabla_{x2}^2 + \nabla_{x3}^2) = (1/a_0^2)\nabla_r^2$$

$$\therefore \quad \hat{H} = \left(-\frac{\hbar^2}{2m}\frac{1}{a_0^2}\right)(\nabla_1^2 + \nabla_2^2) - \left(\frac{Ze^2}{a_0^2}\right)\left(\frac{1}{R_1}+\frac{1}{R_2}\right) + \left(\frac{e^2}{a_0^2}\right)\frac{1}{|R_1 - R_2|} \quad (19.5.3)$$

$$\therefore \quad \hat{H} = \left(\frac{e^2}{a_0^2}\right)\left\{-\tfrac{1}{2}(\nabla_1^2 + \nabla_2^2) - Z\left(\frac{1}{R_1}+\frac{1}{R_2}\right) + \frac{1}{|R_1 - R_2|}\right\} \quad (19.5.4)$$

19.5.2 The PERTURBATION TERM
According to the perturbation theory, one sets
$$\hat{H} = \hat{H}^{(0)} + \hat{H}^{(1)}$$
where
$$\hat{H}^{(0)} = \left\{-\tfrac{1}{2}(\nabla_1^2 + \nabla_2^2) - Z\left(\frac{1}{R_1}+\frac{1}{R_2}\right)\right\}, \text{ in } (e^2/a_0) \text{ units.} \quad (19.5.5)$$

$$\hat{H}^{(1)} = \frac{1}{|R_1 - R_2|}, \quad \text{in } (e^2/a_0) \text{ units} \quad (19.5.6)$$

19.5.3 GROUND STATE ENERGY of helium
The zeroth order eigen functions $|\psi_n^0\rangle$ (i.e. unperturbed states) are the solutions of the eigen value equation of the operator, $\hat{H}^{(0)}$.
$$\hat{H}^{(0)}|\psi_n^0\rangle = E_n^0|\psi_n^0\rangle \quad (19.5.7)$$
If one now sets this eigen function of the two electron system as
$$|\psi_n^0\rangle = |\psi_n^0(1)\rangle \cdot |\psi_n^0(2)\rangle \quad (19.5.8)$$
then the eigen values are
$$E_n^0 = E_n^0(1) + E_n^0(2) \quad (19.5.9)$$
Equation (19.5.7) becomes immediately separable into two equations, so that the GROUND STATE OF HELIUM IS A NON-DEGENERATE SINGLET STATE,
$$(\hat{H}_1^{(0)} + \hat{H}_2^{(0)})|\psi_0^0\rangle = E_0^0|\psi_0^0\rangle$$
i.e.,
$$-\tfrac{1}{2}\nabla_1^2|\psi_0^0(1)\rangle + \left[E_n^0(1) - Z/R_1\right]|\psi_0^0(1)\rangle = 0 \quad (19.5.10)a$$
$$\tfrac{1}{2}\nabla_2^2|\psi_0^0(2)\rangle + \left[E_n^0(2) - Z/R_2\right]|\psi_0^0(2)\rangle = 0 \quad (19.5.11)b$$
These equations are just those for hydrogen atom but with nuclear charge $Z = 2$. For the ground state of the helium atom $|\psi_0^0\rangle \equiv |0\rangle \equiv |1s^2\rangle$,

$$|1s(1)\rangle = |\psi_{100}^0(1)\rangle = \sqrt{1/4\pi}\, 2\,(Z/a_0)^{3/2} e^{-(Z R_1/a_0)} \quad (19.5.12)$$

$$|1s(2)\rangle = |\psi_{100}^0(2)\rangle = \sqrt{1/4\pi}\, 2\,(Z/a_0)^{3/2} e^{-(Z R_2/a_0)}.$$

$$|0\,(He)\rangle \equiv |1s^2\rangle = |1s(1)\rangle \cdot |1s(2)\rangle = (Z^3/\pi a_0^3)\, e^{-(Z(R_1+R_2)/a_0)} \quad (19.5.13)$$

$$\therefore \quad E_{1s^2}^0(He) = \langle \hat{H}_{00}^{(0)} \rangle = \langle 0|\hat{H}^{(0)}|0\rangle = E_{1s}^0(1) + E_{1s}^0(2) = 2Z^2\, E_{1s}^0(H) \quad (19.5.14)$$

where $E_{1s}^0(H)$ = ionization energy (or ground state energy) of hydrogen.
$$E_{1s}^0(H) = E_{1s}^0(1) = E_{1s}^0(2) = -(\tfrac{1}{2})(e^2/a_0) = -13.6 \ eV.$$
Using (19.5.14),
$$\therefore \quad E_{1s^2}^0(He) = 4Z^2 \ E_{1s}^0(H) = 4(-13.6 \ eV) = -108.8 \ eV \quad (19.5.15)$$
But experimentally for ionized Helium atom,
$$E_{1s^2}^0(He) = -78.62 \ eV. \quad (19.5.16)$$

19.5.4 FIRST-ORDER CORRECTION TO ENERGY

$$E_n \approx E_n^{(1)} \cong E_n^{(0)} + \hat{H}_{nn}^{(1)}. \quad (19.3.17)$$

$$A_n^{(1)} = \hat{H}_{nn}^{(1)} = \langle \hat{\psi}_n^0 | \hat{H}^{(1)} | \hat{\psi}_n^0 \rangle \quad (19.3.16)$$

$$E_{1s^2}^{(1)}(He) \cong E_{1s^2}^{(0)}(He) + A_0^{(1)} \cong 2Z^2 \ E_{1s}^0(H) + \langle 1s^2 | \hat{H}^{(1)} | 1s^2 \rangle \quad (19.5.17)$$

$$\therefore \quad A_0^{(1)} \equiv A_{1s^2}^{(1)} = \hat{H}_{00}^{(1)} = \langle 1s^2 | \hat{H}^{(1)} | 1s^2 \rangle = \langle 1s^2 | (1/|R_1 - R_2|) | 1s^2 \rangle \quad (19.3.18)$$

$$= \langle 1s^2 | ((e^2/a_0)/|R_{12}|) | 1s^2 \rangle = \langle ((e^2/a_0)/|R_{12}|) \rangle \quad (19.5.19)$$

Using (19.5.6)

$$A_{1s^2}^{(1)} = \left(\frac{e^2 Z^6}{a_0 \pi^2}\right) \iint \frac{1}{|R_{12}|} e^{2(Z(R_1+R_2)/a_0)} \ d\tau_1 \ d\tau_2. \quad (19.5.20)$$

To evaluate $\left\langle \left(\frac{e^2}{a_0}\right) \frac{1}{|R_{12}|} \right\rangle$

One can expand $|R_{12}|^{-1}$ by using the Associated Legendre Polynomials, $P_\ell^m(\cos\theta)$.

$$P_\ell^m(\rho) = (1-\rho^2)^{m/2} (d^m/d\rho^m) P_\ell(\rho); \quad \ell = |m| = 0,1,2,3,4,....; m \geq 0. \quad (8.4.11)$$

This permits one to derive any $P_\ell^m(\rho)$.

$$P_\ell^m(\rho) = P_\ell(\rho) = [(-1)^\ell/2^\ell \ell!] \ (d^\ell(1-\rho^2)^\ell/d\rho^2) \quad (8.4.13)$$

which is Rodrigue's Formula.

$$\hat{Y}_\ell^m(\theta,\varphi) = (-1)^m \left[(\ell+\tfrac{1}{2})/4\pi)((\ell-m)!/(\ell+m)!)\right]^{1/2} P_\ell^m(\cos\theta) \ e^{i|m|\varphi} \quad (8.7.2)$$

$$|R_{12}|^{-1} = \sum\sum \{((\ell-m)!/(\ell+m)!)\} (R_<^\ell/R_>^{\ell+1}) \cdot P_\ell^m(\cos\theta_1) \cdot P_\ell^m(\cos\theta_2) \cdot e^{i|m|(\varphi_1-\varphi_2)} \quad (19.5.21)$$

$$= (1/R_>) \sum (R_</R_>)^\ell \cdot \left[(4\pi/(\ell+\tfrac{1}{2})\right] \cdot \sum \langle \psi_{n\ell}^0 \rangle \cdot \hat{Y}_\ell^m(\theta_1,\varphi_1) \cdot \hat{Y}_\ell^m(\theta_2,\varphi_2) \quad (19.5.22)$$

where $R_<$ = the smaller of the quantities R_1 and R_2 (i.e. minimum of r_1 and r_2);
$R_>$ = the larger of R_1 and R_2 (i.e. maximum of r_1 and r_2).

$Cos\theta = (r_1 \cdot r_2 / |r_1 \ r_2|)$.

There is no explicit dependence of the angles by the wave functions. Since the *Associated Legendre Polynomials* are *orthogonal*, all the terms in the summation will vanish except those for $\ell = 0$, $m = 0$.

For these $P_0^0(\cos\theta) = 1$.

Thus equation (19.5.20) reduces to

$$A^{(1)}_{1s^2} = \left(\frac{e^2 Z^6}{a_0 \pi^2}\right) \iint \frac{1}{|R_>|} e^{-2(Z(R_1+R_2)/a_0)} d\tau_1\, d\tau_2. \tag{19.5.23}$$

Since $e^{-2(Z(R_1)/a_0)}$ = a potential = constant for $R_> < |R_{12}|$, but varies as $1/R_>$, for $R_> > |R_{12}|$.

$$A^{(1)}_{1s^2} = \left(\frac{16\, e^2\, Z^6}{a_0}\right) \int e^{-2(Z R_1)} \left\{ R_1^{-1} \int e^{-2(Z R_2)} R_2^2\, dR_2 \right\} R_1^2\, dR_1 \tag{19.5.24}$$

$$= \left(\frac{16\, e^2\, Z^6}{a_0}\right)(2ZR_1)^{-1} \left\{ [2 - e^{-2(Z R_1)}((2ZR_1)^2 + (2ZR_1) + 2)] + (2ZR_1)e^{-2(Z R_1)} \right\}$$

$$= \left(\frac{16\, e^2\, Z^6}{a_0}\right)\left\{\left[\frac{5}{128}Z^{-5}\right]\right\} = \left(\frac{16\, e^2\, Z^6}{a_0}\right)\left\{\frac{5}{128}Z^{-5}\right\} = \frac{5}{4} Z\, E^0_{1s}(H) \tag{19.5.25}$$

19.5.5 APPROXIMATE ENERGY of the Normal State of Helium atom

Substituting this value in equation (19.5.17),

$$E^{(1)}_{1s^2}(He) \cong E^{(0)}_{1s^2}(He) + A^{(1)}_0 = E^{(0)}_{1s^2}(He) - \frac{5}{4} Z\, E^0_{1s}(H)$$

$$= 2Z^2\, E^0_{1s}(H) - \frac{5}{4} Z\, E^0_{1s}(H) \tag{19.5.26}$$

$$= [2(2)^2 - \frac{5}{4} 2](-13.6\, eV) = -74.8\, eV. \tag{19.5.27}$$

19.5.6 OUTCOME OF THE STUDY - FAILURE of the First Order Perturbation Theory applied to the Normal State of Helium atom

When the result of (19.5.15) is compared with (19.5.27), the following is found. The $E^{(1)}_{1s^2}(He)$ is of around 27% correction to $E^{(1)}_{1s^2}(He)$, which is ~ 5.5% greater than the experimental value of 2.904 = -2.705 atomic units.

The energy of the ground state of He^+ ion = $Z^2\, E^0_{1s}(H) = -54.4\, eV$.

The first ionization energy, *i.e.* the energy required to remove an electron from the He atom is thus

$$= (Z^2 - \frac{5}{4} Z)\, E^0_{1s}(H) = [2(2)^2 - \frac{5}{4} 2](-13.6\, eV) = -20.4\, eV$$

$$= 1.500\, E^0_{1s}(H) \equiv 1.500\, R_y.$$

The observed (experimental value is 24.58 eV (1.807 R_y), *i.e.* there is an error of 4.18 eV, or ~ 16%.

For the two electron atoms, Li^+ ion, etc. the % error is considerably less than for He, since the interactions between the electrons and the nucleus becomes relatively more important than the interaction between the electrons.

Equation (19.5.26) shows that

$$A^{(1)}_{1s^2} = (2Z^2 / \frac{5}{4} Z) = \frac{5}{16}.$$

This is NOT a *small perturbation*. Therefore the failure of the First Order Perturbation Theory applied to the Normal State of Helium atom is not a surprise! It is in such cases that the 'variation methods', another approximation procedure, gives any degree of accuracy.

19.5.7 LINEAR STARK EFFECT: NORMAL STATE OF HYDROGEN ATOM

It was in 1913 that J. Stark showed the splitting of the spectral lines of the Balmer series (transitions to $n = 2$ state) in hydrogen atom, into a greater multiplicity when excited in an electric field. This phenomenon is known as the Stark Effect. One may recall that the magnitude of an atomic electric field is $E_{ea} \approx 10^{10} Vm^{-1}$. If the electric field to be relatively weak, i.e. < $E < 10^6 Vm^{-1}$, the First Order Perturbation theory will be relevant. The theory of the Stark Effect in hydrogen was the first application of the perturbation theory in Quantum Mechanics. The electric field is said to be STRONG when the splitting of the energy levels is greater then the <u>fine structure</u> splitting

Worked out Example 19.9

Consider the hydrogen atom placed in a perturbing electric field. Calculate the first order energy correction to the ground state, using the perturbation theory.

Solution: $\boxed{\text{STEP \# 1}}$ The Hamiltonian of the H-atom is

$$\hat{H} = \frac{\hat{p}_x^2}{2\mu} - \frac{Ze^2}{\vec{r}_1} = -\frac{\hbar^2}{2\mu}\left(\nabla_x^2\right) - \frac{Ze^2}{\vec{r}_1},$$

whose eigen functions are denoted by $\left|\hat{\psi}^0_{n\ell\, m}\right\rangle = \left|n\ell m_\ell\right\rangle$.

Let $\vec{E}$ = the electric field, $\vec{E}_z$,

applied along the positive Z-axis, so that the electron is acted upon by a force in the –ve Z-axis.

$\boxed{\text{STEP \# 2}}$ The electric dipole moment generated, $\vec{d} = -e.\vec{r}$.

The perturbation term for potential is $\hat{H}^{(1)} = -\vec{d}\cdot\vec{E}_z = -(-e.\vec{r})\cdot\vec{E}_z = +e\vec{E}_z z$.

In non-relativistic approximation, the electric field, $\vec{E}_z$ does not interact with the electron's magnetic moment and for this reason the spin, and the fine structure resulting from S-L coupling can be neglected.

$\boxed{\text{STEP \# 3}}$ $\hat{H}\left|\hat{\psi}^0_{n\ell\, m}\right\rangle = (\hat{H}^{(0)} + \hat{H}^{(1)})\left|n\ell m_\ell\right\rangle = E_n^0\left|n\ell m_\ell\right\rangle$

$(\hat{H}^{(0)} + \hat{H}^{(1)})|1s\rangle = E_{1s}^0|1s\rangle$, for the normal (ground) state,

where $E_1^0 = E_{1s}^0 = -\left((e^2/a_0)/2\right) = -13.6 eV$.

$\boxed{\text{STEP \# 4}}$ $|1s\rangle = \sqrt{1/\pi}\,(1/a_0)^{3/2}\,e^{-(Z r/a_0)}$.

$\ell = 0, ie., even$; means $|1s\rangle \equiv |100\rangle$ has EVEN parity.(an even function of z).

$\boxed{\text{STEP \# 5}}$ Since $\hat{H}^{(1)} = +e\vec{E}_z z$ = an ODD function of z, it changes sign as z is reflected.

$\boxed{\text{STEP \# 6}}$ $A_0^{(1)} \equiv A_{1s}^{(1)} = \langle 1s|\hat{H}^{(1)}|1s\rangle = \langle 1s|(+e\vec{E}_z z)|1s\rangle$

$= (+e\vec{E}_z)\langle 1s|(z)|1s\rangle = (+e\vec{E}_z)\langle z\rangle_{1s} = 0$

$\boxed{\text{STEP \# 7}}$ $E_1 \equiv E_{1s}^1 = (E_{1s}^0 + A_{1s}^{(1)}) \approx (E_{1s}^0 - 0) = -13.6 eV$.

Classically, a system has an electric dipole moment, d, will experience an energy shift of magnitude, $\vec{d}\cdot\vec{E}_z$. Thus an ATOM IN THE GROUND STATE HAS NO PERMANENT DIPOLE MOMENT. In general, systems in non-degenerate states cannot have electric dipole moments; they can have only induced dipole moments. The statement of non-degeneracy is important: it is only then that the states are also eigen states of the parity operator, and then $|\psi^0_{nm\ell}|^2 = 0$ and $<z> = 0$. The all the diagonal elements of $\hat{H}^{(0)}$ with $|\psi^0_{nm\ell}\rangle$ will be zero. The only matrix elements of $\hat{H}^{(1)}$, which will be on-vanishing will have opposite parity states involved. *i.e.* to the first order, perturbation energy of the ground state of a hydrogen atom = unperturbed energy of the H-atom in its ground state. Thus all the NON-DEGENERATE STATES of H-atom will have NO STARK EFFECT. But many MOLECULES ARE WITH PERMANENT DIPOLE MOMENT, WHICH IS DUE TO THEIR GROUND STATES BEING DEGENERATE.

19.5.7 DEGENERATE CASE (The spectrum of $\hat{H}^{(0)}$ is discrete)

19.6.1 DEGENERACY of States and Perturbation:

Thus far it was assumed that the perturbation eigen function differs slightly from a given function $|\hat{\psi}_k^0\rangle$, solution of the unperturbed (ideal) Schrödinger Equation. Most atomic states are degenerate. Therefore, perturbation theory of the non-degenerate case is almost useless. To an experimental physicist, the perturbation theory of degenerate states is of special interest.

A energy eigen value, $E_n^{(0)}$ is said to be *k-fold degenerate* if there are k linearly independent eigen functions $\{|\psi_{n_i}^{(0)}\rangle\}$, so that

$$\hat{H}^{(0)}|\hat{\psi}_{n_i}^0\rangle = E_n^{(0)}|\hat{\psi}_{n_i}^0\rangle, \quad (i=1,2,\ldots,k) \tag{19.2.5}$$

and
$$\langle \hat{\psi}_{n_i}^0 | \hat{\psi}_{n_j}^0 \rangle = \delta_{ij} \tag{19.2.4}$$

The fact that all the $|\psi_{n_i}^{(0)}\rangle$ have the same energy eigen value $E_n^{(0)}$ with the $\hat{H}^{(0)}$ *does not mean* that they also have the same eigen value with some other Hamiltonian, $\hat{H}$.

Even if $\hat{H} = (\hat{H}^{(0)} + \hat{H}^{(1)})$, \hfill (19.2.3)

where $\hat{H}^{(1)}$ is very small, $\hat{H}^{(1)} \ll \hat{H}^{(0)}$, then if $\hat{H}^{(1)}$ acts differently on the various $|\psi_{n_i}^{(0)}\rangle$, it can very well happen that the perturbation $\hat{H}^{(1)}$ removes or partially removes the degeneracy of the degenerate energy levels.

$$|\psi_{n_i}\rangle = (\hat{H}^{(0)} + \hat{H}^{(1)})|\psi_{n_i}\rangle \tag{19.2.6}$$

where $|\psi_{n_i}\rangle = |\hat{\psi}_{n_i}^{(0)}\rangle + \alpha|\hat{\psi}_{n_i}^{(1)}\rangle + \alpha^2|\hat{\psi}_{n_i}^{(2)}\rangle + \cdots$ \hfill (19.3.4)

$E_n = E_{n_i}$, where all the E_{n_i} are DIFFERENT.

In this case one says that THE DEGENERACY IS LIFTED BY PERTURBATION.

Experimentally, this is delightful, as *the difference* $(E_n - E_n^{(0)})$ *can be measured to a very high degree of accuracy* than $E_n^{(0)}$ itself.

19.6.1 PERTURBATION THEORY OF DEGENERATE STATE

For states of definite parity, it might be so that first order correction is zero for all states. Thus a second order correction is to be obtained. It might be desirable to proceed to second order correction in a case where the first order correction is non-zero but where degeneracy remains. But the second order theory discussed earlier to a system where there is degeneracy in the first order correction is doomed to fail, as in (19.3.15)

$$\hat{H}^{(1)}_{kn} + a^{(1)}_{nk}\left(E^{(0)}_k - E^{(0)}_n\right) = A^{(1)}_n \, \delta_{kn} \qquad (19.3.15)$$

where if k^{th} and n^{th} states are degenerate, then for $k \neq n$, both $\delta_{kn} = 0$ and $(E^o_k - E^o_n) = 0$, simultaneously. This gives a trivial result that $\hat{H}^{(1)}_{kn} = 0$, which may or may not be true. This difficulty lies in the fact that $\hat{H}_{kn}$ is not to be found actually by means of the non-degenerate theory.

The treatment of the non-degenerate case till and including equation (19.3.12), *i.e.* till section 19.3.2. is all right.

$$\left[\hat{H}^{(o)} - E^o_n\right]\left|\hat{\psi}^o_n\right\rangle = 0 \qquad (19.3.10)$$

$$\hat{H}^{(1)}\left|\hat{\psi}^o_n\right\rangle + \hat{H}^{(o)}\left|\hat{\varphi}^{(1)}_n\right\rangle - A^{(1)}_n\left|\hat{\psi}^o_n\right\rangle - E^o_n\left|\hat{\varphi}^{(1)}_n\right\rangle = 0, \qquad (19.3.11)$$

$$\hat{H}^{(o)}\left|\hat{\varphi}^{(2)}_n\right\rangle + \hat{H}^{(1)}\left|\hat{\varphi}^{(1)}_n\right\rangle - E^{(o)}_n\left|\hat{\varphi}^{(2)}_n\right\rangle - A^{(1)}_n\left|\hat{\varphi}^{(1)}_n\right\rangle - A^{(2)}_n\left|\hat{\psi}^o_n\right\rangle = 0 \qquad (19.3.12)$$

But $\quad A^{(1)}_n = \hat{H}^{(1)}_{kn} \qquad (19.3.16)$

cannot be true generally if $\left|\psi^{(0)}_{n_i}\right\rangle$ is degenerate.

If one wants to get the $E_{n_1}, E_{n_2}, E_{n_3}, \ldots$, that develop under the influence of $\hat{H}^{(1)}$, one has to pick up those linear combinations

$$\left|\psi^o_i\right\rangle = \sum_{i=1 \text{ to } k} a^{(1)}_{n_i}\left|\hat{\psi}^o_{n_i}\right\rangle$$

that are eigen functions of $\hat{H}^{(1)}$. In matrix language, this means, those linear combinations of $\left|\psi^{(0)}_{n_i}\right\rangle$ that make $\mathcal{H}^{(1)}$ diagonal.

To find this assume that $\hat{H}^{(1)}$ exists and it *lifts the degeneracy completely*. Imagine that the perturbation is turned off gradually, so that the k non-degenerate eigen functions will continuously switch over into exactly k of the infinitely many linear combinations. These k combinations are sometimes called the *Adapted eigen functions* of the perturbation. Once these are found then the first order perturbation theory gives

$$A^{(1)}_n = \hat{H}^{(1)}_{kn} = \left\langle \psi^{(0)}_{n_i}\left|\hat{H}^{(1)}\right|\psi^{(0)}_{n_i}\right\rangle\bigg|_{i=1 \text{ to } k} \qquad (19.3.16)$$

and $\quad a^{(1)}_{n_1 k} = \hat{H}^{(1)}_{k n_i} / [E^o_{n_i} - E^o_k]. \qquad (19.3.18)$

These are valid. If $i = n_j$, the denominator still vanishes in the case of degeneracy, but the numerator also vanishes as $\mathcal{H}^{(1)} = $ diagonal. Then $a^{(1)}_{n_1 k}$ can be found.

19.6.2.1 To show that matrix $\hat{H}^{(1)}$ can always be made diagonal:

Consider for simplicity $k = 5$, *ie* a 5-fold degeneracy for the state $\left|\hat{\psi}^o_n\right\rangle$ in the absence of perturbation. The diagonal matrix $\hat{H}^{(o)}$ will look as follows:
The energy eigen value (E^0_n) is different from
$(E^0_{n<n-1})$, (E^0_{n-1}), (E^0_n) to (E^0_{n+4}), (E^0_{n+5}), $(E^0_{n>n+5})$, *etc*.
So these states will not give any trouble in equation (19.3.19).

$$|\hat{\psi}_n\rangle \cong |\hat{\psi}_n^0\rangle + \left\{\sum \hat{H}_{kn}^{(1)} /(E_n^0 - E_k^0)\right\} |\hat{\psi}_k^0\rangle \tag{19.3.19}$$

The only states one has to worry are the 5 states (E_n^0), (E_{n+1}^0), (E_{n+2}^0), (E_{n+3}^0), and (E_{n+4}^0), all having value (E_n^0). That is the Hamiltonian $\hat{H}^{(1)}$ has to be made diagonal with respect to these 5 states only. If $\hat{H}^{(0)}$ and $\hat{H}^{(1)}$ have to be simultaneously diagonal, within the box shown, they must commute (at least within the box).

$$[\hat{H}^{(0)}, \hat{H}^{(1)}] = 0 \tag{19.6.2}$$

$$\hat{H}^{(0)} = \begin{pmatrix}
E_{n-2}^0 & 0 & 0 & 0 & 0 & 0 & 0 & 0 & 0 & 0 & 0 & 0 & etc \\
0 & E_{n-1}^0 & 0 & 0 & 0 & 0 & 0 & 0 & 0 & 0 & 0 & 0 & etc \\
0 & 0 & E_n^0 & 0 & 0 & 0 & 0 & 0 & 0 & 0 & 0 & 0 & etc \\
0 & 0 & 0 & E_n^0 & 0 & 0 & 0 & 0 & 0 & 0 & 0 & 0 & etc \\
0 & 0 & 0 & 0 & E_n^0 & 0 & 0 & 0 & 0 & 0 & 0 & 0 & etc \\
0 & 0 & 0 & 0 & 0 & E_n^0 & 0 & 0 & 0 & 0 & 0 & 0 & etc \\
0 & 0 & 0 & 0 & 0 & 0 & E_n^0 & 0 & 0 & 0 & 0 & 0 & etc \\
0 & 0 & 0 & 0 & 0 & 0 & 0 & E_{n+5}^0 & 0 & 0 & 0 & 0 & etc \\
0 & 0 & 0 & 0 & 0 & 0 & 0 & 0 & E_{n+6}^0 & 0 & 0 & 0 & etc \\
0 & 0 & 0 & 0 & 0 & 0 & 0 & 0 & 0 & E_{n+7}^0 & 0 & 0 & etc \\
0 & 0 & 0 & 0 & 0 & 0 & 0 & 0 & 0 & 0 & E_{n+8}^0 & 0 & etc \\
0 & 0 & 0 & 0 & 0 & 0 & 0 & 0 & 0 & 0 & 0 & E_{n+9}^0 & etc \\
etc & & & & & & & & & & & &
\end{pmatrix} \tag{19.6.3}$$

The box indicated in $\hat{H}^{(0)}$ in equation (19.6.1) is proportional to the unit matrix, $\hat{H}$, and the unit matrix commutes with any matrix, and so with $\hat{H}^{(1)}$.

$$[\hat{H}^{(1)}, \hat{H}] = 0 \tag{19.6.3}$$

Hence one can always diagonalize $\hat{H}^{(1)}$, in the region of degeneracy.

Thus for the 5-fold degenerate state the procedure is:
(1) Pick 5 different, linearly independent, linear combinations

$$|\psi_{n_1}^{(0)}\rangle, |\psi_{n_2}^{(0)}\rangle, |\psi_{n_3}^{(0)}\rangle, |\psi_{n_4}^{(0)}\rangle, |\psi_{n_5}^{(0)}\rangle$$ of the unperturbed eigen functions,

(2) Form the $(5 \times 5) = 25$ matrix elements of $\hat{H}^{(1)}$ with them,
(3) Diagonalize the (5×5) matrix so obtained using a Unitary Transformation, $\hat{U}$,
(4) The diagonalized form of $\hat{H}^{(1)}$ gives the energy eigen values. One may find the Adapted linear combinations by applying the Unitary Transformation $\hat{U}$ to $|\psi_{n_1}^{(0)}\rangle, |\psi_{n_2}^{(0)}\rangle, |\psi_{n_3}^{(0)}\rangle, |\psi_{n_4}^{(0)}\rangle, |\psi_{n_5}^{(0)}\rangle$, these under the influence of the perturbation Hamiltonian represented in the system of unperturbed eigen function $|\psi_{n_4}^{(0)}\rangle$ becomes as shown in equation (19.6.1).

19.6.2.2 The MATRIX OF THE HAMILTONIAN, $\hat{H}$

$$\hat{H}^{(1)} = \begin{pmatrix} \begin{array}{cc} (\hat{H}^{(1)}_{11} + E^0_1) & \hat{H}^{(1)}_{12} \\ \hat{H}^{(1)}_{21} & (\hat{H}^{(1)}_{22} + E^0_2) \end{array} & \text{Non-zero Region} & \text{Region Non-zero} \\ \hline \text{Non-zero Region} & \begin{array}{c} (\hat{H}^{(1)}_{33} + E^0_3) \\ (\hat{H}^{(1)}_{44} + E^0_4) \quad \text{Non-zero} \\ (\hat{H}^{(1)}_{55} + E^0_5) \\ \text{Non-zero} \quad (\hat{H}^{(1)}_{66} + E^0_6) \\ (\hat{H}^{(1)}_{77} + E^0_7) \end{array} & \text{Non-zero Region} \\ \hline \text{Non-zero Region} & \text{Non-zero Region} & (\hat{H}^{(1)}_{88} + E^0_8) \; etc \end{pmatrix}$$

(19.6.4)

The UNITARY MATRIX, $\hat{U}$

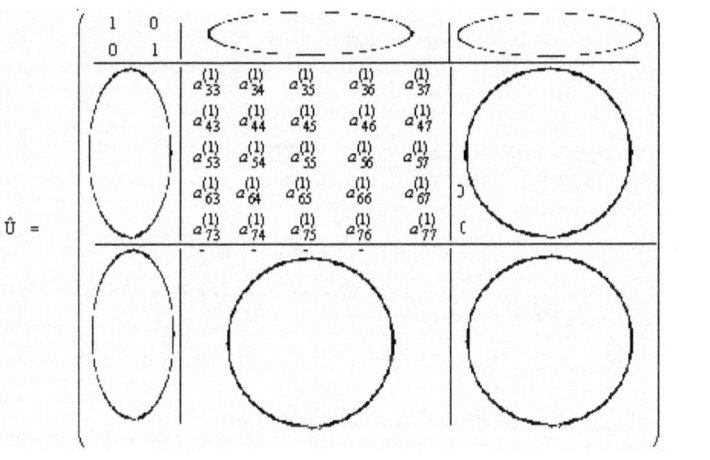

(19.6.5)

19.6.2.4. The DIAGONALIZED HAMILTONIAN, $\hat{H}_d$

As a result of Unitary Transformation

$$\hat{U}^\dagger \hat{H} \hat{U} = \hat{H}_d .\qquad (19.6.6)$$

In the Representation shown (19.6.7) for $\hat{H}_d$, the unperturbed Hamiltonian in the system of the unperturbed eigen functions $\left| \hat{\psi}^o_n \right\rangle$, if instead of arbitrary linear combination the Adapted eigen functions have been used for the state with the energy $(E^0_3 + A^{(1)}_{31})$.

In obtaining the desired values of the five values of the perturbation energy, $A^{(1)}_{31}$, one need not know actually the Adapted eigen functions

$$\hat{H}_d = \begin{pmatrix} \begin{matrix} (\hat{H}^{(1)}_{11}+E^0_1) & \hat{H}^{(1)}_{12} \\ \hat{H}^{(1)}_{21} & (\hat{H}^{(1)}_{22}+E^0_1) \end{matrix} & \text{Non-Zero Region} & \text{Non-Zero Region} \\ \text{Non-Zero Region} & \begin{matrix} (E^0_3+A^{(1)}_{31}) & 0 & 0 & 0 & 0 \\ 0 & (E^0_3+A^{(1)}_{32}) & 0 & 0 & 0 \\ 0 & 0 & (E^0_3+A^{(1)}_{33}) & 0 & 0 \\ 0 & 0 & 0 & (E^0_3+A^{(1)}_{34}) & 0 \\ 0 & 0 & 0 & 0 & (E^0_3+A^{(1)}_{35}) \end{matrix} & \text{Non-Zero Region} \\ \text{Non-Zero Region} & \text{Non-Zero Region} & \text{Non-Zero Region} \end{pmatrix}$$

(19.6.7)

Putting $\hat{H} = (E^0_3 + A^{(1)}_{31})$ in the equation (19.6.6)

$$\hat{U}^\dagger (E^0_3 + A^{(1)}_{31}) \hat{U} = \hat{U}^\dagger (E^0_3) \hat{U} + \hat{U}^\dagger (A^{(1)}_{31}) \hat{U} = (E^0_3 + A^{(1)}_{31}) \tag{19.6.8}$$

or $\qquad \hat{U}^\dagger (A^{(1)}_{31}) \hat{U} = (A^{(1)}_{31})$ (19.6.9)

This is an equation that directly relates the perturbation part of the Hamiltonian, $\hat{H}^{(1)}$ to the First Order Perturbation energy, and it is in this sense that one shall henceforth interpret the matrices $\hat{U}$ and $\hat{H}^{(1)}$.

19.6.3 SUMMARY OF THE DEGENERATE PERTURBATION
In matrix language Degenerate Perturbation Theory is:
1) $\quad \hat{H}^{(0)}$ is diagonal means $\hat{H}^{(1)}$ is not diagonal.

If $\hat{H}^{(1)}$ is diagonalized $\hat{H}^{(0)}$ ceases to be a diagonal matrix, since $[\hat{H}^{(0)}, \hat{H}^{(1)}] \neq 0$.

If one works with $\hat{H} = \hat{H}^{(0)}, \hat{H}^{(1)}$ and with a subset of degenerate states, which are all eigen states with the same eigen value of $\hat{H}^{(0)}$, then for these states $\hat{H}^{(0)}$ is not only diagonal, but $\hat{H}^{(0)}$ is proportional to $\mathcal{I}$. Since $[\hat{H}^{(0)}, \hat{H}] = 0$, one may diagonalize $\hat{H}^{(1)}$ by itself, without affecting $\hat{H}^{(0)}$.

19.7 THE STARK EFFECT: Application of the Perturbation Theory

19.7.1 Particle in a Box
This is an example given as an exercise R.Q. 19.10

19.7.2 LINEAR STARK EFFECT of the First Excited State of HYDROGEN Atom

19.7.2.1 The Stark Effect for $n = 2$ Hydrogen (The Four Degenerate States for $n = 2$)

The Stark effect for the $n = 2$ states of hydrogen requires the use of degenerate state perturbation theory since there are four states with (nearly) the same energies. For this first ignore the hydrogen fine structure and assume that the four states are exactly degenerate, each with unperturbed energy of $\vec{E}_0$.

$$\hat{H}^{(1)} = +e\,\vec{E}_0\,z$$

$$\hat{H}^{(0)} Y_{2\ell}^m = E_2^{(0)} Y_{2\ell}^m$$

It is known from Chapter 8 and Chapter 9 tha

$$Y_\ell^m(\theta,\varphi) = \Theta_\ell^m(\theta) \cdot \Phi_m(\varphi)$$

$$\hat{Y}_\ell^m(\theta,\varphi) = (-1)^m \left[\left(\frac{\ell + \frac{1}{2}}{4\pi} \frac{(\ell-m)!}{(\ell+m)!} \right) \right]^{1/2} P_\ell^m(\cos\theta) \, e^{\,i|m|\varphi}$$

$$\mathfrak{R}_{n\ell}(r) = C_n(\alpha r)^\ell \cdot L_{n+\ell}^{2\ell+1}(\alpha r) \cdot e^{-Zr/na_o}$$

$$\psi(\vec{r}) = \psi(\vec{r},\theta,\varphi) = \mathfrak{R}_{n\ell}(r) \cdot Y_\ell^m(\theta,\varphi) \qquad (9.7.1)$$

$$\mathfrak{R}_{10} = 2(Z/a_o)^{3/2} e^{-Zr/a_o}$$

$$\mathfrak{R}_{20} = (Z/2a_o)^{3/2}[2 - Zr/a_o] \cdot e^{-Zr/2a_o}$$

$$\mathfrak{R}_{21} = \sqrt{(1/3)} \, (Z/2a_o)^{3/2}[Zr/a_o] \cdot e^{-Zr/2a_o}$$

For a hydrogen atom, $n = 2$ (first excited state), in the case of unperturbed system, there are four $n = 2$ states, $Y_{20}^0, Y_{21}^1, Y_{21}^0$ and Y_{21}^{-1}, that are degenerate, having the same energy,

19.7.3.2 To FORM ADAPTED EIGEN FUNCTIONS

To study the perturbation of the $n = 2$ level, in the first order approximation, the following linear combination must be formed:

$$|\Psi_2^0\rangle = \sum_{j=1 \, to \, k} a_{2j}^{(1)} |\varphi_{2_j}^0\rangle \qquad (19.7.13)$$

$$= a_{200}^{(1)} |200\rangle + a_{211}^{(1)} |211\rangle + a_{210}^{(1)} |210\rangle + a_{21-1}^{(1)} |21\text{-}1\rangle$$

19.7.3.3 The PERTURBATION TERM, $\hat{H}^{(1)}$

Let the electric field $\vec{E}_0 \parallel z$, so that angle $\theta = 0$.

$$\hat{H}^{(1)} = +e\,\vec{E}_0 \, r\cos\theta, \qquad (19.7.14)$$

The perturbation due to an electric field in the z direction is $\hat{H}^{(1)} = +e\,\vec{E}_0\,z$

19.7.3.4   H- Atom in an Electric Field FIRST-ORDER CORRECTION, $A_2^{(1)} = \hat{H}_{22}^{(1)}$

So the first order degenerate state perturbation theory equation is

$$\sum_i a_i \left\langle \hat{Y}_\ell^{(j)} \left| \hat{H}^{(0)} + e\vec{E}_0\,z \right| \hat{Y}_\ell^{(i)} \right\rangle = (E_2^{(0)} + E_2^{(1)}) \, a_j$$

This is esentially a (4 x 4) matrix eigenvalue equation. There are 4 eigenvalues $(E_2^{(0)} + E_2^{(1)})$, distinguished by the index n.

Because of the exact degeneracy $[\hat{H}^{(0)} \hat{Y}_\ell^{(j)} = E_2^{(0)} \hat{Y}_\ell^{(j)}]$, $\hat{H}^{(0)}$ and $E_2^{(0)}$ can be eliminated from the equation.

i.e., $$E_2^{(0)} a_j + \sum_i a_i \left\langle \hat{Y}_\ell^{(j)} \left| +e\vec{E}_0\,z \right| \hat{Y}_\ell^{(i)} \right\rangle = E_2^{(0)} a_j + E_2^{(1)} a_j$$

$$\sum_i a_i \left\langle \hat{Y}_\ell^{(j)} \left| +e\vec{E}_0\,z \right| \hat{Y}_\ell^{(i)} \right\rangle = E_2^{(1)} a_j$$

This is just the eigenvalue equation for $\hat{H}^{(1)}$ which we can write out in (pseudo) matrix form

$$\left(\hat{H}^{(1)}\right)\begin{pmatrix} a_1 \\ a_2 \\ a_3 \\ a_4 \end{pmatrix} = E_2^{(1)} \begin{pmatrix} a_1 \\ a_2 \\ a_3 \\ a_4 \end{pmatrix}$$

Now, in fact, most of the matrix elements of $\hat{H}^{(1)}$ are zero. We will show that because $[\hat{L}_z, \hat{z}] = 0$, that all the matrix elements between states of unequal m are zero. Another way of saying this is that the operator z doesn't ``change'' m. Here is a little proof.

$$\langle \hat{Y}_\ell^m |[\hat{L}_z, \hat{z}]| \hat{Y}_{\ell'}^{m'} \rangle = 0 = (m - m') \langle \hat{Y}_\ell^m | \hat{z} | \hat{Y}_{\ell'}^{m'} \rangle$$

This implies that $\langle \hat{Y}_\ell^m | \hat{z} | \hat{Y}_{\ell'}^{m'} \rangle = 0$ unless $(m - m')$.

Lets define the one remaining nonzero (real) matrix element to be $\hat{H}^{(1)}_{\substack{\ell\, m \\ \ell'\, m'}}$

$$\hat{H}^{(1)}_{\substack{\ell\, m \\ \ell'\, m'}} = e\vec{E}_0 \langle \hat{Y}_\ell^m | \hat{z} | \hat{Y}_{\ell'}^{m'} \rangle = e\vec{E}_0 \langle \hat{Y}_{20}^0 | \hat{z} | \hat{Y}_{21}^0 \rangle.$$

The equation (labeled with the basis states to define the order) is.

$$\begin{matrix} |\hat{Y}_{20}^0\rangle \\ |\hat{Y}_{21}^0\rangle \\ |\hat{Y}_{21}^1\rangle \\ |\hat{Y}_{21}^{-1}\rangle \end{matrix} \rightarrow \begin{pmatrix} 0 & 0 & \hat{H}^{(1)}_{\substack{200\\210}} & 0 \\ \hat{H}^{(1)}_{\substack{210\\200}} & 0 & 0 & 0 \\ 0 & 0 & 0 & 0 \\ 0 & 0 & 0 & 0 \end{pmatrix} \begin{pmatrix} a_1 \\ a_2 \\ a_3 \\ a_4 \end{pmatrix} = E_2^{(1)} \begin{pmatrix} a_1 \\ a_2 \\ a_3 \\ a_4 \end{pmatrix}$$

We can see by inspection that the eigenfunctions of this operator are $|\hat{Y}_{21}^1\rangle$, $|\hat{Y}_{21}^{-1}\rangle$, and $\sqrt{1/2}\{|\hat{Y}_{20}^0\rangle \pm |\hat{Y}_{21}^0\rangle\}$ with eigenvalues (of $(\hat{H}^{(1)})$) of 0, 0, and $\pm\hat{H}^{(1)}_{\substack{200\\210}}$.

What remains is to compute $\hat{H}^{(1)}_{\substack{200\\210}}$. Recall $|\hat{Y}_{20}^0\rangle = \sqrt{1/4\pi}$ and $|\hat{Y}_{21}^0\rangle = \sqrt{3/4\pi}\, Cos\theta$.

$$\hat{H}^{(1)}_{\substack{200\\210}} = \left\{ e\vec{E}_0 \int (2a_0)^{-3/2}[2 - (r/a_0)] \cdot \right.$$
$$\left. \cdot e^{-r/2a_0}|\hat{Y}_{20}^0\rangle z(2a_0)^{-3/2}(1/\sqrt{3})(r/a_0)\cdot e^{-r/2a_0}|\hat{Y}_{21}^0\rangle d\tau \right\}$$

$$= 2e\vec{F}_0(2a_0)^{-3}(1/\sqrt{3})\int r^3 dr[1 - (r/2a_0)](r/a_0)\cdot e^{-r/a_0}\int \sqrt{3/4\pi}\, Cos\theta\, |\hat{Y}_{21}^0\rangle d\Omega$$

$$= 2e\vec{E}_0(2)^{-3}(1/\sqrt{3})(1/\sqrt{3})\int_0^\infty \left(\frac{r^4}{a_0^4} - \frac{r^5}{2a_0^5}\right)\cdot e^{-r/a_0}d\tau$$

$$= \frac{a_0 e\vec{E}_0}{12}\left\{\int_0^\infty x^4 \cdot e^{-x}dx - \int_0^\infty x^5 \cdot e^{-x}dx\right\}$$

$$= \frac{a_0 e\vec{E}_0}{12}\{\lfloor 4 - \lfloor 5/2\} = \frac{a_0 e\vec{E}_0}{12}(-36)$$

$$= -3\, e\, a_0\, \vec{E}_0$$

whence $\boxed{E_2^{(1)} = \mp 3\, e\, a_0\, \vec{E}_0}$

This is first order in the electric field, as we would expect in first order (degenerate) perturbation theory.

If the states are not exactly degenerate, we have to leave in the diagonal terms of $\hat{H}^{(0)}$. Assume that the energies of the two (mixed) states are $(E_2^{(0)} \pm \Delta)$, where Δ comes from some other perturbation, like the hydrogen fine structure. (The $|\hat{Y}_{21}^{+1}\rangle$ and $|\hat{Y}_{21}^{-1}\rangle$ are still not mixed by the electric field.)

$$\begin{pmatrix} (E_2^{(0)} - \Delta) & \hat{H}_{200\,210}^{(1)} \\ \hat{H}_{200\,210}^{(1)} & (E_2^{(0)} + \Delta) \end{pmatrix} \begin{pmatrix} a_1 \\ a_2 \end{pmatrix} = E \begin{pmatrix} a_1 \\ a_2 \end{pmatrix}$$

$$E = \vec{E}_0 \pm \sqrt{\left(\hat{H}_{200\,210}^{(1)}\right)^2 + \Delta^2}$$

This is OK in both limits, $\Delta \gg \hat{H}_{200\,210}^{(1)}$, and $\hat{H}_{200\,210}^{(1)} \gg \Delta$. It is also correct when the two corrections are of the same order.

The first order induced energy shifts and corrected energies that one arrives at are:

1) $A_{21+1}^{(1)} = 0$, i.e., $2p_1$ state, $E_{211}^{(1)} = -\frac{1}{8}(e^2/a_o)$. (19.7.23)a

2) $A_{21-1}^{(1)} = 0$, i.e., $2p_{-1}$ state, $E_{21-1}^{(1)} = -\frac{1}{8}(e^2/a_o)$. (19.7.23)b

2) $A_{210}^{(1)} = +3\, e\, a_o\, \vec{E}_0$; (19.7.23)c

3) $A_{200}^{(1)} = -3\, e\, a_o\, \vec{E}_0$. (19.7.23)d

4) $E_{210}^{(1)} = -\frac{1}{8}(e^2/a_o) + 3\, e\, a_o\, \vec{E}_0$, for $2p_0$ state.

5) $E_{200}^{(1)} = -\frac{1}{8}(e^2/a_o) - 3\, e\, a_o\, \vec{E}_0$, for $2s$ state.

The diagonalized matrix $\hat{H}_d$ of the Hamiltonian, $\hat{H}$,

$$(\hat{H}^{(0)} + \hat{H}^{(1)})_d = \begin{pmatrix} (E_2^{(0)} - 3\, e\, a_o\, \vec{E}_0) & 0 & 0 & 0 \\ 0 & (E_2^{(0)} + 3\, e\, a_o\, \vec{E}_0) & 0 & 0 \\ 0 & 0 & E_2^{(0)} & 0 \\ 0 & 0 & 0 & E_2^{(0)} \end{pmatrix} \quad (19.7.24)$$

The states $|210\rangle$ and $|200\rangle$ behave thus as if they have a permanent electric dipole moment, of magnitude ($3\, a_0\, e$), which can be oriented in an electric field $\vec{E}_0$. Hence the name '*Linear Stark Effect*' for the $n = 2$ states of atomic hydrogen when an external electric field is applied. $E_2^{(0)}$ state is split into three levels, one of which is <u>doubly-degenerate</u>.

It is possible to show that from the unitary matrix, which diagonalizes the Hamiltonian, the eigen vectors for $(\pm 3\, e\, a_o\, \vec{E}_0)$ and eigen values 0 and 0 are: $\sqrt{\frac{1}{2}}\begin{pmatrix}1\\1\\0\\0\end{pmatrix}$, $\sqrt{\frac{1}{2}}\begin{pmatrix}1\\-1\\0\\0\end{pmatrix}$, $\begin{pmatrix}0\\0\\1\\0\end{pmatrix}$, and $\begin{pmatrix}0\\0\\0\\1\end{pmatrix}$.

The eigen states of both the $\hat{H}^{(0)}$ and $\hat{H}^{(1)}$ are

$|\,1\,\rangle = \dfrac{|\,2,0,0\,\rangle + |\,2,1,0\,\rangle}{\sqrt{2}}$, $|\,2\,\rangle = \dfrac{|\,2,0,0\,\rangle - |\,2,1,0\,\rangle}{\sqrt{2}}$,

$|\,3\,\rangle = |\,2,1,1\,\rangle$, and $|\,4\,\rangle = |\,2,1,-1\,\rangle$

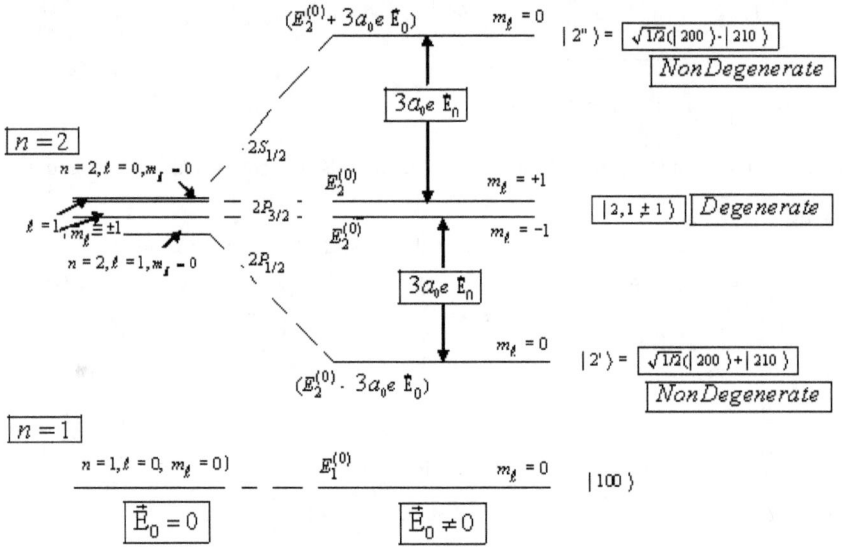

Fig. 19.2. Linear Stark Effect for the n = 2 level of Hydrogen atom.

Fig. 19.2 Linear Stark Effect for the n =2 level of Hydrogen atom.
(Δ (cm^{-1}) = 6.42 x 10^{-5} F (V/cm))

The energy level scheme of the Stark effect in atomic hydrogen, $n-2$ level, ($\vec{E}_0 \approx 10^7\, Vm^{-1}$ used for experimental demonstration) is as shown in Fig. 19.2.

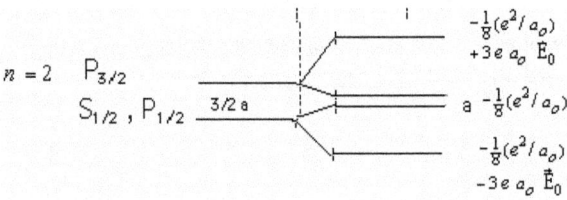

Fig 19.3 The Linear Stark Effect for Hydrogen n =2

19.7.3.5 RESULTS

(1) The states in the presence of an electric field are no longer eigen states of $\hat{L}^2$ but are eigen states of $\hat{L}_z$ as $[\hat{L}^2, \hat{L}_z] \neq 0$, i.e. $\hat{L}_z$ is a *constant of motion*, but $\hat{L}^2$ is not.

(2) When $\hat{H}^{(1)}$ does not conserve $\hat{L}^2$, then the states which diagonalize the new Hamiltonian in an approximation are Superposition of States with different values of the previously conserved quantum number.

(3) System in non-degenerate states cannot have permanent electric dipole moments.

(4) The first-order perturbed energies of the first excited state of hydrogen atom are

$$E^{(1)}_{210} = -\tfrac{1}{8}(e^2/a_0) + 3\, e\, a_0\, \vec{E}_0 \text{ for } 2p_0 \text{ state.}$$

$$E^{(1)}_{200} = -\tfrac{1}{8}(e^2/a_0) - 3\, e\, a_0\, \vec{E}_0 \text{ for } 2s \text{ state.}$$

These two states are NON-DEGENERATE.

The other two states, degenerate when $\vec{E}_0 = 0$, remain degenerate at $\vec{E}_0 \neq 0$,

$2p_1$ state, $E^{(1)}_{211} = -\tfrac{1}{8}(e^2/a_0)$.

$2p_{-1}$ state, $E^{(1)}_{21\text{-}1} = -\tfrac{1}{8}(e^2/a_0)$.

(5) Foster has shown that for very large n, the Stark Effect is <u>linear</u> in $\vec{E}_0$.

(6) For extremely high, $\vec{E}_0$, the Stark Effect is capable of ionizing the atom.

19.7.4 THE LINEAR STARK EFFECT FOR THE $n = 3$ LEVEL IN HYDROGEN

In the presence of a Strong Electric Field, $\vec{E}_0$, the $n = 3$ level of Hydrogen atom can be shown to split into 5 sublevels with

$$E^{(1)}_3 = -\tfrac{1}{8}(e^2/a_0) \pm \tfrac{9}{2}\, e\, a_0\, \vec{E}_0, \qquad \text{19.7.25)a}$$

$$-\tfrac{1}{8}(e^2/a_0) \pm 9\, e\, a_0\, \vec{E}_0, \text{ and} \qquad \text{19.7.25)b}$$

$$-\tfrac{1}{8}(e^2/a_0)) \qquad \text{19.7.25)c}$$

The Selection Rules for the $n = 3$ to $n = 3$ transitions are:

$\Delta m_s = 0$ transitions are said to correspond to π - component,

$\Delta m_s = \pm 1$ transitions are said to correspond to σ - component,

$\Delta\, (\text{cm}^{-1}) = 6.42 \times 10^{-5}\, F\, (V/cm)$

19.7.4 QUDRATIC STARK EFFECT and the NORMAL STATE of HYDROGEN atom

It was seen earlier that the first order correction is zero for all states of definite parity, as in the case of linear Stark Effect., so that perturbation theory fails unless a correction is obtained in the second order perturbation calculation. It might also be desirable to proceed to <u>second order</u> theory in the context of the first order correction is not zero, but where the degeneracy prevails. It is known that

$$\hat{H}^{(0)} = \frac{p^2}{2\mu} - \frac{e^2}{r}$$

$$\hat{H}^{(0)}\left|\psi_{n\ell m}(r)\right\rangle = E^{(0)}_{n\ell m}\left|\psi_{n\ell m}(r)\right\rangle$$

$$E_{n\ell m}^{(0)} = -\frac{e^2}{2a_0^2 n^2}$$

19.7.4.1 The PERTURBATION term, $\hat{H}^{(1)}$:

It is required to find the correction to that solution if an Electric field is applied to the atom. Choosing the axes so that the Electric field $\vec{E}_0 \parallel z$ - direction, so that angle $\theta = 0$. The perturbation is then.

$$\hat{H}^{(1)} = e\,\vec{E}_0\, rCos\theta = e\,\vec{E}_0\, z \qquad (19.7.14)$$

It is typically a small perturbation.

19.7.4.2 FIRST-ORDER CORRECTED ENERGY

$$E_n \approx E_n^{(2)} \cong E_n^{(0)} + A_n^{(1)} + E_n^{(2)}$$

$$\cong E_n^{(0)} + \hat{H}_{nn}^{(1)} + \sum_{i \neq n}^{\infty} \frac{|\hat{H}_{nn}^{(1)}|^2}{E_n^{(0)} - E_i^{(0)}} \qquad (19.4.4)$$

19.7.4.3 To evaluate $A_{1s}^{(1)}$

For non-degenerate states, the first order correction to the energy is zero because the expectation value $\langle z \rangle$ is an odd function.

$$A_n^{(1)} = \hat{H}_{nn}^{(1)} = A_{1s}^{(1)} = \langle 1s \,|\, \hat{H}^{(1)} \,|\, 1s \rangle$$

It can be shown that

$$E_{n\ell m}^{(1)} = e\,\vec{E}_0 \langle Y_{n\ell}^m \,|\, z \,|\, Y_{n\ell}^m \rangle = 0 \qquad (19.7.8)$$

19.7.4.4 SECOND-ORDER CORRECTION, $A_{1s}^{(2)}$

It is thus needed to calculate the *second order* correction. This involves a sum over all the other states.

The relevant equation is (19.7.26), viz

$$A_{1s}^{(2)} = \sum_{i \neq n}^{\infty} \frac{|\hat{H}_{nn}^{(1)}|^2}{E_n^{(0)} - E_i^{(0)}}$$

$$\langle Y_{n\ell}^m \,|\, z \,|\, Y_{10}^0 \rangle = \int d\tau\, \Re_{n\ell}^* (rCos\theta) \Re_{10} Y_\ell^{m*} Y_0^0$$

$$E_{n\ell m}^{(1)} = e^2\,\vec{E}_0^{\,2} \sum_{n\ell m \neq 100} \frac{|\langle Y_{n\ell}^m \,|\, z \,|\, Y_{n\ell}^m \rangle|^2}{E_1^{(0)} - E_n^{(0)}}$$

$$\left|\hat{H}_{n\ell m \atop 100}^{(1)}\right|^2 = \tfrac{1}{3} e^2\, (\vec{E}_0^{\,2})\, \frac{2^8 n^7 (n-1)^{(2n-5)}}{(n+1)^{(2n+5)}}\, a_0^2 \delta_{\ell 0} \delta_{m 0} \equiv f(n) \cdot a_0^2 \delta_{\ell 0} \delta_{m 0}$$

$$E_{100}^{(2)} = e^2\,\vec{E}_0^{\,2} \sum_{n=2}^{\infty} \frac{|\hat{H}_{n\ell m,100}^{(1)}|^2}{E_1^{(0)} - E_n^{(0)}} = e^2\,\vec{E}_0^{\,2} \sum_{n=2}^{\infty} \frac{2 f(n)}{[-1+ (1/n^2)]}$$

$$= -2 e^3\,\vec{E}_0^{\,2} \sum_{n=2}^{\infty} \frac{n^2 f(n)}{[n^2 - 1]}$$

Since the ground state wave function is spherically symmetric $\sum_{i, n \neq 1} f(n) = 1$.

$$\boxed{E_{100}^{(2)} = -2a_0^3 \vec{E}_0^2 (\tfrac{4}{3}) = -\tfrac{8}{3} a_0^3 \vec{E}_0^2} \qquad (19.7.36)$$

This indicates why the energy shift is second order.
Therefore the phenomenon is known as the *quadratic Stark effect*
The electric dipole moment of the atom which is induced by the electric field

$$\vec{d} = -\frac{\partial \Delta E}{\partial E_0} = 4 \, (\tfrac{4}{3}) \, a_0^3 \vec{E}_0 \qquad (19.7.8)$$

The dipole moment is proportional to $\vec{E}$ the Electric field, indicating that it is induced. Then the dipole moment interacts with the E field causing a energy shift.

19.7.4.5 THOMAS-REICHE-KUHN SUM RULE

The relation $\quad \sum_{n\ell m} \langle Y_{n\ell}^m | z | Y_{10}^m \rangle = \sum_{n\ell m} \langle Y_{n\ell}^m | z^2 | Y_{10}^m \rangle \qquad (19.7.33)$

is called 'Thomas-Reiche-Kuhn Sum Rule'.

19.8 Weak-field STARK Effect in Hydrogen

In presence of very weak field, F (V/cm), the Stark splitting is given by
There is a weak-field when the Stark splitting is small compared with the FINE STRUCTURE splitting.
The various levels are: $3S_{1/2}, 3P_{1/2}; 3P_{3/2}, 3D_{3/2}$; and $3D_{5/2}$

19.8 CONCLUSIONS

The theory of first order and second order perturbation of non-degenerate and degenerate systems were dealt with in a simple and in sufficient detail so as to understand the technique well.

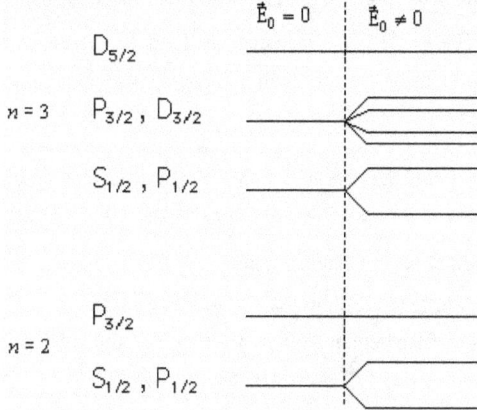

Fig 19.4 The Weak Field Stark Effect in Hydrogen

The sound treatment of the theory was made use of in solving a number of simple to complex problems as worked out examples. These examples were from particle in a box, anharmonic oscillator, hydrogen atom, and hydrogen atom in an electric field and Helium atom.

REVIEW QUESTIONS

R.Q. 19.1 Consider the particle moving in a 1-D box with a linear potential gradient added as perturbation. Calculate the first order correction to the wave function, using the perturbation theory. (Answer: $V_1/E_n^{(0)} = 0.06$, Mixing ratio, $a_{12}^{(1)} = \frac{0.18}{3}(V_1/E_1^{(0)})$, $a_{13}^{(1)} = 0$,

$a_{14}^{(1)} = 0.0096(V_1/E_1^{(0)})$, $\left|\hat{\psi}_n\right> \cong \left|\hat{\psi}_n^{(o)}\right> + \sum_{n\,i} a_{n\,i}^{(1)}\left|\hat{\psi}_i^{(o)}\right>$

$\left|\hat{\psi}_n\right> \cong \left|\hat{\psi}_n^{(o)}\right> + (0.06)\,(V_1/E_1^{(0)})\left|\hat{\psi}_2^{(o)}\right> + 0.0096(V_1/E_1^{(0)})\left|\hat{\psi}_4^{(o)}\right>$).

R.Q. 19.2 A particle moves in a one-dimensional box with a linear potential gradient added as perturbation. Calculate the second order energy correction applying the perturbation theory. (Answers: $\hat{H}_{21}^{(1)} = -0.18\,V_1$, $\hat{H}_{31}^{(1)} = 0$, $\hat{H}_{41}^{(1)} = -0.0144\,V_1$, $\hat{H}_{51}^{(1)} = 0$,

$A_1^{(2)} = \frac{(-0.18\,V_1)^2}{(-3E_1^{(0)})} + \frac{(-0.0144\,V_1)^2}{(-15E_1^{(0)})} \approx -\frac{(0.0109\,V_1)^2}{E_1^{(0)}}$,

$E_1 \approx E_1^{(2)} \cong E_1^{(0)} + A_1^{(1)} + A_1^{(2)} \cong E_1^{(0)} + 0.500\,V_1 - \frac{(0.0109\,V_1)^2}{E_1^{(0)}}$)

R.Q. 19.3 Given a system whose potential has a cubic term, αx^3, added to the parabolic potential (α is a constant, with small value). Calculate the first order energy correction to the ground state, using the perturbation theory. (Answer: ution:

$A_0^{(1)} = \hat{H}_{00}^{(1)} = 0$; $E_0^{(1)} \cong E_0^{(0)} + 0 \cong \frac{1}{2}\hbar\omega_c$)

R.Q. 19.4 Consider a one-dimensional anharmonic oscillator whose Hamiltonian, contains a quartic term, is given by $\hat{H} = \hat{H}^{(0)} + \hat{H}^{(1)} = [\frac{\hat{p}_x^2}{2m} + \frac{1}{2}kx^2] + \beta x^4$. Calculate the Find the energy spectrum of ground state, to the first order correction, using the perturbation theory. β is a small constant. (Answer: $E_0^{(1)} \cong E_0^{(o)} + A_0^{(1)} \cong \frac{1}{2}\hbar\omega_c + (3\,\beta/4)\,(\hbar/k\omega_c)^2$).

R.Q. 19.5 Using the perturbation theory find the first order correction to the energy spectrum of a system whose Hamiltonian is $\hat{H} = \hat{H}^{(0)} + \hat{H}^{(1)} = \left\{[\frac{-\hbar^2}{2m}(\partial^2/\partial x^2) + \frac{1}{2}kx^2] + \alpha x^3 + \beta x^4\right\}$. Calculate the energy spectrum of ground state. α and β are small anharmonic constants. (Answer: $E_n = E_n^{(1)} \cong E_n^{(o)} + A_n^{(1)} \cong [n+\frac{1}{2}\hbar\omega_c] + [0+(3\,\beta/4)\,\hbar\omega_c(2n^2 + 2n+1)]$).

R.Q. 19.6 Find the ground state energy of the helium atom by treating the Coulomb interaction between the two electrons as a perturbation. ($[2(Z)^2 - \frac{5}{4}Z\,](-13.6\,eV) = -74.8\,eV$).

R.Q. 19.7 Evaluate the ground state energy of the helium ion by treating the Coulomb interaction between the two electrons as a perturbation. (;The energy of the ground state of He$^+$ ion $= Z^2\,E_{1s}^0(H) = -54.4\,eV$).

R.Q. 19.8 Consider the ($n_x = 2$, $n_y = 1$, $n_z = 1$), and cyclic permutation) degenerate case of the particle in a 3-D box bounded by an infinite potential barrier and treat it for the first order perturbation theory. Suppose that a perturbing field of the form βx^2 is applied, where $\beta = a$ constant. (Answer: $E_1 \approx E_1^{(1)} = E_1^{(0)} + \hat{H}_{11}^{(1)} = \frac{6\hbar^2 \pi^2}{2ma^2} + \beta a^2 \{\frac{1}{3} - \frac{1}{8\pi^2}\}$;

$E_2 \approx E_2^{(1)} = E_2^{(0)} + \hat{H}_{22}^{(1)} = \frac{6\hbar^2 \pi^2}{2ma^2} + \beta a^2 \{\frac{1}{3} - \frac{1}{2\pi^2}\}$).

R.Q. 19.9 When an electric field, $\vec{E}_0$ is applied to a hydrogen atom, the total Hamiltonian, $\hat{H}$ matrix for the first excited state, that is, n = 2, can be written approximately as

$$(\hat{H}^{(0)} + \hat{H}^{(1)})_d = \begin{pmatrix} (E_2^{(0)} - 3e\, a_0\, \vec{E}_0) & 0 & 0 & 0 \\ 0 & (E_2^{(0)} + 3e\, a_0\, \vec{E}_0) & 0 & 0 \\ 0 & 0 & E_2^{(0)} & 0 \\ 0 & 0 & 0 & E_2^{(0)} \end{pmatrix}$$

where $E_2^{(0)}$ is the energy eigen value, i.e. energy level, of the first excited state for $\vec{E}_0 = 0$, and $a_0 = \hbar^2/me^2$. Find the energy levels of the hydrogen atom for $\vec{E}_0 \neq 0$, but strong). Obtain the Unitary Transformation matrix operator that will diagonalize $\hat{H}$.

$$\text{(Answers: } |u_2\rangle = \begin{pmatrix} \sqrt{1/2} \\ -\sqrt{1/2} \\ 0 \\ 0 \end{pmatrix}, \hat{U} = \begin{pmatrix} \sqrt{1/2} & \sqrt{1/2} & 0 & 0 \\ \sqrt{1/2} & -\sqrt{1/2} & 0 & 0 \\ 0 & 0 & 1 & 0 \\ 0 & 0 & 0 & 1 \end{pmatrix} \text{)}.$$

R.Q. 19.10 A hydrogen atom is exposed to a strong electric field with strength $\vec{E}_0$. Calculate the Stark Effect for the $2s_{1/2}$ and $2p_{1/2}$ levels using perturbation theory. Calculate the first order energies. Show that the energy shifts vary quadratically with the field when $(e\, a_0\, \vec{E}_0) \ll$ and linearly when $(e\, a_0\, \vec{E}_0) \gg$.

R.Q. 19.11 i) Calculate the fine structure for the n=2 level in H. Consider the energies and wavefunctions corresponding to the unperturbed Hamilton operator $\hat{H}^{(0)}$, which is the sum of the kinetic energy operator and the Coulomb potential. The electron has a $\frac{1}{2}$ spin. ii) What is the degeneracy of the n=2 level? Which orthonormal basis can be used to describe level 2? Which quantum numbers are necessary to characterize the wavefunctions?

The fine structure is described with the Hamilton operator $\hat{H} = \hat{H}^{(0)} + \hat{H}_{mv} + \hat{H}_{SO} + \hat{H}_D$ corresponds to the $\hat{H}_{mv}$ relativistic mass increase, $\hat{H}_D$ is the Darwin term, and $\hat{H}_{SO}$ describes the spin-orbit interaction. $\hat{H}_{mv}$ is proportional to mv, $\hat{H}_D$ depends only on r, and $\hat{H}_{SO} = V_{LS}(r)$. The hyperfine interaction is neglected.

iii) Show that $\hat{H}_{SO}$ does not commute with $\hat{L}_z$. What does it imply for the matrix that describes $\hat{H}_{SO}$ in the basis for n = 2?

iv) Show that $\hat{H}_{SO}$ commutes with and that the matrix consists of a (2x2) matrix for 2s and a (6x6) matrix for 2p.

v) To simplify the problem, consider the total angular momentum operator. Use the triangle rule to determine which j, and s are possible for $n=2$. These states are labelled.

Show that and $\hat{J}_z$ commute with $\hat{H}^{(0)}$ and $\hat{H}_{SO}$.

vi) In which basis are , $\hat{J}_z$, $\hat{H}^{(0)}$ and $\hat{H}_{SO}$, diagonal?

vii) Write down $\hat{H}_{SO}$ in the above matrix using the integrals where is the radial wavefunction for 2. How many energy levels are there?

viii) It can be shown that $\hat{H}_{mv}$ and $\hat{H}_D$ commute with . What is the effect of $\hat{H}_{mv}$ and $\hat{H}_D$ on the $n=2$ shell?

R.Q. 19.12 A Hydrogen atom is exposed to a strong electric field $\vec{E}_0$. Obtain using the perturbation theory the effect on the levels $2s_{1/2}, 2p_{1/2}$ and $2p_{3/2}$ and draw the energy level diagram. (Answers: $(-\frac{1}{8}(e^2/a_0) \pm 3 e a_0 \vec{E}_0)$ and $-\frac{1}{8}(e^2/a_0)$).

R.Q. 19.13 Show that in the Linear Stark Effect the $n=3$ state of H-atom is split into 5 equally spaced components. (Answers: $E_3^{(1)} = -\frac{1}{8}(e^2/a_0) \pm \frac{9}{2} e a_0 \vec{E}_0$, $-\frac{1}{8}(e^2/a_0) \pm 9 e a_0 \vec{E}_0$, and $-\frac{1}{8}(e^2/a_0)$)

R.Q. 19.14 Calculate the second order correction to the energy of the ground state of the Hydrogen atom, using perturbation theory (Answer: $E_{100}^{(2)} = -\frac{8}{3} a_0^3 \vec{E}_0^2$).

R.Q. 19.14 Using the perturbation theory find the first order correction to the energy spectrum of a harmonic oscillator with anharmonic perturbation ax^4 (Answer: $\Delta E_n^1 = a<n|x^4|n> = \frac{a\hbar^2}{4m^2\omega^2} <n|(\hat{a}+\hat{a}^\dagger)^4|n> = = \frac{3a\hbar^2}{4m^2\omega^2}(2n^2+2n+1)$).

R.Q. 19.15 Illustrate the method of time-independent perturbation theory for degenerate states by applying it to explain the Stark Effect in Hydrogen atom.

R.Q. 19.16 A system has three states. The perturbed Hamiltonian Matrix is $\begin{pmatrix} E_1 & 0 & a \\ 0 & E_1 & b \\ a^* & b^* & E_2 \end{pmatrix}$ when $E_2 > E_1$. Diagonalize the matrix to find the exact eigen values.

R.Q. 19.17 Develop te time independent perturbation theory for non-degenerate states and obtain eigen functions and eigen values up to second order.

R.Q. 19.18 Show that a Hydrogen atom in its first excited state behaves asthough it has a permanent electric dipole moment of magnitude $3ea$ that can be oriented in three different ways.

&&*&*&*&*&*&*&*&*&*

Chapter 20

VARIATIONAL PRINCIPLE
AND JKB APPROXIMATION

Chapter 20

VARIATIONAL PRINCIPLE AND JKB APPROXIMATION

"If we did all the things that we are capable of doing, we would literally astound ourselves".
— Thomas Edison

20.1 INTRODUCTION

It has been stated in Chapter 19 that the Schrödinger Equation cannot be solved exactly for any atom or molecule more complicated than the ideal Hydrogen atom. In that Chapter the perturbation methods when used were seen to solve the Schrödinger Equation of a complicated system to any desired accuracy. It frequently happens that a physical system cannot be solved directly is also not amenable to the perturbation method. This may be due to the fact that the system either does not resemble a system which can be solved exactly, or that the perturbation term, $\hat{H}^{(1)}$ is too large in relation to the exactly solvable part of the Hamiltonian (*i.e.* $\hat{H}^{(o)}$). This situation was seen in the case of the Normal State of Helium atom. $\hat{H}^{(o)}$ of an atom with Z electrons has the form

$$\hat{H}^{(o)} = \sum_{i=1} \frac{\hat{p}_i^2}{2m} - \sum_{i=1} Ze^2 \left(\bar{r}^{-1}\right) + \sum_{i>j} e^2 / |\bar{r}_{ij}|^{-1} \quad (20.1.1)$$

$$= \sum_{i=1} \left(-\hbar^2/2m\right)\nabla_i^2 - \sum_{i=1} Ze^2 \left(\bar{r}^{-1}\right) + \sum_{i>j} e^2 / |\bar{r}_i - \bar{r}_j|^{-1} \quad (20.1.2)$$

where m is the mass of the electron.

It is a partial differential equation in $3Z$- Dimensions. Perturbation Theory turned out to be adequate for $Z = 2$, but as the number of electrons increase, the SHIELDING EFFECTS, not taken into consideration by the First Order Perturbation Theory, becomes more important.

In this Chapter, I will present the other most widely used method, *viz.* the <u>Methods of Variation</u>. The basic equation of the method will be presented first and then it will be applied to a variety of problems. The Variational Methods have the virtue of maintaining the single-particle picture; while at the same time yielding single-particle function that incorporate the screening corrections.

20.2 THE RITZ MTHOD OF VARIATION

20.2.1 The VARIATIONAL METHOD provides an UPPER BOUND to the GROUND STATE ENERGY of a System

To illustrate the Variational Method, consider a general system. Let it possesses a Hamiltonian operator $\mathcal{H}$ and a set of orthonormal eigen functions $\{|\hat{\psi}_n\rangle\}$, such that

$$\langle \hat{\psi}_m || \hat{\psi}_n \rangle = \delta_{mn} \quad (20.2.1)$$

$$\hat{H} |\hat{\psi}_i\rangle = E_i |\hat{\psi}_i\rangle,$$

or $\quad \hat{H} | \hat{i} \rangle = E_i | \hat{i} \rangle \quad (20.2.2)$

i.e. the system has a non-degenerate set of eigen values

$$E_0 < E_1 < E_2 < E_3 < \ldots < E_n, \qquad (20.2.3)$$

corresponding to the orthonormal set $\{|\hat{i}\rangle\}$.

Consider an ARBITRARY or TRIAL WAVE FUNCTION, $|\psi\rangle$ such that

$$|\Psi\rangle = |\Psi(\alpha_1, \alpha_1, \alpha_1, \ldots)\rangle \qquad (20.2.4)$$

α_i forms a set of parameters to be determined.

This function can be expanded in a complete set of eigen states of $\hat{H}$,

$$|\Psi\rangle = \sum_i a_i |\hat{i}\rangle \qquad (20.2.5)$$

20.2.2 BASIC IDEA of the Variation Method

The expectation value of the energy gives the average energy of the system, in a state corresponding to the particular function used, viz. $|\Psi\rangle$.

i.e. $\quad E = \langle \hat{H} \rangle = \dfrac{\langle \Psi | \hat{H} | \Psi \rangle}{\langle \Psi \| \Psi \rangle} = \dfrac{\langle \Psi | \hat{H} | \Psi \rangle}{\sum_i |a_i|^2}$

$$\langle \Psi | \hat{H} | \Psi \rangle = \sum_i \sum_j a_i^* a_j \langle \hat{i} | \hat{H} | \hat{j} \rangle = \sum_i |a_8|^2 E_i \qquad (20.2.6)$$

$$\langle \Psi | \hat{H} | \Psi \rangle = \sum \sum a_i^* a_j \langle i | \hat{H} | j \rangle = \sum_i |a_i|^2 E_i .$$

$$E = \langle \Psi | \hat{H} | \Psi \rangle = \left[\dfrac{\sum_i |a_i|^2 E_i}{\sum_i |a_i|^2} \right] = \sum_i |a_i|^2 E_i \qquad (20.2.7)$$

These equations determine the parameters and energy of the system.

If each of the eigen value on the right is replaced by the lowest eigen value of $\hat{H}$, viz. E_0, one obtains the inequality:

$$E_0 \leq \dfrac{\langle \Psi | \hat{H} | \Psi \rangle}{\langle \Psi \| \Psi \rangle} \qquad (20.2.8)$$

Thus $\quad \langle \hat{H} \rangle \geq E_0 \sum_i |a_i|^2 \geq E_0 \qquad (20.2.9)$

This is the basis of the Rayleigh-Ritz Variational Method for approximate calculation of E_0.

20.2.3 UPPER BOUND ON THE EXPECTATION VALUE of the ground state of the system

Thus the use of the equation (20.2.9) provides an upper bound on the expectation value of the system, E_0, although not a very useful one. That is, it entitles one to choose a $|\Psi\rangle$ that depends on a number of free parameters, $\alpha_1, \alpha_2, \alpha_3, \ldots$; as in

$$|\Psi\rangle \equiv |\Psi(\alpha_1, \alpha_2, \alpha_3, \ldots)\rangle \qquad (20.2.4)$$

20.2.3.1 CALCULATION of $\langle \Psi | \hat{H} | \Psi \rangle$ and its MINIMIZATION with respect to α_i

To minimize the value of $\langle \Psi | \hat{H} | \Psi \rangle$ with respect to the parameters,

find $\quad \dfrac{\partial \langle \hat{H} \rangle}{\partial \alpha_i} = 0, \qquad (20.2.10)$

so that the upper bound so obtained might well be a good approximation to the actual energy. Hence the name *Variational Method*, The usefulness of this method lies in the fact that a trial wave function, $|\Psi(\alpha_1, \alpha_2, \alpha_3, \ldots)\rangle$, can be chosen such that it contains one or more *variational parameters* for the minimization procedure. *The judicious choice of this trial function is very important for an effective application of the method.* Once the ground state energy and wave functions are obtained, the method can be applied to the first excited state by choosing a trial function that is orthogonal to the ground state.

20.2.3.2 PROOF OF equation (20.2.9), viz. $\langle \hat{H} \rangle \geq E_0 \sum_i |a_i|^2 \geq E_0$

Let $|\Psi(\alpha_1, \alpha_2, \alpha_3, ...)\rangle$ be varied by an incremental value, $|\delta\psi\rangle$, say, then

$$\langle \hat{H} \rangle + \langle \delta \hat{H} \rangle = \frac{\langle \Psi+\delta\psi | \hat{H} | \Psi+\delta\psi \rangle}{\langle \Psi+\delta\psi || \Psi+\delta\psi \rangle} \quad (20.2.11)$$

where $\langle \delta \hat{H} \rangle$ = change in the average value of $\hat{H}$.

Ignoring terms such as $\langle \delta\psi | \hat{H} | \delta\psi \rangle$,

$$= \langle \Psi | \hat{H} | \Psi \rangle - \langle \hat{H} \rangle \langle \psi || \psi \rangle + \langle \delta\psi | \hat{H} | \Psi \rangle - \langle \hat{H} \rangle \langle \delta\psi || \psi \rangle$$
$$+ \langle \psi | \hat{H} | \delta\psi \rangle - \langle \hat{H} \rangle \langle \psi || \delta\psi \rangle$$
$$= \langle \delta\hat{H} \rangle \{\langle \psi || \psi \rangle + \langle \delta\psi || \psi \rangle + \langle \psi || \delta\psi \rangle\} \quad (20.2.12)$$

But $\langle \Psi | \hat{H} | \Psi \rangle = \langle \hat{H} \rangle \langle \psi || \psi \rangle$

$$\langle \delta\psi | \hat{H} | \Psi \rangle - \langle \hat{H} \rangle \langle \delta\psi || \psi \rangle + \langle \psi | \hat{H} | \delta\psi \rangle - \langle \hat{H} \rangle \langle \psi || \delta\psi \rangle$$
$$= \langle \delta\hat{H} \rangle \{\langle \psi || \psi \rangle + \langle \delta\psi || \psi \rangle + \langle \psi || \delta\psi \rangle\}$$

However, if $\langle \hat{H} \rangle$ = a stationary value,

$$\langle \delta\hat{H} \rangle = 0$$

So $\langle \delta\psi | \hat{H} | \Psi \rangle - \langle \hat{H} \rangle \langle \delta\psi || \psi \rangle + \langle \psi | \hat{H} | \delta\psi \rangle - \langle \hat{H} \rangle \langle \psi || \delta\psi \rangle = 0$

$\langle \delta\psi | |\hat{H} - \langle \delta\hat{H} \rangle| | \Psi \rangle - \langle \psi | |\hat{H} - \langle \delta\hat{H} \rangle| | \delta\psi \rangle = 0$.

This is satisfied when

$$\langle \delta\psi | |\hat{H} - \langle \delta\hat{H} \rangle| | \Psi \rangle = \langle \psi | |\hat{H} - \langle \delta\hat{H} \rangle| | \delta\psi \rangle = 0.$$

Or, $\hat{H} | \Psi \rangle = \langle \hat{H} \rangle | \Psi \rangle \quad (20.2.13)$

Since $|\delta\psi\rangle$ is an arbitrary variable function, comparing equation (20.2.13) with the Schrödinger Equation for a stationary state, equation (20.2.2), one notes that

$$E = \langle \hat{H} \rangle.$$

Hence, if one finds a completely flexible function $|\Psi\rangle$, such that a small change has zero first order effect on $\langle \hat{H} \rangle$, then $|\Psi\rangle$ is the eigen function and $\langle \hat{H} \rangle$ is the eigen value required from the solution of the Schrödinger Equation (20.2.2.). Therefore the Variational Principle is another way of expressing the Schrödinger wave Equation.

When $\langle \delta\hat{H} \rangle = 0$,

$\langle \hat{H} \rangle$ = any one of the three possible values:
(i) a maximum,
(ii) a point of inflexion, or
(iii) a minimum.

E_0 = energy of the ground state, which is the lowest possible $\langle \hat{H} \rangle$ for a completely flexible wave function.

Therefore, for an arbitrary ground state wave function, $|\Psi\rangle$,

$$\frac{\langle \Psi | \hat{H} | \Psi \rangle}{\langle \Psi || \Psi \rangle} = \langle \hat{H} \rangle \geq E_0$$

This aspect of the Variational Principle is the most valuable in practice.

Worked out Example 20.1

Apply the Variational Principle and find the ground state energy of a Harmonic Oscillator.

Solution: Step # 1 Given $\hat{H}^{(0)} = [\frac{\hat{p}_x^2}{2m} + \frac{1}{2} k x^2]$, (6.2.4), variable x extends over the range $-\infty$ to $+\infty$.

Hence a function $|\Psi(x)\rangle$ has to satisfy the following conditions,

$\langle \Psi | \hat{H} | \Psi \rangle =$ a minimum, (20.2.14), subject to

$$\langle \Psi || \Psi \rangle = \int_{-\infty}^{+\infty} \Psi(x) * \Psi(x) \, dx = 1 \qquad (20.2.15)$$

The normalizing condition for real $|\Psi\rangle$ means $|\Psi(x)\rangle$ must be an even function of x.

Step # 2 Therefore let the function be x^2.

Furthermore, since $\langle \Psi | \hat{H} | \Psi \rangle$ must converge, it is clear that

$\langle \Psi || \Psi \rangle = |\Psi|^2 \to 0$, as $x \to \infty$

These properties suggest the simplest function

$$|\Psi(x)\rangle = A e^{-\alpha x^2}, (\alpha > 0) \qquad (20.2.16)$$

Step # 3 Consider A = a constant, and $\alpha =$ a variational parameter.

Then by the relation $\int_{-\infty}^{+\infty} \Psi(x) * \Psi(x) \, dx = 2 \int_0^{+\infty} \Psi(x) * \Psi(x) \, dx$

$$= 2 \int_0^{+\infty} |A e^{-\alpha x^2}|^2 \, dx = A^2 \left(\frac{1}{2} \frac{\pi}{\alpha}\right)^{1/2} = 1 \qquad (20.2.17)$$

Step # 4 $\hat{H}^{(0)} | \Psi(x) \rangle = [\frac{\hat{p}_x^2}{2m} + \frac{1}{2} k x^2] | \Psi(x) \rangle. = -\frac{1}{2} A^2 \hbar^2 \{(4\alpha^2 - \hbar^2 \omega^2) x^2 - 2\alpha\} e^{-\alpha x^2}$

$\langle \Psi | \hat{H}^{(0)} | \Psi \rangle = \langle \Psi | [\frac{\hat{p}_x^2}{2m} + \frac{1}{2} k x^2] | \Psi \rangle$

$$= -A^2 \hbar^2 \int_{-\infty}^{+\infty} \{(4\alpha^2 - \hbar^2 \omega^2) x^2 - 2\alpha\} e^{-\alpha x^2} dx \qquad (20.2.18)$$

Step # 5 To evaluate (20.2.18), using integration by parts gives

$$\int_0^{+\infty} x^2 e^{-2\alpha x^2} dx = \left(\frac{1}{8} \frac{1}{\alpha}\right)\left(\frac{1}{2} \frac{\pi}{\alpha}\right)^{1/2} \qquad (20.2.19)$$

Then (20.2.17) and (20.2.19) give (20.2.14)

$$\langle \Psi | \hat{H}^{(0)} | \Psi \rangle = \frac{\omega^2}{8\alpha} + \left(\frac{1}{2} \hbar^2 \alpha\right) \qquad (20.2.20)$$

Step # 6 The equation $\frac{d}{d\alpha} \langle \Psi | \hat{H}^{(0)} | \Psi \rangle = 0$ gives $\alpha = \frac{1}{2} \frac{\omega}{\hbar}$ (20.2.21)

Step # 7 Hence the required condition is $|\Psi(x)\rangle = \left(\frac{\pi \hbar}{\omega}\right)^{1/4} e^{-(\omega/2\hbar) x^2}$ (20.2.22)

and the lowest energy of the HO is $E_0 = \langle \Psi | \hat{H} | \Psi \rangle_{minimum} = \frac{1}{2} \hbar \omega$ (20.2.33)

CONCLUSION: The Variational Method gave the energy of the SYSTEM without solving the Schrödinger Equation.

20.3 VARIATIONAL METHOD APPLIED TO THE SPECTRUM OF HELIUM ATOM

Consider the nucleus of the helium atom as a fixed centre of the central force. (Fig. 20.1).

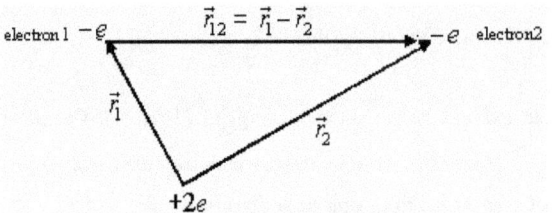

Fig. 20.1   The Helium atom

Neglect
  (i)   SPIN-ORBIT Interaction terms and
  (ii) The interaction between the MAGNETIC MOMENTS of the two electrons of the Helium atom.

$$\hat{H} = \frac{\hat{p}_1^2}{2m} + \frac{\hat{p}_2^2}{2m} - \frac{Ze^2}{r_1} - \frac{Ze^2}{r_2} + \frac{e^2}{|r_1 - r_2|}. \tag{19.5.1}$$

$$\hat{H} = \left(-\frac{\hbar^2}{2m}\right)\left(\nabla_1^2 + \nabla_2^2\right) - \left(\frac{Ze^2}{r_1} + \frac{Ze^2}{r_2}\right) + \frac{e^2}{|\vec{r}_1 - \vec{r}_2|} \tag{19.5.2}$$

where m is the mass of the electron.

$$\hat{H} = \hat{H}^{(0)} + \frac{e^2}{|\vec{r}_1 - \vec{r}_2|}$$

Step # 1 Choose the simplest trial function:

The simplest trial function for the variational calculation for the helium atom is the product of two ground-state wave functions of a hydrogenic atom,

$$|1s(1)\rangle = \sqrt{1/\pi}\,(Z^*/a_0)^{3/2}\,e^{-(Z^* r_1/a_0)} \tag{20.3.1a}$$

$$|1s(2)\rangle = \sqrt{1/\pi}\,(Z^*/a_0)^{3/2}\,e^{-(Z^* r_2/a_0)} \tag{20.3.1b}$$

where Z^* is the arbitrary charge on the nucleus. Z^* is now taken as the free parameter. In this case the normalized wave function is given by

$$|0\,(He)\rangle \equiv |1s^2\rangle = |1s(1)\rangle \cdot |1s(2)\rangle$$

$$= \left(Z^{*3}/\pi a_0^3\right) e^{-\left(Z^*(r_1 + r_2)/a_0\right)} \tag{20.3.2}$$

Step # 2 Let the variational parameter is

$$\alpha = Z^*/a_0. \tag{20.3.3}$$

$$|0\,(He)\rangle = \left(\alpha^3/\pi\right) e^{-(\alpha(r_1 + r_2))} \tag{20.3.4}$$

Let Z^* takes into consideration, the effect of the partial screening of the nucleus by electron 2. This wave function $|1s^2\rangle$ becomes the exact wave function of the interaction between the electrons (in the Hamiltonian, $\hat{H}^{(0)}$ tends to zero, and $Z^* \Rightarrow Z = 2$.

Now $\left[(1/2m)p^2 - Ze^2/r \right] | 1s \rangle = E_{1s} | 1s \rangle$ (20.3.5)

$E_{1s} = \left(-\frac{1}{2} e^2 / a_0 \right) Z^{*2} = -13.6 \, Z^{*2} \, eV$.

Neglect (i) Spin-Orbit interaction terms and
(ii) The interaction between the magnetic moments of the two electrons of the helium atom.

$$\hat{H} = \left(-\frac{\hbar^2}{2m} \right) (\nabla_1^2 + \nabla_2^2) - \left(\frac{Ze^2}{\vec{r}_1} + \frac{Ze^2}{\vec{r}_2} \right) + \frac{e^2}{|\vec{r}_1 - \vec{r}_2|}$$

$$\hat{H} = \hat{H}^{(0)} = \langle 1s^2 | \hat{H}^{(0)} | 1s^2 \rangle + \langle 1s^2 | \frac{e^2}{|\vec{r}_1 - \vec{r}_2|} | 1s^2 \rangle \quad (20.3.6)$$

Step # 3   To find the expectation of the Hamiltonian, i.e. $\langle \hat{H} \rangle$:

$$\hat{H} = \hat{H}^{(0)} + \frac{e^2}{|\vec{r}_1 - \vec{r}_2|}$$

$<E>_{ground} = E_{|1s^2\rangle} = \langle 1s^2 | \hat{H}^{(0)} | 1s^2 \rangle + \langle 1s^2 | \frac{e^2}{|\vec{r}_1 - \vec{r}_2|} | 1s^2 \rangle$ (20.3.7)

$<\hat{H}^{(0)}>_{ground} = <\hat{H}_1^{(0)}>_{ground} + <\hat{H}_2^{(0)}>_{ground}$ (20.3.8)

Step # 4 :   With electron 1:

$<\hat{H}_1^{(0)}>_{gr} = \int_{-\infty}^{+\infty} d\tau_1 \int_{-\infty}^{+\infty} d\tau_2 \psi_{1S}(r_1)^* \psi_{1S}(r_1)^* \left[\frac{\hat{p}_1^2}{2m} - \frac{Ze^2}{\vec{r}_1} \right] \psi_{1S}(r_1) \, \psi_{1S}(r_2)$ (20.3.9)

Splitting Z into $Z^* - (Z^* - Z)$,

$<\hat{H}_1^{(0)}>_{gr} = \int_{-\infty}^{+\infty} d\tau_1$

$\times \int_{-\infty}^{+\infty} d\tau_2 \psi_{1S}(r_1)^* \psi_{1S}(r_1)^* \left[\frac{\hat{p}_1^2}{2m} - \frac{Z^* e^2}{\vec{r}_1} - \frac{(Z-Z^*)e^2}{\vec{r}_1} \right] \psi_{1S}(r_1) \, \psi_{1S}(r_2)$ (20.3.10)

By using the VIRIAL Theorem for atoms, viz.

$<V>_{n\ell m} = -2 <T>_{nlm}$ (20.3.11)

$<T>_{1S} = \left\langle \frac{\hat{p}_1^2}{2m} \right\rangle_{1S} = Z^{*2} \frac{e^2}{2 a_0}$. (20.3.12)

$<V>_{1S} = \left\langle \left| \frac{1}{r_1} \right| \right\rangle_{1S} = \int_{-\infty}^{+\infty} d\tau_1 \, \psi_{1S}(r_1)^* \left| \frac{1}{r_1} \right| \psi_{1S}(r_1) = \int_{-\infty}^{+\infty} d\tau_1 | \psi_{1S}(r_1) |^2 \left| \frac{1}{r_1} \right|$

$= \left(Z^{*3} / \pi a_0^3 \right) \int_{-\infty}^{+\infty} d\tau_1 | \psi_{1S}(r_1) |^2 (r_1)^{-1}$

$= \left(Z^{*3} / \pi a_0^3 \right) \int_0^{+\infty} \int_0^{\pi} \int_0^{2\pi} (r_1)^{-1} e^{-2(Z^* r_1 / a_0)} \sin\theta \, d\theta_1 \, d\varphi_1 dr$

$= \left(4Z^{*3} / a_0^3 \right) \int_0^{+\infty} (r_1) \, e^{-2(Z^* r_1 / a_0)} dr_1 = \left(Z^* / a_0 \right)$ (20.3.13)

$<V>_{1S} = \left\langle Z^* \, e^2 (r_1)^{-1} \right\rangle_{1S} = -Z^* \, e^2 [Z^* / a_0] = -Z^{*2} [e^2 / a_0]$ (20.3.14)

$\left\langle \hat{H}_1^{(0)} \right\rangle_{gr} = \{ \frac{1}{2} Z^{*2} [e^2 / a_0] - Z^{*2} [e^2 / a_0] \}$

$$+(Z^* - Z) e^2 \langle (r_1)^{-1} \rangle \{\tfrac{1}{2} Z^{*2}[e^2/a_0] - Z^{*2}[e^2/a_0]\} + (Z^* - Z)(-Z^* e^2/a_0)$$

$$= [e^2/a_0]\{-\tfrac{1}{2} Z^{*2} + (Z^* - Z)Z^*\} \qquad (20.3.15)\text{a}$$

An identical term comes from the Hamiltonian of electron 2, i.e.

$$\langle \hat{H}_2^{(0)} \rangle_{gr} = [e^2/a_0]\{-\tfrac{1}{2} Z^{*2} + (Z^* - Z)Z^*\} \qquad (20.3.15)\text{b}$$

$$<E>_{gr} = E_{1S^2} = \langle 1s^2 \mid \hat{H}_i^{(0)} \mid 1s^2 \rangle + \langle 1s^2 \mid \frac{e^2}{\|\vec{r}_{12}\|} \mid 1s^2 \rangle \qquad (20.3.7)$$

$$= 2 \frac{e^2}{a_0}\{-\tfrac{1}{2} Z^{*2} + (Z^* - Z)Z^*\} + \langle 1s^2 \mid \frac{e^2}{\|\vec{r}_{12}\|} \mid 1s^2 \rangle \qquad (20.3.16)$$

The expectation value of the electron-electron repulsion (as calculated from the first order perturbation theory) is $\langle 1s^2 \mid \frac{e^2}{\|\vec{r}_{12}\|} \mid 1s^2 \rangle = -\tfrac{5}{8} Z^* \frac{e^2}{a_0}$ (20.3.17)

Adding up these terms,

$$<E>_{gr} = E_{1S^2} = 2 \frac{e^2}{a_0}\{-\tfrac{1}{2} Z^{*2} + (Z^* - Z)Z^*\} - \tfrac{5}{8} Z^* \frac{e^2}{a_0}$$

$$= \left(-\frac{e^2}{2a_0}\right)\{2Z^{*2} + 4Z^*(Z^* - Z) - \tfrac{5}{4} Z^*\}$$

$$\boxed{<E>_{gr} = E_{1S^2} = \left(-\frac{e^2}{2a_0}\right)\{2Z^{*2} + 4Z^*(Z^* - Z) - \tfrac{5}{4} Z^*\}} \qquad (20.3.18)$$

Step # 5 Minimizing the value of $\langle \hat{H} \rangle_{gr}$ with respect to Z^* (the <u>variational parameter</u>)

$$\left(\frac{d}{dZ^*}\right)\left(-\frac{e^2}{2a_0}\right)\{2Z^{*2} + 4Z^*(Z^* - Z) - \tfrac{5}{4} Z^*\}$$

$$= \left(-\frac{e^2}{2a_0}\right)\{4Z - 4Z^* - \tfrac{5}{4}\} = 0 .$$

i.e., $\left. Z^* \right|_{\langle \hat{H} \rangle_{gr} = \text{Minimum}} = (Z - \tfrac{5}{16}) = \tfrac{27}{16} \approx 1.7 \qquad (20.3.19)$

for the screened Helium nucleus.

i.e. the Hydrogen wave function gives the best energy when Z is replaced by $(Z - \tfrac{5}{16})$, i.e. each electron screens the nucleus from the other electron.

Step # 6 Substituting this value (20.3.19) of Z^* in the expression for (20.3.18)
The binding energy of the Helium atom is

$$<E>_{gr} = E_{1S^2} \leq 2E_{1S} = 2\left(\frac{Z^* e^2}{2a_0}\right)$$

$$\leq 2\left(-\frac{e^2}{2a_0}\right)(Z - \tfrac{5}{16}) = 5.7(-13.6 eV) \leq -77 eV \qquad (20.3.20)$$

This is the energy necessary to obtain doubly ionized Helium, He^{++}. This value is in excellent agreement with the experimental value of $E_{1S^2} = -78.62 \, eV \, (= 2.9[e^2/a_0])$. It should be recalled that the value of E_{1S^2} of -78.62 eV, estimated by the Variational Method, is very much better than the First Order Perturbation result of -74.8 eV.

The *first ionization energy* of He atom is given by

$$[E_{1S^2} - E_{1S}(He)] = 5.7(-13.6eV) - Z^2[-e^2/2a_0] = -1.7[-e^2/2a_0] = -23eV,$$

which is comparable with the experimental value of -24.575 eV. (The First Order Perturbation Theory gives for this -20.40 eV)..

TABLE 20.1   [1 $R_y = \{-½ e^4/a_0\}$]

	H	He	Li$^+$	Be^{++}
Theory	-0.055 Ry	1.695 Ry	5.445 Ry	11.195 Ry
Experiment	0.055 Ry	1.807 Ry	5.560 Ry	11.112 Ry
Difference	0.110 Ry	0.112 Ry	0.115 Ry	0.117 Ry

20.4   THE JWKB METHOD (or WKB Approximation)

20.4.1   Introduction

This is a semi-classical approximation used in 1923 by H. Jeffreys, and in 1926 by G. Wentzel, H.A. Kramers and L. Brillouin. The motion of a particle in a region of constant potential can be represented by a superposition of plane waves traveling to and fro. Each such wave has the form

$$|\psi(x)\rangle = e^{\pm ikx}, \qquad (20.4.1)$$

with $\quad \hbar k = \sqrt{2m[E-V(x)]} \qquad (20.4.2)$

But in practical cases, the potential is not constant. Provided that the potential is *slowly varying* in space, the JWKB approximation method is particularly useful, especially in 1-D cases and in problems with radial symmetry.

To apply the WKB approximation to an arbitrary potential $V(x)$, simply find the phase integral S as a function of the energy W. This can always be done by computer integration, where all the tedium is eliminated by the automatic calculations. Then the energy eigenvalues are the values of W for which $S/\pi = n + \frac{1}{2}$. This can be done for anharmonic oscillators, such as vibrating diatomic molecules, or for the radial function in atomic spectra or in atomic collision problems, where the centrifugal potential is added to the self-consistent potential $V(r)$. A large number of such calculations were done in the 30's and 40's, before great computing power was available for more elaborate schemes, and the results were quite satisfactory.

The WKB approximation is second only to perturbation theory as a fruitful method of calculation.

20.4.2   The WKB Method applied to GENERAL POTENTIAL

Calculation of the tunneling probability for a square barrier potential was seen earlier. How can one adopt this analysis to an irregularly shaped barrier? In general, one has to rely on approximation methods. It was seen that equation (5.4.42)

$$\tau = \omega_3/\omega_1 = |F/A|^2 (k_3/k_1) \qquad (5.4.42)$$

Equation (5.4.40) can be written as

$$\frac{F}{A} = (4i k_1 k_2) e^{i k_1 a \cdot k_3 b} \{(k_2^2 - k_1 k_3) + i(k_1 + k_3) k_2\}^{-1} e^{-k_2 c} \qquad (5.4.43)$$

for $V(x) = \begin{cases} 0, & \text{for } x < a \\ V_1, & \text{for } a < x < b, \\ V_2, & \text{for } x < b \end{cases}$ (5.4.21)

$k_1^2 = 2m[V_1 - E]/\hbar^2$. (5.4.27)

If the barrier width c is so small that $k_2 c \gg 1$

$$\tau \triangleq \left|\frac{F}{A}\right|^2 = G(k_1, k_2, k_3) \, e^{-2k_2 c} \qquad (5.4.44)$$

If $k_2 c$ = large enough,

$$\text{Ln } \tau \triangleq -2 k_2 c$$

i.e. linear in $k_2 c$.

If one follows the first potential barrier by another one for which the corresponding parameters are k' and c'

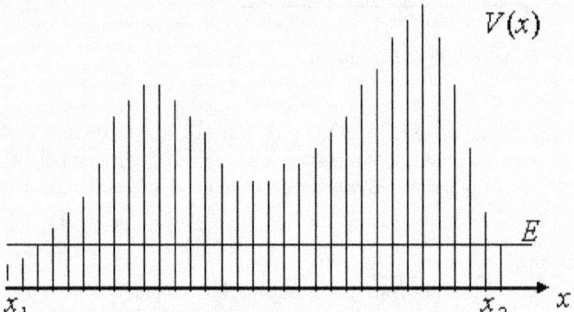

Fig. 20.2   Arbitrarily shaped potential barrier V(x).

then $Ln \, \tau$ will be modified by the addition to it of $-2\,k'c'$. A smooth arbitrarily shaped adjacent rectangular barriers, each of infinitesimal width Δx, (Fig. 20.2). In the region $x_1 < x < x_2$, one may use

$k_1^2 = 2m[V_1 - E]/\hbar^2$ and $Ln \, \tau \triangleq -2 k_2 c$

to write the contribution to $Ln \, \tau$ from one of these infinitesimal barriers.

$$-2\left(2m[V(x)-E]/\hbar^2\right)\Delta x, \qquad (20.4.3)$$

and $\tau \propto e^{-2\int dx \left\{2\sqrt{(2m[V(x)-E])}/\hbar\right\}}$. (20.4.4)

This is the WKB APPROXIMATION FORMULA, which was derived more rigorously by them.

20.4.3   The WKB Approximation applied to ALPHA DECAY

Now one can replace

$$k_2 c \text{ by } \int_{r_1}^{r_2} k_2 \, dr, \qquad (5.6.11)$$

The Coulomb force can now be included, so that for $r > R$, the potential becomes

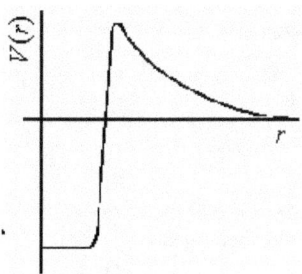

Fig. 20.3 Nuclear Potential

$$V(r) = k\,(2\,Ze^2/r)$$

Z e is the charge of the *daughter* nucleus. The potential is assumed to be unaltered for r > R.
Equation (5.6.2) becomes

$$\propto e^{-\frac{2}{\hbar}\int\{2\sqrt{(2m[V_1-E])}\}\,dr} \qquad (5.6.12)$$

20.4.3.1 To evaluate the integral

$$E = V(r_2) + T = (2\,Ze^2/r_2), \qquad (5.6.13)$$

Assuming T as negligibly small compared to E.

$$\int k_2\,dr = -\frac{2}{\hbar}\int\{2\sqrt{(2m[V_1-E])}\}\,dr$$

if λ = radio-active decay constant, the *Radio-active Law* is

$$\frac{dN(t)}{dt} - \lambda N(t) \rightarrow N(t) = N_0 e^{-\lambda t}(t). \qquad (5.6.5)$$

$$\therefore \quad Ln\,\tau_{1/2} = 2\,k_2 c + B \text{ (a constant)} \qquad (5.6.10)$$

where $\tau_{1/2}$ = half-life of the radioactive material.
Or, $\quad Ln\,\lambda = -2\,k_2 c + B_0$ (a constant).
Following Section 5.6, one gets

$$\therefore \quad k_2 c \approx \frac{\pi\,Z\,e^2}{\hbar}\{\sqrt{(2m/E)}\} \approx \frac{\pi\,e^2}{\hbar}\{\sqrt{4m}\,\frac{Z}{v}\} \qquad (5.6.16)$$

$$k_2 c \propto \{\frac{Z}{v}\} = A\,\frac{Z}{v}. \qquad (5.6.17)$$

Substituting this equation (5.6.17) in equation (5.6.10), results in the equation:
$\quad Ln\,\tau_{1/2} = 2\,k_2 c + B$ (a constant) $\qquad (5.6.10)$

$$\therefore \quad Ln\,\tau_{1/2} = A\,\frac{Z}{v} + B. \qquad (5.6.18)$$

where both A and B are constants.

$$Ln\,\lambda = -A\,\frac{Z}{v} - B \qquad (5.6.19)$$

20.4.4 The WKB method in the H-atom

The WKB Method is found useful in solving Schrödinger Equation for highly excited states. For example, the Radial Equation for a particle moving in a spherically symmetric potential V*(r)* satisfies such a 1-D equation.

$$\nabla_r^2 \Re(r) + \frac{2\mu}{\hbar^2}[E - V(r) - \frac{\ell(\ell+1)}{2\mu\,r^2}]\,\Re(r) = 0 \qquad (9.4.8)$$

where $\quad \psi(r) \equiv \psi(r,\theta,\varphi) = \Re(r)\cdot Y(\theta,\varphi) \qquad (9.4.1)$

One is interested in solutions in the region where the quantity

$$\varnothing(r) = \frac{2\mu}{\hbar^2}[E - V(r) - \frac{\ell(\ell+1)}{2\mu r^2}] = +\text{ve.} \quad (20.4.5)$$

This is shown in Fig. 20.4.

In this region the WKB solution is

$$\Re(r) = A\sqrt{[\varnothing(r)]}\ \text{Cos}\int_a^r \left(\varnothing(r)^{1/2}\,dr' - \frac{\pi}{4}\right) \quad (20.4.6)$$

where A is the normalization constant,
a = the only one classical turning point; viz. it is zero of $\varnothing(r)$.
For $r > a$, $\varnothing(r) = +\text{ve}$, and the WKB solution holds good.

Frequently, it is usual to replace the centrifugal potential, $\frac{\ell(\ell+1)}{2\mu r^2}$, occurring in $\varnothing(r)$,

by $\frac{(\ell+\frac{1}{2})^2}{2\,r^2}$.

This improves the WKB solution. When there are two turning points, a and a_1
For $a < r < a_1$, $\varnothing(r) = +\text{ve}$ and the WKB solution holds good (Fig 20.2). The phase

$$\Theta(r) = \int_a^r \left(\varnothing(r)^{1/2}\,dr' - \frac{\pi}{4}\right) \quad (20.4.7)$$

must satisfy the quantization condition
$\Theta(a_1) = (n+\frac{1}{2})\pi$.

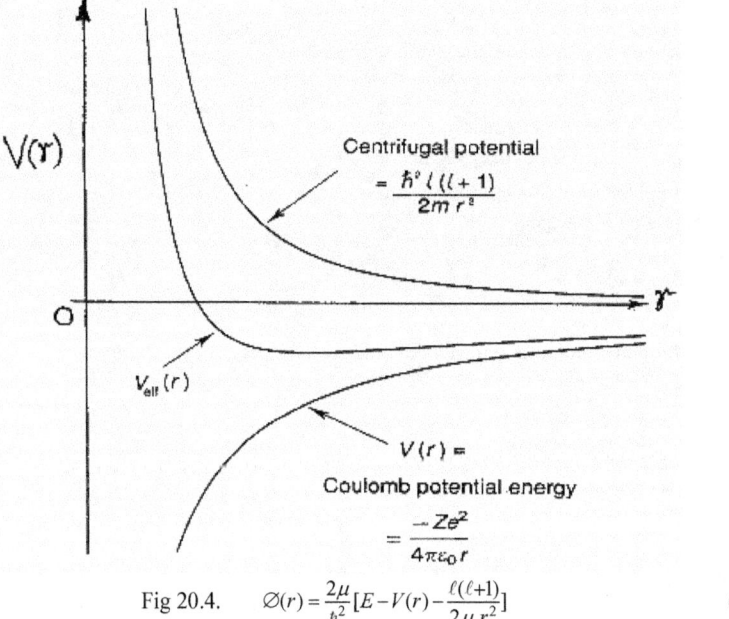

Fig 20.4. $\quad \varnothing(r) = \frac{2\mu}{\hbar^2}[E - V(r) - \frac{\ell(\ell+1)}{2\mu r^2}] \quad (20.4.8)$

20.5 CONCLUSIONS

The Ritz Method can give very accurate results for the ground state energy of two electron atoms, and with sufficient labour can be extended to yield good results for some excited states. The variational wave function is also determined, although it is not expected to be as accurate as the energy eigen value. For complex atoms this approach would become prohibitorily cumbersome if carried out to a satisfactory degree of accuracy.

Therefore the simplest approximation is to ignore the electron –correlation, which has a number of facets like

(1) Because of the equality of charges they avoid being in the same region of space, or avoid being at the same distance along a radius - radial correlation,

(2) They will be found o opposite sides of the nucleus, i.e. they avoid being at the same angle to the nucleus – angular correlation-

(3) Spin correlation, has its explanation in the form of the *Pauli's Exclusion Principle*; electrons with the same spin are unlikely to be found in the same region of space.

A respectable approximate wave function for an atom should take into account of each of the three types of correlation.

20.5.1 STO Orbitals

John Clarke Slater, in 1930, introduced a type of Atomic Orbitals, known as *Slater type Orbitals* (STO). He proposed a set of rules for taking into account the influence of shielding. The H-atom angular wave functions (of the unperturbed Hamiltonian) are preserved as such, but the Radial wave Functions are replaced by a new set. The *Slater wave Function*

$$\mathfrak{R}_S(r) = N_n \, r^{n-1} \, e^{-\xi r/a_0} \tag{20.5.1}$$

where n = Principal Quantum number, ξ takes into account the screening. Slater has published rules for determining ξ-values.

$$\xi = Z_{\text{eff}} / n. \tag{20.5.2}$$

Z_{eff} = effective nuclear charge.

20.5.2 HCF Orbitals

For these complex atoms there exist, *Hartree Focks Self-Consistent Field* (HF SCF) Method. This method is useful for obtaining energy levels and wave functions for atoms. Douglas Hartree introduced the computer method of calculating the ζ values using this method.

REVIEW QUESTIONS

R.Q. 20.1 Determine the best possible ground state energy of the Hydrogen atom. Given,

$$\hat{H} = \frac{\hat{p}_x^2}{2\mu} - \frac{Ze^2}{r} = -\frac{\hbar^2}{2\mu}(\nabla_x^2) - \frac{Ze^2}{r}, \text{ (Answer: } E_0 \text{ (minimum)} = -\frac{4}{3\pi}\left(\frac{k^2 \mu e^4}{\hbar^2}\right) = -0.424\left(\frac{k^2 \mu e^4}{\hbar^2}\right),$$

compared to the exact value $\left(-\frac{e^2}{2a_0}\right)$).

R.Q. 20.2 Estimate the ground state energy for an atom with two electrons and a nuclear charge Ze. Use the Variational Principle. Given the Trial wave function of the form

$$|\psi(\vec{r}_1, \vec{r}_2)\rangle = \left(Z^{*3} / \pi \, a_0^3\right) e^{-\left(Z^*(r_1 + r_2)/a_0\right)}, \text{ where } \vec{r}_1, \vec{r}_2 \text{ are the distances of the two electrons}$$

from the nucleus; $a_0 = \hbar^2 / m \, e^2$, and Z^* an adjustable parameter.

R.Q. 20.3   Estimate the ground state energy for the anharmonic oscillator, with
$$\hat{H} = [\frac{\hat{p}_x^2}{2m} + \frac{1}{2}kx^2] + \beta x^4$$
Use the Variational Principle. (Answer: Step1: Consider $|\psi\rangle = |\psi_o\rangle + \lambda|\psi_2\rangle$ where $|\psi_o\rangle$ and $|\psi_2\rangle$ are the oscillator . eigen functions. Step 2: Choose λ and β as the variational parameters. Step 3. Find $<\hat{H}>_{gr}$ and minimize with respect to the variational parameter. Step 4: Evaluate $<\hat{H}_2^{(0)}>_{gr}$)

R.Q. 20.4   Write down a Trial function for the Variational calculation of the nth excited state of a system.

R.Q. 20.5   Describe the Variation method. Discuss how it is applied to find ground state energy of the Helium atom.

&&*&*&*&*&*&*

Chapter 21

FINE STRUCTURE OF HYDROGEN –
(Application of Perturbation Theory to Real Hydrogen),
AND ZEEMAN EFFECT

Chapter 21

FINE STRUCTURE OF HYDROGEN –
(Application of Perturbation Theory to Real Hydrogen), AND ZEEMAN EFFECT

"Strength does not come from physical capacity. It comes from an indomitable will"
Mohandas Karamchand Gandhi

21.1 INTRODUCTION

The general structure of the spectrum of Hydrogen atom was calculated to first approximation in Chapter 9. Often an exact mathematical solution of the Schrödinger Equation cannot be found for problems other than the ideal cases, as seen in Chapter 19. It was shown that approximate solutions were obtained to the eigen values and eigen functions for interacting potentials that do not lead to exactly solvable equations. In fact no physical system can be studied without approximation. The Time Independent Perturbation Theory was covered up in Chapter 19.

The discussions of H-like atoms are based on the ideal Hamiltonian of the H-atom (19.7.1) is

$$\boxed{\hat{H}^{(0)} = \frac{\hat{p}_x^2}{2\mu} - \frac{Ze^2}{r_1} = -\frac{\hbar^2}{2\mu}\left(\nabla_x^2\right) - \frac{Ze^2}{r_1}}$$
(21.1.1)

whose eigen functions are denoted by $\left| \psi^0_{n\ell m} \right\rangle \equiv \left| n, \ell, m_\ell \right\rangle$.

In a more realistic treatment, several corrections must be taken into account. In the Centre of Mass Frame the original electron-proton Hamiltonian, viz.

$$\frac{\hat{p}_x^2}{2\mu} = \frac{\hat{p}_m^2}{2\mu} + \frac{\hat{p}_M^2}{2M}$$
(21.1.2)

is replaced, when the kinetic energy of the electron is altered, when relativistic corrections are taken into account.

$$\frac{\hat{p}_m^2}{2\mu} + \frac{\hat{p}_M^2}{2M} = c\sqrt{(mc^2 + p^2)} + \frac{\hat{p}_M^2}{2M}$$

$$\cong (mc^2 + \frac{\hat{p}_m^2}{2m}) - \frac{(\hat{p}_m^2)^2}{2\,m^2\,c^2} + \frac{\hat{p}_M^2}{2M} \cong$$
(21.1.3)

$$\cong (mc^2 + \frac{\hat{p}^2}{2\mu}) - \frac{(\hat{p}_m^2)^2}{2\,m^2\,c^2}$$

$$\boxed{\hat{H} = (mc^2 + \frac{\hat{p}^2}{2\mu}) - \frac{(\hat{p}_m^2)^2}{2\,m^2\,c^2}}$$
(21.1.4)

where mc^2 = electron rest mass term may be neglected.

Thus the relativistic correction term, $\mathcal{H}^{(1)}$, appears in the Hamiltonian.

$$\boxed{\hat{H}^{(1)} = -\frac{(\hat{p})^2}{2\,m^2\,c^2}}$$
(21.1.5)

$$<\hat{H}^{(1)}> / <\hat{H}^{(0)}> \quad = \frac{<\hat{p}^2>}{m^2\,c^2} \approx \frac{(m\,c\,Z\,\alpha)^2}{m^2\,c^2} \approx (Z\alpha)^2 .$$
(21.1.6)

$\approx 10^{-5}$, for Hydrogen,
This is less than the μ-effects; which is why α is called the *Fine Structure Constant*.
The Schrödinger Equation is set up for the motion, (9.3.1) viz.

$$\nabla_T^2 \psi_T(r) + \frac{2\mu}{\hbar^2} E_T \psi_T(r) = 0 \tag{21.1.7}$$

i.e.
$$\left[\nabla_{CM}^2 + \nabla_{rel}^2\right]\psi_T(r) + \frac{2\mu}{\hbar^2} E_T \psi_T(r) = 0 \tag{21.1.8}$$

Another correction $\hat{H}^{(2)}$, that is of the same order as $\hat{H}^{(1)}$, is due to the existence of the electron spin. Consider the electron is at rest relative to the nucleus. Then the proton is moving, so that there is a current present, and the electron 'sees' a magnetic field, $\vec{B}$ which would be $\vec{B} = -\vec{v} \wedge \vec{E}/c$, assuming the electron due to its spin.

$$\hat{H}^{(2)} = -\vec{\mu}_i \cdot \vec{B} = \frac{e}{mc}(\vec{s} \cdot \vec{B}) = \frac{e}{mc}\left[\vec{s} \cdot (-\vec{v} \wedge \vec{E}/c)\right]$$

$$= -\left(\frac{e}{m^2 c^2}\right)\left[\vec{s} \cdot (\vec{v} \wedge \vec{E})\right] = -\left(\frac{e}{m^2 c^2}\right)\left[\vec{s} \cdot (\vec{p} \wedge \vec{v})\right]\left\{r^{-1} \vec{\nabla}\psi(r)\right\}$$

$$= -\left(\frac{1}{m^2 c^2}\right)\left[\vec{s} \cdot (\vec{r} \wedge \vec{p})\right]\left\{r^{-1} \vec{\nabla} e\, \psi(r)\right\}$$

$$= -\left(\frac{1}{m^2 c^2}\right)\left[\vec{s} \cdot \vec{L}\right]\left\{r^{-1} \vec{\nabla} e\, \psi(r)\right\}$$

$$\boxed{\hat{H}^{(2)} = -\left(\frac{1}{m^2 c^2}\right)\left[\vec{s} \cdot \vec{L}\right]\left\{r^{-1} \vec{\nabla} e\, \psi(r)\right\}} \tag{21.1.9}$$

where the factor ½ has appeared as a result of the electron does not move in a straight line (THOMAS PRECESSION EFFECT).

$$<\hat{H}^{(2)}>/<\hat{H}^{(0)}> \cong \left(\frac{\hbar}{mca_0}\right)^2 \approx (\alpha)^2. \tag{21.1.10}$$

The real Hamiltonian of a real Hydrogen atom is thus
$$\hat{H} = \hat{H}^{(0)} + \hat{H}^{(1)} + \hat{H}^{(2)}. \tag{21.1.11}$$
Quantum Electrodynamics (QED) shows that all other corrections are smaller than $\hat{H}^{(1)}$ and $\hat{H}^{(2)}$ by further powers of α.

21.2 PERTURBATION CALCULATION WITH EFFECTS OF $\hat{H}^{(1)}$

To calculate the effect of $\hat{H}^{(1)}$ in the spectrum of a H-like atom,
$$\hat{H}^{(1)} = \frac{(\hat{p}^2)^2}{2 m^3 c^2}. \tag{21.1.5}$$

Rewriting, $\hat{H}^{(1)} = -\frac{1}{2 m c^2}\left(\frac{\hat{p}^2}{2m}\right)^2$

$$= -\frac{1}{2mc^2}\left(\hat{H}^{(0)} + \frac{Ze^2}{r_1}\right)^2 \tag{21.2.1}$$

Here the reduced mass effect has been neglected.
So one has to diagonalize $\hat{H}^{(1)}$, in the degenerate subspaces belonging to a degenerate value of n.
$\hat{H}^{(1)}$ commutes with all the angular momentum operators.
First order correction to energy to the nth level,

$$E_n \approx E_n^{(1)} \equiv E_n^{(0)} + \hat{H}_{nn}^{(1)} \tag{19.3.17}$$

$$A_n^{(1)} = \hat{H}_{nn}^{(1)} \tag{19.3.16}$$

$$\left| \psi^0{}_{n\ell m} \right\rangle \equiv \left| n, \ell, m_\ell \right\rangle$$

The matrix element of the operator $\hat{H}^{(1)}$,

$$\hat{H}_{nn}^{(1)} = \left\langle \psi^0{}_{n\ell m} \left| \hat{H}^{(1)} \right| \psi^0{}_{n\ell m} \right\rangle$$

$$= -\frac{1}{2mc^2} \left\langle n, \ell, m_\ell \left| \left(\hat{H}^{(0)} + \frac{Ze^2}{\tilde{r}_1} \right) \left(\hat{H}^{(0)} + \frac{Ze^2}{\tilde{r}_1} \right) \right| n, \ell, m_\ell \right\rangle$$

$$= -\frac{1}{2mc^2} \left(E_n^{(0)2} + 2 E_n^{(0)} Ze^2 <r^{-1}>_{n\ell} + (Ze^2)^2 <r^{-2}>_{n\ell} \right)$$

$$= -\frac{1}{2mc^2} \left(\frac{mc^2(Z\alpha)^2}{2n^2} \right)^2$$

$$- \left(\frac{2Z e^2 m c^2 (Z\alpha)^2}{2n^2} \right) \left(\frac{Z}{n^2 a_o} \right) + (Ze^2)^2 \left(\frac{Z^2}{n^2 a_o (\ell+\frac{1}{2})} \right)$$

with $\alpha = e^2 / \hbar c$

$$A_{n\ell}^{(1)} = -\frac{1}{2mc^2} (Z\alpha)^2 \left\{ \left(\frac{(Z\alpha)^2}{n^2 a_o (\ell+\frac{1}{2})} \right) - \left(\frac{3(Z\alpha)^2}{4n^4} \right) \right\} \tag{21.2.1}$$

$$\text{Using } <r^{-1}>_{n\ell} = \left(\frac{Z}{n^2 a_o} \right). \tag{9.9.5}$$

$$<r^{-2}>_{n\ell} = \left(\frac{Z^2}{n^3 a_o^2 (\ell+\frac{1}{2})} \right) \tag{9.9.11}$$

$$\boxed{A_{n\ell}^{(1)} = -\frac{1}{4mc^2} (Z\alpha)^2 \left(\frac{3}{4n^4} \right)} \tag{21.2.2}$$

It can be seen that the spin of the electron does not enter into this first order correction of energy.

21.3 SECOND ORDER PERTURBATION CORRECTION TO ENERGY, $\hat{H}^{(2)}$:(Perturbation in the Hamiltonian of isolated atoms, the S-L. Interaction term of the magnetic moment of an electron with its surroundings)

$\hat{H}^{(2)}$ does depend on the spin, and for the unperturbed wave functions one must take two component wave functions since the expectation value, $<\hat{H}^{(2)}>$ is required.

$$\hat{H}^{(2)} = -\left(\frac{1}{2m^2 c^2} \right) [\vec{S} \cdot \vec{L}] \left\{ r^{-1} \vec{\nabla} e \, \psi(r) \right\} \tag{21.1.9}$$

$\hat{H}^{(1)}$ is only compatible with $\{\hat{L}^2, \hat{J}^2, \hat{J}_z\}$ set.

$$\therefore \quad \hat{H}^{(2)} = -\left(\frac{Ze^2}{2m^2 c^2} \right) [\vec{S} \cdot \vec{L}] \left\{ r^{-3} \right\} \tag{21.3.1}$$

$$= \hat{H}_{LS}$$

Here again one has an example of the use of *Degenerate* Perturbation Theory. For a given s and ℓ, there are $2(2\ell+1)$ degenerate eigen states of $\hat{H}^{(0)}$, with the additional factor of 2 coming from the two spin states. Thus the calculation of the shift in energy involves a diagonalization of a sub-matrix. One may simplify the procedure by noting that
$$J = L + S \tag{21.3.2}$$
Implies that $\hat{J}^2 = \hat{L}^2 + \hat{S}^2 + 2\,\hat{S}\cdot\hat{L}$

i.e. $\quad \hat{S}\cdot\hat{L} = \tfrac{1}{2}(\hat{J}^2 - \hat{L}^2 - \hat{S}^2) \tag{21.3.3}$

Thus if the degenerate eigen functions are combined into linear combinations that are eigen functions of $\hat{J}^2$ (They are already eigenfunctions of $\hat{J}_z = \hat{S}_z + \hat{L}_z$), then these liner combinations will diagonalize $H^{(2)}$. The appropriate linear combinations are obtained.

Since $\hat{S}$ does not depend on any external coordinates and $\hat{L}$ depends on spatial coordinates, $[\hat{L}, \hat{S}] = 0$.

Therefore the components of $\hat{J} = \hat{L} + \hat{S}$ satisfy the angular momentum Commutation Relations. The appropriate linear combinations yield

$$\hat{S}\cdot\hat{L}\left| j = (\ell+\tfrac{1}{2}), m_j = (m+\tfrac{1}{2})\right\rangle = \tfrac{1}{2}(\hat{J}^2 - \hat{L}^2 - \hat{S}^2)\left| j = (\ell+\tfrac{1}{2}), m_j = (m+\tfrac{1}{2})\right\rangle$$

$$= \tfrac{1}{2}\hbar^2 (\ell+\tfrac{1}{2})[\ell+3\cdot\tfrac{1}{2}] - [\ell(\ell+1) - \tfrac{3}{4}]\left| j = (\ell+\tfrac{1}{2}), m_j = (m+\tfrac{1}{2})\right\rangle$$

$$= \tfrac{1}{2}\hbar^2 \ell \left| j = (\ell+\tfrac{1}{2}), m_j = (m+\tfrac{1}{2})\right\rangle \tag{21.3.4}$$

and $\quad \hat{S}\cdot\hat{L}\left| j = (\ell-\tfrac{1}{2}), m_j = (m+\tfrac{1}{2})\right\rangle$

$$= \tfrac{1}{2}\hbar^2 (\ell-\tfrac{1}{2})[\ell+\tfrac{1}{2}] - [\ell(\ell+1) - \tfrac{3}{4}]\left| j = (\ell-\tfrac{1}{2}), m_j = (m+\tfrac{1}{2})\right\rangle$$

$$= -\tfrac{1}{2}\hbar^2 (\ell+1)\left| j = (\ell-\tfrac{1}{2}), m_j = (m+\tfrac{1}{2})\right\rangle \tag{21.3.5}$$

For a given ℓ-value, there are $[2(\ell+\tfrac{1}{2})+1)]+[2(\ell-\tfrac{1}{2})+1)]$ states. What has happened is that the degenerate states have merely been re-arranged, but the two groups that they have been split into behave differently under the action of $\hat{H}^{(2)}$. If the linear combinations are denoted by $\left| j = (\ell-\tfrac{1}{2}), m_j = (m+\tfrac{1}{2})\right\rangle$

$$\left\langle j = (\ell-\tfrac{1}{2}), m_j = (m+\tfrac{1}{2}); \ell \right| \hat{H}^{(2)} \left| j = (\ell-\tfrac{1}{2}), m_j = (m+\tfrac{1}{2}); \ell \right\rangle$$

$$= \left(\frac{Ze^2}{2m^2 c^2}\right)\cdot\left[\tfrac{1}{2}\hbar^2\right]\left\{\begin{array}{c}\ell \\ -\ell-1\end{array}\right\}\langle n, \ell\,|\{r^{-3}\}|\,n, \ell\rangle \tag{21.3.6}$$

for $j = (\ell\pm\tfrac{1}{2})$, respectively.

Here the result used is (9.9.11),

viz. $\quad \langle r^{-3}\rangle_{n\ell} = Z^3 /[n^3 a_0^3 \ell(\ell+\tfrac{1}{2})(\ell+1)]$

Then one gets the shift in energy,

$$\boxed{E_{n\ell}^{(2)} = \left[\tfrac{1}{4} m c^2 (Z\alpha)^4\right]\left\{\begin{array}{c}\ell \\ -\ell-1\end{array}\right\}\cdot\frac{1}{[n^3 \ell(\ell+\tfrac{1}{2})(\ell+1)]}}$$

(21.3.7)

Worked out Example 21.1

Evaluate the relativistic fine structure correction term, due to

$$\hat{H}^{(1)} = -\left(\frac{Z^4 e^8}{8\hbar^4 c^2}\right) = -\left(\frac{(p^2)^2}{8 m_0^3 c^2}\right).$$

Solution: **Step #1** Given $\hat{H}^{(1)} = -\left(\frac{Z^4 e^8}{8\hbar^4 c^2}\right) = -\left(\frac{(p^2)^2}{8 m_0^3 c^2}\right).$; $A_n^{(1)} = \hat{H}_{nn}^{(1)}$

$<r^{-1}>_{n\ell} = \left(\frac{Z}{n^2 a_o}\right)$ (9.9.5); $<r^{-2}>_{n\ell} = \left(\frac{Z^2}{n^3 a_o^2 (\ell+\frac{1}{2})}\right)$ (9.9.11)

$E_n^{(0)} = -\left[\frac{1}{2n^2} m_0 c^2 (Z\alpha)^2\right]$

Step #2 $A_{n\ell m}^{(1)} = \langle n, \ell, m | \left(-\frac{(p^2)^2}{8 m_0^3 c^2}\right) | n, \ell, m \rangle$

$= \langle n, \ell, m | \left(-\frac{1}{2 m_0 c^2}\right)\left(\frac{p^2}{2 m_0}\right)^2 = <n\,\ell\,m|\,(-1/2\,m_0 c^2)\,(\mathcal{H}_0+Z\,e^2/r)^2\,|n\,\ell\,m>$

$= \langle n, \ell, m | \left(-\frac{1}{2 m_0 c^2}\right)\left(\hat{H}^{(0)} + \frac{Ze^2}{r}\right)^2 | n, \ell, m \rangle$

$= \left(-\frac{1}{2 m_0 c^2}\right)\{[E_n^{(0)}]^2 + 2E_n^{(0)} Ze^2\}\langle n, \ell, m | \{r^{-1}\} | n, \ell, m \rangle$

$+(Ze^2)^2 \langle n, \ell, m | \{r^{-2}\} | n, \ell, m \rangle$

Step #3 Using the values of $<r^{-1}>, <r^{-2}>$, and $E_n^{(0)}$

$$A_{n\ell m}^{(1)} = \left[\frac{1}{2n^2} m_0 c^2 (Z\alpha)^2\right]\left(\frac{Z\alpha^2}{2 n^2}\right)\frac{n}{[(\ell+\frac{1}{2})-\frac{3}{4}]} \quad (20.3.8)$$

The spin of the electron does not enter into this Energy Shift.

21.4.1 TOTAL ENERGY SHIFT, $\Delta E_{n\ell}$

Total Energy Shift, $\Delta E_{n\ell}$ is obtained from (21.2.2) and (21.3.7)

$$\Delta E_{n\ell} = -\left[\frac{1}{2 n^3} m_0 c^2 (Z\alpha)^4\right]\left\{\frac{1}{(j+\frac{1}{2})} - \frac{3}{4n}\right\} \quad (21.4.1)$$

for both $\ell = (j \mp \tfrac{1}{2})$.

This result will agree perfectly with the result when worked with the *Relativistic Dirac Equation* (Chapter 25), when $\ell = 0$.
The splitting of the Hydrogen energy levels is schematically shown in Fig. 9.3.
The final degeneracy of the $^2S_{½}$ and $^2P_{½}$ states is also the result of the Relativistic Dirac Theory; this degeneracy is actually lifted by QED effects. The upward shift of the $^2S_{1/2}$ state is called the *Lamb Shift*, to be dealt with in Chapter 25 and Chapter 26.

21.4.2 THE ZEEMAN EFFECT IN HYDROGEN ATOM: Application of Degenerate Perturbation Theory

21.5.1 Equation of Motion of a Charged particle in a Magnetic Field

In hydrogenic atoms with *constant* magnetic field, $\vec{B}$, which can be taken to be uniform over atomic dimensions, the vector potential $\vec{A}$ can be written as

$$\vec{A}(\vec{r}.t) = \frac{1}{2}(\vec{B} \wedge \vec{r}) \qquad (21.5.1)$$

which satisfies $\vec{B}(\vec{r}.t) = [\vec{\nabla} \wedge \vec{A}(\vec{r}.t)] \qquad (21.5.2)$

Let $\vec{B} \parallel Z - axis$.

$\therefore \quad \vec{A}(\vec{r}.t)$ has components $(-\frac{1}{2}\hat{j} B_y, +\frac{1}{2}\hat{i} B_x, 0)$.

A particle of mass m, charge q moving with a velocity, $\vec{v}$ in an EM Field, $\vec{E}(\vec{r}.t)$ and $\vec{B}(\vec{r}.t)$ when subject to a LORENTZ FORCE, $\vec{F}$

$$\vec{F} = q[\vec{E}(\vec{r}.t) + \vec{v} \wedge \vec{B}(\vec{r}.t)] \qquad (21.5.3)$$

$$= q\{-\nabla\Phi - \partial \vec{A}/\partial t + \vec{v} \wedge (\nabla \wedge \vec{A})\} \qquad (21.5.4)$$

where $\Phi = \Phi(\vec{r}.t)$, the scalar EM potential,

$\vec{A} = \vec{A}(\vec{r}.t)$, the vector EM potential.

This can be achieved if one takes the Lagrangian, L as

$$L = \frac{1}{2}m v^2 - q\Phi - q\vec{v} \cdot \vec{A}. \qquad (21.5.5)$$

and works in the Cartesian Coordinate System.

$$m\frac{d^2\vec{r}}{dt^2} = \vec{F}. \qquad (21.5.6)$$

The *generalized momenta*, $\vec{p}_i$, defined as $(\partial L / \partial \dot{q}_i)$, with $\dot{q}_i = \partial q_i / \partial t$, are

$$\vec{p} = m\vec{v} + q\vec{A} \qquad (21.5.7)$$

The Hamiltonian H of the system is

$$H = \frac{1}{2m}[\vec{p} - q\vec{A}]^2 + q\Phi. \qquad (21.5.8)$$

The Schrödinger equation for such a particle in the Hydrogen atom is

$$i\hbar(\partial/\partial t)|\psi(r,t)\rangle = \left\{\frac{1}{2m}[(-i\hbar\vec{\nabla}) - q\vec{A}]^2 + e\vec{A}]^2 - kZe^2/r^2\right\}|\psi(r,t)\rangle \qquad (21.5.9)$$

where $k = (1/4\pi\varepsilon_0)$

$$\vec{E}(\vec{r}.t) = -\vec{\nabla}\Phi(\vec{r}.t) - \partial\vec{A}(\vec{r}.t)/\partial t, \qquad (21.5.1)$$

$$\vec{B}(\vec{r}.t) = [\nabla \wedge \vec{A}(\vec{r}.t)] \qquad (21.5.2)$$

The potentials are not completely defined by equations (21.5.1) and (21.5.2). It must satisfy the GAUGE INVARIANCE, i.e.

$$\vec{\nabla} \cdot \vec{A}(\vec{r}.t) = 0. \qquad (21.5.10)$$

i.e., use the *Coulomb Gauge*,

$$[\vec{\nabla} \cdot \vec{A}(\vec{r}.t)]|\psi(r,t)\rangle = \vec{A} \cdot \{\vec{\nabla}|\psi(r,t)\rangle\} + (\vec{\nabla} \cdot \vec{A})|\psi(r,t)\rangle = \vec{A} \cdot \{\vec{\nabla}|\psi(r,t)\rangle\} \qquad (21.5.11)$$

so that $\vec{\nabla}$ and $\vec{A}$ commute. Making use of this fact equation (21.5.9) may be written as

$$i\hbar(\partial/\partial t)|\psi(r,t)\rangle$$

$$= \left\{-\frac{1}{2m}[\hbar\vec{\nabla}^2 - kZe^2/r^2\right\}|\psi(r,t)\rangle - i\hbar e\vec{A}\cdot\vec{\nabla}/m + (e\vec{A})^2/2m]|\psi(r,t)\rangle \qquad (21.5.12)$$

For WEAK FIELD, the *term in $\vec{A}$ is less than the term linear in $\vec{A}$*.

Consider the Non-relativistic Schrödinger Equation, viz.,

$$\left\{-\frac{1}{2m}[\hbar\vec{\nabla}^2 - kZe^2/r^2 - i\hbar e\vec{A}\cdot\vec{\nabla}/m + (e\vec{A})^2/2m] + e\vec{A}]^2\right\}|\psi(r,t)\rangle$$
$$= E|\psi(r,t)\rangle \qquad (21.5.12)$$

where reduced mass effects are neglected.

21.5.2 The LINEAR TERM in $\vec{A}$

$$\vec{A}(\vec{r}.t) = \tfrac{1}{2}(\vec{B}\wedge\vec{r}) \qquad (21.5.1)$$

The linear term in $\vec{A}$ appeared in equation (21.5.12) is
$$-i\hbar e\vec{A}\cdot\vec{\nabla}/m = (i\hbar e/2m)(\vec{B}\wedge\vec{r})\cdot\vec{\nabla}$$
$$= (-i\hbar e/2m)((\vec{B}\wedge\vec{r})\cdot\vec{\nabla} = (e/2m)(\vec{B}\cdot\vec{L}). \qquad (21.5.13)$$
where $\quad \vec{L} = -i\hbar\{\vec{r}\wedge\vec{\nabla}\}$ $\qquad\qquad\qquad\qquad\qquad\qquad\qquad\qquad\qquad\qquad (21.5.14)$

21.5.3 The QUADRATIC TERM in $\vec{A}$

The quadratic term in $\vec{A}$ appeared in equation (21.5.12) is
$$(e\vec{A})^2/2m = (e^2/8m)(\vec{B}\wedge\vec{r})^2 = (e^2/8m)(\vec{B}^2\ \vec{r})^2 - (\vec{B}\cdot\vec{r})^2 \qquad (21.5.15)$$

21.5.4 Why is the quadratic term in $\vec{A}$ may be neglected?

The ratio of equations (21.5.15) to (21.5.13) is
$$(e^2 a_o \vec{B})/4\hbar \approx 10^{-6}\vec{B} \qquad (21.5.16)$$

where $\vec{B}$ is in units of Tesla (T).
In the Laboratory, magnetic fields encountered do not exceed 10T, so that for most purposes the quadratic term is negligible.

21.5.5 The interaction energy of the magnetic field, $\vec{B}$

The linear term in $\vec{A}$ in equation (21.5.12) corresponds to the interaction energy of the magnetic field $\vec{B}$ with an orbital *magnetic dipole moment*, $\vec{\mu}_L$, which is defined as (Fig. 21.1)

$$\boxed{\vec{\mu}_L = -(e/2m)\vec{L} = -\mu_B\cdot\vec{L}/\hbar} \qquad (21.5.17)$$

where μ_B = Bohr Magneton.

The interaction energy is given by (21.5.13)
$$-(e/2m)\vec{B}\cdot\vec{L} = -\vec{\mu}_L\cdot\vec{B}$$

$$\boxed{\hat{H}_1^{(1)} = -\vec{\mu}_L\cdot\vec{B}} \qquad (21.5.18)$$

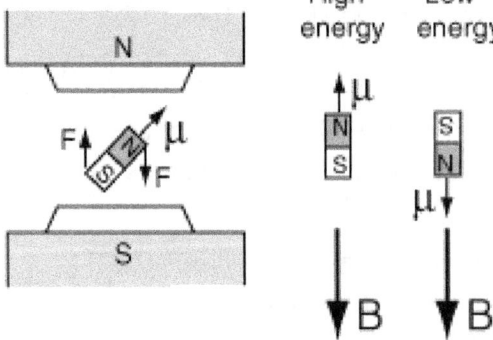

Fig.21.1. Interaction of the atomic magnet and external field $\vec{B}_{ext} = \vec{B}$

Intrinsic magnetic moment of the electron is

$$\vec{\mu}_S = -g_e(e/2m)\vec{S} = -g_e\mu_B \cdot \vec{S}/\hbar \tag{21.5.19}$$

where $g_e = 2$, the *Spin Gyro-Magnetic Ratio* for Electron.

This *spin magnetic moment* gives additional interaction energy (Fig. 21.2.)

$$\boxed{\hat{H}_2^{(1)} = -\vec{\mu}_S \cdot \vec{B} = -g_e\mu_B \cdot \vec{S}/\hbar}. \tag{21.5.20}$$

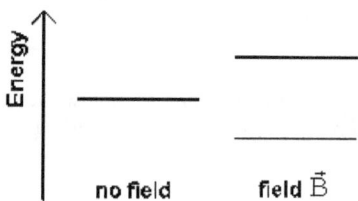

Fig. 21.2 Interaction energy $\hat{H}_2^{(1)} = -\vec{\mu}_S \cdot \vec{B} = -g_e\mu_B \cdot \vec{S}/\hbar$ splits the energy level.

21.5.6 The Perturbation Term in the Non-relativistic Hamiltonian of a Hydrogen atom in a Magnetic field (S-L interaction and spin magnetic interaction)

The Schrödinger Equation for the present case is

$$\left\{-\frac{1}{2m}[\hbar\vec{\nabla}^2 - kZe^2/r - \xi(r)\,\vec{L}\cdot\vec{S} + (\mu_B/\hbar)(\vec{L}+2\vec{S})\cdot\vec{B}\right\}|\,\psi(r,t)\rangle = E|\,\psi(r,t)\rangle \tag{21.5.21}$$

where $\xi(r) = \dfrac{1}{2m^2c^2}(kZe^2/r^3)$. $\tag{21.5.22}$

as $\xi(r) = \dfrac{1}{2m^2c^2}r^{-1}(\dfrac{dV}{dr})$,

with $V(r) = -kZe^2/r$.

$|\,\psi(r,t)\rangle$ is spin orbital.

21.5.7 STRONG MAGNETIC FIELD: THE NORMAL ZEEMAN EFFECT

The first person to study the effects of magnetic fields on the optical spectra of atoms was Zeeman in1896. He observed that the transition lines split when the field is applied. Further work showed that the interaction between the atoms and the field can be classified into two regimes:
1) Strong or Weak fields: the Zeeman Effect, either **Normal** or **Anomalous**;
2) Very Strong fields: the **Paschen-Back** Effect.

The "Normal" Zeeman Effect is so-called because it agrees with the classical theory developed by Lorentz.

The "Anomalous" Zeeman Effect is caused by electron spin, and is therefore a completely quantum result.

21.5.7.1 Multiplets: LANDE INTERVAL RULE

It is known that a perturbation $\hat{H}_{SL}^{(1)}$ would split the energy level (L, S) into multiplets labeled by the quantum number J:

$$E_J^{(1)} = \langle L\,S\,J\,M_J |\, \hat{H}_{SL}^{(1)} \,| L\,S\,J\,M_J \rangle$$
$$= \xi(L,S)\,\tfrac{1}{2}\,[J(J+1)-L(L+1)-S(S+1)] \quad (21.5.23)$$

where $\xi(L,S)$ is a function of L, S, ξ (ξ measures the strength of coupling).
This equation has the following meaning:

a) Each level in a multiplet, i.e. a sublevel $^{2S+1}L_J$ is $(2J+1)$-fold degenerate with respect to the quantum number, M_J.

b) The energy interval between levels differing by unity in their J-value is proportional to the higher J-value of the pair. This is known as the <u>Lande Interval Rule</u>,

$$\boxed{E_J^{(1)} - E_{J-1}^{(1)} = J\,\xi(L,S)}. \quad (21.5.24)$$

c) Since J may take values from $L+S$ to $L-S$ in integral steps, a Term would split into $2S+1$ or $2L+1$ levels, depending on whether $L > S$ or $S > L$ (when $L < S$ the multiplicity is said to be not fully developed).

21.5.7.2 HUND's Rules and the Relative Energies of Terms

A set of rules (empirical) was devised by Frederick Hund for identifying the lowest energy term

$$\left|^{2S+1}\boxed{L}_J\right.$$
Term Symbol

of a configuration with the minimum of calculation,

i) Select first the Terms of largest S. i.e. Terms with maximum multiplicity lies lowest in energy. Example: For the configuration, $2p^2$, one expects the ^3P Term to lie lowest in energy.

ii) Among these Terms, the Term of largest L is lowest in energy, i.e. for a given multiplicity, the term with highest value of L lies lowest in energy. Example: Between ^3P Term and ^3F Term in a particular configuration, then ^3P Term will have the lowest energy.

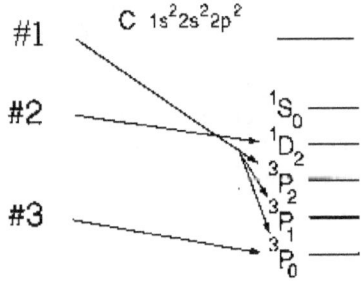

Hund's Rules in Nutshell
Fig 21.3 Hund's Rules in a Nutshell

iii) Multiplets formed from equivalent electrons are regular when less than half the shell is occupied, but inverted when more than half the shell is occupied. A Term is regular if the lowest J-value of the multiplet has the lowest energy, and is inverted if the largest J-value has the lowest energy. (Fig 21.4). The origin of this rule is the **spin-orbit coupling**.

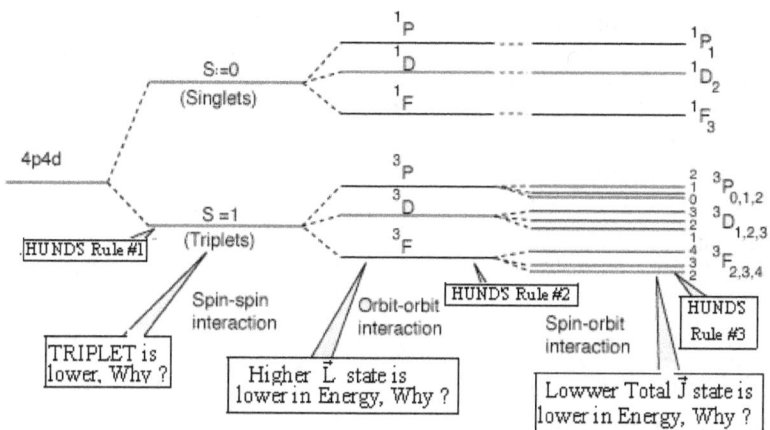

HUND'S RULES Illustrated
Fig 21.4 Illustration of the application of Hund's Rules to energy levels

Worked out Example 21.2

Predict the lowest energy level for a system whose configuration is $2p^2$.

Solution: Step#1 Given the state of the system $2p^2$ and Hund's Rules

Step#2 2p shell can accommodate maximum of 6 electrons. $2p^2$ corresponds to the shell that is less than half-filled.

Step #3 The ground term is $2\,^3P$, it has three levels, with J = 2, 1, 0.

Step #4 Applying the Hund's Rules #1 and #3, i.e. because of the spin-orbit coupling, the lowest energy level is $2\,^3P_0$.

21.5.8 EXTERNAL MAGNETIC FIELDS AND ATOMIC ENERGY LEVELS

P. Zeeman, in 1896, discovered experimentally that when a sodium flame is placed between the poles of a powerful electromagnet the two lines of the first *principal doublet* are considerably broadened, and that their outside edges were polarized as predicted. T. Preston (1898) using an instrument with greater dispersion and resolving power was able to show not only that certain line were split up into *triplets* when viewed perpendicular to the field, but that others were split into as many as four and even six components. He also pointed out that the pattern of all lines, usually called *Zeeman patterns*, belonging to the same series of spectrum lines was the same and was characteristic of that series. This is now known as <u>Preston's Law</u>.

The Normal Zeeman Effect is observed in <u>atoms with no spin</u>. The total spin of an N-electron atom is given by: $S = \sum_{j=1}^{N} s_j$.

21.5.8.1 Zeeman Splitting; LANDE g-factor, g_L

Thus each level in a multiplet, i.e. a sublevel $^{2S+1}\boxed{L}_J$, is (2J +1)- fold degenerate with respect to the quantum number, M_J. The removal of degeneracy by magnetic field, $\vec{B}_{ext}$. is called the **ZEEMAN EFFECT**. It is the basis of both microwave PMR (Paramagnetic Resonance) phenomena and magnetic frequency tuning of optical masers. The spectra produced by this Zeeman Effect are called the **Fine Structure** in Spectroscopy. The Zeeman Effect is expressed as an additional perturbation term in the Hamiltonian, written as

$$\hat{H}^{(1)} = \vec{B}_{ext} \cdot (\vec{L} + 2\vec{S})\mu_B / \hbar \tag{21.5.25}$$

where μ_B = *Bohr Magneton*.

$\hat{H}^{(1)}$ can be considered as a perturbation to the sublevels if the Zeeman splitting is smaller than the sublevel splitting..

In the absence of a magnetic field, the atoms (or ions) have no preferred spatial orientation. Treating the S-L interaction as a perturbation, and $\vec{B} \parallel Z$ - axis (Fig. 21.5. and Fig 21.6),

$$\hat{H}_B^{(1)} = \vec{B}_Z \cdot (\vec{L}_z + 2\vec{S}_z)\mu_B / \hbar \tag{21.5.26}$$

$$\left\{-\frac{1}{2m}[\hbar^2 \vec{\nabla}^2 - kZe^2/r]\right\}| \psi(r,t) \rangle = \left\{ E - (\mu_B/\hbar)(\vec{L}_z + 2\vec{S}_z) \cdot \vec{B}_z \right\}| \psi(r,t) \rangle \tag{21.5.27}$$

Strong magnetic field, $\vec{B}_{ext}$ means

Magnetic interaction > S-L interaction

From 1st Order Perturbation Theory, the perturbed 1st order energies are:

$$E_J^{(1)} = \langle LSJM_J | \left\{ \mu_B (\vec{L}_z + 2\vec{S}_z) \cdot \vec{B}_z \right\} | LSJM_J \rangle$$

$$\vec{L}_z + 2\vec{S}_z = \vec{J}_z + \vec{S}_z \tag{21.5.28}$$

Therefore $\quad \vec{J}^2 = \vec{L}^2 + \vec{S}^2 - 2\vec{L}\,\vec{S}\,Cos\theta_J \tag{21.5.29}$

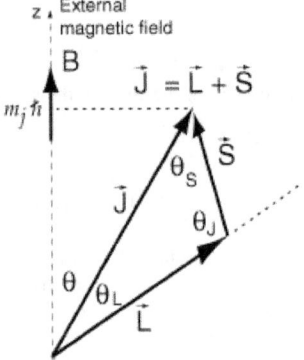

Fig. 21.5 Lande vector Coupling: Slow precession of $\vec{J}$ around $\vec{B}_{ext}$ in the Anomalous Zeeman effect. The Spin-Orbit Coupling / Magnetic interaction causes $\vec{L}$, and $\vec{S}$ to process much more rapidly around $\vec{J}$.

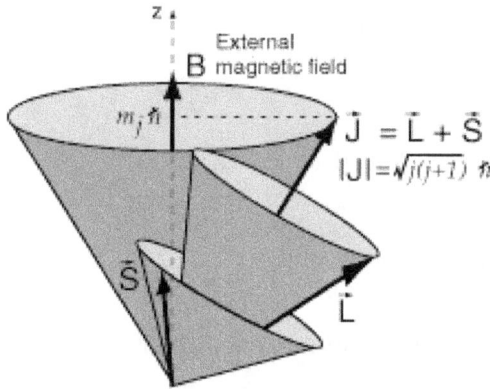

Fig. 21.6. The vector diagram to calculate the Lande g_J-factor. In a weak field, $\vec{J}$ precesses slowly about $\vec{B}_{ext}$ with J_z fixed (quantized), while $\vec{L}$, and $\vec{S}$ precess rapidly about $\vec{B}_{ext}$.

$$E_J^{(1)} = \vec{B}_z M_J \mu_B + \vec{B}_z \mu_B \langle L\,S\,J\,M_J | \{\vec{S}_z\} | L\,S\,J\,M_J \rangle \qquad (21.5.30)$$

$\langle L\,S\,J\,M_J | \{\vec{S}_z\} | L\,S\,J\,M_J \rangle$ can be calculated again from the properties of angular momentum operators (for example, by means of the *Clebsch-Gordan* coefficients; See J.C. Slater, "*Quantum Theory of Atomic Structure*" (Vol. II, MGH, NY. 1960, Sections 24-25) to be

$$\langle L\,S\,J\,M_J | \{\vec{S}_z\} | L\,S\,J\,M_J \rangle = M_J \; \frac{1}{2} \frac{[J(J+1)-L(L+1)+S(S+1)]}{[J(J+1)]} \qquad (21.5.31)$$

Hence $\quad E_J^{(1)}(L\,S\,J\,M_J) = g(L\,S\,J\,M_J)\,\vec{B}_z M_J \mu_B \qquad (21.5.32)$

where $\quad g_L \equiv g_J \equiv g(L\,S\,J\,M_J) = 1 + \frac{[J(J+1) + S(S+1) - L(L+1)]}{[2J(J+1)]} \qquad (21.5.33)$

$$= \frac{[3J(J+1) + S(S+1) - L(L+1)]}{[2J(J+1)]}$$

g_L is the Lande g-factor.

21.5.8.2 EQUALLY SPACED ZEEMAN levels

This means that each individual multiplet is split by the magnetic field into $(2J+1)$ EQUALLY SPACED Zeeman levels. The energy difference between adjacent Zeeman levels

$$\delta E_J^{(1)} = g_J\,\vec{B}_z \mu_B \qquad (21.5.34)$$

In the case of *NORMAL Zeeman Effect* the spin has no part to play in the splitting, and one has g = 1.

21.5.8.3 THE NORMAL ZEEMAN EFFECT IN HYDROGEN $2p \leftrightarrow 1s$

The Normal Zeeman Effect is observed in atoms with no spin. The total spin of an N-electron atom is given by: $S = \sum_{j=1}^{N} s_j$

Consider $n = 2$ for Hydrogen atom, the fine structure splitting is

$$\delta E_J^{(1)}\,(J = \ell + \tfrac{1}{2}, j = \ell - \tfrac{1}{2}) = E_{n\ell}\,(Z\alpha)^2 [n\,\ell(\ell+1)]^{-1}.$$

$$= (Z^2\alpha^4)[2n^3 \ell(\ell+1)]^{-1} \dot{A}. \qquad (21.5.35)$$

$$= (0.365) Z^4 cm^{-1}.$$

21.5.8.4 LARMOR Frequency, v_L; LORENTZ Triplet

The *Selection Rules* for electric dipole transitions require

$$\Delta m_\ell = 0, \text{ and } \Delta m_\ell = 0, \pm 1. \qquad (21.5.37)$$

Thus <u>(n - n') line is split into *three* (3) components</u> (Fig. 21.7).

$v_{n n'}$ = original frequency (Linearly polarized).

$$v_{\Delta m_\ell = 0} \equiv v_{n n'} = \pi - \text{Component, (Circularly polarized)} \qquad (21.5.38)$$

$$v_{\Delta m_\ell = \pm 1} \equiv (v_{n n'} \pm v_L) = \sigma - \text{Component (Circularly polarized)}, \qquad (21.5.39)$$

v_L is Larmor Frequency,

$$\boxed{v_L = \mu_B \vec{B}_z / h} \qquad (21.5.40)$$

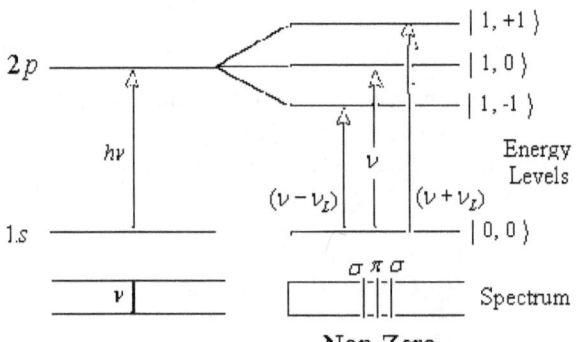

Normal Zeeman Effect in Hydrogen
Fig. 21.7 The Normal Zeeman Effect in Hydrogen, $2p \leftrightarrow 1s$

This splitting is called the <u>*Normal Zeeman Effect*</u>. The three lines are said to form a Lorentz Triplet. The entire phenomenon of the Normal Zeeman Effect is self explanatory from the schematic diagram of Fig. 21.7.

21.5.8.5 The case of Cadmium

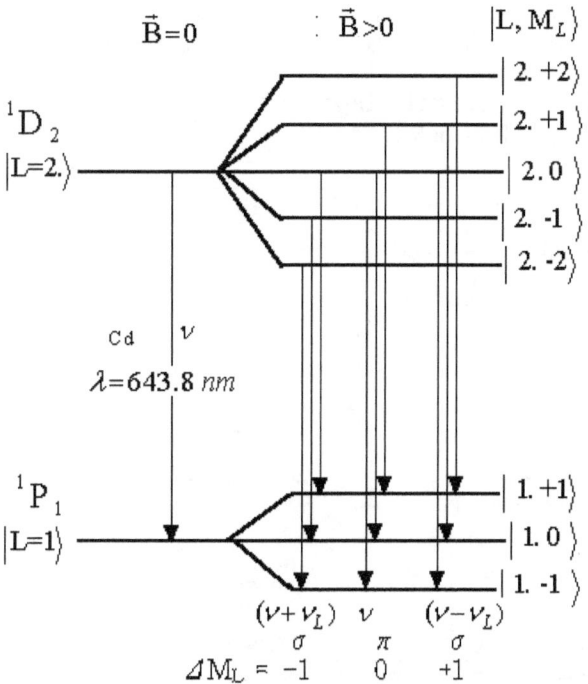

NORMAL ZEEMAN EFFECT in Cadmium

Fig. 21.8 The splitting of the p($\ell = 2$) to d ($\ell = 1$) transition energy levels and the splitting of the transitions into three groups of coincident lines. $\Delta m_\ell = 0$ transition is unshifted, but $\Delta m_\ell = \pm 1$ occur at $(h\nu \mp \mu_B \vec{B}_z)$.

TABLE 21.1   The Normal Zeeman Effect

Δm_ℓ	Energy	Longitudinal Observation	Transverse Observation
+1	$(h\nu - \mu_B \vec{B}_z)$	σ^+ (Circular Polzn)	E ⊥ B
0	$h\nu$	π (Linear Polzn)	E ∥ B
-1	$(h\nu + \mu_B \vec{B}_z)$	σ^- (Circular Polzn)	E ⊥ B

One can observe the Normal Zeeman Effect, for example, in the red spectral line of cadmium ($\lambda = 643.8$ nm). It corresponds to the transition $^1D_2$ ($J = 2$, $S = 0$) → $^1P_1$ ($J = 1$, $S = 0$) of an electron of the fifth shell (Fig 21.8).

21.5.8.6 The Normal Zeeman Effect in the case of D-lines of Sodium.
The pattern is schematically shown in the self-explanatory diagram in Fig 21.9.

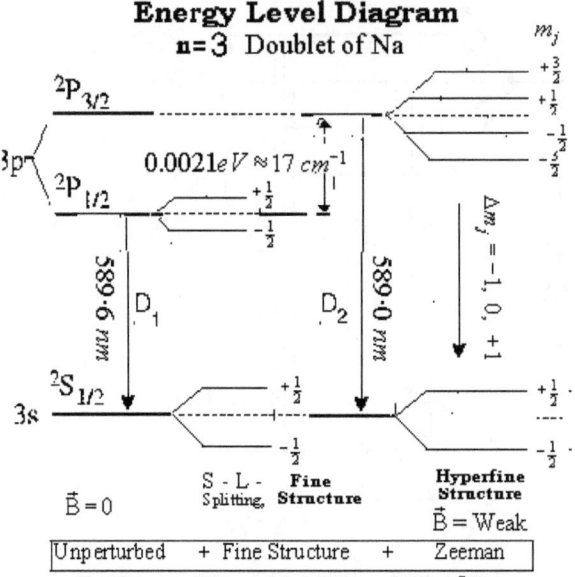

Fig 21.9 The Normal Zeeman Effect of the $3p \to 3s$ doublets of Sodium

21.5.8.7 ANOMALOUS ZEEMAN EFFECT (WEAK FIELD)

More generally, one observes the ``Anomalous Zeeman Effect''. In this case, not only the interaction between the external field and the magnetic moment of the orbital must be taken into account, but also the interaction with the intrinsic magnetic moment of the electron. This leads to a more complex splitting of the energy levels. If the magnetic field is weaker than 10^5 Gauss, the Zeeman term, $\vec{B}_z$ can be regarded as a perturbation compared to the relativistic corrections. The Zeeman splitting in First Order Perturbation Theory becomes

$$\Delta E = (\mu_B \vec{B}_z) m_j \left[1 \pm \frac{1}{2\ell + 1} \right] \qquad (21.5.43)$$

All degenerate levels are split due to the magnetic field. The levels $(\ell \pm \tfrac{1}{2})$, split by the magnetic field, become $(2\ell + 2)$ and 2ℓ levels. Specifically, examine the energy level scheme of a p-electron, $\ell = 1$ ($P_{3/2}$, $P_{1/2}$. In the weak $\vec{B}_z$ case the energy shifts are linear in $\vec{B}_z$ with slopes determined by $m_j \left[1 \pm \frac{1}{2\ell+1} \right]$ (equation (21.5.43)). As one increases $\vec{B}_z$, mixing becomes possible between states with the same m value, for example, $P_{3/2}$ with $m_s = \pm \tfrac{1}{2}$ and $P_{1/2}$ with $m_s = \pm \tfrac{1}{2}$. The sodium D lines correspond to the $3p \to 3s$ transition. At $\vec{B}_z = 0$, the spin-orbit interaction splits the upper 3p ^2P Term into the $^2P_{3/2}$ and $^2P_{1/2}$ levels separated by $17\ cm^{-1}$.

The lower $^2S_{1/2}$ level has no spin-orbit interaction. The Landé g_J-factors of the levels are given in Table below:

Landé g_J factors for Levels of Sodium D lines

Level	J	L	S	g_J
$^2P_{3/2}$	$\frac{3}{2}$	1	$\frac{1}{2}$	$\frac{4}{3}$
$^2P_{1/2}$	$\frac{1}{2}$	1	$\frac{1}{2}$	$\frac{2}{3}$
$^2S_{1/2}$	$\frac{1}{2}$	0	$\frac{1}{2}$	2

$$E_{n;\,Weak\,\vec{B}_{ext}} = E_n^{(0)} + \frac{e}{2m} g_J m_j \hbar \vec{B}_{ext}$$

Worked out Example 21.4
Account for the form of the Zeeman Effect when a magnetic field is applied to the transition $3\,^2D_{3/2} \to 2\,^2P_{1/2}$.

Solution: Step #1 It is known that $g_J \equiv g(L\,S\,J\,M_J) = 1 + \frac{[J(J+1) + S(S+1) - L(L+1)]}{[2J(J+1)]}$;
Selection Rule is $\Delta M_J = 0, \pm 1$

Step #2 For the level $3\,^2D_{3/2}$, $L = 2$, $S = \frac{1}{2}$ and $J = \frac{3}{2}$. It follows that $g_J(3\,^2D_{3/2}) = 4/5$

Step #3 For the level $2\,^2P_{1/2}$, $L = 1$, $S = 1/2$ and $J = 3/2$; so $g_J(2\,^2P_{1/2}) = 2/3$.

Step #4 The splittings are therefore of magnitude $\Delta E(^2D_{3/2}) = \frac{4}{5}(\mu_B \vec{B}_z)$ Term and $\Delta E(2\,^2P_{1/2}) = \frac{2}{3}(\mu_B \vec{B}_z)$ Term. Step #5 Apply the selection rule, $\Delta M_J = 0, \pm 1$. There are thus six allowed transitions, where it is seen that they form three **doublets**.

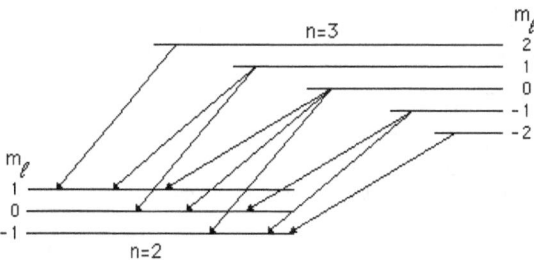

21.5.7.8 The Zeeman Effect and Quantum Electronics
Zeeman Effects are important in Quantum Electronics because they make possible a method by which the values of energy levels can be slightly altered, thereby causing a fine tuning of emission and absorption frequencies in the optical and infrared ranges. They also provide the

distinct energy levels whose separations are in the microwave or radio frequency ranges ideal for masers.

Worked out Example 21.3

Examine if the spins play any part in Zeeman splitting of Atomic levels.

Solution. Step #1 The Lande splitting factor, $g(L\,S\,J\,M_J) = 1 + \dfrac{[J(J+1) + S(S+1) - L(L+1)]}{[2J(J+1)]}$

Step #2 To examine the role of spin, put S = 0, which means $J = L + S = L + 0 = L$, i.e., $g_J = 1$. Hence spins play no role in Zeeman Effect. Such is the case in the Normal Zeeman Effect.

21.5.9 The PASCHEN-BACK EFFECT (VERY STRONG FIELD)

The criterion for observing the Paschen-Back Effect is that the interaction with the external magnetic field should be very much stronger than the spin-orbit interaction:

Magnetic interaction ≫ S-L interaction

In sodium, the field strength equivalent to the spin-orbit interaction for the D-lines is given by:

$\Delta E = \vec{\mu}_B g \cdot \vec{B}_{ext} = (5.9 \times 10^{-5}\, eV/T)\, 2\, \vec{B}_{ext} = 0.0021\, eV$

$\Delta E = \vec{\mu}_B g \cdot \vec{B}_{int} = (9.27 \times 10^{-24}\, J/T)\, 2\, \vec{B}_{int} = 17\, cm^{-1}$

$\vec{B}_{int} = 18\, Tesla$

This is a very large field, and is not achievable in normal laboratory conditions. On the other hand, since the spin-orbit interaction decreases with the atomic number Z, the splitting for the equivalent transition in lithium is only $0.3\, cm^{-1}$. This means that we can reach the very strong field regime for fields $\approx 0.6 - T$. This is readily achievable, and allows the Paschen-Back Effect to be observed.

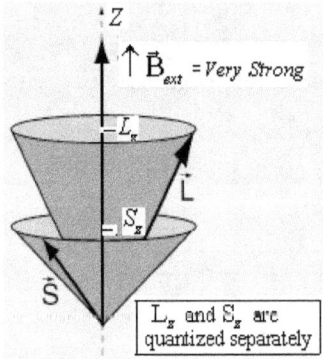

Fig 21.10 S-L interaction broken and L and S are have quantized energies independent of each other.

If the magnitude of Zeeman perturbation is much larger than the S-L.Interaction it is commonly known as the PASCHEN-BACK Effect. When $\vec{B}_{ext}$ = very strong, the S-L interaction is broken; and the $\vec{\mu}_S$ and $\vec{\mu}_L$ have independent orientations, quantized according to independent quantization of S and L along the $\vec{B}_{ext}$. This is totally different from the Zeeman case. One can show that

$$\boxed{E_{n.\,VeryStrong\,\vec{B}_{ext}} = E_n^{(0)} + \frac{e}{2m}(m_\ell + 2m_s)\hbar\,\vec{B}_{ext}}$$

The good quantum numbers in the case are m_ℓ and m_s.

21.6 THE 21-cm (1420 MHz) LINE OF NEUTRAL HYDROGEN

Both spin-orbit coupling and relativity effects combine to produce fine structure shifts of the order of $\approx 10^{-3}$ to $\approx 10^{-6}\,eV$. These can be detected by a good constant deviation prism spectrometer. By when the energy shift is below $\approx 10^{-6}\,eV$, then one has to resort to Fabry-Perot etalon or Lummer-Gehrcke plate spectroscopy. This is where the Hyperfine Structure of a spectral line falls. Hyperfine Structure is produced by the interaction of electron and proton magnetic moments. The magnitude of the shift becomes very small (*i.e.* hyperfine) because, unlike the S.L interaction of electrons, the magnetic moment of a particle is given by

$$\mu = g\left(\frac{e\hbar}{2m}\right). \tag{21.6.1}$$

the g-factor is of order of unity, and the proton mass, $M_p \approx 2000\,m_e$. This is the manifestation in Hydrogen as line at about 1420 *MHz* that radio astronomers use to map Hydrogen clouds in the galaxy. This radiation is emitted as a result of the spins of the electron and proton spontaneously decays from a Triplet state with spins parallel to the Singlet one.

$$\Delta E = 5.876 \times 10^{-6}\,eV = 1420\,MHz = 21\,cm. \tag{21.6.2}$$

Further slight corrections come from relativity, from QED (*i.e.*, the Lamb shift) and from the finite size and orbital motion of the proton. This yields a value of

$$\Delta E = 1420.405751786\,MHz \pm 0.01\,Hz \tag{21.6.3}$$

This is one of the most accurately measured constants in physics.

The receiver on a radio telescope that detects radio waves focused by the dish will be tuned so that it is sensitive to a range frequencies centered on 1420 *MHz*, or, equivalently a wavelength of 21 *cm*. This wavelength has been chosen because it corresponds to an energy transition of single neutral hydrogen. Since this emission occurs in the quietest part of the EM spectrum and traces the most abundant element in the Universe its frequency may serve as a galactic WATER HOLE around which different species may congregate.

21.7 CONCLUSIONS

1) It is known that an atomic system, say the simple hydrogen atom, the energy is given by

$$E_n^{(0)} \propto \frac{1}{n^2}$$

2) When the spin of electron is accounted by S-L interaction, then

$$\vec{\mu}_S = -g_e(e/2m)\vec{S} = -g_e\vec{\mu}_B \cdot \vec{S}/\hbar$$

$$\vec{\mu}_L = -(e/2m)\vec{L} = -\vec{\mu}_B \cdot \vec{L}/\hbar$$

$$\vec{B}_{internal} = \frac{\mu_o e}{4\pi\,m\,r^3}\vec{L} = 0.28\text{-T for 2p state of hydrogen}$$

$$E_{nS.L} = -\vec{\mu}_S \cdot \vec{B}_{internal} = g_e\frac{\mu_o e^2}{4\pi\,m^2\,r^3}\vec{S}\cdot\vec{L}$$

$$A_{n\ell}^{(1)} = -\frac{1}{4\,m\,c^2}(Z\alpha)^2\left(\frac{3}{4n^4}\right)$$

Because spin is relativistic in origin, the s-L interaction may be viewed as a relativistic effect. $\equiv 1/137$, the fine structure constant appears here.

3) Additional (correction) energy takes place and depends on total angular momentum, j.
$$J = L + S$$

4) When an external magnetic field $\vec{B}_{ext}$ is applied, i.e. $\vec{B}_{ext} \neq 0$, to the system, states degenerate when there is no external field, $\vec{B}_{ext} = 0$, will split into different sublevels. How they split depends on the $\vec{B}_{ext}$.

a) The Zeeman Effect is when $\vec{B}_{ext}$ = weak compared to $\vec{B}_{internal}$. In this case the S-L interaction "locks" S and L vectors into a total angular momentum J, and the quantization of J along the external Z-axis gives a quantized interaction energy, $\Delta E_{\mu.B} = -\vec{\mu}_J \cdot \vec{B}_{ext}$, where $\vec{\mu}_J$ locked together $\vec{\mu}_S$ and $\vec{\mu}_L$ (Fig 21.6). $\vec{J}$ will precess about the Z-axis at a speed much slower than that $\vec{L}$, and $\vec{S}$ do independently.

$$E_{n. \text{ Weak } \vec{B}_{ext}} = E_n^{(0)} + \frac{e}{2m} g_J \, m_j \hbar \, \vec{B}_{ext}$$

b) The Paschen-Back Effect is when the $\vec{B}_{ext}$ = very strong. Here the spin-orbit coupling is broken, and $\vec{\mu}_S$ and $\vec{\mu}_L$ have independent orientation energies, $\vec{L}$, and $\vec{S}$ are quantized independently along the external field (Fig 21.10). They precess around the field independently.

$$E_{n. \text{ [VeryStrong } \vec{B}_{ext}]} = E_n^{(0)} + \frac{e}{2m} (m_\ell + 2m_s) \, \hbar \, \vec{B}_{ext}$$

REVIEW QUESTIONS

R.Q. 21.1   Estimate electron's spin magnetic moment in a hydrogen atom Given:
$e = 1.6021 \times 10^{-19} C$, $\hbar = (h/2\pi) = 1.054 \times 10^{-34} J-s$, $m_e = 9.1094 \times 10^{-31}$ kg, (Answer: $|\vec{\mu}_S| = g_e(e/2m)\sqrt{s(s+1)} \, \hbar = 1.6 \times 10^{-23} \, A-m^2$).

R.Q. 21.2   Determine the intrinsic magnetic field of the orbiting electron in a hydrogen atom, and hence estimate the orientation energy produced by the S-L interaction. Given:
$e = 1.6021 \times 10^{-19} C$, $m_e = 9.1094 \times 10^{-31}$ kg, $\hbar = (h/2\pi) = 1.054 \times 10^{-34} J-s$,
$\mu_0 = 4\pi \times 10^{-7} T \, m \, A^{-1}$, $a_0 = \hbar^2/m_e e^2 = 0.5292$ nm (Answers:
$\vec{B}_{\text{due to } L} = [\mu_0 e/4\pi m_e (4a_0)^3] \sqrt{\ell(\ell+1)} \, \hbar] = 0.28 \, T$; $|\vec{\mu}_S| = g_e(e/2m)\sqrt{s(s+1)} \, \hbar = 1.6 \times 10^{-23} A - m^2$,
$|\vec{\mu}_S \cdot \vec{B}| \cong 10^{-23} J \sim 10^{-4} eV$)

R.Q. 21.3   Determine the Spin-Orbit interaction energy of the orbiting electron in a hydrogen atom, Given: $\hbar = (h/2\pi) = 1.054 \times 10^{-34} J-s$, $\mu_B = 9.2740 \times 10^{-24} \, A \, m^2 \, (= JT^{-1})$, (Answers:;
$|g_e \mu_B \cdot \vec{S}/\hbar| \cong 10^{-23} J \sim 10^{-4} eV$).

R.Q. 21.4   Consider sample consisting atomic hydrogen is immersed in an externally applied 0.05-T magnetic field. a) Find out the number of sublevels appearing as a result of splitting of the $1s_{1/2}$.

b) Into how many is the $2p_{3/2}$ level splitting ?, c) In each case of a) and b) how large is the energy splitting d) Into how many lines is the $2p_{3/2} \to 1s_{1/2}$ spectral line split by the magnetic field? e) What is the energy splitting between the lines? f) By how much about is the wavelength of a spectral line shifted? Given The estimated internal magnetic field is 0.3-T, and the selection rule is $\Delta m_J = 1$, $e = 1.6021 \times 10^{-19} C$, $m_e = 9.1094 \times 10^{-31}$ kg, $\hbar = (h/2\pi) = 1.054 \times 10^{-34} J-s$

(Answers: a) 2 for $1s_{1/2}$; b) 4 for $2p_{3/2}$, c) $g(2p_{3/2}) = 4/3, g(1s_{1/2}) = 1$,

$E(2p_{3/2}) = \begin{Bmatrix} \pm \frac{3}{2} \\ \pm \frac{1}{2} \end{Bmatrix} (3.9x10^{-6} eV)$, $E(1s_{1/2}) = \begin{Bmatrix} \pm \frac{1}{2} \end{Bmatrix} (5.8 \times 10^{-6} eV)$; d) Six eually spaced transitions are allowed, and the energy splitting is $[E(2p_{3/2}) - E(1s_{1/2})] + E = E_{n\ell} + (\frac{e}{2m}\vec{B}_z\hbar) \begin{Bmatrix} \pm \frac{5}{3} \\ \pm 1 \\ \pm \frac{1}{3} \end{Bmatrix}$;

e) $(\frac{e\hbar}{2m}\vec{B}_z)\frac{2}{3} = 1.9 \times 10^{-6} eV$, f) $\lambda_{Lyman} = 122$ nm, $\Delta\lambda_{Lyman} = 2.5 \times 10^{-5}$ nm).

R.Q. 21.5   Account for the form of the Zeeman Effect when a magnetic field of 1-T is applied to the transition $3\,^2D_{3/2} \to 2\,^2P_{1/2}$. Given $\mu_B = 9.2740 \times 10^{-24} A\,m^2 (= JT^{-1})$ (Answer: Anomalous Zeeman Three doublet pattern,
$\Delta E = M_J(\mu_B \vec{B}_z) = 5.79 \times 10^{-5} eV$)

R.Q. 21.6 a) Differentiate clearly the quantum mechanics behind the Zeeman Effect and Paschen-Back Effect., b) Why is it that Paschen-Back Effect can not be observed experimentally in D-lines of Sodium $3p \to 3s$, whereas it is possible for equivalent transitions in Lithium ? Given, ΔE (3p doublet states) $= 17\,cm^{-1}$, for sodium, and for lithium $0.3\,cm^{-1}$.

R.Q, 21.7   Calculate the fine structure for the $n = 2$ level in H. Consider the energies and wavefunctions corresponding to the unperturbed Hamilton operator $H^{(0)}$, which is the sum of the kinetic energy operator and the Coulomb potential. The electron has a $\frac{1}{2}$ spin. 1) What is the degeneracy of the $n = 2$ level? 2) Which orthonormal basis can be used to describe level 2? Which quantum numbers are necessary to characterize the wave functions? 3) The fine structure is described with the Hamilton operator $W_f = W_{mv} + W_{SO} + W_D$, where W_{mv} corresponds to the relativistic mass increase, W_D is the Darwin term, and W_{SO} describes the spin-orbit interaction. W_{mv} is proportional to , W_D depends only on r, and $W_{SO} = V_{LS}(r)$. The hyperfine interaction is neglected. Show that W_{SO} does not commute with $\hat{L}_z$. What does it imply for the matrix that describes W_{SO} in the basis for $n = 2$? 4) Show that W_{SO} commutes with and that the matrix consists of a 2x2 matrix for 2s and a 6x6 matrix for 2p.
5) To simplify the problem, consider the total angular momentum operator . Use the triangle rule to determine which j, and s are possible for $n = 2$. These states are labeled. 6) Show that can be written as . 7) Show that and $\hat{J}_z$ commute with $H^{(0)}$ and W_{SO}.
8) In which basis are , $\hat{J}_z$, $H^{(0)}$, W_{SO}, diagonal? 9) Write down W_{SO} in the above matrix using the integrals where is the radial wave function for 2. How many energy levels are there? 10) It can be shown that W_{mv} and W_D commute with $H^{(0)}$. What is the effect of W_{mv} and W_D on the n=2 shell? (Answer: 4) =0,1, j=1/2, 3/2 ; 8) 3 energy levels).

R.Q. 21.8 a) A hydrogen atom is exposed to a very strong magnetic field, so that the interaction with the field is larger than the spin-orbit interaction. Show that this problem can be solved exactly. Express the energies and wave functions.

b) Consider the spin-orbit interaction as a small perturbation. Calculate approximately the energy levels. (Paschen Back effect)

c) Assume that the interaction with the magnetic field is a small perturbation of the spin-orbit Hamiltonian. What are the energies and wave functions?

R.Q. 21.9 Show that the correction due to the Electron Spin appears in the expression of the Hamiltonian as $\hat{H}^{(2)} = -\vec{\mu}_l \cdot \vec{B} = \frac{e}{mc}(\vec{s} \cdot \vec{B})$ Obtain an expression for the Energy Shift, $\Delta E_{n\ell m}$.

Given: $E_n^{(0)} = [-m_o c^2 (Z\alpha)^2]/2n^2$, $<r^{-1}>_{n\ell} = z/n^2 a_o$, $<r^{-2}>_{n\ell} = z/[n^2 a_o(\ell+\frac{1}{2})]$ (Answer: $[\frac{1}{4} m c^2 (Z\alpha)^4] \begin{Bmatrix} \ell \\ -\ell-1 \end{Bmatrix} \cdot [n^3 \ell(\ell+\frac{1}{2})(\ell+1)]^{-1}$, i.e. $<\mathcal{H}_{LS}>$ = finite, in s-states.)

R.Q. 21.10 Obtain an expression for the energy shift due to the Darwin Term,

$\hat{H}^{(3)} = -\left(\frac{\hbar^2}{8 m_o^2 c^2}\right) \nabla^2 V = \left(\frac{\pi \hbar^2 Z e^2}{2 m_o^2 c^2}\right) \delta^3(x)$, and the correction (to the perturbed energy), $\Delta E_{n\ell m}$.

(Answer: $<\hat{H}^{(3)}>$ is formally identical to $<\hat{H}^{(2)}>_{n,j=1,\ell=0}$).

%&*%&*%&*%&*%&*

Chapter 22

QUANTUM MECHANICS – 4:
SCATTERING (COLLISION OR REACTION) THEORY

Chapter 22

QUANTUM MECHANICS – 4:
SCATTERING (COLLISION OR REACTION) THEORY

"People are just about as happy as they make up their minds to be". - Abraham Lincoln

" I have yet to see a problem however complicated, which when you looked at it in the right way, did not become still more complicated" - Paul Anderson

22.1.1 INTRODUCTION

So far the matter dealt with in this book have been concerned mostly with particles held in potential wells, central potentials, or other entrapments. It was the 'discrete energy' spectrum of the operator that characterized this. What happens to a particle traveling through matter? Much of what one knows about the forces and interactions in atoms and nuclei has been learnt from scattering experiments. The particles are 'scattered' by a target atom and the scattered particles are detected, experimentally. Usually one knows the nature of the particles used as projectiles, their momentum and perhaps their polarization; and the detector may simply give the intensity as a function of angle.

22.1.1 Comparison of studies involving 'discrete' and 'continuous' parts of the energy spectra
(1) The most significant aspect of scattering processes is that one is concerned with the 'continuous' part of the energy spectrum. If the zero of energy is chosen in the conventional manner, this corresponds to the positive eigen values of the Schrödinger Equation and to eigen functions of the unbound states.

(2) In the case of bound states the interest was to solve the Schrödinger Equation for potentials and on the discrete energy eigen values; this enables to compare directly the calculated discrete eigen values of theory and measured spectral frequencies of experiment. On the other hand, in scattering, where the continuous spectrum is involved, the incident beam gives the energy, and the intensities are the object of measurement and prediction. These measure the probability of finding the particle at a certain location, and they are related to the *eigen functions* rather than the *eigen values*.

(3) In the case of bound state problems, one assumes the conservation of energy, *i.e.* given by equation (1.22),
viz. $\quad E_i - E_f = \hbar\omega$ \hfill (22.1.1)

Scattering data can be compared with theoretical predictions only if one elucidates carefully the various stages of scattering process. Relating observed intensities to the calculated wave functions is the first problem of Scattering Theory.

22.1.2 Time Development of a System and the S-matrix
The time development of the wave function Ψ of a system is described by the Schrödinger Equation
$$\hat{H}(\vec{r},\vec{p})\,\psi(r,t) = i\hbar\,(\partial/\partial t)\,\psi(r,t) \quad (22.1.2)$$

for the Hamiltonian operator, $\hat{H}$ of the system which develops in time.
a) In non-relativistic Quantum Mechanics,
$$\hat{H} \equiv \hat{H}(\vec{r}, \vec{p}), \text{ where } \hat{p} \to -i\hbar \hat{\nabla}$$
The applications of Schrödinger Equation fall into two broad classes:
(i)   Bound states, and
(ii)   Scattering processes.
Knowledge of elementary particles comes from scattering process. A typical reaction/scattering process between elementary particles is
$$a + b \to c + d + e.$$

| 1st Stage | : Particles a & b are prepared with definite momentum and definite spin orientations.

| 2nd Stage | : Particles a and b interact in a region of dimension $\sim 10^{-13}$ cm.

| 3rd Stage | : The reaction products c, d, e emerge, counter detects giving the probability distribution $W(p_c, pd, p_e)$ of the momenta (and spins) of the products relative to a & b.
(b) In non-relativistic Quantum Mechanics
Schrödinger Equation allows calculation of W for elastic scattering: $a + b \to a + b$.
(c) For Relativistic Quantum Mechanics
Synthesis of Quantum Mechanics and Special Relativity leads to *Quantum Field Theory* (QFT) and a framework for the description of annihilation and creation of particles. In this case a Schrödinger Equation of the form $\hat{H}\, \psi(r,t) = i\hbar\, (\partial/\partial t)\, \psi(r,t)$, still exists, but ψ is no longer a simple function of particle positions, and solutions are possible only in perturbation theory. Here the '*scattering or reaction amplitudes*', called the S-matrix element.

22.1.3 S-MATRIX ELEMENT DEFINED

The initial and final particles in a scattering experiment are *free* particles.
Free particle states = function (momentum, spin projection, and other internal quantum numbers such as iso-spin).
Let Φ_a = initial state of the system,
After the interaction,
Ψ = final state of the system. It is complicated, as it will have components describing various inelastic final states as well as components describing elastic scattering.
Because of the requirements of principle of superstition and Schrödinger Equation being linear, one can write
$$\Psi = \hat{S}\Phi_a. \qquad (22.1.3)$$
where $\hat{S}$ is called the *Scattering Operator*.
From Scattering Theory it will be seen that
$$\hat{S} = \text{a function of } \hat{H}.$$
If Φ_b = state of the particle detected in the detector counter, the probability amplitude to find Φ_b in the actual state Ψ, is (Φ_b, Ψ).
$$\hat{S}_{ba} = (\Phi_b, \hat{S}\Phi_a) = \text{matrix element}.$$
Φ_a are the states which form a complete set of ortho-normal functions,
i.e. $(\Phi_b, \Phi_a) = \delta_{ba}$
It is known that
$$\hat{S}^\dagger \hat{S} = \hat{I} = \hat{S}\, \hat{S}^\dagger.$$
The unitary nature of $\hat{S}$ expresses the Conservation of Probability.

(c) For non-interacting particles:
Final state = initial state, $\hat{S} = \hat{I}$.
It is convenient to introduce the Transition Operator, T defined by
$$\hat{S} = \hat{I} + \hat{T}_i.$$
The matrix elements of $\hat{T}$ describe the interaction between the constituents of the system. Invariance Principles lead to conditions on both $\hat{S}$ and $\hat{T}$ matrix elements, much is established by experimental tests. In *Quantum Electro Dynamics* (QED) and weak interaction theory, it is shown how to derive S-matrix elements from more fundamental quantities: the interaction $\hat{H}$ and the quantized field operators occurring in it.

22.1.4 SCATTERING FUNDAMENTALS

22.2.1 SCATTERING AMPLITUDE: Potential Scattering
In all the work, on assumes that the force vanishes as $r \to \infty$. By choosing the origin of the energy scale appropriately, one can always arrange the potential V(r), such that
$$\underset{r \to \infty}{Lt}\ V(r) = 0;\ |r| \leq r_o,\ \text{say},\ V(r) \neq 0. \tag{22.2.1}$$
For the particle the spectrum of energy E now breaks up into two distinct parts: (1) the continuum ($E \geq 0$), and (2) Solutions for $E < 0$.

22.2.2 THE CONTINUUM ($E \geq 0$)
These solutions describe scattering; the wave functions are not localized about r = 0. For a short-range potential,
$$\underset{r \to \infty}{Lt}\ rV(r) = 0 \tag{22.2.2}$$
defines the short range forces.

22.2.3 SOLUTIONS for $E < 0$
Condition (22.2.1) implies that there is only a discrete set of such solutions, in general, as they correspond to bound states because the wave functions are localized about r.
The wave equation is
$$\frac{\hbar^2}{2\mu} \hat{\nabla}^2\ \psi_k(r) + V\ \psi_k(r) = E\ \psi_k(r) \tag{22.2.3}$$
Putting $k^2 = \frac{2\mu}{\hbar^2} E$ \hfill (22.2.4)
$$U(r) = \frac{2\mu}{\hbar^2} V(r) \tag{22.2.5}$$
where E is the energy of the particle of mass μ, momentum $\vec{p} = \hbar\ \vec{k}$.
$$r = r(r, \theta, \varphi),$$
$$r^2 = x^2 + y^2 + z^2 \tag{9.3.5}$$
in the relative r coordinates, the Spherical Polar Coordinates, (r, θ, φ), are defined earlier in Chapter 8, Section 8.2.3.
φ is the angle between the projection of r in the xy plane and the +z- axis, measured in the direction shown in Fig 4.4,
$$\varphi = \text{Azimuth angle} = \tan^{-1}(y/x)\ . \tag{9.3.6}$$
$x = r\ Sin\theta\ Cos\varphi;\ y = r\ Sin\theta\ Sin\varphi;\ z = r\ Cos\theta$.
$$d\tau = dx\ dy\ dz = (r^2 dr)\ (Sin\theta\ d\theta)\ d\varphi = (r^2 dr)\ d\rho\ d\varphi \tag{9.3.7}$$

$$(\hat{\nabla}^2 + k^2)\, \psi_k(r) = U(r)\, \psi_k(r) \tag{22.2.6}$$

These two situations (1) and (2) envisaged above correspond to the following boundary conditions, which must be stipulated in addition to equation (22.2.6).

22.2.4 BASIC INTEGRAL FORMULA for Scattering

In analogy with the phenomenon of diffraction in optics and acoustics, one may visualize
(i) The incident flux as a superposition of PLANE waves,

$$\Phi_{inc} \equiv \Phi_k = A\, e^{i\vec{k}\cdot\vec{r}} \tag{22.2.7}$$

incident on the scattered, $|A|^2$ = unity.

(ii) The scattering centre itself as a weak source of SPHERICAL waves,

$$\Psi_{Sph} = e^{i\vec{k}\cdot\vec{r}}/r \tag{22.2.8}$$

$\vec{r}$ in the denominator insures that the intensity obeys the **Inverse Square Law.**, and
(iii) As a result of scattering event near the scattering centre, may be determined by the superposition of the incident plane wave and the newly generated spherical wave appears an outgoing scattered wave, $\Psi_{Sc}(r)$.

Using the boundary condition for 22.2.1, *i.e.* for $E \geq 0$, these solutions are normalized to δ-function, *i.e.*

$$\int d\tau\, \Phi_k{}^*(r)\, \Phi_{k'}(r) = \delta(k-k') \tag{22.2.9}$$

Applying the boundary condition for 22.2.2, *viz.*, for $E > 0$, the solutions fall off sufficiently as $r \to \infty$, so as to yield normalizable wave functions.

The standard way of incorporating these boundary conditions into a differential equation (22.2.6) is to convert the equation into an integral equation. This is illustrated in Worked out Example 22.1.

Green's Functions (alternative names are Impulse Response, Influence Function, and System Function) are of great use in many areas of physics.

Worked out Example 22.1

Derive Green's Function for the equation $(\nabla^2 + k^2)\, \psi_k(\vec{r}) = U(r)\, \psi_k(\vec{r})$.

Solution:

$\boxed{\text{Step \# 1}}$ It is known that $\psi_k(\vec{r}) = \Phi_{inc} + \Psi_{Sph}(\vec{r})$; $(\nabla^2 + k^2)\, \psi_k(\vec{r}) = U(r)\, \psi_k(\vec{r})$

$\boxed{\text{Step \# 2}}$ $(\nabla^2 + k^2)[\Phi_{inc} + \Psi_{Sph}(\vec{r})] = U(r)\,[\Phi_{inc} + \Psi_{Sph}(\vec{r})]$

$(\nabla^2 + k^2)[A\, e^{i\vec{k}\cdot\vec{r}}] + (\nabla^2 + k^2)\, \Psi_{Sph}(\vec{r}) = U(r)\,[\Phi_{inc} + \Psi_{Sph}(\vec{r})]$

i.e. $0 + (\nabla^2 + k^2)\, \Psi_{Sph}(\vec{r}) = U(r)\, \Psi_k(\vec{r})]$

$\boxed{\text{Step \# 3}}$ Applying Green's Function Method,

$$\Psi_{Sph}(\vec{r}) = -\left(\frac{1}{4\pi}\right) \int \left\{ e^{i\vec{k}\cdot|\vec{r}-\vec{r}'|} / |\vec{r}-\vec{r}'| \right\} U(r)\, \Psi_k(\vec{r})\, d\tau',$$

$\vec{r}'$ = position vector of the wave scattered from the region.

$\boxed{\text{Step \# 4}}$ The *outward moving wave* (scattered wave) Green's Function

$$\Psi_k^{(+)}(r) \equiv \Psi_{Sc}(\vec{r}) = \Phi_k(\vec{r}) - \left(\frac{1}{4\pi}\right) \int \left\{ e^{i\vec{k}\cdot|\vec{r}-\vec{r}'|} / |\vec{r}-\vec{r}'| \right\} U(r)\, \Psi_k(\vec{r})\, d\tau'$$

This is the <u>Basic Integral Equation of Scattering</u> Theory,

$$\boxed{\Psi_k^{(+)}(\vec{r}) \equiv \Psi_{Sc}(\vec{r}) = \Phi_k(\vec{r}) - (\tfrac{1}{4\pi}) \int \left\{ e^{i\vec{k}\cdot|\vec{r}-\vec{r}'|}/|\vec{r}-\vec{r}'| \right\} U(r)\, \Psi_k(\vec{r})\, d\tau'} \quad (22.2.10)$$

where since $z = r\cos\theta$,
$$\vec{k}\cdot\vec{r} = kr\cos\theta,$$
$$\Phi_k(\vec{r}) = e^{i\vec{k}\cdot\vec{r}} = e^{ikz}. \quad (22.2.11)$$
is the plane wave and satisfies the homogeneous equation
$$(\nabla^2 + k^2)\,\Phi_k(\vec{r}) = \hat{H}_o \Phi_k(\vec{r}). \quad (22.2.12)$$
is an Eigen Equation.

22.2.5 To EXPAND $e^{i\vec{k}\cdot|\vec{r}-\vec{r}'|}$ IN POWERS of $\vec{r}'$

When $r \to \infty$, and $U(r)$ decreases rapidly so that
$\vec{r} \gg \vec{r}'$, everywhere.
Then introduce the unit vector $\hat{n} \parallel \vec{r}$, and
$$|\vec{r}-\vec{r}'| = k(\vec{r}-\vec{r}')\cdot\hat{n} = \cos\pi = -1.$$
$$k|\vec{r}-\vec{r}'| = k\sqrt{\{(\vec{r}-\vec{r}')^*(\vec{r}-\vec{r}')\}} = k\{r - (\hat{n}\cdot\vec{r}') + \tfrac{1}{2}(\tfrac{1}{r})(\hat{n}\wedge\vec{r}')^2 + \cdots\}$$
$$\approx k\{r - (\hat{n}\cdot\vec{r}')\} \approx k\{r - \vec{r}'\cos\theta\}$$
$$\therefore \quad k|\vec{r}-\vec{r}'| \approx \{kr - \vec{k}\cdot\vec{r}'\} \quad (22.2.13)$$

and as $\hat{n} = \tfrac{\vec{r}}{r}$ is a unit vector $\parallel \vec{r}$.
$$\vec{k}' = k\hat{n} = k\tfrac{\vec{r}}{r}. \quad (22.2.13)a$$

Similarly, $|\vec{r}-\vec{r}'|^{-1} = \{r - (\hat{n}\cdot\vec{r}') + \tfrac{1}{2}(\tfrac{1}{r})(\hat{n}\wedge\vec{r}')^2 + \cdots\}^{-1}$.
$$\approx [r - (\hat{n}\cdot\vec{r}')]^{-1}$$

Recalling that $V(r)$, and hence $U(r)$, has a limited range, i.e. falls off faster with r than $1/r$, one recognizes that for $r \to \infty$, it is consistent to neglect in the exponential the quadratic term of
$$\{r - (\hat{n}\cdot\vec{r}') + \tfrac{1}{2}(\tfrac{1}{r})(\hat{n}\wedge\vec{r}')^2 + \cdots\}$$
and simultaneously neglect the r' term in the denominator of the expansion of
$$|\vec{r}-\vec{r}'|^{-1} \approx [r - (\hat{n}\cdot\vec{r}')]^{-1}.$$
The Centre of Mass (CM) System is assumed.
Let the collision be ELASTIC.
i.e., $\quad |\vec{k}| = |\vec{k}'| \quad (22.2.14)$

Therefore, the outward moving part of the wave function $\Psi_k^{(+)}(r)$ has the ASYMPTOTIC FORM, i.e. the true solution of the Schrödinger Equation becomes

$$\boxed{\Psi_k^{(+)}(r) \sim \Phi_k(\vec{r}) - (\tfrac{1}{4\pi})[e^{i\vec{k}\cdot\vec{r}}/r] \int \left\{ e^{-i\vec{k}\cdot\vec{r}'} \right\} U(r')\, \Psi_k^+(\vec{r}')\, d\tau'}$$

plane wave, $e^{i\vec{k}\cdot\vec{r}}$ ↗ and is a ↖ outgoing spherical wave
beam of particles with unit density. with time dependence, $e^{-iEt/\hbar}$

i.e., $\quad \boxed{\Psi_k^{(+)}(r) \equiv \Psi_{Sc}(r) \sim (\tfrac{1}{2\pi})^{3/2}\left\{ e^{i\vec{k}\cdot\vec{r}} + [e^{i\vec{k}\cdot\vec{r}}/r] \right\} f(\vec{k}',\vec{k})} \quad (22.2.16)$

for $r \gg r_0$, or as $kr \to \infty$, i.e. at large distances from the scattering centre, where $f(\vec{k}',\vec{k}) \equiv A(E,\theta,\varphi) = -(\frac{1}{4\pi}) \int \{e^{-i\vec{k}\cdot\vec{r}'}\} U(r') \Psi_k^+(\vec{r}') d\tau'$

$= -(\frac{1}{4\pi}) \langle \Phi_k(\vec{r}) \mid U(r') \mid \Psi_k^+(\vec{r}') \rangle$

$= -(\frac{1}{4\pi}) (\Phi_k(\vec{r}), U(r') \Psi_k^+(\vec{r}'))$ (22.2.17)

is the <u>Scattering Amplitude</u>.

and $\vec{k}' = k \, (\vec{r}'/r)$ (22.2.13)

22.2.6 SCATTERING AMPLITUDE, $f(\vec{k}',\vec{k}) \equiv A(E,\theta,\varphi)$

$f(\vec{k}',\vec{k}) \equiv A(E,\theta,\varphi) = -(\frac{1}{4\pi}) (\Phi_k(\vec{r}), U(r') \Psi_k^+(\vec{r}'))$ (22.2.17)

is known as the Scattering Amplitude. It plays a key role in the Scattering Theory, because $|f(\vec{k}',\vec{k})|^2$ is the <u>Differential Scattering Cross-section</u>, which is related to the potential energy function. The angle factor in f is included to account for some angular asymmetry ne may note the following points: A schematic diagram of a typical scattering experiment is given in Fig.22.1.

(1) $\hbar \vec{k}$ = a vector pointing from the scattering centre to the observation point, $\vec{r}$.

           = the momentum of the particles scattered into the direction, $\vec{k}'$.

(2) $|\vec{k}| = |\vec{k}'|$, i.e. the collision be *elastic*, since the particles were assumed to be structureless.

(3)   The incident beam's cylindrical symmetry cannot be destroyed by scattering off a spherically symmetric target. Consequently, f can only depend on $\vec{k}$ and $\vec{k}\cdot\vec{k}$, or equivalently, on k and θ.

(4)   The incident (plane) wave is considered as a superposition of plane waves. The scattering centre is a source of spherical wave. The scattered wave is a superposition of plane wave and spherical wave giving Asymptotic Form.

(5)   Incident intensity, $|\Phi_{inc}(r)|^2 \frac{\hbar k}{m} = \frac{\hbar k}{m}$, as

$\vec{j} = \frac{\hbar}{2m}[\Psi^*(\nabla\varphi) - (\nabla\varphi)^*\Psi]$.

(6)   Intensity of the outgoing beam scattered at angle θ,

$N(\theta) = J(\theta) = \frac{\hbar}{i2m}[\Psi_{Sc}^*(\nabla\Psi_{Sc}) - (\nabla\Psi_{Sc})^*\Psi_{Sc}]$

$= |f(\vec{k}',\vec{k},\theta)[e^{i\vec{k}\cdot\vec{r}}/\vec{r}]|^2 \frac{\hbar k}{m} r_o - i\frac{\hbar}{mr^3} f(\vec{k}',\vec{k},\theta) \frac{\partial f}{\partial \theta} \theta_o$

$\quad\quad -i\frac{\hbar}{mr^3 Sin\theta} Im \left[f(\vec{k}',\vec{k},\theta) \frac{\partial f}{\partial \varphi} \right] \varphi_o$

where r_o, θ_o, φ_o are unit vectors in the direction of increasing $\vec{r}, \vec{\theta}, \vec{\varphi}$. At very large r, for which

$\Psi_k^{(+)}(r) \equiv \Psi_{Sc}(r) \sim \{[e^{i\vec{k}\cdot\vec{r}}/\vec{r}]\} f(\vec{k}',\vec{k})$

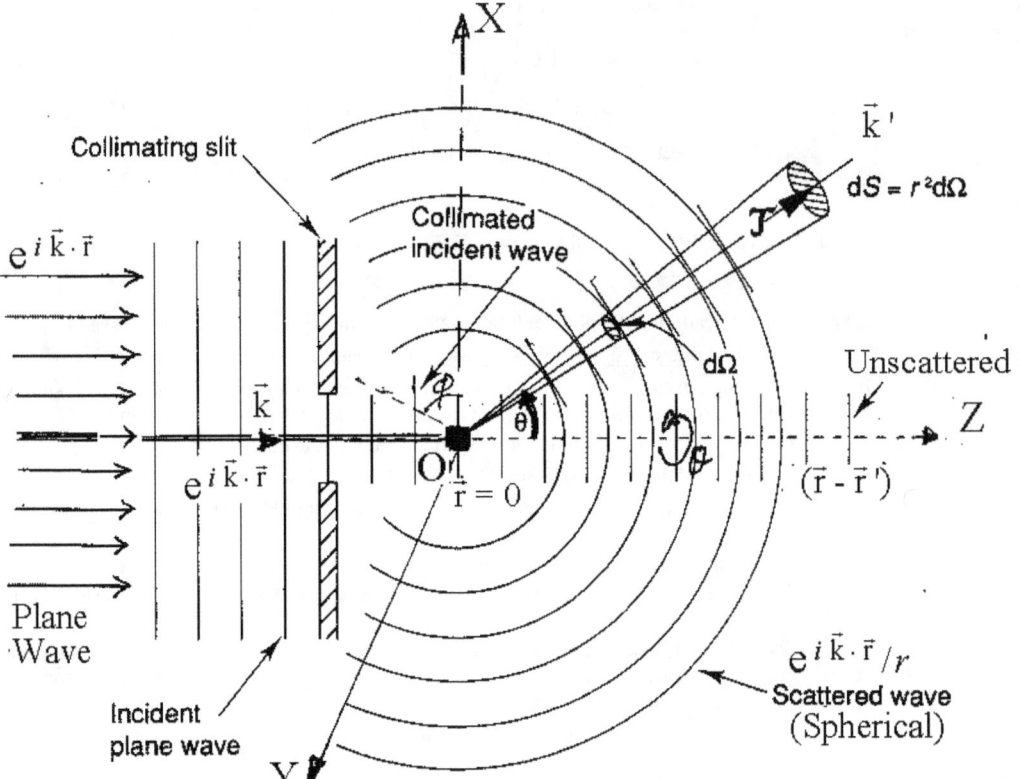

Schematic diagram of scattering of a plane wave incident on a scattering centre at O. θ is the angle of scattering.

Fig. 22.1 Scattering Experiment illustrated

is valid, only the radial component is important.

$$\frac{\text{Scattered Intensity}}{\text{Area, }A} = |f(\hbar \vec{k}', \vec{k})|^2 \frac{\hbar \vec{k}}{m r^2}$$

22.2.7 SCATTERING CROSS SECTION, $\sigma(\vec{k}) = |f(\vec{k}', \vec{k}, \theta)|$
    is directly proportional to the experimentally observed Cross-Section

22.2.7.1 DIFFERENTIAL SCATTERING CROSS-SECTION, $\sigma(\vec{k},\theta) \, d\Omega$

Experimental data in a scattering experiment are commonly expressed in terms of the *Differential Scattering Cross-Section*, defined as $\sigma(\vec{k},\theta) \, d\Omega$. Fig. 22.2 illustrates the various quantities used in getting an expression for $\sigma(\vec{k},\theta) \, d\Omega$

$$\sigma(\vec{k},\theta)\,d\Omega = \frac{\text{\# of Particles scattered into } d\Omega \text{ at angle } \theta}{\text{\# of Particles incident per Unit area}}$$

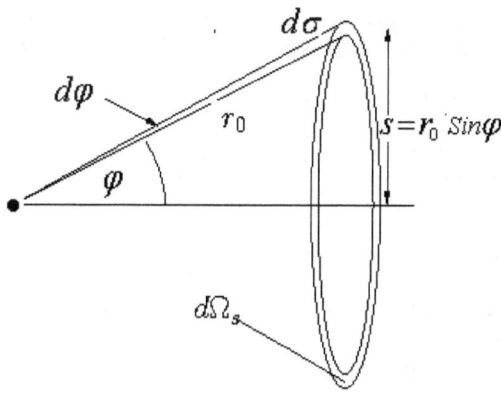

Fig.22.2 The small part of the beam incident falls on the detector, defines the element of cross section, dσ.

$$\sigma(\vec{k},\theta) = \frac{d\sigma(\vec{k},\theta)}{d\omega} = \frac{dN_{Sc}/d\Omega}{dN_{inc}/d\Omega} \quad (22.2.18)$$

where dN_{inc} = # of particles that traverse elemental area dA normal to **k** in the incident packet, dN_{Sc} = # of particles scattered into the cone subtended by the solid angle, $d\Omega$,

$$d\Omega = \sin\theta\, d\theta\, d\varphi. \quad (22.2.19)$$

22.2.7.2 TOTAL SCATTERING CROSS-SECTION

If the potential has spin dependence, *i.e.* azimuthal dependence, the *Total Cross-Section*, σ_T, is by definition,

$$\sigma_T \equiv \sigma(\vec{k}) = \int_{-1}^{+1} d\cos\theta \int_0^{2\pi} d\varphi\, \frac{d\sigma}{d\Omega} \quad (22.2.20)$$

$$= \frac{N(\theta)}{(\hbar k/m)A} = |f(\hbar\vec{k}',\vec{k})|^2\, \frac{\hbar \vec{k}}{m\, r^2}$$

It can be shown that

$$dN(\theta) = |f(\hbar\vec{k}',\vec{k})|^2\, \frac{\hbar\vec{k}}{m}\, \frac{dA}{r^2} = |f(\vec{k}',\vec{k})|^2\, \frac{\hbar\vec{k}}{m}\, d\Omega.$$

Equation (22.2.16) yields

$$\frac{d\sigma}{d\Omega} = |f(\vec{k}',\vec{k})|^2. \quad (22.2.21)$$

$$\sigma_T \equiv \sigma(\vec{k}) = \int \frac{d\sigma}{d\Omega}\, d\Omega = \int |f(\vec{k}',\vec{k})|^2 d\Omega. \quad (22.2.22)$$

i.e. $|f(\vec{k}',\vec{k},\theta)|$ is directly proportional to the experimentally observed Cross-Section.

22.2.8 SCATTERING CROSS SECTIONS in the Centre-of-Mass and Laboratory Frames

22.2.8.1 The L- and the CM-frames Differentiated

The L-system (Laboratory system) is fixed in space, with the target particles (before recoil) at rest. This is the system in which all the angles actually measured in Laboratory in performing an experiment are referred. In analyzing scattering measurements in terms of an interaction potential between beam and target particles, however, the Center-of-Mass system (CM System) is more convenient. As the name implies; in this system the CM of the INCIDENT PARTICLE-TARGET PARTICLE SYSTEM is stationary. In the CM system; the scattering can be described as taking place from a fixed scattering centre (CM), with the two particles (*viz.*, incident and target) remaining collinear with this centre with both moving either toward or away from this centre with equal momentum.

Since scattering experiments are performed in the L-frame, while the theory is conveniently worked out in the CM-frame, it is necessary to be able to transform scattering results from the L- frame to the CM-frame and the vice-versa.

22.2.8.2 To OBTAIN A RELATION BETWEEN the L- and CM-SYSTEMS

Consider *non-relativistic* particles,
i.e., $v \ll c$ in the L-frame. (22.2.26)
m_1 = mass of the projectile particle,
v_1 = velocity of the particle,
and it moves toward a particle of
m_2 = mass of the second particle, i.e. target
$v_2 = 0$, velocity of the second particle,

(*a*) **BEFORE COLLISION**
(i) For motion In the L-frame, the Conservation of Momentum requires
The momentum of the CM = the sum of the momenta of the two particles.
i.e. If the CM is located at position X,
$(m_1 + m_2)X = m_1 x_1 + m_1 x_2$
giving the CM, X as
$(m_1 + m_2)\dfrac{dX}{dt} = m_1 \dfrac{d}{dt} x_1 + m_1 \dfrac{d}{dt} x_2,$
$(m_1 + m_2) V = m_1 v_1 + m_1 v_2,$

The CM velocity, $V = \dfrac{m_1}{(m_1+m_2)} v_1 = \dfrac{\mu}{m_2} v_1.$ (22.2.27)

Total kinetic energy, in the L-system,
$T_L = \tfrac{1}{2} m_1 v_1^2,$ (22.2.28)

(ii) For motion in the CM-frame;
An observer located at the CM and moving with it, experiences momentum,
$\vec{p} = m_1 (v_1 - V) - m_2 V$
$= m_1 v_1 - \dfrac{m_1}{(m_1+m_2)} v_1 - m_2 \dfrac{m_1}{(m_1+m_2)} v_1,$
$= m_1 \dfrac{m_2}{(m_1+m_2)} v_1 - m_2 \dfrac{m_1}{(m_1+m_2)} v_1 = 0.$

Total kinetic energy, in the CM-system,
$T_C = \tfrac{1}{2} m_1 (v_1 - V)^2 + \tfrac{1}{2} m_2 V^2,$ (22.2.29)

On manipulation,
$T_C = \tfrac{1}{2} m_1 v_1^2 + \tfrac{1}{2} (m_1 + m_2) V^2 - (m_1 + m_2) V^2.$

$$T_C = T_L - \tfrac{1}{2}(m_1 + m_2) V^2. \tag{22.2.30}$$

$$\frac{T_C}{T_L} = 1 - \left\{ \left(\tfrac{1}{2}(m_1 + m_2) V^2\right) / \left(\tfrac{1}{2} m_1 v_1^2\right) \right\}$$

$$= 1 - \frac{m_1}{(m_1 + m_2)} = \frac{\mu}{m_1} \tag{22.2.31}$$

(b) **AFTER COLLISION**

In the CM-frame, each particle is scattered at the same angle, θ; thus conservation of kinetic energy requires conservation of velocities and each particle has the same kinetic energy after the collision that it had before the collision. Using single quote symbol to the quantities before the collision,

Total initial momentum = total final momentum.

i.e., $\quad m_1 v_1^2 = m_1 v_2^2$. $\tag{22.2.32}$

Total initial k. E. = Total final k. E.

$$\tfrac{1}{2} m_1 (v_1 - V)^2 + \tfrac{1}{2} m_2 V^2 = \tfrac{1}{2} m_1 v_1'^2 + \tfrac{1}{2} m_2 v_2'^2 \tag{22.2.33}$$

The only solution, which satisfies both these equations, is

$$v_1' = (v_1 - V)$$

and $\quad v_2' = V$. $\tag{22.2.34}$

Thus the velocity of each particle in the L-system is obtained by adding V *vectorially* to the velocity of each in the CM-frame. On the other hand, the speeds of the particles in the CM-frame will be unchanged by elastic collisions, but the observer sees the direction of each changes by the same amount φ.

$$(v_1 - V) \sin\theta_C = v_1' \sin\theta_L,$$

$$V + (v_1 - V) \cos\theta_C = v_1' \cos\theta_L,$$

and $\quad \tan\theta_L = \dfrac{\sin\theta_C}{[V/(v_1-V)] + \cos\theta_C} \tag{22.2.35}$

This relation holds good for both Classical and Quantum Mechanics, since the conservation of energy and momentum are both fundamental to Classical as well as Quantum Mechanics.

$$\tan\theta_L = \frac{\sin\theta_C}{\eta + \cos\theta_C} \tag{22.2.36}$$

$$\boxed{\eta = m_1 / m_2}. \tag{22.2.37}$$

If $\quad m_1 \ll m_2, \ \theta_L \approx \theta_C$

Table 22.1

	L-Frame	CM-frame
Momentum of m_1	$m_1 v_1$	μv_1
Momentum of m_2	0	$-\mu v_1$
Total kinetic energy	$T_L = \tfrac{1}{2} m_1 v_1^2$,	$T_C = T_L - \tfrac{1}{2}(m_1 + m_2) V^2$.

22.2.8.3 RELATION BETWEEN SCATTERING CROSS-SECTIONS in L- and CM-frames

It is convenient to draw an auxiliary triangle whose sides may be labeled in accordance with the quantities in the relations (22.2.36).
Using the fact that the total cross-section is the same in each frame, one may write

$$\iint_{\varphi = 0-2\pi} \sigma_L(\theta,\varphi) \, Sin\theta_L \, d\theta_L \, d\varphi_L$$

$$= \iint_{\varphi = 0-2\pi} \sigma_C(\theta,\varphi) \, Sin\theta_C \, d\theta_C \, d\varphi_C$$

$$2\pi \int \sigma_L(\theta,\varphi) \, Sin\theta_L \, d\theta_L = 2\pi \int \sigma_C(\theta,\varphi) \, Sin\theta_C \, d\theta_C \, .$$

Then $\quad \theta_L = \theta_C \dfrac{Sin\theta_C \, d\theta_C}{Sin\theta_L \, d\theta_L}$ \hfill (22.2.38)a

Differentiating $\quad Tan\theta_L = \dfrac{Sin\theta_C}{\eta + Cos\theta_C}$,

$$Sec^2\theta_L \, d\theta_L = [\dfrac{1 + \eta \, Cos\theta_C}{(\eta + Cos\theta_C)^2}] \, d\theta_C ,$$

Or, $\quad \dfrac{d\theta_C}{d\theta_L} = \dfrac{(\eta + Cos\theta_C)^2 \, Sec^2\theta_L}{(\eta + Cos\theta_C)^2 \, Sin\theta_C}$ \hfill (22.2.38)b

From the triangle mentioned above,

$$\dfrac{Sec^2\theta_L}{Sin\theta_L} = \dfrac{(1 + \eta^2 + 2\eta \, Cos\theta_C)^{1/2}}{(1 + \eta \, Cos\theta_C)} \quad (22.2.38)c$$

Combining equations (22.2.38)a, (22.2.38)b, (22.2.38)c,

$$\boxed{\theta_L = \theta_C \dfrac{(1 + \eta^2 + 2\eta \, Cos\theta_C)^{1/2}}{(1 + \eta \, Cos\theta_C)}} \quad (22.2.39)$$

An isotropic Scattering Cross-Section in the C-frame will thus have a peak in the forward direction in the L-frame by the factor, $(1+\eta)$.

When the target is much more massive than the projectile, i.e.
$m_2 \gg m_1$, $\eta = m_1/m_2$ becomes a very small fraction,
there is no distinction between the angles θ_C and θ_L in the C-frame and L-frame, respectively.

22.3 PARTIAL WAVE TREATMENT TO SCATTERING AT LOW ENERGIES (E < 10 MeV)

There can be no difference fundamentally between the interaction between two bound particles and that of two free particles. A useful quantum mechanical approximation for the case of elastic scattering is to be discussed now. A typical example is the n - p scattering at low energies. The METHOD OF PARTIAL WAVES (i.e. Partial Wave Treatment of Scattering) is used to analyze such a scattering experiment. For this it is necessary to be able to calculate the phase shifts, $\delta_\ell(k)$. Lord Rayleigh originally developed the method for analysis of the scattering of sound waves from obstacles; H. Faxen and J Holtsmark (1927) applied it in quantum mechanical scattering problems.

Consider the simplest problem where V(r) = 0 (i.e. no scattering at all) and that the particles are point (structure less) particles, as follows:
Then the cross-section can be calculated from an energy eigen function, $|\psi_k\rangle$, if

$$(\nabla^2 + k^2)|\psi_k\rangle = 0.$$

22.3.1 Solution of the Schrödinger Equation in the force-free (V(r) = 0) region
Putting $k^2 = 2\mu E/\hbar^2$ \hfill (22.3.1)

where E is the energy of the incident particle, $\Phi_{inc}(r)$ of mass μ, having momentum, $\vec{p} = \hbar \vec{k}$.
(V(r) = 0, is assumed).
The Schrödinger Equation is

$$(\nabla^2 + k^2)\Phi_{inc}(r) = 0 \qquad (22.3.2)$$

This has the solution, which is a plane wave,

$$\Phi_{inc}(r) \equiv \Phi_k(r) = A\, e^{i\vec{k}\cdot\vec{r}} = A\, e^{ikz}. \qquad (22.3.3)$$

It is known for central force potential (from H-atom) that $|\psi_k(r,\theta,\varphi)\rangle$ are simultaneous eigen functions
$\hat{H}_o$, $\hat{L}^2$ and $\hat{L}_z$;

i.e. $[\hat{H}_o, \hat{L}_z] = 0$, $[\hat{L}^2, \hat{L}_z] = 0$, $[\hat{H}_o, \hat{L}^2] = 0$.

The complete solution shows that, as in the case of a bound particle, the angular part of the $|\psi_{n\ell m}\rangle$ is independent of E or V.

$$|\Phi_{n\ell m}\rangle = \tfrac{1}{kr}\sum\sum \{A_{\ell m}(kr) + B_{\ell m}\,\eta_\ell(kr)\}\, P_\ell^m(Cos\theta)\, e^{i|m|\varphi} \qquad (22.3.4)$$

$$\Phi_k(r) = A\, e^{ikz} = \sum \{a_\ell\, \chi_\ell(r)\}\, P_\ell(Cos\theta) \qquad (22.3.5)$$

which is obtained by

(1) Expanding e^{ikz} in Spherical Polar Coordinates,

(2) Exclude the possibility of φ dependence as $\vec{r} \parallel Z$, i.e. only contribution from $m = 0$ occurs,

(3) The angular part (θ) of the solution is Legendre Polynomials, $P_\ell(Cos\theta)$,

(4) $\eta_\ell(kr)$, Neumann Function, does not contribute, as it is not regular at the coordinate centre.

where $\chi_\ell(r) = j_\ell(kr)$,

ℓ^{th} order Spherical Bessel's Function = the regular solution of the Radial part of the Schrödinger Equation.

Thus $\Phi_k(r) = A\, e^{ikz} = \sum a_\ell(k)\, j_\ell(kr)\, P_\ell(Cos\theta) \qquad (22.3.6)$

22.3.2 BAUER'S FORMLA

$$\Phi_k(r) = A\, e^{ikz} = \sum (2\ell+1)(i)^\ell\, j_\ell(kr)\, P_\ell(Cos\theta) \qquad (22.3.7)$$

22.3.3 $\Psi_{Sc}(\vec{r})$ - Introduction of ℓth partial phase shift, $\delta_\ell(k)$

But $\Psi_{Sc}(\vec{r}) = \sum a_\ell(k)\, \Re_{k\ell}(r)\, Y_\ell^m(\theta,\varphi) = \sum a_\ell(k)\, \Re_{k\ell}(r)\, P_\ell(Cos\theta) \qquad (22.3.6)$

where $m = 0$ alone is taken.

The scattering integral equation is (22.2.16), viz.,

$$\Psi_{Sc}(r) \sim (\tfrac{1}{2\pi})^{3/2} \left\{ e^{i\vec{k}\cdot\vec{r}} + [e^{i\vec{k}\cdot\vec{r}}/r] \right\} f(\vec{k}',\vec{k}) \qquad (22.2.16)$$

$$\sim \left\{ \sum (2\ell+1)(i)^\ell\, j_\ell(kr)\, P_\ell(Cos\theta) + [e^{i\vec{k}\cdot\vec{r}}/r] \right\} f(\vec{k}',\vec{k}) \qquad (22.3.12)$$

as $r \gg r_o$.

When V(r) = 0, i.e. for large distances ($r \gg r_o$) from the scattering centre, then

$$\Re_{k\ell}(r) \to j_\ell(kr) \qquad (22.3.13)$$

It is known that

$$J_\ell(kr) = \sqrt{\tfrac{\pi k r}{2}}\, J_{\ell+1}(kr) \qquad (22.3.14)$$

$$= (-1)^\ell\, (\tfrac{r}{k})^\ell\, (\tfrac{1}{r})\, \tfrac{d^\ell}{dr^\ell} J_o(kr) = (-1)^\ell\, (\tfrac{r}{k})^\ell\, (\tfrac{1}{r})\, \tfrac{d^\ell}{dr^\ell}[\tfrac{Sinkr}{kr}] \qquad (22.3.15)$$

Further $J_m(y) = \sqrt{\frac{2}{\pi y}} \cos\{y - \frac{\pi}{4} - \frac{\pi m}{2}\}$ (22.3.16)

As $V(r) \xrightarrow{r \gg r_0} 0$, $J_\ell(kr)$ has Asymptotic Form,

$$J_\ell(kr) \xrightarrow{kr \to \infty} \frac{\sin\{kr - \ell\pi/2\}}{kr}$$ (22.3.17)

But $V(r) \to 0$ faster than the Coulomb Potential, which recommends modification in equation (22.3.17), leading to

$$\Re_{k\ell}(r) \to J_\ell(kr) \xrightarrow{kr \to \infty} \frac{\sin\{kr - \ell\pi/2 + \delta_\ell(k)\}}{kr}$$ (22.3.18)

where $\delta_\ell(k) = \ell^{th}$ partial phase shift.
is a measure of the degree to which $\Re_{k\ell}(r)$ and $J_\ell(kr)$ differ at infinity.
Since V(r), the scattering potential, is responsible for the difference between $\Re_{k\ell}(r)$ and $J_\ell(kr)$, $\delta_\ell(k)$ determines, in part, the form of the Scattering Cross-Section.

22.3.4 The RADIATION CONDITION

Substituting for $\Re_{k\ell}(r)$ and $J_\ell(kr)$, the asymptotic form, the integral equation (22.3.12) becomes

$$\Psi_{Sc}(\vec{r}) = \sum a_\ell(k) P_\ell(\cos\theta) \frac{\sin\{kr - \ell\pi/2 + \delta_\ell(k)\}}{kr}$$

$$\sim \sum (2\ell+1)(i)^\ell P_\ell(\cos\theta) \frac{\sin\{kr - \ell\pi/2\}}{kr} + [e^{i\vec{k}\cdot\vec{r}}/r] f(\vec{k}',\vec{k})$$ (22.3.19)

This is not a plane wave, but is a DISTORTED PLANE WAVE.
or, from equation (22.3.12)

$$\sum \frac{P_\ell(\rho)}{2ik} \left[e^{-i\ell\pi/2} a_\ell e^{i\delta_\ell(k)} - (2\ell+1)(i)^\ell e^{i\vec{k}\cdot\vec{r}}/r \right] = [e^{i\vec{k}\cdot\vec{r}}/r] f(\vec{k}',\vec{k}).$$ (22.3.20)

because $(i)^\ell = e^{i\ell\pi/2}$, as $a^x = e^{x \log a}$
This is possible only if

$$\boxed{a_\ell = (2\ell+1)(i)^\ell e^{i\delta_\ell(k)}}$$ (22.3.21)

This is called the Radiation condition.
Substituting this result in (22.3.20),

$$\sum \frac{P_\ell(\rho)}{2ik} \left[e^{-i\ell\pi/2} (2\ell+1)(i)^\ell [e^{i\delta_\ell(k)} - 1] \{e^{i\vec{k}\cdot\vec{r}}/r\} \right] = [e^{i\vec{k}\cdot\vec{r}}/r] f(\vec{k}',\vec{k})$$ (22.3.22)

$$f(\vec{k}',\vec{k}) = \sum \frac{P_\ell(\rho)}{2ik} \left[e^{-i\ell\pi/2} (2\ell+1)(i)^\ell [e^{i\delta_\ell(k)} - 1] \right]$$

or $\quad f(\vec{k}',\vec{k}) = \frac{1}{k}\sum (2\ell+1) e^{i\delta_\ell(k)} \sin\delta_\ell(k) P_\ell(\rho)$ (22.3.21)

$$\sigma(k,\theta) = |f(\vec{k}',\vec{k})|^2$$

$$= (\frac{1}{k})^2 \sum (2\ell+1)(2\ell'+1) e^{i(\delta_\ell(k) - \delta_{\ell'}(k))} \sin\delta_\ell(k) \sin\delta_{\ell'}(k) P_\ell(\rho) P_{\ell'}(\rho).$$

Case 1: When V(r) = 0, because of the definition of $\delta_\ell(k)$ given after equation (22.3.18), all $\delta_\ell(k) \to 0$;
i.e,. $\quad \sigma(k,\theta) = 0$

Case 2: Because of the **orthogonality relation** of $P_\ell(\rho)$,
For $\ell = \ell'$, $\quad \sigma_T(k) = \int \frac{d\sigma}{d\Omega} d\Omega = 2\pi \int \sigma(\rho) d\rho$.

$$\boxed{\sigma_T(k) = \frac{4\pi}{k^2} \sum (2\ell+1) \operatorname{Sin}\delta_\ell^2(k)} \quad (22.3.22)$$

Thus maximum value for $\sigma_T(k)$ occurs when the scattering phase shift is

$$\pm\tfrac{1}{2}\pi,\ \pm\tfrac{3}{2}\pi,\ \pm\tfrac{5}{2}\pi,\ \ldots\ldots,\pm(n+\tfrac{1}{2})\pi$$

i.e., to this sum each angular momentum contributes at most the ℓth partial wave cross-section, given by

$$\sigma_\ell(k)_{Max} = \frac{4\pi}{k^2}(2\ell+1) = \frac{\lambda^2}{\pi}(2\ell+1) \quad (22.3.23)$$

This value is of the same order of magnitude as the maximum classical scattering cross-section per unit ℏ.

22.3.5 The OPTICAL THEOREM

$$\operatorname{Im} f(\vec{k}',\vec{k}) = (\tfrac{1}{k})\sum (2\ell+1)\ \operatorname{Sin}\delta_\ell(k)\ P_\ell(\rho)$$

and for $\theta = 0$, $\rho = 1$, $P_\ell(\rho) = 1$;

$f(\theta = 0)$ = forward scattering amplitude,

i.e. $\quad = f(0) = (\tfrac{1}{k})\sum (2\ell+1)\ \operatorname{Sin}^2\delta_\ell(k) \quad (22.3.24)$

$$\boxed{\sigma_T(k) = \frac{4\pi}{k}\operatorname{Im} f(0) = \frac{4\pi}{k^2}\sum (2\ell+1)\operatorname{Sin}^2\delta_\ell(k)} \quad (22.3.25)$$

is called the OPTICAL THEOREM in Scattering Theory.

From the notation of the conservation of probability, it is evident that when scattering takes place, the removal of particles from the beam should affect the forward scattering amplitude, f(0). Actually, $\operatorname{Im} f(0) \propto$ # of scattered particles removed from the incident beam.

The Optical Theorem is analogous to $T + R = 1$, in 1-D scattering.

22.3.6 To APPLY PARTIAL WAVE PHASE SHIFT APPROXIMATION:

In order to apply the partial phase shift approximation to a scattering experiment, it is necessary to be able to calculate the phase shifts, $\delta_\ell(k)$. In general,

<u>Case (i)</u> all $\delta_\ell(k) = $ +ve, for an attractive potential.

This means that the phase is advanced and the wave is pulled in toward the scatterer in the interaction region. The peak of the outgoing wave is thus spatially behind the free particle wave. (Fig.).

<u>Case (ii)</u> all $\delta_\ell(k) = $ -ve, for a repulsive potential. Just the opposite to that of case (i) occurs.

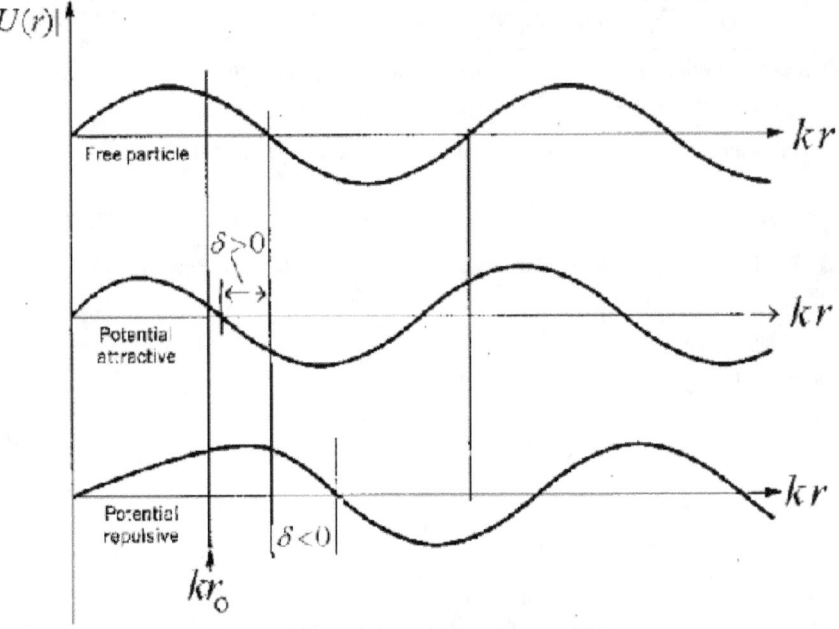

Phase Shifts $\delta_\ell(k)$ and Scattering Potentials

Fig. 22. 3 Positive and negative phase shifts, $\delta_\ell(k)$ produced by attractive and repulsive scattering potentials.

The scattering cross-section, however, does not depend upon the sign of $\delta_\ell(k)$. $\delta_\ell(k)$ are obtained by matching the Radial Solutions of the Radial Schrödinger Equation, for V(r) = 0 to the Radial Functions which include the effect of the scattering potential. However, the partial wave method is frequently applied to situations where only the first one or two phase shifts are required, as the argument given below will show.

a) s-wave scattering:

It was seen that

$$\sigma_\ell(k)_{Max} = \frac{4\pi}{k^2}(2\ell+1) = \frac{\lambda^2}{\pi}(2\ell+1) \qquad (22.3.23)$$

i.e. each partial wave corresponds to a definite value of angular momentum, $\ell = \hbar k$, since there will be essentially no scattering for values $r > r_o$ (range of potential).

$$\ell < \hbar k\, r_o, \qquad (22.3.26)$$

This determines how many values of ℓ will be required to represent scattering. $\ell < k\, r_0$, is required for scattering to occur.

Worked out Example 22.2

Discuss the elastic scattering of a beam of low energy by an impenetrable (or perfect) rigid sphere of radius, r_0.

Solution: Step # 1 Given, low energy particles, k = small, or $k r_0 \ll 1, E \ll \hbar^2/2m$, for a short-range potential (r_0 = small), $kr_0 \ll 1$, so that $\ell = 0$ is the only angular momentum state that need be considered. Such an event is called S-wave scattering, in analogy with the terminology of spectroscopy. Here the distribution of the scattered particles is nearly independent of the scattering angle (*i.e.* isotropic) in the CM-system.

$$\text{Im } f(\vec{k}',\vec{k}) = (\tfrac{1}{k})\sum (2\ell+1)\ \text{Sin}\delta_\ell(k)\ P_\ell(\rho)$$

Step # 2 $\text{Im } f(\vec{k}',\vec{k}) = (\tfrac{1}{k})\ e^{i\ \delta_0(k)}\ \text{Sin}\delta_0(k)$ when all $\delta_\ell(k)$, except $\delta_0(k)$ vanish.

Step # 3 For S-wave scattering, (22.3.25) gives

$$\sigma^{(S)}(k) = \frac{4\pi}{k^2}\sum (2\ell+1)\ \text{Sin}^2\delta_\ell(k) \tag{22.3.27}$$

The de Broglie wavelength of the incident low energy particle is $\lambda \gg r_0$, the detector must be able to resolve $\delta_0(k)$ of the order of $2\pi r_0/\lambda = kr_0$, if anisotropies are to be involved. Hence if both k = small and r_0 = small, $\delta_\ell(k) = 0$ in the detection. Applying the S-wave approximation to the classical collision of two hard spheres, the maximum phase shift is $\delta_0(k) = kr_0$, r_0 = the sum of the radii of the two spheres.

Step # 4 For $kr_0 \ll 1$, $\text{Sin}\delta_0(k) \sim \delta_0(k)$, and $\sigma^{(S)}(k) = \frac{4\pi}{k^2}[\delta_0(k)]^2 \sim 4\pi r_0^2 = 4$ times the

classical value of geometrical cross-section. The additional cross-section, *i.e.* the factor 4, in the quantum mechanical case is due to interference and diffraction effects.

22.3.6 SUMMARY
The various functions associated with elastic scattering
a) The phase shifts, $\delta_\ell(k)$
b) The eigen values of the S-matrix,

$$S_\ell(k) = e^{i\ 2\ \delta_\ell(k)} \tag{22.3.28}$$

c) Known as the "S-functions" for each value of the orbital angular momentum ℓ of the partial waves,
d) Partial wave amplitudes,

$$f_\ell(k) = (1/2ik)[S_\ell(k)-1] = (1/k)\{e^{i\ \delta_\ell(k)}\ \text{Sin}^2\delta_\ell(k)\}, \tag{22.3.29}$$

e) The scattering amplitude,

$$f(\vec{k}',\vec{k}) = (\tfrac{1}{k})\sum (2\ell+1)\ e^{i\ 2\ \delta_\ell(k)}\ \text{Sin}\delta_\ell(k)\ P_\ell(\rho) = \sum (2\ell+1)\ f_\ell(k)\ P_\ell(\rho) \tag{22.3.21}$$

f) The differential cross-section for elastic scattering,

$$d\sigma(k,\theta) = \sigma(k,\theta)\ d\Omega = |\ f(\vec{k}',\vec{k})|^2\ d\Omega \tag{22.2.18}$$

g)   The total cross-section for elastic scattering,

$$\sigma_T(k) = \frac{4\pi}{k}\ \text{Im } f(0) = \frac{4\pi}{k^2}\sum (2\ell+1)\ \text{Sin}^2\delta_\ell(k)\ .$$

22.3.9 RAMSAUER-TOWNSEND EFFECT

It will be seen that when electrons are scattered by a spherical/ square well potential, (R.Q. 22.7)

$$\sigma_0(k) \cong 4\pi r_0^2 \{\ \frac{\tan k_{mc}\ r_0}{k_{mc}\ r_0} - 1\ \}^2. \tag{22.3.39}$$

i.e., $\sigma_0(k) = 0$, when $\tan k_{mc}\ r_0 = k_{mc}\ r_0$.   (22.3.42)

and $\delta_0 = \pi, 2\pi, ..., n\pi$.

Nearly complete transmission of electrons is observed for scattering by rare gas atoms at ~ 0.7 eV. For energies other than this value a scattering $\sigma_o(k) = 0$, is measured. This is the phenomenon of Ramsauer – Townsend Effect.

22.4 ELASTIC SCATTERING AT HIGHER ENERGIES –THE FIRST BORN APPROXIMATION (THE PERTURBATION TREATMENT OF STATONARY SCATTERING)

Max Born in 1926 derived an expression for the scattering amplitude in his quantum theory of scattering. If particles with sufficiently high average momentum, $\hbar k$, (high), *i.e.* higher energy contribution by many partial waves to the scattering are present, and it is therefore preferable to avoid the angular momentum decomposition, one has, in the non-relativistic regime,

$$[E - V(r)] \gg V(r). \tag{22.4.1}$$

i.e., $\quad E \gg 2V(r)$, means

$$V(r)\,a \ll \hbar v.$$

i.e., $\quad Z \ll 137\,v/c \tag{22.4.2}$

where $V(r)$ is a weak potential. A procedure that leads to a very useful approximation both when the $V(r)$ is weak and E = very high, is the *Born Approximation*. In this approximation one considers the scattering process as a transition, just like radiation transitions, with the difference that here one considers the transitions:

Continuum → *Continuum*. (22.4.2)

The differential cross-section,

$$d\sigma(k,\theta) = \sigma(k,\theta)\,d\Omega = |\,f(\vec{k}\,',\vec{k}\,)|^2\,d\Omega \tag{22.2.18}$$

where $f(\vec{k}\,',\vec{k}\,)$, the scattering amplitude, appears as the coefficient of the outgoing wave,

$$\Psi_k^{(+)}(r) \sim \Phi_k(\vec{r}) - (\tfrac{1}{4\pi})\,[e^{i\vec{k}\cdot\vec{r}}/r\,]\int\left\{e^{-i\vec{k}\cdot\vec{r}\,'}\right\}U(r')\,\Psi_k^+(\vec{r}\,')\,d\tau' \tag{22.2.15}$$

in the asymptotic solution of the Schrodinger equation

$$(\nabla^2 + k^2)\,\Psi_k(r) = U(r)\,\Psi_k(r) \tag{22.2.6}$$

22.4.1 ZEROTH APPROXIMATION, $\Psi_k^{(0)}(r)$

In the zeroth approximation, one neglects the integral term (*i.e.* the 2nd term in (22.2.15)), and set

$$\Psi_k^{(0)}(\vec{r}) = [e^{ikz} - 0] = e^{ikz} \tag{22.4.3}$$

the inhomogeneity term in (22.2.15), as $\vec{k} \parallel Z$.

22.4.2 FIRST BORN APPROXIMATION, $\Psi_k^{(0)}(r)$

One uses the zeroth approximation, $\Psi_k^{(0)}(r)$, given by (22.4.3), on the right- hand side of (22.2.15) to compute the first approximation, $\Psi_k^{(1)}(r)$..

$$\Psi_k^{(1)}(r) \sim (\tfrac{1}{4\pi})\,[e^{ikz} - \tfrac{1}{4\pi}\,]\,[e^{i\vec{k}\cdot\vec{r}}/r\,]\int\left\{e^{-i\vec{k}\cdot\vec{r}\,'}\right\}U(r')\,\Psi_k^{(0)}(r')\,d\tau' \tag{22.4.4}$$

22.4.3 nth APPROXIMATION, $\Psi_k^{(n)}(r)$

Like-wise one may insert $\Psi_k^{(1)}(r)$ on the right-hand side of (22.2.15) to compute $\Psi_k^{(2)}(r)$, etc.

$$\Psi_k^{(n)}(r) \sim (\tfrac{1}{4\pi})\,[e^{ikz} - \tfrac{1}{4\pi}\,]\,[e^{i\vec{k}\cdot\vec{r}}/r\,]\int\left\{e^{-i\vec{k}\cdot\vec{r}\,'}\right\}U(r')\,\Psi_k^{(n-1)}(r')\,d\tau' \tag{22.4.5}$$

22.4.4 ELASTIC SCATTERING AMPLITUDE IN FIRST BORN APPROXN

Since $\vec{k}'$ (final momentum) depends only on the direction but not on the magnitude of $\vec{r}$, so that the integral in $\psi_k^{(1)}(\mathbf{r})$ given in (22.4.4) is indeed an angular amplitude. $[e^{i\vec{k}\cdot\vec{r}}/r]$ is a spherical outgoing wave. Thus equation (22.4.4) is equivalent to (22.2.16), viz.

$$\Psi_k^{(+)}(r) \equiv \Psi_{Sc}(r) \sim (\tfrac{1}{2\pi})^{3/2} \left\{ e^{i\vec{k}\cdot\vec{r}} + [e^{i\vec{k}\cdot\vec{r}}/\vec{r}] \right\} f(\vec{k}',\vec{k}) \qquad (22.2.16)$$

Thus $f(\vec{k}',\vec{k}) \equiv A(E,\theta,\varphi) = -(\tfrac{1}{4\pi}) \int \left\{ e^{-i\vec{k}\cdot\vec{r}'} \right\} U(r') \Psi_k^{(0)}(r) d\tau'$

$$= -(\tfrac{1}{4\pi}) \langle \Phi_{k'} | U(r') | \Psi_k^{(0)} \rangle \qquad (22.4.6)$$

$$f(\vec{k}',\vec{k}) = -(\tfrac{1}{4\pi}) \left(\Phi_{k'}, U(r') \Psi_k^+ \right) \qquad (22.2.17)$$

Thus the scattering potential V(r) is considered as a perturbation on $|\Phi_k\rangle$ state, the incident plane wave, and $|\Phi_k\rangle$ state (particle which experienced the V(r) but asymptotically behaves like a spherical wave), which are both free equal energy eigen function of the V(r) = 0 potential problem. That is, the scattering process is a transition from the initial and final free particle states $|\Phi_k\rangle$ and $|\Psi_k^+\rangle$. States with equal energy but different k-directions of momentum, have the $\sigma(k,\theta) = d\sigma(k,\theta)/d\Omega$ is related to the transition probability.

Substituting $\quad \left| \Psi_k^{(0)} \right\rangle = e^{ikz|} \qquad (22.4.3)$

in $f(\vec{k}',\vec{k})$ in (22.4.6),

$$f(\vec{k}',\vec{k}) = -(\tfrac{1}{4\pi}) \langle \Phi_{k'} | U(r') | \Psi_k^{(0)} \rangle$$

$$= -(\tfrac{1}{4\pi}) \int e^{-i\vec{k}\cdot\vec{r}'} U(r') \Psi_k^{(0)}(r) d\tau'$$

$$= -(\tfrac{1}{4\pi}) \int e^{-i\vec{k}\cdot\vec{r}'} U(r') e^{-i\vec{k}\cdot\vec{r}'} d\tau'$$

$$f(\vec{k}',\vec{k}) = (\tfrac{m}{2\pi\hbar^2}) \int V(r) e^{-i(\vec{k}'-\vec{k})\cdot\vec{r}} d\tau' \qquad (22.4.7)$$

Equation (22.4.7) is known as the *elastic Scattering Amplitude in the First Born Approximation*. Hence, contributions to the scattering arise only where $V(r) \neq$ small, regardless of the limits of integration on r.

22.4.5 ALTERNATIVE FORM OF THE EXPRESSION for $f(\vec{k}',\vec{k})$

Alternately, introducing

$\vec{K} = (\vec{k} - \vec{k}') =$ momentum transfer vector $\qquad (22.4.8)$

$\vec{K}^2 = |(\vec{k}-\vec{k}')|^2 = \vec{k}^2 + \vec{k}'^2 - 2\vec{k}\cdot\vec{k}' = k^2 + k'^2 - 2kk'\cos\theta$

where $\vec{k} \| Z$ and $|\vec{k}\cdot| \equiv k = |\vec{k}'| \equiv k'$, $\cos\theta = \vec{k}\cdot\vec{k}'/\vec{k}'^2$,

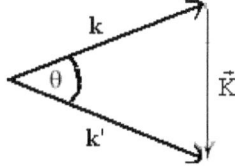

for <u>elastic scattering</u>, and the concerned triangle is isosceles, as shown..

$$\vec{K}^2 = 2k^2(1-Cos\theta) = (2k\,Sin\tfrac{1}{2}\theta)^2.\qquad(22.4.9)$$

Equation (22.4.7) becomes

$$f(\vec{k}',\vec{k}) = -\left(\tfrac{1}{4\pi}\right)\int e^{-i\vec{K}\cdot\vec{r}'}\,U(r')\,d\tau' \qquad(22.4.10)$$
$$\propto \text{F.T. of } V(\mathbf{r}),\text{ in the k-space.}$$

22.4.5 VALIDITY CONDITION for BORN APPROXIMATION:

It is seen that the Born Approximation is likely to be valid for weak potentials and high incident energies, *i.e.* $V(r)\,a \ll h\,v$. This is because Time-Dependent Perturbation Theory has been used in this scattering, *i.e.* scattering process is regarded as a time –dependent transition for an initial plane wave state $|\,i\,\rangle \equiv |\,\varphi_i\,\rangle$ to a final plane wave state $|\,f\,\rangle \equiv |\,\varphi_f\,\rangle$. In the CM-frame it was effectively a one-particle problem, and the particle makes transition from $|\,\varphi_k\,\rangle \to |\,\varphi_k^{(+)}\,\rangle$. This presupposes that the amplitude of the scattering wave is small compared to that of the incident wave and that $V(r) \ll E$, the energy of the incident particles. Thus E must be large enough so that they may be still represented by plane waves after the scattering occurs. The Born Approximation affords a rapid estimate of scattering cross-section and is accurate for reasonably high energies, In comparison with the interaction potential. Because of its simplicity it has enjoyed enormous popularity among atomic and nuclear physicists.

22.4.5 STANDARD FORM OF BORN APPROXIMATION – Born Approximation applied to a Central potential:
This aspect is understood through the Example below:

Worked out Example 22.3
Consider scattering of high energy particles by a central potential. Obtain an expression for the scattering amplitude in the standard form of the Born Approximation.

Solution: Step # 1 It is known that Central potential $V(\vec{r})$ means $V(\vec{r})$ depends only on the distance $|\vec{r}|$ and not on the direction of $\vec{r}$. $V(\vec{r}) \equiv V(r)$. The elastic Scattering Amplitude,(22.4.10)

$$f(\vec{k}',\vec{k}) = -\left(\tfrac{1}{4\pi}\right)\int e^{-i\vec{K}\cdot\vec{r}'}\,U(r')\,d\tau\,;\ U(r) = \tfrac{2\mu}{\hbar^2}V(r);\ \vec{K}^2 = 2k^2(1-Cos\theta) = (2k\,Sin\tfrac{1}{2}\theta)^2\,;$$

$$V(r') \propto \tfrac{1}{r'}$$

Step # 2 Define an auxiliary set of spherical polar coordinates with the polar axis $\vec{k} \parallel Z$, for the purpose of integration.

$$f(\vec{k}',\vec{k}) = -\left(\tfrac{1}{4\pi}\right)\int e^{-i\vec{K}\cdot\vec{r}'}\,U(r')\,d\tau$$

i.e., $\quad V(K) = \int V(r')\,e^{-i\vec{K}\cdot\vec{r}'}\,d\tau' = -\left(\tfrac{1}{K^2}\right)\int V(r')\,\nabla^2 e^{i\vec{K}\cdot\vec{r}'}\,d\tau'$

$$= -\left(\tfrac{1}{K^2}\right)\int e^{-i\vec{K}\cdot\vec{r}'}\,\nabla^2 V(r')\,d\tau'$$

$$= \int_0^\infty V(r')\,r^2\,dr' \cdot \int_0^\pi e^{iK\,r'\,Cos\theta}\,Sin\,\theta\,d\theta \cdot \int_0^{2\pi}d\varphi$$

$$= 2\pi\int_0^\infty V(r')\,r^2\,dr' \cdot \int_0^\pi e^{iK\,r'\,Cos\theta'}\,Sin\,\theta'\,d\theta'$$

Step # 3 $\int_0^\pi Cos\,Kr'\,Cos\theta\,Sin^2\theta\,d\theta = K\,r'\,CosKr'\,\{Cos\theta\}_0^\pi$

Step # 4 $V(K) = 2 \text{ Cos}Kr' \int_0^\pi e^{iKr'\cos\theta'} \sin\theta\, d\theta$

$= \left[-\frac{1}{Kr'} e^{iKr'\cos\theta'}\right]_0^\pi = \frac{2\cos Kr'}{Kr'} \int_0^\pi \sin Kr\, dr$

$= \underset{Kr \to 0}{Lt} \int_0^\pi \sin Kr\, dr\, e^{-i\vec{K}\cdot\vec{r}'} = (\frac{1}{K})$

Step # 5 The potential being Central $V(r') \propto \frac{1}{r'}$

$V(K) = (\frac{4\pi}{K}) \int_0^\infty V(r') \frac{\sin Kr'}{Kr'} r'^2\, dr' = -(\frac{2m}{\hbar^2 K}) \int_0^\infty V(r') \frac{\sin Kr'}{Kr'} r'^2\, dr'$

$\boxed{f(\vec{k}',\vec{k}) = -(\frac{2m}{\hbar^2 K}) \int_0^\infty V(r') \frac{\sin Kr'}{Kr'} r'^2\, dr'}$ \hfill (22.4.13)

This equation is referred to as the *Scattering Amplitude in the Standard Form of the Born Approximation*.

Step # 6 $\sigma(k,\theta) = d\sigma(k,\theta)/d\Omega = |f(\vec{k}',\vec{k})|^2$

$= |\int_0^\infty U(r') \frac{\sin Kr'}{Kr'} r'^2\, dr'|^2$

REVIEW QUESTIONS

R.Q. 22.1 In a scattering experiment with $A\, e^{i\vec{k}\cdot\vec{r}}$ as incident beam find an expression for J of the scattered beam.

Answer: $J = \frac{\hbar \vec{k}}{m} + i\, \frac{\hbar \vec{k}}{m} |f(\vec{k}',\vec{k})|^2 \frac{1}{r^2}$).

R.Q. 22.2 Discuss the elastic scattering of a beam of low energy by an impenetrable (or perfect) rigid sphere of radius, r_0. (Answer: $\sigma^{(S)}(k) = \frac{4\pi}{k^2}[\delta_0(k)]^2 \sim 4\pi r_0^2 = 4$ times the classical value of geometrical cross-section. The additional cross-section, i.e. the factor 4, in the quantum mechanical case is due to interference and diffraction effects.= 4 times the classical value of geometrical cross-section. The additional cross-section, i.e. the factor 4, in the quantum mechanical case is due to interference and diffraction effects.)

R.Q. 22.3 Find the scattering cross-section of slow particles in the repulsive field, $V(r) = V_0\, (r < r_0)$; $V(r) = 0\, (r > r_0)$. (Answer: $\sigma^{(S)}(k) = \frac{4\pi}{k^2}[\delta_0(k)]^2 \sim 4\pi r_0^2 = 4$ times the classical value of geometrical cross-section.)

R.Q. 22.4 Show that the relation that holds good between the elastic Scattering Amplitude $f(\theta = 0)$ and the total Scattering Cross-Section, $\sigma_T(k,\theta)$ is $\sigma_T(k) = \frac{4\pi}{k} \text{Im}\, f(0)$.

R.Q. 22.5 Obtain an expression for Optical theorem in Scattering Theory.
(Answer: $\sigma_T(k,\theta) = \frac{4\pi}{k} \text{Im}\, f(0) = \frac{4\pi}{k^2} \sum (2\ell+1) \sin^2 \delta_\ell(k)$)

R.Q. 22.6 Find the Total Cross-Section for scattering of low energy particles by the potential energy function (attractive square well potential), $V(r) = -V_0$ $(r < r_0)$, $V(r) = 0$ $(r > r_0)$. (Answer: $\sigma_0(k) \cong 4\pi r_0^2 \{k_m^4 r_0^4 / 9\}$, $k_m^2 = 2\mu(E - V_0)/\hbar^2$)

R.Q. 22.7 Find the total cross-section for scattering of low energy particles by the following spherical potential well,
$V(r) = -V_0$ $(0 < r < r_0)$; $V(r) = 0$ $(r > r_0)$
(Answer: $\sigma_0(k) \cong 4\pi r_0^2 \{k_m^4 r_0^4 / 9\}$, $k_m^2 = 2\mu(E - V_0)/\hbar^2$)

R.Q. 22.8 Apply the Born approximation to a beam of electrons is scattered elastically by Yukawa potential (i.e., a screened Coulomb Potential, or a potential with spherical symmetry),
$V(r) = (-Ze^2/r) e^{-r/a} = (V_0/\alpha r) e^{-\alpha r}$, where a is screening radius. Find the Differential Scattering Cross-section. (Answer: $f(\vec{k}', \vec{k}) = (\frac{2mZe^2}{\hbar^2 K^2})[1 - \frac{1}{1+(Ka)^2}]$; $= (\frac{-2mV_0}{\alpha \hbar^2})[1 - \frac{1}{K^2 + \alpha^2}]$

$= (\frac{-2mV_0}{\alpha \hbar^2})[1 - \frac{1}{4k^2 Sin^2 \frac{1}{2}\theta + \alpha^2}]$; $\sigma(k,\theta) = [16\pi \ a^2] \left\{ \frac{m Z e^2 a/\hbar^2}{1 + (Ka)^2} \right\}^2$

R.Q. 22.9 Apply Born Approximation to scattering of particles of mass m by the Coulomb Potential between two charges q_1 and q_2 and show that the result is identical to the classical Rutherford scattering. Answer; $V(r) = (V_0/\alpha r) e^{-\alpha r} = (-Z_1 Z_2 e^2/r) e^{-\alpha r} = (q_1 q_2/r) e^{-\alpha r}$;
$\sigma(k, \theta) = (Z_1^2 Z_2^2 e^4 / 16 E^2 Sin^4 \frac{1}{2}\theta)$
This is the result of both the classical Rutherford scattering cross-section and the exact quantum mechanical evaluation of the case for pure Coulomb Potential).

R.Q. 22.10 Find the Differential and Total Cross-Sections for the elastic scattering of fast electrons by a Hydrogen atom, by using the Born Approximation. Given: $|100\rangle = 2(a_0)^{-3/2} e^{-r/a_0}$ (Answer: .
$\sigma_T(k) = (\pi \ a_0^3/3)[7k^4 a_0^4 + 18k^2 a_0^2 + 12]/(k^2 a_0^2 + 1)^3 \xrightarrow{k a_0 \gg 1} [7\pi/3k^2]$)

R.Q. 22.11 Using the Born Approximation, find the Differential and Total Cross-Sections for the elastic scattering of fast electrons by a Helium atom. Given: $|100\rangle = (2/\pi b^3) e^{-2r/b}$, $b = (16/27)a_0$
(Answer: $\sigma_T(k) = \{(\pi \ b^3/3)[7k^4 b^4 + 18k^2 b^2 + 12]/(k^2 b^2 + 1)^3\}Z^2$
$\xrightarrow{k a_0 \gg 1} [28\pi/3k^2]$)

R.Q. 22.12 Calculate the Differential and Total Cross-Sections for the elastic scattering of fast electrons by an attractive Gaussian potential, $V(r) = -V_0 \ e^{-r^2/r_0^2}$. Apply the Born Approximation.
(Answer: $\sigma_T(k) = \{(\pi/2k^2)[m r_0^2 \ V_0/\hbar^2]^2 \{1 - e^{-2k^2 r_0^2}\}$).

R.Q. 22.13 Find $\sigma_k(\theta)$ and σ_k for the scattering of a particle from a perfectly rigid sphere (an infinitely repulsive potential) of radius a. Choose the energy of the particle such that $ka \ll 1$ (Answer: $\sigma_k(\theta) = |f_k(\theta)|^2 = (1/k^2) Sin^2 \delta_0$; $\sigma_k = (4\pi/k^2) Sin^2 \delta_0$ $\sigma_k = (4\pi a^2)$.

R.Q. 22.14 A slow particle is scattered by a spherical potential well of the form $V(r) = -V_0$, for $r < a$, $V(r) = 0$, for $r > a$.
(a) Write down the radial wave equation for this potential and boundary conditions that apply at $r = 0$, $r = a$, and $r = \infty$.
(b) Assume that the de Broglie wavelength exceeds the dimension of the well, so that s-wave scattering dominates and write solutions both inside and outside the $r = a$ sphere. Using the continuity conditions at $r = a$, calculate the phase shift that occurs at this boundary. (Answers: a) $[\partial^2/\partial r^2 + k^2 - U(r) - \ell(\ell+1)/r^2]\Re(r) = 0$; $V(r) = \hbar^2 U(r)/2m$, $E = \hbar^2 k^2/2m$;
$\delta_o = n\pi + \tan^{-1}(C) - ka$; $\psi_{k_0}(r) = C_1 Sin(kr + \delta_o)$)

R.Q. 22.15 What must $V_0 a^2$ be for a 3-dimensional square well potential in order that the scattering cross section be zero at zero bombarding energy (Ramsauer-Townsend effect)? (Answer:
$\sigma_k(\theta) = |f(k(\theta))|^2 = (1/k^2) Sin^2 \delta_o$ whwn $\delta_o = \pi$, $\sigma_k(\theta) = \sigma_k = 0$;
$ka \to 0$ means $\sqrt{U_0}a = \tan^{-1}\sqrt{U_0}a$, $n=1, b =1.352$, $\tan^{-1}\sqrt{U_0}a = 4.497$;
$V_0 a^2 = \hbar^2 U_0 a^2 / 2m = 20.2\hbar^2 / 2m$)

R.Q. 22.16 Determine the differential scattering cross section $\sigma_k(\theta)$ in units of cm^2/Sr for a particle of mass $m = 9.1 \times 10^{-31} kg$ incident on a spherically symmetric potential $V(r) = 0$, $0 < r < a$, $V(r) = V_0$, $a < r < b$, $V(r) = 0$, $r > b$, with $a = 0.05nm$ and $b = 0.1nm$. Let $E = 1eV$ and $V_0 = 0.8eV$.
(Answer: $\delta_o = -2.8 \times 10^{-2}$, $\sigma_k(\theta) = 3 \times 10^{-19} cm^2 / Sr$

R.Q. 22.17 An electron of incident momentum $\vec{k}_i$ is scattered elastically by the electric field of an atom of atomic number Z. The potential due to the nucleus is of the form $V_0(r) = -Ze^2/r$. This potential is screened by the atomic electron cloud. As a result, the **total** potential energy of the incident electron is $V(r) = (-Ze^2/r) e^{-r/a}$, where a is the radius of the atom. Let $\vec{k}_f$ be the final momentum of the electron and $\vec{K} = \vec{k}_i - \vec{k}_f$ be the momentum transfer.
(a) Calculate the differential cross section, $d\sigma(k,\theta)/d\Omega$, in the Born approximation for the scattering of the electron by theatom (using the screened potential V(r).)(b) Now calculate the differential cross section, $d\sigma(k,\theta)/d\Omega$, in the Born approximation for the scattering of the electron by the nucleus only (using the unscreened potential $V_0(r)$.)
(c) Plot the ratio of the two cross sections as a function of $x = a|\vec{K}|$ and briefly discuss the limits $x \to 0$ and $x \to \infty$. (Answers: a) $\sigma(k,\theta,\varphi) = (4\mu^2 Z^2 e^4 / \hbar^4)[1/(a^{-2} + 4k^2 Sin^2 \frac{1}{2}\theta)]^2$, b)
$\sigma(k,\theta,\varphi) = (4\mu^2 Z^2 e^4 / \hbar^4)/[16k^4 Sin^4 \frac{1}{2}\theta)]$

R.Q. 22.18 Evaluate, in the Born Approximation, the differential cross section for the scattering of a particle of mass m by a delta-function potential $V(r) = B \delta(r)$. Comment on the angular and velocity dependence. Find the total cross section.(Answers: $\sigma_T(k) = B\mu^2/(\pi\hbar^4)$)

R.Q. 22.19 What is the Born approximation? Write down the Born approximation scattering amplitude for the case of a Central potential. Deduce the criterion for the validity of the approximation.

R.Q. 22.20 Show that differential scattering cross section has the dimension of area.

R.Q. 22.21 Show that the scttering amplitude in the case of scattering by Square-well potential of depth V_0 and range a is given by $f(\theta) = \frac{2m}{\hbar^2} \frac{V_0}{k^3} [Sinka - ka\, Coska]$.

*&^*&^*&^*&^*&^*&^*

Chapter 23

IDENTICAL PARTICLES (MANY- BODY SYSTEMS)

Chapter 23

IDENTICAL PARTICLES
(MANY- BODY SYSTEMS)

"People are just about as happy as they make up their minds to be".   Abraham Lincoln.
"If we did all the things that we are capable of doing, we would literally astound ourselves." – Thomas A.Edison.
"We cannot solve the problem that we have created with same thinking that created them" - A. Einstein
"Whatever you see as duality is unreal"   Adi Shankara
"The plurality that we perceive is only an appearance. it is not real " -   Erwin Schrödinger.

23.1   INTRODUCTION

It was seen already in earlier Chapters how to write down the wave functions and wave equations for systems consisting of two or more particles. What happens is that the wave function contains an x, y, and z spatial component corresponding to each individual particle. Even in the case of a simple system like the H-atom, the wave function of six spatial variables and t. The identity of particles plays a more important role in Quantum Mechanics than in Classical Physics. It is always possible in Classical Physics to minutely alter each particle of a system so as to make it identifiable without affecting appreciably its mechanical behavior. This is, however, quite impossible in the Quantum World with particles such as electrons. Such particles do not have enough degrees of freedom so that each particle can be marked differently. The trajectory of each particle cannot be followed closely in a quantum mechanical system. Such observations disturb the system in an uncontrollable manner.

By IDENTICAL PARTICLES it is meant that the particles possess exactly the same properties, and they cannot be distinguished from one another by physical measurements of their properties. When the wave functions of two or more identical particles overlap, then, it is not possible in principle to keep track of the trajectories. The situation is also different from the two-particle system discussed earlier; for there is only one matter field, say that of photon or that of electron. And the existence of several different quanta of the field, photons or electrons, is best described by mentioning that the entire field is in one of its higher states of quantum excitation. The construction of the wave function of a system of identical particles is the subject matter of this Chapter, using the new features, viz. the nature of symmetry requirements.

The problem of a MANY-ELECTRON system (an atom, a molecule or a solid) is the problem of taking proper account of the interaction of the electrons. Calculation of the properties of systems containing many electrons becomes almost impossibly complicated when one attempts to include the interactions of the electrons. The difficulty can be avoided either by neglecting this interaction altogether, or by considering only its average effect (i.e. independent electron approximation). It is of concern to us to calculate the total energy of a many-electron system, or more specifically, that part of it known as the CORRELATION ENERGY. The important techniques include Perturbation Theory and the use of Green's Functions

23.2 SYMMETRY AND ANTISYMMETRY

23.2.1 A SYSTEM WITH MANY-IDENTITICAL PARTICLES AND HIGH DEGENERACY

Consider a system consisting of n identical particles. The wave function of the system is given by

$$|\Psi(1,2,3,4,\ldots,n)\rangle = \langle 1,2,3,4,\ldots,n|\Psi\rangle \tag{23.2.1}$$

where 1,2,3, , n are symbols used to denote the eigen values of a complete set of observables of the particles. For instance, for a set of particles without spin, the eigen values may be $1 = x'_1$, $2 = x'_2$, and so on, or they may be $1 = p'_1$, $2 = p'_2$, *etc.* For a set of particles of spin-$\frac{1}{2}$, they may be $1 = (x'_1, \sigma_{z1})$, $2 = (x'_2, \sigma_{z2})$, *etc.*, where $\sigma_{z1} = \pm 1 \pm 1$ is the eigen value of the Pauli matrix σ_z for the particle. If the particle has additional quantum numbers, such as is 0-spin, these are to be included also. To simplify the discussion let 1, 2, 3, 4,..., n refer to the coordinates of the particles. Then

$$|\Psi(1,2,3,4,\ldots,n)\rangle = \text{probability amplitude}$$

for finding particle 1 *at* x'_1, particle 2 *at* x'_2, and so on, If one interchanges two particles, say particles 1 and 2, to obtain $|\Psi(2,1,3,4,\ldots,n)\rangle$ = the probability amplitude for finding particle 1 *at* x'_2, , and so on.

If the particles are really identical, there are no interactions that can distinguish them. It follows that any observable operator must treat all identical particles in exactly the same way. There it must be a symmetrical function of the particle operators (coordinates, momenta, spins, etc.). For instance, the Hamiltonian, $\hat{H}$ of a system of Z spin less electrons interacting with a nucleus of charge +Ze and with each other through Coulomb forces is

$$\hat{H} = \sum_{i=1}^{Z}[\tfrac{1}{2m}p_i^2 - \tfrac{Ze^2}{r_i}] + \sum_i^Z\sum_j^Z[\tfrac{e^2}{|r_i-r_{ij}|}] \tag{23.2.2}$$

It is a symmetric function of the operators $\{p_i, x_i\}$ of the particles.

It is convenient to introduce the PERMUTATION OPERATOR, $\hat{P}_{ij}$, which interchange the particles *i* and *j*. Thus

$$P_{ij}|\Psi(1,2,3,i,\ldots,j,\ldots,n)\rangle = |\Psi(1,2,3,j,\ldots,i,\ldots,n)\rangle \tag{23.2.3}$$

Because the Hamiltonian is symmetric in the particles

$$\hat{P}_{ij}\hat{H}|\Psi(1,2,3,i,\ldots,j,\ldots,n)\rangle\ |\Psi(1,2,3,j,\ldots,i,\ldots,n)\rangle$$

$$= \hat{H}|\Psi(1,2,3,i,\ldots,j,\ldots,n)\rangle\ \hat{P}_{ij}|\Psi(1,2,3,j,\ldots,i,\ldots,n)\rangle \tag{23.2.4}$$

Therefore $\qquad \hat{P}_{ij}\hat{H} = \hat{H},\hat{P}_{ij},$ (23.2.5)

i.e., $[\hat{H},\hat{P}_{ij}] = 0$.

$\hat{P}_{ij}$ commutes with $\hat{H}$ means $\hat{P}_{ij}$ is a CONSTANT OF THE MOTION. Further, the eigen functions of $\hat{P}_{ij}$ are simultaneously eigen functions of $\hat{H}$. However, not all of the $\hat{P}_{ij}$ s commute

with one another. For example consider two such operators, $\hat{P}_{12}$ and $\hat{P}_{13}$, such that
$|\Psi(1,2,3,i,.....,j,.....,n)\rangle$.

$$\hat{P}_{12}\hat{P}_{13}|\Psi(1,2,3)\rangle = |\Psi(3,1,2)\rangle$$
$$\hat{P}_{13}\hat{P}_{12}|\Psi(1,2,3)\rangle = |\Psi(2,3,1)\rangle,$$

So $\hat{P}_{12}\hat{P}_{13} \neq \hat{P}_{13}\hat{P}_{12}$

It follows that, in general, not all of the $\hat{P}_{ij}$, can be diagonalized simultaneously.

Let that $\hat{P}_{ij}|\Psi(1,2,3,j,.....,i,.....,n)\rangle$ is an eigen function of $\hat{H}$ with eigen value E. Because $\hat{P}_{ij}$ commutes with $\hat{H}$, $\hat{P}_{ij}|\Psi\rangle$ must be another eigen function of $\hat{H}$ with the same eigen value E. This means a set of degenerate states with the same eigen value E can be generated from $|\Psi(1,2,3,i,.....,j,.....,n)\rangle$ by operating with all the permutation operators. It would appear that a many-identical particle system would be highly degenerate.
However, in practice, not all of the degenerate states are physically permitted.

SYMMETRIC and ANTI-SYMMETRIC STATES:
There are two states having rather special properties:
(1) A state of a system with many identical particles is said to be COMPLETELY SYMMETRIC, if

$$\hat{P}_{ij}|\Psi^S(1,2,3,i,.....,j,.....,n)\rangle = |\Psi^S(1,2,3,j,.....,i,.....,n)\rangle$$
$$= +1|\Psi^S(1,2,3,i,.....,j,.....,n)\rangle \quad (23.2.7)$$

for all i and j.

i.e., $|\Psi^S(2,1)\rangle = +1|\Psi^S(1,2)\rangle$ (23.2.7)a

(2) The COMPLETE ANTI-SYMMETRIC state, for which

$$\hat{P}_{ij}|\Psi^a(1,2,3,i,.....,j,.....,n)\rangle = |\Psi^a(1,2,3,i,.....,j,.....,n)\rangle$$
$$= -1|\Psi^S(1,2,3,j,.....,i,.....,n)\rangle \quad (23.2.8)$$

for all i and j.

i.e., $|\Psi^a(2,1)\rangle = -1|\Psi^a(1,2)\rangle$ (23.2.9) b

These two types of states are eigen functions of all the permutation operators $\hat{P}_{ij}$ with the eigen values ±1. Since $[[\hat{H},\hat{P}_{ij}] = 0$, a system which is initially in a completely Symmetric $|\Psi^S(1,2)\rangle$ (or, completely Anti-symmetric, $|\Psi^a(1,2)\rangle$) State remains in a completely symmetric (or anti-symmetric) state for all time. It is an experimental fact that the completely symmetric and completely Anti-symmetric states are the only two states found in nature.

Further, particles with integral spin always have *completely Symmetric States*: they are said to obey <u>Bose-Einstein Statistics</u>. They are called *bosons*. Examples of bosons are pions, photons, and He nuclei (*i.e.* α – particle). Particles with half-integral spin always have *completely Anti-symmetric States*: They are said to obey the <u>Fermi-Dirac Statistics</u>. They are known as *Fermions*. Examples of fermions are electrons, protons, and neutrinos.

No reason can be given at this point for the occurrence only completely Symmetric or completely Anti-symmetric States in nature. Some insight is gained if one understands *Second Quantization* (discussed in Chapter 26). Pauli first proved the connection between spin and statistics on the basis of a general assumption about the structure of Relativistic Quantum Field Theory.

Any function of r_1 and r_2, viz., $|\Psi(1,2)\rangle$ can be split up into a symmetrical and an anti-symmetric part,

$$|\Psi(1,2)\rangle = +\frac{1}{2}\left[|\Psi^S(1,2)\rangle + |\Psi^a(1,2)\rangle\right] \qquad (23.2.10)$$

with $\quad |\Psi^S(1,2)\rangle = |\Psi(1,2)\rangle + |\Psi(2,1)\rangle \qquad (23.2.11)$

and $\quad |\Psi^a(1,2)\rangle = |\Psi(1,2)\rangle - |\Psi(2,1)\rangle \qquad (23.2.12)$

Suppose that there are three particles described by an arbitrary function $|\Psi(1,2,3)\rangle$. The operator that interchanges the ith and j^{th} labels of the particles, viz. $\hat{P}_{ij}$,

For example, $\hat{P}_{12}|\Psi(3,1,2)\rangle = |\Psi(1,3,2)\rangle$, etc. then

$$\hat{P}_{12}\hat{P}_{13}|\Psi(1,2,3)\rangle = |\Psi(2,3,1)\rangle$$
$$\hat{P}_{13}\hat{P}_{12}|\Psi(1,2,3)\rangle = |\Psi(3,1,2)\rangle,$$

and so $[\hat{P}_{12}, \hat{P}_{13}] \neq 0.$, as before.

One can arrive at the result,

$$|\Psi^S(1,2,3)\rangle = |\Psi(1,2,3)\rangle + |\Psi(2,3,1)\rangle + |\Psi(3,1,2)\rangle$$
$$+ |\Psi(2,1,3)\rangle + |\Psi(1,3,2)\rangle + |\Psi(3,2,1)\rangle \qquad (23.2.13)$$

and $\quad |\Psi^a(1,2,3)\rangle = |\Psi(1,2,3)\rangle + |\Psi(2,3,1)\rangle + |\Psi(3,1,2)\rangle$
$$- |\Psi(2,1,3)\rangle - |\Psi(1,3,2)\rangle - |\Psi(3,2,1)\rangle \qquad (23.2.14)$$

23.2.2 NON-INTERACTING PARTICLES; The CENTRAL FIELD APPROXIMATION

Consider that the identical particles that compose a system are non-interacting. This is a usual approximation. Let it be so that in many electron atomic system that each electron is assumed to in a common potential, which represents the Coulomb potential, of the nucleus plus the average potential of all the other electrons. The Hamiltonian of the many-electron atom is then written as

$$\hat{H}(1,2,....,n) = \sum_{i=1 \text{ to } n} \hat{H}(i) \qquad (23.2.15)$$

where $\hat{H}(i)$ = single-particle Hamiltonian for the ith particle. Since the particles are identical, all the single-particle Hamiltonians have the same functional form. This is the CENTRAL FIELD APPROXIMATION of the many-body problems. Assume that the single-particle eigen value equation is solved.

$$\hat{H}(i)|\varphi_a(i)\rangle = E_a|\varphi_a(i)\rangle \qquad (23.2.16)$$

The function $\quad |\Psi(1,2,3,.....,n)\rangle = |\varphi_a(1)\rangle|\varphi_b(2)\rangle|\varphi_c(3)\rangle........|\varphi_f(n)\rangle .(23.2.17)$

is easily seen to be a solution of

$$\hat{H}(1,2,....,n)|\Psi(1,2,3,.....,n)\rangle = E|\Psi(1,2,3,.....,n)\rangle \qquad (23.2.18)$$

with the eigen value

$$E = E_a + E_b + E_c + ... + E_f \qquad (23.2.19)$$

However, it does not have the symmetry of either equation (23.2.7) or (23.2.8).

23.2.3 Case: for FERMIONS

One can easily construct an eigen function of $\mathcal{H}(1, 2,\ldots, n)$ which has the symmetry of (23.2.8). It is

$$|\Psi(1,2,3,\ldots,n)\rangle = \sqrt{\frac{1}{n!}}\sum_{p}(-1)^{P}\hat{P}\Big[|\varphi_{a}(1)\rangle|\varphi_{b}(2)\rangle|\varphi_{c}(3)\rangle\ldots|\varphi_{f}(n)\rangle\Big] \quad (23.2.20)$$

where $\hat{P}$ = the Permutation Operator that permutes the arguments 1, 2, 3, , n among functions $|\varphi_{a}(1)\rangle|\varphi_{b}(2)\rangle|\varphi_{c}(3)\rangle\ldots|\varphi_{f}(n)\rangle$. There are $n!$ possible permutations and all of these are to be included in the sum. The factor $(-1)^{P}$ means that this factor is to be multiplied by -1, if the permutation is ODD, and by +1, if the permutation is EVEN. Each term in the sum is an eigen function of $\hat{H}(1,2,\ldots,n)$ with the same eigen value, viz. $E = E_a + E_b + \ldots E_f$.

Therefore, $\quad \hat{H}|\Psi\rangle = E|\Psi\rangle$

It is easily seen that

$\quad \hat{P}_{ij}|\Psi\rangle = -|\Psi\rangle$, for all i and j.

If there are only two particles, evidently, being Fermions,

$$\left|\Psi^{a}(1,2)\right\rangle = \frac{1}{\sqrt{2!}}\Big\{\left|\Psi(1,2)\right\rangle - \left|\Psi(2,1)\right\rangle\Big\} \quad (23.2.21)$$

where the relation (23.2.17) gives

$$\left|\Psi(1,2)\right\rangle = |\varphi_{a}(1)\rangle\cdot|\varphi_{b}(2)\rangle \quad (23.2.22)$$

With three fermions, the form is

$$\left|\Psi^{a}(1,2,3)\right\rangle = \frac{1}{\sqrt{3!}}\left\{\begin{array}{l}\left|\Psi(1,2,3)\right\rangle + \left|\Psi(2,3,1)\right\rangle + \left|\Psi(3,1,2)\right\rangle \\ -\left|\Psi(2,1,3)\right\rangle - \left|\Psi(1,3,2)\right\rangle - \left|\Psi(3,2,1)\right\rangle\end{array}\right\} \quad (23.2.23)$$

On substituting for $\left|\Psi(1,2,3)\right\rangle$, etc. from the relation (23.2.20),

$$\left|\Psi^{a}(1,2,3)\right\rangle = \frac{1}{\sqrt{3!}}\left\{\begin{array}{l}|\varphi_{a}(1)\rangle\cdot|\varphi_{b}(2)\rangle\cdot|\varphi_{c}(3)\rangle + |\varphi_{a}(2)\rangle\cdot|\varphi_{b}(3)\rangle\cdot|\varphi_{c}(1)\rangle + |\varphi_{a}(3)\rangle\cdot|\varphi_{b}(1)\rangle\cdot|\varphi_{c}(2)\rangle \\ -|\varphi_{a}(2)\rangle\cdot|\varphi_{b}(1)\rangle\cdot|\varphi_{c}(3)\rangle - |\varphi_{a}(3)\rangle\cdot|\varphi_{b}(2)\rangle\cdot|\varphi_{c}(1)\rangle - |\varphi_{a}(1)\rangle\cdot|\varphi_{b}(3)\rangle\cdot|\varphi_{c}(2)\rangle\end{array}\right\}$$
$$(23.2.24)$$

23.2.4 SLATER DETERMINANT and DETERMINANTAL Wave functions

There is an elegant way to construct such a wave function, which was introduced by J.C. Slater. The method makes use of the property of determinants.

For N Fermions, $\left|\Psi^{a}(1,2,\ldots N)\right\rangle$ can be written as a (NxN) determinant, as follows

$$\left|\Psi^{a}(1,2,\ldots N)\right\rangle = \frac{1}{\sqrt{N!}}\begin{vmatrix}|\varphi_{a}(1)\rangle & |\varphi_{a}(2)\rangle & |\varphi_{a}(3)\rangle & \cdots & |\varphi_{a}(N)\rangle \\ |\varphi_{b}(1)\rangle & |\varphi_{b}(2)\rangle & |\varphi_{b}(3)\rangle & \cdots & |\varphi_{b}(N)\rangle \\ |\varphi_{c}(1)\rangle & |\varphi_{c}(2)\rangle & |\varphi_{c}(3)\rangle & \cdots & |\varphi_{c}(N)\rangle \\ & & & & \\ & & & & \\ |\varphi_{f}(1)\rangle & |\varphi_{f}(2)\rangle & |\varphi_{f}(3)\rangle & \cdots & |\varphi_{f}(N)\rangle\end{vmatrix} \quad (23.2.25)$$

This is popularly called a SLATER DETERMINANT. This is characteristic of the Central Field Approximation. In the preceding equations, clearly the interchange of two particles involves the interchange of two columns in the determinant, and this changes the sign. If two electrons are in the same energy eigen state and if they are in the same spin state, then the determinant vanishes. Thus $|\Psi^a(1,2,...N)\rangle = 0$, if any two sets of quantum numbers, say a and b, are the same, since two rows of the determinant are the same..

Since the state functions, $|\varphi_a(1)\rangle, |\varphi_b(2)\rangle, |\varphi_c(3)\rangle, \cdots, |\varphi_f(N)\rangle$ are orthogonal, different terms in equation (23.2.25) are also orthogonal.

i.e. $\langle \varphi_a()||\varphi_b()\rangle = 0,$, etc.

The factor $\frac{1}{\sqrt{N!}}$ is needed to normalize the $|\Psi^a\rangle$ to unity, when $|\varphi\rangle$ are normalized to unity.

i.e., $\langle \varphi_a||\varphi_a\rangle = \langle \varphi_b||\varphi_b\rangle = \cdots = 1,$

$\langle \Psi^a||\Psi^a\rangle = 1.$

The Normalizing factor in the **N-electron**_determinant wave function_ (23.3.25) is $\frac{1}{\sqrt{N!}}$ because an (N x N) determinant will have N! elements. A Slater determinant is found written in abbreviated form by only writing the diagonal elements:

$$|\varphi_a(1) \quad \varphi_b(2) \quad \varphi_c(3) \quad \cdots \quad \varphi_f(N)|$$

23.2.5 PAULI EXCLUSION PRINCIPLE Emerges

Since $|\Psi^a()\rangle$ vanishes if any two Fermions have the same set of quantum numbers, one may say that it is impossible for two Fermions to be in the same single-particle state. This is the famous *Pauli Exclusion Principle*, and was discovered (1924) by the study of atomic spectra before the advent of Quantum Mechanics. The general statement is: NO TWO ELECTRONS CAN BE IN THE SAME QUANTUM STATE.

For Anti-Symmetrized wave function, it is possible to show that the OVERLAP INTEGRAL,

$$C^2 \int |\Psi^a_1()\rangle^* |\Psi^a_2()\rangle dx = C^2 \langle \Psi^a_1()| \Psi^a_1()\rangle = 1 \qquad (23.2.26)$$

for a two-electron system, with the electron with 'a' label and the other with label 'b', in some spatial region R.

For example, if in the case of the two-electron system, 'a' in the GAUSIIAN WAVE PACKET, say, $C e^{-\beta x^2/2}$ at the origin ($x = 0$) and 'b' at some location $x = L$, and represented by $C e^{-\beta(x-L)^2/2}$, leads to the OVERLAP INTEGRAL $\propto e^{-\beta L^2/2}$. It is evident that as L becomes large the overlap integral vanishes very rapidly. This simple calculation shows that the wave function of an electron under consideration need not be anti-symmetrized with any or all other distant electrons. That is, the Pauli Exclusion Principle must be applied in atoms and molecules, but not in crystalline lattices, where the spacing between the atoms is several times the *Bohr Radius*. In other words, ANTI-SYMMETRIZATION for ELECTRONS IS NOT NECESSARY WHENEVER THEIR SEPARATION IS LARGE.

23.2.6 Case: BOSONS

Equation (23.2.20) is replaced by

$$|\Psi^S(1,2,3,....,n)\rangle = \sqrt{\frac{n_a! n_b!... n_f!}{n!}} \sum_P \hat{P} |\varphi_a(1)\rangle \cdot |\varphi_b(2)\rangle \cdots \cdot |\varphi_c(n)\rangle \qquad (23.2.27)$$

It is easily verified that

$$\hat{H}|\Psi\rangle = E|\Psi\rangle.$$

and $\quad \hat{P}_{ij}|\Psi\rangle = +|\Psi\rangle, \quad$ for all i and j.

This means there is no restriction on the number of bosons that have the same quantum numbers. Pauli's Exclusion Principle is **violated** by Bosons.

In the normalization factor, n_a is the number of times that the state $|\varphi_a\rangle$ appears in a term of equation (23.2.25).

Worked out Example 23.1

Write out the eigen function of $\hat{H}(1,2,3)$ in the case of Lithium atom. Express the same in the form of Slater Determinant.

Solution: Step # 1 A Lithium atom has atomic number $Z = 3$.

$$|\Psi(1,2,3,.....,n)\rangle = \sqrt{\frac{1}{n!}}\sum_p (-1)^p \hat{P}\left[|\varphi_a(1)\rangle|\varphi_b(2)\rangle|\varphi_c(3)\rangle.........|\varphi_f(n)\rangle\right] \quad (23.2.20)$$

$$|\Psi^a(1,2,3)\rangle = \frac{1}{\sqrt{3!}}\left\{\begin{array}{l}|\Psi(1,2,3)\rangle + |\Psi(2,3,1)\rangle + |\Psi(3,1,2)\rangle \\ -|\Psi(2,1,3)\rangle - |\Psi(1,3,2)\rangle - |\Psi(3,2,1)\rangle\end{array}\right\} \quad (23.2.23)$$

Step # 2 On substituting for $|\Psi(1,2,3)\rangle$, etc. from the relation (23.2.20),

$$|\Psi^a(1,2,3)\rangle = \frac{1}{\sqrt{3!}}\left\{\begin{array}{l}|\varphi_a(1)\rangle\cdot|\varphi_b(2)\rangle\cdot|\varphi_c(3)\rangle + |\varphi_a(2)\rangle\cdot|\varphi_b(3)\rangle\cdot|\varphi_c(1)\rangle + |\varphi_a(3)\rangle\cdot|\varphi_b(1)\rangle\cdot|\varphi_c(2)\rangle \\ -|\varphi_a(2)\rangle\cdot|\varphi_b(1)\rangle\cdot|\varphi_c(3)\rangle - |\varphi_a(3)\rangle\cdot|\varphi_b(2)\rangle\cdot|\varphi_c(1)\rangle - |\varphi_a(1)\rangle\cdot|\varphi_b(3)\rangle\cdot|\varphi_c(2)\rangle\end{array}\right\}$$
$$(23.2.24)$$

Step # 3 As a determinant, $\quad |\Psi^a(1,2,.3)\rangle = \frac{1}{\sqrt{3!}}\begin{vmatrix}|\varphi_a(1)\rangle & |\varphi_a(2)\rangle & |\varphi_a(3)\rangle \\ |\varphi_b(1)\rangle & |\varphi_b(2)\rangle & |\varphi_b(3)\rangle \\ |\varphi_c(1)\rangle & |\varphi_c(2)\rangle & |\varphi_c(3)\rangle\end{vmatrix}$

$$\langle\varphi_a()|\varphi_b()\rangle = \delta_{ab}$$

23.2.7 ERTURBATIONS TO THE CENTRAL FIELD APPROXIMATUION

A detailed analysis showed that the energy levels of the atoms and ions, except those levels, which have closed-shell configurations, are degenerate in the Central Field Approximation. This is because, in the incomplete shells, one can assign electrons to various values os quantum numbers m_ℓ and m_s, without getting identical sets of quantum numbers. In the closed-shell configurations, all the m_ℓ and m_s quantum numbers are occupied and the energy levels are non-degenerate. The total perturbation Hamiltonian is added to the Central Field Approximation to give the exact Hamiltonian of the system. The procedure consists of two parts. (1) the difference between the actual Electro-Static Interactions and the Central Field, and (2) the Spin-Orbital Interaction.

The important information one gets, from a detailed study, is that

a) Each degenerate configuration energy level is split by $\hat{H}_{es}$ into sub-energy levels,

b) The number of sub-levels ina particular configuration equals the terms available in that configuration, and can be calculated by M_L and M_S onsiderations, and

c) The energy values of the sub-levels are expressible using $\hat{H}_{es}$

23.8 The T0TAL HAMILTONIAN of a Many-Electron Atom:

The analysis of the energy levels can be approximated by successive orders of perturbation, as indicated in the expression for the Hamiltonian, as follows:

$$\hat{H} = \sum_{k=1}^{k=N} \frac{1}{2m} p_k^2 - \sum_1^N Ze^2 [r_k]^{-1} + \sum_{i>j} e^2 [r_{ij}]^{-1} + \sum_1^{'N} \xi(r_k) \vec{\ell}_k \cdot \vec{s}_k$$

$$+ \sum_1^N e^2 |\vec{r}_e - \vec{r}_k|^{-1} + \xi(\vec{r}_e) \vec{\ell}_e \cdot \vec{s}_e \quad (23.2.28)$$

where $\sum'$ means summations over all the electrons except the excited electron, and the subscript e refers to the excited electron, and.

$\sum_{k=1}^{k=N} \frac{1}{2m} p_k^2 - \sum_1^N Ze^2 [r_k]^{-1} + \sum_{i>j} e^2 [r_{ij}]^{-1}$ = Non-relativistic part.

$\sum_1^{'N} \xi(r_k) \vec{\ell}_k \cdot \vec{s}_k$ = 1st order perturbation, (S.L. interaction)

$\sum_1^N e^2 |\vec{r}_e - \vec{r}_k|^{-1}$ = 2nd order perturbation,

$\xi(\vec{r}_e) \vec{\ell}_e \cdot \vec{s}_e$ = 3rd order perturbation.

23.3 THE HELIUM ATOM
(The treatment has already been described in Section 19.5)

The Hamiltonian of the atom is (19.5.1) and (19.5.2)

$$\hat{H}^{(0)} = \frac{\hat{p}_1^2}{2m} + \frac{\hat{p}_2^2}{2m} + Ze^2 \left(\frac{1}{r_1} + \frac{1}{r_2} \right) + \frac{e^2}{|\vec{r}_{12}|}$$

$$= \left(-\frac{\hbar^2}{2m} \right) (\nabla_1^2 + \nabla_2^2) - \left(\frac{Ze^2}{r_1} + \frac{Ze^2}{r_2} \right) + \frac{e^2}{|\vec{r}_1 - \vec{r}_2|}$$

where m is the mass of the electron.

For the purpose of simplifying the calculation of the integrals involved in problems of this type, it is convenient to use atomic units, *i.e.*, in terms of the Bohr Radius, a_0, (19.5.3) one gets (19.5.4)

$a_0 = \hbar^2 / me^2$,

$\therefore \quad \hat{H} = \left(\frac{e^2}{a_0} \right) \left\{ -\frac{1}{2}(\nabla_1^2 + \nabla_2^2) - Z\left(\frac{1}{R_1} + \frac{1}{R_2} \right) + \frac{1}{|R_1 - R_2|} \right\}$

23.3.1 The PERTURBATION TERM
According to the Perturbation Theory, (19,5,5) and (19.5.6) one sets
$\hat{H} = \hat{H}^{(0)} + \hat{H}^{(1)}$,

where $\hat{H}^{(0)} = \left\{ -\frac{1}{2}(\nabla_1^2 + \nabla_2^2) - Z\left(\frac{1}{R_1} + \frac{1}{R_2} \right) \right\}$ in (e^2 / a_0) units.

$\hat{H}^{(1)} = \frac{1}{|R_1 - R_2|}$, in (e^2 / a_0) units.

23.3.2 GROUND STATE ENERGY of HELIUM
The zeroth order eigen functions $|\psi_n^0\rangle$ (*i.e.* unperturbed states) are the solutions of the Eigen Value Equation of the operator, $\hat{H}^{(0)}$.

$$\hat{H}^{(0)}|\hat{\psi}_n^0\rangle = E_n^0 |\hat{\psi}_n^0\rangle \quad (19.5.7)$$

If one now sets this eigen function of the two electron system as
$$|\hat{\psi}_n^0\rangle = |\hat{\psi}_n^0(1)\rangle \cdot |\hat{\psi}_n^0(2)\rangle \quad (19.5.8)$$
then the eigen values are
$$E_n^0 = E_n^0(1) + E_n^0(2) \quad (19.5.9)$$
For the ground state of the Helium atom $|0\rangle \equiv |1s^2\rangle$,

$$|1s(1)\rangle = |\hat{\psi}_{100}^0(1)\rangle = \sqrt{1/4\pi}\ 2\ (Z/a_0)^{3/2}\ e^{-(Z R_1/a_0)} \quad (19.5.12)$$

$$|1s(2)\rangle = |\hat{\psi}_{100}^0(2)\rangle = \sqrt{1/4\pi}\ 2\ (Z/a_0)^{3/2}\ e^{-(Z R_2/a_0)}$$

$$|0(He)\rangle \equiv |1s^2\rangle = |1s(1)\rangle \cdot |1s(2)\rangle = (Z^3/\pi a_0^3)\ e^{-(Z(R_1+R_2)/a_0)} \quad (19.5.13)$$

$$E_{1s^2}^0(He) = \langle \hat{H}_{00}^{(0)} \rangle = \langle 0|\hat{H}^{(0)}|0\rangle = E_{1s}^0(1) + E_{1s}^0(2) = 2Z^2\ E_{1s}^0(H)$$
(19.5.14)

where $E_{1s}^0(H)$ = ionization energy (or ground state energy) of Hydrogen.

$$E_{1s}^0(H) = E_{1s}^0(1) = E_{1s}^0(2) = -(\tfrac{1}{2})(e^2/a_0) = -13.6\ eV$$

Using (19.5.14),
$$E_{1s^2}^0(He) = 4Z^2\ E_{1s}^0(H) = 4(-13.6\ eV) = -108.8\ eV \quad (19.5.15)$$

But experimentally for ionized Helium atom,
$$E_{1s^2}^0(He) = -78.62\ eV. \quad (19.5.16)$$

23.3.3 FIRST ORDER Correction to Energy for Helium

$$E_n = E_n^{(1)} \approx E_n^{(0)} + H_{mn}^{(1)}. \quad (19.5.17)$$

$$A_n^{(1)} \approx H_{mn}^{(1)} = \langle \psi_n^{(0)}|H^{(1)}|\psi_n^{(0)}\rangle \quad (19.5.16)$$

$$E_{1s^2}^{(1)}(He) = E_{1s^2}^{(0)}(He) + A_n^{(1)} \approx 2Z^2 E^{(0)}(H) + \langle 1s^2|H^{(1)}|1s^2\rangle \quad (19.3.17)$$

$$\therefore A_n^{(1)} \approx +A_{1s^2}^{(1)} = H_{00}^{(1)} = \langle 1s^2|H^{(1)}|1s^2\rangle \quad (19.5.18)$$

$$= (e^2/a_0)\hat{H}^{(1)}\frac{1}{|R_{12}|} \quad (19.5.19)$$

$$A_{1s^2}^{(1)} = \left(\frac{e^2}{a_0}\right)\left(\frac{Z^6}{\pi^2}\right) \iint |R_{12}|^{-1}\ e^{-2Z(R_1+R_2)}\ d\tau_1\ d\tau_2. \quad (19.5.20)$$

23.3.3.1 TO EVALUATE $\langle (e^2/a_0)|R_{12}|^{-1}\rangle$

One can expand $|R_{12}|^{-1}$ by using the associated Legendre Polynomials, $P_\ell^m(\rho)$.

Thus equation (19.5.20) reduces to
$$A_{1s^2}^{(1)} = \left(\frac{e^2}{a_0}\right)\left(\frac{Z^6}{\pi^2}\right) \iint |R_>|^{-1}\ e^{-2Z(R_1+R_2)}\ d\tau_1\ d\tau_2 \quad (19.5.23)$$

Since e^{-2ZR_1/a_0} = a potential = constant for $R_> < |R_{12}|$, but varies as R_2^{-1}, for $R_2 > R_{12}$.

$$A_{1s^2}^{(1)} = 16Z^6(e^2/a_0)\{(5/128)Z^{-5}\} = (5/8)Z(e^2/a_0) = -(5/4)Z\ E_{1s}^{(0)}(H) \quad (19.5.25).$$

23.3.3 APPROXIMATE ENERGY of the NORMALSTATE of HELIUM ATOM

Substituting this value (19.1.25) in equation (19.5.17), approx. energy

$$E^{(1)}_{1s^2}(He) = E^{(0)}_{1s^2}(He) + A^{(1)}_{1s^2} = 2Z^2 E^{(0)}(H) - (5/4)Z\, E^{(0)}_{1s}(H)$$

$$= \left[2Z^2 - (5/4)Z\right] E^{(0)}_{1s}(H). \tag{19.5.26}$$

$$= \left[2Z^2 - (5/4)Z\right] (-13.6\ eV) = -74.8\ eV. \tag{19.5.27}$$

.23.4 ORTHO- AND PARA - HELIUM

Except when both electrons are in the ground state their spins need not be in opposite directions. The total spin momentum for the atom will therefore be either 0 or 1. There will thus be two possible spectra, which will behave differently, in an applied magnetic field. This is attributed to two types of helium, so distinguished, called Parahelium and Orthohelium. Two levels of Parahelium are all Singlets, whereas those of Orthohelium are found under high resolution to be Tiplets except for the S levels, which are Snglets.
Let the non-relativistic spin wave functions are used.

Introduce spin wave functions $\chi_1(\sigma_3)$ and $\chi_2(\sigma_3)$ one can form an Anti-symmetric atomic wave function in two possible ways

$$\Psi_O = \Psi_+ = \{\Psi_a(1)\Psi_b(2) - \Psi_a(2)\Psi_b(1)\} \times \{\chi_1(1)\chi_2(2) + \chi_1(2)\chi_2(1)\} \tag{23.4.1}$$

and $\quad \Psi_P = \Psi_- = \{\Psi_a(1)\Psi_b(2) + \Psi_a(2)\Psi_b(1)\} \times \{\chi_1(1)\chi_2(2) - \chi_1(2)\chi_2(1)\} \tag{23.4.2}$

The first ($\Psi_O = \Psi_+$) corresponds to PARALLEL SPINS (ORTHOHELIUM) while the second ($\Psi_P = \Psi_-$) corresponds to the *anti-parallel spins* (PARAHELIUM). If the electrons have quantum numbers (ℓ_1, λ_1) and (ℓ_2, λ_2) the momenta can be combined by the LANDE VECTOR MODEL. By writing

$$L = M_1 + M_2,$$
$$S = \sigma_1 + \sigma_2,$$

One has the normal (or Russell-Saunders) Coupling whereby

$$\vec{J} = \vec{L} + \vec{S}.$$

The case of Parahelium corresponds to $S = 0$ so that $J = L$, and the eigen values of J in an assigned direction are merely those of L, *viz.* s ℏ. Thus the Parahelium terms are SINGLETS in an applied magnetic field.

In the Orthohelium S ha the eigen values ± ℏ in an assigned direction. For the vectors M_1, M_2 have eigen values $S_1\hbar, S_2\hbar$ the resulting quantum number for $M_1 + M_2$ will be any one of the integers lying between $|S_1 + S_2|$ and $|S_1 - S_2|$. Denote any one of these by s. Then for each s, J will have values K ℏ, where

$$K = (s - 1),\ 0,\ (s + 1).$$

Thus the Orthohelium terms will be TRIPLETS in a magnetic field. The absence of terms corresponding to transitions to the state s = 0 in Orthohelium provided initial inspiration for the Exclusion Principle.

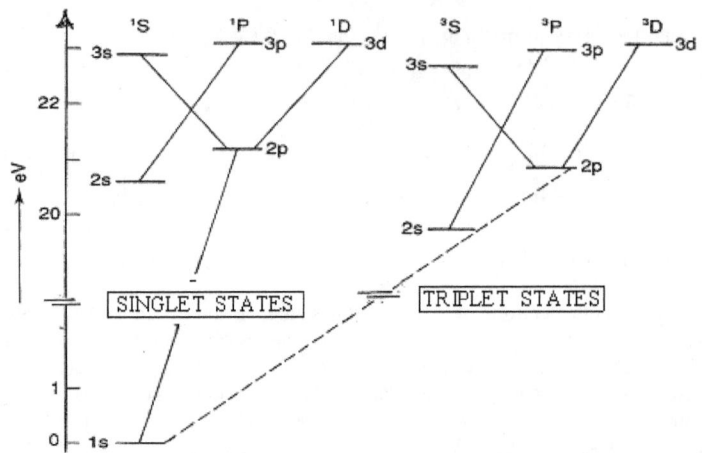

Fig. 23.1. The lowest-lying energy levels and some their electric dipole transitions for Helium.

23.4.1 SPIN CORRELATION

It was seen that two electrons, in a He atom, tend to avoid each other if they are described by an Anti-symmetric spatial wave function. However, if the two electrons are described by such a wave function, it follows that their spins must be in a Symmetrical state, and hence $S = 1$. To summarize it is said that PARALLEL SPINS TEND TO AVOID ONE ANOTHER. This effect is called SPIN CORRELATION. That is, if the spins are parallel, then the *Pauli Principle* requires them to have an anti-symmetric wave function, which implies that they cannot be found at the same point simultaneously.

A consequence of the Spin Correlation is that the triplet term arising from a configuration lies lower in energy than the Singlet term of the same configuration.

23.5 MANY-ELECTRON ATOMS

23.5.1 EFFECTIVE NUCLEAR CHARGE:

A crude description of the ground state of the Helium atom is configuration $1s^2$ with both electrons in hydrogenic 1s orbitals with $Z = 2$. An improved description takes into account the repulsion between the electrons, and the effect of the repulsion on the orbitals occupied can be simulated to some extent by replacing the true nuclear charge, Ze, by an EFFECTIVE NUCLEAR CHARGE, $Z_{eff} e$. As per the **Variation Principle** used,

$$Z^*\big|_{<H>gr=Minimum} = (Z - \tfrac{5}{16}) = \tfrac{27}{16} \approx 1.7 \qquad (20.3.19)$$

This approach to the description of atomic structure can be extended to other many-electron atoms; I shall give a brief description of what is involved.

The ORBITAL APPROXIMATION is the basis of most descriptions of atomic structure, where it is supposed that each electron occupies its own atomic orbital, resembling the hydrogenic orbital. The structure of an atom is expressed in terms of a configuration, such as $1s^2 2s^2 2p^6$ for Neon. –The wave function for this atom as the approximation

$|\psi\rangle = \sqrt{\frac{1}{10!}} \; \det | 1s^\alpha(1) \; 2s^\beta(2) \cdots 2p^\beta(10) |$ for the Neon atom.

23.5.2 NUCLEAR SCREENING CONSTANT, σ

But the actual many-electron wave function $|\psi^a(1,2,.,.,.,N)\rangle$ can be written as a (N x N) determinant (23.2.25).

As a result of the CENTRAL FIELD APPROXIMATION the order of sub shells is explained; and the nuclear charge is reduced to $(Z-\sigma)e$, and hence one can write

$$Z_{eff} \; e = (Z-\sigma) \; e \qquad (23.5.1)$$

The quantity σ is called the NUCLEAR SCREENING CONSTANT, and is characteristic of the orbital that the electron (labeled 1) occupies. Thus σ is different for 2 s_ and 2 p_ orbitals. It also depends n the configuration of the atom, and σ for a given orbital has different values on the ground and excited states. Strictly speaking the shielding constant varies with distance, and an electron does not have the single value of σ However, in the next approximation one replaces the varying value of σ by its average value, and hence treats Z_{eff} as a constant, typical of the atom, and of the orbital occupied by the electron of interest. Thus Z = 2 is replaced Z_{eff} = 1.3 for each electron of a He atom, for the average value of σ = 0.7 is ascribed to each electron.

23.5.3 SHIELDING and PENETRATION:

The reduction of the nuclear charge due to the presence of the other electrons in an atom is called SHIELDING; and its magnitude is determined by the extent of PENETRATION of core regions of the atom, the extent to which the electron of interest will be found close to the nucleus and inside the spherical shells of charge.

23.5.4 PERIODICITY and BUILDING-UP PRINCIPLE

The ground state electron configurations are determined experimentally by an analysis of their spectra or, by magnetic measurements. Theses configurations show a periodicity that mirrors the block, group, and period structure of the **Periodic Table**. The rationalization of the observed configurations is normally expressed in terms of the BUILDING-UP PRINCIPLE. Accordingly, electrons are allowed to occupy atomic orbitals in an order that mirrors the structure of the Periodic Table. (Fig. 23.2) and subject to the *Pauli Exclusion Principle* that no more than two electrons can occupy any one orbital, and if two do occupy an orbital, then their spins must be paired.

THE PERIODIC TABLE OF ELEMENTS

GROUP →	I	II											III	IV	V	VI	VII	VIII
PERIOD ↓																		
1	Z→1 H		Metals					Non-Metals										2 H
2	3 Li	4 Be							Metalloids				5 B	6 C	7 N	8 O	9 F	10 Ne
3	11 Na	12 Mg											13 Al	14 Si	15 P	16 S	17 Cl	18 Ar
4	19 K	20 Ca	21 Sc	22 Ti	23 V	24 Cr	25 Mn	26 Fe	27 Co	28 Ni	29 Cu	30 Zn	31 Ga	32 Ge	33 As	34 Se	35 Br	36 Kr
5	37 Rb	38 Sr	39 Y	40 Zr	41 Nb	42 Mo	43 Tc	44 Ru	45 Rh	46 Pd	47 Ag	48 Cd	49 In	50 Sn	51 Sb	52 Te	53 I	54 Xe
6	55 Cs	56 Ba	71 Lu	72 Hf	73 Ta	74 W	75 Re	76 Os	77 Ir	78 Pt	79 Au	80 Hg	81 Tl	82 Pb	83 Bi	84 Po	85 At	86 Rn
7	87 Fr	88 Ra	103 Lr	104 Rf	105 Db	106 Sg	107 Bh	108 Hs	109 Mt	110 Uun	111 Uuu	112 Uub	113 Uut	114 Uuq	115 Uup	116 Uuh	117 Uus	118 Uuo

*Lanthanoids

57 La	58 Ce	59 Pr	60 Nd	61 Pm	62 Sm	63 Eu	64 Gd	65 Tb	66 Dy	67 Ho	68 Er	69 Tm	70 Yb

**Actinoids

89 Ac	90 Th	91 Pa	92 U	93 Np	94 Pu	95 Am	96 Cm	97 Bk	98 Cf	99 Es	100 Fm	101 Md	102 No	103 Lr

Fig. 23.2 The Periodic Table of Elements.

The order of occupation largely follows the order of energy levels as determined by penetration and shielding, with ns-orbitals being occupied before np-orbitals. The lowering of energy of ns-orbitals is so great that in certain regions of the Periodic Table they lie below the $(n-1)$ d-orbitals of an inner shell: the occupation of 4s-orbitals before 3d-orbitals is a well-known example of this phenomenon, and it accounts for the intrusion of the d-block into the structure of the Periodic Table of Elements.

23.5.5 SLATER ATOMIC ORBITALS (STO):

No definitive form can be given for the atomic orbitals of many-electron atoms because the orbital approximation is very primitive. Nevertheless, it is often helpful to have available a set of approximate AO- s which model the actual wave functions found by using the more sophisticated numerical techniques. These **Slater type Orbitals** (STO) are constructed as follows:

(1) An orbital with quantum number n, ℓ, m belonging to a nucleus of an atom of atomic number Z is written as

$$\Psi(r) \equiv \Psi_{n\ell m}(r, \theta, \varphi) = \Re_{n\ell}(r) \cdot Y_\ell^m(\theta, \varphi) \quad (9.4.1)$$

$$= N \, [r^{(n_{eff})}]^{-1} \, e^{[-Z_{eff} \, \rho / n_{eff}]} \cdot Y_\ell^m(\theta, \varphi) \quad (23.5.2)$$

where N is a normalization constant, $Y_\ell^m(\theta, \varphi)$ is spherical harmonic, and $\rho = (r/a_0)$.

(2) The effective principal quantum number, n_{eff}, is related to the true principal quantum number, n, by the following mapping:

$n \to n_{eff}$: $1 \to 1$ $2 \to 2$ $3 \to 3$ $4 \to 3.7$ $5 \to 4.0$ $6 \to 4.2$

(3) The effective atomic number, Z_{eff}, is taken for the neutral ground state atoms.

23.5.5.1 Deficiencies on using STOs

Care should be taken since STOs of different n values, but of the same values of ℓ are no orthogonal. Again ns-orbitals with $n>1$ have zero amplitude at the nucleus.

23.5.6 SELF-CONSISTENT FIELDS (SCFs):

The best AOs are found by numerical solution of the Schrödinger Equation. The original procedure was introduced by D.R. Hartree and is known as the method of SELF-CONSISTENT FIELDS (SCF). V. Fock and J.C. Slater to include the effects of electron exchange improved the procedure, and the orbitals obtained by their methods are called HARTREE-FOCK ORBITALS.

The assumption behind the technique is that any one electron moves in a potential, which is a spherical average of the potential due to all the other electrons and the nucleus. And which can be expressed as a single charge centred on the nucleus. The Schrödinger Equation is numerically integrated for that electron, etc. The Hartree-Fock Equations on which the procedure is based are slightly tricky to derive, but they are reasonably easy to interpret.

Worked out Example 23.2

Construct the Term Symbols that can arise from the configurations (a) $2p^1 3p^1$ and (b) $2p^5$

Solution: Step # 1 Construct the possible values of L by using the Clebsch-Gordan (C.G.) Series, and identify the corresponding letters.

Step # 2 Construct the possible values of S similarly, and

Step # 3 : Work out the multiplicities.

Step # 4 Construct the values of J from the value of L and S for each term by using the C.G. series again.

(a) $\ell_1 = 1$ and $\ell_2 = 2$, so L = 2, 1, 0, and the configuration gives rise to D, P, and S Terms. Two electrons have S =1,0, giving rise to a Triplet and Singlet Terms, respectively, so the complete set of Terms is: $^3D, ^1D, ^3P, ^1P, ^3S, ^1S$. The values of J that can arise are formed from J = (L + S), (L+ S -1), …,|L – S|, and so the complete list of Term Symbols is
$^3D_3, ^1D_2, ^3D_1, ^1D_2, ^3P_3, ^3P_1, ^3P_0, ^1P_1, ^3S_1, ^1S_0$.

(b) The configuration is equivalent to a single hole in a shell. So $L = \ell = 1$, corresponding to a P term. Because $S = s = \frac{1}{2}$ for the hole the Term Symbol is ^2P. The two levels of these Terms are $^2P_{3/2}, ^2P_{1/2}$,

23.5.7 HUND's RULES and the RELATIVE ENERGIES of TERMS

A set of rules (empirical) was devised by Frederick Hund for identifying the lowest energy term of a configuration with the minimum of calculation,

Rule # 1 : Select first the Terms of largest S. *i.e.* Terms with maximum multiplicity lies lowest in energy.

Example: For the configuration, $2p^2$, one expects the ^3P Term to lie lowest in energy.

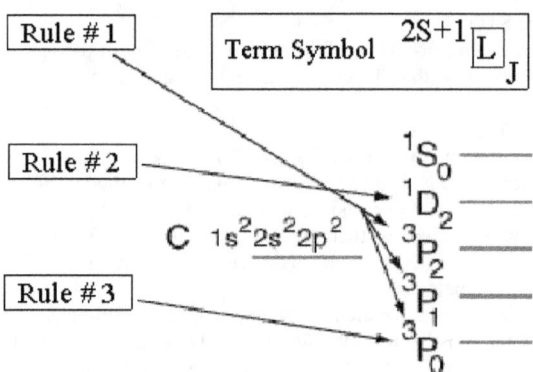

Fig. 23.3 Hund's Rule # 1 illustrated

Rule # 2: Among these Terms, the Term of largest L is lowest in energy, *i.e.* for a given multiplicity, the term with highest value of L lies lowest in energy.
Example: Between ^3P Term and ^3F Term in a particular configuration, then ^3P Term will have the lowest energy.

Rule # 3: Multiplets formed from *equivalent electrons* are regular when less than half the shell is occupied, but inverted when more than half the shell is occupied. A Term is regular if the lowest J-value of the multiplet has the lowest energy, and is inverted if the largest J-value has the lowest energy. The origin of this rule is the spin-orbit coupling.
Example: Since p^2 is less than half-filled the three levels ^3P are expected to lie in the order $^3P_0 < {^3P_1} < {^3P_2}$.

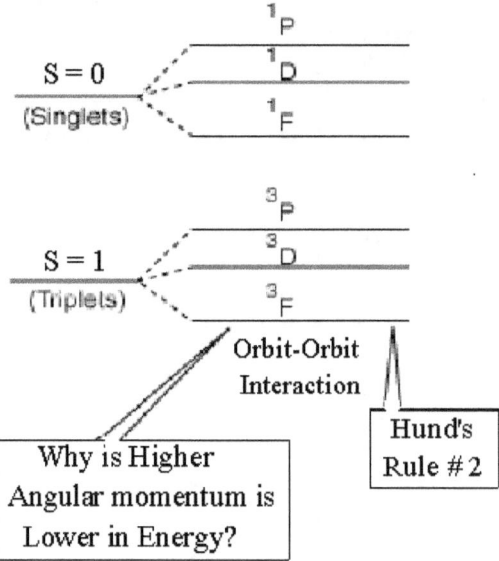

Fig. 23.4 Illustration of Hund's Rule # 2

23.6 THE HARTREE-FOCK SCF METHOD

A primary objective of Quantum Mechanics of the molecules is the solution of the non-relativistic, time-independent Schrödinger Equation, and in particular the calculation of the electronic structure of atoms and molecules. There are two principal methods of molecular structure, viz.,
1) Molecular Orbital theory and
2) Valence Bond theory.

These are of most interest to Chemistry; and it is not my intention to deal with them in this book. This Chapter is intended to introduce some of the techniques that are used to solve the Schrödinger Equation for electrons in molecules. All such techniques make heavy use of Personal Computers, and so it is only proper to establish the equations that are the basis of such computations and describe some of the approximations to make the computations feasible. All the material here is based on the Born-Oppenheimer Approximation.

23.6.1 The BORN-OPPENHEIMER APPROXIMATION

Even in the case of H_2^+ ion, the simplest of the 3-particle system, its Schrödinger Equation cannot be solved analytically. To overcome this difficulty the *Born-Oppenheimer Approximation* is adopted, which takes note of the great difference in masses of electrons and nuclei. Because of this difference, the electrons can respond almost instantaneously to displacement of the nuclei. Therefore instead of trying to solve the Schrödinger Equation for all the particles simultaneously, it is possible to regard the nuclei as fixed in position and to solve the Schrödinger Equation for the electrons in the static electric potential arising from the nuclei in that particular arrangement. This means that the electronic energy can be found at a set of fixed inter-nuclear distances, and then plotted as a function of R_{AB} (One nucleus at point A and the other at B).

$$\hat{H} = \left\{ -\tfrac{1}{2}\hat{\nabla}^2 - \tfrac{m}{M_A}\tfrac{1}{2}\hat{\nabla}_A^2 - \tfrac{m}{M_B}\tfrac{1}{2}\hat{\nabla}_B^2 - \tfrac{1}{r_A} - \tfrac{1}{r_B} + \tfrac{1}{R_{AB}} \right\}, \text{ in}(e^2/a_0) \text{ units.} \quad (23.6.1)$$

where the terms represent the kinetic energies of the electron, nucleus A and B, the potential energy of attraction between the electron and the two nuclei, and nuclear repulsion energy, of the H_2^+ ion.

Mathematically, the *Born-Oppenheimer Approximation* allows one to drop the nuclear kinetic energy terms from the Hamiltonian $\hat{H}$ in equation (23.6.1).

$$\hat{H} = -\tfrac{1}{2}\hat{\nabla}^2 - \sum_{i=A}^{B}\tfrac{1}{r_i} + \tfrac{1}{R_{AB}}, \text{ in } (e^2/a_0) \text{ units.} \quad (23.6.2)$$

When this procedure is followed for the H_2^+, the plot is seen to have a minimum. A minimum in the curve corresponds to the formation of a **stable** molecule, and the depth of the minimum is DISSOCIATION or Binding Energy.

The problem of H_2^+ is the exactly solvable because it is reduced to a one-electron problem.

If elliptical coordinates are used then it is possible to separate variables and find a solution to equation

$$\hat{H}\Psi_i = E_i\Psi_i \quad . \quad (23.6.3)$$

of the form $\quad \Psi = U(\mu) \cdot V(\nu) \cdot \Phi(\varphi) \quad (23.6.4)$

where the three functions U, V, and Φ are functions of $\mu, \nu,$ and φ, respectively. It is not the concern about the exact solution to the H_2^+ problem. This has rotational symmetry about the AB axis. The $\Phi(\varphi)$ part of the eigen value equation leads to

$$\frac{d^2\Phi}{d\varphi^2} = -m^2\Phi \qquad (23.6.5)$$

which has the familiar solution

$$\Phi = A\, e^{i\, m\, \varphi}, \quad \text{where } m = 0, \pm 1, \pm 2, \ldots \qquad (23.6.5)$$

These solutions have physical interpretation based on the fact that they are eigen functions of the operator $\hat{L}_z$, where now the Z-axis is defined by the AB bond. In the case of H_2^+, the energy will depend on m, except that states for $\pm m$ will have the same energy. Thus considering only angular degeneracy, one gets

    (a) A singly degenerate level for $m = 0$,
    (b) Doubly degenerate levels for $m = \pm 1, \pm 2, \ldots, \text{etc.}$

The discussion of the Φ function for H_2^+ has been given so as to introduce the nomenclature to describe the states of diatomic molecules. This nomenclature is similar to the s, p, d, f, g, nomenclature used in atomic orbitals except that Greek letters are used. Thus for various values of m, the orbitals are labeled as follows:

m	Orbital designation
0	σ
± 1	π
± 2	δ

Orbitals analogous to the atomic p_x and p_y orbitals can be constructed by taking appropriate linear combinations of the $\pm m$ functions in equation (23.6.5).

23.6.2 The HARTREE SCF POTENTIAL

A second method for computing the potential for the atom is due to Hartree (1928). In this method, it is assumed that each electron moves in a central field that can be calculated from the nuclear potential and the wave functions of all the other electrons, where the charge density for each electron is given by the squared magnitude of its wave function. The Schrödinger Equation is solved for each electron moving in its own central field in a *self-consistent* fashion. Calculating the Coulomb *interaction* between the chosen electron and all other electrons is the method to get the Hartree potential, but that the coordinates of the other (N -1) electrons are integrated out of the problem thus providing an average potential seen by the electron due to the average distance away of the other (N-1) electrons. Here the wave function mentioned is the $\left| \Psi^a(1,2,\ldots N) \right\rangle$, *i.e.* Slater Determinantal type, described by equation (23.2.25). Since the position of each electron changes as the other electrons move, iteration to self-consistency is required. In essence, one wants to solve the N equations for the N electrons ($i = 1, 2, 3, \ldots N$)

$$\hat{H}\Psi(r_j) = \frac{-\hbar^2}{2m}\hat{\nabla}_i^2 + \left[-\frac{Z\,k\,e^2}{r} + \sum_{j \neq i} \int |\Psi(r_j)|^2 \frac{k\,e^2}{|r_j - r_i|} d^3 r_j \right] \Psi r_j) = E_i \Psi_j \qquad (23.6.6)$$

To begin, a potential that approximates that represented by

$$\left[-\frac{Z\,k\,e^2}{r} + \sum_{j \neq i} \int |\Psi(r_j)|^2 \frac{k\,e^2}{|r_j - r_i|} d^3 r_j \right]$$

is assumed and the charge distributions calculated. These are then inserted into *Poisson's Equation* to calculate the next iteration for the potential. This is then used to calculate a new charge distribution, and the process is repeated until consistency is achieved. Hence the method is known by the name SELF-CONSISTANT FIELD (SCF) method. In some sense it is an iterative approach to the *Variational Method* introduced in Chapter 20. In a self-consistent procedure, a trial set of spin orbitals, say $|\varphi_a(i)\rangle$, is formulated and used to formulate the FOCK OPERATOR, $\hat{f}_i$, then the H F EQUATIONS

$$\hat{f}_i |\varphi_a(i)\rangle = \varepsilon_a |\varphi_a(i)\rangle \tag{23.6.7}$$

are solved to obtain a new set of spinorbitals which are used to construct a revised Fock operator, and so on. The cycle of formulation and reformulation is repeated until a convergence criterion is satisfied.

There is one more assumption suppressed until now, and this is that the angular effects in the third term, viz.,

$$\sum_{j \neq i} \int |\Psi(r_j)|^2 \frac{k e^2}{|r_j - r_i|} d^3 r_j ,$$

are averaged over to give a spherically symmetric potential.

A further modification was introduced to include the 'Exchange Interaction' as a modification to the third term of the $\hat{H}$, a form that is called the <u>Hartree-Fock Approximation</u>.

However, the interaction between the orbital angular momentum and the spin angular momentum of the electron must also be included, and this effect leads to an observable '<u>Fine Structure' splitting</u> of the energy levels. The Spin-Orbit Interaction can be written in terms of an additional energy

$$\sum_i \xi_i(r) \vec{L}_i \cdot \vec{S}_i , \tag{23.6.8}$$

where $\quad \xi_i(r) = \frac{1}{2m^2c^2} r^{-1}(\frac{dV}{dr}) \tag{23.6.9}$

with $\quad V(r) = -kZe^2 / r$.

The presence of the factor c^2 clearly indicates that this correction is due to a relativistic one.

So $\left\{ \frac{-\hbar^2}{2m} \hat{\nabla}_i^2 + \left[-\frac{Z k e^2}{r} + \sum_{j \neq i} \int |\Psi(r_j)|^2 \frac{k e^2}{|r_j - r_i|} d^3 r_j \right] - \sum_i \xi_i(r) \vec{L}_i \cdot \vec{S}_i \right\} \Psi r(\vec{r}, t) = E_i \Psi(\vec{r}, t)$

$$\tag{23.6.10}$$

23.7 CORRELATION ENERGY of a MANY-ELECTRON SYSTEM
(An Atom, a Molecule or a Solid)

23.7.1 The Many-electron problem is the problem of taking proper account of the interaction of the electrons.

Calculation of the properties of systems containing many electrons becomes almost impossibly complicated when one attempts to include the interactions of the electrons. The difficulty can be avoided either by neglecting this interaction altogether, or by considering only its average effect (i.e. independent electron approximation). It is of concern to us to calculate the total energy of a many-electron system, or more specifically, that part of it known as the CORRELATION ENERGY. The important techniques include perturbation theory and the use of GREEN'S FUNCTIONS.

From Section 23.2 it is known that the Determinantal functions form the orthonormal set, and on shall assume that this set is complete, in that any N-electron wave function $|\Psi\rangle$ can be expanded in an infinite series of the type the coefficients b_a being suitably chosen constant. This function is correctly Anti-symmetric, because interchange of the

$$|\Psi\rangle = \sum b_a |\varphi_a\rangle \qquad (23.7.1)$$

coordinates of two electrons changes the sign of every determinant.

23.7.2 MATRIX ELEMENTS

The problem is to find the eigen functions and eigen values of the Schrödinger Equation

$$\hat{H}(1,2,....,n)|\Psi(1,2,3,.....,n)\rangle = E|\Psi(1,2,3,.....,n)\rangle \qquad (23.2.18)$$

One can then write the matrix element of $\mathcal{H}$ between the states $|\varphi_b\rangle$ and $|\varphi_a\rangle$ as

$$\hat{H}_{ba} = \langle \varphi_b | \hat{H} | \varphi_a \rangle \qquad (23.7.2)$$

where $\langle \varphi_b \| \varphi_a \rangle = \delta_{ba}$ (27.3.3)

Owing to the orthonormality of $|\varphi_a\rangle$, one gets

$$\sum b_a(\hat{H}_{ba} - E\,\delta_{ba})|\varphi_a\rangle = 0 \qquad (23.7.4)$$

There is an infinite number of such equations, corresponding to all the states $|\varphi_b\rangle$, and each equation has an infinite number of terms. In other words, there one gets an infinite set of simultaneous linear equations in the b_a. The determinant of the coefficients of the b_a must vanish, for consistency requirement.

i.e., $\det|(\hat{H}_{ba} - E\,\delta_{ba})| = 0 \qquad (23.7.5)$

The determinant is of infinite order, E appearing only in the elements of the leading diagonal.

If the states $|\varphi_a\rangle$ are the eigen states of a Hamiltonian which differs from $\hat{H}$ by only a small perturbing term, then the approximate evaluation of the roots of the equation (23.7.5) is very much simpler. However, it is important to note that one has not made any such assumption here, so that equation (23.7.5) is correct, and in principle one gets the correct values for the energy levels for one complete orthonormal set of N-electron determinantal functions $|\varphi_a\rangle$.

The matrix elements, therefore, constitute a complete representation of $\hat{H}$ Any operator equivalent to $\hat{H}$ will have the same matrix elements of $\hat{H}$ (Chapter 11).

For the N-electron system, the analysis of the energy levels can be approximated by successive orders of perturbation, as indicated in the expression for the Hamiltonian, as follows:

where $\hat{H} = \sum_{k=1}^{k=N} \frac{1}{2m} p_k^2 - \sum_1^N Ze^2 [r_k]^{-1} + \sum_{i>j}' e^2 [r_{ij}]^{-1} + \sum_1^{'N} \xi(r_k)\vec{\ell}_k \cdot \vec{s}_k$

$$+ \sum_1^N e^2 |\vec{r}_e - \vec{r}_k|^{-1} + \xi(\vec{r}_e)\ell_e \cdot s_e \qquad (23.2.28)$$

where $\sum'$ means summations over all the electrons except the excited electron, and the subscript e refers to the excited electron, and.

$\sum_{k=1}^{k=N} \frac{1}{2m} p_k^2 - \sum_1^N Ze^2 [r_k]^{-1} + \sum_{i>j}' e^2 [r_{ij}]^{-1}$ = Non-relativistic part.

$\sum_1^{'N} \xi(r_k)\vec{\ell}_k \cdot \vec{s}_k$ = 1st order perturbation, (S.L. interaction)

$\sum_1^N e^2 |\vec{r}_e - \vec{r}_k|^{-1}$ = 2nd order perturbation,

$\xi(\vec{r}_e)\ell_e \cdot s_e$ = 3rd order perturbation.

Putting $\hat{H} = \hat{H}^{(0)} + \hat{H}^{(1)}$,

where $\hat{H}^{(0)} = \sum_{k=1}^{k=N} \frac{1}{2m} p_k^2 - \sum_1^N Ze^2 [r_k]^{-1} + \sum_{i>j}' e^2 [r_{ij}]^{-1} + \sum_1^{'N} \xi(r_k)\vec{\ell}_k \cdot \vec{s}_k$

$$\hat{H}^{(1)} = \sum_{1}^{N} e^2 \, | \vec{r}_e - \vec{r}_k |^{-1} \qquad (23.7.7)$$

23.7.2.1 MATRIX ELEMENTS of $\hat{H}^{(0)}$

For brevity, one may summarize:

$\hat{H}^{(0)}$ can have non-zero matrix elements only between two Determinantal Functions which are either the same or differ in a single one-electron function only, and these matrix elements are given by

$$\langle \varphi_a | \hat{H}^{(0)} | \varphi_a \rangle \qquad (23.7.8)$$

and $\quad \langle \varphi_b | \hat{H}^{(0)} | \varphi_a \rangle \qquad (23.7.9)$

23.7.2.2. MATRIX ELEMENTS of $\hat{H}^{(1)}$

$| \Psi^a(1,2,3,.....,n) \rangle$ is given by equation (23.2.24) extended to N-electrons.

If $\quad | \varphi_b \rangle = | \varphi_a \rangle$

it is known that becomes

$$\langle \varphi_a | \hat{H}^{(1)} | \varphi_a \rangle = \tfrac{1}{2} \sum_{j \neq 1}^{N} \sum_i \left\{ \begin{array}{l} \langle \varphi_a(i)\,\varphi_a(j) | \hat{H}^{(1)} | \varphi_a(i)\varphi_a(j) \rangle \\ - \{ \langle \varphi_a(j)\,\varphi_a(i) | \hat{H}^{(1)} | \varphi_a(i)\varphi_a(j) \rangle \} \end{array} \right\} \quad (23.7.10)$$

Suppose the $| \varphi_a \rangle$ and $| \varphi_b \rangle$ differ in only one function, and

$| \varphi_a(g) \rangle = | \varphi_b(g) \rangle$ for $g \neq k$,

but $\quad | \varphi_a(g) \rangle \neq | \varphi_b(k) \rangle$.

Then $\quad \langle \varphi_a | \hat{H}^{(1)} | \varphi_a \rangle$

$$= \{ \langle \varphi_b(k)\,\varphi_a(j) | \hat{H}^{(1)} | \varphi_a(k)\varphi_b(j) \rangle - \{ \langle \varphi_a(j)\,\varphi_b(k) | \hat{H}^{(1)} | \varphi_a(k)\varphi_a(j) \rangle \} \}$$

if $i \neq k$, $\qquad (23.7.11)$a

$$= \{ \langle \varphi_a(i)\,\varphi_b(k) | \hat{H}^{(1)} | \varphi_a(i)\varphi_a(k) \rangle - \{ \langle \varphi_b(k)\,\varphi_a(i) | \hat{H}^{(1)} | \varphi_a(i)\varphi_a(k) \rangle \} \},$$

if $j = k$, $\qquad (23.7.11)$b

$= 0$, otherwise. $\qquad (23.7.11)$c

The higher-order corrections can be found in the same way, but they are very complicated and Rayleigh-Schrodinger Perturbation Theory is rarely taken beyond the 2nd order.

23.7.3 SECOND QUANTIZATION (Occupation Number Representation)

A new mathematical formalism, known as SECOND QUANTIZATION (described in detail in Chapter 26, section 26.3) is convenient for further developments in the theory quantum systems. The Determinantal Function is

$$| \Psi(1,2,3,.....,n) \rangle = \sqrt{\tfrac{1}{n!}} \sum_P (-1)^P \hat{P} \big[| \varphi_a(1) \rangle | \varphi_b(2) \rangle | \varphi_c(3) \rangle .... | \varphi_f(n) \rangle \big] \quad (23.2.20)$$

Introduce now the destruction (annihilation) operator, $\hat{c}_{ak}$ which removes a state $| \varphi_a(k) \rangle$ from a determinantal function containing this state, that is to say, it converts an N-electron function containing $| \varphi_a(k) \rangle$ into an (N −1)-electron function not containing $| \varphi_a(k) \rangle$.

$$\hat{c}_{ak} | \Psi(1,2,3,.....,N) \rangle = \pm \sqrt{\frac{1}{(N-1)!}} \sum_P (-1)^P \hat{P} \Big[| \varphi_a(1) \rangle | \varphi_b(2) \rangle | \varphi_c(3) \rangle .... | \varphi_f(N-1) \rangle \Big]$$
(23.7.12)

Similarly defining the creation operator, $\hat{c}_{ak}^\dagger$ which adds a state $| \varphi_a(k) \rangle$ to a Determinantal Function not containing this state, that is to say, it converts an $(N-1)$-electron function not containing $| \varphi_a(k) \rangle$ into an N-electron function containing $| \varphi_a(k) \rangle$,

$$\hat{c}_{ak}^\dagger | \Psi^a(1,2,3,.....,N) \rangle = 0,$$
(23.7.13)

23.7.4.. The VACUUMSTATE $| 0,0,0,....,0 \rangle$

If one operates with the destruction operator c_1 upon the 1st order Determinantal Function $| \Psi(1,0,0,.....,0) \rangle$, in which only one-electron state $| \varphi_1(1) \rangle$ is occupied, the result is a fictitious 'zeroth order determinant' in which no one-electron state is occupied. This is called the VACUUMSTATE or EMPTYSTATE, denoted as $| \Psi(1,0,0,.....,0) \rangle$ or $| 0,0,0,....,0 \rangle$.

23.7.4.2. MATRIX ELEMENTS $\langle \varphi_b | \hat{H}^{(0)} | \varphi_a \rangle$ for a ONE-ELECTRON ATOM

For a on-electron system has only a single occupancy number and has value unity. Suppose that
$$| \Psi_a(,,,.k,....,) \rangle = | \varphi_k(1) \rangle$$
$$| \Psi_b(,,,.1,....,) \rangle = | \varphi_1(1) \rangle$$
$$\langle \varphi_a | \hat{H}^{(0)} | \varphi_b \rangle = \langle \varphi_a \Big| \sum_{1,1} \langle i | \hat{H}^{(0)} | j \rangle \hat{c}_1^\dagger \hat{c}_1 \Big| \varphi_b \rangle$$
$$= \Big| \sum_{1,1} \langle i | \hat{H}^{(0)} | j \rangle \langle \varphi_a \Big\| \hat{c}_1^\dagger \hat{c}_1 \Big| \varphi_b \rangle$$
(23.7.15)

23.4.3 MATRIX ELEMENTS $\langle \varphi_a | \hat{H}^{(1)} | \varphi_b \rangle$ for a TWO-ELECTRON ATOM

It can be shown that
$$\langle \varphi_a | \hat{H}^{(1)} | \varphi_b \rangle = \langle \varphi_a \Big\| \frac{1}{2} \sum_{1i,j,k,l} \langle ij | \hat{H}^{(1)} | kl \rangle \Big| \hat{c}_i^\dagger \hat{c}_j^\dagger \hat{c}_l \hat{c}_k \Big| \varphi_b \rangle$$
$$= \Big| \frac{1}{2} \sum_{1i,j,k,l} \langle ij | \hat{H}^{(1)} | kl \rangle \Big| \hat{c}_i^\dagger \hat{c}_j^\dagger \hat{c}_l \hat{c}_k$$
(23.7.16)

23.7.4.4 The CORRELATION ENERGY, W

The effect of the Coulomb repulsion between electrons is to correlate the electronic motions in such a way as to reduce the probability of two electrons closely approaching each other. Such correlations among the electronic motions may be called Coulomb correlations, to distinguish from another type of correlations, which are due to the *Pauli Principle*. The HF method, for example, takes account of the latter, but takes no account at all of Coulomb correlations.

The CORRELATION ENERGY, W is defined as the total energy, calculated with proper allowance for Coulomb correlations, minus the HF energy.

23.7.5 The TIME-DEVELOPMENT OPERATOR, $\hat{T}$

While dealing with the S-, H-, and Interaction Pictures the Time Development Operator, $\hat{U}(t,t_0)$ was introduced.

$$\hat{U}(t,t_0) = e^{\{-i\hat{H}(t-t_0)/\hbar\}}$$
(12.2.17)

$\hat{U}^\dagger(t, t_0) \hat{U}(t, t_0) = \Im$.

It can be shown that

$$\hat{U}(t, t') = 1 + \sum_{n=1}^{\infty} (-i)^n \int_{t'}^{t} \hat{H}^{(1)}(t_1) \, dt_1 \int_{t'}^{t_2} \hat{H}^{(1)}(t_2) \, dt_2 \cdots \int_{t'}^{t_{n-1}} \hat{H}^{(1)}(t_n) \, dt_n \quad (23.7.17)$$

23.7.6 The CHRONOLOGICAL OPERATOR, $\hat{P}$

In the application of Perturbation Theory to a many-electron system, one would like to make use of the expression (23.7.17) for U(t, t') as it stands. However a more compact expression may be obtained with the aid of an operator, due to Dyson, which is called the CHRONOLOGICAL OPERATOR, denoted by $\hat{P}$ (not to be confused with the Permutation Operator, $\hat{P}$ introduced in Section 23.2), to distinguish it from the similar Time-ordering Operator, due to Gian-Carlo Wick.

This operator, when applied to a product of time-dependent operators $A(t_1) B(t_2) \ldots$ taken at different times, t_1, t_2, ….arranges the operators so that the times decrease from left to right. Thus the product $\hat{A}(t_1) \hat{B}(t_2)$ one has

$$\hat{P}\{\hat{A}(t_1) \hat{B}(t_2)\} = \hat{A}(t_1) \hat{B}(t_2), \quad \text{if } t_1 > t_2$$
$$= \hat{B}(t_2) \hat{A}(t_1), \quad \text{if } t_2 > t_1. \quad (27.3.18)$$

Thus $\int_{t'}^{t} \hat{H}^{(1)}(t_1) \, dt_1 \int_{t'}^{t_1} \hat{H}^{(1)}(t_2) \, dt_2 = \frac{1}{2} \int_{t'}^{t} \int_{t'}^{t_1} \hat{P}\left[\hat{H}^{(1)}(t_1) \hat{H}^{(1)}(t_2)\right] dt_1 dt_2 \quad (23.7.19)$

Accordingly, if $t > t'$; the regions of integration for (23.7.19) is obviously,

$$\int_{t'}^{t} \int_{t'}^{t_1} \hat{P}\left[\hat{H}^{(1)}(t_1)\hat{H}^{(1)}(t_2)\right] dt_1 dt_2 = \int_{t'}^{t} dt_1 \int_{t'}^{t_1} \hat{P}\left[\hat{H}^{(1)}(t_1) \hat{H}^{(1)}(t_2)\right] dt_2 + \int_{t_1}^{t} \hat{P}\left[\hat{H}^{(1)}(t_1) \hat{H}^{(1)}(t_2)\right] dt_2$$

$$= \int_{t'}^{t} dt_1 \int_{t'}^{t_1} \hat{H}^{(1)}(t_1) \hat{H}^{(1)}(t_2) \, dt_2 + \int_{t_1}^{t} \hat{H}^{(1)}(t_2) \hat{H}^{(1)}(t_1) \, dt_2 \quad (23.7.20)$$

It can be shown that

$$\int_{t'}^{t} \int_{t'}^{t_1} \hat{P}\left[\hat{H}^{(1)}(t_1)\hat{H}^{(1)}(t_2)\right] dt_1 dt_2 = 2 \int_{t'}^{t} dt_1 \int_{t'}^{t_1} \hat{H}^{(1)}(t_1) \hat{H}^{(1)}(t_2) \, dt_2$$

$$= 2 \int_{t'}^{t} \hat{H}^{(1)}(t_1) \, dt_1 \int_{t'}^{t_1} \hat{H}^{(1)}(t_2) \, dt_2 \quad (23.7.21)$$

Generalization of this gives

$$\hat{U}(t, t') = 1 + \sum_{n=1}^{\infty} \left[(-i)^n / n!\right] \int_{t'}^{t} dt_1 \int_{t'}^{t} dt_2 \cdots \int_{t'}^{t} dt_n \, \hat{P}\left[\hat{H}^{(1)}(t_1)\hat{H}^{(1)}(t_2) \cdots \hat{H}^{(1)}(t_n)\right]$$

(23.7.22)

23.7.6.1 COMPACT FORM of Equation (23.7.22):

$$\hat{U}(t, t_0) = \hat{P} \, e^{-i \int_{t'}^{t} \hat{H}(t_1) dt_1} \quad (23.7.23)$$

Since $\left\{\int_{t'}^{t_1} \hat{H}^{(1)}(t) \, dt_1\right\}^n = \int_{t'}^{t} \hat{H}^{(1)}(t_1) \, dt_1 \cdots \int_{t'}^{t_1} \hat{H}^{(1)}(t_n) \, dt_n$

23.8 FEYNMAN DIAGRAMS

23.8.1 REVOLUTIONS IN QUANTUM MECHANICS

Processes in which the absorption or emission of a *single photon* are spontaneous and stimulated ones. Both, in the fundamental theory of radiation processes as well as in many practical applications of non-linear optics, it will be necessary to study effects in which several or even *many photons* play a role. High order Perturbation Theory is useful and necessary for qualitative and quantitative discussions. The modern view of interactions of elementary charged particles and the EM Field is embodied in a Quantum Field Theory. R.P. Feynman (1948), Sin-Itiro Tomonaga (1948) and Julian Schwinger (1948) independently developed Quantum Electro-Dynamics (QED).

In the relative antiquity of the 1930s and 1940s books on Quantum Mechanics looked difficult because their pages covered with integral signs and complicated sum over states of particles. The introduction by Paul Dirac of his bracket notation are pages of much of their inhospitality. However, another simplification took place in the 1950s when physicists began to draw little squiggles like those shown in the margin, and were still able to claim that they are doing calculations. In this second revolution in Quantum Mechanics, to get an overview of the possible processes, Feynman (1918 - 88) showed how squiggles could be used to portray mathematical expressions systematically. These diagrams go by his name. They provide a picture of the mechanisms operating in particle processes and I will describe them in the context QED. This technique of graphs is particularly useful to view EM processes. (Fig. 23.5). Some of its basic ideas are called *Feynman Rules*.

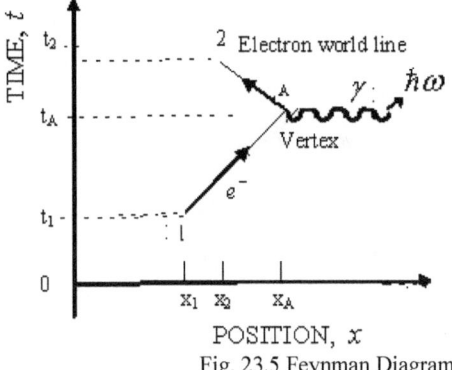

Fig. 23.5 Feynman Diagram

The Fig 23.5 says that the electron e^- is moving from $(x_1, t_1) \rightarrow (x_2, t_2)$ interacts with the EM Field at (x_A, t_A). The vertex shows the basic interaction between the electron and the EM Field, $\gamma(\hbar\omega)$.

23.8.2 FEYNMAN RULES

The rules for the construction and interpretation of Feynman Diagrams are as follow:

The basic element of a Feynman Diagram is a VERTEX representing the interaction of a fundamental fermion with a mediating Boson. The axes are space and time. Straight lines indicate the paths of Fermions, whereas the mediating Bosons are by wavy lines. That is

(1) Energy and momentum are conserved at the VERTEX,
(2) Electric charge is conserved; there are other conservation rules that apply in other circumstances and that will be described later.
(3) Solid STRAIGHT LINES with arrowheads pointing in the direction of increasing time are used to represent PARTICLE Fermions propagating toward in time.

Arrowheads pointing in the reverse direction represent ANTI-PARTICLE of Fermions propagating toward in time.

Is technically a propagator (since the line for a Fermion represents the mathematical expression for the propagation of the particle through time).
(4) BROKEN, **wavy**, or CURLY lines are used to represent BOSONS,

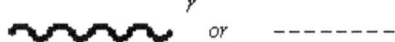

(5) Lines having one end at the boundary of the diagram represent free (that is real) particle approaching or leaving a reaction (normally a boundary is drawn only when it is explicitly required).
(6) LINES that JOIN TWO VERTICES (internal lines) normally represent VIRTUAL particles. There are exceptions to this when an internal line represents a real but unstable particle, which is a compound state of the initial particles, for example the diagram for $e^- e^+ \to Z^0 \to n\, n$.
(7) The time ordering of the vertices connected by an internal line is not determined so that two diagrams having an internal line apparently oriented differently with respect to time, but otherwise not the same, are NOT different diagrams.

(8) Every particle at the boundary should be LABELED with a momentum; If this is done two diagrams, which might otherwise appear to be the same, become different diagrams. However, one does not include momentum labels unless necessary. Time increases from left to right. One can define a time axis, and orient the other three lines in any way one wishes; say time runs from left to right across the page.
Some Feynman Diagrams for typical processes are shown in Fig. 23.6 to Fig. 23.17.

23.8.2.1 EMISSION OF A PHOTON BY AN ELECTRON: Fig.23.6 indicates the emission of a photon, γ ☐☐by an electron, e⁻. The vertex represents EM coupling.

Fig. 23.6

23.8.2.2. ABSORPTION OF A PHOTON BY AN ELECTRON:
Fig. 23.7 indicates an electron e⁻ absorbing a photon γ. The vertex is for EM coupling.

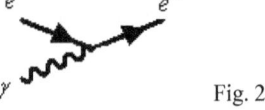

Fig. 23.7

23.8.2.3. EMISSION OF A PHOTON BY A POSITRON:
A positron e^+ emits a photon, γ in Fig 23.8.

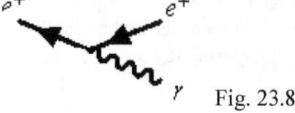

Fig. 23.8

23.8.2.4 ABSORPTION OF A PHOTON BY A POSITRON:
A positron e^+ absorbs a photon γ, in Fig23.9.

Fig. 23.9.

23.8.2.5 PAIR PRODUCTION:
A photon γ converting to an electron-positron (e^- - e^+) pair, where the arrowhead points backward in time for the positron, (Fig. 23.10).

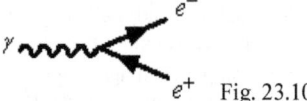

Fig. 23.10

23.8.2.6. A TWO PARTICLE COLLISION

This is the simplest process of "scattering" event. Fig.23.11 indicates that the process is started and ended with one electron and one photon. Only the momentum and energies change in the process

Fig. 23.11.

23.8.2.7. DIFFERENT ORIENTATIONS (or, TIME-ORDERING):

Consider Fig.23.11. Feynman tells one to draw all possible diagrams. First add one intermediate photon line to Fig. 23.11. The one finds a three time-ordered diagram, as shown in Fig. 23.12, or

Fig. 23.12   Fi g. 23. 13

It is to be understood that Fig. 23.12 and Fig. 23.13. are two different orientations of the same event.

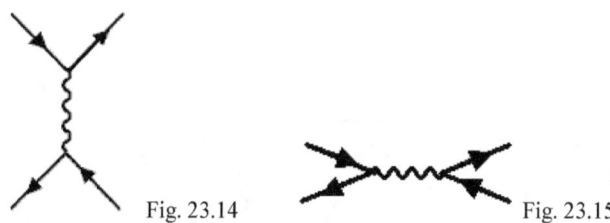

Fig. 23.14   Fig. 23.15

Feynman diagram, Fig. 23.15, is really a quite different process. It is an intermediate stage with only a photon (virtual photon) present.

23.8.2.8 A COMPLICATED FEYNMAN DIAGRAM:

One can also draw more complicated diagrams by adding more photons, for example, as in Fig. 23.16. and Fig. 23.17.

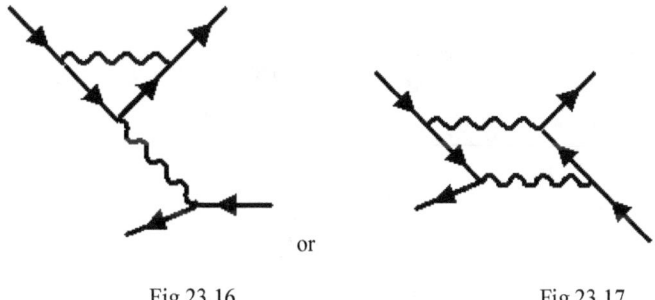

Fig.23.16.   or   Fig.23.17

In fact one could have any number of photons. Each diagram has definite "complex amplitude" – related to a set of Feynman rules. One part of these rules is that there is a multiplication factor of $\alpha = \frac{1}{137} = ke^2/hc$, for each photon, so amplitudes for diagrams with many photons are smaller, compared with processes with one. 'e' factor in ke^2/hc shows it is EM coupling or electronic charge.

In all these diagrams, one arrow points towards the vertex, and one away. This reflects the conservation of electric charge. Momentum and angular momentum both are also assumed to be conserved.

23.8.2.9 Deeply virtual Compton scattering

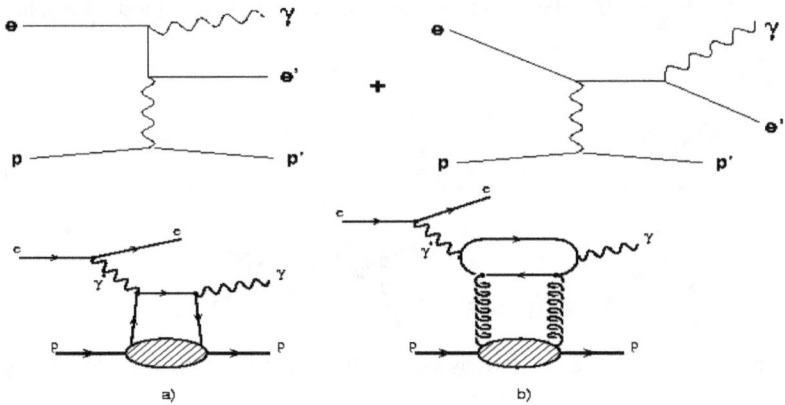

Fig. 23.18 Deeplyvirtual Compton scattering

Worked out Example 23.10

Give the Feynman Diagram for $e^- - \gamma$ scattering.

Solution. $e^- - \gamma$ scattering is Compton Effect.

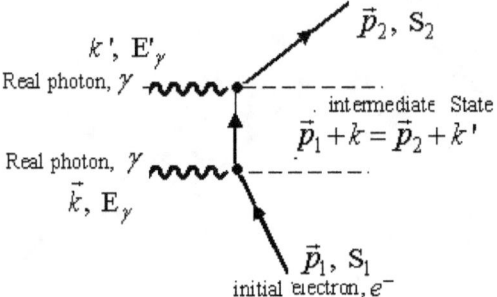

Fig. 23.21 $e^- - \gamma$ scatteing

Virtual photon whose 4-momenum is determined by conservation of 4-momenta of the vertices.

23.8.2.10 Feynman Diagram and the Lamb Shift
(See Chapter 26)

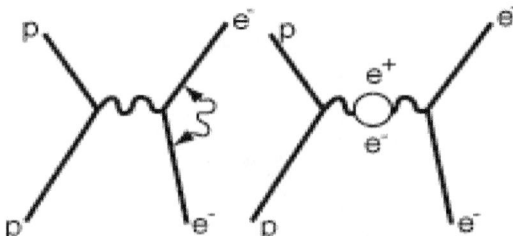

Fig. 23. 22: Feynman graph of the Lamb Shift

23.8.3. FEYNMAN DIAGRAMS for ATOMIC STRUCTURE

23.8.3.1. REAL and VIRTUAL PROCESSES

Energy conservation is violated in all the basic processes. They are called the *virtual* processes to emphasize that they cannot occur in isolation in free space. They violate the relation, $E^2 = p^2c^2 + m_0^2 c^4$. To make a REAL process two or more virtual processes must be combined in such a way that energy conservation is only violated for a short period of time, τ compatible with the energy-time Uncertainty Principle: $\tau \cdot \Delta E = \hbar$. In particular the initial and final states must have the same energy.

Some Feynman Diagrams for typical processes are shown in Fig. 23.6 to Fig. 23. 21. They illustrate the emission of a *virtual* photon; emission of additional virtual photons and re-absorption by the same electron; the exchanged photon converts to a pair, which later annihilate; exchange of two photons ; a molecule emitting and absorbing a photon (Fig 23.22 and Fig. 23. 23); the Raman effect (Fig. 23.24); Bremstrahlung (Fig. 23.25).

Worked out Example 23.11

Give the Feynman Diagram for the emission of a γ by a molecule.

Solution

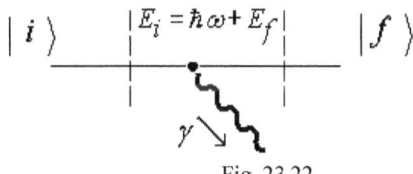

Fig. 23.22

Worked out Example 23.12

Give the Feynman Diagram associated with the Raman Effect.

Solution:

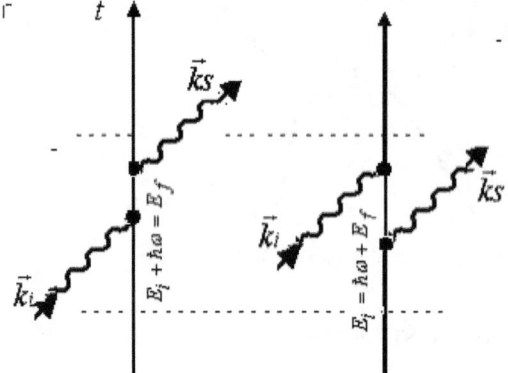

Fig. 23.23 The Raman Effect

Once the Feynman Rules are understood for translating the arrangement of vertices into mathematical expressions, one can interpret any of these processes mathematically. One can use them to predict the intensities of spectral transitions and strengths of the interacting forces involved.

23.8.4. IMPORTANCE

23.8.4.1   The AMPLITUDE OF A DIAGRAM:

Worked out Example 23.14
Give the Feynman graph of the *Bremstrahlung* of an electron in a Coulomb Field
Solution:

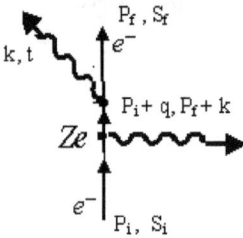

Fig 23.25 Bremstrahlung

Feynman Diagrams are important in the QED calculation of cross sections and of transition rates. Every diagram represents a contribution to the total probability amplitude for the process. The technique for calculating such amplitudes is not the subject matter here, but I can mention one useful thing. The amplitude of a given diagram contains several factors, including an 'e' (the charge on an electron) for every vertex; thus the amplitude for $e^+ - e^-$ elastic scattering is proportional to e^2, or in dimensionless term, in $\alpha = \frac{1}{137} = ke^2/\hbar c$. It follows that the cross section is proportional to α^2. The '*bremstrahlung*' process has amplitude, which contains e for each vertex and Ze for the virtual photon attaching to the nuclear charge; the cross section is therefore proportional to $Z^2 \alpha^3$.

Another property is that the virtual photon has to transfer momentum, $\vec{q} = \vec{p} - \vec{p}\,'$, and this makes the amplitude proportional to q^{-2}.

Worked out Example 23.15
Draw the Feynman Diagram for the emission of a Boson by a fundamental Fermion, any one in the processes involving Weak Interactions.

Solution:

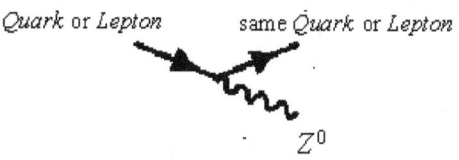

Fig. 23.26 Weak interaction

OR

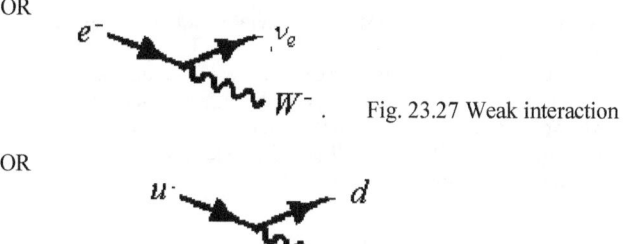

Fig. 23.27 Weak interaction

OR

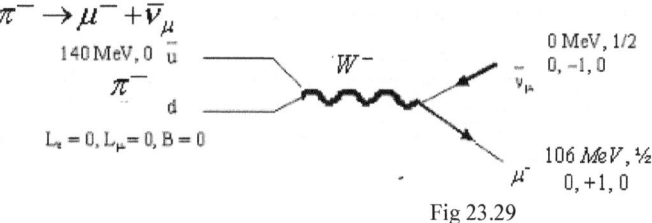

Fig. 23.28 Weak interaction

Worked out Example 23.16 Draw the Feynman Diagram for the Spontaneous Decay of π^-.
Solution:

$\pi^- \to \mu^- + \bar{\nu}_\mu$

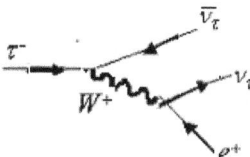

140 MeV, 0 $\bar{u}$

π^-

d

$L_e = 0, L_\mu = 0, B = 0$

W$^-$

0 MeV, 1/2
0, –1, 0
$\bar{\nu}_\mu$

106 MeV, ½
0, +1, 0
μ^-

Fig 23.29

Worked out Example 23.17
Draw the Feynman diagram for the proposed decay $\tau^- \to e^+ + \nu_\tau + \bar{\nu}_\tau$

Solution:

τ^-

$\bar{\nu}_\tau$

ν_τ

W$^+$

e^+

Fig. 23.35 Weak interaction

23.8.4.2. LEADING-ORDER DIAGRAM

This is a diagram which represents a given physical process and which has the minimum possible number of vertices. There may be more than one such diagram.

The calculations of cross sections and transition rates depend on the use of perturbation theory. The Feynman diagrams are a geometrical way of representing the terms in the perturbation calculation. The leading-order diagrams correspond to the LOWEST -ORDER PERTURBAION calculation. It is given by

$$A_1 = \langle \varphi_0 | \hat{U}_1 | \varphi_0 \rangle = \langle \varphi_0 | \{(-i)\int_{-\infty}^{t} \hat{H}^{(1)}(t_1) dt_1 \} | \varphi_0 \rangle \qquad (23.8.1)$$

23.8.4.3 HIGHER-ORDER DIAGRAM

For the second –order term one gets

$$A_2 = \langle \varphi_0 | \hat{U}_2 | \varphi_0 \rangle = \langle \varphi_0 | \{(-i)^2 \int_{-\infty}^{t} \hat{H}^{(1)}(t_1) dt_1 \int_{-\infty}^{t} \hat{H}^{(1)}(t_2) dt_2 \} | \varphi_0 \rangle \qquad (23.8.2)$$

The nth - order term may be easily deduced from (23.8.1) and (23.8.2) using $A_n = \langle \varphi_0 | \hat{U}_n | \varphi_0 \rangle$.

One may also define the creation and destruction operators for holes and particles (excited states) in terms of the previous operators.

In addition to the leading-order diagrams there are higher- order diagrams, each of which contribute extra amplitudes to the total. Since they can only be drawn by adding internal lines they must always involve two more vertices for each step on increasing order. It is the smallness of α (=1/137) relative to 1, which normally makes all but next to leading-order contributions negligible and the calculations manageable.

Worked out Example 23.18

What is the limitation of Feynman technique in application to processes?

Solution:

Feynman technique is only useful when the type of coupling involved in the event is small, i.e., for EM or weak interactions, and not for strong interactions, except for very high energies.

In addition to the leading-order diagrams there are higher- order diagrams, each of which contribute extra amplitudes to the total. Since they can only be drawn by adding internal lines they must always involve two more vertices for each step on increasing order. It is the smallness of α (=1/137) relative to 1, which normally makes all but next to leading-order contributions negligible and the calculations manageable.

REVIEW QUESTIONS

R.Q. 23.1   Write down the Determinantal wave function for a Normal Lithium atom. Prove that the s-state can at the most accommodate two electrons. Answer: Since $|\varphi_c(m)\rangle = a|\varphi_a(m)\rangle + b|\varphi_a(m)\rangle$,etc., $\left| \Psi^a (1,2,.3) \right\rangle = \frac{1}{\sqrt{3!}} x(0) = 0$. It shows that it is impossible for the three electrons to be in the ground state, because the s-state ($\ell = 0$) can at the most accommodate two electrons).

R.Q. 23.2 Define a tensor operator. Deduce the Wigner-Eckart theorem and give a proof for the theorem.

R.Q. 23.3 What do you mean by identical particles? Define Symmetric and Anti-symmetric wave functions. Show how to construct Symmetric and Anti-symmetric wave functions for a system of identical particles.

R.Q. 23.4 Evaluate the Coulomb integral for the configuration $1s^2$ of a hydrogenic atom given the following:

$$|R_{12}|^{-1} = (1/R_>)\sum (R_</R_>)^\ell \cdot \left[4\pi/(\ell+\tfrac{1}{2})\right] \cdot \sum |\psi^o_{n\ell}\rangle \cdot \hat{Y}^m_\ell(\theta_1,\varphi_1) \cdot \hat{Y}^m_\ell(\theta_2,\varphi_2)$$

when $R_1 < R_2$, and with R_1 and R_2 interchanged when $R_1 < R_2$. What is this value for a Helium atom? Given:

$$|\psi\rangle_{1s^2} = \pi^{-1/2}\left(\frac{Z}{a_o}\right)^{3/2} e^{-\left(\frac{ZR}{a_o}\right)}$$

(Answer: $J = \tfrac{5}{8}\left(\tfrac{e^2}{4\pi\varepsilon_o}\right)\left(\tfrac{Z}{a_o}\right)$). For He atom $\approx 34eV$).

R.Q. 23.5 Predict the lowest energy level for the configuration, $2p^2$ (Answer: 2p shell can accommodate maximum of 6 electrons. $2p^2$ shell is less than half-filled. The ground term is ^3P, it has three levels, with $J = 2, 1, 0$. Prediction, as per Hund's Rules, & of the L-S Coupling, of the lowest energy level is $^3P_0$).

R.Q. 23.6. In the case of N = 3, what state does the operator c_2 gives on the given state

$$|\Psi(1,2,3,....,N)\rangle = \sqrt{\tfrac{1}{3!}}\sum_p (-1)^p \hat{P}\left[|\varphi_a(1)\rangle |\varphi_b(2)\rangle |\varphi_c(2)\rangle\right].$$

How will it be affected if $\hat{c}_2^\dagger$ were to operate on it? (Answer: $\hat{c}_2 |\Psi(1,2,3)\rangle = \sqrt{\tfrac{1}{3!}}\begin{pmatrix}\varphi_1(1) & \varphi_1(2)\\ \varphi_2(1) & \varphi_2(2)\end{pmatrix}$; $\hat{c}_2^\dagger |\Psi(1,2,3)\rangle = 0$)

R.Q. 23.6 Give the Feynman Diagram to show that a positron (e^+) and an electron (e^-) meet to annihilate each other producing a photon, γ. (:Answer:

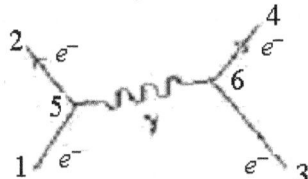

)

R.Q. 23.7 Draw the Feynman diagram for $e^- - e^-$ scattering. (Answer: One of the e^- goes from 1 to 3 and the other from 2 to 4 reroute the $1^{st} e^-$ emits a γ at 5, which the 2nd absorbs at 6. The e^- lines have arrows as they move forward in time.

R.Q. 23.8 In the case of N = 3, what state does the operator c_2 gives on the given state

$$|\Psi(1,2,3,....,N)\rangle = \sqrt{\tfrac{1}{3!}}\sum_p (-1)^p \hat{P}\left[|\varphi_a(1)\rangle |\varphi_b(2)\rangle |\varphi_c(3)\rangle\right].$$

How will it be affected if $\hat{c}_2^\dagger$ were to operate on it?

(Answer: $\hat{c}_2 |\Psi(1,2,3)\rangle = \sqrt{\tfrac{1}{3!}}\begin{pmatrix}\varphi_1(1) & \varphi_1(2)\\ \varphi_2(1) & \varphi_2(2)\end{pmatrix}$; $\hat{c}_2^\dagger |\Psi(1,2,3)\rangle = 0$)

R.Q. 23.9 Find the effect of operating $\hat{c}_1^\dagger$ on $|0,0,0,....,0\rangle$.

(Answer: $\hat{c}_1^\dagger | 0,0,0,.....,0 \rangle = | 1,0,0,.....,0 \rangle$)

R.Q. 23.10 Prove that

$$\hat{U}(t,t') = 1 + \sum_{n=1}^{\infty}\left[(-i)^n/n!\right]\int_{t'}^{t} dt_1 \int_{t'}^{t} dt_2 \cdots \int_{t'}^{t} dt_n\, \hat{P}\left[\hat{H}^{(1)}(t_1)\hat{H}^{(1)}(t_2)\cdots\hat{H}^{(1)}(t_n)\right]$$

where $\hat{P}$ is the Chronological Operator and $\hat{U}$ is the Time Development operator.

R.Q. 23.11 Identify the process indicated by the Feynman Diagram

(Answer:

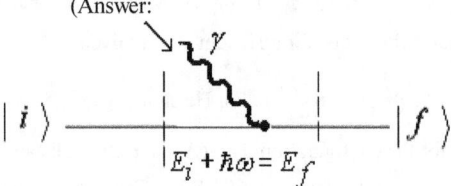

Atomic or molecular absorption).

R.Q. 23.12 Draw the Feynman Diagram for the emission of a Boson by a fundamental Fermion, in the process involving Electro Magnetic interactions.

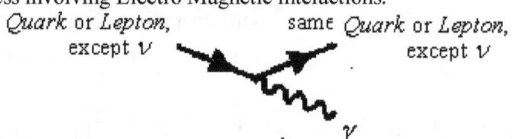

(Answer:).

R.Q. 23.13 Draw the Feynman Diagram for the emission of a Boson by a fundamental Fermion, one in the process involving Strong Interactions

(Answeer: Strong interaction)

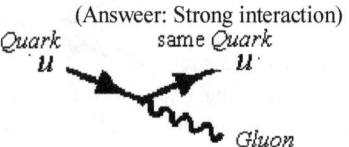

R.Q. 23.14 Draw the Feynman Diagram for the creation of a particle and an anti-particle. (Answer:

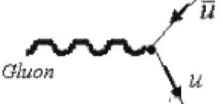

R.Q. 23.15 Draw the Feynman diagram for the destruction of a particle and anti-particle results.
(Answer:

R.Q. 23.16 Draw the Feynman diagram for the Spontaneous Decay described by $n \to p + e^- + \bar{\nu}_e$
(Answer:

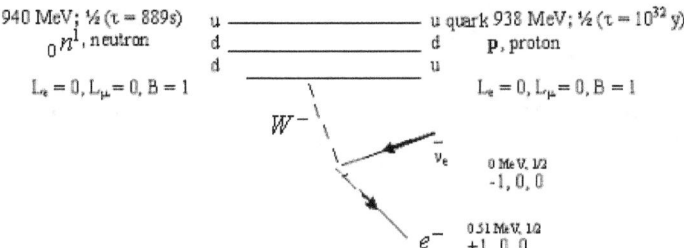

R.Q. 23.17   Show the Feynman diagram of a Weak interaction.

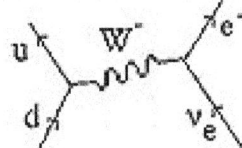

(Answer Weak interaction).

R.Q. 23.18   Illustrate the Feynman Diagram for the Electron Capture (EC) process in any standard nuclear decay.

(Answer: Consider the radioactive decay of $_{27}Co^{57}$ nuclide, as shown, as a form (of EC)

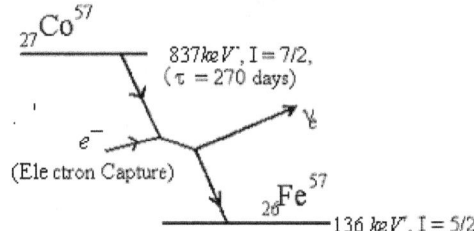

of β-decay..

*&^*&^*&^*&^*&^*

Chapter 24

TIME DEPENDENT PERTURBATION THEORY

Chapter 24

TIME DEPENDENT PERTURBATION THEORY

"Before telling me anything, I'd like you to pass a little test.
It's called the *Triple Filter Test*.(Truth, Goodness & Usefulness)":   Socrates
"Fate determines who comes into our lives. The heart determines who stays... " Socrates

24.1.1 INTRODUCTION

It was seen earlier in Chapters 19 and 20 that both the approximations viz. the Stationary State Perturbation Theory (Time Independent Perturbation Theory) and the Variational Principle are powerful, and the two may be applied to the calculation of molecular properties. Perturbation Theory is particularly useful when one is interested in responses of atoms and molecules to electric and magnetic fields. When the Hamiltonian, $\hat{H}$ of a system is time-dependent it is generally impossible to obtain exact solutions of the Schrödinger Equation.

Three approximation schemes are known, and all these are based on the following assumption: $\hat{H}$ of the system can in some way be approximated to a time-independent $\hat{H}_0$. There are three types of time dependences. They are:

1) Time dependent part of $\hat{H}$ is less than $\hat{H}_0$. In this case the PERTURBATION THEORY can be applied.

2) $\hat{H}$ changes very slowly on a time-scale set by the periodic times associated with the approximate stationary state solutions. For the rate of change of change of $\hat{H}$ small, this is known as the ADIABATIC APPROXIMATION.

3) In the third case, $\hat{H}$ changes rapidly, where SUDDEN APPROXIMATION is used.

There is another approximation technique, known as the *Self-Consistent Fields*, and it contains an iterative procedure for solving the Schrödinger Equation for systems of many particles (Chapter 23).

The Stationary State Perturbation Theory is quite successful because molecular response is so rapid that for all practical purposes the systems forget as though the perturbed states were not occurred due to switching. There is a much more important reason for studying the Time-Dependent Perturbation Theory than to understand quantum mechanics. Many important perturbations never 'settle down' to a constant value; a molecule exposed to EM radiation is especially an important example.

24.2 PERTURBATIVE EXPANSION

24.2.1 The SOLUTION to Schrödinger equation:
A very important type of time-dependent problem is where the Hamiltonian

$$\hat{H}(\vec{r},t) = \hat{H}_0(\vec{r}) + \hat{H}^{(1)}(\vec{r},t) \qquad (24.2.1)$$

As an example, for the perturbing part $\hat{H}^{(1)}(\vec{r},t)$ is the case of an atomic system subject to an oscillating electric field (such as that associated with an EM wave) which can cause transitions to occur from one energy state to another of the system (as a result of interaction between the oscillating EM Field and the system's induced instantaneous dipole moment).
<u>Case (a)</u>: $t \le 0$,

Let $\hat{H}^{(1)}(\vec{r},t) = 0$, for $t \le 0$, i.e.
$$\hat{H}^{(1)}(\vec{r},t) = \hat{H}_0(\vec{r}).$$
Then $\hat{H}_0 |\Psi_n^0, t\rangle = E_n^0 |\Psi_n^0, t\rangle$ (24.2.2)

which has solutions
$$|\Psi_n^0, t\rangle = |\Psi_n^0\rangle e^{-iE_n t/\hbar}.$$ (24.2.3)

where E_n^0 are the eigen energies corresponding to the eigen states $|\Psi_n^0, t\rangle$.

Case (b): for $t > 0$,
one has the actual Hamiltonian given by equation (24.2.1).
As in the case of time-independent perturbation, let
$$\hat{H}^{(1)}(\vec{r},t) \ll \hat{H}_0(\vec{r})$$ (24.2.4)

Therefore one has to deal with the TDSE (Time-Dependent Schrödinger Equation)
$$\hat{H}|\Psi_n, t\rangle = [i\hbar \, \partial/\partial t]|\Psi_n, t\rangle$$ (24.2.5)

For the systematic development of perturbation theory, it is convenient to separate off the time evolution occurring due to $\hat{H}_0$ or $t < 0$. For this *Interaction Picture* is used.

$$|\Psi_n^0(t)\rangle_I = e^{-i\hat{H}_0 t/\hbar} |\Psi_n(t)\rangle_S.$$ (24.2.6)

and $\hat{H}_I(t)|\Psi_n^0(\vec{r},t)\rangle_I = [i\hbar \, \partial/\partial t]|\Psi_n(\vec{r},t)\rangle_S$ (24.2.7)

where $\hat{H}_I(t) = e^{i\hat{H}_0 t/\hbar} \hat{H}(t) e^{-i\hat{H}_0 t/\hbar}$. (24.2.8)

and $|\Psi_n(t)\rangle_I = |\Psi_n(t=0)\rangle_I + [1/i\hbar] \int_0^t dt' \, \hat{H}_I(t') |\Psi_n(t')\rangle_I$. (24.2.9)

24.2.2. A TWO=LEVEL PERTURBED SYSTEM

24.2.2.1. UNPERTURBED state

Denoting the two values of energy as E_1 and E_2 corresponding to the states $|\Psi_1\rangle$ and $|\Psi_2\rangle$, these are the solutions of (24.2.2.).

Here $|\Psi_n^0, t\rangle = |\Psi_n^0\rangle e^{-iE_n t/\hbar}$ (24.2.3)

These basis functions satisfy the relation
$$\langle \Psi_1^0 | \Psi_2^0 \rangle = \delta \Psi_{12}$$ (24.2.10)

and general unperturbed state
$$|\Psi^0\rangle = \sum_1^2 c_n^0 |\hat{\psi}_n^0\rangle$$ (24.2.11)

and $|c_n^0|^2$ = probability of finding the system in the state $|\Psi_n^0\rangle$.

24.2.2.2. PERTURBED System

In the presence of the perturbation, $\hat{H}^{(1)}(\vec{r},t)$ the perturbed state $|\Psi_n\rangle$, of the two level system, has to be found. It is described by the linear combination

$$|\Psi(t)\rangle = \sum_{1}^{2} c_i(t)|\hat{\psi}_i^0(t)\rangle \qquad (24.2.12)$$

where $c_i(t)$ are the expansion coefficients. They are themselves time-dependent, because of time dependence of the Hamiltonian. The significance of this time dependence is of fundamental importance. So the over all time dependence of $|\Psi(t)\rangle$ arises both from

(i)   the oscillating time dependence of the base functions, $|\Psi_n^0, t\rangle$, and

(ii) the time dependence of the coefficients, $c_i(t)$.

$c_i(t)$ represents the fact that, the system evolves under the influence of the perturbation. If it starts from $|1\rangle$ to $|2\rangle$,

the probability p that at any instant the system is in state $|2\rangle$ is

$$p|2\rangle = |c_i(t)|^2,$$

the probability p that at any instant the system is in state $|1\rangle$ is

$$p|1\rangle = 1 - |c_i(t)|^2,$$

From equation (24.2.5)

$$\hat{H}|\Psi_n(\vec{r},t)\rangle = [i\hbar\, \partial/\partial t]|\Psi_n(\vec{r},t)\rangle \qquad (24.2.5)$$

and the general solution is

$$|\Psi(t,t)\rangle = \sum_{1}^{2} a_i(t)|\hat{\psi}_i^0(\vec{r},t)\rangle = \sum_{1}^{2} a_i(t)|\hat{\psi}_{0i}^0\rangle e^{-iE_it/\hbar} \qquad (23.2.12)$$

24.2.2.3 To FIND THE COEFFICIENTS, $a_i(t)$

Substitution of (24.2.12) and (24.2.1) in (24.2.5)

$$[\hat{H}_0(\vec{r}) + \hat{H}^{(1)}(\vec{r},t)]\sum_{1}^{2} a_i(t)|\hat{\psi}_i^0(\vec{r},t)\rangle = [i\hbar\, \partial/\partial t]\sum_{1}^{2} a_i(t)|\hat{\psi}_i^0(\vec{r},t)\rangle \qquad (24.2.13)$$

L.H.S.$= a_1(t)\hat{H}_0|\hat{\psi}_1^0\rangle + a_2(t)\hat{H}_0|\hat{\psi}_2^0\rangle + a_1(t)\hat{H}^{(1)}|\hat{\psi}_1^0\rangle + a_2(t)\hat{H}^{(1)}|\hat{\psi}_2^0\rangle$ (24.2.14)a

R.H.S. $= [i\hbar\, \partial/\partial t]a_1(t)|\hat{\psi}_1^0\rangle + [i\hbar\, \partial/\partial t]a_2(t)|\hat{\psi}_2^0\rangle$

$= [i\hbar\, a_1(t)(\partial/\partial t)|\hat{\psi}_1^0\rangle + i\hbar|\hat{\psi}_1^0\rangle (\partial/\partial t)a_1(t)] + [i\hbar\, a_2(t)(\partial/\partial t)|\hat{\psi}_2^0\rangle$

$+ [i\hbar\, a_2(t)(\partial/\partial t)|\hat{\psi}_2^0\rangle + i\hbar|\hat{\psi}_2^0\rangle (\partial/\partial t)a_2(t)].$ (24.2.14)b

To find the rate of change, $(\partial/\partial t)a_1(t)$ and $(\partial/\partial t)a_2(t)$ one must use the orthonormality property of the basis functions,

$$\langle \hat{\psi}_2^0 | \hat{\psi}_1^0 \rangle = \delta_{12} \qquad (24.2.10)$$

Expressing (24.2.14) a and b with time-independent and time-dependent factors, explicitly

$$a_1(t)\hat{H}^{(1)}|\hat{\psi}_{01}^0\rangle e^{-iE_1 t/\hbar} + a_2(t)\hat{H}^{(1)}|\hat{\psi}_2^0\rangle e^{-iE_2 t/\hbar}$$

$$= i\hbar\,(\partial/\partial t)a_1(t)|\hat{\psi}_{101}^0\rangle e^{-iE_1 t/\hbar} + i\hbar\,(\partial/\partial t)a_2(t)|\hat{\psi}_{02}^0\rangle e^{-iE_2 t/\hbar} \qquad (24.2.11)$$

Multiplying throughout by $\langle \hat{\psi}_{01}^0 |$ and applying $\langle \hat{\psi}_2^0 | \hat{\psi}_1^0 \rangle = \delta_{12}$,

$$a_1(t)\hat{H}_{11}^{(1)}(t)e^{-iE_1 t/\hbar} + a_2(t)\hat{H}_{12}^{(1)}(t)e^{-iE_2 t/\hbar} = i\hbar\, a_1(t)e^{-iE_1 t/\hbar}.$$

where $\hat{H}_{ij}^{(1)} = \langle \hat{\psi}_{0i}^0 | \hat{H}^{(1)}(t) | \hat{\psi}_{0j}^0 \rangle \equiv \langle i | \hat{H}^{(1)}(t) | j \rangle$ (24.2.12)

= the matrix element.

$(\partial/\partial t) a_1(t) = (1/\hbar) \, a_2(t) \hat{H}_{12}^{(1)}(t) \, e^{-i(E_1-E_2)t/\hbar} = (1/\hbar) \, a_2(t) \hat{H}_{12}^{(1)}(t) \, e^{-i\omega_{12}t/\hbar}$ (24.2.12) a

Similarly, $(\partial/\partial t) a_2(t) = (1/\hbar) \, a_1(t) \hat{H}_{21}^{(1)}(t) \, e^{+i(E_1-E_2)t/\hbar} = (1/\hbar) \, a_1(t) \hat{H}_{21}^{(1)}(t) \, e^{-i\omega_{21}t/\hbar}$

(24.2.12) b

The solution of these two equations, say $a_1(t)$ depends on the time dependence of $a_2(t)$, and the vice-versa.

$$\omega_{12} = (E_1 - E_2)/\hbar$$ (24.2.14)

is the **Bohr frequency** corresponding to $|\Psi_1^0\rangle \to |\Psi_2^0\rangle$.

24.3. CONSTANT PERTURBATION: (Time - Independent Perturbation)

Let $\hat{H}^{(1)}(t) =$ a constant perturbation, and let it is switched on at time $t = 0$.

24.3.1 COUPLED DIFFERENTIAL EQUATIONS

Case (a) For $t \geq 0$:
The system has matrix elements

$\hat{H}_{12}^{(1)}(t) \neq 0$, (24.3.1)

$\hat{H}_{21}^{(1)}(t) = 0$, (24.3.2)

between states $|\hat{\psi}_1^0\rangle$ to $|\hat{\psi}_2^0\rangle$.

Case (b): For $t < 0$.
i.e. before $\hat{H}^{(1)}(t)$ is switched on,

$(\partial/\partial t)a_1(t=0) = (\partial/\partial t)a_1(0) = 0$. (24.3.3)

$\therefore (\partial/\partial t)a_1(0) = 0$ (24.3.4)

Case (c): for $t > 0$,

$(\partial/\partial t)a_1(t) = \frac{1}{\hbar}(\partial/\partial t)a_2(t) \cdot \hat{H}_{12}^{(1)}(t) \, e^{-i\omega_{12}t}$. (24.2.12)a

$\therefore (\partial/\partial t)a_2(t) = \frac{1}{i\hbar} a_1(t) \cdot \hat{H}_{21}^{(1)}(t) \, e^{-i\omega_{21}t}$ (24.2.12)b

These are *coupled differential equations*.

24.3.2 Solving of the Coupled Differential Equations:

Of the several methods to solve (23.2.13), Laplace Transforms technique is the best one.
Differentiating equations (24.2.12)b with respect to time t,

$(\partial^2/\partial t^2)a_1(t) = \frac{1}{i\hbar}[(\partial/\partial t)a_1(t)] \cdot \hat{H}_{21}^{(1)}(t) \, e^{-i\omega_{21}t} + i\omega_{12} \frac{1}{i\hbar} a_1(t) \cdot \hat{H}_{21}^{(1)}(t) \, e^{-i\omega_{21}t}$

$(\partial^2/\partial t^2)a_2(t) = \left(\frac{1}{i\hbar}\right)^2 a_2(t)] \cdot \hat{H}_{12}^{(1)}(t) \cdot \hat{H}_{21}^{(1)} + i\omega_{12}(\partial/\partial t)a_2(t)$

i.e. $\hat{H}_{12}^{(1)}(t) \cdot \hat{H}_{21}^{(1)} = \hbar^2 \nabla^2$ (24.3.5)

Then $(\partial^2/\partial t^2)a_2(t) = \left(A \cdot e^{i\Omega t} + e^{-i\Omega t}\right) e^{-i\omega_{12}t}$ (24.3.6)

$\Omega = \frac{1}{2}\sqrt{\left(\omega_{12}^2 + 4\nabla^2\right)}$ (24.3.7)

Similarly, $(\partial^2/\partial t^2)a_1(t) = \left(A \cdot e^{i\Omega t} + e^{-i\Omega t}\right) e^{+i\omega_{12}t}$ (24.3.8)

Now, if t = 0, the system was definitely at state $|\hat{\psi}_1^0\rangle$,

$a_1(0) = 1$, and $a_2(0) = 0$.
These determine A and B.

$$a_1(t) = \{Cos\Omega t + i(\omega_{12}/2\Omega)\cdot Sin\Omega t\} e^{-i\frac{1}{2}\cdot\omega_{12}\cdot t}. \qquad (24.3.9)$$

$$a_2(t) = \{-(V/\Omega)\cdot Sin\Omega t\} e^{+i\frac{1}{2}\cdot\omega_{12}\cdot t}. \qquad (24.3.10)$$

are the values of $a_1(t)$ at any time after $t = 0$.

24.3.3 The Rabi Formula

The Probability of finding the system in the state $|\hat{\psi}_1^0\rangle$, at any time

$$P|1\rangle = \|a_1(t)\|^2.$$

For state $|2\rangle$, the initially unoccupied state, one finds

$$P|2\rangle = \|a_1(2)\|^2 - 1 - \|a_1(t)\|^2,$$

$$P|2\rangle = \{4|V|^2/(\omega_{12}^2 + 4|V|^2)\}\cdot Sin^2\left[\tfrac{1}{2}\sqrt{(\omega_{12}^2 + 4|V|^2)}\cdot t\right] \qquad (24..3.11)$$

This is the RABI Formula.

24.3.3.1 DISCUSSIONS - Degenerate System:

Consider the case of a system in which $E_1 = E_2$, i.e. $\omega_{12} = 0$.
Initially, for $t < 0$,

$$P|1\rangle = \|a_1(t)\|^2 \text{ .in state } |1\rangle$$

so that $P|2\rangle = Sin^2|V|t$ \qquad 24.3.12)

This means the system oscillates between states $|1\rangle$ and $|2\rangle$. So the system periodically may be found with certainty in $|2\rangle$. The frequency with the system appears in $|2\rangle$ is 2V. So the system makes transition from $|1\rangle$ to $|2\rangle$ and vice-versa in a time

$$t = \pi/2V. \qquad (24.3.13)$$

which is known as *wait time*. This is the property of a degenerate system. Thus the weakest perturbation $\hat{H}^{(1)}(t)$ can drive the populations (occupation probabilities) completely between degenerate states $|1\rangle$ and $|2\rangle$..

24.3.3.2. Non-degenerate case:

If $(E_1-E_2) \gg \hat{H}^{(1)}(t)$, i.e. $(E_1-E_2) = $ large, that is $\omega_{12}^2 \gg 4|V|^2$,

$$P|2\rangle \geq \{4|V|^2/(\omega_{12}^2 + 4|V|^2)\}\cdot Sin^2\left[\tfrac{1}{2}\sqrt{(\omega_{12}^2 + 4|V|^2)}\cdot t\right] \qquad (24.3.11)$$

$$P|2\rangle \approx \{2|V|/\omega_{12}\}^2 \cdot Sin^2\left[\tfrac{1}{2}\omega_{12}\cdot t\right]. \qquad (24.3.14)$$

The population still oscillates between 0 and $\{2|V|/\omega_{12}\}^2$ ($\ll 1$).

A weak perturbation during $|1\rangle$ and $|2\rangle$ is only a small value.

Now as (E_1-E_2) increases ω_{12} also increases, but the maximum value of $P|2\rangle$ decreases. In the limit, as $\hat{H}^{(1)}(t) = 0$, i.e. $4|V|^2/\omega_{12}^2 \to 0$, the population $P|2\rangle$ never rises from 0. Thus the important criterion for the effectiveness of the perturbation is not its absolute strength but its strength relative to (E_1-E_2).

24.3.2 MIXED STATES:

For the degenerate case one saw that for doubly degenerate levels, the wait time is $t = \pi/2V$. On the other hand, if one waits for $t = \pi/2V$,

$$P|2\rangle = \tfrac{1}{2} \text{ and } P|1\rangle = \tfrac{1}{2}.$$

If $\hat{H}^{(1)}(t) = 0$, at the end of $t = \pi/2V$, it is possible to prepare systems in 50% in state $|1\rangle$ and 50% in state $|2\rangle$, because now

$$(\partial/\partial t)a_1(t) = (\partial/\partial t)a_2(t) = 0.$$

This is a general way of preparing mixed states, and is widely used in NMR (Nuclear Magnetic Resonance) spectroscopic method, where nuclei are exposed to magnetic field for exactly specified lengths of time to cause them to evolve from a pure spin state, $|\alpha\rangle$ into a superposition of $|\alpha\rangle$ and $|\beta\rangle$ states. This is the underlying Quantum Mechanics of *Pulse methods* in NMR.

Worked out Example # 24.1

Prepare a degenerate two-level system in a mixed state in which there is equal probability of finding it in either case. Explain how you could do this.

Solution:

A state is known to persist in the same composition once prepared, under constant perturbation.

| Step # 1 | Given: Rabi Formula

$$P|2\rangle = \left\{4|V|^2 / (\omega_{12}^2 + 4|V|^2)\right\} \cdot Sin^2\left[\tfrac{1}{2}\sqrt{(\omega_{12}^2 + 4|V|^2)} \cdot t\right] \quad (24.3.11)$$

| Step # 2 | $P|1\rangle = 0.5, P|2\rangle = 0.5$.

and immediately turn off the perturbation. $\hat{H}^{(1)}(t) = 0$

| Step # 3 |: $P|2\rangle = \left\{4|V|^2 / (\omega_{12}^2 + 4|V|^2)\right\} \cdot Sin^2\left[\tfrac{1}{2}\sqrt{(\omega_{12}^2 + 4|V|^2)} \cdot t\right]$

$= 0.5$, when $t = \pi/4V$.

| Step # 4 |: Apply to the state $|1\rangle$ initially the perturbation, and

| Step # 5 |: turn off after a time $t = \pi/4V$.

Though the wave function of the system will oscillate, the probability of finding the system in either state will remain at 0.5, until another perturbation is applied..

24.3.4 GENERAL PERTURBATION: THE VARIATION OF $a_i(t)$:

It is known from section (24.2.2.) that

(i) a two-level system requires solution of a 2nd order differential equation; while for a 3-level system a solution of 3rd order differential equation is required, and so on.

(ii) The perturbation $\hat{H}^{(1)}(t)$ = constant, when it was on. So the differential equation becomes complicated as $\hat{H}^{(1)}(t)$ has a complicated time dependence. Even the case of Cos ωt is very difficult.

Dirac invented a technique known as *variation of constants*. It is a generalization of the two-level problem. The method is as follows:

$$\hat{H}(\vec{r},t) = \hat{H}_0(\vec{r}) + \hat{H}^{(1)}(\vec{r},t) \quad (24.2.1)$$

$$\hat{H}_0 \left| \Psi_n^0, t \right\rangle = E_n^0 \left| \Psi_n^0, t \right\rangle \quad (24.2.2)$$

which have solutions

$$|n,t\rangle \equiv |\Psi^0_n, t\rangle = |\Psi^0_n\rangle e^{-iE_n t/\hbar}. \qquad (24.2.3)$$

$$\hat{H}|\Psi_n(r,t)\rangle = [i\hbar\, \partial/\partial t]|\Psi_n(r,t)\rangle \qquad (24.2.5)$$

$$|\Psi(t,t)\rangle = \sum_1^2 a_i(t)|\hat{\psi}^0_i(\bar{r},t)\rangle \qquad (24.2.12)$$

$$\langle \hat{\psi}^0_k | \hat{\psi}^0_n \rangle = \delta_{kn}.$$

where $a_i(t)$ are the coefficients which evolve with time, and a subsequent measurement of energy, say at any time t, may yield E_k with probability, $p(k) = |a_i(t)|^2 =$ probability of finding the system in the state,$|k\rangle$, given that at time t = 0 it occupies the state $|n,0\rangle$. To find how the linear combination (24.2.12) of $|\Psi(t)\rangle$ evolves with time t, solve the Schrödinger Equation (24.2.5), i.e. find the differential equation for the mixing coefficients, $a_k(t)$, and then solve it.

24.3.4.1. The BASIC LAW in Time-Dependent Perturbation Theory:

As in the case of the 2-level system, one gets the generalized version of equation (34.2.12), viz.,

$$(\partial/\partial t)a_k(t) = [1/i\hbar\,]\partial/\partial t] \sum_{l\&n} a_n(t)\,\hat{H}^{(1)}_{kn}(t)\, e^{-i\omega_{kn} t}. \qquad (24.3.15)$$

where $\omega_{kn} = (E^{(0)}_k - E^{(0)}_n)/\hbar$. (24.3.16)

the matrix element,

$$\hat{H}^{(1)}_{kn} = \langle k|\hat{H}^{(1)}|n\rangle \qquad (24.3.17)$$

Equation (24.3.15) is the basic law in Time Dependent Perturbation Theory.

As only the term n = k survives (since $\langle \hat{\psi}^0_k | \hat{\psi}^0_n \rangle = \delta_{kn}$, and as $\delta_{kn} \to 1, |n\rangle \to |k\rangle$; and $e^{-i\omega_{kn} t} \to 1$. Therefore the equation (24.3.15) is **exact**. It is a coupled differential equation, and is impossible to solve exactly since it relates $(\partial/\partial t)a_k(t)$, the kth coefficient, to all the others, $a_n(t)$ (i.e. other than $a_k(t)$).

24.3.4.2 To solve the equation (24.3.15):

The way of solving a differential equation is to integrate it. In the present case the limits of integration are:
(i) from initial time, t = 0, when

$$|\Psi(t)\rangle = \sum_{l\&n} a_i(t)|\hat{\psi}^0_i(t)\rangle$$

with $a_n(0) = 0$. (24.3.18)

II) to the time of interest, $t = 0$ to t_0, when the coefficient is $a_n(t_0)$.

However, for stationary states $|k\rangle$ and $|n\rangle$, equation (24.3.15) becomes

$$(\partial/\partial t)a_k(t) = [-i/\hbar\,]] \sum_{i\&n} a_n(t)\,\hat{H}^{(1)}_{kn}(t)\, e^{-i\omega_{kn} t}. \qquad (24.3.19)$$

For small perturbing potential, $\hat{H}^{(1)}(t)$ it is not a bad step to assume the $(\partial/\partial t)a_k(t)$ = constant.

i.e., $$[-i/\hbar\,]] \sum_{i\&n} a_n(t)\,\hat{H}^{(1)}_{kn}(t)\, e^{-i\omega_{kn} t} = \text{constant}. \qquad (24.3.20)$$

At t = 0, the system is in its initial state, $|i\rangle$ and so

$$a_i(0) = 1 \quad . \tag{24.3.21}$$

All other coefficients are zero.

At $t = t_0 \neq$ long, i.e. finite for which $\hat{H}^{(1)}(t)$ acts and since $\hat{H}^{(1)}(t) =$ weak, the system is in some state $|i\rangle$ which is not the initial state, $|i\rangle$.

$$a_f(0) = 0 . \tag{24.3.22}$$

such that $a_i(t_0) \approx 1$. \hfill (24.3.23)

$$(i\hbar)(\partial/\partial t)a_f(t_0) \approx \hat{H}^{(1)}_{fi}(t) e^{-i\omega_{fi} t_0} . \tag{24.3.24}$$

The probability that the system is in any state other than $|i\rangle$ is always low that all the terms in the sum on the rhs of (24.3.15) can be set equal to zero, with the exception of the term with $n = i$.

$$a_f(t_0) - a_f(0) = (-i/\hbar) \int_0^{t_0} a_i(t) \hat{H}^{(1)}_{fi}(t) e^{-i\omega_{fi} t} dt \tag{24.3.25}$$

Similarly, $a_i(t)$ barely changes from $a_i(t_0) \approx 1$, over the period t_0 over which the $\hat{H}^{(1)}(t)$ is active.

$$a_i(t_0) = 1. \tag{24.3.26}$$

Equation (24.3.25) becomes

$$a_f(t_0) - a_f(0) = (-i/\hbar) \int_0^{t_0} \hat{H}^{(1)}_{fi}(t) e^{-i\omega_{fi} t} dt$$

i.e., $\quad a_f(t) = (-i/\hbar) \int_0^{t_0} \hat{H}^{(1)}_{fi}(t) e^{-i\omega_{fi} t} dt$, for $f \neq i$. \hfill (24.3.27)

and $\quad a_i(t) = 1 - (i/\hbar) \int_0^{t_0} \hat{H}^{(1)}_{ii}(t) dt$, \hfill (24.3.28)

or, $\quad a_i(t) \approx e^{\left\{-(i/\hbar) \int_0^{t_0} \hat{H}^{(1)}_{ii}(t) dt\right\}}$ \hfill (24.3.29)

Equation (24.3.27) is an EXPLICIT expression for the value of the coefficient of a state $|f\rangle$ that was unoccupied initially.

24.3.4.3 DRAWBACKS of the Approximation:

(a) One ignored the possibility that the perturbation can take the system from its initial state $|i\rangle$ to some selected state $|f\rangle$ by an indirect route of excitation, in which it induces several transitions in sequence.

(b) $\hat{H}^{(1)}(t)$ is allowed to act only once; i.e. only 1st Order Perturbation Theory is dealt with.

24.4. TRANSITIONS INTO A CONINUOUS SPECTRA, AND CONSTANT PERTURBATION:

Quantum mechanical theory of scattering was seen using *Born's Approximation*, in Chapter 22. Examples of transitions into continuous spectra appear in scattering, Radioactivity by means of Alpha Decay and Optical Transitions. The kind of problem that arises in this connection is the discussion of three quantities, viz.

    (i)   transition probability,
    (ii)   transition rate, and
    (iii) intensities of spectral lines.

24.4.1 QUANTITATIVE FORM of time dependence of the probability of a state at time t:

In order to obtain the quantitative form of time dependence of the probability of a state $|\Psi_n\rangle$, occupied by the system at time t, consider a *Constant* (time independent) Perturbation,

$$\hat{H}^{(1)}(t) \equiv \hat{H}^{(1)}. \tag{24.4.1}$$

Let $\hat{H}^{(1)}$ is turned on at $t \geq 0$.

$\hat{H}^{(1)}$ acts over the period $t = 0$ to $t = t_o$.

Let $|i\rangle$ = initial state, and
        $|f\rangle$ = selected final state.

a) At $t = 0$:

The system is definitely in $|i\rangle$,

$$a_i(0) = 1, \tag{24.3.21}$$
$$a_f(0) = 0. \tag{24.3.22}$$

From (24.3.28) $a_i(t) = 1 - (i/\hbar) \int_0^{t_o} \hat{H}^{(1)}_{ii}(t)\, dt = 1 - (i/\hbar)\, \hat{H}^{(1)}_{ii}(t_o - 0)$

$$= 1 - (i/\hbar)\, \hat{H}^{(1)}_{ii}(t_o). \tag{24.4.2}$$

b) At $t \geq 0$, such that $0 < t < t_o$,

Equation (24.3.27) is

$$a_f(t) = (-i/\hbar) \int_0^{t_o} \hat{H}^{(1)}_{fi}(t)\, e^{-i\omega_{fi} t}\, dt \tag{24.3.27}$$

$$= \{(-i/\hbar)\, \hat{H}^{(1)}_{fi}\} \left[e^{-i\omega_{fi} t} \Big/ i\omega_{fi} \right]_0^{t_o}$$

$$= \{(1/i\omega_{fi})\, \hat{H}^{(1)}_{fi}\} \left(1 - e^{i\omega_{fi} t_o}\right) \tag{24.4.3}$$

This is exact. The probability of perturbation causing a transition of the system from $|i\rangle \to |f\rangle$, from $t = 0 \to t = t_o$, i.e. in time t_o is $P|f\rangle$.

$$P|f\rangle = |a_f(t)|^2 = \left| \{(1/i\omega_{fi})\, \hat{H}^{(1)}_{fi}\} \left(1 - e^{i\omega_{fi} t_o}\right) \right|^2.$$

Put $x = \omega_{fi} t_o$; $Sin A = 2 Sin \tfrac{1}{2} A\, Cos \tfrac{1}{2} A$; $Cos A = 1 - 2 Sin^2 \tfrac{1}{2} A$

$$\left(1 - e^{i\omega_{fi} t_o}\right) = [1 - e^{ix}] = 1 - (Cos\, x - i Sin\, x) = [(1 - Cos\, x) - i Sin\, x)]$$

$$= 2 Sin^2 \tfrac{1}{2} x - i\, [2 Sin \tfrac{1}{2} x\, Cos \tfrac{1}{2} x] = 2 Sin \tfrac{1}{2} x\, [Sin \tfrac{1}{2} x - i\, Cos \tfrac{1}{2} x]$$

$$= (2 \sin\tfrac{1}{2} x)(-i) e^{i\tfrac{1}{2}x}$$

$$[1 - e^{ix}] = 2 \sin\tfrac{1}{2} x \quad = \quad |1 - e^{ix}|^2 = (1 - e^{ix})(1 - e^{-ix})$$

$$= (1 + 1 - e^{ix} - e^{-ix}) = \{2 - (\cos x - i \sin x) - (\cos x + i \sin x)\}$$

$$= 2(1 - \cos x) = 4 \sin^2 \tfrac{1}{2} x$$

To first order approximation,

$$P|f\rangle = |a_f(t)|^2 = \left| \{(1/i\omega_{fi})\, \hat{H}^{(1)}_{fi}\} \left(1 - e^{i\omega_{fi} t_o}\right) \right|^2 .$$

$$\therefore \quad P|f\rangle = \{(2/i\omega_{fi})\, \hat{H}^{(1)}_{fi}\}^2 \sin^2[(E^{(0)}_f - E^{(0)}_i)/2\hbar]\, t_o . \qquad (24.4.4)$$

$$= \{\hat{H}^{(1)}_{fi}/\hbar\}^2 \sin^2 \tfrac{1}{2}\beta\, t_o / (\tfrac{1}{2}\beta)^2 , \text{ for } f \neq i$$

where $\beta = (E^{(0)}_f - E^{(0)}_i)/\hbar$.

24.4.1.1. Proof of $a_f(t)$ resembling a δ-function, $\delta(E - E^{(0)}_i)$:

This expression (24.4.4) shows that the probability, $P|f\rangle$ oscillates at angular frequency given by $\tfrac{1}{2}\omega_{fi}$, but that its effect is appreciable only for values of $f \approx i$, where $a_f(t)$ becomes maximum.

The derivation of equation (24.4.4) assumes that $a_f(t)$ remains small, implying that $\hat{H}^{(1)}_{fi} \ll (E^{(0)}_f - E^{(0)}_i)$. For large values of t_o, in equation (24.4.4),

$$\left(\sin^2 \tfrac{1}{2}\beta\, t_o\right) / (\tfrac{1}{2}\beta) \to 2\pi\, t_o\, \delta(\tfrac{1}{2}\beta). \qquad (24.4.5)$$

This means that $a_f(t)$ resembles a δ-function, $\delta(E - E^{(0)}_i)$.

24.4.1.2 TRANSITIONS to a group of final states:

Consider that there is a quasi-continuum of closely-spaced states, the area under the $|a_f(t)|^2$ versus E graph (if plotted) represents the Transition Probability, from $|m\rangle$ to a group of states clustered about state $|k\rangle$ (which is part of the continuum), where $E_k > E_m$.
If $g(E) = $ density of states

$$g(E) = \left(\sin^2 \tfrac{1}{2}(\omega_{kmi} - \omega)\, t\right) / \left(\tfrac{1}{2}(\omega_{kmi} - \omega)^2\right) \qquad (24.4.6)$$

$g(E_k)\, dE = $ # of final states $|f\rangle$ in the energy range E_k to $(E_k + dE)$,

$t_o = $ time that $\hat{H}^{(1)}$ acts on the system.

Total probability of transition to all states lying in a band near E_k out of the initial state E_m is

$$P(t) = \sum P(f) \qquad (24.4.7)$$

$$= \int_{-\infty}^{+\infty} |a_f(t)|^2 g(E_k)\, dE = [\hat{H}^{(1)}_{km}]^2 g(E_k)\, (t_o/2\hbar) \int_{-\infty}^{+\infty} \left(\sin^2 x / x^2\right) dx$$

where $x = i\, \omega_{kmi}\, t_o$.

$$\int_{-\infty}^{+\infty} \left(\sin^2 x / x^2\right) dx, x = \pi \qquad (24.4.8)$$

is a standard integral, or evaluating bt the method of contour integration.

$$P(t) = \sum P(f) = (2\pi/k) |\hat{H}_{km}^{(1)}|^2 g(E_k) \, t_o. \qquad (24.4.9)$$

Thus $P(t) \propto t_o$ that $|\hat{H}_{km}^{(1)}|^2$ acts on the system,

$$P(t) \propto |\hat{H}_{km}^{(1)}|^2 \text{ that acts on the system.}$$

Both $g(E_k)$ and $|\hat{H}_{km}^{(1)}|$ are assumed to be slowly varying and are regarded as constants for the integration o of equation (24.4.9).

24.4.2. TRANSITION RATES - to first order approximation – FERMI's GOLDEN RULE

24.4.2.1 Between two states:

Let $|m\rangle$ and $|n\rangle$ are non-degenerate states.

Defining the *Transition Rate per unit time, R* as the time rate of change of the probability of being in the initially empty state, $|k\rangle$:

$$R_{n \to k} = dP_k(t)/dt \qquad (24.4.10)$$

The transition rate per unit time to the band is
$$R = dP(t)/t$$

In equation (24.4.9) is

$$R_{n \to k} = (2\pi/k) |\hat{H}_{kn}^{(1)}|^2 g(E_k) = (2\pi/k) |\hat{H}_{kn}^{(1)}|^2 g(E_k \pm \hbar\omega) \qquad (24.4.11)$$

= Independent on time,

Because of the multiple applications of the equation (24.4.11) Fermi coined the phrase, known as the **_Fermi's Golden Rule._**

24.4.2.1 Covering a band of frequencies:

When a system is exposed to a Constant Perturbation, $\hat{H}^{(1)}$ covering a band of frequencies, the same procedure above gives R. Consider here the perturbation as a sum (integral) of different frequency constants,

$g(\omega)$ = frequency density of states (Radiation density), *i.e.*, # of components with frequencies in the range ω to $\omega + d\omega$ is $g(\omega) \, d\omega$

$$P_k(t) = \int_{-\infty}^{+\infty} p|f\rangle g(\omega) \, d\omega$$

Using $g(\omega) \approx g(\omega_{km}) \approx \hbar^{-1} g(E_{km})$

$$R_{m \to k} = (1/\hbar^2) |\hat{H}_{km}^{(1)}|^2 g(\nu_{km}). \qquad (24.4.12)$$

This is the 2nd form of the **_Fermi's Golden Rule_**.

It states that the Transition Rate, $R_{n \to k}$, for $|n\rangle$ and $|k\rangle$ is

(a) $\neq 0$, between continuum states of the same energy,
(b) $\propto |\hat{H}_{km}^{(1)}|^2$ that connects $|n\rangle$ and $|k\rangle$ in the system,
(c) $\propto g(E_k)$ of $|k\rangle$.

Fermi's Golden Rule has a wide range of application in the field of atomic transitions as well as in Scattering Theory. In Radiation Theory,

$$R_{n \to k} = (d/dt)|a_f(t)|^2 = B_{km} \, g(\omega_{km}) \qquad (24.4.13)$$

24.4.2.1 VALIDITY OF THE GOLDEN RULE:

The delta-function, $\delta(E - E_m^{(0)})$ has to be understood properly for this.

i.e., $\quad \delta_k(\alpha) = Sin^2 \alpha t / \pi \alpha^2 t$,

with $\quad \alpha = (E_n - E_m)/2\hbar$,

has to be considered.

For every finite t, the width of the function is $4\pi\hbar/t$.

Condition 1:
In order that this function is to be replaceable by a δ – function, the width of the energy distribution of the final states

$\Delta E > 2\pi\hbar/t$. (24.4.14)

Condition 2:
Many states must lie within this δ – like function, because only then on can characterize this set of states by a density of states, $g(E)$. Denoting the separation of the energy levels by $\delta\varepsilon$, one gets

$\Delta E \gg 2\pi\hbar/t \gg \delta\varepsilon$,

or, $\quad 2\pi\hbar/\Delta E \ll 1 \ll 2\pi\hbar/\delta\varepsilon$, (24.4.15)

24.4.2.2 INTENSITY of Spectral Lines:
Spectral intensities are proportional to transition rates, because they depend upon the rate at which energy is pumped into the EM Field or absorbed from it.

24.5. EINSTEIN TRANSITION PROBABILITIES:

The 2nd form of the *Fermi's Golden Rule* is

$R_{m \to k} = (1/\hbar^2) |\hat{H}^{(1)}_{k\,m}|^2 \, g(v_{km})$. (24.4.12)

with $g(v_{km})$ as the Radiation density. One has the downward / upward transition rates per atom

$R_{n \to k} = B_{km} \, g(\omega_{km})$. (24.4.13)

Einstein studied the equilibrium between matter and radiation, and distinguished several processes:

24.5.1 STIMULATED ABSORPTION (upward transition rate):
Let $E_k > E_m$ i.e. state $|k\rangle$ lies higher in energy than $|m\rangle$.

$|m\rangle \to |k\rangle$ corresponds to **absorption**.

The rate of absorption of EM radiation is Stimulated absorption
(Upward transition rate): $\quad R_{k \leftarrow m} = B_{mk} \, g(\omega_{km})$. (24.5.1)

where B_{mk} is EINSTEIN'S STIMULATED *(or induced) absorption coefficient*.

24.5.2 INDUCED EMISSION (Downward transition rate / atom):

$|k\rangle \to |m\rangle$ corresponds to **emission**.

Einstein introduced the COEFFICIENT OF STIMULATED (induced) EMISSION, B_{km}.

The rate of emission of EM radiation is Stimulated (induced) emission
(Downward transition rate): $R_{k \to m} = B_{km} \, g(\omega_{km})$. (24.5.2)

Since $|\hat{H}^{(1)}_{k\,n}|^2 =$ constant, when m and k exchange places,

$B_{mk} = B_{km}$. (24.5.3)

$$R_{k \to m} = R_{k \leftarrow m}. \qquad (24.5.4)$$

These two coefficients alone are unable to account for the existence of thermal equilibrium between matter and radiation, according to Einstein. There must be another process in operation, which follows emission to occur even in the absence of stimulated radiation. This is the process known as **spontaneous emission**.

24.5.3 SPONTANEOUS EMISSION:

Einstein introduced the coefficient of spontaneous emission, A_{km}.
The rate of emission of EM radiation is Spontaneous emission
(Downward transition rate): $R_{k \to m} = A_{km}$ \qquad (24.5.5)
According to Einstein's postulate, TOTAL EMISSION RATE
$$W_{k \leftarrow m} = A_{km} + B_{km} g(\omega_{km}). \qquad (24.5.6)$$
TOTAL RATE OF ABSORPTION,
$$W_{k \to m} = 0 + B_{mk} g(\omega_{km}). \qquad (24.5.7)$$
$g(\omega_{km}) d\omega$ = spectral radiation density
= energy / unit volume of radiation at ω_{km} and $(\omega_{km} + d\omega)$.

Consider a system of molecules in thermal equilibrium with radiation of a black body at some temperature, T. At equilibrium, let
\# of molecules in state $| m \rangle = N_m$ = constant with time.
\# of molecules in state $| k \rangle = N_k$ = constant with time,
Total absorption of all molecules = $N_m R_{k \leftarrow m}$.
Total emission of all molecules = $N_k R_{k \to m}$.
Since N_k and N_m are both constants in time, means $N_k < N_m$,
requiring the condition $R_{k \to m}$ per atom $> R_{k \leftarrow m}$ per atom.
At equilibrium temperature, T, according to the BOLTZMANN CRITERION,

$$\frac{N_m}{N_k} = e^{\hbar \omega_{km}/k_B T} \qquad (24.5.8)$$

$$g(\omega_{km}) = \frac{A_{km}}{B_{km}} \left[e^{\hbar \omega_{km}/k_B T} - 1 \right]^{-1} . \qquad (24.5.9)$$

But the radiation density of states $g(\omega)$, at thermal equilibrium at T, is also given by the Planck formula

$$g(\omega_{km}) = \left(\frac{8\pi \hbar v_{km}^3}{c^3} \right) \left[e^{\hbar \omega_{km}/k_B T} - 1 \right]^{-1} .$$

$$= \left(\frac{2 \hbar \omega_{km}^3}{\pi c^3} \right) \left[e^{\hbar \omega_{km}/k_B T} - 1 \right]^{-1} = \left(\frac{2 \hbar \omega_{km}^3}{\pi c^3} \right) \left[e^{\hbar \omega_{km}/k_B T} - 1 \right]^{-1}. \qquad (24.5.10)$$

Comparing equations (24.5.9) and (24.5.10)

$$\frac{A_{km}}{B_{km}} = \left(\frac{2 \hbar \omega_{km}^3}{\pi c^3} \right) \qquad (24.5.11)$$

This indicates that

$$A_{km} \propto \omega_{km}^3. \qquad (24.5.12)$$

The explanation as to **why X-ray lasers are not possible** lies in this result. A_{km} increases with the third power of X-ray energy, causing difficulty to maintain population inversion.

Therefore, the **spontaneous transitions have their rate independent of the field intensity**.

Worked out Example 24.2.

Calculate the rates of stimulated and spontaneous emission of the $|n=3, \ell=1\rangle \rightarrow |n=2, \ell=0\rangle$, i.e. $3p \rightarrow 2s$ transition, in hydrogen (red H_α radiation) when it is inside a cavity at a temperature of 1000 K.. Given:

$$|n, \ell\rangle \equiv \Re_{n\ell}(r) = C_n(\alpha r)^\ell \cdot L_{n+\ell}^{2\ell+1}(\alpha r) \cdot e^{-Zr/na_o}. \qquad (9.6.11)$$

$$|n, \ell\rangle = \sqrt{[-(n-l-1)!]/[2n(n+l)!]^3} \, \rho^{\ell+3/2} i^\ell e^{-\rho/2}.$$

$$\rho = \alpha r = (2Z/na_o)r.$$

Solution: **Step # 1**   Given $|n, \ell\rangle \equiv \Re_{n\ell}(r) = C_n(\alpha r)^\ell \cdot L_{n+\ell}^{2\ell+1}(\alpha r) \cdot e^{-Zr/na_o}$.,

$|3p\rangle \equiv |n=3, \ell=1\rangle$, $|2s\rangle \equiv |n=3, \ell=1\rangle$, $Z=1$

Step # 2 : $\hat{\Re}_{31}(r) = -\left(\frac{1}{a_o}\right)^{3/2} \left(\frac{1}{9x\sqrt{6}}\right)(4-\rho)\rho\, e^{-\rho/2} Y_3^{\pm 1}$.

$\hat{\Re}_{20}(r) = -\left(\frac{1}{a_o}\right)^{3/2} \left(\frac{1}{2x\sqrt{2}}\right)(2-\rho) e^{-\rho/2} \sqrt{\frac{1}{2\pi}}$.

Step # 3 The transition dipole moment, for $3p \rightarrow 2s$ transition

$\mu_z = e\langle 3p | z | 2s \rangle = -(3^3 x 2^{10}/5^0)\, e\, a_o = -18\, e\, a_o = 1.500 \times 10^{-29}$ Cm.

$|\mu|^2 = |\mu_x|^2 + |\mu_y|^2 + |\mu_z|^2 = 9.393\, e^2\, a_o^2 = 1.500 \times 10^{-58}\, C^2 m^2$.

Step # 4 (1) Stimulated emission coeff., $B_{km} = [1/6\, \varepsilon_0\, \hbar^2]\, |\mu_{km}|^2 = 1.143 \times 10^{21}\, J^{-1} m^3 \cdot s^{-2}$.

(2) $\nu_{3p \rightarrow 2s} = cR\{\frac{1}{4} - \frac{1}{3\,3}\} = cR\, \frac{5}{36} = 4.567 \times 10^{14}$ Hz.

(3)   Spontaneous emission coeff., $A_{km} = \{[2\hbar\, \omega_{km}^3]/(\pi c^3)\}\, |\mu_{km}|^2 = 6.728 \times 10^7\, s^{-1}$.

(4)   At T = 1000K, and for the transition frequency, density of states $g(\omega_{km})$

$= \{(2\hbar\, \omega_{km}^3(\,))/(\pi c^3)\}[e^{\hbar\omega_{km}/k_B T} - 1]^{-1} = 1.782 \times 10^{-23}\, J\, Hz^{-1} m^{-3}$.

(5)   Stimulated emission rate, $W_{k \rightarrow m}^{St} = +B_{km}\, g(\omega_{km}) = 2.036 \times 10^{-2}\, s^{-1}$.

(6)   Spontaneous emission rate, $W_{k \leftarrow m}^{Sp} = A_{km} = 6.728 \times 10^7\, s^{-1}$.

24.6. LIFE-TIME AND ENERGY UNCERTAINTY:

It is now obtained from (24.4.4) that, to first order,

$$P|f\rangle = |\hat{H}^{(1)}(t)|^2\, \{\hat{H}_{f\,i}^{(1)}/\hbar\omega_{f\,i}\}^2 [2\, Sin\tfrac{1}{2}\omega_{f\,i}t_o]^2.$$

$$\therefore\ P|f\rangle = \{2\, \hat{H}_{f\,i}^{(1)}/\hbar\omega_{f\,i}\}^2\, Sin^2[(E_f^{(0)} - E_i^{(0)})/2\hbar]\, t_o \qquad (24.6.1)$$

During the transition given by (24.4.4), ENERGY IS CONSERVED to within the limits of the Uncertainty Principle,

$$\Delta E \cdot \Delta t \approx \hbar \qquad (24.6.2)$$

Although the system can oscillate between states of widely separated energies, the greater this separation, the shorter is the lifetime, τ, of the excited state. However, as the resonance condition is approached, $\Delta E \to 0$, and $\Delta t \to \infty$, implying long-lived transitions can occur. In equations (24.4.9) and (24.4.11) energy is conserved as the transitions occur between states which are close to $E_i^{(0)}$

It is known that if a state possesses a definite energy it has a

$$|\psi\rangle = |\psi(t)\rangle = |\psi_o(0)\rangle e^{-iE\,t/\hbar}, \qquad (24.6.3)$$

and $|\psi|^2 = |\psi(t)|^2$ If a system $|\psi_i\rangle$ makes a transition to another state $|\psi_f\rangle$, $|\psi_i\rangle$ is described by a wave function that is decaying in amplitude $|\psi_{oi}\rangle$

If $|\psi_{oi}\rangle$ decays exponentially so that

$$|\psi_i\rangle = |\psi_{oi}\rangle e^{(-iE\,t/\hbar)-(t/2\tau)}, \qquad (24.6.4)$$

and $\qquad |\psi|^2 = |\psi_{oi}|^2 \, e^{-t/\tau}, \qquad (24.6.5)$

where $\tau =$ time constant for the decay.

To find its energy:

One has to express the result as a product of $e^{-iE\,t/\hbar}$.

Consider the decaying function $|\psi_i\rangle$ is modeled by a superposition of oscillating functions. In fact, using the technique of Fourier transforms, to write

$$e^{(-iE\,t/\hbar)-(t/2\tau)} = \int_{-\infty}^{+\infty} g(E')\, e^{-iE'\,t/\hbar}\, dE',$$

$$g(E') = \frac{(\hbar/\tau)}{[(E-E')^2 + (\hbar/\tau)^2]} \qquad (24.6.6)$$

Thus the decaying $|\psi_i\rangle$ corresponds to a range of energies (all the values of E' in the integral). Therefore, any $|\psi\rangle$ with $\tau =$ finite has to be regarded as having an imprecise energy.

One can arrive at a quantitative relation between τ and energy range by considering the shape of the weighting function g(E'). It can be shown that the Half-Width-at-Half-Maximum (HWHM)

HWHM $= (\hbar/\tau) = \delta E$ (24.6.7)
Lifetime broadening, $\tau \cdot \delta E \approx \hbar/2$ (24.6.8)
SPREAD OF ENERGY $\propto (1/\tau)$ (24.6.9)

Worked out Example 24.3.

Calculate the spontaneous lifetime for $|n=3, \ell=1\rangle \to |n=2, \ell=0\rangle$ in atomic hydrogen.

Given: $|n, \ell\rangle \equiv \Re_{n\ell}(r) = C_n(\alpha r)^\ell \cdot L_{n+\ell}^{2\ell+1}(\alpha r) \cdot e^{-Zr/na_o}$,

$|n, \ell\rangle = \sqrt{[-(n-l-1)!]/[2n(n+l)!]^3}\, \rho^{\ell+3/2} i^\ell e^{-\rho/2}$, and $\rho = \alpha r = (2Z/na_o)r$.

Solution: Step # 1 Given $|n, \ell\rangle \equiv \Re_{n\ell}(r) = C_n(\alpha r)^\ell \cdot L_{n+\ell}^{2\ell+1}(\alpha r) \cdot e^{-Zr/na_o}$,

$|n, \ell\rangle = \sqrt{[-(n-l-1)!]/[2n(n+l)!]^3}\, \rho^{\ell+3/2} i^\ell e^{-\rho/2}$, $\rho = \alpha r = (2Z/na_o)r$, $Z=1$,

$|3p\rangle \equiv |n=3, \ell=1\rangle$, $|2s\rangle \equiv |n=3, \ell=1\rangle$.

Step # 2 .|2p> state: degeneracy $= 2\ell+1 = 3$. These are characterized by $m_\ell = +1, 0, -1$.

These three upper states are: $|z\rangle = |2, 1, 0\rangle = \sqrt{(1/32\pi)}(1/a_o)^{3/2} \cdot z \cdot e^{-r/2a_o}$

$|x\rangle = \sqrt{(1/2)} \, (|\, 2, 1, +1\,\rangle + |\, 2, 1, -1\,\rangle) = \sqrt{(1/32\pi)}(1/a_o)^{3/2} \cdot x \cdot e^{-r/2a_o}$.

$|y\rangle = \sqrt{(1/2)} \, (|\, 2, 1, +1\,\rangle - |\, 2, 1, -1\,\rangle) = \sqrt{(1/32\pi)}(1/a_o)^{3/2} \cdot y \cdot e^{-r/2a_o}$.

Step # 3 The ground state is $|\, 1, 0, 0\,\rangle = |\, 1s\,\rangle$:

Step # 4 Spontaneous emission rate, $W^{Sp}_{k \leftarrow m} = (1/\tau_{Sp}) = \int_0^{+\infty} W^{(\ell)} dv_\ell$

$= [(2\pi/\hbar)e^2 \hbar \omega_\ell / 2V\varepsilon)] \int_0^{+\infty} g(v_\ell) \, dv_\ell$ \hfill (24.6.10)

$\int_0^{+\infty} W^{(\ell)} dv_\ell = [(2\pi/\hbar)e^2 \hbar \omega_\ell / 2V\varepsilon)] \, \{8\pi v_\ell^2 n_\ell^3 V/c^3\} \, dv_\ell$

$= \{16\pi^3 e^2 v_o^2 n_\ell^3 / 3\,\varepsilon\hbar c^3\}$

$= \{16\pi^3 e^2 v_o^2 n_\ell^3 / 3\,\varepsilon\hbar c^3\}\{|\langle 1||x|2\rangle|^2 + |\langle 1||y|2\rangle|^2 + |\langle 1||z|2\rangle|^2\}$.

where $n_\ell = W^{(\ell)Ind} / W^{(\ell)Sp}$.

$|\langle 1||y|2\rangle| \propto \int xy \, e^{-3r/2a_o} d\tau = 0$.

$|\langle 1||z|2\rangle| \propto \int zx \, e^{-3r/2a_o} d\tau = 0$.

$|\langle 1||x|2\rangle| \propto \sqrt{(1/32)}(1/\pi)(1/a_o)^4 \int r^4 \, e^{-3r/2a_o} dr$

$\int_0^\pi \sin^3\vartheta \, d\vartheta \int_0^{2\pi} \cos^2\varphi \, d\varphi = (128/243)\sqrt{2} \, a_o = 0.7450 \, a_o$.

Step # 5 $\hbar\omega_{21} = (-13.6 eV)(1^{-2} - 2^{-2}) = 3.288 \times 10^{15}$

$W^{(\ell)Sp}_{2 \to 1} = 6.27 \times 10^8 \, s^{-1}$.

$\tau_{Sp} = \{W^{(\ell)Sp}_{2 \to 1}\}^{-1} = 1.60 \times 10^{-9} \, s$.

Worked out Example 24.4.

Calculate the life-time of the upper state due to spontaneous emission and the upper state due to stimulated emission of the $3p \to 2s$, transition, in Hydrogen (red H_α radiation) when it is inside a cavity at a temperature of 1000 K. Given:

$|n, \ell\rangle \equiv \Re_{n\ell}(r) = C_n(\alpha r)^\ell \cdot L^{2\ell+1}_{n+\ell}(\alpha r) \cdot e^{-Zr/na_o}$,

$|n, \ell\rangle = \sqrt{[-(n-\ell-1)!]/[2n\,(n+\ell)!]^3} \, \rho^{\ell+3/2} i^\ell e^{-\rho/2}$, and $\rho = \alpha r = (2Z/na_o)r$

Solution: Step # 1   Given $|n, \ell\rangle \equiv \Re_{n\ell}(r) = C_n(\alpha r)^\ell \cdot L^{2\ell+1}_{n+\ell}(\alpha r) \cdot e^{-Zr/na_o}$,

$|3p\rangle \equiv |\, n=3, \ell=1\,\rangle, \, |2s\rangle \equiv |\, n=3, \ell=1\,\rangle, \, Z=1$

Step # 2: $\hat{\Re}_{31}(r) = \left(\dfrac{1}{a_o}\right)^{3/2} \left(\dfrac{1}{9x\sqrt{6}}\right) (4-\rho)\rho \, e^{-\rho/2} Y_3^{\pm 1}$.

$\hat{\Re}_{20}(r) = -\left(\dfrac{1}{a_o}\right)^{3/2} \left(\dfrac{1}{2x\sqrt{2}}\right) (2-\rho) \, e^{-\rho/2} \sqrt{\dfrac{1}{2\pi}}$.

Step # 3 The transition dipole moment, for $3p \to 2s$ transition

$\mu_z = e\langle\, 3p\,|\,z\,|\,2s\,\rangle = -(3^3 x 2^{10}/5^0) \, e \, a_o = -18 \, e \, a_o = 1.500 \times 10^{-29} \, Cm$.

$|\mu|^2 = |\mu_x|^2 + |\mu_y|^2 + |\mu_z|^2 = 9.393 \, e^2 \, a_o^2 = 1.500 \times 10^{-58} \, C^2 m^2$.

Step # 4 (1)   Stimulated emission coeff., $B_{km} = [1/6 \, \varepsilon_0 \, \hbar^2] \, |\mu_{km}|^2 = 1.143 \times 10^{21} \, J^{-1} m^3 . s^{-2}$.

(2) $\nu_{3p \to 2s} = cR\{\dfrac{1}{4} - \dfrac{1}{3\,3}\} = cR \, \dfrac{5}{36} = 4.567 \times 10^{14} \, Hz$.

(3) Spontaneous emission coeff., $A_{km} = \{[2 \hbar \omega_{km}^3]/(\pi c^3)\} |\mu_{km}|^2 = 6.728 \times 10^7 s^{-1}$.

(4) At T = 1000K, and for the transition frequency, density of states $g(\omega_{km})$

$= \{(2 \hbar \omega_{km}^3 ( ))/(\pi c^3)\} [e^{\hbar\omega_{km}/k_BT} -1]^{-1} = 1.782 \times 10^{-23} J\ Hz^{-1} m^{-3}$.

(5) Stimulated emission rate, $W_{k \to m}^{St} = +B_{km}\ g(\omega_{km}) = 2.036 \times 10^{-2} s^{-1}$.

(6) Spontaneous emission rate, $W_{k \leftarrow m}^{Sp} = A_{km} = 6.728 \times 10^7 s^{-1}$.

Step # 5 1) The lifetime of the upper state due to spontaneous emission,

$\tau_{Sp} = (A_{km})^{-1} = 1.49 \times 10^{-9} s = 1.49\ ns$

2) The life-time of the upper state due to stimulated emission,

$\tau_{Sp} = (A_{km})^{-1} = 1.49 \times 10^{-9} s = 1.49\ ns$.

24.7. PERIODIC (or HARMONIC) PERTURBATION

24.7.1 Introduction:

A most common way of inducing transitions between stationary states of quantum systems is by applying Harmonic (i.e. periodic) Perturbation. This is extremely important in physics because this model may approximate nearly all of the weak interactions of EM radiation with matter. Further, the phenomenon of NMR in solids and resonance absorption in optical spectroscopy are described n this harmonic approximation.

If λ = turning on or off parameter,

$$\hat{H}(\vec{r},t) = \hat{H}_0(\vec{r}) + \lambda \hat{H}^{(1)}(\vec{r},t) \qquad (24.7.1)$$

Let the harmonic perturbation inducing transitions $|i\rangle \to |j\rangle$ of discrete operators of the form given by.

$$\hat{H}^{(1)}(t) = \hat{H}^{()}(t)\ e^{-i\omega t} + (\hat{H}^{()}(t))^\dagger\ e^{i\omega t}, \qquad (24.7.2)$$

i.e. $\lambda \hat{H}^{(1)}(\vec{r},t) = \lambda \hat{H}^{(1)}(r)\{e^{+-i\omega t} + e^{-i\omega t}\} \qquad (24.7.3)$

where $\hat{H}^{(1)}(r) = \hat{H}_0(\vec{r})\ e^{i \vec{k}\cdot\vec{r}}$.

Let the perturbation acts from time t = 0 to $t = t_o$, i.e. $0 < t < t_o$.
At $t \geq 0$, such that $0 < t < t_o$,
Equation (24.3.27) is

$$a_f(t) = (-i/\hbar) \int_{0 \to t_o} \hat{H}_{fi}^{(1)}\ e^{i\omega_{fi}t}\ dt, \qquad (24.3.27)$$

Using equation (24.7.3),

$$a_k(t) = (-i/\hbar) \int_{0 \to t_o} \hat{H}_{ki}^{(1)} \{e^{i(\omega_{ki}+\omega)t} + e^{i(\omega_{ki}-\omega)t}\} dt \qquad (24.7.4)$$

$$a_k(t) = -\hat{H}_{ki}^{(1)} \left[(e^{i(\omega_{ki}+\omega)t_o} -1)/\hbar(\omega_{ki}+\omega) \right] + \left[(e^{i(\omega_{ki}-\omega)t_o} -1)/\hbar(\omega_{ki}-\omega) \right] \qquad (24.7.5)$$

STIMULATED (Induced) Transitions:

$\omega \sim -\omega_{ki}$, i.e. stimulated emission,

$\hbar\omega \sim [E_k^{(0)} - E_i^{(0)}] = +ve$, \qquad (24.7.6) the 1st term > 2nd term in equation (24.7.5).

An incident photon $\hbar\omega$ induces a transition, i.e. $E_i^{(0)} \to E_k^{(0)}$.

This maximum probability occurs when

$$\hbar\omega = E_i^{(0)} - E_k^{(0)}.\qquad(24.7.7)$$

In resonant absorption:

$$\hbar\omega \approx \left(E_i^{(0)} - E_k^{(0)}\right)\qquad(24.7.8)$$

In this case the 2nd term in equation (24.7.5) dominates the 1st. Here the photon excites the system to a high energy state, $E_k^{(0)} > E_i^{(0)}$.

For ω which are not close to $|\omega_{k\,i}|$, neither the 1st term nor the 2nd term gets large, and

$a_k(t)$ = negligibly small.

$\lim_{\omega \to 0}$ (Equation 24.7.5) tends to the case of Constant Perturbation

Thus it can be seen that only one term in equation is important at any time, Probability of absorption / emission,

$$P(|k\rangle) = |a_k(t)|^2 = 4|\hat{H}_{ki}^{(1)}/\hbar(\omega_{k\,i}\pm\omega)|^2 \sin^2\tfrac{1}{2}(\omega_{k\,i}\pm\omega)t_o.\qquad(24.7.9)$$

For large values of t_o, $|i\rangle \to |k\rangle$.

$$[\sin^2(\beta/2)t_o/(\beta/2)] \to 2\pi t_o\,\delta(\beta)\qquad(24.4.5)$$

TRANSITION RATE, $R_k = (2\pi/\hbar)|\hat{H}_{ki}^{(1)}|^2 g(E_k)\,\delta(E_{k\,i}\mp E)\qquad(24.7.10)$

where $E = \hbar\omega$.

24.7.2 FORBIDDEN TRANSITIONS:

If the matrix element, $\hat{H}_{ki}^{(1)}$, coupling the states vanishes,

i.e. $\qquad \hat{H}_{ki}^{(1)} \to 0 \qquad(24.7.11)$

the transition does not occur; $|i\rangle \to |k\rangle$ is *forbidden*, i.e. $|i\rangle \leftarrow // \to |k\rangle$.

Non-occurring transitions in a given system are forbidden, and the allowed ones are known as *Selection Rules*. It is important in spectroscopy to know how the atomic quantum number may change during atomic transitions. The rules governing these have been known as selection rules.

24.7.3 SELECTION RULES – The ELECTRIC DIPOLE TRANSITIONS (E1 Radiation), (Radiative Transitions in Atomic Hydrogen), (Harmonic Oscillator in a periodic Electric Field) - Radiation Theory, illustrated:

Consider transitions in Hydrogen induced by thermal (black-body) EM radiation. To understands this let such an atom is subject to a plane EM wave of angular frequency, ω whose electric field $\vec{E} \parallel \vec{z}$, and has a magnitude

$$\vec{E}(\vec{r},t) = \vec{E}_o\,e^{i\vec{k}\cdot\vec{r}}[e^{+i\omega t} + e^{-i\omega t}]\qquad(24.7.12)$$

24.7.4.1. DIPOLE-DIPOLE transition:

The electric field of a photon can be expressed as

$$\vec{E}(\vec{r},t) = \vec{E}_o\,e^{i(\vec{k}\cdot\vec{r} - \omega t)}\qquad(24.7.13)$$

This can be expanded in power series.

$$\vec{E}(\vec{r},t) = \vec{E}_o e^{-i\omega t}[1 + i\vec{k}\cdot\vec{r} + (i\vec{k}\cdot\vec{r})^2 + \cdots] \quad (24.7.14)$$

If the wavelength of the radiation, λ is greater than the dimension of the atom, a_o, the spatial variation of the E may be ignored in a 1st approximation, is called the DIPOLE APPROXIMATION. This expansion is valid in the visible region of EM spectrum; where λ is such that $(\lambda/a_o) \sim 10^4$, i.e., $(i\vec{k}\cdot\vec{r}) < 1$, ..and $e^{i\vec{k}\cdot\vec{r}} \approx 1$.

$$\vec{E}(\vec{r},t) = \vec{E}(t) = \vec{E}_o [e^{+i\omega t} + e^{-i\omega t}] \quad (24.7.15)$$
$$= \vec{E}_o \cos\omega t$$

(a) DIPOLE MOMENT, $\vec{D}$:

The energy of interaction, W of the electric dipole, symbolized by letter $\vec{D}$, with this field $\vec{E}(t)$ is

$$W(r,t) = -\vec{E}(t)\cdot\vec{D}. \quad (24.7.16)$$

where the electric dipole moment,

$$\vec{D} = e\cdot\vec{r}, \quad (24.7.17)$$

for the one-electron atom.

The Hamiltonian, $\hat{H}$ is

$$\hat{H}(\vec{r},t) = \hat{H}_0(\vec{r}) + \hat{H}^{(1)}(\vec{r},t) \quad (24.7.1)$$
$$= \hat{H}_0(\vec{r}) + V(\vec{r},t)$$

where $\hat{H}_0(\vec{r}) = \{(-\hbar^2/2m)\nabla^2 - Ze^2/(4\pi\varepsilon_o r^2)\}$

$$\hat{H}^{(1)}(\vec{r},t) \equiv V(\vec{r},t) = -\vec{E}(t)\cdot\vec{D} = -\vec{E}(t)\cdot(-e\cdot\vec{r}) = e\cdot\vec{r}\cdot\vec{E}(t)$$
$$= e\,\vec{r}\cdot\vec{E}_0\{e^{i\omega t} + e^{-i\omega t}\} = -\vec{D}\cdot\vec{E}_0\{e^{i\omega t} + e^{-i\omega t}\} \quad (24.7.18)$$

where $\vec{E}_o = (\hat{i}\vec{E}_{ox} + \hat{j}\vec{E}_{oy} + \hat{k}\vec{E}_{oz}) = \hat{k}\vec{E}_{oz} \quad (24.7.18)\text{a}$

since $\vec{E}_o \parallel \vec{E}_{oz}$, i.e. polarized.

(b) The induced emission and absorption rates, R_k

Rates of electric dipole transitions, from $|j\rangle \to |k\rangle$ is given by equation (24.7.10), viz.

$$R_k = (2\pi/h)|\hat{H}^{(1)}_{ki}|^2 g(E_k)\delta(E_{k\,i} \mp E). \quad (24.7.10)$$

$$R_k = (2\pi/h)|\langle k|\vec{D}\cdot\vec{E}_{oz}|j\rangle|^2 g(E_k)\delta(E_{k\,i} \mp E) \quad (24.7.19)$$

If the incident radiation is *unpolarized*, and isotropic,

$$E_{ox} = E_{oy} = E_{oz}.$$

Since $\vec{E}_o \parallel \vec{E}_{oz}$, and

$$E_{ox}^2 + E_{oy}^2 + E_{oz}^2 = E_o^2$$

$$E_{ox}^2 = E_{oy}^2 = E_{oz}^2 = \tfrac{1}{3}E_o^2 \quad (24.7.20)$$

The energy density $\rho_{rad}(\omega)$ of an EM wave is given by

$$\rho_{rad}(\omega) = \tfrac{1}{4\pi} <E_o^2> \quad (24.7.21)$$
$$= \tfrac{3}{4\pi} <E_{oz}^2> \quad (24.7.22)$$

$$|\hat{H}^{(1)}_{kj}|^2 = |\langle k|\vec{D}\cdot\vec{E}_{oz}|j\rangle|^2 = \tfrac{4\pi}{3}\rho_{rad}(\omega)\langle k|\vec{D}|j\rangle|^2.$$
$$= \tfrac{4\pi}{3}\rho_{rad}(\omega)|\vec{D}_{kj}|^2 \quad (24.7.23)$$

where $|\vec{D}_{x,kj}|^2 = |\vec{D}_{y,kj}|^2 = |\vec{D}_{kj}|^2 = |\vec{D}_{z,kj}|^2$.

This value (24.7.23) is substituted in (24.7.19) to get

$$R_k = (2\pi/\hbar)|\tfrac{4\pi}{3}\rho_{rad}(\omega)|\vec{D}_{kj}|^2 \delta(E_{kj} \mp E)$$

$$R_k = (2\pi/\hbar)|\tfrac{4\pi}{3}\rho_{rad}(\omega)|\vec{D}_{kj}|^2 \delta(E_{kj} \mp E)$$

$$\hbar\omega \approx (E_k^{(0)} - E_j^{(0)}) \tag{24.7.8}$$

$$R_k = (2\pi/\hbar)|\tfrac{4\pi}{3}\rho_{rad}(\omega)|\vec{D}_{kj}|^2 \delta(E_k^{(0)} - E_j^{(0)} \mp E) \tag{24.7.24}$$

(c) **EINSTEIN COEFFICIENTS**

For stimulated (induced) emission, B_{kj} [or absorption, B_{jk}] of photons is

$$R_{k \to m} = B_{km} g(\omega_{km}) \tag{24.5.2}$$
$$= B_{kj} \rho_{rad}(\omega) \tag{24.7.25}$$

where $B_{kj} = B_{jk}$. (24.5.3)

$$B_{kj} = \tfrac{8\pi^2}{3\hbar}|\vec{D}_{kj}|^2 \tag{24.7.26}$$

On the other hand, for spontaneous emission, A_{kj} **can not be calculated directly** from the electric dipole approximation; it can only be deducted from B_{kj}. From (24.5.11) and (24.7.26)

$$A_{kj} = (\tfrac{2\hbar\omega_{kj}^3}{\pi c^3}) B_{kj} \tag{24.7.27}$$

$$= (\tfrac{2\hbar\omega_{jk}^3}{\pi c^3}) \tfrac{8\pi^2}{3\hbar}|\vec{D}_{kj}|^2$$

$$A_{kj} = (16\pi/3c^3)\, \omega_{jk}^3 |\vec{D}_{kj}|^2 \tag{24.7.28}$$

(d) RADIATIVE TRANSITIONS occur in *FIRST ORDER* only, if:
From (24.7.24)

$$R_k = \tfrac{8\pi^2}{3\hbar}\rho_{rad}(\omega)|\vec{D}_{kj}|^2 \delta(E_k^{(0)} - E_j^{(0)} \mp E) = 0, \tag{2.7.29}$$

as $|\vec{D}_{kj}|^2 = 0$.

Thus it is seen that *radiative transitions* occur in *first order* only, if
(i) Radiation of the exciting frequency, ω is present, and
(iii) The matrix element, $|\vec{D}_{kj}|^2 \neq 0$. This leads to the selection rules.

24.7.4.2 SELECTION RULES for electric dipole transitions (E1), HYDROGEN ATOM:
Consider the hydrogen atom.
The matrix elements are

$$\hat{H}_{kn}^{(1)}(t) = \langle k | \hat{H}^{(1)}(t) | n \rangle \tag{24.3.17}$$

$$\vec{D}_{kj} = \langle n'\ell'm' | \vec{D} | n\ell m \rangle$$

In Cartesian coordinates,

$$\vec{D}_{xkj} = \langle n'\ell'm' | \vec{D}_x | n\ell m \rangle = e\langle n'\ell'm' | (r\, \text{Sin}\vartheta\, \text{Cos}\varphi) | n\ell m \rangle$$
$$= e\langle n' | (r) | n\ell m \rangle \cdot \langle \ell' | (\text{Sin}\vartheta) | \ell \rangle \cdot \langle m' | (\text{Cos}\varphi) | m \rangle$$
$$= e\langle n' | (r) | n\ell m \rangle \cdot \delta_{\ell',\ell\pm 1} \cdot \delta_{m',m\pm 1} \tag{24.7.30}$$

$$\vec{D}_{ykj} = e\langle n' | (r) | n\ell m \rangle \cdot \langle \ell' | (\text{Cos}\vartheta) | \ell \rangle \cdot \langle m' | (\text{Sin}\varphi) | m \rangle$$
$$= e\langle n' | (r) | n\ell m \rangle \cdot \delta_{\ell',\ell\pm 1} \cdot \delta_{m',m\pm 1} \tag{24.7.31}$$

$$\vec{D}_{zkj} = e\langle n' | (r) | n \ell m \rangle \cdot \langle \ell' | (\cos\vartheta) | \ell \rangle \cdot \langle m' | 1 | m \rangle$$
$$= e\langle n' | (r) | n \ell m \rangle \cdot \delta_{\ell',\ell\pm 1} \cdot \delta_{m',m} \quad (24.7.32)$$

If spin is also included,
$$\vec{D}_{zkj} \propto \delta_{m\ell',m\ell} \cdot \delta_{ms',ms}$$

Thus the selection rules for ELECTRIC DIPOLE transition (**E1**),
$\langle n \ell m | \leftrightarrow | n \ell m \rangle$ for arbitrary polarization of EM radiation, $\vec{D}_{kj}$. vanishes unless

$$\boxed{\Delta\ell = \ell' - \ell = \pm 1}, \quad (24.7.33)\text{ a}$$

$\vec{D}$ operator connects only states with different parity.
$$\boxed{\Delta m_\ell = m_{\ell'} - m_\ell = 0, \pm 1}, \quad (24.7.33)\text{ b}$$

only either $\vec{D}_{xkj} = 0$, or $\vec{D}_{ykj}$ and $\vec{D}_{zkj}$ are 0

$$\boxed{\Delta m_s = m_{s'} - m_\ell = 0}, \quad (24.7.33)\text{ c}$$

$\vec{D}$ operator has the effect on the spin of the electron.
All other transitions are forbidden in first order.
Or, Selection rules for $\langle n \ell m | \leftrightarrow | n \ell m \rangle$ for arbitrary polarization of EM radiation, $\vec{D}_{kj}$ vanishes unless

$$\boxed{\begin{array}{l} \Delta\ell = \ell' - \ell = \pm 1 \\ \Delta j = j' - j = 0, \pm 1, \quad j = 0 \leftarrow / \rightarrow j' = 0 \\ \Delta m_j = m_{j'} - m_j = 0, \quad m_j = 0 \leftarrow / \rightarrow m_{j'} = 0 \end{array}} \quad (24.7.34)$$

There are no such rules governing the change in n, the principal quantum number.

Thus these are the statements concerning **vanishing** or possibly **non-vanishing matrix elements** $\vec{D}_{kj}$, are called **Selection Rules**.

24.8. MULTI-POLE RADIATIONS (E2, M1):

However, other types of transition can be induced by EM radiation.

$$\vec{E}(\vec{r}, t) = \vec{E}_o \, e^{i(\vec{k}\cdot\vec{r} - \omega t)}. \quad (24.7.13)$$

which, on expansion, in power series is

$$\vec{E}(\vec{r}, t) = \vec{E}_o \, e^{-i\omega t}[1 + i\vec{k}\cdot\vec{r} + (i\vec{k}\cdot\vec{r})^2 + \cdots] \quad (24.7.14)$$

The first term gives the electric dipole transition (E1).

24.8.1. The second order approximation

$$\vec{E}(\vec{r}, t) = \vec{E}_o \, e^{-i\omega t}[1 + i\vec{k}\cdot\vec{r}] \quad (24.8.1)$$

This introduces 2nd- ordered multipole and a corresponding new set of selection rules. A FORBIDDEN TRANSITION in the dipole approximation merely means that OTHER KINDS OF POLES L (i.e. 2^L) are playing a role in the process. The 2nd term ($\vec{k} \cdot \vec{r}$) and 3rd term ($i\vec{k} \cdot \vec{r})^2$ lead to transitions known as ELECTRIC QUADRUPOLE (E2) and ELECTRIC OCTUPOLE (E3), etc.

24.8.2. The Selection Rules for **E2 radiation** are:

Neglecting spin,

$$\Delta\ell = \ell' - \ell = 0, \pm 2, \tag{24.8.2}$$

$$\Delta m_s = m_{s'} - m_\ell = 0, \pm 1, \pm 2 ; \tag{24.8.3}$$

With spin, $\Delta j = j' - j = 0, \pm 1, \pm 2$ \hfill (24.8.4)

$$\Delta m_j = m_{j'} - m_j = 0, \pm 1, \pm 2 ; \tag{24.8.5}$$

No parity change.

24.8.3. The atomic electron interacts with the magnetic field associated with the EM wave as well as the E.

This leads to the transitions known as MAGNETIC DIPOLE (M1) and MAGNETIC QUADRUPOLE (M2), etc.

The Selection Rules for **M1 radiation** are:

Neglecting spin,

$$\Delta\ell = \ell' - \ell = 0, \tag{24.8.6}$$

$$\Delta m_s = m_{s'} - m_\ell = 0, \pm 1, \tag{24.8.7}$$

With spin,

$$\Delta j = j' - j = 0, \tag{24.8.8}$$

$$\Delta m_j = m_{j'} - m_j = 0, \pm 1, \tag{24.8.9}$$

24.8.4. One can in a similar way apply harmonic perturbation to obtain selection rules for Vibrational IR and RAMAN TRANSITIONS in molecules, in the phenomena of Magnetic Resonance (NMR, etc) and MOSSBAUER Spectroscopy.

REVIEW QUESTIONS

R.Q. 24.1 Write down a formal expression for a harmonic perturbation. Justify the terms in the expression.

R.Q. 24.2 Discuss the dipole approximation, the induced emission and absorption in te case of an atom placed in an EM Field.

R.Q. 24.3 Apply time dependent perturbation method o obtain Fermi's Golden Rule for the case of harmonic perturbation.

&^*^&^*(^&^*

Chapter 25

QUANTUM MECHANICS 5 -
RELATIVISTIC QUANTUM MECHANICS

Chapter 25

QUANTUM MECHANICS 5 - RELATIVISTIC QUANTUM MECHANICS

*"Perseverance is a golden trait of scientists. No matter what happens...
Do not shrink back, but move forward. Life is a constant challenge, but it is worthwhile to mostly accept it.
And never stop singing"- by unknown*

25.1.1 INTRODUCTION

Up to this point the Quantum Theory dealt with has concerned itself with the behaviour of particles that inhabit a Galilean space-time. The theories of both Schrödinger and Pauli are non-relativistic Quantum Mechanics. Accordingly, the fact that point-mass particles (the structure less ones) having mass, m_0 can move in space at a velocity $v \approx c$ is ignored; and analyses of results of experiments performed at energies higher than m_0c^2 energies of the particles are involved.

For many purposes, in atomic, molecular and condensed matter physics, the Galilean Theory is quite adequate. The framework needed to describe the properties of elementary particles involves relativity and quantum mechanics. Instead of Euclidean structure, the actual space-time has a structure that is much closer to that of the Minkowski space-time of special relativity. A relativistic generalization of Quantum Mechanics requires
i) Introducing new particle concepts, and
ii) Introducing, besides spin, a new degree of freedom, for the electron. This is difficult because one cannot interpret this degree of freedom within the limits of the one-body problem. However, it is possible to formulate the problem of one body (electron in a given external EM Field in accordance with the theory of relativity. Dirac formulated this problem when he suggested his equation of motion.

Particle creation – usually in the form of particle - anti-particle pairs – is an essential feature of high-energy physics. The traditional wave equation formalism (even in the relativistic one) would seem inappropriate since one introduces one wave function for each particle degree of freedom. So the natural formalism would seem one embodying the possibility of particle creation and destruction from the beginning leading to Quantum Field Theory.
Non-relativistic Schrödinger Equation is

$$\hat{H}\Psi = E\Psi = ih\frac{d\Psi}{dt} \qquad (25.1.1)$$

It is formed based on the non-relativistic relationship between energy E and momentum, p;

$$\hat{H} = E = (T+V) = \frac{p^2}{2m} + V \qquad (25.1.2)$$

Therefore it can be applied only to electrons whose speed $v \ll c$, and kinetic energy,

$$T = [(m_o c^2)/\sqrt{(1-v^2/c^2)} - m_o c^2] \qquad (25.1.3)$$

i.e., $\quad T = (p^2/2m) \ll m_o c^2$.

25.1.1. WHAT IS TO BE SATISFIED BY A CORRECT Quantum Theory?

A correct Quantum Theory should satisfy the requirement of RELATIVITY. Relativistic Quantum Theory must be formulated in a *Loremtz covariant form* form, since the laws of motion valid in one inertial frame must be true in all inertial frames of reference (Einstein's postulate). All the *postulates of non-relativistic* Quantum Mechanics are retained in Relativistic Quantum Mechanics.

25.1.2. PRELIMINARIES
25.1.2.1. LORENTZ TRANSFORMATION:

If one reference frame, S' moves relatively to the other, S at constant speed, v along +X axis, the classical (Galilean) view point is as follows:
Let the origins, O and O', of S and S', are superimposed at time $t = 0$.
Let an event occurs at point $P(x', y', z')$ at time t'.
Since in classical physics **Time is absolute**, $t = t'$. the GALILEAN Transformation Equations for *position coordinates (i.e.,* Classical Transformation for position) are:

$$x' = x + vt'; \; y = y'; \; z = z' \text{ ; and } t = t'. \tag{25.1.4}$$

Similarly, the GALILEAN VELOCITY Transformation equations are:

$$u_x = u_x' + v', \; u_y = u_y'; \text{ and } u_z = u_z'. \tag{25.1.5}$$

Again frames S' moves relatively to S, at constant speed v along the +ve x-axis with space-time coordinates $\bar{r}' = (x', y', z', t')$ and $\bar{r} = (x, y, z, t)$, respectively. Then one has

$$x^2 + y^2 + z^2 + c^2 t^2 = 0$$
$$x'^2 + y'^2 + z'^2 + c^2 t'^2 = 0.$$

In the case of *Special Relativity*, an important concept is that of an *inertial frame of reference*; and Einstein's two fundamental postulates, viz., (1) "*the form of each physical law is the same in all inertial frames*" and (2) "*light moves with the same speed relative to all observers*". An event is anything with a location in space and time. There arise three consequences:

(1). Relative Simultaneity is that two events at different locations, simultaneous In one frame of reference will not be simultaneous in a frame of reference moving relative to the first.
The transformation satisfying $\bar{r} \rightarrow \bar{r}'$ is

$$x' = \beta(x - vt) \; ; \; y' = y; \; z' = z; \; t' = \beta(t - vx/c^2)$$

with the factor $\beta = 1/(1 - v^2/c^2)$. (25.1.6)

In matrix form these equations may be written as follows:

$$\begin{bmatrix} x' \\ y' \\ z' \\ ct' \end{bmatrix} = \begin{bmatrix} \beta & 0 & 0 & -\beta v/c \\ 0 & 1 & 0 & 0 \\ 0 & 0 & 1 & 0 \\ -\beta v/c & 0 & 0 & \beta \end{bmatrix} \begin{bmatrix} x \\ y \\ z \\ ct \end{bmatrix} \tag{25.1.6)a}$$

The set of equations (25.1.6) is called *a priori* LORENTZ TRANSFORMATION EQUATIONS (a similar set of equations was first written by H.A. Lorentz, 1904; to explain results of Michelson & Morley experiment and Maxwell's Equations) derived by Einstein in 1905.

(2). Thus the space and time coordinates mix. It can be shown that an object of length L_o and L in the frames S and S' (Fig. 25.1) are such that there is FITZGERALD (LENGTH) CONTRACTION (The length of an object in a frame through which it moves, determined by viewing its ends at the same time (two events) in that frame, is smaller that the length of the object in the frame in which it is at rest),

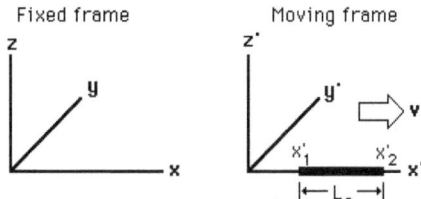

Fig. 25.1. Moving Frame OX'Y'Z' with respect to the Fixed Frame OXYZ.

$$L = L_o \sqrt{(1-v^2/c^2)}. \qquad (25.1.7)$$

(3). Similarly the time Δt_o between two events occurring at the same place in S and S' (Fig. 25.2) are related by

$$\Delta t = \Delta t_o / \sqrt{(1-v^2/c^2)}. \qquad (25.1.8)$$

in the TIME DILATION *Formula* (i.e., Two events occurring at the same location in one frame will be separated by a longer time interval in a frame moving relative to the first).

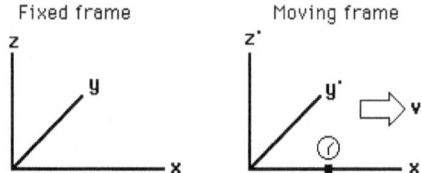

Fig. 25.2. Time dilation illustrated.

Worked out Example 25.1.

Rama and Gopi are on identical spaceships of length $\ell = 12\sqrt{3}$ m. Rama's frame (S') moves to the right relative to Gopi's (S). If clocks are glued at the nose, centre and tail of each ship, at $t = 0$ the tails of the ships are aligned. Rama's tail –end clock reads zero, and the nose of Rama's ship is aligned exactly with the centre of Gopi's. Determine the Lorentz transformation matrix fro Gopi's frame to Rama's frame. Using matrix multiplications verify the initial time on the nose-end clock on Rama's ship.

Solution: Step # 1 At $t = 0$, Rama's ship length, $L = \frac{1}{2}\ell$. But in the frame in which it is at rest, S'), Ship of Rama is of length $L_o = \ell$. Thus $L = L_o / \beta$, whence $\beta = 2$.

Step # 2 $1/(1-v^2/c^2) = 2$, i.e. $v = \sqrt{3/2}\ c$. The Lorentz transformation equation is

$$\begin{bmatrix} 2 & 0 & 0 & -\sqrt{3} \\ 0 & 1 & 0 & 0 \\ 0 & 0 & 1 & 0 \\ -\sqrt{3} & 0 & 0 & 2 \end{bmatrix}$$

at $t = 0$ of S,

$$\begin{bmatrix} x' \\ y' \\ z' \\ ct' \end{bmatrix} = \begin{bmatrix} 2 & 0 & 0 & -\sqrt{3} \\ 0 & 1 & 0 & 0 \\ 0 & 0 & 1 & 0 \\ -\sqrt{3} & 0 & 0 & 2 \end{bmatrix} \begin{bmatrix} 6\sqrt{3}\ m \\ 0 \\ 0 \\ 0 \end{bmatrix}$$

$x' = 2 (2\sqrt{3} \, m)$

25.1.2.2 MAXWELL'S EQUATIONS:

The classical equations of Maxwell are:

$\hat{\nabla} \cdot \vec{E} = 0$; $\hat{\nabla} \wedge \vec{B} - (\partial/\partial t)\vec{E} = \vec{j}$;

$\hat{\nabla} \cdot \vec{B} = 0$; and $\hat{\nabla} \wedge \vec{E} + (\partial/\partial t)\vec{B} = 0$.   (25.1.9)

The source, in turn, obeys the Conservation Laws

$(\partial \rho / \partial t) + \hat{\nabla} \cdot \vec{J} = 0$,   (25.1.10)

Because (i) $\hat{\nabla} \cdot (\hat{\nabla} \wedge \vec{E} \text{ or } \vec{B}) = 0$,

(ii) $\hat{\nabla} \wedge (\hat{\nabla}\phi) = 0$.

One may replace $\vec{B}$ and $\vec{E}$ with A° and $\mathbf{A}$ as follows:

$\vec{E} = \hat{\nabla}A^\circ - (\partial/\partial t)A$; $\vec{B} = \hat{\nabla}A$.   (25.1.11)

These equations do NOT transform according to STANDARD GALILEAN TRANSFORMATION. So it led to the discovery of SPECIAL THEORY OF RELATIVITY.

25.1.2.3 VECTOR EQUATIONS:

There are systems with many degrees of freedom. The three spatial dimensions with time are referred to as a **four-vector**. Any four quantities [$x^\mu = (t, x)$)] that transform from one frame S to another S' via Lorentz Transformation constitute a 4-vector. Lagrangian Formalism can be extended to the EM Field. It is convenient to use a compact notation for vector, tensor, etc, fields that exhibits relativistic covariance. With Greek indices running from 0 to 3:

$x^\mu = (t, x)$)   (25.1.12) a

To generalize to 4-D space-time, use the 4-vector notation and introduce CONTRA-VARIANT 4-vector,

x^μ (where $\mu = 0, 1, 2, 3$) = $(x^o, x^1, x^2, x^3) \equiv (ct, x, y, z)$ (25.1.12) b

and CO-VARIANT 4-vector (formal invariance),

$x_\mu = (x_0, x_1, x_2, x_3) \equiv (ct, -x, -y, -z)$ (25.1.12) c

To see this RELATIVISTIC INVARIANCE, one defines the quantities

$A^\mu = (A^\circ \text{ and } A)$ (25.1.12) d

$J^\mu = (\rho \text{ and } j)$ (25.1.12) e

where ρ and **j** are charge and current densities, and A is potential.

$\partial_\mu = \partial / \partial x^\mu$ (25.1.12) f

Then the current CONSERVATION LAW can be written as

$\partial_\mu J^\mu = 0$.   (25.1.13)

and MAXWELL'S EQUATIONS can be summarized as

$\partial_\mu F^{\mu\nu} = J^{\mu\,\nu}$.   (25.1.14)

where $F_{\mu\nu}$ = Maxwell tensor

$F_{\mu\nu} = \partial_\mu A_\nu - \partial_\nu A_\mu$.   (25.1.15) a

$F^{\mu\nu} = \partial^\mu A^\nu - \partial^\nu A^\mu$.   (25.1.15) b or $F^{\mu\nu} = \begin{bmatrix} 0 & -E & -E^2 & -E^3 \\ E^1 & 0 & -B^3 & +B^2 \\ E^2 & B^3 & 0 & -B^1 \\ -E^3 & -B^2 & +B^1 & 0 \end{bmatrix}$ (25.1.16)

and $F^{0i} = -E^i$
$F^{ij} = -\varepsilon^{ijk} B^k$; (25.1.17)
$F^{\mu\nu} = -F^{\nu\mu}$.

LEVI-CIVITA'S anti-symmetric symbol, $\varepsilon^{\mu\nu\rho\sigma}$:
Defining $\varepsilon^{\mu\nu\rho\sigma} = 1$, if $\mu\,\nu\,\rho\,\sigma$ is *even* permutation of (0, 1, 2, 3).
    = -1, if $\mu\,\nu\,\rho\,\sigma$ is *odd* permutation of (0, 1, 2, 3).
    = 0, if $\mu\,\nu\,\rho\,\sigma$ of (0, 1, 2, 3) is otherwise.
    $= -\varepsilon_{\mu\nu\rho\sigma}$.

$\varepsilon_{\mu\nu\rho\sigma}$ is used transform an anti-symmetric tensor into its dual.
$$F^{\mu\nu} = -F^{\nu\mu} = \frac{1}{2}\varepsilon^{\mu\nu\rho\sigma} F_{\rho\sigma} .$$
It can be shown that
$$F_{\mu\nu} F^{\nu\mu} = -2(E^2 - B^2) \quad . \qquad (25.1.18)$$
To derive Maxwell's Equations,
Action, $L = -\frac{1}{4} F_{\mu\nu} F^{\nu\mu} = -\frac{1}{2}(E^2 - B^2)$ (25.1.19)
The EULER-LAGRANGE EQUATION, describing a particle motion, in generalized coordinates is,
$$\partial/\partial q^i = (\frac{d}{dt})[\delta L(q^i, q^{\cdot i})/\delta q^{\cdot i}] \qquad (25.1.20)$$
with $q^{\cdot i} = (\frac{d}{dt}) q^i$.

Insert (25.1.18) into (25.1.19) to get the Maxwell's equation,
$$\partial_\mu F^{\nu\mu} = 0 . \qquad (25.1.14)$$

25.1.3 RELATIVISTIC NOTATION:
Any theory of fundamental nature must be consistent with
    i) Relativity, and
    ii) Quantum Theory.
Consider two events in SPACE-TIME (r - t).
Define ds = the distance between two points in space.

Lorentz Transformations (25.1.6) are:
Length Contraction: $x' = \beta (x - vt)$, (25.1.6)
i.e., $x' = (x - vt)/\sqrt{(1 - v^2/c^2)}$ (25.1.21)
Time Dilation: $t' = \beta (t - vx/c^2)$, (25.1.8)
i.e., $t' = (t - vx/c^2)/\sqrt{(1 - v^2/c^2)}$. (25.1.22)
In order that ds be the same for all inertial observers ds must be invariant under Lorentz Transformations and rotations, i.e. for convenience,

$$ds^2 = c^2 dt^2 - (dx^2 + dy^2 + dz^2) \tag{25.1.23}$$

$ds^2 = (dx^2 + dy^2 + dz^2) - c^2 dt^2$. can also be taken). With this definition,

$ds^2 > 0$, for events separated by time-like interaction,

$ds^2 < 0$, for events separated by space-like interaction,

$ds^2 = 0$, for events separated by null (*i.e.*, light-like) interaction,

In 3-vector notation:

In 3-D space, $\mathbf{r} \equiv (x, y, z) =$ a 3-vector,

$$dr^2 = (dx^2 + dy^2 + dz^2)$$

= invariant under rotation,
= sum of squared quantities
= positive definite.

25.1.4 The METRIC TENSORS, $g_{\mu\nu}$ and $g^{\nu\mu}$:

$$g_\mu = (x_0, x_1, x_2, x_3) \equiv (ct, -sx, -y, -z) \tag{25.1.25}$$

$$ds^2 = \sum dx^\mu dx_\mu = c^2 dt^2 - (dx^2 + dy^2 + dz^2).$$

i.e., $\quad ds^2 = c^2 dt^2 \tag{25.1.26}$

$\tau =$ PROPER TIME along the trajectory,

$$d\tau = \sqrt{(1 - v^2/c^2)} \, dt \tag{25.1.27}$$

i.e., the scalar product $\sum (x^\mu, x_\mu) =$ constant. $\tag{25.1.28}$

or, the summation convention (*i.e.* an index once appearing in an upper and once in a lower position is automatically summed from 0 to 3) simplifies this as

$$x^\mu x_\mu = \text{constant.} \tag{25.1.29}$$

Further, from (25.1.26), $ds^2 = c^2 dt^2$.

i.e., $\quad ds = \sqrt{c^2 dt^2} = c\sqrt{(1 - v^2/c^2)} \, dt = \sqrt{(c^2 - v^2)} \, dt \tag{25.1.30}$

Introducing the *metric tensor*, $g_{\mu\nu}$ that is used to lower or raise the Lorentz indices

$$x_\mu = g_{\mu\nu} x^\nu, \tag{25.1.31}$$

Operator transcription, $\hat{H} \to i\hbar(\partial/\partial t) =$ Lorentz invariant,

since it is a correspondence between two contra-variant 4-vectors.

$$p^\mu \to i\hbar \, (\partial/\partial x_\mu) \tag{25.1.32}$$

the *metric tensor*, $g_{\mu\nu}$ which is used to LOWER or RAISE the Lorentz indices (25.1.31)

$$x_\mu = g_{\mu\nu} x^\mu = g_{\mu 0} x^o, + g_{\mu 1} x^1, + g_{\mu 2} x^2, + g_{\mu 3} x^3,$$

i.e., $\quad x_o = x^o, x_1 = -x^1, x_2 = -x^2, x_3 = -x^3,$

i.e., $\quad g_{\mu\nu} =$ a diagonal matrix.

$$g_{\mu\nu} = \begin{bmatrix} 1 & -0 & 0 & 0 \\ 0 & -1 & 0 & 0 \\ 0 & 0 & -1 & 0 \\ 0 & 0 & 0 & -1 \end{bmatrix} \tag{25.1.33}$$

$$g^{\mu\nu} = \begin{bmatrix} 1 & -0 & 0 & 0 \\ 0 & -1 & 0 & 0 \\ 0 & 0 & -1 & 0 \\ 0 & 0 & 0 & -1 \end{bmatrix} \qquad (25.1.34)$$

as in Minkowski space (*i.e.* Cartesian coordinates). $g_{\mu\nu}$ contains, therefore, all information about the geometry of space-time, which in this case is the Minkowski space. In SPECIAL RELATIVITY, $g_{\mu\nu}$ plays only a PASSIVE role. in GENERAL RELATIVITY, $g_{\mu\nu}$ plays an ACTIVE role, since the geometry of space depends on what matter is in it.
Defining, CO-VARIANT

$$\nabla_\mu \equiv \partial_\mu \equiv (\partial/\partial x^\mu) = (\partial_o, \partial_1, \partial_2, \partial_3). \qquad (25.1.35)\text{ a}$$

$$= \{\tfrac{1}{c}(\partial/\partial t), (\partial/\partial x), (\partial/\partial y), (\partial/\partial z)\} = \{\tfrac{1}{c}(\partial/\partial t), \hat{\nabla}\} \qquad (25.1.35)\text{ b}$$

CONTRA-VARIANT $\nabla^\mu \equiv \partial^\mu = g^{\mu\nu}\partial_\nu = \{\tfrac{1}{c}(\partial/\partial t), -\hat{\nabla}\} \qquad (25.1.36)$

25.1.6. D' Alembertian operator, $\Box^2$:

Define the differential 4-SPACE GRADIENT operator, $\Box$ such that

$$\Box = (\partial_x, \partial_y, \partial_z, \tfrac{1}{ic}\partial_t) \qquad (25.1.37)\text{ a}$$

D' Alembertian operator,

i.e., $\hat{\Box}^2 \equiv \hat{\Box}\cdot\hat{\Box} = \partial^\mu \partial_\mu = \sum \partial_\mu{}^2 = [\nabla^2 - \tfrac{1}{c^2}(\partial^2/\partial t^2)] \qquad (25.1.37)\text{ b}$

25.1.7 The ENERGY-MOMENTUM 4-vector of a particle:

The 4-momentum of a free particle,

$$p^\mu = m\, u^\mu \qquad (25.1.38)$$

Contra-variant vector

$$p^\mu = (E/c,\, \mathbf{p}) \qquad (25.1.39)$$

Co-variant momentum 4-vector

$$p_\mu = (E/c, -\mathbf{p}) \qquad (25.1.40)$$

Invariant, $p^2 = p^\mu p_\mu = (E^2/c^2 - \mathbf{p}\cdot\mathbf{p}) = (E^2/c^2 - p^2) = m^2 c^2 \qquad (25.1.41)$

25.1.8 The 4-velocity, u^μ

$$d\tau = \sqrt{(1 - v^2/c^2)}\, dt \qquad (25.1.27)$$

$$u^\mu = (dx^\mu/d\tau) = \{c\, dt/d\tau,\, dx/d\tau\} = [(dt/d\tau)(c, \mathbf{v})] \qquad (25.1.42)$$

$$\mathbf{p}\cdot\mathbf{x} = \mathbf{p}\cdot(\mathbf{p}\,t) = p_\mu x^\mu = [E\,t/c^2 - \mathbf{p}\cdot\mathbf{r}] \qquad (25.1.43)$$

25.2. SINGLE PARTICLE RELATIVICTIC EQUATIONS:

25.2.1 THE KLEIN-GORDON EQUATION: (Klein- Fock Equation)
(Schrödinger- Gordon Equation)
Consider two events in space-time, (r, t) and (r + dr, t + dt).
Consider a fast free particle of rest mass, m_o, moving with velocity, v.

Lagrangian, $\quad L = T - V = m\,c^2\,(d\tau/dt)$

$$L_o = m_o\, c^2 \sqrt{(1-v^2/c^2)} \qquad (25.2.1)$$

where $v \approx c$. in field-free space (*i.e.* v = 0).
This is RELATIVISTICALLY INVARIANT.

25.2.1.1 TRADITIONAL WAVE EQUATION Approximation:
By definition, the momentum of the particle is

$$p = (\partial L/\partial v) = mv = m_o v/\sqrt{(1-v^2/c^2)} \qquad (25.2.2)$$

Energy,
$$E = v\,(dL/\partial v) - L = vp + m_o c^2/\sqrt{(1-v^2/c^2)}.$$
$$= (1/c)\left\{p\,v\,c + m_o c^2 \sqrt{(c^2-v^2)}\right\}$$
$$= (1/c)\left\{p\,v\,c + \sqrt{m_o^2 c^4 - m_o c^2\, v^2}\right\}.$$
$$= \left\{p\,v + \sqrt{m_o^2 c^2 - m_o\, v^2}\right\}.$$

$$m = m_o/\sqrt{(1-v^2/c^2)}.;\quad m_o^2 = m^2\,(1-v^2/c^2).$$
$$m_o^2 c^2 = m^2 c^2 - m^2 v^2;\quad m_o^2 c^4 = m^2 c^4 - m^2 v^2 c^2.$$
$$m_o^2 c^4 = m^2 c^4 - (mv)^2 c^2.;\quad m^2 c^4 = m_o^2 c^4 + p^2 c^2.$$
$$m\,c^2 = \sqrt{(m_o^2 c^4 + p^2 c^2)};$$

$$\therefore E = m\,c^2 = \sqrt{(m_o^2 c^4 + p^2 c^2)} \qquad (25.2.3)$$

For a relativistic wave equation, the correct E - p relation must be used.
Schrödinger, and later Klein & Gordon, used

$$\hat{H} = E = (T+V) = (p^2/2m) + V \qquad (25.1.2)$$

Non-relativistic, $\hat{H} = E = (p^2/2m) + 0$.

Relativistic: $\hat{H} = E = \left\{p\,v + \sqrt{m_o^2 c^2 - m_o\, v^2}\right\} = \left\{m_o^2 c^2 + \sqrt{1+p^2/m_o\, c^2}\right\}$

$$\boxed{\hat{H} = E = \sqrt{m_o^2 c^4 + p^2 c^2}} \qquad (25.2.3)$$

If one replaces the momentum p by
$$\hat{p}_x \rightarrow i\hbar(\partial/\partial x) \qquad (25.2.4)\,a$$
$$\hat{p}^2 = -\hbar^2 \hat{\nabla}^2 ; \qquad (25.2.4)\,b$$

one will have to face the difficulty that the square root of a linear operator, $\hat{p}_x$ i.e. (linear operator, $\hat{p}_x)^{1/2}$ = not uniquely defined. This is avoided by taking the square, $\hat{p}^2$. It is seen that

$$E = \sqrt{(m_o^2 c^4 + p^2 c^2)} \qquad (25.2.3)$$

Thus this becomes the Energy-Momentum-Mass relationship.

$$E^2 = (m_o^2 c^4 + p^2 c^2) \qquad (25.2.5)$$

Describe the classical particle as a wave packet
$$|\Psi(r,t)\rangle = \text{state function} = \int d^3k\, C(k) e^{-i\,p\cdot r/\hbar} e^{-i\,E\,t/\hbar}.$$

where E is given by equation(25.2.5). This wave packet would move with the Group Velocity,

$$(\partial\omega/\partial k) = (\partial E/\partial p) = c^2 p/\sqrt{(m_o^2 c^4 + p^2 c^2)}.$$

which is the velocity, v of the CLASSICAL particle.

On substituting, $\hat{E} \to i\hbar(\partial/\partial t)$; (25.2.6)

$$\hat{p}^2 \to -\hbar^2 \hat{\nabla}^2 \ ;$$

in $E^2 = (m_o^2 c^4 + p^2 c^2)$.

$$-\hbar^2(\partial^2/\partial t^2)|\Psi(r,t)\rangle = (m_o^2 c^4 - \hbar^2 \hat{\nabla}^2 c^2)|\Psi(r,t)\rangle \quad (25.2.7)$$

Rearranging, one gets

$$-\frac{1}{c^2}(\partial^2/\partial t^2)|\Psi(r,t)\rangle = -\hat{\nabla}^2|\Psi(r,t)\rangle + (m_o^2 c^4/\hbar^2)|\Psi(r,t)\rangle \quad (25.2.8)$$

De Broglie first proposed this equation in 1924, but is now more usually called the KLEIN-GORDON EQUATION. Setting the D' Alembertian, a scalar operator, $\Box^2$

$$\Box^2 \equiv \left\{ \hat{\nabla}^2 - \frac{1}{c^2}(\partial^2/\partial t^2) \right\} \quad (25.1.37)$$

and $p = m_o c \equiv \hbar k$ (25.2.9)

one gets the customary form of the more used K G EQUATION (Klein-Gordon Equation)

$$\boxed{\{\Box^2 - k^2\}|\Psi(r,t)\rangle = 0} \quad . \quad (25.2.10)$$

25.2.1.2. Other forms of the KG Equation:

Using $p = m_o c \equiv \hbar k$

Equation (25.2.10) can be written as

$$\{\Box^2 - m_o c^2/\hbar^2\}|\Psi(r,t)\rangle = 0 \quad . \quad (25.2.11)$$

Putting $\hat{\Box} \equiv \left\{ \frac{1}{c^2}(\partial^2/\partial t^2) - \hat{\nabla}^2 \right\}$ (25.2.12)

$$\{\hat{\Box} + m_o c^2/\hbar^2\}|\Psi(r,t)\rangle = 0 \quad . \quad (25.2.13)$$

This is also a form of K G EQUATION.
When one deals with everyday physical situations, it is convenient to use the SI system of units, based upon *metre*, *kilogram* and *second*. In fundamental (particle) physics, it is much more convenient system of units, known as NATURAL UNITS, is used in which
(i) Velocity of light, c = 1 means 299,792,458 *m* is equivalent to 1 SECOND.
(ii) Planck Constant, $h = 1$, means ERG and SECOND are inverses of each other.
This also means that GRAM AND CM are inversely related.
(iii) The third basic unit being energy 1 M*eV*.
Thus the KG Equation (25.2.13) becomes

$$\left\{ -\hat{\nabla}^2 + \frac{1}{c^2}(\partial^2/\partial t^2) + m_o c^2/\hbar^2 \right\} \Psi(r,t)\rangle = 0 \ .$$

$$\left\{ -\hat{\nabla}^2 + (\partial^2/\partial t^2) + m_o \right\} \Psi(r,t)\rangle = 0 \ . \quad (25.2.14)$$

This is yet another form of the K G EQUATION. Further it assumes the form

$$\boxed{\{\Box^2 + m_o^2\}|\Psi(r,t)\rangle = 0} \quad . \quad (25.2.15)$$

Worked out Example 25.3.
   Show that the KG equation is Lorentz invariant.
Solution: $\boxed{Step\ \#\ 1}$ Given: $\{\Box^2 - k^2\}|\Psi(r,t)\rangle = 0$ (25.2.10);
   $ds = \sqrt{(c^2 - v^2)}\ dt$ (25.1.30)

Step # 2 $p_x = m v_x = m_o v_x / \sqrt{(1-v^2/c^2)} = m_o (dx/dt) / \sqrt{(1-v^2/c^2)}$.

$= m_o (dx/dt) / \sqrt{(1-v^2/c^2)} = m_o c (dx/ds)$

Similarly, $p_y = m_o c (dy/ds)$, and $p_z = m_o c (dz/ds)$,

$$E = mc^2 = m_o c^2 / \sqrt{(1-v^2/c^2)} = (dt/dt(c^2-v^2))/c = m_o c^2 (dt/ds).$$

These quantities m_o, c and ds are invariants,

i.e., the components p_x, p_y, p_z are transformed similar to dx, dy, dz;

i.e., similar to x, y, z. Further, E transforms like t;

$p_x \sim x, p_y \sim y, p_x \sim z$; and $E \sim c^2 t$.

Substituting energy and momentum in the Lorentz transformation,

$x' = \beta(x - vt); y' = y; z' = z; t' = \beta(t - vx/c^2)$,

with $\quad \beta = 1/\sqrt{(1-v^2/c^2)}$. $\hfill$ (25.1.6)

It is seen that the Lorentz transformation yield the correct limiting transition to a Galilean transformation, it is necessary to include $m_o c^2$ in the energy E, for then $p_x' = p_x - m_o V$, and T does not give a correct limiting transition.

25.2.2 PROBABILITY DENSITY:

For the Non-Relativistic Schrodinger Equation (NRSE), the probability density,

$\rho = \Psi^* \Psi \neq$ a scalar $\hfill$ (25.2.16)

PROBABILITY CURRENT DENSITY,

$j = (i\hbar/2m)[\Psi^* \nabla \Psi - \Psi \nabla \Psi^*]$ $\hfill$ (25.2.17)

They obey the continuity equation

$(\partial/\partial t)\rho + \nabla j = (\partial/\partial t)\Psi^* \Psi + \nabla(-i\hbar/2m)[\Psi^* \nabla \Psi - \Psi \nabla \Psi^*] = 0$ $\hfill$ (25.2.18)

To be purely relativistic,

$\rho = \Psi^* \Psi \neq$ a scalar. $\hfill$ (25.2.19)

But $\rho = \Psi^* \Psi$ transforms as the time component of a 4-vector whose space component is **j**.

So $\quad \rho = (i\hbar/2m)[\Psi^* (\partial/\partial t)\Psi - \Psi(\partial/\partial t)\Psi^*]$ $\hfill$ (25.2.20)

and with $j^\mu = (\rho, j) = (i\hbar/m)\Psi^*[\partial_o, \nabla]\Psi = (i\hbar/m)\Psi^* \partial^\mu \Psi$

(25.2.21)

where $\quad A \partial^\mu B^{def} = \frac{1}{2}[A \partial^\mu B^{def} - (\partial^\mu A) B^{def}]$ $\hfill$ (25.2.22)

and Contra-variant $\partial^\mu = g^{\mu\nu} \partial_\nu = (1/c)[(\partial/\partial t) - \nabla]$ $\hfill$ (25.1.36)

The continuity equation is $\quad (\partial/\partial t)\rho + \nabla j = 0$, becomes

$\partial^\mu j^\mu = (i\hbar/2m)[\Psi^* \nabla \Psi - \Psi \nabla \Psi^*] = 0$ $\hfill$ (25.2.23)

since $\quad \{\Box^2 + m_o^2\} | \Psi(r,t) \rangle = 0$. $\hfill$ (25.2.15)

To show that the spatial current, $j_{KG} \equiv j_{SE}$:

But ρ_{KG} is not identical with ρ_{SE}, since ρ_{KG} contains $(\partial/\partial t)$, i.e. the K G Equation is a second order in $(\partial/\partial t)$, Ψ and $(\partial/\partial t)\Psi$ can be fixed arbitrarily at a given time.

i.e., $\quad \rho_{SE} = +$ve definite.

So $\quad \rho_{KG} \neq +$ve definite.

25..2.2.1 SOLUTIONS TO THE K G EQUATION in Coordinate Space:
The K G Equation is
$$\{\Box^2 - m_o c^2 / \hbar^2\} | \Psi(r,t) \rangle = 0 \qquad (25.2.11)$$
The PLANE WAVE SOLUTION has the form
$$\Psi(r,t) = N\, e^{(-iE\,t\, +i\,\mathbf{p}\,\bullet\,\mathbf{r})/\hbar}. \qquad (25.2.24)$$
where $\mathbf{p} \bullet \mathbf{r}$ = a 4-vector scalar product = $p_\mu x^\mu\ p_\mu x^\mu = E\,t - \mathbf{p}\bullet\mathbf{x}$ $\qquad$ (25.2.25)

In order this $\Psi(r,t)$ be a solution of the KG Equation, one finds by direct substitution that E must be related to p by the condition (25.2.5), viz,
$$E^2 = (m_o^2 c^4 + p^2 c^2). \qquad (25.2.5)$$
That is, $E^2 = (m^2 + p^2)$. $\qquad (25.2.26)$

This implies that for a given 3- momentum, p there are
$$E = \pm\sqrt{(p^2 + m^2)}. \qquad (25.2.27)$$
This means there are positive and negative energy solutions. Thus the most striking feature of the Klein-Gordon Equation is the existence of negative energy solutions.

25.2.3. The KG Equation, in itself, is not a sufficient foundation of Relativistic Quantum Mechanics: The INTERPRETATION of the K G EQUATION
$$\{\Box^2 - k^2\} | \Psi(r,t) \rangle = 0. \qquad (25.2.10)$$
i.e., $\quad -\dfrac{1}{c^2}(\partial^2/\partial t^2)| \Psi(r,t) \rangle = -\hat{\nabla}^2 | \Psi(r,t) \rangle + \left(m_o^2 c^4 / \hbar^2\right)| \Psi(r,t) \rangle \quad (25.2.8)$

as the equation for $\Psi(r,t)$ of a relativistic free particle meets with great difficulties, because

(1) EQUATION (25.2.10) is a second order differential equation, i.e., it contains $(\partial^2/\partial t^2)$; so its solutions require initial conditions on $\Psi(r,t)$ as well as $(\partial/\partial t)\Psi$, and $(\partial/\partial t)\Psi$ has no direct physical interpretation. $(\partial/\partial t)\Psi$ is present in the ρ and so negative values of ρ.,

(2) Klein-Gordon Equation (25.2.10) gives rise to both +ve and –ve energy solutions, as
$$E^2 = (m_o^2 c^4 + \hbar^2 c^2 k^2) = \hbar^2 \omega^2, \qquad (25.2.5)$$

(3) This equation is not applicable to electrons, as it is not a single-particle equation, with wave function Ψ,

(4) This is a single wave equation and does not include the spin of electron, and so Spin-Orbit Interaction (a relativistic effect) cannot be entered.

(5) This equation can be employed for only spin-zero particles like mesons.

(6) $\rho = \Psi^* \Psi \neq$ probability density of the state over the space as it is time dependent, and so not +ve definite,

(7) Stationary state solutions of the K G Equation are not in general orthogonal to one another.

It will be seen later that the interpretation of Ψ as a quantum field clears up these problems. It was only in1934, Pauli and Weisskopf, pointed out the validity of the K G Equation as a field equation that is to be quantized, as the EM Field Equation are.

25.3 HAMILTONIAN OF A CHARGED PARTICLE IN EM FIELD:

25.3.1 CLASSICAL PICTURE:

The effect of EM field on a classical particle (having a charge, q) is specified by the vector EM potential, A(r, t) and scalar EM field potential, Φ(r, t) is described by

Non-relativistic: $\hat{H} = \{[p-(q/c)A]^2/2m\} + q\Phi$ (25.3.1) a

i.e., $[\gamma_\mu p_\mu + c\gamma_\mu A_\mu - m]\Psi = 0$ (25.3.1) b

In the operator form
$[i\gamma_\mu \partial_\mu \bullet m]\Psi = -e\gamma_\mu A_\mu \Psi$ (25.3.1) c

Relativistic: $\hat{H} = \sqrt{m_o^2 c^4 - (cp - qA)^2} + q\Phi$. (25.3.2)

25.3.2 QUANTUM MECHANICAL PICTURE:

Consider the use of the prescription in a 4-D Minkowskii space of events. A(r, t) and Φ(r, t) modify the canonical momentum of the particle with charge q from $p \to p_\mu$ and $E \to E$.

where $p_\mu = (p, iE/c)$ = 4-momentum, (25.3.3)

$A_\mu = (A, i\Phi)$ = 4-potential, (25.3.4)

(a) Free particle motion:
Then the energy of a relativistic free particle is

$E = \sqrt{m_o^2 c^4 + p^2 c^2}$. (25.2.3)

$E = \hbar\omega \to i\hbar(\partial/\partial t)$; (25.2.6)

$\hat{p}_x = \hbar k \to -i\hbar(\partial/\partial x)$; (25.2.4) a

$\hat{p}^2 \to -\hbar^2 \nabla^2$; (25.2.4) b

∴ $\omega \to i\ (\partial/\partial t)$; (25.3.5)

$k \to -i(\partial/\partial x) = -i\nabla_x$ (25.3.6)

Equation (25.2.3) becomes

$\hbar\omega = \sqrt{(m_o^2 c^4 - \hbar^2 k^2 c^2)}$. (25.3.7)

$i\hbar(\partial/\partial t) = \sqrt{(m_o^2 c^4 - c^2 \hbar^2 \nabla^2)}$ (25.3.8)

$i\hbar(\partial/\partial t)\Psi(r,t) = \sqrt{(m_o^2 c^4 - c^2 \hbar^2 \nabla^2)}\ \Psi(r,t)$. (25.3.9)

Equivalently, $\omega \to i\ [(\partial/\partial t) + (iq/\hbar)\ \Phi]$; (25.3.10)

$\nabla_x \to [\nabla_x - (iq/\hbar)\ A]$. (25.3.11)

a) MOTION IN AN EM FIELD:
Using equation (25.2.3)

$(E - q\Phi)^2 = c^2[p_x - (iq/c)A_\mu]^2 + m_o^2 c^4$. (25.3.12)

Equation (25.3.12) is replaced by

$[i\ (\partial/\partial t) - q\ \Phi]^2\ \Psi(r,t) = m_o^2 c^4\ \Psi(r,t) + c^2[(-\hbar/i)\nabla_x - (q/c)A_\mu]^2\ \Psi(r,t)$ (25.3.13)

or $[\Box - k^2 - (2q/ich)A\nabla - (q/ch)^2 A^2 + (2q/ihc)\Phi\ (\partial/\partial t)\ \Psi(r,t)] = 0$ (25.3.14)

Originally considered by Schrödinger himself equation (25.3.13); and he subsequently disregarded it as it led to negative probabilities.

For static electric and magnetic fields,
$(\partial/\partial t)A_\mu = (\partial/\partial t)\Phi = 0$. (25.3.15)

The Klein-Gordon Equation given by (25.2.10) and (25.3.15) reduces to the Schrödinger Equation as $(v/c) \to 0$.

Relativistically, for a system,

$E =$ (non-relativistic energy) $+ m_o c^2$.

Therefore, the rest mass energy, $m_o c^2$, must be reflected in the quantum mechanical wave function.

$$\Psi(r,t)] = e^{-m_o c^2 t/\hbar} \cdot \Psi \quad \text{(kinetic, non-relativistic)} \tag{25.3.16}$$

Substituting (25.3.16) in (25.3.13)

$$[-\hbar^2(\partial^2/\partial t^2) + 2i\hbar(m_o c^2 - q\Phi)(\partial/\partial t) - [q\Phi - (m_o c^2 - q\Phi)]\Psi(\text{kinetic})$$

$$= \{c^2 - [p - (q/c)A]^2\}\Psi(\text{kinetic}) \tag{25.3.17}$$

When $v \ll c$, i.e. non-relativistic,

$$[i(\partial/\partial t)\Psi(\text{kinetic}) = \{[p-(q/c)A]^2/2m + q\Phi\}\Psi(\text{kinetic}) \tag{25.3.18}$$

is the Schrödinger Equation.

Since spin properties do not arise in the KG equation (25.3.17), it governs the behavior of spin-less bosons *eg.*, Pi mesons.

$$\Psi(r,t)] = e^{-iEt/\hbar} \cdot \psi_i(r) = e^{-iEt/\hbar} e^{i\,p\bullet x} \tag{25.3.19}$$

Substituted in the K G Equation (25.3.13), gives the energy eigen value KG equation.

25.3.3 .EM FIELD as an Infinite Dynamical System: 4-GRADIENT: & GAUGE TRANSFORMATION:

Maxwell's Equations, in tensor version, are

$$\partial_\mu F^{\mu\nu} = 0. \tag{25.2.28 a}$$

and $\quad \partial_\mu F^{\mu\nu} = j^\nu$. $\tag{25.2.28 b}$

where $F^{\mu\nu}$ represents the field-strength tensor.

If $\quad A^\mu = $ 4-potential,

current conservation is given by

$$\partial_\mu j^\nu = 0.$$

$$F^{\mu\nu} = \partial^\mu A^\nu - \partial^\nu A^\mu \tag{25.1.15 b}$$

i.e., $\quad E = \nabla A^0 - (\partial/\partial t)A$;

$$B = \nabla \wedge A. \tag{15.1.11}$$

This has a particular solution, in the vicinity of a point taken as origin of the form

$$A^\mu(x) = \int_0^1 d\lambda \, \lambda F^{\mu\nu}(\lambda x) x, \tag{25.3.20}$$

This potential is not uniquely defined by (25.1.15) b. This can be modified by the addition f a 4-gadient. This is called a GAUGE Transforma5tion.

$$A^\mu(x) \to A^\mu(x) + \partial^\mu \Phi \tag{25.3.21}$$

For regular fields satisfying Maxwell's Equations (25.2.28) a & b, one has to define A^μ over all space-time. Putting $\quad \square = \square = \{(1/c^2)(\partial^2/\partial t^2) - \nabla^2\}$

(25.2.12)

as $\quad \square = \{(\partial^2/\partial t^2) - \nabla^2\}$ $\tag{25.2.31}$

the equivalent form of Maxwell's Equations become
$$\Box A^{\mu} = \partial^{\mu} A^{\nu} (\partial_{\nu} A^{\nu}) = j^{\mu}. \tag{25.3.22}$$
This form of Maxwell's Equations is clearly not affected by a Gauge Transformation.

25.3.4 THE DIRAC EQUATION FOR A FREE PARTICLE:

P.A.M. Dirac in 1928 developed an alternative to the K G Equation. Because in the latter the (r - t) space-time derivatives enter in the second order, satisfying the requirements of a relativistic wave equation (where the derivatives with respect to r and t must enter in the same order). The K G Equation is not satisfactory; in the non-relativistic limit it does not imply the existence of intrinsic spin for the particle, so as to abandon its interpretation as a SINGLE PARTICLE EQUATION. To combine the relativistic invariance with Quantum Mechanics, consider the Correspondence Principle. According to Dirac, if an equation in the second derivatives with respect to r and t, to get the charge density, $p = \Psi^* \Psi \geq 0$, an equation in the first derivative in time, t could be valid. To keep space-time symmetry, this equation must be of first order in space derivatives also. So Dirac chose to preserve

(i) The linear from in $(\partial / \partial t)$ of the Schrödinger Equation
$$\hat{H}\Psi = i\hbar(\partial/\partial t)\Psi. \tag{25.4.1}$$

(iii)   The relativistic Hamiltonian, $\hat{H}$ had to be linear in ∇, i.e. **p**.

25.3.5 The DIRAC HAMILTONIAN, $\hat{H}_D$:

For a free massive particle, m_0 the energy E is given in terms of momentum, p by
$$E = \sqrt{(m_0^2 c^4 + p^2 c^2)}, \tag{25.2.3}$$
in the relativistic case.
Using the convenient system of units, $\hbar = 1, c = 1$,
$$\hat{H} = E = \sqrt{(m_0^2 + p^2)}, \tag{25.4.2}$$
Dirac modified $\hat{H}$ to
$$\boxed{\hat{H}_D = \{c\, \alpha \cdot p + \beta\, m_0 c^2\}} \tag{25.4.3 a}$$
i.e.,
$$\boxed{\hat{H}_D = \{\alpha \cdot p + \beta\, m_0\}} \tag{25.4.3 b}$$

It is known that $\hat{H} = \hat{E} \to i\hbar(\partial/\partial t)$;   (25.2.6)
$p_\mu = \hbar k \to -i\hbar(\partial/\partial x_\mu)$;   (25.2.4) a
$p_\mu p^\mu = E^2 - p^2 = m_0^2 \to -\hbar^2 \nabla^2$;   (25.2.4) b
∴ $E \sim \omega \to i(\partial/\partial t)$;   (25.3.5)
$p_o \sim k \to i(\partial/\partial x_\mu) = -i\nabla_\mu$.   (25.3.6)
$\hat{H}_D = \{c\, \alpha \cdot p + \beta\, m_0 c^2\}$ (25.4.3) a
$\hat{H}_D = \{\alpha_\mu \cdot (\nabla^\mu / i) + \beta\, m_0\}$ (25.4.4) a
$\mathcal{H}_D = [\alpha_\mu (\nabla^\mu / i) + \beta\, m_0]$ (25.4.4) b
$\hat{H}_D = \left\{ (c\hbar/i) \sum_{k=1}^{3} \alpha_k (\partial/\partial x_k) + \beta\, m_0 c^2 \right\}$.   (25.4.4) c

Equations (25.4.3) a & b and (25.4.4) a, b & c are all different forms of the DIRAC HAMILTONIAN.

α_μ and β are hermitian matrices.

25.3.6 The DIRAC EQUATION:

∴ $\boxed{\hat{H}_D \Psi = E \Psi}$ (25.4.5)

becomes the Dirac equation, viz.

$\boxed{\left\{ \alpha_\mu \cdot (\nabla^\mu / i) + \beta\, m_o \right\} \Psi = i(\partial / \partial t)\Psi}$ (25.4.6) a

or, in CANONICAL FORM,

$\boxed{\left\{ (\hbar/i)(\partial/\partial t) + \hat{H} \right\} \Psi(x,t) = 0}$ (25.4.6) b

$\left\{ (\hbar/i)(\partial/\partial t) + (c\hbar/i) \sum_{k=1}^{3} \alpha_k (\partial/\partial x_k) + \beta\, m_o c^2 \right\} \Psi(x,t) = 0$ (25.4.6) c

$\left\{ \alpha_\mu \cdot (\nabla^\mu / i) + \beta\, m_o \right\} \Psi = i(\partial / \partial t)\Psi$ (25.4.6) d

$\left\{ (\partial/\partial t) + \alpha_\mu \nabla^\mu + i\beta\, m_o \right\} \Psi = 0$ (25.4.6) e

where Ψ is a VECTOR WAVE FUNCTION.

To make it possible Dirac proposed that α_μ and β matrices be anti-commuting matrices of square equal to unity.

$\boxed{\alpha_\mu = i\beta\, r_\mu}$. (25.4.7)

with the Greek indices running from 1 to 3, the ANTI-COMMUTING RELATIONS are:

$\{\alpha_\mu, \alpha_\nu\} = 0, \quad \mu \neq \nu$. (25.4.8) a

$\{\alpha_\mu, \beta\} = 0$. (25.4.8) b

$\alpha_\mu^2 - \beta^2 = \hat{I}$. (25.4.8) c

with the bracket $\{A, B\}$ of two operators A and B standing for the symmetric combination, $AB + BA$, called the Anti-commutator

$\{A, B\} = AB + BA$.

Squaring (25.4.6) a or b,

$\left\{ c\alpha_\mu \cdot (\hbar \nabla^\mu / i) + \beta\, m_o c^2 \right\}^2 \Psi = -\hbar^2 (\partial^2 / \partial t^2)\Psi$ (25.4.9) a

or $\left\{ \alpha_\mu \cdot (\nabla^\mu / i) + \beta\, m_o \right\}^2 \Psi = -(\partial^2 / \partial t^2)\Psi$ (25.4.9) b

25.4 The DIRAC EQUATION FOR A FREE PARTICLE:

Writing explicitly by expanding (25.4.9) a, is

$-\frac{1}{\hbar}\Big\{ -c^2 \left\{ \left(\alpha_1^2 \nabla_x^2 + \alpha_2^2 \nabla_y^2 + \alpha_3^2 \nabla_z^2 \right) \hbar^2 \right) + \beta^2\, m_o^2 c4^4$

$-c^2 (\alpha_1 \alpha_2 + \alpha_2 \alpha_1) \nabla_x \nabla_y \hbar^2$

$-c^2 (\alpha_1 \alpha_3 + \alpha_3 \alpha_1) \nabla_x \nabla_z \hbar^2$

$-c^2 (\alpha_2 \alpha_3 + \alpha_3 \alpha_2) \nabla_y \nabla_z \hbar^2$

$$\left.\begin{array}{l}+ m (\alpha_1 \beta + \beta \alpha_1) \nabla_x \hbar c^3 \\ + m (\alpha_2 \beta + \beta \alpha_2) \nabla_y \hbar c^3 \\ + m (\alpha_3 \beta + \beta \alpha_3) \nabla_z \hbar c^3\end{array}\right\} \Psi = -\hbar^2 (\partial^2 / \partial t^2) \Psi \qquad (25.4.10)$$

This is the DIRAC EQUATION OF A FREE PARTICLE. This equation cannot be accomplished if α_μ and β were mere numbers and are independent of r, p, t and E; but it is possible if they satisfy the anti-commuting properties, *viz.* equation (25.4.8). This equation is 4-D matrix equation.

25.4.1 The DIRAC REPRESENTATION the $\gamma - s$:

There are three representations of the gamma matrices found commonly. The Dirac notation is probably accepted as the standard one. The new notation uses the symbol γ^μ.

$\gamma^0 = \beta$.

$\gamma^i = \beta \cdot \alpha^i$, where $i = 1, 2, 3$. (Anti-hermitian)

$$\{\gamma^\mu, \gamma^\nu\} = 2g^{\mu\nu} = \gamma^\mu \gamma^\nu + \gamma^\nu \gamma^\mu, \qquad (25.4.11)$$

and the FEYNMAN SLASH, defined as

$$\partial\!\!\!/ = \alpha_\mu \gamma^\mu \qquad (25.4.12)$$

This enables one to write the Dirac equation (25.4.6) a in the new form

$$(i \gamma^\mu \partial_\mu - m_0) \Psi \equiv (i (\partial\!\!\!/) - m_0) \Psi = 0 \qquad (25.4.13)$$

The K G Equation is then obtained by multiplying Dirac equation (25.4.13) by $[\partial\!\!\!/(\partial\!\!\!/) + m]$.

$$((\partial\!\!\!/)^2 - m\beta) \Psi = 0 \qquad (25.4.14)$$

Note: The matrices α_μ and β commute with p and r, only if

$$\{\gamma_\mu, \alpha_\nu\} = 2\delta_{\mu\nu} I. \qquad (25.4.15)\text{ a}$$
$$\{\alpha, \beta\} = 0 \qquad (25.4.15)\text{ b}$$
$$\beta^2 = I \qquad (25.4.15)\text{ c}$$

It can be shown that the lowest order of matrices that satisfy these relations are (4x4). Four is the smallest Dimension in which matrices fulfilling (25.4.11) can be found. Dirac showed that there does not exist any set of four (2 x 2) or four 3-by-3 matrices that meet the requirements in the α_μ and β given by (25.4.11) and/ or (25.4.15).

25.4.2 An EXPLICIT REPRESENTATION OF DIRAC'S MATRICES, γ^D

$$\gamma^0 = \begin{pmatrix} \hat{I} & 0 \\ 0 & -\hat{I} \end{pmatrix} \qquad (25.4.17)$$

$$\gamma^i = \begin{pmatrix} 0 & \sigma^i \\ -\sigma^i & 0 \end{pmatrix} \qquad (25.4.18)$$

in terms of the (2 x 2) UNIT Matrix and PAULI Matrices.

$$\hat{I} = \begin{pmatrix} 1 & 0 \\ 0 & 1 \end{pmatrix} \qquad (25.4.19)$$

$$\sigma^1 = -\begin{pmatrix} 0 & 1 \\ 1 & 0 \end{pmatrix} \qquad (25.4.20)\text{ a}$$

$$\sigma^2 = -\begin{pmatrix} 0 & -i \\ i & 0 \end{pmatrix} \quad (25.4.20)\ b$$

$$\sigma^3 = \begin{pmatrix} 1 & 0 \\ 0 & -1 \end{pmatrix} \quad (25.4.20)\ c$$

Equation (25.4.18) is abbreviated for the following matrices:

$$\alpha^1 = \begin{pmatrix} 0 & \sigma^1 \\ \sigma^1 & 0 \end{pmatrix} \equiv \begin{bmatrix} 0 & 0 & 0 & 1 \\ 0 & 0 & 1 & 0 \\ 0 & 1 & 0 & 0 \\ 1 & 0 & 0 & 0 \end{bmatrix} \quad (25.4.21)\ a$$

$$\alpha^2 = \begin{pmatrix} 0 & \sigma^2 \\ \sigma^2 & 0 \end{pmatrix} \equiv \begin{bmatrix} 0 & 0 & 0 & -i \\ 0 & 0 & -i & 0 \\ 0 & -i & 0 & 0 \\ -i & 0 & 0 & 0 \end{bmatrix} \quad (25.4.21)\ b$$

$$\alpha^3 = \begin{pmatrix} 0 & \sigma^3 \\ \sigma^3 & 0 \end{pmatrix} \equiv \begin{bmatrix} 0 & 0 & 1 & 0 \\ 0 & 0 & 0 & -1 \\ 1 & 0 & 0 & 0 \\ 0 & -1 & 0 & 0 \end{bmatrix} \quad (25.4.21)\ c$$

$$\beta = \gamma^0 = \begin{pmatrix} \hat{I} & 0 \\ 0 & -\hat{I} \end{pmatrix} \equiv \begin{bmatrix} 1 & 0 & 0 & 0 \\ 0 & 1 & 0 & 0 \\ 0 & 0 & -1 & 0 \\ 0 & 0 & 0 & -1 \end{bmatrix} \quad (25.4.22)$$

Worked out Example 25.4.

Use the explicit Representation of α^i and β to verify the anti-commuting properties.

Solution:

$$\alpha^1 = \begin{pmatrix} 0 & \sigma^1 \\ \sigma^1 & 0 \end{pmatrix} \equiv \begin{bmatrix} 0 & 0 & 0 & 1 \\ 0 & 0 & 1 & 0 \\ 0 & 1 & 0 & 0 \\ 1 & 0 & 0 & 0 \end{bmatrix} \quad (25.4.21)\ a,$$

$$\beta = \gamma^0 = \begin{pmatrix} \hat{I} & 0 \\ 0 & -\hat{I} \end{pmatrix} \equiv \begin{bmatrix} 1 & 0 & 0 & 0 \\ 0 & 1 & 0 & 0 \\ 0 & 0 & -1 & 0 \\ 0 & 0 & 0 & -1 \end{bmatrix} \quad (25.4.22)$$

Substituting, the matrices, $\{\alpha^1, \beta\} = (\alpha^1\beta + \beta\alpha^1) = 0$

$$\alpha^1 = \begin{pmatrix} 0 & \sigma^1 \\ \sigma^1 & 0 \end{pmatrix} \equiv \begin{bmatrix} 0 & 0 & 0 & 1 \\ 0 & 0 & 1 & 0 \\ 0 & 1 & 0 & 0 \\ 1 & 0 & 0 & 0 \end{bmatrix}, \quad (25.4.21)\ a,\ ,$$

$$\alpha^2 = \begin{pmatrix} 0 & \sigma^2 \\ \sigma^2 & 0 \end{pmatrix} \equiv \begin{bmatrix} 0 & 0 & 0 & -i \\ 0 & 0 & -i & 0 \\ 0 & -i & 0 & 0 \\ -i & 0 & 0 & 0 \end{bmatrix} \qquad (25.4.21)\,b$$

Substituting, the matrices,
$$\{\alpha^1, \alpha^2\} = (\alpha^1 \alpha^2 + \alpha^2 \alpha^1) = 0.$$
$$\{\alpha^1, \alpha^1\} = (\alpha^1 \alpha^1 + \alpha^1 \alpha^1) = 2\,\hat{I}.$$

Similarly, substituting, the matrices,
$$\beta^2 = \hat{I}$$

All the 4 Dirac Matrices are seen to Anti-commute one another.

25.4.3 The MAJORANA REPRESENTATION of γ^μ

It was seen that the hermitian $\gamma^0 = \beta$, and $\gamma^1 = \beta\,\alpha^1$.

Now defining $\gamma_5 = \gamma^5 = i\,\gamma^0\gamma^1\gamma^2\gamma^3 = -(1/4!)\,\varepsilon_{\mu\nu\rho\sigma} = \gamma^\mu\gamma^\nu\gamma^\rho\gamma^\sigma$

$$\varepsilon_{\mu\nu\rho\sigma}\,\gamma^\mu\,\gamma^\nu\gamma^\rho\,\gamma^\sigma, \qquad (25.4.22)\,a$$
$$\gamma_5 = -i\,\gamma_0\gamma_1\gamma_2\gamma_3 = i\,\gamma^3\gamma^2\gamma^1\gamma^0 = \gamma_5^\dagger.$$
(25.4.22) b

$$\therefore \quad \gamma_5^2 = \hat{I} \qquad (25.4.23)$$

$$\{\gamma_5, \gamma^\mu\} = 0.$$

In Majorana Representation,

$$\gamma^0 = \beta = \sigma^1 \otimes \sigma^1 = \begin{pmatrix} 0 & \sigma^2 \\ \sigma^2 & 0 \end{pmatrix} = \alpha_2 \text{ of Dirac Representation.}$$

$$\alpha^1 = -\sigma^1 \otimes \sigma^1 = \begin{pmatrix} 0 & -\sigma^1 \\ -\sigma^1 & 0 \end{pmatrix} = -\alpha_1 \text{ of Dirac Representation.}$$

$$\alpha^2 = \sigma^3 \otimes \hat{I} = \begin{pmatrix} \hat{I} & 0 \\ 0 & -\hat{I} \end{pmatrix} = \beta \text{ of Dirac Representation.}$$

$$\alpha^3 = \sigma^1 \otimes \sigma^3 = \begin{pmatrix} 0 & -\sigma^3 \\ -\sigma^3 & 0 \end{pmatrix} = -\alpha_3 \text{ of Dirac Representation.}$$

$$\gamma^1 = i\,\hat{I} \otimes \sigma^3 = \begin{pmatrix} i\sigma^3 & 0 \\ 0 & i\sigma^3 \end{pmatrix} \qquad (25.4.24)$$

$$\gamma^2 = -i\,\hat{I} \otimes \sigma^2 = \begin{pmatrix} 0 & -\sigma^3 \\ -\sigma^3 & 0 \end{pmatrix}$$

$$\gamma^3 = -i\,\hat{I} \otimes \sigma^1 = \begin{pmatrix} -i\sigma^1 & 0 \\ 0 & -i\sigma^1 \end{pmatrix}$$

$$\gamma_5 = \gamma^5 = \sigma^3 \otimes \sigma^2 = \begin{pmatrix} \sigma^2 & 0 \\ 0 & -\sigma^2 \end{pmatrix}$$

$$C = -i\,\sigma^1 \otimes \sigma^2 = \begin{pmatrix} 0 & -i\sigma^2 \\ -i\sigma^2 & 0 \end{pmatrix}$$

$$\gamma^\mu{}_{Majorana} = U\,\gamma^\mu{}_{Dirac}\,U^\dagger. \tag{25.4.25}$$

$$U = U^\dagger = \sqrt{1/2}\begin{pmatrix} \hat{I} & \sigma^2 \\ \sigma^2 & -\hat{I} \end{pmatrix} \tag{25.4.26}$$

Among all equivalent representations obtained by a NON-SINGULAR TRANSFORMATION (25.4.25), the MAJORANA REPRESENTATION plays a crucial role. It is designed to make the *Dirac Equation real*. This is an achievement obtained by interchanging α^2 and β, and changing the sign of α^1 and α^3 as in (25.4.24). Then β only is imaginary.
The Dirac Equation

$$\boxed{[\,(\partial/\partial t)+\alpha_\mu\cdot\nabla^\mu+i\beta\,m_o\,]\Psi = 0} \tag{25.4.6 c}$$

becomes real. Its solutions are linear combinations of real solutions. The matrix U that enables changes to such a Representation is given by (25.4.25).

25.4.4 SYMMETRIC Form of the Dirac Equation:

Dirac $\hat{H}_D = [c(\hbar/i)\sum_{k=1}^{3}\alpha_k(\partial/\partial x_k)+\beta\,m_o c^2]$.

$$[(\hbar/i)(\partial/\partial t)+\hat{H}_D]\,\Psi(x,t)=0 \tag{25.4.6 b}$$

is the CANONICAL form of the Dirac Equation.
Multiply equation (25.4.6) b by $(1/\hbar c)\beta$, and introduce the new operator, γ_k,

$$\gamma_k \equiv -i\beta\,\alpha_k \tag{25.4.27}$$

with $k = 1, 2, 3$.

$$\gamma_4 \equiv \beta, \tag{25.4.28}$$

then one gets the complete SYMMETRIC FORM of the Dirac Equation,

$$\boxed{(\gamma_\mu\,\partial_\mu+k)\Psi = 0} \tag{25.4.29}$$

with $\mu = 1, 2, 3, 4$.
$k = m_o c/\hbar$.

$$\gamma_\mu\partial_\mu = \sum_{\mu=1\text{ to }4}\gamma^\mu(\partial/\partial x^\mu) \tag{25.4.30}$$

using (25.1.12) f $\chi_\mu\gamma_\nu + \gamma_\nu\chi_\mu = 2\delta_{\mu\nu}$ \tag{25.4.31}

25.4.5 RELATIVISTIC INVARIANCE of The Dirac Equation:

The requirement of a Relativistic Wave Equation, like the Dirac Equation, is covariance

under Lorentz Transformation.

Worked out Example 25.6.
Demonstrate that the Dirac Equation is covariant under Lorentz Transformation.
Solution:

$$[(\hbar/i)(\partial/\partial t)+c(\hbar/i)\sum_{k=1}^{3}\alpha_k(\partial/\partial x_k)+\beta\,m_o c^2]\,\Psi(x,t)=0 \tag{25.4.6 b}$$

where (Dirac $\hat{H}_D$) $\hat{H}_D = [c(\hbar/i)\sum_{k=1}^{3}\alpha_k(\partial/\partial x_k)+\beta\,m_o c^2]$

$$[(\hbar/i)(\partial/\partial t) + \hat{H}_D]\Psi(x,t) = 0.$$ (25.4.6) b

is the Dirac Equation in the Canonical Form, which is the famous Schrödinger Form, $\hat{H}\Psi = E\Psi$, with the difference that $\hat{H}$ is different. Consider two Coordinate Lorentz Frames S and S', x_μ and x'_μ

Lorentz frame x'_μ is obtained from x_μ according to

$$x'_\mu = \alpha_{\mu\nu} x_\nu.$$

$$x_\mu = \alpha_{\nu\mu} x'_\nu.$$ (25.4.32)

Since under a Lorentz Transformation (L),
∂_μ behaves in the same way as x_μ do as per equation (25.4.32).

$$\partial_\mu = \alpha_{\nu\mu} \partial'_\nu.$$ (25.4.33)

Then $(\gamma_\mu \partial_\mu + k)\Psi = 0$. (25.4.29)

becomes $\{\alpha_{\nu\mu} \gamma_\mu \partial'_\nu + k\}\Psi = 0$. (25.4.34)

The 4 quantities $\gamma_\nu = \alpha_{\nu\mu} \gamma_\mu$,

obey the same algebraic relations as the Dirac Matrices themselves, *i.e.*,

$$\gamma_\mu \gamma_\nu + \gamma_\nu \gamma_\mu = 2\delta_{\mu\nu}$$ (25.4.31)

Now the fundamental theorem of Dirac algebra has only one Irreducible Representation (IR), and this is 4-dimensional. I.e. there must exist a non-singular (4 x 4) matrix S(L) such that

$$\alpha_{\nu\mu}\gamma_\mu = S(L)^{-1}\gamma_\nu S(L)$$ (25.4.35)

Substituting this in the previous equation, and multiplying from left by S(L), and using the notation

$$\Psi'(x') = S(L)\Psi(x)$$ (25.4.36)

where $\Psi(x)$ and $\Psi'(x')$ are wave functions in the two frames,

X' = L x .

One gets from $(\alpha_{\nu\mu}\gamma_\mu \partial'_\nu + k)\Psi = 0$. (25.4.34)

$[S(L)^{-1}\gamma_\nu S(L)\partial'_\nu + k]\Psi = 0$.

$S(L)\{S(L)^{-1}\gamma_\nu S(L)\partial'_\nu \Psi(x) + S(L)\} + k\Psi(x) = 0$.

i.e., $\gamma_\nu S(L)\partial'_\nu S(L)\Psi(x) + k\Psi(x) = 0$.

$[[\gamma_\nu \partial'_\nu + k]\Psi(x) = 0$. (25.4.37)

This Dirac Equation in the primed Lorentz frame is identical to the one in the original (unprimed) frame; $\Psi(x) \rightarrow \Psi'(x')$. That establishes the covariance of the Dirac Equation under Lorentz Transformation.

25.4.6 **INVARIANCE OF THE DIRAC EQUATION** under Rotation Transformation; invariant under Space Inversion Transformation:
This is not to be discussed in this book.

25.4.7 The **DIRAC EQUATION** as a **FOUR COUPLED DIFFERENTIAL EQUATION**:
The Dirac Wave Equation for a Free Particle:
The Dirac (energy eigen value) Equation for a free particle is given by

$$\{c\alpha_\mu \cdot (\hbar\nabla^\mu/i) + \beta m_o c^2\}\Psi_P = i\hbar(\partial/\partial t)\Psi_P$$ (25.4.6) a

where Ψ_P is the momentum eigen function.

$$\Psi_P = \begin{bmatrix} \Psi_P(1) \\ \Psi_P(2) \\ \Psi_P(3) \\ \Psi_P(4) \end{bmatrix} \quad (25.4.38)$$

The Dirac Equation (25.4.10) as well as (25.4.13) can be regarded as 4-component wave equation, *i.e.* a (4x4) matrix differential equation for the 4-component spinor (state vector), Ψ_P in the spin space. This, therefore, accounts for many properties of spin-½ particles. Alternately, the Dirac Equation can be considered as a set of 4-component differential equations for the components of Ψ. Each component of Ψ separately satisfies the K G Equation.
Equation (25.4.6) a can be written in full,

$$(-c\hbar/i)\begin{bmatrix} 0 & 0 & 0 & 1 \\ 9 & 9 & 1 & 0 \\ 0 & 1 & 0 & 0 \\ 1 & 0 & 0 & 0 \end{bmatrix} \begin{pmatrix} (\partial/\partial x)\Psi_P(1) \\ (\partial/\partial x)\Psi_P(2) \\ (\partial/\partial x)\Psi_P(3) \\ (\partial/\partial x)\Psi_P(4) \end{pmatrix} + (-c\hbar/i)\begin{bmatrix} 0 & 0 & 0 & -i \\ 9 & 9 & i & 0 \\ 0 & -i & 0 & 0 \\ i & 0 & 0 & 0 \end{bmatrix}\begin{pmatrix} (\partial/\partial y)\Psi_P(1) \\ (\partial/\partial y)\Psi_P(2) \\ (\partial/\partial y)\Psi_P(3) \\ (\partial/\partial y)\Psi_P(4) \end{pmatrix}$$

$$+ (-c\hbar/i)\begin{bmatrix} 0 & 0 & 1 & 9 \\ 9 & 9 & 0 & -1 \\ 1 & 0 & 0 & 0 \\ 0 & -1 & 0 & 0 \end{bmatrix}\begin{pmatrix} (\partial/\partial z)\Psi_P(1) \\ (\partial/\partial z)\Psi_P(2) \\ (\partial/\partial z)\Psi_P(3) \\ (\partial/\partial z)\Psi_P(4) \end{pmatrix}$$

$$+ (-m_o c^2)\begin{bmatrix} 1 & 0 & 0 & 9 \\ 9 & 1 & 0 & 0 \\ 0 & 0 & -1 & 0 \\ 0 & 0 & 0 & -1 \end{bmatrix}\begin{pmatrix} \Psi_P(1) \\ \Psi_P(2) \\ \Psi_P(3) \\ \Psi_P(4) \end{pmatrix} - E\begin{pmatrix} \Psi_P(1) \\ \Psi_P(2) \\ \Psi_P(3) \\ \Psi_P(4) \end{pmatrix} = 0 \quad .(25.4.39)$$

Now incorporate each of the matrix operators on the column symbol to its right, this equation becomes

$$-(c\hbar/i)\begin{pmatrix} (\partial/\partial x)\Psi_P(4) \\ (\partial/\partial x)\Psi_P(3) \\ (\partial/\partial x)\Psi_P(2) \\ (\partial/\partial x)\Psi_P(1) \end{pmatrix} -(c\hbar/i)\begin{pmatrix} (-i\,\partial/\partial y)\Psi_P(4) \\ (+i\,\partial/\partial y)\Psi_P(3) \\ (-i\,\partial/\partial y)\Psi_P(3) \\ (+i\,\partial/\partial y)\Psi_P(1) \end{pmatrix} -(c\hbar/i)\begin{pmatrix} (+\partial/\partial z)\Psi_P(3) \\ (-\partial/\partial z)\Psi_P(4) \\ (+\partial/\partial z)\Psi_P(1) \\ (-\partial/\partial z)\Psi_P(2) \end{pmatrix}$$

$$-(m_o c^2)\begin{pmatrix} +\Psi_P(1) \\ +\Psi_P(2) \\ -\Psi_P(3) \\ -\Psi_P(4) \end{pmatrix} - E\begin{pmatrix} \Psi_P(1) \\ \Psi_P(2) \\ \Psi_P(3) \\ \Psi_P(4) \end{pmatrix} = 0 \quad (25.4.40)$$

This is just like the vector equation,
$\quad A + B + C = 0$,
which means $\quad A_x + B_x + C_x = 0$,
$A_y + B_y + C_x = 0$, etc.

An equation on column symbols would therefore mean that the same relationship holds for 'each' of the four components.

Written in order, the four equations obtained by equating corresponding components, in (25.4.40), are:

$$[E+ m_oc^2]\Psi_P(1) + 0 + (c\hbar/i)(\partial/\partial z)\Psi_P(3)+(c\hbar/i)[(\partial/\partial x)-i(\partial/\partial y)]\Psi_P(4) = 0$$
$$0 + [E+ m_oc^2]\Psi_P(2) + (c\hbar/i)[(\partial/\partial x) + i(\partial/\partial y)]\Psi_P(3) -(c\hbar/i)(\partial/\partial y)\Psi_P(4)=0$$
$$(c\hbar/i)[(\partial/\partial z)\Psi_P(1) +(c\hbar/i)[(\partial/\partial x)-i(\partial/\partial y)]\Psi_P(2) +[E- m_oc^2]\Psi_P(3) + 0 = 0$$
$$c\hbar/i)[(\partial/\partial x)+i(\partial/\partial y)]\Psi_P(1)-(c\hbar/i)(\partial/\partial z)\Psi_P(2)+ 0 +[E- m_oc^2]\Psi_P(4) = 0 .$$

(25.4.41)

This *set of linear, partial differential equations is* the DIRAC WAVE EQUATION for a free particle.

(I) It tells how the four components of Ψ_P must vary in space,
(II) If the new Ψ_P is to be well behaved, 'each' of its components must be well behaved, etc.,
(III) For Ψ_P to be a non-trivial solution of equation (25.4.41) at least one of its components must be non-zero,
(IV) The above 4 equations form a set of four coupled differential equations.

25.5.1 HOLE THEORY-NEGATIVE ENERGY STATES AND FREE RELATIVISTIC PARTICLE (PLANE WAVE) SOLUTIONS:

25.5.2 PLANE WAVE APPROXIMATION:

To find the solutions of the Dirac wave Equation (25.4.41) is generally more difficult than to find solutions to the non-relativistic Schrödinger Equation,

$$\hat{H}\Psi = E\Psi = i\hbar(\partial/\partial t)\Psi \qquad (25.1.1)$$

So, for simplicity, let the solution to (25.4.41) has the form of a plane wave propagating along the x-axis, viz.,

The WAVE FUNCTION, $\left|\Psi_{P_\mu}^+(p,t)\right\rangle$ of a free electron with 4-momentum, p_μ (and 3-momentum,p) is the product of a **plane wave**, $e^{i\,p\cdot r/\hbar}e^{-i\,E\,t/\hbar}$ and a **Spinor**, $\left|U_{P\downarrow}^+(p,r)\right\rangle$.

Momentum wave function,

$$\left|\Psi_{P_\mu}^+(p,t)\right\rangle - \left|U_{P\downarrow}^+\right\rangle e^{i\,p\cdot r/\hbar}e^{-i\,E\,t/\hbar} \qquad (25.5.1)$$

with $\quad p\cdot r \equiv k\cdot r = x\,k_x + y\,k_y + z\,k_z$

$$\left|\Psi_{P_\mu}^+(p,t)\right\rangle = \begin{bmatrix} U_P(1) \\ U_P(2) \\ U_P(3) \\ U_P(4) \end{bmatrix} = \begin{bmatrix} A_1 e^{-i\,p_1\,x/\hbar} \\ A_2 e^{-i\,p_1\,x/\hbar} \\ A_3 e^{-i\,p_1\,x/\hbar} \\ A_4 e^{-i\,p_1\,x/\hbar} \end{bmatrix}. \qquad (25.5.2)$$

This form for $\left|\Psi_P^+(p,t)\right\rangle$ implies that the waves are at any instant t the same for all values of y and z (out to $-\infty$ to $+\infty$) and that they are sinusoidal in form, along the x-axis, from $-\infty$ to $+\infty$. A-s are constants.

The amplitude, $\left|U_{Ps}^+\right\rangle$ of the 4-component Dirac spinor, *i.e.* the bi-spinor (or 4-spinors) satisfies the eigen value equation

$$\hat{H}_D\left|U_{P\downarrow}^+\right\rangle = [c\alpha \cdot p + \beta m_o c^2]\left|U_{P\downarrow}^+\right\rangle \tag{25.5.4}$$

where $\hat{H}_D = [c\alpha \cdot p + \beta m_o c^2]$ (25.4.3) a

does not commute with the components of angular momentum, **L**.

$$[\hat{H}_D, L_i] \neq 0,$$

$$[\hat{H}_D, L_i] = -c\hbar^2[\alpha_2(\partial/\partial x_3) - \alpha_3(\partial/\partial x_2)]$$

∴ L is not a constant of the motion.

25.5.3 CONSERVATION OF ANGULAR MOMENTUM VIOLATION not allowed:

To save the angular momentum Conservation Law, one must amend L by an additional factor, S, the spin of the Dirac particle.

$$\mathbf{J} = \mathbf{L} + \mathbf{S}. \tag{25.5.3}$$

such that $[S_1, S_2] = i\hbar S_3$.

(1) $[L, S] = 0$.
(2) $[J, \hat{H}_D] = 0$.
(3) It can be seen easily that $S = \frac{1}{2}\hbar\sigma$.

$$\sigma_1 = -i\,\gamma_2\,\gamma_3\,.$$
$$\sigma_2 = -i\,\gamma_3\,\gamma_1\,.$$
$$\sigma_3 = -i\,\gamma_1\,\gamma_2$$

satisfy all these. Substituting these in the equation (25.4.3) gives the 4 x 4 matrix eigen value equation,

$$[c\alpha \cdot p + \beta m_o c^2]\left|U_{P\downarrow}^+\right\rangle = E_P\left|U_{P\downarrow}^+\right\rangle \tag{25.5.4}$$

where $\alpha_1\left|U_{Ps}^+\right\rangle = \begin{bmatrix} U_P(4) \\ U_P(3) \\ U_P(2) \\ U_P(1) \end{bmatrix}$

$$\alpha_2\left|U_{Ps}^+\right\rangle = \begin{bmatrix} +U_P(4) \\ -U_P(3) \\ +U_P(1) \\ -U_P(2) \end{bmatrix}$$

$$\alpha_3\left|U_{Ps}^+\right\rangle = \begin{bmatrix} +U_P(3) \\ -U_P(4) \\ +U_P(1) \\ -U_P(2) \end{bmatrix}$$

$$\beta\left|U_{Ps}^+\right\rangle = \begin{bmatrix} +U_P(1) \\ -U_P(2) \\ -U_P(3) \\ -U_P(4) \end{bmatrix}$$

Each component of (25.5.1) is a periodic waveform, of wavelength, λ, propagating in the +ve x-direction.

25.5.4 4 HOMOGENEOUS LINEAR EQUATIONS:

Substituting these one gets the equation (25.4.41), viz.,

$$[E + m_o c^2]\Psi_P(1) + \quad 0 \quad +(ch/i)(\partial/\partial z)\Psi_P(3) + (ch/i)[(\partial/\partial x) - i(\partial/\partial y)]\Psi_P(4) = 0$$
$$\quad 0 \quad + [E + m_o c^2]\Psi_P(2) + (ch/i)[(\partial/\partial x) + i(\partial/\partial y)]\Psi_P(3) - (ch/i)(\partial/\partial y)\Psi_P(4) = 0$$
$$(ch/i)[(\partial/\partial z)\Psi_P(1) + (ch/i)[(\partial/\partial x) - i(\partial/\partial y)]\Psi_P(2) + [E - m_o c^2]\Psi_P(3) + \quad 0 \quad = 0$$
$$ch/i)[(\partial/\partial x) + i(\partial/\partial y)]\Psi_P(1) - (ch/i)(\partial/\partial z)\Psi_P(2) + \quad 0 \quad +[E - m_o c^2]\Psi_P(4) = 0 \,.$$
$$\hspace{10cm}(25.4.41)$$

These are 4 homogeneous linear equations for $U_P(1), U_P(2), U_P(3)$, and $U_P(4)$. For NON-TRIVIAL solutions, $|\text{Determinant}| = 0$, and one finds the eigen values E, as usual.
Replacing the differential operators by their corresponding momentum components

$$[E + m_o c^2] U_P(1) + \quad 0 \quad +c p_z U_P(3) +c [p_x + ip_y] U_P(4) = 0$$
$$\quad 0 \quad + [E + m_o c^2] U_P(2) + c(c_x - ip_y) U_P(3) \quad -c p_z U_P(4) = 0$$
$$c p_z U_P(1) + c (p_x + ip_y) U_P(2) + [E - m_o c^2] U_P(3) + \quad 0 \quad = 0$$
$$c(p_x - ip_y) U_P(1) - c p_z U_P(2) + \quad 0 \quad +[E - m_o c^2] U_P(4) = 0$$
$$\hspace{10cm}(25.5.5)$$

i.e., $$(\hat{H}_D - E)|U_P\rangle = \begin{bmatrix} (m_o c^2 - E) & 0 & c p_z & c(p_x - ip_y) \\ 0 & (m_o c^2 - E) & c(p_x + ip_y) & c p_z \\ c p_z & c(p_x - ip_y) & (-m_o c^2 - E) & 0 \\ c(p_x + ip_y) & -c p_z & 0 & (-m_o c^2 - E) \end{bmatrix} \begin{bmatrix} U_P(1) \\ U_P(2) \\ U_P(3) \\ U_P(4) \end{bmatrix} = 0$$
$$\hspace{10cm}(25.5.6)$$

Setting the determinant of the coefficients equal to zero, this equation reduces to
$$(m_o c^2 - E)^2 (m_o c^2 + E)^2 + 2(m_o c^2 - E)(m_o c^2 + E)c^2 p^2 + c^4 p^4 = 0$$
i.e., $\quad [m_o^2 c^4 - E^2 + c^2 p^2]^2 = 0 \,.$ $\hspace{5cm}(25.5.7)$

25.5.4 EIGEN VALUES, E:

Thus from (25.5.7) the eigen values, or allowed energies E, are
$$E = E_\pm = \pm \sqrt{c \, ([m_o^2 c^2 + p^2]} \,. \hspace{4cm}(25.5.8)$$

i.e., $E_+ = \sqrt{c([m_0^2 c^2 + p^2])}$. Twice

and $E_- = -\sqrt{c([m_0^2 c^2 + p^2])}$. Twice.

For each p, E has not only twice the positive value, *i.e.* two linearly independent solutions for $E_+ = +\sqrt{c([m_0^2 c^2 + p^2])}$, but also twice the NEGATIVE VALUES $E = -(m_0^2 c^4 + c^2 p^2)^{1/2}$.

This means that to every positive energy state of the particle, there corresponds a negative energy state (as seen in the K G Equation). Equation (25.5.8) is exactly the basic relationship between E, p and rest mass required by the Theory of Relativity,

i.e., $E = +\sqrt{c([m_0^2 c^2 + p^2])}$.

25.5.5 Dirac's Interpretation of the POTENTIAL CATASTROPHE:

At first sight, equation (25.5.8)

vi.z, $E = E_\pm = \pm \sqrt{c([m_0^2 c^2 + p^2])}$,

gives the *potential catastrophe* might suggest that there are states of negative mass (rest energy, $m_0 c^2$). But the other interpretation that led Dirac to triumph is as follows:

When there is no observed field one supposes that the states E_- are completely filled. These ELECTRONS cannot therefore produce an external field. One takes as the 'null field' that which is observed when all the states E_+ are empty and the entire E_- states are filled. If one further assumes that an EM Field can affect these particles of negative energy then these can be transferred to the states E_+. In this case one observes a field due to the HOLE, which appears in the states, E_-.

Energy of a HOLE $= -(E_-) = +$ ve energy.

That is, the properties of the HOLE will be negative sign put to those of the electron when in its negative energy state. Thus a HOLE will behave like a particle of positive energy, positive rest mass and positive charge, +e, because $-(-e) = +e$. One calls this a HOLE a positron, and it differs from an ORDINARY electron solely by its positive charge.

C.D. Anderson confirmed the existence of the positron later (in 1932). It is clear that the positron can only be produced in the association of an electron, since the former is the 'hole', which remains after the latter, has appeared (in a state E_+). This is the phenomenon of PAIR PRODUCTION.

Since E_- have the range $-\infty < E_- < -m_0 c^2$,

E_+, have the range $+m_0 c^2 < E_+ < +\infty$,

It follows that energy of at least $2 m_0 c^2$ is required to produce a positron-electron pair.

E_- - *Energy particle* solutions propagate BACKWARD in time

$\equiv E_+$ energy 'hole' (*anti-particle*) solutions propagate FORWARD IN TIME.

R. Feynman in 1962 gave an alternative interpretation called the WAVE FUNCTION INTERPRETATION to the potential catastrophe.

25.5.6 EIGEN SPINORS or the negative energy states:

The eigen vectors are

(a)   for $E_+ = +\sqrt{c([m_0^2 c^2 + p^2])}$,

there are two linearly independent solutions,

$$|U_P^I\rangle \approx |U_{P\downarrow}^+\rangle = N_P \begin{bmatrix} 1 \\ 0 \\ cp_z/(E_+ + m_oc^2) \\ c(p_x + ip_y)/(E_+ + m_oc^2) \end{bmatrix} \qquad (25.5.9)$$

and

$$|U_P^{II}\rangle \approx |U_{P\downarrow}^+\rangle = N_P \begin{bmatrix} 0 \\ 1 \\ c(p_x + ip_y)/(E_+ + m_oc^2) \\ -cp_z/(E_+ + m_oc^2) \end{bmatrix} \qquad (25.5.9)b$$

where $N_P = \sqrt{[(E_+ + m_oc^2)/(2E_+)]}$, $\qquad (25.5.10)$

In the total state, $|U_P^+\rangle$ of a Dirac particle, the momentum is denoted by p, its spin orientation is indicated by arrows up and down, and its energy type by +ve or -ve. A plane wave has no orbital angular momentum, and so spin is a Constant of the Motion; and $[S, \hat{H}_D] = 0$.

(b)   The eigen vectors for the $E_- = -\sqrt{c([m_o^2c^2 + p^2]}$
there are another two linearly independent solutions

$$|U_P^{III}\rangle \approx |U_{P\downarrow}^-\rangle = N_P \begin{bmatrix} -cp_z/(-E_- + m_oc^2) \\ -c(p_x + ip_y)/(-E_- + m_oc^2) \\ 1 \\ 0 \end{bmatrix} \qquad (25.5.11)a$$

and $\quad |U_P^{IV}\rangle \approx |U_{P\downarrow}^-\rangle = N_P \begin{bmatrix} c(p_x + ip_y)/(-E_- + m_oc^2) \\ 0 \quad cp_z/(-E_- + m_oc^2) \\ 0 \\ 1 \end{bmatrix} \qquad (25.5.11)b$

All the 4 spinors may be normalized for N_P.
It is easily checked that

$$\langle U_P^{(\mu)} || U_P^{(\nu)} \rangle = \delta_{\mu\nu} \qquad (25.5.11)c$$

Thus one can write the FREE-PARTICLE STATE FUNCTION as

$$|\Psi_{P(\lambda)}(x,t)\rangle = |U_P^{(\lambda)}\rangle e^{iPx/\hbar} e^{-iEt/\hbar}. \qquad (25.5.1)b$$

The index λ takes on the four values λ =I, II, III, IV. Solutions to λ =I, II, correspond to the positive sign of the energy in equation (25.5.8); solutions with λ = III, IV, to the negative sign.

25.5.7   The 4-fold DEGENERACY of the Dirac state:

The 4-fold degeneracy, as seen from the discussions earlier, viz. +ve energy spin-up and spin-down, and the other two -ve energy spin-up and spin-down, is the explanation of the mathematical fact that the Dirac particle state $|\Psi(p,t)\rangle$ given by (25.5.1) has four components, $|\Psi_P^-(p,t)\rangle$, $|\Psi_P^-(p,t)\rangle$, $|\Psi_P^+(p,t)\rangle$ and $|\Psi_P^+(p,t)\rangle$. The wave function, $|\Psi_P^+(p,t)\rangle$ of a free electron with 4-momentum, p_μ (and 3-momentum,p) is the product of a plane wave, $e^{i\mathbf{p}\cdot\mathbf{r}/\hbar}e^{-iEt/\hbar}$ and a spinor, $|U_{P\downarrow}^+\rangle$.

Observe!

For $|p| < m_oc$, the 3rd and 4th component of the +ve energy solutions, $|U_{P\downarrow}^I\rangle$ and $|U_{P\downarrow}^{II}\rangle$, are very small; and the 1st and 2nd component of the –ve energy solutions, $|U_{P\downarrow}^{III}\rangle$ and $|U_{P\downarrow}^{IV}\rangle$, are very small (of order $(p|/m_oc)$).

Worked out Example 25.7.

What kind of degeneracy is present for a Dirac free particle? Give explicit expressions for the degenerate states. What is the energy?

Solution:

The 4-fold degeneracy; +ve energy spin-up and spin-down, and the other two are -ve energy spin-up and spin-down.

The degenerate states are: $|U_{P\downarrow}^I\rangle, |U_{P\downarrow}^{II}\rangle, |U_{P\downarrow}^{III}\rangle$ and $|U_{P\downarrow}^{IV}\rangle$, whose explicit expressions are given below:

(1) $\quad |U_P^I\rangle \approx |U_{P\downarrow}^+\rangle = N_P \begin{bmatrix} 1 \\ 0 \\ cp_z/(E_+ + m_oc^2) \\ c(p_x + ip_y)/(E_+ + m_oc^2) \end{bmatrix}$ \hfill (25.5.9) a

(2) $\quad |U_P^{II}\rangle \approx |U_{P\downarrow}^+\rangle = N_P \begin{bmatrix} 0 \\ 1 \\ c(p_x + ip_y)/(E_+ + m_oc^2) \\ -cp_z/(E_+ + m_oc^2) \end{bmatrix}$ \hfill (25.5.9) b

where $\quad N_P = \sqrt{[(E_+ + m_oc^2)/(2E_+)]}$, \hfill (25.5.10)

(4) $\quad |U_P^{III}\rangle \approx |U_{P\downarrow}^-\rangle = N_P \begin{bmatrix} -cp_z/(-E_- + m_oc^2) \\ -c(p_x + ip_y)/(-E_- + m_oc^2) \\ 1 \\ 0 \end{bmatrix}$ \hfill (25.5.11) a

and

(5) $\quad \left|U_P^{IV}\right\rangle \approx \left|U_{P\Downarrow}^{-}\right\rangle = N_P \begin{bmatrix} c(p_x + ip_y)/(-E_- + m_o c^2) \\ cp_z/(-E_- + m_o c^2) \\ 0 \\ 1 \end{bmatrix}$ (25.5.11) b

The energy $E = E_\pm = \pm \sqrt{c([m_o^2 c^2 + p^2])}$.; i.e. $E_+ = +\sqrt{c([m_o^2 c^2 + p^2])}$. Twice,
and $\quad E_- = -\sqrt{c([m_o^2 c^2 + p^2])}$. Twice.

Worked out Example 25.8.

A Dirac particle is moving along the z-axis. It has negative energy and has spin up. Give its Representation, $\left|\Psi_{\overline{P}_\mu}^{-}(p,t)\right\rangle$.

Solution:
It is known that
$$\left|\Psi_{\overline{P}_\mu}^{-}(p,t)\right\rangle = \left|U_{P\downarrow}^{-}\right\rangle e^{-i\,\mathbf{p}\cdot\mathbf{r}/\hbar} e^{-iEt/\hbar}. \qquad (25.5.1)$$

$\left|U_P^{III}\right\rangle \approx \left|U_{P\downarrow}^{-}\right\rangle = N_P \begin{bmatrix} -cp_z/(-E_- + m_o c^2) \\ -c(p_x + ip_y)/(-E_- + m_o c^2) \\ 1 \\ 0 \end{bmatrix}$ (25.5.11) a

$N_P = \sqrt{[(E_+ + m_o c^2)/(2E_+)]}$; (25.5.10); $E_- = -\sqrt{c([m_o^2 c^2 + p^2])}$ (25.5.8)

So along the z-axis,
$$\left|\Psi_{\overline{P}\uparrow}^{-}(p,t)\right\rangle = \left|U_{P\uparrow}^{-}\right\rangle e^{-i\,\mathbf{p}\cdot\mathbf{r}/\hbar} e^{-iEt/\hbar}. \qquad (25.5.12)$$

where $\left|U_P^{III}\right\rangle \approx \left|U_{P\Downarrow}^{-}\right\rangle = N_P \begin{bmatrix} c(p_x + ip_y)/(-E_- + m_o c^2) \\ 0 \\ 1 \\ 0 \end{bmatrix}$ (25.5.13)

Worked out Example 25.9.

Show that in the non-relativistic limit, the relativistic spinors, $\left|\Psi_{P\uparrow}^{+}(p,t)\right\rangle$ and $\left|\Psi_{P\downarrow}^{+}(p,t)\right\rangle$, reduce to he two component spinors.

Solution:

The wave function, $\left|\Psi_{P\uparrow}^{+}(p,t)\right\rangle$ of a free electron with 4-momentum, p_μ (and 3-momentum,p) is the product of a plane wave, $e^{+i\,\mathbf{p}\cdot\mathbf{r}/\hbar} e^{-iEt/\hbar}$ .and a spinor, $\left|U_{P\uparrow}^{+}\right\rangle$, viz.

$$\left|\Psi_{P\uparrow}^{+}(p,t)\right\rangle = \left|U_{P\uparrow}^{+}\right\rangle e^{-i\,\mathbf{p}\cdot\mathbf{r}/\hbar} e^{-iEt/\hbar}$$

$$|U_P^I\rangle \equiv |U_{P\uparrow}^+\rangle = N_P \begin{bmatrix} 1 \\ 0 \\ cp_z/(E_+ + m_0c^2) \\ c(p_x + ip_y)/(E_+ + m_0c^2) \end{bmatrix} \quad (25.5.9)\,a$$

and

$$|U_P^{II}\rangle \approx |U_{P\downarrow}^+\rangle = N_P \begin{bmatrix} 0 \\ 1 \\ c(p_x + ip_y)/(E_+ + m_0c^2) \\ -cp_z/(E_+ + m_0c^2) \end{bmatrix} \quad (25.5.9)\,b$$

$$N_P = \sqrt{[(E_+ + m_0c^2)/(2E_+)]} \quad (25.5.10)$$

$$E_+ = +\sqrt{c([m_0^2c^2 + p^2]}. \quad (25.5.8)$$

In the non-relativistic limit, $(v/c) \to 0$,

The lower two components (*i.e.* 3rd and 4th) of (25.5.9) a & b do not decide the properties of the Dirac particle; as they are of order (v/c) times the upper two components. So in the non-relativistic limit,

$$|\Psi_{P\uparrow}^-(p,t)\rangle = |U_{P\uparrow}^-\rangle e^{-i\,p\cdot r/\hbar} \quad |\Psi_{P\uparrow}^+(p,t)\rangle = |u_{P\uparrow}^+\rangle e^{i\,p\cdot r/\hbar}.$$

and $\quad |\Psi_{P\downarrow}^+(p,t)\rangle = |U_{P\downarrow}^+\rangle e^{+i\,\cdot r/\hbar}$;

and (25.5.9) a & b reduce to

$$|U_P^I\rangle \approx |U_{P\uparrow}^+\rangle = N_P \begin{bmatrix} 1 \\ 0 \end{bmatrix} \quad (25.5.14)$$

and $\quad |U_P^{II}\rangle \approx |U_{P\downarrow}^+\rangle = N_P \begin{bmatrix} 0 \\ 1 \end{bmatrix} \quad (25.5.15)$

These expressions represent the free spin-½ particle, with spin-up and spin-down, respectively. That is the expressions (25.5.9) a & b reduce to the two component spinors. The two solutions with energy $E_+ = +\sqrt{c([m_0^2c^2 + p^2]}$ are represented fully by

$$|U_P^I\rangle \approx |U_{P\uparrow}^+\rangle = N_P \begin{bmatrix} 1 \\ 0 \\ 0 \\ 0 \end{bmatrix} \quad (25.5.16)$$

and

$$|U_P^{II}\rangle \approx |U_{P\downarrow}^+\rangle = N_P \begin{bmatrix} 0 \\ 1 \\ 0 \\ 0 \end{bmatrix} \quad (25.5.17)$$

$$N_P = \sqrt{[(E_+ + m_0c^2)/(2E_+)]} \to 1 \quad (25.5.10)$$

These are simultaneously eigen spinors of S_z; with eigen values $+\frac{1}{2}\hbar$ and $-\frac{1}{2}\hbar$, respectively.

Note: Similarly, the non-relativistic eigen spinors of S_z; with energy

$$E_- = -\sqrt{c([m_0^2c^2 + p^2]} \quad \text{are:}$$

$$\left|U_P^{III}\right\rangle \approx \left|U_{P\uparrow}^{-}\right\rangle = N_P \begin{bmatrix} 0 \\ 0 \\ 1 \\ 0 \end{bmatrix} \qquad (25.5.18)$$

and $\left|U_P^{IV}\right\rangle \approx \left|U_{P\downarrow}^{-}\right\rangle = N_P \begin{bmatrix} 0 \\ 0 \\ 0 \\ 1 \end{bmatrix}$ (25.5.19)

Again these are simultaneously eigen spinors of S_z; with eigen values $+\frac{1}{2}\hbar$ and $-\frac{1}{2}\hbar$, respectively.

Worked out Example 25.10.
List the impressive body of 'truth' in the Dirac Equation.
Solution:
The following list the impressive body of 'truth' in the Dirac Equation:
i) it predicts the correct H-atom energy spectrum,
ii) correctly describes electrons of spin $\frac{1}{2}$ and g = 2, The strength of the magnetism of a spinning particle is measured by its g-factor. It is really only the 'simplest' Dirac Equation which gives g = 2.
Iiii) at first predicts positrons; negative energy solutions leads to the existence anti-particles. As well as particles by positive energy solutions. The work of Weyl, Oppenheimer and others convinced Dirac that the 'hole' is positron (and not proton).

25.6 THE DIRAC PARTICLE IN AN EM FIELD
(THE INTERACTION OF A DIRAC PARTICLE WITH AN EM FIELD):

To obtain the relativistic equation of motion of a charged particle in an EM Field, one may use the prescription in 4-D Minkowskii - Lorentz space of events.

25.6.1. The PRESENCE OIF an EM FIELD:
The presence of an EM Field, acting on the matter field of the particle, is as usual taken into account by the replacement $\mathbf{p}^\mu$, 4-momentum (25.3.3)
This prescription defines a minimal interaction with the field and is specified by the Gauge potential,

$A^\mu = (A_k, i\Phi) = $ 4- vector potential, (25.3.4)
$= (A_1, A_2, A_3, A_4)$

where A_k, etc. and Φ completely specify an EM Field.
The 4-vector EM potential, $A_k = (A_1, A_2, A_3)$, (25.6.1)
and Scalar EM Potential, $A_+ = i\Phi$. (25.6.2)
Defining 4-momentum, $p^\mu \to p\mu - (q/c)A^\mu$, (25.3.3)
Canonical momentum of the particle = p
Charge of the particle = q
A_k and Φ completely define the EM Field.
Defining a 4-D Curl of A_k,

$$F_{ik} = [(\partial/\partial x_i)A_k - (\partial/\partial x_k)A_i] \qquad (25.6.3)$$
$$= \partial_i A_k - \partial_k A_i$$
= anti-symmetric tensor → '*EM Field tensor*'.

F_{ik} has 6 components: 3 electric + 3 magnetic fields.

Magnetic field, $B = (F_{12}, F_{13}, F_{31})$ \qquad (25.6.4)

Electric field, $iE = (F_{41}, F_{42}, F_{43})$ \qquad (25.6.5)

where $\quad E = -\nabla\Phi - (1/c)(\partial/\partial t)A_k$; \qquad (25.6.6)

$\quad B = \nabla \wedge A$. \qquad (25.6.7)

25.6.2. DIRAC HAMILTONIAN OF A CHARGED PARTICLE IN AN EM FIELD:

Consider the Dirac particle as point charge.

$$\hat{H}_D = [c\,\alpha \cdot p + \beta\, m_0 c^2] \qquad (25.4.3)\,a$$

Canonical replacements of p and E are required to get the $\mathcal{H}_D$ in an EM Field is as follows:

$$p_\mu \to p_\mu - (q/c)A_\mu, \qquad (25.3.3)$$

and $\quad E \to x p_x - L = E - q\Phi$. \qquad (25.6.8)

$\therefore \quad \hat{H}_D \Psi = E\Psi$ \qquad (25.4.5)

viz., $\quad [c\,\alpha \cdot p + \beta\, m_0 c^2]\Psi = E\Psi$

becomes $[c\,\alpha \cdot (p_\mu - (q/c)A_\mu) + \beta\, m_0 c^2]\Psi = (E - q\Phi)\Psi$

which is $\quad [(c\,\alpha \cdot p_\mu + \beta\, m_0 c^2) + (q\Phi - q\,\alpha \cdot A_\mu)]\Psi = E\Psi$ \qquad (25.6.9)

which is $\hat{H}_D \Psi = E\Psi$.

The Dirac Hamiltonian of a charged particle in an EM Field

$\hat{H}_D$ = Dirac Hamiltonian of free particle, $\hat{H}_D^{(0)}$ + interaction Hamiltonian, $\hat{H}_D^{(1)}$.

$$\hat{H}_D = \hat{H}_D^{(0)} + \hat{H}_D^{(1)}. \qquad (25.6.10)$$
$$\hat{H}_D^{(0)} = (c\,\alpha \cdot p_\mu + \beta\, m_0 c^2) \qquad (25.6.11)$$
$$\hat{H}_D^{(1)} = (q\Phi - q\,\alpha \cdot A_\mu) \qquad (25.6.12)$$

25.6.3 The Four Coupled Equations:

The state Ψ in (25.6.9) is expressed as

$$\Psi = \begin{bmatrix} \psi_p(1) \\ \psi_p(2) \\ \psi_p(3) \\ \psi_p(4) \end{bmatrix} \qquad (25.6.13)$$

Thus the Dirac equation for the particle in an EM Field (25.6.9) is equivalent to four equations similar to the ones in the theory of free electron (25.4.39).

$$(E - q\Phi - \beta\, m_0 c^2)\psi_p(1) = (c\hbar/i)\{(\partial/\partial x) - i(\partial/\partial y)\psi_p(4) + (\partial/\partial z)\psi_p(3)$$
$$-q\{(A_x - iA_y)\psi_p(4) + A_z \psi_p(3)$$

$$(E - q\varphi - m_0 c^2)\Psi_p(2) = (c\hbar/i)\{[(\partial/\partial x) - i(\partial/\partial y)]\Psi_p(3) - (\partial/\partial z)\}\Psi_p(4)$$
$$-q\{(A_x - iA_y)\Psi_p(3) - A_z \Psi_p(4)\}.$$

$$(E - q\varphi + m_0 c^2)\Psi_p(3) = (c\hbar/i)\{[(\partial/\partial x) - i(\partial/\partial y)]\Psi_p(2) + (\partial/\partial z)\}\Psi_p(1).$$

$$-q\{(A_x - iA_y)\Psi_p(2) + A_z\Psi_p(1)\}.$$

$$(E - q\varphi + m_0c^2)\Psi_p(4) = (c\hbar/i)\{[(\partial/\partial x) + i(\partial/\partial y)]\Psi_p(1) - (\partial/\partial z)\Psi_p(2)$$
$$-q\{(A_x + iA_y)\Psi_p(1) - A_z\Psi_p(2)\} \qquad (25.6.14)$$

25.6.4 DICOTONIC VARIABLES:

Introducing the two component column matrix (i.e. dicotonic variables),

$$|\psi_p\rangle = |U_p\rangle = \begin{bmatrix} u^+ \\ u^- \end{bmatrix} \qquad (25.6.15)$$

$$|u^+\rangle = \begin{bmatrix} \psi_p(1) \\ \psi_p(2) \end{bmatrix}$$

$$|u^-\rangle = \begin{bmatrix} \psi_p(3) \\ \psi_p(4) \end{bmatrix} \qquad (25.6.16)$$

and the Pauli matrices, $\sigma = (\sigma_x, \sigma_y, \sigma_z,)$ (25.4.20), the equation (25.6.9) becomes a pair of coupled equations;

$$(i/c\hbar)(E - m_0c^2 - q\Phi)|u^+\rangle = \sigma[\nabla - (iq/c\hbar)A]|u^-\rangle \qquad (25.6.17)\text{ a}$$

$$(i/c\hbar)(E + m_0c^2 - q\Phi)|u^-\rangle = \sigma[\nabla - (iq/c\hbar)A]|u^+\rangle \qquad (25.6.17)\text{ b}$$

which can be written as

$$(1/c)(E - m_0c^2 - q\Phi)|u^+\rangle = \sigma \cdot [p - (q/c)A]|u^-\rangle \qquad (25.6.18)\text{ a}$$

$$(1/c)(E + m_0c^2 - q\Phi)|u^-\rangle = \sigma \cdot [p - (q/c)A]|u^+\rangle \qquad (25.6.18)\text{ b}$$

25.6.4 ESTIMATION of $|u^+\rangle$ and $|u^-\rangle$:

$$(1/c)(E + m_0c^2 - q\Phi)|u^-\rangle = \sigma \cdot [p - (q/c)A]|u^+\rangle \qquad (25.6.17)\text{ b}$$

gives $$|u^-\rangle = \frac{c\,[p - (q/c)A \cdot \sigma]}{(E + m_0c^2 - q\Phi)}|u^+\rangle \qquad (25.6.19)$$

L.H.S. of (25.6.18)a is

$$(1/c^2)(E - m_0c^2 - q\Phi)(E + m_0c^2 - q\Phi)|u^+\rangle = (1/c^2)(E - q\Phi)^2 - m_0^2c^4)|u^+\rangle$$

$$= (1/c)(E + m_0c^2 - q\Phi)\sigma \cdot [p - (q/c)A]|u^-\rangle$$

R.H.S. of (25.6.18) a

$$= \{\sigma \cdot [p - (q/c)A](1/c)(E + m_0c^2 - q\Phi) - (q/c^2)\sigma \cdot [E, A] - (q/c)\sigma[\Phi, p]\}|u^-\rangle$$

$$= \{[\sigma \cdot p - (q/c)A]^2|u^+\rangle = (q\hbar/i\,c^2)\sigma(\partial/\partial t)A + (q\hbar/i\,c)\sigma \cdot \nabla\Phi]\}|u^-\rangle$$

$$= \{[p - (q/c)A]^2|u^+\rangle + i\,\sigma \cdot [p - (q/c)A]\wedge[p - (q/c)A]|u^+\rangle - (q\hbar/i\,c)[\sigma, E]\}|u^-\rangle$$

in $[\sigma, E]$, $E = \nabla\Phi - (1/c)(\partial/\partial t)A$.

$[p-(q/c)A] \wedge [p-(q/c)A = -(q/c)+(A \wedge p)$.

as $[p, A] \neq 0$; $= -(q\hbar/i\ c)(\nabla \wedge A) = -(q\hbar/i\ c)B$

Find the

$$\{(1/c^2)(E-q\Phi)^2 - m_0^2 c^2) - [p-(q/c)A]^2\}\left| u^+ \right\rangle$$
$$= -(q\hbar/i\ c)B \cdot \sigma \left| u^- \right\rangle - (q\hbar/i\ c)(\sigma \cdot r)\left| u^- \right\rangle \qquad (25.6.20)$$

L.H.S. of equation (25.6.20) only would yield the K G Equation.

25.6.6 Even in the Non-RelativisticLimit, a DIRAC PARTICLE has SPIN-MAGNETIC Properties

Reducing further, after neglecting terms with $(1/c^3)$,

$$E = (m_0 c^2 + T) \qquad (25.6.21)$$

$$(1/c)(E - m_0 c^2 - q\Phi)\left| u^- \right\rangle = \sigma \cdot [p-(q/c)A]\left| u^+ \right\rangle \qquad (25.6.18)\ b$$

gives, in the lowest approximation,

$$\left| u^- \right\rangle = \frac{1}{m_0 c} \sigma \cdot p \left| u^+ \right\rangle$$

Using for the R.H.S. of equation (25.6.20) the operator identities,

$(\sigma \cdot r)(\sigma \cdot p) = (\sigma \cdot p) + i \sigma \cdot E \wedge p)$

provided that $[\sigma, r][\sigma, p] = 0$

Equation (25.6.20) becomes $T\left| u^+ \right\rangle = \hat{H}_D \left| u^+ \right\rangle \qquad (25.6.22)$

where $\hat{H}_D = \left\{ \frac{1}{2m_0}[p-(q/c)A_\mu]^2 + q\Phi \right\} - \left[\frac{1}{8m_0^3 c^2}[p-(q/c)A_\mu]^4 - \frac{q\hbar}{4\ i\ m_0^2 c^2} E \cdot p \right]$

$-\left\{ \frac{q\hbar}{4\ m_0^2 c^2}(\sigma \cdot E \wedge p) - \frac{q\hbar}{2\ m_0 c} B \wedge \sigma \right\} \qquad (25.6.23)$

(i) The first two terms (within curly brackets) are the **classical Hamiltonian**, $\hat{H}$,
(ii) The next two terms (within square brackets) are **spin-independent relativistic corrections**,
(iii) The interesting terms are the last two terms (within curly brackets); viz.
(a) $-(q\hbar/2\ m_0 c)\ \sigma \wedge B$, $\qquad (25.6.24)$

which is the **energy of magnetic moment**, $\bar{\mu}_{z\ s}$, given by

$\bar{\mu}_z = (q\hbar/2\ m_0 c) = \bar{\mu}_0 \sigma = (g_s q\ S/m_0 c)$

in the magnetic field **B**.

$S = \frac{1}{2}\hbar \sigma$,

$g_s = 2$, **is Spin-gyro-magnetic Ratio**.

(b) $-(q\hbar/4\ i\ m_0^2 c^2)\ \sigma \cdot E \wedge p$, $\qquad (25.6.25)$

is the **mutual energy** of $\bar{\mu}_0 \sigma$ in apparent magnetic field,

$E \wedge \frac{v}{c} \approx \frac{1}{m_0 c} r \wedge p$ divided by 2.

(THOMAS CORRECTION).

Thus, even in the non-relativistic limit, a Dirac particle has spin-magnetic properties. It is quite remarkable that this feature, which originated from relativistic considerations, remains in the limit $\frac{v}{c} \to 0$!

25.6.7. SPIN OF A PARTICLE is a Consequence of RELATIVISTIC NATURE of the Particle:

Spin has been shown to be a consequence of the relativistic nature of the electron. The starting point of the discussion on μ should be thus the Dirac Equation, where the relativistic electron wave function has 4 – components. In the non-relativistic limit, the electron and positron parts of this equation may be separated by means of *Foldy – Wouthuysen* Transformation to give an equation for the 2-component wave function describing the electron alone. The fact that the electron wave function does have two components is consistent with the electron possessing a degree of freedom corresponding to a spin angular momentum $s = \frac{1}{2}\hbar$, that can point either up or down.

The most important terms from a physicist's point of view that are contained in this reduction of the Dirac Equation are given by the Hamiltonian,

$$\hat{H}_D = (1/2\,m)[p - (q/c)A_\mu]^2 - (1/8\,m^2 c^2)\,p^4 + V(r)$$

$$- \{(q/mc)(s \cdot \nabla \wedge A_\mu) - (1/2\,m^2 c^2)\, s \cdot [\nabla V(r) \wedge (p - (q/c)A_\mu)]\}$$

$$+ (\hbar^2/8\,m^2 c^2)\nabla^2 V(r). \tag{25.6.21}$$

This differs from expression (25.3.1)

$$\hat{H} = (1/2\,m)[p - (q/c)A_\mu]^2 + V(r) \tag{25.3.1}$$

in having

(a)   a term in p^4, *i.e.* $(1/8\,m^2 c^2)\,p^4$ a relativistic correction to the kinetic energy,

(b)   two terms involving the spin angular momentum **s**.

Since $\nabla \wedge A_\mu = H$,

the presence of the first of the terms containing **s** shows that the electron to have a magnetic moment of

$$\mu_s = (e/mc)\,s, \text{ due to its spin.} \tag{25.6.22}$$

$\mu_s / s =$ can be determined by experiment,

$= (e/mc)$, for many ferromagnetic substances.

This indicates that it is the spin of the electron rather than its orbital motion, which would have led to a value $(e/2mc)$ for this ratio that is principally responsible for the magnetic properties of these materials. The operator **S** has the same commutative properties as the operator L. This shows that in the limit of small applied magnetic fields, $H \to 0$, the μ due to a spin less electron is proportional to the orbital angular momentum $L = r \wedge p$.

(c)   the term, $(1/2\,m^2 c^2)\,s \cdot [\nabla V(r) \wedge (p - (q/c)A_\mu)]$, may be written as

$$\boxed{\Delta E(\text{spin-orbit}) = (1/2\,m^2 c^2)\,(1/r)\,(dV/dr)\,S \cdot L} \tag{25.6.23}$$

(d)   the term, $\boxed{(\hbar^2/8\,m^2 c^2)\nabla^2 V(r)}$, is the DARWIN term,   (25.6.24)

produces 1st order energy correction.

Worked out Example 25.11

Estimate the Darwin term for Hydrogen, and show that fine structure effects in Hydrogen are thus 1st order relativistic effects of the Dirac Equation. Given: Darwin term = $(\hbar^2 / 8\, m^2 c^2)\, \nabla^2 V(r)$

Solution:

Darwin term, ΔE (Darwin) = $(\hbar^2 / 8\, m^2 c^2) \nabla^2\, V(r)$

i.e., ΔE (Darwin) $\propto \langle n\ell\, |\nabla^2\, V(r)|\, n\ell \rangle$

$$\langle n\ell\, |(d\,V/dr)|\, n\ell \rangle = \int_0^\infty R_{n\ell}(r)\, \frac{dV}{dr}\, \frac{dR_{n\ell}}{dr}\, r^2\, dr$$

For Hydrogen, $V(r) = e^2/r$,

$$\Delta E\,(\text{Darwin}) \propto e^2 \int_0^\infty R_{n\ell}(r)\, \frac{1}{r^2}\, \frac{dR_{n\ell}}{dr}\, r^2\, dr$$

$$= e^2 \int_0^\infty R_{n\ell}(r)\, dR_{n\ell} = \frac{1}{2} e^2 |R_{n\ell}(r)|^2$$

$R_{n\ell}(r)$ vanishes at ∞, for all bound states.
It also vanishes at the origin for all but the s-states; the energy corrections exist only for these states.
This means Fine Structure effects in Hydrogen are 1st order relativistic effects of the Dirac Equation. One may therefore identify a **Dirac Particle with a relativistic electron**!

25.6.8 The FOLDY-WOUTHUYSEN TRANSFORMATION:

A systematic procedure was developed by Foldy and Wouthuysen, *viz.* a Canonical Transformation, which decouples the Dirac Equation into two-component wave function equations, one reduced to the Pauli description in the non-relativistic limit, and the other describes the negative energy states (The fact that electron wave function does have two components as consistent with the electron possessing a degree of freedom corresponding to a spin and of spin angular momentum exactly half of the Planck Constant, $\frac{1}{2}\hbar$).

Worked out Example 25.12.
Apply the Foldy-Wouthuysen Approximation to a Dirac free particle Transformation.
Solution:

Consider $\hat{H}_D = [\{c\hbar/i\} \sum_{k=1}^{3} \alpha_k (\partial/\partial x_k) + \beta\, m_o c^2]$

$\{(\hbar/i)(\partial/\partial t) + \hat{H}_D\}\Psi(x,t) = 0$. \hfill (25.4.6) b

Let $\hat{U}_F = e^{+iS}$ = an operator (unitary) that removes from the equation operators such as α_k which couple the large to the small components. Such an operator is 'odd' (eg. $\alpha_1, \gamma, \gamma_5$, *etc.*) and operators that do not couple large and small components are 'even' (example. $\hat{I}, \beta$, etc.).
S = a hermitian, and explicitly not t-dependent.

$\Psi'(x,t) = e^{+iS}\Psi(x,t)$.

$(\hbar/i)(\partial/\partial t))\Psi'(x,t) = e^{+iS}\, \hat{H}_D\, \Psi(x,t) = e^{+iS}\, \hat{H}_D\, \Psi(x,t)\, e^{-iS}\, \Psi(x,t)$.

where $\hat{H}_D = [\alpha \cdot p + \beta\, m_o]$

Now find a $\hat{U}_F$ which change $\hat{H}_{Spin} = \sigma_x B_x + \sigma_z B_z$ into a form which contains only 'even' operators is (θ_0 is rotation about y-axis),

$$\frac{1}{e^2} \sigma_y B_o = \frac{1}{e^2} \sigma_z \sigma_x \theta_o$$

with $\theta_o = \tan^{-1}(B_x / B_z)$,

i.e., $e^{+iS} = e^{+iS} \beta \, \alpha \cdot p \theta_o(p) = Cos|p|\theta_o(p) + \{\beta[\alpha, p]/|p|\} Sin p |\theta_o(p)$

where the RHS is established by expansion of the exponential in powers of θ.

So $\hat{H}'_D = \{Cos|p|\theta_o(p) + \{\beta [\alpha, p]/|p|\} Sin p |\theta_o(p)\}$

$[\alpha \cdot p + \beta \, m_o]\{Cos|p|\theta(p) - \{\beta [\alpha, p]/|p|\} Sin p |\theta_o(p)$

$= [\alpha \cdot p + \beta \, m_o]\{Cos|p|\theta(p) - \{\beta [\alpha, p]/|p|\} Sin p |\theta_o(p)\}^2$.

$= [\alpha \cdot p + \beta \, m_o] \, e^{\, 2\beta \, [\alpha, p]\theta}$.

$= [\alpha, p]\{Cos \, 2|p|\theta - (m/|p|) \, Sin \, 2|p|\theta + \beta \, \{m_o Cos \, 2|p|\theta - |p| \, Sin \, 2|p|\theta\}$

In order to eliminate the odd operator, one chooses

$\tan[2|p|\theta] = |p|/m$

and the transformed Hamiltonian is $\quad \hat{H}' = \beta \sqrt{(m^2 + p^2)}$.

Worked out Example 25.13.

Apply the FW Transformation for the motion of a Dirac electron in an EM Field, to obtain S.L.interaction term as well as the Zitterbewegung motion giving the Darwin term.

Solution:

(a) S.L. interaction:

For spherically symmetric static potential, V(r),

$\hat{H}^{(2)} \equiv \hat{H}_{LS} = -(e/4m^2) \, \sigma \cdot E \wedge p - (e/8m^2) \, \sigma \cdot \nabla \wedge E$.

$\sigma \cdot E \wedge p = -(1/r)(\partial V/\partial r) \, \sigma \cdot r \wedge p = -(1/r)(\partial V/\partial r) \, \sigma \cdot L$,

$\hat{H}_{LS} = (e/4m^2) \, (1/r)(\partial V/\partial r) \, \sigma \cdot L$ \hfill (25.4.7)

If $B' = -\nabla \wedge E$, experienced by a moving electron,

$-(e/2m) \, \sigma \cdot \nabla \wedge E = -(e/2m) \, \sigma \cdot B' = -(e/2m^2) \, \sigma \cdot (p \wedge E)$.

This factor is reduced by a factor 2 owing to the THOMAS PRECESSION Effect and indicates that the L of the electron has the $g_e = 1$.

(b) ZITTERBEWEGUNG of the electron:

The perturbation, known as the Darwin term,

$$\boxed{\begin{array}{l} \hat{H}^{(3)} - (\hbar^2/8m_o^2 c^2) \, \nabla^2 V = (\hbar^2/m_o c^2)\{4\pi c Q_{Nuc}(x)\} \\ = (\pi \hbar^2 Ze^2 / 2 \, m_o^2 c^2) \, \delta^{(3)}(x) \end{array}}$$

with nuclear charge density, $Q_{Nuc}(x)$. This term can be understood from the ZITTERBEWEGUNG of the electron. According to the Relativistic Theory the position of a localized electron fluctuates with

$$\boxed{\delta(r) = (\hbar/m_o c) = \text{Compton Wavelength}}.$$

The electron feels, therefore, on the average δ – potential

$$<[\langle V(x+\delta x)]\rangle> = V(x) + <(\delta x \; \nabla V)> + \tfrac{1}{2}<(\delta x \nabla)(\delta x \; \nabla)V(x)>$$
$$= V(x) + \tfrac{1}{6}(\delta r)^2 \nabla^2 V(x). \quad (25.6.25)$$

The correction thus calculated is in qualitative agreement in form, sign, and magnitude with the Darwin term.

Because of the δ-function, a contribution

$$<\hat{H}^{(3)}> = (\pi \hbar^2 Z e^2 / 2 m_o^2 c^2)|\psi(0)|^2 = m_o c^2 [(Z\alpha)^4 / 2n^3]\,\delta_{\ell 0}. \quad (25.6.26)$$

This arises for s-wave states only.

$<\hat{H}^{(3)}>$ is formally identical to $<\hat{H}^{(2)}>_{n,\,j=1,\,\ell=0}$.

Worked out Example 25.14.

Show that the 'Zitterbewegung' term in the Dirac Hamiltonian of an electron in an EM Field vanishes if the wave packet for a free electron is a superposition of only positive or only negative energy value.

Solution:

The projections onto the positive and negative energy solutions are:

$$A_+ = \tfrac{1}{2}[1+\hat{H}_D/E_+] = \tfrac{1}{2}\{1+[c\,\alpha\cdot p + \beta\, m_o c^2]/E_+\}$$

$$A_- = \tfrac{1}{2}[1-\hat{H}_D/E_+] = \tfrac{1}{2}\{1-[c\,\alpha\cdot p + \beta\, m_o c^2]/E_+\}$$

$$E_+ = \sqrt{(m_o^2 c^4 + p^2 c^2)}\,,$$

$A_\pm$ are respective projection operators since

$$A_\pm^2 = \left\{\tfrac{1}{2}[1\pm\hat{H}_D/E_+]\right\}^2 = A_\pm,$$

i.e., $A_\pm$ is idempotent.

Now $\displaystyle A_+ \begin{bmatrix} U^I \\ U^{II} \end{bmatrix} = A_+ \left\{ \begin{bmatrix} U^I \\ 0 \end{bmatrix} + \begin{bmatrix} 0 \\ U^{II} \end{bmatrix} \right\}$

$$= \left\{\tfrac{1}{2}[1+\hat{H}_D/E_+]\right\}\left\{\begin{bmatrix} U^I \\ 0 \end{bmatrix} + \begin{bmatrix} 0 \\ U^{II} \end{bmatrix}\right\}$$

$$= \left\{\tfrac{1}{2}\{1+\hat{H}_D/E_+\}\right\}\begin{bmatrix} U^I \\ 0 \end{bmatrix} + \left\{\tfrac{1}{2}\{1+\hat{H}_D/E_+\}\right\}\begin{bmatrix} 0 \\ U^{II} \end{bmatrix}$$

$$= \begin{bmatrix} U^I \\ 0 \end{bmatrix}$$

Similarly, $\displaystyle A_- \begin{bmatrix} U^I \\ U^{II} \end{bmatrix} = \begin{bmatrix} 0 \\ U^{II} \end{bmatrix}$

If wave packet = superposition of only positive or only negative energy waves,

$$<ci\hbar\,[\alpha(0)-cp/\hat{H}_D)][(e^{-i\hat{H}t/\hbar})-1]/2\hat{H}_D>$$

$$=<A_\pm U | \left(ci\hbar\,[\alpha(0)-cp/\hat{H}_D)][(e^{-i\hat{H}t/\hbar})-1]/2\hat{H}_D\right) | A_\pm U>$$

Since $A_\pm$ is hermitian,

$$A_\pm\,[\alpha(0)-cp/\hat{H}_D)][(e^{-i\hat{H}t/\hbar}-1)/2\hat{H}_D]A_\pm$$

$$= A_\pm\,[\alpha(0)-cp/\hat{H}_D)A_\pm\,[(e^{-i\hat{H}t/\hbar}-1)/2\hat{H}_D]$$

since $[A_\pm, \hat{H}_D] = 0$.

$$= A_\pm [\alpha(0) - cp/\hat{H}_D) A_\pm$$
$$= \tfrac{1}{4} [1\pm\hat{H}_D/E_+)[\alpha\pm\alpha\,\hat{H}_D/E_+) - \tfrac{1}{4} [1\pm\hat{H}_D/E_+)\tfrac{1}{4}[cp/\hat{H}_D + cp/\hat{H}_D)$$
$$= A_\pm[\alpha(0) \cdot c\,p/\hat{H}_D]A_\pm$$
$$= \tfrac{1}{4}[\alpha + [(\alpha\,\hat{H}_D + \hat{H}_D\alpha \cdot 2c\,p)/\hat{H}_D] + (\hat{H}_D/E_+^2)(\alpha\,\hat{H}_D - c\,p) - c\,p/\hat{H}_D]$$

But $[\alpha, \hat{H}_D] = 2c\,p$.

$$A_+[\alpha \cdot c\,p/\hat{H}_D]A_+ = \tfrac{1}{4}[\alpha \cdot c\,p/\hat{H}_D] + (\hat{H}_D/E_+^2)(\alpha\,\hat{H}_D - c\,p)$$
$$= \tfrac{1}{4}\hat{H}_D[\hat{H}_D\alpha - c\,p + (\hat{H}_D^2/E_+^2)(\alpha\,\hat{H}_D - c\,p)]$$
$$= \tfrac{1}{4}\hat{H}_D[\alpha\,\hat{H}_D + \hat{H}_D\alpha - 2c\,p = 0.$$

Hence the result is obtained.

Worked out Example 25.13.

Obtain the eigen values of the operator, $\hat{k} = (\beta/\hbar)(\sigma'\cdot L + \hbar)$

Solution:

$$\hat{k} = (\beta/\hbar)(\sigma'\cdot L + \hbar).$$
$$\hat{k}^2 = (1/\hbar^2)(\beta/\hbar)(\sigma'\cdot L + \hbar)(\beta/\hbar)(\sigma'\cdot L + \hbar)$$
$$\hat{k}^2 = (1/\hbar^2)\beta^2[(\sigma'\cdot L)^2 + 2\hbar\sigma'\cdot L + (L+\hbar)^2]$$
$$= (1/\hbar^2)[L^2 + i\,\sigma'\cdot(L\Lambda L)^2 + 2\hbar\sigma'\cdot L + \hbar^2]$$
$$= (1/\hbar^2)[L^2 + \hbar\,\sigma'\cdot L + \hbar^2\sigma'^2/4 - \hbar^2\sigma'^2/4 + \hbar^2]$$
$$= (1/\hbar^2)[(L + \tfrac{1}{2}\hbar\,\sigma')^2 - \hbar^2\sigma'^2/4 + \hbar^2)] = (1/\hbar^2)[J^2 + \hbar^2/4].$$

The eigen values of $\hat{k}^2$ are $J(J+1) + \tfrac{1}{4}$, i.e. $(j+\tfrac{1}{2})^2$, $j = \tfrac{1}{2}$, $j = \tfrac{3}{2}$, etc.

Worked out Example 25.16.

Transform the Dirac Hamiltonian using the transformation e^{iN}, where
$N = -(i/2mc)\beta\,\alpha\cdot p\,f\,(|p|/mc)$ with $f = (mc/|p|)\tan^{-1}(|p|/mc)$

Solution: $\hat{H}_D - [c\,\alpha\cdot p + \beta\,m_0 c^2]$.

FW transformed Hamiltonian is $\hat{H}_D' = e^{iN}\hat{H}_D\,e^{-iN}$
$$= e^{iN\beta}[c\beta\,\alpha\cdot p + \beta\,m_0 c^2]\,e^{-iN}.$$
$$= e^{iN\beta}(e^{-iN\beta})[c\,\alpha\cdot p + \beta\,m_0 c^2]\,e^{-iN}.$$
$$= e^{iN\beta}(1 - iN - N^2/\underline{2}\text{-}, \cdot, \cdot,)\beta\,\hat{H}_D = 2\,iN\,\hat{H}_D,$$

because $[N, \beta\alpha\cdot p] = 0$ and $[N, \beta]_+ = 0$.

So $\hat{H}_D' = e^{(\beta\,\alpha\cdot p.f/m_0 c)}$
$$= [1 + (1/m_0 c)(\beta\,\alpha\cdot p.f) + (1/m_0\,c)^2\underline{|2}\,(\beta\,\alpha\cdot p.f)^2]$$

One can prove that $(\beta\,\alpha\cdot p.f)^n = \pm P^n$, for $n = $ even. ;
+ve sign for $n = 0, 4, 8, ..$; –ve sign for $n = 2, 6, 10,..$

$(\beta\,\alpha\cdot p.f)^n = \pm P^{n-1}(\beta\,\alpha\cdot p.f)$, for $n = $ odd.
+ve sign for $n = 1, 5, 9, ..$; –ve ssign for $n = 3, 7, 11, ..$

One has $e^{2iN} = Cos(p.f/m_o c) + (\beta \alpha \cdot p)/p \, Sin(p.f/m_o c)$.

$$\hat{H}_D{}' = [(\alpha \cdot p)/p][c p \, Cos(p.f/m_o c) - m_o c^2 \, Sin(p.f/m_o c)]$$
$$+ \beta[m_o c^2 \, Cos(p.f/m_o c) - c p \, Sin(p.f/m_o c)].$$

with $f = (mc/|p|) \tan^{-1}(|p|/mc)$

$$Sin(p.f/m_o c) = p/\sqrt{(p^2 + m_o^2 c^2)}.$$
$$Cos(p.f/m_o c) = m_o c/\sqrt{(p^2 + m_o^2 c^2)}.$$
$$\hat{H}_D{}' = \beta[m_o c^2) m_o c)/\sqrt{(p^2 + m_o^2 c^2)} + c p p/\sqrt{(p^2 + m_o^2 c^2)}\,]\,].$$

Thus $\hat{H}_D{}'$ has the simple form, $\hat{H}_D{}' = \beta E'$.

Using $\beta = \gamma^o = \begin{pmatrix} \hat{I} & 0 \\ 0 & -\hat{I} \end{pmatrix} \equiv \begin{bmatrix} 1 & 0 & 0 & 0 \\ 0 & 1 & 0 & 0 \\ 0 & 0 & -1 & 0 \\ 0 & 0 & 0 & -1 \end{bmatrix}$ \hfill (25.4.22)

One sees the eigen values of $\hat{H}_D{}'$ are $\pm|E'|$ and the eigen vectors are

$$\begin{bmatrix} u(1) \\ u(2) \\ 0 \\ 0 \end{bmatrix} \text{ and } \begin{bmatrix} 0 \\ 0 \\ u(3) \\ u(4) \end{bmatrix}, \text{ respectively.}$$

25.7. DIRAC PARTICLE IN A CENTRAL FORCE POTENTIAL – HYDOGEN ATOM

25.7.1 Dirac Hamiltonian for the electron in the H-atom:

This section turns to a discussion of the bound state solutions of the Dirac Equation of a bound particle. It is interesting to consider the leading relativistic corrections to a Dirac Particle in a Coulomb field, *i.e.* due to a charge of $+Ze$. This can be described by

$$-q\Phi = V(r) = \frac{1}{4\pi \varepsilon_o} \frac{Z e q}{r}, \hfill (25.7.1)$$

$$A_\mu 0 \hfill (25.7.2)$$

The Dirac Hamiltonian appropriate is given by equation (25.6.10), *viz.*

$$\hat{H}_D = \hat{H}_D{}^{(0)} + \hat{H}_D{}^{(1)} \hfill (25.6.10)$$
$$\hat{H}_D{}^{(0)} - [c \, \alpha \cdot p_\mu + \beta \, m_o c^2] \hfill (25.6.11)$$
$$\hat{H}_D{}^{(1)} = [q\Phi - q \, \alpha \cdot A_\mu] \hfill (25.6.12)$$

For an electron in the Hydrogen atom,

$$\hat{H}_D = [c \, \alpha \cdot p_\mu + \beta \, m_o c^2] + V(r). \hfill (25.7.3)$$

As in section (25.6.4), introducing the two component column matrix (*i.e.* dicotonic variables),

$$|\Psi_P\rangle = |U_P\rangle = \begin{bmatrix} u^+ \\ u^- \end{bmatrix}. \hfill (25.6.15)$$

$$|u^+\rangle = \begin{bmatrix} \psi_P(1) \\ \psi_P(2) \end{bmatrix}$$

$$|u^-\rangle = \begin{bmatrix} \psi_p(3) \\ \psi_p(4) \end{bmatrix} \qquad (25.6.16)$$

and the Pauli matrices, $\sigma = (\sigma_x, \sigma_y, \sigma_z)$, the equation (25.6.9) becomes a pair of coupled equations;

$$(1/c\hbar)(E - m_o c^2 - q\varphi)|u^+\rangle = \sigma \cdot [\nabla - (iq/c\hbar)A]|u^-\rangle \qquad (25.6.17)\text{ a}$$

$$(1/c\hbar)(E + m_o c^2 - q\varphi)|u^-\rangle = \sigma \cdot [\nabla - (iq/c\hbar)A]|u^+\rangle \qquad (25.6.17)\text{ b}$$

which can be written as

$$(1/c)(E - m_o c^2 - q\varphi)|u^+\rangle = \sigma \cdot [p - (q/c)A]|u^-\rangle \qquad (25.6.18)\text{ a}$$

$$(1/c)(E + m_o c^2 - q\varphi)|u^-\rangle = \sigma \cdot [p - (q/c)A]|u^+\rangle \qquad (25.6.18)\text{ b}$$

25.7.2 HELICITY Operator:

Equation (25.6.18) becomes

$$(1/c)(E - m_o c^2 + V(r))|u^+\rangle = (\sigma \cdot p)|u^-\rangle \qquad (25.7.4)\text{ a}$$

$$(1/c)(E + m_o c^2 + V(r))|u^-\rangle = (\sigma \cdot p)|u^+\rangle \qquad (25.7.4)\text{ b}$$

$$(\sigma \cdot p) = \text{pseudo-scalar} \qquad (25.7.5)$$

$(\sigma \cdot p)$ is called the *Helicity Operator*.

Angular Momentum is given by

$$\hbar(L + S) = \hbar J = r \wedge p + \tfrac{1}{2}\hbar\sigma', \qquad (25.7.6)$$

commutes with $\hat{H}_D$.

$$[\hat{H}_D, J] = 0 \qquad (25.7.7)$$

but $\quad [\hat{H}_D, L] \neq 0 \qquad (25.7.8)$

$$J^2|u^+\rangle = j(j+1)\hbar^2|u^+\rangle$$

and $\quad J_z|u^+\rangle = m_j \hbar |u^+\rangle, \quad -j \leq m_j \leq +j. \qquad (25.7.9)$

J^2 and J_z are diagonal.

Then σ' has the same commutative properties as that of σ (the Pauli matrices), viz.

$$\sigma' \wedge \sigma' = 2i\,\sigma', \qquad (17.2.32)$$

$$\sigma'^2 = \sigma'^2_x + \sigma'^2_y + \sigma'^2_z.$$

$$\sigma'^2_x = \sigma'^2_y = \sigma'^2_z = \begin{pmatrix} 1 & 0 \\ 0 & 1 \end{pmatrix} = \hat{I} \qquad (17.2.37)$$

The from equations (25.7.6) and (25.7.9)
Allowable values of ℓ and ℓ_z are

$$\ell = j \pm \tfrac{1}{2}$$

$$\ell_z = m_j \pm \tfrac{1}{2}.$$

From (25.7.4), because of the helicity operator, $(\sigma \cdot p)$, it results in $|u^+\rangle$ and $|u^-\rangle$ to have *opposite parity*. From this one gets two types of solutions, as (25.5.9) & (25.5.11).

For $\ell = j - \frac{1}{2}$,
the Dirac components are:

$$|u^+\rangle = \{\mathfrak{R}(r)/\sqrt{2j}\} \begin{bmatrix} \sqrt{j+m}\ Y_{j-\frac{1}{2},\ m-\frac{1}{2}} \\ \sqrt{j-m}\ Y_{j-\frac{1}{2},\ m+\frac{1}{2}} \end{bmatrix} \quad (25.7.10)$$

$$\equiv \mathfrak{R}(r)\ \mathbb{Z}_{j,\ j-\frac{1}{2},\ m}.$$

$$|u^-\rangle = \{S(r)/\sqrt{2(j+1)}\} \begin{bmatrix} +\sqrt{j+1-m}\ Y_{j+\frac{1}{2},\ m-\frac{1}{2}} \\ -\sqrt{j+1+m}\ Y_{j+\frac{1}{2},\ m+\frac{1}{2}} \end{bmatrix} \quad (25.7.11)$$

$$\equiv iS(r)\ \mathbb{Z}_{j,\ j+\frac{1}{2},\ m}.$$

$\mathbb{Z}_{j,\ j+\frac{1}{2},\ m}$ are the *dicotonic* functions, the properties of which are:

(i)   They play the role of the spherical harmonics, for problems with spin,
(ii) They have $\ell = j \pm \frac{1}{2}$

$$(\sigma \cdot r)[f(r)\ \mathbb{Z}_{j,\ j\pm\frac{1}{2},\ m}] = rf(r)\ \mathbb{Z}_{j,\ j\pm\frac{1}{2},\ m}. \quad (25.7.12)$$

$$(\sigma \cdot p)\ [f(r)\ \mathbb{Z}_{j,\ j\pm\frac{1}{2},\ m}] = (\hbar/i)[f'(r) + (1 \pm j \pm \frac{1}{2})(f/r)] \quad (25.7.13)$$

Substituting (25.7.11) in (25.7.4),

$$(1/c\hbar)(E - m_o c^2 + V(r))|\mathfrak{R}(r)\rangle = [|S'(r)\rangle + (j + \frac{3}{2})(S'(r)/r)] \quad (25.7.14)$$

$$(1/c\hbar)(E + m_o c^2 + V(r))|S(r)\rangle = -[|\mathfrak{R}'(r)\rangle + (j - \frac{1}{2})(\mathfrak{R}(r)/r)] \quad (25.7.15)$$

These two first order equations (25.7.14) and (25.7.15) are the equivalent of the single-relativistic radial equation of the second order.
In this solution,
(a)   First type:
      $\mathfrak{R}(r) =$ large, $S(r) =$ small, for $\ell = j \pm \frac{1}{2}$.
(b)   Second type:
      $\ell = j + \frac{1}{2}$

$$|u^+\rangle \equiv \mathfrak{R}(r)\ \mathbb{Z}_{j,\ j+\frac{1}{2},\ m}. \quad (25.7.16)$$

$$|u^-\rangle \equiv iS(r)\ \mathbb{Z}_{j,\ j-\frac{1}{2},\ m}. \quad (25.7.17)$$

The two coupled radial equations are, instead of (25.7.14) & (25.7.15),

$$(1/c\hbar)(E - m_o c^2 + V(r))[|\mathfrak{R}(r)\rangle = -|S'(r)\rangle + (j - \frac{1}{2})[S(r)/r] \quad (25.7.18)$$

$$(1/c\hbar)(E + m_o c^2 + V(r))|S(r)\rangle = |\mathfrak{R}'(r)\rangle + (j + \frac{3}{2})[\mathfrak{R}(r)/r] \quad (25.7.19)$$

25.8. KEPLER PROBLEM IN DIRAC THEORY – FINE STRUCTURE LEVELS of the H-atom:

25.8.1 For a Dirac Particle in a Central Potential:
If V(r) is the Central Potential, it was seen that, for a Dirac Particle

$$(1/c\hbar)(E-m_oc^2+V(r))|\Re(r)\rangle=[|S'(r)\rangle+(j+\tfrac{3}{2})(S'(r)/r)] \qquad (25.7.14)$$

$$(1/c\hbar)(E+m_oc^2+V(r))|S(r)\rangle=-[|\Re'(r)\rangle+(j-\tfrac{1}{2})(\Re(r)/r)] \qquad (25.7.15)$$

and
$$(1/c\hbar)(E-m_oc^2+V(r))[|\Re(r)\rangle=-|S'(r)\rangle+(j-\tfrac{1}{2})[S(r)/r)] \qquad (25.7.18)$$

$$(1/c\hbar)(E+m_oc^2+V(r))|S(r)\rangle=|\Re'(r)\rangle+(j+\tfrac{3}{2})[\Re(r)/r)] \qquad (25.7.19)$$

For the Coulomb potential,
$$V(r) = +(1/4\pi\varepsilon_o)\frac{Zeq}{r}, \qquad (25.7.1)$$

Where $q = -e$. and $V(r) = -(1/4\pi\varepsilon_o)\frac{Ze^2}{r}$, $\qquad (25.8.1)$

Then equations (25.7.14), (25.7.15), (25.7.18) and (25.7.19) can be **solved exactly**.

25.8.2 GROUND STATE of H-like atom:

Consider, for example, the ground state of H-like atom.
$$j = +\tfrac{1}{2}, \quad \ell = 0.$$

Equation (25.7.14) and (25.7.15) reduces to
$$(E_o - \mu + Z_o/r)|\Re(r)\rangle = |S'(r)\rangle + (2/r)|S(r)\rangle \qquad (25.8.2)\,a$$

$$(E_o + \mu + Z_o/r)|S(r)\rangle = -|\Re'(r)\rangle \qquad (25.8.2)\,b$$

with the abbreviations,
$$E_o = (1/c\hbar)E, \quad \mu = m_o c/\hbar$$

$$Z_o = Ze^2/c\hbar = Z/137 = Z/\alpha. \qquad (25.8.3)$$

It will be convenient if one tries
$$|\Re(r)\rangle = r^\beta e^{-\lambda t}, \qquad (25.8.4)$$

where β and λ are constants.
On substitution in (25.8.2) one gets
$$\beta = \{-1 + \sqrt{(1-Z_o^2)}\} \qquad (25.8.5)\,a$$

$$\lambda = Z_o \mu = Z_o m_o e/\hbar^2 \qquad (25.8.5)\,b$$

and
$$\frac{|S(r)\rangle}{|\Re'(r)\rangle} = \{-1 + \sqrt{(1-Z_o^2)}\}/Z = \text{a constant}. \qquad (25.8.5)\,c$$

$$Z_o = \mu\sqrt{(1-Z_o^2)}.$$

Or
$$E = m_o c^2 \sqrt{(1-(Ze^2/c\hbar^2)}$$
$$= m_o c^2 - (Ze^4 m_o/2\hbar^2) - (Z^4 e^8/8\hbar^4 c^2)$$
$$+ (1/2\,m_o^2\,c^2)\,S \cdot L\,(1/r)\,(d/dr)\,V(r)$$
$$+ (\pi\hbar^2\,Z\,e^2/2\,m_o^2 c^2)\,\delta^{(3)}(x). \qquad (25.8.6)$$

where 1st term, $m_o c^2$ = the Rest energy of electron, $\qquad (25.8.7)$
2nd term is = non-relativistic one, $\qquad (25.8.8)$
3rd term is = relativistic Fine Structure, a correction due to relativistic kinetic energy,
$$= m_o\,c^2\,[\frac{(Z\alpha)^2}{2n^2}]\,[\frac{(Z\alpha)^2}{n^2}]\,[\frac{n}{\ell+\tfrac{1}{2}} - \tfrac{3}{4}]. \qquad (25.8.9)$$

4th term is = Spin-Orbit Coupling correction,

$$= m_o c^2 \left[\frac{(Z\alpha)^4}{4n^3(\ell(\ell+\frac{1}{2})(\ell+1)} \right] \begin{Bmatrix} \ell \\ -\ell-1 \end{Bmatrix} \tag{25.8.10}$$

5th is Darwin term.$= (R_y \ Z/n^2)(Z\alpha/n)^2 \{\frac{3}{4} - [n/(j+\frac{1}{2})]\}$. (25.8.11)

NORMALIZED SOLUTION is

$|\Re(r)\rangle = 2Z_o \mu\sqrt{(1-Z_o^2)} \{2\mu\sqrt{[1+(1-Z_o^2)}/2\sqrt{(1-Z_o^2)}/2. \sqrt{[-1+v(1-Z_o^2)}\ e^{-Z_o\mu r}$.

(25.8.7) a

$|S(r)\rangle = \{\sqrt{[1-(1-Z_o^2)]}/Z_o\}|\Re(r)\rangle$ (25.8.7) b

These have to be substituted in the Dirac components (25.7.10) and (25.7.11), with $j=+\frac{1}{2}, m=\pm\frac{1}{2}$, to obtain the two normalized ground state solutions with electron spin up or down.

Worked out Example 25.17.

The Helicity Operator is defined as (σ', p). Obtain its eigen values.

Solution:

$(\sigma', p)^2 = 1$.

The eigen values of $(\sigma', p) = \pm 1$.

So $(\sigma'. p)^2/\sigma''^2 = (\sigma'. p)^2 = 1$.

$(\sigma'. p)^2 = \sigma''^2 = 3$.

The eigen values of $(\sigma'. p)$ are $\pm\sqrt{3}$.

Worked out Example 25.18.

Evaluate the Relativistic Fine Structure correction term, due to

$\hat{H}^{(1)} = - Z^4 e^8 / 8\hbar^4 c^2 = -(p^2)^2 / 8 m_o^3 c^2$

Solution: $\hat{H}^{(2)} = - Z^4 e^8 / 8\hbar^4 c^2 = -(p^2)^2 / 8 m_o^3 c^2$

$A^{(1)}_{n,\ell m} = \langle n\ell m|-(p^2)^2/8m_o^3 c^2|n\ell m\rangle$

$= \langle n\ell m|-(1/2 m_o c^2)(p^2/2 m_o)^2|n\ell m\rangle$

$= \langle n\ell m|-(1/2 m_o c^2)(\hat{H}_o + Ze^2/r)^2|n\ell m\rangle$

$= -(1/2 m_o c^2)\langle n\ell m|(\hat{H}_o + Ze^2/r)^2|n\ell m\rangle$

$= (1/2 m_o c^2)\{E_n^{o2} + 2E_n^o Ze^2\}\langle n\ell m|(1/r)|n\ell m\rangle$

$+ Ze^2 \langle n\ell m|(1/r^2)|n\ell m\rangle$

Use $\langle n\ell m|(1/r)|n\ell m\rangle = (Z/a_o n^2)$ (9.9.11)

$\langle n\ell m|(1/r^2)|n\ell m\rangle = [Z^2/a_o^2 n^3(\ell+\frac{1}{2})]$ (9.9.21)

$E_n^o = m_o c^2 (Z\alpha)^2/2n^2$

$A^{(1)}_{n,\ell m} = (1/2 m_o c^2)\{E_n^{o2} + 2E_n^o Ze^2(Z/a_o n^2)\} + Ze^2[Z^2/a_o^2 n^3(\ell+\frac{1}{2})]$

$= m_o c^2\left[(Z\alpha)^2/2n^2\right][(Z\alpha)^2/2n^2]\{[n/(\ell+\frac{1}{2})]-\frac{3}{4}\}$ (25.8.8)

The spin of the electron does not enter into this energy shift.

Worked out Example 25.19.

Show that the correction due to the electron spin appears in the expression $-\vec{\mu} \cdot \vec{B} = (e/mc) \vec{S} \cdot \vec{B}$. Obtain an expression for the energy shift, $\Delta E_{n,\ell m}$.

Solution: $\hat{H}^{(2)} = -\vec{\mu} \cdot \vec{B} = (e/mc) \vec{S} \cdot \vec{B} = (e/mc^2) \vec{S} \cdot \vec{v} \wedge \vec{E}$

$= -(e/m^2c^2) \vec{S} \cdot \vec{p} \wedge \nabla \psi = -(e/m^2c^2) \vec{S} \cdot \vec{p} \wedge \vec{v} (1/r)(d/dr)\psi(r)$

$= (1/m^2c^2) \vec{S} \cdot \vec{r} \wedge \vec{p} (1/r)(d/dr)[e\psi(r)]$

$= \tfrac{1}{2}(1/m^2c^2) \vec{S} \cdot \vec{L} (1/r)(d/dr)[e\psi(r)]$ (25.8.9)

Consider two component wave functions for the unperturbed state function, $\Psi(r)$. It is required to find the expectation value of $\hat{H}^{(2)}$.

$\hat{H}^{(2)} = \tfrac{1}{2}(1/m^2c^2) \vec{S} \cdot \vec{L} (1/r)(d/dr)[e\Psi(r)]$

$= \tfrac{1}{2}(Ze^2/m^2c^2) \vec{S} \cdot \vec{L} (1/r^3) \equiv \hat{H}_{LS}$.

$\hat{H}^{(2)}$ is compatible with the set $\{L^2, J^2, J_z\}$.

This is a case of Degenerate Perturbation Theory. For a given s and ℓ, there are $2(2\ell+1)$ degenerate eigen states of $\hat{H}^{(o)}$, with the additional factor 2 coming from the two spin states. To calculate the energy shift one has to diagonalize a sub-matrix. To simplify it is to be noted that $J = S + L$.

Implying $\quad J^2 = S^2 + 2 S \cdot L + L^2$,

$S \cdot L = \tfrac{1}{2}[J^2 - L^2 - S^2]$ (25.8.10)

Combining the degenerate eigen states into linear combinations that are eigen states of J^2 to diagonalize $\hat{H}^{(2)}$,

$[L, S] = 0$,

since L depends on spatial coordinates and S does not.

The appropriate linear combinations yield

$S \cdot L \, |\ell+\tfrac{1}{2}, m+\tfrac{1}{2}\rangle = \tfrac{1}{2}[J^2 - L^2 - S^2] \, |j = \ell + \tfrac{1}{2}, m_j = m+\tfrac{1}{2}\rangle$

$= \tfrac{1}{2} \hbar^2 [(\ell+\tfrac{1}{2})(\ell + \tfrac{3}{2}) - \ell(\ell+1) - \tfrac{3}{4}] \, |(\ell+\tfrac{1}{2}), (m+\tfrac{1}{2})\rangle$,

$= \tfrac{1}{2} \hbar^2 \ell \, |(\ell+\tfrac{1}{2}), (m+\tfrac{1}{2})\rangle$,   (25.8.11)

$S \cdot L \, |\ell-\tfrac{1}{2}, m+\tfrac{1}{2}\rangle = = \tfrac{1}{2} \hbar^2 [(\ell-\tfrac{1}{2})(\ell + \tfrac{1}{2}) - \ell(\ell+1) - \tfrac{3}{4}] \, |(\ell-\tfrac{1}{2}), (m+\tfrac{1}{2})\rangle$,

$= \tfrac{1}{2} \hbar^2 (\ell+1) \, |(\ell-\tfrac{1}{2}), (m+\tfrac{1}{2})\rangle$ (25.8.12)

For a given ℓ value,

The # of states $= [2(\ell+\tfrac{1}{2})+1] + [2(\ell-\tfrac{1}{2})+1]$ (25.8.13)

$\hat{H}^{(2)} = \tfrac{1}{2}(Ze^2/m^2c^2) \vec{S} \cdot \vec{L} (1/r^3) = \tfrac{1}{2}(Ze^2/m^2c^2) \tfrac{1}{2} \hbar^2 \ell (1/r^3)$ (25.8.14)

If the linear combinations are $|\varphi_{\ell m}\rangle$

$\langle j, m_j, \ell | \hat{H}^{(2)} | j, m_j, \ell \rangle = \langle j, m_j, \ell | \{\tfrac{1}{2}(Ze^2/m^2c^2) \tfrac{1}{2}\hbar^2 \ell (1/r^3)\} | j, m_j, \ell \rangle$

$= \{\tfrac{1}{2}(Ze^2/m^2c^2)\tfrac{1}{2}\hbar^2\}\ell \int_0^\infty dr \, r^2 < n, \ell |(1/r^3)| n, \ell >$, and $= \{\frac{1}{2} (Ze^2/m^2c^2) \frac{1}{2} \hbar^2 \ell\}[-\ell-1] \int_0^\infty dr \, r^2 <n, \ell |(1/r^3)|n, \ell>,$

for $j = \ell \pm \frac{1}{2}$, respectively.

$<n, \ell |(1/r^3)|n, \ell> = [Z^2/a_o^3][1/n^3(\ell+\frac{1}{2})(\ell+1)].$

$$A^{(2)}_{n,\ell m} = \frac{1}{2} E_n^o (Z\alpha)^2 [1/n^3(\ell+\frac{1}{2})(\ell+1)] \begin{Bmatrix} \ell \\ -\ell-1 \end{Bmatrix} \qquad (25.8.15)$$

One would think that $<\hat{H}_{LS}> = 0$ in the s-states, since $<L, S>_{\ell=0} = 0$.
However, $<n, \ell = 0|(1/r^3)|n, \ell = 0>$ diverges.

But $<(1/r^3)> <L, S>_{j=\ell+\frac{1}{2}} = \frac{1}{2}\ell$

$<(1/r^3)>_{n,\ell} = [Z^2/a_o^3][1/n^3(\ell+\frac{1}{2})(\ell+1)\ell] \qquad (9.9.22)$

i.e. $<\hat{H}_{LS}> = $ finite, in s-states.

Worked out Example 25.20.
Obtain an expression for the energy shift due to the Darwin term, $\hat{H}^{(3)} = (\hbar^2/8 m_0^2 c^2) \nabla^2 V$
$= (\pi \hbar^2 Z e^2 / 2 m_0^2 c^2) \delta^{(3)}(x)$. The correction (to the perturbed energy), $\Delta E_{n,\ell m}$.

Solution:
The perturbation, known as the Darwin term,
$\hat{H}^{(3)} = (\hbar^2/8 m_0^2 c^2) \nabla^2 V = (\hbar^2/8 m_0^2 c^2)[4\pi e Q_{nucl}(x)] = (\pi \hbar^2 Z e^2 / 2 m_0^2 c^2) \delta^{(3)}(x)$, with
nuclear charge density, $Q_{nucl}(x)$. This term can be understood from the *Zitterbewegung* of the electron. According to the relativistic theory the position of a localized electron fluctuates with $\lambda = \hbar / m_0 c = $ Compton Wavelength. The electron feels, therefore, on the average a potential
$<[V(x+\delta x)]> = V(x) + <\delta x \, \nabla V> + \frac{1}{2}<(\delta x \, \nabla)(\delta x \, \nabla) V(x)>$
$= V(x) + (1/6)(\delta r)^2 \nabla^2 V(x). \qquad (25.8.16)$

The correction thus calculated is in qualitative agreement in form, sign, and magnitude with the Darwin term.
Because of the δ-function, a contribution
$\hat{H}^{(3)} = (\pi \hbar^2 Z e^2 / 2 m_0^2 c^2)|\Psi(0)|^2 = m_0 c^2 [(Z\alpha)^4/2n^3 \delta_{\ell 0}. \qquad (25.8.17)$

arises for s-wave states only. $<\hat{H}^{(3)}>$ is formally identical to $<\hat{H}^{(2)}>_{n,j=1,\ell=0}$,

25.9. BEYOND THE FINE STRUCTURE CORRECTIONS-; RELATIVISTIC CORRECTIONS & THE LAMB SHIFT; HYPERFINE STRUCTURE:

25.9.1 The FINE STRUCTURE terms in Hydrogen:

The fine structure terms in Hydrogen provide corrections of the order of 10^{-4} with respect to electrostatic Bohr levels. Additional corrections exist which produce much smaller corrections than those due to Fine Structure, as given explicitly in equation (25.8.6) and in Chapter 21:-
(i)   The proton, which has spin magnetic moment, μ_P, can interact with the dipole associated with the electron spin, producing *Hyperfine Structure* in the spectrum,
(ii)   The hydrogenic electron experiences a Coulomb potential only outside the proton. Inside the proton ($r < 1$ nm), $V(r) \neq (1/r)$-type. This produces a minor shift in the hydrogenic energy spectrum,

(iii) Another factor of importance is related to the interaction between an electron and the radiation vacuum. In the absence of external radiation, the $<E>$ and $<B>$ both vanish. However, the non-vanishing fluctuations of the quantized field can still interact with the electron to cause what is known as the *Lamb Shift*.

Quantization of the EM Field not only ensures consistency with Heisenberg's Uncertainty Principle, but it also justifies the notion of a photon as a field quantum.

For example, in the case of a H-atom, $V(r) = -(Z\alpha/r)$,

Table 25.1

Term symbol	n	ℓ	j	$E_{n,j}$
$1S_{\frac{1}{2}}$	1	0	½	$m\sqrt{(1-Z^2\alpha^2)}$
$2S_{\frac{1}{2}}$	2	0	½	$m\sqrt{\{[1+\sqrt{1-Z^2\alpha^2}]/2\}}$
$2P_{\frac{1}{2}}$	2	1	½	$m\sqrt{\{[1+\sqrt{1-Z^2\alpha^2}]/2\}}$
$2P_{\frac{3}{2}}$	2	1	3/2	$(m/2)\sqrt{(4-Z^2\alpha^2)}$

25.9.2 The LAMB SHIFT

25.9.2.1. The Experimental Results

The experimental results of the fine structure of atomic Hydrogen as well as H-like atoms are in broad agreement with the Dirac's relativistic predictions. However, the agreement is not perfect. The largest discrepancy is observed in the Fine Structure of the n = 2 levels of atomic Hydrogen. Non-relativistically, $2S_{1/2}, 2P_{1/2}, 2P_{3/2}$ are all degenerate; it is found that relativistically, $2S_{1/2}$ and $2P_{1/2}$ are still degenerate. It is now known that virtual emission processes involve CHANGE IN THE MASS OF THE ELECTRON. The resulting EM SELF-ENERGY for free electrons of momentum p is proportional to p^2. This mechanism leads to an energy shift in the $2S_{1/2}$ and $2P_{1/2}$ Fine Structure levels of Hydrogen (Fig. 25.3), called LAMB SHIFT.

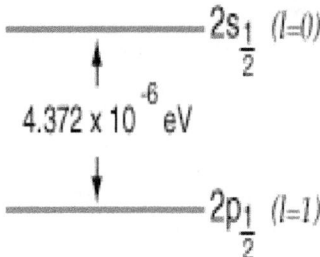

Fig. 25. 3. Energy splitting due to Lamb Shift, $2S_{1/2} \rightarrow 2P_{1/2}$

Willis E. Lamb (and R.C. Retherford and colleagues) from 1940 carried out a series of experiments in which the Fine Structure was studied by RF spectroscopy of the transitions $2S_{1/2} \rightarrow 2P_{1/2}, 2P_{3/2}$ with a resolution of $\approx 10^{-4}$ GHz. He confirmed that the Fine Structure of the 2P state is accurately given by the Dirac Theory, and in fact used his measurements to obtain a more precise value for the Fine Structure constant, α. A more dramatic finding in 1947 (Lamb &

Retherford, Phys. Rev. Vol.72, pp241), for which Lamb received the Nobel Prize for Physics in 1955, was that $2S_{1/2}$ and $2P_{1/2}$ levels for n =2 are not, in fact, degenerate. The $2S_½$ state was found shifted upward by 1040 MHz. This displacement is called the Lamb shift, is due to radiative coupling of the electron with the Vacuum Field. In 1972, Hansch was able to measure the splitting by optical high-resolution spectroscopy. Accordingly, he found that the $2S_{1/2}$ state was shifted upward (above the degenerate level) by 1060 MHz. Erickson (1971) theoretically obtained 1057.9 MHz; and Mohr (1975) got 1057.86 MHz. Experimentally in 1975, Lundeen and Pipkin found 1057.89 MHz. Andrews and Newton (1976) experimentally measured this shift as 1057.86 MHz.

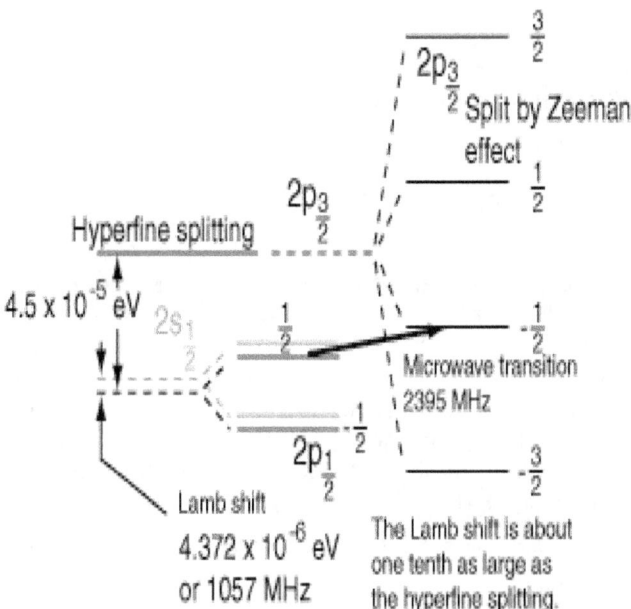

Fig. 25.4. : Energy level splitting of the hydrogen $2S_{1/2}, 2P_{1/2}, 2P_{3/2}$; and the Lamb Shift is compared with Zeeman splitting and Hyperfine splitting.

The discovery of the Lamb Shift (1947) in Hydrogen initiated rapid and amazing progress in QED.

25.9.2.2 THEORY of the LAMB SHIFT:
Feynman Diagram of the Lamb Shift is shown in Fig 25.5.

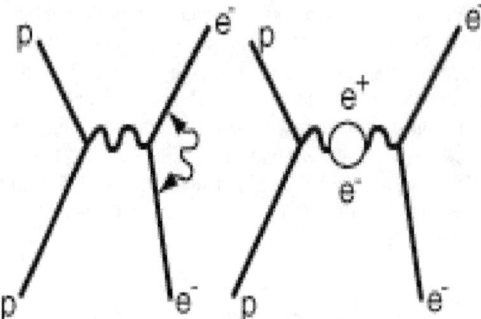
Fig.25.5: Feynman diagram of the Lamb Shift

25.10. HYPERFINE STRUCTURE:

25.10.1. THEORY:

In addition to the Lamb Shift, there is a correction, which contributes the Hyperfine Structure. The nuclear spin, I leads to the nuclear magnetic moment, $\mu_N = Ze\, g_N /(2c\, nm)\, I$. where nm = Nuclear Magneton. It generates a vector potential,

$$A = -\mu_N /\nabla(1/r) = \mu_N \wedge (i/r^3). \qquad (25.10.1)$$

and a magnetic field $\quad \vec{B} = \nabla \wedge A = -\{\mu_N \nabla^2(1/r) - \nabla(\mu_N \cdot \nabla)(1/r)\} \qquad (25.10.2)$

First consider the s-electrons. The interaction of the magnetic moment of the electrons μ with the magnetic field B of the nucleus gives rise to the HYPERFINE INTERACTION, $\mathcal{H}_{h,f}$.

$$\hat{H}_{h,f} = -\vec{\mu}_N \cdot \vec{B} = (e/m_0 c)\, \vec{S} \cdot \vec{B}..$$

$$= (Z\, e^2\, g_N /2\, m_0\, nm\, c^2)\, S\, \{-\hat{I}\, \nabla^2(1/r) - \nabla(\hat{I}\cdot\nabla)(1/r)\} \qquad (25.10.3)$$

Since $\quad \nabla^2(1/r) = -4\pi\, \delta^{(3)}(x)$),

and $\quad \int_0^\infty d^{3(x)} \{\hat{I}\cdot \nabla^2(1/r)\}\, [\Psi_{n,0}(r)]^2 = (\hat{I}/3) \int_0^\infty d^{3(x)}\{\nabla^2(1/r)\}\, [\Psi_{n,0}(r)]^2$.

For readily symmetric s states, as a first step in the 1st Order Perturbation Theory, the spatial expectation value in the states $|n, j=½, \ell=0>$ becomes

$$<\hat{H}_{h,f}>_{|n, j=½, \ell=0>} = (4/3)\, (m_0/nm)\, [(Z\,\alpha)^4\, m_0\, c^2]\, (1/n^3)\, [S\cdot\hat{I}/\hbar^2] \quad (25.10.4)$$

Therefore $\hat{H}_{h,f} < \hat{H}_{fine}$, by a factor, (m_0/nm). Analogous to the S.L coupling, J is introduced to diagonalize S. L, here the total spin F = S + I has to be introduced for S.I interaction. Then

$$\tfrac{1}{2}\hbar^2\, S \cdot I = (\tfrac{1}{2}/\hbar^2)\,(F^2 - S^2 - I^2) = \tfrac{1}{2}F(F+1) - \tfrac{3}{4} - I(I+1)$$

$$= \tfrac{1}{2}I, \text{ for } F = I + \tfrac{1}{2},$$

or, $\qquad = \tfrac{1}{2}(-I-1), \text{ for } F = I - \tfrac{1}{2}.$

For the Hydrogen atom,

$$g_N = g_P = 5.56, I = \tfrac{1}{2}.$$

The s-wave states in the Hydrogen atom are, therefore, either in SingletState (F = 0, ground state), or a TripletState (F = 1, excited state). The splitting of the nth level for s-electrons is, therefore,

$<\hat{H}_{h,f.}> |n, j=½, \ell=0> = \Delta E_{n,\ell m}(hf)\,]$
$= (4/3)(m_0/nm)[(Z\alpha)^4 m_0 c^2](1/n^3)[S\cdot\hat{I}/\hbar^2]$ (25.10.5)

For $1S_{1/2}$, $\Delta E(1S_{1/2}) = 5.876 \times 10^{-6} eV = 1421 MHz = 21 cm$

$2S_{1/2}$ $\Delta E(2S_{1/2}) = (\frac{1}{8}) 5.876 \times 10^{-6} eV = 177 MHz$.

$2P_{1/2}$, $\Delta E(2P_{1/2}) = (\frac{1}{24}) 5.876 \times 10^{-6} eV = 59 MHz$.

25.10.2.1 The 21 cm (1421 MHz) Line of Hydrogen:

Both Spin-Orbit (S.L) Coupling and relativity effects combine to produce Fine Structure shifts of the order of $\approx 10^{-3}$ to $\approx 10^{-6} eV$. These can be detected by a good constant deviation prism spectrometer. By when the energy shift is below $\approx 10^{-6} eV$, then one has to resort to Fabry-Perot Etalon or Lummer-Gehrcke Plate spectroscopy. This is where the Hyperfine Structure of a spectral line falls. Hyperfine Structure is produced by the interaction of electron and proton magnetic moments. The magnitude of the shift becomes very small (*i.e.* hyperfine) because, unlike the S.L interaction of electrons, the magnetic moment of a particle is given by
$\mu = g(e\hbar/2m)$. (25.10.6)
the g-factor is of order of unity, and the proton mass, $m_p \approx 2000\, m_e$. This is the manifestation in Hydrogen as line at about 1421 MHz that radio astronomers use to map Hydrogen clouds in the Galaxy. This radiation is emitted as a result of the spins of the electron and proton spontaneously decays from a Triplet state with spins parallel to the Singlet one.
$\Delta E(1S_{1/2}) = 5.876 \times 10^{-6} eV = 1421 MHz = 21 cm$ (25.10.7)
Further slight corrections, to $1S_{1/2}$, come from relativity, from QED (*i.e.* the Lamb Shift) and from the finite size and orbital motion of the proton. This yields a value of
$\Delta E = 1420.405751786\, MHz \pm 0.01\, Hz$. (25.10.8)
This is one of the MOST ACCURATELY measured constants in physics.
The 21 cm radiation is very important in astronomy. From its intensity, Doppler broadening and Doppler Shift, one gets the information concerning the density, temperature, and motion, respectively, of interstellar and inter-galactic Hydrogen clouds. Due to the Hyperfine Interaction, all of the levels in Hydrogen are split into Doublets.

25.10.3. HYPERFINE INTERACTIONS including the Orbital part:

The orbital moment of the electron, L and the magnetic moment of the nucleus,
$\mu_N = (Zeg_N)/(2c\, nm)\, I$ interact.
$\hat{H}_{h,f,L}$ = interaction of L with $\vec{\mu}_N$,
$= (i\hbar e/m_0 c) A\cdot \nabla = (-e/m_0 c)(\mu_N/r^3)(x \wedge p)$
$= (-e/m_0 c)(1/r^3)(\vec{\mu}_N \cdot \vec{L})..$ (25.10.9)
The magnetic field, $\vec{B}$ of the nucleus,
$\vec{B} = -\{\mu_N \delta^{(3)}(x)(-8\pi/3) + (\mu_N/r^3) - 3(\mu_N .x) x/r^5\,)\}$ (25.10.10)
Total hyperfine interaction

$$\hat{H}_{h.f} = \hat{H}_{hfL} + (e/m_0 c)\vec{S}\cdot\vec{B} = Z e^2 g_N / (2\mu_N m_0 c^2) \wedge \{(1/r^3)\vec{I}\cdot\vec{L}$$
$$+ (8\pi/3)\delta^{(3)}(x)\vec{I}\cdot\vec{S} - (\vec{I}\cdot\vec{S})(1/r^3) + [3(I\cdot x)(S\cdot x)/r^5]\} \quad (25.10.11)$$

Fig.25.4. shows also the energy level splitting due to Hyperfine Interactions in Hydrogen.

25.11. CONCLUSIONS:

To modify the Quantum Mechanics presented in the earlier 24 Chapters so that it could account for the behaviour of charged particle moving in an EM Field, two main problems needed to be solved. First, the particles, especially electrons, that were to be the subjects of the Quantum Field Theory, often had to be supposed to move at speeds approaching that of light, so some relativistic extension of Quantum Theory was made. Schrödinger Equation is not covariant; and so it was appropriately extended to get the Dirac Equation, which is relativistically covariant. Second, the EM Field is classical; a quantum treatment of the EM Field was not made in this Chapter.

The experimental results of the Fine Structure of atomic Hydrogen as well as H-like atoms are in broad agreement with the Dirac's relativistic predictions. However, the agreement is not perfect. The largest discrepancy is observed in the fine structure of the n = 2 levels of atomic hydrogen. The discovery of the Lamb Shift played a central role in the development of Quantum Electro Dynamics (QED). The first relativistic QED, involving *spin-less* particles, was by Heisenberg and Pauli (1929). The development of QED involving charged relativistic particles of *spin -½* required the knowledge of the Dirac's Relativistic Equation (1928), which predicted the existence of the positron. A set of *Anti-commuting operators* (creation and destruction operators) would ensure the validity of the Pauli's Exclusion Principle by providing *Anti-symmetric wave functions*. These symmetry properties could now be imposed on the Dirac wave function by requiring it to satisfy anti-commuting relations. The electron wave function thus becomes the ELECTRON OPERATOR FIELD (Second Quantization). Dirac & Heisenberg (1934) developed first QED of relativistic quantum mechanical electrons. The discovery of the Lamb Shift (1947) in Hydrogen initiated rapid and amazing progress in QED.

REVIEW QUESTIONS

R.Q. 25.1 Apply the K.G. Equation to the Hydrogen atom. Comment on the result'

R.Q. 25.2 Summarize the arguments in favour of regarding the Dirac Equation as the relativistic wave equation of the electron.

R.Q. 25.3 Show how the Dirac Equation accounts for the Fine structure of the Hydrogen atom.

R.Q. 25.4 Obtain the free particle solutions to the Dirac's Equation.

R.Q. 25.5 Illustrate that the Dirac's Equation constrains the spin of the particle to $\frac{1}{2}$.

R.Q. 25.6 It takes light 40 years to reach Planet Z from Earth. If Govind has just born, what speed is required for him so that he reaches Planet Z by the time he is 30 years old? (Answer: $v = 0.8\,c$, $\beta = 1/\sqrt{(1-0.8)} = 5/3$., To Govind, the "Earth – Planet Z" is only 24 *ly* away (long)).

R.Q. 25.7. Show that the KG equation is Lorentz invariant.

R.Q. 25.8 Use the explicit Representation of α^i and β to verify the anti-commuting properties.

R.Q. 25.9 Demonstrate that the Dirac Equation is covariant under Lorentz Transformation.

R.Q. 25.10 Show that in the non-relativistic limit, the relativistic spinors, $\left|\Psi^+_{P\uparrow}(p,t)\right\rangle$ and $\left|\Psi^+_{P\downarrow}(p,t)\right\rangle$, reduce to the two component spinors. (Answer: $\left|U_P^I\right\rangle \approx \left|U_{P\uparrow}^+\right\rangle = N_P \begin{bmatrix}1\\0\end{bmatrix}$, and $\left|U_P^{II}\right\rangle \approx \left|U_{P\downarrow}^+\right\rangle = N_P \begin{bmatrix}0\\1\end{bmatrix}$).

R.Q. 25.11 Estimate the Darwin term for Hydrogen, and show that fine structure effects in Hydrogen are thus 1st order relativistic effects of the Dirac Equation. Given: Darwin term ΔE (Darwin) $= (\hbar^2/8\, m^2 c^2)\nabla^2 V(r)$. (Answer: ΔE **(Darwin)** $= e^2 \int_0^\infty R_{n\,\ell}(r)\, dR_{n\,\ell} = \frac{1}{2}e^2\,|R_{n\,\ell}(r)|^2$, $R_{n\,\ell}(r) = 0$ for bound states.)

R.Q. 25.12 Obtain the eigen values of the operator $\hat{k} = (\beta/\hbar)(\sigma'\cdot L + \hbar)$. (Answer: The eigen values of $\hat{k}^2$ are $J(J+1)+\frac{1}{4}$, i.e., $(j+\frac{1}{2})^2$, $j=\frac{1}{2}$, $j=\frac{3}{2}$, etc).

R.Q. 25.13 Transform the Dirac Hamiltonian using the transformation e^{iN}, where $N = -(i/2\,mc)\beta\,\alpha\cdot p\,f\,(|p|/mc)$ with $f = (mc/|p|)\tan^{-1}(|p|/mc)$.

R.Q. 25.14 The Helicity Operator is defined as (σ', p). Obtain its eigen values. (Answer: $\pm\sqrt{3}$).

R.Q. 25.15 Show that the correction due to the electron spin appears in the expression $-\bar{\mu}\cdot\vec{B} = (e/mc)\,\vec{S}\cdot\vec{B}$. Obtain an expression for the energy shift $\Delta E_{n,\ell m}$.

R.Q. 25.16 Describe with theory the Lambshif.t.:

R.Q. 25.17 Give an account on Hyperfine structure.

R.Q. 25.18 What is the origin of 21 *cm* line of hydrogen? Briefly describe.

R.Q. 25.19 Obtain an expression for the energy shift due to the Darwin term,
$$\hat{H}^{(3)} = (\hbar^2/8\,m_0^2\,c^2)\,\nabla^2 V = (\pi\hbar^2\,Z e^2/2\,m_0^2\,c^2)\,\delta^{(3)}(x)$$
The correction (to the perturbed energy), $\Delta E_{n,\ell m}$.

&&*&*&*&*&*&*&*&*

Chapter 26

QUANTUM MECHANICS -6:
[INTERACTING FIELDS]
QUANTUM FIELD THEORY (QFT),
QUANTUM ELECTRODYNAMICS (QED)

Chapter 26

QUANTUM MECHANICS -6: [INTERACTING FIELDS] QUANTUM FIELD THEORY (QFT), QUANTUM ELECTRODYNAMICS (QED)

Education is the manifestation of perfection already in man- Swami Vevekananda

26.1. INTRODUCTION:

26.1.1 GOAL OF LOCAL QUANTUM FIELD THEORY (QFT)

To account for the properties of sub-atomic particles that make up matter and for the forces between these particles, Quantum Field Theory (QFT) was introduced. . This theory is a natural extension of Quantum Mechanics and, indeed, originally grew out of attempts to provide a quantum mechanical description of the way charged particles like electrons interact with the EM Field. This Chapter presents the general structure of Field Theory, Path Integrals, Second Quantization, quantized K.G Field, the Dirac Field, and the EM Field, Feynman Rules, Lamb Shift and discusses in brief other developments, like QCD, in Quantum Mechanics.

Field Theory was developed in the late 1920s and early 1930s to describe the interactions of electrons and photons. It was a natural synthesis of Quantum Mechanics and Relativistic Wave Equations like Maxwell's Equations (that describe the electrical and magnetic properties) and the Dirac Equation (an attempt to describe the properties of the field associated with the electron). LOCAL QUANTUM FIELD THEORY (QFT) is the only way to combine a quantum mechanical theory of particles with Special Relativity consistent with causality. (CAUSALITY is the general principle that causes should always happen before their effects). The word LOCAL here means the theory in which the interactions that cause scattering or production or destruction of particles take place at single SPACE-TIME POINTS.

To modify the Quantum Mechanics, presented from Chapter 2 through chapter 24, so that it could account for the behaviour of charged particle moving in an EM Field, two main problems were in need of solutions. First, the particles, especially electrons, that were to be the subjects of the Quantum Field Theory, often had to be supposed to move at speeds approaching that of light, so some relativistic extension of Quantum Theory was made. Schrödinger Equation is not covariant; and so it was appropriately extended to get the Dirac Equation, which is relativistically covariant. This was dealt with in Chapter 25. Second, the EM Field is classical; a quantum treatment of the EM Field was made in this Chapter. These difficulties, and some others that rose in the course of developing QED, were surmounted and an enormously successful theory emerged. One of the successes of this theory was the fact that particles, light quanta, came out of the theory in a natural way as a consequence of a quantum treatment of the EM Field. The obvious idea, then, was that all particles arise from their own field in the same way; viz. photons from the EM Field, electrons from an electron field, mesons from a meson field, *etc.*

26.1.2 PROBLEMS THAT CONFRONT THEORY:

One of the serious problems that the theory must confront is that the number and nature of particles in the microphysical world is not fixed. Sub-atomic particles are rather created and destroyed, and this must become the pivotal fact to any relevant theory of sub-atomic particles. The traditional Quantum Mechanics, treated so far in this book, did not take into account the number of particles present and it determines the number of variables included in the wave equation. A new wave equation would have to be calculated for every creation and destruction event.

26.1.3 Why is it necessary to extend the TRADITIONAL THEORY?

The immediate goals for the extension of traditional Quantum Mechanics, therefore, are: (1) to include classical fields, like EM Field and gravitational fields, within the quantum mechanical framework, (2) to generalize Quantum Mechanics to include the domain of relativity theory, as in Chapter 25, and (3) to take into account of the creation and annihilation of particles. Finally, it is aimed at to devise a Unified Theory that will explain all of the forces of nature and particles on which they interact.

26.1.4 RENORMALIZATION:

The calculation of quantities from QED means using Perturbation Theory. The accuracy of measurements was such that the results of 1st order perturbation calculations were good by are not quite precise. But higher order calculations give results that were infinite. The cure of such difficulties is called RENORMALIZATION; and it involves redefinition of parameters like charge and mass of the electron. Renormalization is considered to be an essential property of any theory of elementary particle and their interactions.

26.1.5. FEYNMAN DIAGRAMS:

The Quantum Field Theory of EM processes is called QED and is the simplest of the Gauge Theories. It has been used to make precise predictions, which have been checked to the limit of experimental accuracy. Feynman Diagrams are helpful in visualizing EM processes and in calculating reaction amplitudes. They are dealt with in detail in Chapter 23. The process of e^+e^- annihilation provides an application of the techniques, which is of relevance to the study of quark properties. The underlying significance of the Gauge Principle is considered at length.

26.1.6 REAL AND VIRTUAL PARTICLES:

The Heisenberg UNCERTAINTY PRINCIPLE allows having an energy uncertainty ΔE for a time Δt, given by $\Delta E.\Delta t \simeq \hbar/2$. A particle which is free and stable (not interacting with external fields or with other particles) has $\Delta t = \infty$, so that $\Delta E = 0$. Translating the idea to an inertial frame in which the particle at rest will have rest energy $m_o c^2$. In any other inertial frame, the total energy, E and momentum, p must satisfy the relation, $E^2 = p^2 c^2 + m_o c^4$. Such a particle is said to be REAL.

However, a particle may have only a brief existence and not become truly free of sources; and as per the Uncertainty Principle, does not satisfy the relation, $E^2 = p^2 c^2 + m_o c^4$. Such a particle is said to be VIRTUAL.

26.1.7 GAUGE INVARIANCE:

According to theories in NEW PHYSICS, interactions between fundamental fields (like electrons, quarks, weak vector bosons, so on) are dictated by a Gauge Principle. This principle arises from the requirement that the quantities, which are conserved, are conserved LOCALLY and not merely GLOBALLY. Invariances are related to conservation laws in both Classical and Quantum Mechanics. Conservation Laws – such as charge conservation in electromagnetism- play a central role in Gauge Theories in that they are closely related to the dynamics.

Gauge Theories are theories characterized by a close inter-relation between three conceptual elements: Symmetries, Conservation laws and Dynamics. Noether's Theorem (1918) connects symmetries and conservation laws, using a Lagrangian Formulation of Field Theory (It states that if Lagrangian, $\mathcal{L}$ = invariant under a continuous Transformation, then there will be an associated symmetry current). Gauge theories have a particular type of symmetry, called GAUGE INVARIANCE.

LOCAL gauge Invariance is the common feature underlying all the forces of nature: EM, Weak, Strong and Gravitational forces. It has been possible to unify the description of the EM and weak force using the principles of local gauge invariance. Gauge Invariance means that it is possible to express a theory in a form, which is invariant under a group of Transformations made on internal coordinates (for the electro-weak and strong forces), which vary smoothly from point to point in space-time.

GLOBAL gauge Invariance refers to independence of the theory to changes of the internal coordinates, which are the same in all points in space-time. Each force is related to a particular symmetry of nature and hence to a particular group of Transformations and to particular conservation laws.

Gauge invariance has very important consequences. This is important because QED, QCD and the UNIFIED electro-weak theory are all examples of Gauge Theories with somewhat different forms of gauge invariance in each case. The simplest type of gauge invariance is that found in electromagnetism, and is connected to to charge conservation. The so-called 'Gauge Principle' makes gauge invariance the fundamental requirement from which the detailed properties of the interaction are deduced.

26.2 QUANTUM FIELD THEORY (QFT): GENERAL FORMALISM:
26.2.1 OVERVIEW on the Dirac Theory of the free electron:

In Chapter 25, a detailed account on the relativistic electron based on Dirac Theory has been presented. From this the existence of its anti-particle – positron – emerged naturally. The non-relativistic spin-up $|U_{p\uparrow}\rangle$ and spin-down $|U_{p\downarrow}\rangle$ states of the electron are represented by two-component entities called the spinors, $|U_p\rangle$. There are 4 basic states (4-fold degeneracy, of the electron: two with positive energy, one having spin-up, $|U^+_{p\uparrow}\rangle$ (i.e. $|U^I_p\rangle$) and the other spin-down, $|U^+_{p\downarrow}\rangle$ (i.e. $|U^{II}_p\rangle$), and two with negative energy. The 4-fold degeneracy; +ve energy spin-up and spin-down, and the other two are -ve energy spin-up and spin-down, $|U^-_{p\uparrow}\rangle$ (i.e $|U^{III}_p\rangle$) and $|U^-_{p\downarrow}\rangle$ (i.e. $|U^{IV}_p\rangle$). The wave function, $|\Psi^+_{p\uparrow}(p,t)\rangle$ of a free electron with 4-momentum, p_μ (and 3-momentum, p) is the product of a plane wave, $e^{i\,p.r/\hbar}\,e^{-i\,E\,t/\hbar}$ and a spinor, $|U^+_{p\downarrow}\rangle$, viz.,

$$|\Psi^+_{p\uparrow}(p,t)\rangle = |U^+_{p\downarrow}\rangle e^{i\,p.r/\hbar}\,e^{-i\,E\,t/\hbar}. \qquad (25.5.1)$$

The Dirac wave functions are the solutions of the Dirac free particle Equation,

$$(i\,\gamma^\mu \partial_\mu - m_o)|\Psi\rangle = 0 \qquad (25.4.13)$$

where one recognizes $i\partial_\mu$ as the 4-momentum operator discussed earlier in Chapter 25. The component γ^μ–s form a 4-vector in space-time, but have more complex structure than vectors.

Each component is itself a 4x4 matrix acting in the 'spin' space spanned by the 4-component spinors, $|U_{Ps}\rangle$. By implication the mass in the Dirac Equation needs to act on the spinors: m_o is taken to mean $m_o \Im$, where $\Im$ is a 4 x 4 unit matrix. Dirac's idea fails when applied to Bosons; bosons do not obey the Pauli's Exclusion Principle, so the continuum of negative energy states can never be filled. QED gives a more consistent explanation, which removes the need for a continuum of occupied negative energy states. According to R.P. Feynman, a negative energy particle moving backward in time is exactly equivalent to a positive energy anti-particle moving forward in time.

In the presence of an EM Field, specified by the Gauge potential,

$$A^\mu = (A_k, i\Phi) = 4\text{ - vector potential,} \qquad (25.3.4)$$

$$= (A_1, A_2, A_3, A_4)$$

where A_1, etc. and Φ completely specify an EM Field.

The 4-vector EM potential, $A_k = (A_1, A_2, A_3)$, (25.6.1)

and   Scalar EM Potential, $A_4 = i\Phi$. (25.6.2)

In the presence of EM Field the standard classical expression for the energy of a particle with charge q and mass m is:

Non-relativistic: $H = E = \{[p-(q/c)]\}^2 / 2m + q\Phi$ (25.3.1)

i.e., $\qquad (E - q\Phi) = \{[p-(q/c)]\}^2 / 2m$.

The inference can be made that the 4-momentum p is replaced by

$$[p-(q/c)A] \equiv [(E-q\Phi), p-(q/c)A] .$$

This expresses what is called the minimal EM Coupling. The Dirac Equation for an electron in the presence of an EM Field becomes, as a consequence,

i.e., $\qquad (\gamma_\mu p_\mu + e\gamma_\mu A_\mu - m) |\Psi\rangle = 0$ (25.3.1) b

In the operator form $\quad (i\gamma_\mu \partial_\mu - m)|\Psi\rangle = -e\gamma_\mu A_\mu$ (25.3.1)c

Neutrinos are also spin-½ particles but unlike the electrons possess no mass or charge. But the mass-less neutrino only requires a 2-component spinor for its description. Denoting this by $|w(v)\rangle$, and if the neutrino 4-momentum is (E, p) it satisfies the equation

$$E|w(v)\rangle = (\sigma.p)|w(v)\rangle \qquad (26.2.1)$$

Now $\quad E = |p|,$

For a mass-less particle so that $(\sigma.p)/E$ is a component along the direction of motion and has eigen values ± 1; this spin component is called the HELICITY, H. The neutrino has negative helicity ($H = -1$), i.e. it is LEFT-HANDED. Its anti-particle, the anti-neutrino, is also mass less and is RIGHT-HANDED. The sign of the anti-neutrino can be deduced with the help of Dirac Representation of negative energy states. This analysis shows that because neutrinos are mass less they exist in nature and have a definite handedness. Note that all mass-less particles seem to lose spin components, the photon may have helicity +1 (right circularly polarized) or –1 (left circularly polarized) but never zero.

26.2.2. The PHYSICS OF SELECTION OF APPROPRIATE LAGRANGIAN, L to use in the KEY-PATH INTEGRAL:

The extension of Quantum Theory to include the EM Field is conceptually easy and direct: the amplitude for a transition from one field configuration to another is found by using the KEY-PATH INTEGRAL with the classical Lagrangian, $\mathcal{L}$ for the EM Field. The Lagrangian used corresponds to the Maxwell's Equations. This is the case with photons. In the case of other particles, say mesons, one has to replace the Lagrangian for the EM Field to the one for the meson field. All of the physics, then, is in the selection of appropriate Lagrangian to use.

26.2.3 To CALCULATE THE LAGRANGIAN for a 1-D scalar Field, $\Phi(z,t)$:

The Lagrangian for the EM Field is complicated. I will, therefore, adopt the treatment followed by C.F. Stevens. The fussy detail involved (though they are at the heart of the physical problem) in treating the EM Field obscures the essential aspects of the theory. So consider the limitation of a 1-D scalar field, $\Phi(z,t)$, which depends only on a single spatial coordinate z and on time t, rather than a 3-D pair of fields required for electromagnetism. The transition amplitude that describes the field's behaviour is, according to the equation for the KEY-PATH INTEGRAL,

$$K\Phi(z,t)\Phi(\xi,t) = \int D\ \Phi\{(i/\hbar)\iint dzdt L(\Phi,\partial_\mu\Phi)\}. \qquad (26.2.1)$$

Note that the argument in the exponential function is now an integral over both time and space and not just time, as it is usually the case. L is LAGRANGIAN DENSITY – i.e. field Lagrangian – which must be integrated over the space. To simplify, $L(\Phi,\partial_\mu\Phi)$ is expanded in a power series to lowest order in $\partial_\mu\Phi$ (which is second order in this case to keep the symmetry of the $L(\Phi,\partial_\mu\Phi)$ in time and space) to get

$$L(\Phi,\partial_\mu\Phi) = (\partial_\mu\Phi)(\partial_\mu\Phi) \qquad (26.2.2)$$

where $\quad \partial_\mu = (\partial_z, (i/c)(\partial_t))$.

The repeated indices in μ indicate the Einstein summation convention.

This $L(\Phi, \partial_\mu \Phi)$ is the simplest possible one that is covariant and treats space as symmetric. Let this will describe the "Electro Magnetic" Field with a simple wave equation.

26.2.4 The LAGRANGIAN (L) FOR A LINEAR CHAIN (Mechanical Field):

To extend the Feynman Formulation of Quantum Mechanics to a field, one must have an expression for the field's Lagrangian. Consider a linear chain of identical, massless, ideal Hooke's law springs connecting N point masses, each with mass $m = M/N$ (M is the total mass). The chain extends from z = 0 to z = Z. Let ϕ_k is the deviation from its equilibrium position. When all the ϕ_k are 0, the mass points are evenly spaced. For this system, the Lagrangian, L

$$L = \text{Kinetic energy - potential energy}$$

$$L = \sum_{k=1}^{N}\left(m\,\phi_k'^2 / 2\right) - \sum_{k=0}^{N} V_k \tag{26.2.3}$$

where V_k = potential energy of the kth spring, and ϕ_k' is the time derivative of the ϕ_k.

$$L = \sum_{k=1}^{N}\left(m\,\phi_k'^2 / 2\right) - \tfrac{1}{2}\kappa\sum_{k=1}^{N}\left(\phi_{k+1} - \phi_k\right)^2. \tag{26.2.3}$$

Where κ is force constant of the spring.

When $N \to \infty$, $\Delta z = z/(N+1)$, $m = M/N$, $\kappa\,\Delta z = \kappa_o$, and defining

$$\mu = M/Z = Nm/[(N+1)\Delta z]$$

so
$$m = [(N+1)/N]\mu\,\Delta z,$$
$$\kappa = \kappa_o / \Delta z,$$

Equation (26.2.3) becomes

$$L = \frac{1}{2}\sum_{k=1}^{N}\left(\phi_k'^2 [(N+1)/N]\mu\,\Delta z\right) - \frac{1}{2}\sum_{k=0}^{N}\kappa_o[(\phi_{k+1} - \phi_k)/\Delta z]^2 \Delta z \; z \tag{26.2.4}$$

As $N \to \infty$, $\quad L = \frac{1}{2}\int_0^\infty \{\mu(\frac{\partial\phi}{\partial t})^2 - \kappa_o(\frac{\partial\phi}{\partial z})^2\}dz.$ \hfill (26.2.5)

whence $L_d = \{\mu(\frac{\partial\phi}{\partial t})^2 - \kappa_o(\frac{\partial\phi}{\partial z})^2\}$ \hfill (26.2.6)

26.2.5 2-D WAVE EQUATION:

The action S for this mechanical field, ϕ is

$$S[\phi(z,t)] = \int L dt = \iint dz\, dt\, L_d\phi'), \qquad (26.2.7)$$

Where ϕ' = argument of L_d indicates both the temporal and spatial derivatives of ϕ.

It is now desirable to identify the Euler- Lagrange Equation that corresponds to this Lagrangian. Then it turns out to be a wave equation. For this select the constants of the Lagrangian density L_d (26.2.6) so that it now reads

$$L_d = \{(\mu/\kappa_o)(\frac{\partial \phi}{\partial t})^2 - (\frac{\partial \phi}{\partial z})^2\} = \frac{1}{c^2}\{(\frac{\partial \phi}{\partial t})^2 - (\frac{\partial \phi}{\partial z})^2\}, \qquad (26.2.8)$$

The EULER-LAGRANGE EQUATION is

$$(\partial L / \partial \phi) - (\partial / \partial t)(\partial L / \partial_t \phi) - (\partial / \partial z)(\partial L / \partial_z \phi) = 0. \qquad (26.2.9)$$

giving the 2-D wave equation

$$-\frac{1}{c^2}\{(\frac{\partial^2 \phi}{\partial t^2}) + (\frac{\partial^2 \phi}{\partial z^2})\} = 0, \qquad (26.2.10)$$

Thus one can see how the Lagrangian density arises and why the integral that is the argument of the exponential function in the KEY PATH INTEGRAL (26.2.1) must include both space and time variables when one treats the system quantum mechanically. Note that the Lagrangian density that arises from the coupled harmonic oscillators, as described above, is the same as the one for the example below:

26.2.5.1. The FIELD-TRANSITION AMPLITUDE:

Just as with traditional Quantum Mechanics, the key quantity for Field Theory is the transition amplitude.

$$K\varphi(z,t)\varphi(z',t') = \int D\varphi \exp\{(i/\hbar)\int_\varsigma^z \int_\varsigma^{t'} dz\, dt\, L_d(\varphi, \partial_\mu \varphi)\} \qquad (26.2.11)$$

$$= \int D\varphi \exp\{(i/\hbar)\int d\xi\, L_d\}$$

In the last equation of this chain $\xi = (z.t) = $ 4-D, generally.

But the integral in $\exp\{(i/\hbar)\int_\varsigma^z \int_\varsigma^{t'} dz\, dt\, L_d(\varphi, \partial_\mu \varphi)\}$ uses 2-D volume element $d\xi = dz\, dt$

26.2.5.2. $\langle 0| \rightarrow |0\rangle$ transition:

The transition amplitude must be calculated with anew for each initial state and final state. This can be done by starting and ending in a standard BLANK state and the insert the initial state with a driving function. These two states can be chosen to be the VACUUM States, $\langle 0|$ and

$|0\rangle$ of the system, and one can calculate the amplitude of the $<0| \Rightarrow |0>$ transition, denoted as $\langle 0||0\rangle$.

26.2.6.1 GENERATING FUNCTIONAL, W[J]:

The interesting states are produced by the driving function J (as will be described below). The resulting amplitude of the $\langle 0| \Rightarrow |0\rangle$ transition with driving in between is called a GENERATING FUNCTIONAL (as the one used to characterize a Harmonic Oscillator, in Chapter 6). Thus one may start with the generating functional:

$$W[J] = \int D\varphi \exp\left\{(i/\hbar)\int_{-\infty}^{+\infty} L_d(\varphi, \partial_\mu\varphi) + J(\xi)d\xi\right\} \quad (26.2.12)$$

where $J(\xi)$ = Forcing function used to give some desired states.

The system is assumed to start (at $t = -\infty$) and finish (at $t = +\infty$) in a LOWEST ENERGY *vacuum* state, $|0\rangle$, so W[J] = amplitude of $\langle 0| \Rightarrow |0\rangle$ transition, specifically. The trick behind the generating functional is like that used in probability theory for generating functions that produce all of the moments of a probability distribution. Thus the generating functional enables calculation of all the specific transition amplitudes that one wants.

26.2.6.2 GREEN'S FUNCTION for the Field:

The central entity for relating experimentally observable quantities to field –theoretic predictions is the Green's Function for the field. Many experiments measure life times of particles and scattering cross sections, and these can be found from the Green's Function. At present here the goal is to define the Green's Function and relate it to the generating functional given above. The Green's Function is also identified with another function, *viz.* the FEYNMAN PROPAGATOR. The Green's Function for the field, Φ(z, t) is defined as

$$G(\xi,\xi') = \int D\varphi(\xi)\varphi(\xi') \exp\left\{(i/\hbar)\int_{-\infty}^{+\infty} d\xi\, L_d(\varphi, \partial_\mu\varphi)\right\} \quad (26.2.13)$$

with $d\xi = dz\, dt$.

This gives the probability of ending at $\varphi(\xi)$, given that the system started at $\varphi(\xi')$.

26.2.6.3 EVALUATION of Green's Function:

The Green's Function is calculated from the generating functional W[J] by functional differentiation according to the relation

$$G(\xi,\xi') = (\hbar/i)^2 \delta^2 W[J]/[\delta J(\xi)\, \delta J(\xi')]_{J=0} \quad (26.2.13)$$

because the functional differential operators bring down the functions φ that multiply J:
Using (26.2.12),

$$\delta^2 W[J]/[\delta J(\xi) \; \delta J(\xi')]$$

$$= \delta^2/[\delta J(\xi) \; \delta J(\xi')] \int D\varphi \exp\left\{(i/\hbar)\int_{-\infty}^{+\infty}[L_d(\varphi, \partial_\mu \varphi) + J\varphi]d\xi\right\}$$

$$=(i/\hbar)^2 \int D \; \varphi(\xi)\varphi(\xi')\exp\left\{(i/\hbar)\int_{-\infty}^{+\infty}[L_d(\varphi, \partial_\mu \varphi) + J\varphi]d\xi\right\}. \quad (26.2.14)$$

There are really not just one Green's Function but a whole family, whose members are denoted $G^{(n)}$; the nth member of this family is the nth order functional derivative of W[J] with respect to $J(\xi)$ evaluated at J = 0, and what was specified as G earlier is really $G^{(2)}$. Again these family members are like the various moments calculated from a probability-generating function.

26.2.6.4 CONCEPT OF A FUNCTIONAL:

A functional y takes a function $f(x)$ as an argument, i.e. $y = g[f(x)]$. as does a function of a function, viz. $y = g(f(x))$ and assigns it to a number. However, a function of a function just looks at the value of the argument function, but a functional looks at the entire behaviour of its argument function.

Eg.: if $f(x) = x^2$, and $\quad g(x) = e^{-x}$; then

$$y = g(f(x)) = e^{-f(x)} = \exp(-x^2),$$

is a usual function of a function to which, for example, the chain rule could be applied for taking derivatives. For each value of x, we would then assign a value to y by evaluating $y = g(f(x))$ at x. The functional $y = g[f(x)]$ (note that square brackets are used to make it distinct from a usual function of a function) would also have a value that depends on the function $f(x)$. But now y would depend not on the value of $f(x)$ at a particular point x, but on the entire path of $f(x)$ as x varies over some specified range (say, from 0 to 1). i.e., in the example above, if $f(x) = x^2$, and $g(x) = e^{-x}$; then the value of the functional $g[f]$ is defined as

$$y = g[f(x)] = \int_0^{+1} f(x)dx = \int_0^{+1} x^2 dx$$

Here y depends on the entire behaviour of $f(x)$ over the range $x \in [0,1]$.

Note that if g were defined as

$$g[f\{x\}, w] = \int_0^w f(x)dx$$

(note the upper limit of the integration), g would be a functional of f and a function of w.

<u>Functionals</u> can be thought of as extensions of functions of many variables. They are used in a variety of areas of modern physics; for example they play a central role in the Feynman treatment of Quantum Mechanics, presented in this Chapter.

26.2.6.5. The RELATION BETWEEN the GENERATING FUNCTIONAL, W[J] and the PROPAGATOR:

At this point a proper treatment of field theory (followed by C.F. Stevens) is an examination of the meaning of the Green's function and how this function might be explicitly calculated. In order to simplify the quantum mechanical GENERATING FUNCTIONAL, it is usual to express W[J] as a product of two factors. *viz.* W[0] and an exponential function that involves the driving function J. In other words, the result is *the relation between the Generating Functional and the Propagator.*

$$W[J] = W[0] \exp\left\{(i/2\hbar) \int\int d\xi \, d\xi' J(\xi) \Delta(\xi, \xi') J(\xi')\right\} \quad (26.2.14)$$

26.2.7 The FEYNMAN PROPAGATOR, $\Delta_F(\xi, \xi')$:

W[J] may be expressed as a product of two factors. *viz.* W[0] and an exponential function that involves the driving function J., as in equation (26.2.14),

$$W[J] = W[0] \exp\left\{(i/2\hbar) \int\int d\xi \, d\xi' J(\xi) \Delta(\xi, \xi') J(\xi')\right\}$$

W[0] = the amplitude of the vacuum-vacuum transition, $\langle 0| \Rightarrow |0\rangle$ in the absence of driving, and is given by

$$W[0] = \int D\varphi \exp(i/\hbar) \int d\xi \, L_d \quad (26.2.15)$$

The function, $\Delta_F(\xi, \xi') = $ *Feynman Propagator*.

This expression is crucial because the generating functional, the object needed to relate theory to experimental measurements, an be calculated from the constant W[0] = 1 (because the system starting in the VacuumState surely ends there without driving) and an expression that depends only on the driving function. The Green's function $G \equiv G^{(2)}$ is seen (26.2.13) to be just

$$G(\xi, \xi') = (\hbar/i)^2 \delta^2 W[J]/[\delta J(\xi) \, \delta J(\xi')]_{J=0}.$$

$$= (i/\hbar)^2 \delta^2/[\delta J(\xi) \, \delta J(\xi')] \exp\left\{(i/\hbar) \int\int d\xi \, d\xi' J(\xi) \Delta(\xi, \xi') J(\xi')\right\}_{J=0} \quad (26.2.16)$$

$$= (\hbar/i)\Delta_F(\xi,\xi'). \tag{26.2.17}$$

26.2.7.1 DEFINITION OF the FEYNMAN PROPAGATOR:

It is defined to be the negative of the impulse response of the EULER-LAGRANGE Equation that corresponds to the Lagrangian, L.

The Lagrangian for the 'EM' Field gives the wave equation as its related Euler-Lagrange equation.

What is the propagator associated with this wave equation and what is the physical interpretation? To find the impulse response for the wave equation operator, start by applying the Fourier Transform, defined as

$$FT[\Delta_F(z,t)] = \int_{-\infty}^{+\infty} dz \int_{-\infty}^{+\infty} dt\, \Delta_F(z,t) e^{-i\{(pz/\hbar)+\omega t\}} \equiv \Delta'_F(\omega,p) \tag{26.2.18}$$

to the wave equation with δ-function driving:

$$(1/c^2)[\partial^2 \Delta_F/\partial t^2] - [\partial^2 \Delta_F/\partial z^2] = \delta(z).\delta(t), \tag{26.2.19}$$

where $\Delta_F(z,t)$ = the required impulse response (Propagator).

The Propagator, $\Delta_F(z,t)$ summarizes the Quantum Mechanics of the system.

26.2.7.2. PROPAGATOR THEORY:

[A common type of calculation is that of a scattering cross section for a particular process. In the usual formulation of Quantum Mechanics the quantities q and p are replaced by the operators q and p, which obey Heisenberg Commutation Relations].

The Fourier transformed Green Function $\Delta_F(\omega, p)$ depends on frequency ω and momentum, p. Because

$$FT[(\partial^2 \Delta_F/\partial t^2)] = (-\omega^2 \Delta'_F)$$

and $\quad FT[(\partial^2 \Delta_F/\partial z^2)] = -(p/\hbar)^2 \Delta'_F, \tag{26.2.20}$

The equation for Δ'_F becomes

$$\{(-\omega^2/c^2) + (p/\hbar)^2\}\Delta'_F(\omega,p)$$
$$= \{(-\omega^2 \hbar^2 + p^2 c^2)/c^2 \hbar^2\}\Delta'_F(\omega,p) = 1 \tag{26.2.21}$$

Or, solving for

$$\Delta'_F \quad \Delta'_F(\omega,p) = -c^2\hbar^2/(\omega^2\hbar^2 - p^2c^2) = -c^2\hbar^2/(\omega\hbar - pc)(\omega\hbar + pc). \tag{26.2.22}$$

This is the impulse response for the wave equation operator in momentum / frequency space. The Propagator in position / time space is given by the inverse Fourier Transforms

$$\Delta_F(z,t) = (1/2\pi)^2 \hbar \int_{-\infty}^{+\infty} dp\, e^{i(pz/\hbar)} \int_{-\infty}^{+\infty} d\omega\, e^{i\omega t} \{-c^2\hbar^2/(\omega\hbar - pc)(\omega\hbar + pc)\}$$

$$= c^2\hbar^2 (1/2\pi)^2 \hbar \int_{-\infty}^{+\infty} dp\, e^{i(pz/\hbar)} \int_{-\infty}^{+\infty} d\omega\, e^{i\omega t} \{-1/(\omega\hbar - pc)(\omega\hbar + pc)\} \quad (26.2.23)$$

To perform this inverse FT, consider first the integral over ω,

viz.,
$$(1/2\pi)\int_{-\infty}^{+\infty} e^{i\omega t}\{1/(\omega\hbar - pc)(\omega\hbar + pc)\}d\omega. \quad (26.2.24)$$

This integration requires that the poles on the –axis is properly handled; the treatment of these poles is equivalent to deciding on the boundary conditions for the equation. The idea is to place the poles just above and just below the ω-axis so that they make a contribution specified by the residue theorem when a closed integration path that goes through $\pm i\omega$ is selected. This is a standard procedure to insert initial conditions for this situation, and the final solution depends on just how the poles are treated. When the pole is slightly displaced from the ω-axis and the FT is changed into a contour integral by extending ω into the complex plane, one gets

$$(1/2\pi)\int_{-\infty}^{+\infty} e^{i\omega t}\{1/(\omega\hbar - pc)(\omega\hbar + pc)\}d\omega$$

$$= (1/2\pi) \oint e^{i\omega t}\{1/(\omega\hbar - pc)(\omega\hbar + pc)\}d\omega. \quad (26.2.25)$$

because, if the argument, for example, of the exponential is negative, one closes the contour through $+i\infty$. To simplify thus, let the positive pole (the one at $\omega\hbar$) is placed on the negative side of the ω-axis, and the *vice-versa*. When the path for $t > 0$ is closed through $+i\omega$, only the pole at $-\omega\hbar$ contributes ($\omega = -cp/\hbar$), so the residue is $-2\pi i\, e^{-ipz/\hbar}/2cp$, and the integral becomes

$$(1/2\pi) \oint e^{i\omega t}\{1/(\omega\hbar - pc)(\omega\hbar + pc)\}d\omega = -2\pi i\, e^{-ipz/\hbar}/2cp, (26.2.26)$$

Now perform this result to the FT over momentum for $t > 0$ to give the required impulse response.

$$\Delta_F(z,t) = (ic/4\pi)\int_{-\epsilon}^{+\infty} dp\, e^{i(z-ct)p/\hbar}$$

$$G(\xi,\xi') = (\hbar/i)^2 \delta^2 W[J]/[\delta J(\xi)\, \delta J(\xi')]_{J=0}$$

$$= (-c\hbar/2)\, \theta(z-ct) \quad (26.2.27)$$

where $\theta(z-ct)$ = unit step function.

It is defined as
$$\theta(z-ct) = (-1/2\pi)\int_{-\epsilon}^{+\infty} dk\, \{e^{iky}/k\} \quad (26.2.28)$$

A perusal of the derivative of this function

$$\frac{d}{dy}\left[(-1/2\pi)\int_{-\epsilon}^{+\infty} dk \{e^{iky}/k\}\right] = (1/2\pi)\int_{-\epsilon}^{+\infty} dk\, e^{iky}$$

$$= \delta(y), \text{ the Dirac delta function.}$$

The EM Field Propagator is therefore

$$\Delta_F(z,t;z',t') = (-c\hbar/2)\theta\{(z-z')-c(t-t')\} \tag{26.2.29}$$

This event, i.e., the traveling edge of the step function, moves from z' at t' to z at t with the velocity of light, c. Further the energy associated with the transmitted wave is $\omega\hbar$.

Worked out Example 26.1.

Show that the Propagator $\Delta_F(q_f,t_f;q_i,t_i) = \langle q_f,t_f \| q_i,t_i \rangle$.

Solution:

Given the wave function $|q_i,t_i\rangle$ at time t_i, the Propagator, $\Delta_F(q_f,t_f;q_i,t_i)$ permits one to know the motion of a particle in space-time. The Propagator gives the corresponding wave function at a later space-time q_f, t_f. The Propagator is given by

$$|q_f,t_f\rangle = \int_{q_i}^{t_i} \Delta_F(q_f,t_f;q_i,t_i)dq_i$$

It is known that $|q,t\rangle = e^{-iH\,t/\hbar}|q\rangle_H$

is the relation between a state in the S- and H- Pictures.

Define moving frame $|q,t\rangle = e^{-iH\,t/\hbar}|q\rangle$.

$$|q,t\rangle = \langle q,t \| q\rangle_H$$

$$\langle q_f,t_f \| q\rangle = \langle q_f,t_f \| q_i,t_i\rangle\langle q_i,t_i \| q\rangle dq_i$$

$$|q_f,t_f\rangle = \int \langle q_f,t_f \| q_i,t_i\rangle |q_i,t_i\rangle dq_i,$$

$$- \int_{q_i}^{t_i} \Delta_F(q_f,t_f;q_i,t_i)dq_i \quad \Delta_F(q_f,t_f;q_i,t_i)dq_i \quad -\langle q_f,t_f \| q_i,t_i\rangle.$$

26.2.7.3. The TRVALLING EDGE OF THE STEP FUNCTION moves from z' at t' to z at t, why?
Unambiguous identification of the object as the PHOTON of the scalar "EM" Field:

For this it is necessary to find the momentum of the wave; the energy E of a mass less, free-moving (non-interacting) particle is given by $E = cp$, a relation that has origin from the expression (25.2.3), viz., $E^2 = [m_0^2 c^4 + p^2 c^2]$, with mass $m_0 = 0$. It is to be kept in mind that the traveling wave must have no mass, because it moves at c, and a massive object cannot move with c.

It is to be remembered that starting with an expression for the Green's function in momentum/frequency space and then used inverse transforms to find the position-time version. The original momentum/frequency equation contained poles that determined the initial conditions of the wave, and the positive energy pole dictates that

$$E = \omega \hbar = cp.$$

Therefore the wave discussed here constitutes an object that propagates and has a quantized energy $E = \hbar \omega$. Because the Propagator describes the movement at the speed of light, c of an object with energy $E \propto \omega$, its frequency, one identifies unambiguously the object as the 'photon' of the scalar "EM" Field.

26.2.7.4 SUMMARY of what was done in this section to show that the quantal nature of the EM Field arises naturally:

(1) First the started with the Lagrangian for a classical "EM" Field.

(2) Next used the KEY PATH INTEGRAL to find the transition amplitude for an "EM" Field going from the lowest energy Vacuum State at $t = -\infty$. To the same state after a long time ($t = \infty$).

(3) Calculated the transition amplitude

(4) This transition amplitude turned out to provide one with the Green's Function (or equivalently, the Propagator, Δ_F) that is the amplitude for going from one field state to another.

(5) When the Propagator is found, a quantum of energy that propagates with the speed of light, a "photon", emerges. Thus the quantal nature of the EM Field arises naturally from the same sort of process that earlier led to the quantum mechanical nature of particle.

26.2.7.5 To COMBINE QUANTUM MECHANICS OF THE FIELD with that of PARTICLES:

The next step is to combine the quantum mechanics of the field with that of particles. Before that it is necessary to examine further the significance of the Propagator, Δ_F.

The 4th order Green's Function is obtained from the Generating Functional; and it is

$$G(\xi_1, \xi_2, \xi_1', \xi_2') = (\hbar/i)^4 \delta^4 W[J]/[\delta J(\xi_1) \, \delta J(\xi_2), \, \delta J(\xi_1') \delta J(\xi_2')]_{J=0} \quad (26.2.30)$$

If the system Consists of particles, this function specifies the amplitude for particle 1 to go from $\xi_1' \to \xi_1$ and for particle 2 to go from $\xi_2' \to \xi_2$. The 6th order Green's Function would, in turn, describe the amplitude for 3 particles, *etc.* Thus the Generating Functional describes the movement of any number of particles whenever the Lagrangian can be given a particle interpretation.

Because of this, all of the above orders of Green's Functions can be calculated from the Propagator, Δ_F. Thus the entire description of a quantum system with any number of particles is in the hand once the Propagator is known. This shows that the generating functional describes the movement of any number of particles whenever the Lagrangian can be given a particle interpretation. Because of the relation between the Generating Functional and the Propagator, displayed above, all of the various orders of Green's Functions can be calculated from the Propagator, Δ_F. This to say that the entire description of a quantum system with any number of particles is in hand once the Propagator is known.

26.3. SECOND QUANTIZATION:

The term 'Second Quantization' is an unfortunate one, in so far as it suggests a theory that is 'twice as quantum mechanical' as the one was started with in any book on introductory Quantum Mechanics. This is emphatically not the case: all that is done is to develop a convenient mathematical technique for dealing with the original theory. The origin of the term briefly is this. Addition or subtraction of particles to or from the system is represented by creation and annihilation operators, which are closely analogous to the raising and lowering operators of the Harmonic Oscillator. From these one can construct FIELD OPERATORS, which in the absence of interaction, satisfy the same Schrödinger Equation as single particle wave function. By turning a wave function, which is acted on by operators representing physical quantities, into an operator which itself act on state vectors, one might appear to be adding a further layer of quantumness!.

26.3.1 DISCREPANCIES OF THE TRADITIONAL QUANTUM MECHANICS:

The traditional formulation of Quantum Mechanics suffered from two problems:

(1) The Schrödinger Equation is not covariant; a Relativistic Quantum Mechanics is required for rapidly moving particles, and

(2) The traditional QM did not easily describe systems in which the number of particles is changing.

The solution to the first discrepancy is to develop a covariant Lagrangian to replace the one that leads to the Schrödinger Equation. This aspect has been discussed in Section 26.2, where the treatment of the "EM" Field in which the Lagrangian is covariant. The second problem of describing systems in which the particle number is changing as the system evolves; it is simpler to deal with the extension of the traditional non-relativistic Quantum Mechanics.

26.3.2 PROBLEM OF VARYING NUMBERS OF PARTICLES:

It is known from the earlier section that the Generating Functional used above naturally dealt with this problem. The various orders of Green's Functions can all be calculated from just the Propagator, enabling description of systems of varying numbers of particles.

The key is this observation: the Propagator calculated from the Euler-Lagrange Equation describes the amplitude of a particle to go from $z',t' \to z,t$, and this is what the transition amplitude, $K(z,t;z',t')$ does. Further, $K(z,t;z',t')$ is just the Green's Function for the Schrödinger Equation as its Euler-Lagrange equation.

The sequence of steps is:

(i) This Lagrangian is denoted as L_S. It is this L_S finds use in the KEY PATH INTEGRAL.

(ii) The resulting formulation will express the system's behavior in terms of the Propagator that is the transition amplitude in traditional case, for single particle.

(iii) Inclusion of multi-particle numbers is possible since higher order Green's Functions, can be easily calculated from just the Propagator.

26.3.3 To FIND THE LAGRANGIAN APPROPRIATE:

Recall that the Schrödinger Equation for the system with single spatial dimension (1-D) is

$$i\hbar(\partial/\partial t)\psi(z,t) = \{(-i\hbar\nabla)^2/2m\}\psi(z,t) + V(z)\psi(z,t) \qquad (26.3.1)$$

Thus in the Lagrangian Density, L_d, one needs terms that include $\psi(r, t)$ and $\psi^*(r, t)$ in a symmetric way and that through the general Euler-Lagrange Equation,

$$(\partial L_S/\partial \phi) - (\partial/\partial t)(\partial L_S/\partial_t \phi) - (\partial/\partial z)(\partial L_S/\partial_z \phi) = 0. \qquad (26.2.9)$$

will yield each of the terms in Schrödinger Equation. One finds the Lagrangian by examining three terms that, it will be apparent, are the right ones to use for the construction of the Lagrangian. First, suppose that L_S contains a term like $V\psi^*\psi$, and operate on this term with $(\partial \psi^*/\partial)$:

$$(\partial \psi^*/\partial)V\psi^*\psi = V(z)\psi \qquad (26.3.2)$$

This operation gives the potential energy term in the Schrödinger Equation.

Next, consider

$$(\partial/\partial z)(\partial/\partial \psi^*)[(\hbar^2/2m)(\partial_z\psi^*\partial_z\psi)] = (\hbar^2/2m)(\partial^2/\partial z^2)\psi \qquad (26.3.3)$$

which gives the spatial part of the Schrödinger Equation.

Thus the Lagrangian Density,

$$L_d = (-i/\hbar)(\psi^*\partial_t\psi - \psi\partial_t\psi^*) - [(\hbar^2/2m)(\partial_z\psi^*\partial_z\psi)] + V\psi^*\psi \qquad (26.3.4)$$

has Schrödinger Equation (and its complex conjugate, found by using ψ rather than ψ^* in taking derivatives of L_d) as its Euler-Lagrange Equation. The GENERATING FUNCTIONAL,

26.3.4. The GENERATING FUNCTIONAL:

$$W[J, J^*] = \iint D\psi \, D\psi^* \exp(i/\hbar) \iint [L_d + J^*\psi + J\psi^*] dz dt \qquad (26.3.5)$$

is associated with the Propagator, $\Delta_F(z,t;z',t') = K(z,t;z',t')$ used to describe the motion of a single particle in traditional Quantum Mechanics, and the Propagator is just the solution for Schrödinger Equation with delta function driving. Systems with various particle numbers are described by constructing the appropriate-order Green's Functions from this Propagator, as described above.

The procedure just described is usually called *Second Quantization*.

Now the first quantized result is contained within the second quantized formulation because the second quantized Propagator is identical to the first-quantized transition amplitude. But this new formulation extends the theory to systems with variable numbers of particles. All of this worked because of the particular interrelation of Lagrangians, Euler-Lagrange Equations, Green's Functions of various orders, and the propagator that appears in the Generating Functional orders

26.3.5 DIFFERENCE BETWEEN FIRST AND SECOND QUANTIZATION:

First quantization was the subject of traditional Quantum Mechanics. It provides a quantum mechanical description of a single particle's behavior by using the classical Lagrangian in the KEY PATH INTEGRAL. The result of this is a field $\psi(z,t)$ that specifies the particle probabilistically. It promotes classical variables, like position and momentum, to the status of operators that act on Hilbert Space vectors.

For second quantization, a particular Lagrangian L_d is found that replaces the position and its derivatives in the Lagrangian with the first quantized field ψ and its derivatives. In Second Quantization, wave functions are promoted to operator status.

z and its derivatives appear in the Lagrangian of first quantized system, whereas ψ and its derivatives appear in the second quanized Lagrangian.

Now the first quantized result is contained within the second quantized formulation because the second quantized Propagator is identical to the first-quantized transition amplitude.

This new formulation extends the theory to systems with variable numbers of particles.

26.3.5.1. ALTERNATIVE PATH to Second Quantization:

This is done as follows:
Replace wave functions with operators, and use the operator equation of motion in the H-Picture, as the basis for the description.

26.4. INTERACTING FIELDS:

26.4.1. The possibilities of various processes in collisions (scattering) between particles whose interactions may be regarded as small are calculated by means of Perturbation Theory. In its ordinary form (non-relativistic quantum mechanics), however, the formalism of this theory has the defect of not exhibiting explicitly the conditions of relativistic invariance. If extended to relativistic problems, the final result will satisfy these conditions, but the non-invariant form of the intermediates expressions considerably complicates the calculations. A consistent relativistic (invariant) Perturbation Theory was established by R.P. Feynman in 1948-49.

The preceding section has focused on free fields, but of course, the most interesting problems relate to interactions. This section extends method as to how interactions are included in the theory.

26.4.2 HOW TO INCLUDE INTERACTIONS in the Theory?

The basic idea is very easy: replace the free-field Lagrangian in the KEY-PATH INTEGRAL with one that includes the interactions.

Example, the "electromagnetic" (EM)- Field Lagrangian, $L_S(\varphi')$ is

$$L_S(\varphi') = (1/c^2)(\partial_t \varphi)^2 - (\partial_z \varphi)^2, \qquad (26.4.1)$$

and the Schrödinger Lagrangian, $\mathcal{L}_S$ is

$$L_S = (-i/\hbar)(\psi^*\partial_t\psi - \psi\partial_t\psi^*) - [(\hbar^2/2m)(\partial_t\psi^*\partial_t\psi)] + V\psi^*\psi \qquad (26.3.4)$$

So the total Lagrangian, $\mathcal{L}_T$ is

$$L_T = L_S + L_d + \alpha\psi^*\varphi\psi \qquad (26.4.2)$$

Here the Lagrangian density for the interaction has been taken as having the very simple form, $\alpha\psi^*\varphi\psi$, where α = the coupling constant that characterizes the strength of the interaction. Clearly this interaction term couples the two fields, because the changes in one will influence the other, since the interaction term alters the action. In general, to find the right expression for the interaction Lagrangian is where the physics is. As before,

$$W[J, J^*, F] = \iiint D\psi \, D\psi^* D\varphi \exp(i/\hbar) \iint [L_S + L_d + \alpha\psi^*\varphi\psi + J^*\psi + J\psi^* + F\varphi] dz dt$$
(26.4.3)

Again as before, the Green's Functions of all orders are calculated by taking the functional derivatives of W with respect to the forcing functions; these Green's functions describe the system.

The relation between the Generating Functional and the Propagator no longer holds for this situation. In order to surmount the difficulties, define the free-field Generating Functional as

$$W[J, J^*, F] = \iiint D\psi \, D\psi^* D\varphi \exp(i/\hbar) \iint [L_S + L_d + J^*\psi + J\psi^* + F\varphi] dz dt \quad (26.4.4)$$

Initially, assume that λ is so small that the field interaction can be approximated by

$$\exp(i/\hbar) \iint [\alpha\psi^*\varphi\psi] dz dt \approx 1 + (\alpha i/\hbar) \iint [\psi^*\varphi\psi] dz dt \quad (26.4.5)$$

This means that W can be written approximately as

$$= \iiint D\psi \, D\psi^* D\varphi \exp(i/\hbar) \iint [L_S + L_d + J^*\psi + J\psi^* + F\varphi] dz dt \, .$$

$$. \exp\left\{(\alpha i/\hbar) \iint [\psi^*\varphi\psi] dz dt\right\} . \quad (26.4.6)$$

$$= \iiint D\psi \, D\psi^* D\varphi \exp(i/\hbar) \iint [L_S + L_d + J^*\psi + J\psi^* + F\varphi] dz dt \, .$$

$$. \left[1 + \left\{(\alpha i/\hbar) \iint [\psi^*\varphi\psi] dz dt\right\}\right] \quad (26.4.7)$$

$$W_0 + (\alpha i/\hbar) \iiint D\psi \, D\psi^* D\varphi \exp(i/\hbar) \iint [L_S + L_d + J^*\psi + J\psi^* + F\varphi] dz dt \, t \quad (26.4.8)$$

For electrodynamics (with real EM Field, and a relativistic electron described by the Dirac Equation), α = 1/137, so only a few terms in the expansion give a very accurate approximation.

26.4.3.   ANTI - PARTICLES:

Time and space appeared symmetrically in the Lagrangian, in the case of the EM Field. Therefore the Lagrangian is relativistically invariant. The symmetry produced two poles in the momentum / frequency-space representation of the Propagator, $G(\omega, p)$. One of these poles was responsible for the space-time Propagator, $\Delta_F(z, t)$ and arose out of the Fourier Transformation of the wave equation. This is when $t > 0$. The $t < 0$ solution is the same except that it corresponds to the 'photon' that propagates backward in time, and particles with negative energy. This is interpreted first by Feynman as a positive-energy 'antiparticle' propagating forward in time.

The prediction of the anti-particle, called positron, was the greatest triumph of the Dirac Relativistic Theory of the Electron.

26.4.3.1 What is the theoretical basis behind the prediction that
'all particles have their anti-particle counterparts'?

The number and nature of particles in the sub-atomic world is not fixed. Sub-atomic particles are created and destroyed. Therefore the number of variables to be included in the wave function is determined by the number of particles present. Traditional Quantum Mechanics is suited for this. An extended Quantum Mechanics must, therefore, include classical fields, generalize using relativity and take into account the creation and annihilation of particles. This leads to the finding a Lagrangian that gives Schrödinger Equation as its Euler-Lagrange Equation The Lagrangian must depend on $\psi(z,t)$ and on the derivatives of this function. The Lagrangian must be covariant, so the space-time must appear symmetrically in it. This symmetry produces two poles, in the momentum/ frequency-space representation of the Propagator, $G(\omega,p)$; one for the particle and the for the anti-particle. Because of the symmetry introduced by the requirement that LAGRANGIANS FOR THE FIELD BE RELATIVISTICALLY COVARIANT, all particles have their ant-particle counterparts. The construction of satisfactory Lagrangians is at the core of Field Theory, and it turns on exploiting various SYMMETRIES.

26.5. THE QUANTUM FIELD:

26.5.1. CREATION AND DESTRUCTION OPERATORS, $\hat{c}_s^\dagger$, $\hat{c}_s$:

The mathematical statement of this problem (of a physical system that consists of a VARIABLE NUMBER OF charged particles) requires introduction of operators of a new kind, called creation / annihilation operators, $\hat{a}^\dagger$ and $\hat{a}$, discussed in detail in Chapter 7. These operators of a new kind are a formal generalization of the one-particle wave function.

It was seen earlier that to quantize the matter field – Second Quantization, of the Schrödinger wave field (*i.e..* single particle, ψ). The many particle system is analogous to a "system of photons". Photons are quanta of EM (Maxwell) Field; the overall behavior of a "photon system" is characterized by the EM wave equation. Second quantization of these equations brings out the corpuscular aspect of photons. The Bose operators, symbolized by $\hat{b}^\dagger$ and $\hat{b}$, instead of $\hat{a}^\dagger$ and $\hat{a}$, are utilized. Likewise, the overall behavior of particle systems is described by the matter field equation. The quantal properties are brought out by means of 2nd quantization. In the case of fermions, the creation and annihilation operators are symbolized as $\hat{c}_s^\dagger$ and $\hat{c}_s$, to distinguish them from the boson operators. Subscript symbol r is used to indicate the one-particle state to which the operators apply. $\hat{c}_s^{\dagger\dagger}$ and $\hat{c}_s$, obey the Anti-commuting Relations:

$$[\hat{c}_r, \hat{c}_s^{\dagger}] = \delta_{rs} . \quad (26.5.1)$$

As in the case of bosons, a VACUUM STATE is defined as $|0\rangle$, in which no particles are present. Thus particles are 'excitation quanta' and the vacuum is the ground state of a field system.

The fact that particles interact through EM Field, *i.e.* quantization of the EM Field results (the theory is QED) in the observation of individual photons, eg. the photoelectric effect; and Einstein's Equations for the gravitational field predict gravitational radiation, and a possibility of observing individual '*gravitons*'; and the interaction field between quarks is the strong interaction and the quanta are the '*gluons*'; (and the dynamics of quark-gluon system is QCD). The weak field acquires quanta – the W and Z bosons.

26.5.2. QUANTUM THEORY of the MASSIVE K.G. Field:
26.5.2.1. The Massive K.G. Field (Spin-0 Particles):

Now it is quite understood from the previous sections, how Quantum Theory is married to Field Theory.

The K.G. Equation is

$$(\Box + m_0 c^2 / \hbar^2)\psi(z,t) = 0 \quad . \quad (25.2.13)$$

The classical Euler-Lagrange Equation is given by

$$(\partial L_S / \partial \phi) - (\partial / \partial t)(\partial L_S / \partial_t \phi) - \hat{\nabla}_z (\partial L_S / \hat{\nabla}_z \phi) = 0 . \quad (26.2.9)$$

This in the relativistic invariant form is

$$(\partial L_S / \partial \phi) - \partial_\mu (\partial L_S / \partial_\mu \phi) = 0 . \quad (26.3.19)$$

with $\partial_\mu = (\partial / \partial z^\mu)$

The action S for mechanical field, ϕ is

$$S[\varphi(z,t)] = \iint d^4 z \, L_S(\phi') , \quad (26.2.7)$$

where ϕ' = argument of L_S indicates both the temporal and spatial derivatives of ϕ; which is relativistically invariant if L_S = invariant, which means the L_S describes a mass less particle. Then it is given by the expression

L_S = invariant, $L_S = \frac{1}{2}\partial_\mu \varphi\, \partial^\mu \varphi - \frac{1}{2}m_0^2 \varphi^2$. (26.5.2)

The plane wave solutions of the K.G. Equation (the field equation) have frequencies

$$\omega^2 = k^2 + m_0^2.$$ (26.5.3)

This is the correct energy-momentum relation for a massive particle.
To quantize this field,

$$\varphi(z,t) = \int_{-\infty}^{+\infty} \{(1/2\pi)^3 2\omega\}\, d^3k [\hat{a}(k)e^{-ikz} + \hat{a}^\dagger(k)e^{ikz}].$$ (26.5.4)

and conjugate momentum

$$\pi(z,t) = \int_{-\infty}^{+\infty} \{(1/2\pi)^3 2\omega\}\, d^3k(-i\omega)[\hat{a}(k)e^{-ikz} - \hat{a}^\dagger(k)e^{ikz}]_+ \quad .(26.5.5)$$

$$H_{KG} = \int_{-\infty}^{+\infty} d^3k \frac{1}{2}[\pi^2 + \nabla_z\varphi.\nabla_z\varphi + m_0^2 \varphi^2]$$ (26.5.6)

that $\quad [\hat{a}(k), \hat{a}^\dagger(k)] = (2\pi)^3 2\omega\, \delta^3(k\text{-}k')$.

$$H_{KG} = \int_{-\infty}^{+\infty} \{d^3k/(2\pi)^3 2\omega\}\frac{1}{2}[\hat{a}(k)\,\hat{a}^\dagger(k) + \hat{a}^\dagger(k)\,\hat{a}(k)]\omega \ .$$

Dropping he infinite zero-point energy, one gets

$$H_{KG} = \int_{-\infty}^{+\infty} \{d^3k/(2\pi)^3 2\omega\}\, \hat{a}^\dagger(k)\,\hat{a}(k)\omega$$ (26.5.7)

where $\hat{a}^\dagger(k)$ and $\hat{a}(k)$ stand for the creation and destruction operators.

Since $\varphi = \varphi^\dagger$ = a real scalar field, it corresponds to a unique particle state of a given mass, m_0.
But for a given mass there are two distinct states, *viz.* 'particle' and its 'anti-particle' states.
Let φ_1 and φ_2 denote two real scalar fields having distinct types of mode operators, whose

$$L_S = \frac{1}{2}\partial_\mu \varphi_1\, \partial^\mu \varphi_1 - \frac{1}{2}m_0^2 \varphi_1^2 + \frac{1}{2}\partial_\mu \varphi_2\, \partial^\mu \varphi_2 - \frac{1}{2}m_0^2 \varphi_2^2.$$ (26.5.8)

= Invariant, if φ_1 and φ_2 are replaced by φ'_1 and φ'_2.

$\varphi'_1 = (Cos\ \alpha)\varphi_1 - (Sin\ \alpha)\varphi_1$,

$\varphi'_2 = (Sin\ \alpha)\varphi_2 + (Cos\ \alpha)\varphi_2$, (26.5.9)

This is like a rotation of coordinates about the z-axis, *i.e.* this is a "O(2) Transformation", *i.e.* the 2-D rotation group, O(2).

$L_S(\varphi_1,\varphi_2) = L_S(\varphi'_1,\varphi'_2)$ under O(2).

26.5.2.2. SYMMETRY CURRENT, j^{μ}.

If ε = infinitesimal rotation,

$$\varphi'_1 = \varphi_1 - \varepsilon\, \varphi_2\,;\ \varphi'_2 = \varepsilon\, \varphi_1 + \varphi_2, \tag{26.5.10}$$

It can be shown that

$$\partial^{\mu}[-(\partial L_S/\partial_{\mu}\varphi_1)\varepsilon\, \varphi_2 - (\partial L_S/\partial_{\mu}\varphi_2)\varepsilon\, \varphi_1] = 0, \tag{26.5.11}$$

giving $\quad \mathbb{N}^{\mu} = \varphi_1 \partial^{\mu}\varphi_2 - \varphi_2 \partial^{\mu}\varphi_1$ = conserved. $\tag{26.5.12}$

So $\quad \partial_{\mu}\mathbb{N}^{\mu} = 0$. $\tag{26.5.13}$

Such conserved 4-vector operators are called 'SYMMETRY CURRENTS', denoted by J^{μ}.

26.5.2.3. NOETHER'S THEOREM:

According to Emmy Noether (1918) the law of the conservation of momentum implies invariance of the laws of physics to a spatial displacement. It states that if $\mathcal{L}$ = invariant, under a continuous Transformation then there will be an associated symmetry current.

i.e., $\quad \partial_{\mu}\mathbb{N}^{\mu} = 0$, means

$$\partial \mathbb{N}^0/\partial t + \nabla \mathbb{N} = 0. \tag{26.5.14}$$

$$(d/dt)\int_{-\infty}^{+\infty} \mathbb{N}^0.d^3z + \int_{-\infty}^{+\infty} \mathbb{N}.dS = 0 \tag{26.5.15}$$

The 2nd term vanishes since the fields die off sufficiently fast at infinity.

$$\mathbb{N} = \int_{-\infty}^{+\infty} \mathbb{N}^0 d\tau = \text{a constant in time.} \tag{26.5.16}$$

If one considers a complex K.G. Field (charged scalar field), it can be shown that $\mathbb{N}$ would distinguish the 'particle' from its 'anti-particle'

26.5.3. The SPINOR FIELD (Dirac Field): (Fermionic Fields and the Spin statistics connection):

26.5.3.1 FUNDAMENTAL FIELDS:

The fundamental Fields in nature are of two kinds:

(1) SPINOR FIELDS

Examples of matter fields of this type are LEPTONS AND QUARKS.

(2) GAUGE FIELDS.

EM Field quanta, WEAK particles, GLUONS are examples to this type of fields.

26.5.3.2 SPINOR FIELD QUANTIZATION:

Having shown how to quantize massive K.G Field, I turn now to take the case of the Spinor Field.

Using the Lagrangian approach, one may vary $\psi^\dagger$ in the Action Principle with

$$L_D = [i\psi^\dagger(\partial/\partial t) + i\psi^\dagger\alpha.\nabla\psi - m_0\psi^\dagger\beta\psi] \qquad (26.5.17)$$

The classical Euler-Lagrange Equation is given by

$$(\partial L_D/\partial\psi) - (\partial/\partial t)(\partial L_D/\partial_t\psi) - \hat{\nabla}_z(\partial L_D/\hat{\nabla}_z\psi) = 0 \qquad (26.2.9)$$

This in the relativistic invariant form is

$$(\partial L_D/\partial\psi) - \partial_\mu(\partial L_D/\partial_\mu\psi) = 0 \qquad (26.3.19)$$

with $\partial_\mu = \partial/\partial z^\mu$.

$$\partial L_D = (\partial L_D/\partial\psi)\delta\psi + \partial_\mu(\partial L_D/\partial_\mu\psi)\delta(\partial_\mu\psi) + (\psi \to \psi^\dagger).$$

On substitution, the Euler-Lagrange Equation becomes the celebrated Dirac Equation

$$(i\gamma^\mu\partial_\mu - m_0)\psi = 0 \qquad (25.4.13)$$

The relativistic invariance of this is more evident in the γ^μ matrix notation (γ^μ transforms) as vector under the Lorentz group).

i.e., by choosing $L_D = i\psi^*(i\gamma^\mu\partial_\mu)\psi - \psi^* m_0\psi$

$$= i\frac{1}{2}\psi^*\gamma^\mu(\partial_\mu\psi) - (\partial_\mu\psi^*)\gamma^\mu\psi + m_0\psi^*\psi \qquad (26.5.18)$$

In this L_D, ψ, ψ^* are treated as dynamically independent fields. So canonical momentum field (momentum canonically conjugate to the Spinor Field)

$$\pi(z) = [\partial L_D/\partial(\partial_t\psi(z)] = i\psi(z) \qquad (26.5.19)$$

$\therefore \quad H_D = \pi\,\partial_t\psi - L_D,$

$$H_D = \psi^\dagger\gamma^0(-i\gamma^i\partial_i + m_0)\psi = \psi^\dagger\gamma^0(-i\gamma^0\partial_D\psi) = \psi^\dagger i(\partial/\partial t)\psi \qquad (26.5.20)$$

So the negative energy difficulty is not removed by treating the Dirac Equation as a field equation as it was with the K.G. Equation. It is only removed on SECOND QUANTIZATION, quantization of the spinor field.

This attempt to quantize ψ and ψ^* is as follows:

Decompose $\psi(z)$ into Fourier moments, i.e. construct a mode expansion in terms of plane wave solutions.

$$\psi(z) = \int\{d^3k/(2\pi)^3 2\omega\}\sum_{s=1,2}[\hat{c}_s(k)u(k,s)e^{-ikz} + \hat{d}_s^\dagger(k)v(k,s)e^{ikz}]$$

(26.5.21)

$$\psi^*(z) = \int \{d^3k/(2\pi)^3 2\omega\} \sum_{s=1,2} [\hat{c}_s(k) u^*(k,s) e^{ikz} + \hat{d}_s^\dagger(k) v^*(k,s) e^{-ikz}] \quad (26.5.22)$$

where $u(k,s)$ and $v(k,s)$ are +ve and −ve energy spinors.

$\hat{c}_s^\dagger(k)$ = creation operator for a Dirac Particle of spin s and momentum, k, and it multiplies the +ve energy states.

$\hat{d}_s(k)$ = creation operator for a Dirac Anti-particle of spin s and momentum, k, and it multiplies the -ve energy states.(or annihilation operator for Dirac Particles having +ve energies).

$$\hat{c}_s(k)|0\rangle = 0, \text{ for all k; and s} = 1,2. \quad (26.5.23)$$

A two-Fermion State is obtained as

$$|k_1,s_1;k_2,s_2\rangle = \hat{c}_s^\dagger(k_1)\hat{c}_s^\dagger(k_2)|0\rangle \quad (26.5.24)$$

To accommodate the anti-symmetric nature of Fermion States under the exchange of $k_1 \leftrightarrow k_2$, and $s_1 \leftrightarrow s_2$, and the Pauli's Exclusion Principle, the anti-commutation relations are used.

$$\{A,B\} \equiv AB+BA \quad (26.5.25)$$

$$\hat{c}_{s1}^\dagger(k_1)\hat{c}_{s2}^\dagger(k_2) + \hat{c}_{s2}^\dagger(k_2)\hat{c}_{s1}^\dagger(k_1) = 0. \quad (26.5.26)$$

$$|k_1,s_1;k_2,s_2\rangle = -|k_2,s_2:k_1,s_1\rangle$$

$$\{\hat{c}_{s1}^\dagger(k_1)\hat{c}_{s2}^\dagger(k_2)\} = \{\hat{d}_{s1}^\dagger(k_1)\hat{d}_{s2}^\dagger(k_2)\} = \{(2\pi)^3 2\omega\} d^3(k_1-k_2)\delta_{s1,s2} \quad (26.5.27)$$

Jordan and Wigner first proposed this.

$$H_D = \int_{-\infty}^{+\infty} \{d^3z \ \psi^\dagger(z)(i\partial/\partial t)\psi(z)\}.$$

L_D = invariant under the U(1) Transformation.

$$\psi \to \psi^* = e^{-i\alpha}\psi \quad (26.5.28)$$

i.e., $\quad \psi \to \psi^* = \psi - i c \psi$.

Noether symmetry current is $\mathbb{N}^\mu = \psi^* \gamma^\mu \psi$.

and the associated symmetry operator is

$$\hat{\mathbb{N}} = \int_{-\infty}^{+\infty} \{d^3z \ \psi^\dagger(z)\psi(z)\} = \text{clearly a kind of number operator for the fermion case.}$$

Using the properties of a Dirac Spinor,

$$\hat{\mathbb{N}} = \int_{-\infty}^{+\infty} \{d^3k/(2\pi)^3 2\omega \sum_s^{1,2} [\hat{c}_s^\dagger(k)\hat{c}_s(k) + \hat{d}_s(k)\hat{d}_s^\dagger(k)] \quad (26.5.29)$$

Dirac Hamiltonian $H_D = \int_{-\infty}^{+\infty} \{d^3z \ \psi^\dagger(z)(i\partial/\partial t)\psi(z)$.

i.e., $H_D = \int_{-\infty}^{+\infty} \{d^3k/(2\pi)^3 2\omega\} \sum_s^{1,2} [\hat{c}_s^\dagger(k)\hat{c}_s(k) + \hat{d}^\dagger(k)\hat{d}_s^\dagger(k)] \ \omega$. (26.5.30)

This has not used the Commutation Relations of $\hat{c}$, $\hat{c}^\dagger$, $\hat{d}$ and $\hat{d}^\dagger$. If one uses them the it will be seen that $\mathcal{H}_D$ would contain the difference between two number operators for 'c' and 'd' particles; and is not +ve definite, as required. Moreover, the 'd' s ought to be the anti-particles of the 'c's, carrying opposite $\hat{N}$ value; but $\hat{N} \propto$ sum of of 'c's and 'd's number operators, counting +1 for each type, which does not fit this interpretation.

If the Anti-commutation Relations are assumed one gets

$$\hat{N} = \int_{-\infty}^{+\infty} \{d^3k/(2\pi)^3 2\omega \sum_s^{1,2} [\hat{c}_s^\dagger(k)\hat{c}_s(k) - \hat{d}_s(k)\hat{d}_s^\dagger(k)].$$

$$H_D = \int_{-\infty}^{+\infty} \{d^3k/(2\pi)^3 2\omega\} \sum_s^{1,2} [\hat{c}_s^\dagger(k)\hat{c}_s(k) + \hat{d}^\dagger(k)\hat{d}_s^\dagger(k)] \ \omega \quad (26.5.31)$$

which enables interpretation of 'd' quanta as the anti-particles of the 'c' quanta.

26.5.3.3. SPIN STATISTICS THEOREM:

Field Theories defined with integral spin are quantized with commutators are called BOSONS, while theories with half-integral spins are quantized with anti-commutators and are called FERMIONS.

26.5.4.4. SUCCESS of QFT:

Perhaps the most impressive successes of early Quantum Field Theory were the calculations of scattering probabilities for a variety of processes. These were possible because the theory contained a small dimensionless parameter, the 'coupling constant',
$\alpha = (e^2/\hbar c) = 7.2973 \times 10^{-3} = 1/137.036$. Scattering probabilities could be calculated unambiguously to first order in α. This means that the contribution to the probability that is proportional to the small parameter α is calculated while the contributions proportional to α^2 and higher powers of α are ignored. The processes which were studied included (where e^- is an electron and e^+ is a positron):

Electron - photon scattering	Klein and Nishina (1929)
e^+ - e^- - annihilation into photons	Dirac (1930)
Electron-electron scattering	Moller (1932)
$e^- \to e^-$ + photon in nuclear field	Bethe and Heitler (1934)
Photon $\to e^- e^-$ in nuclear field	" " "
$e^+ e^- \to e^+ e^-$.	Homi J. Bhabha, (1936)

They were in reasonable agreement with experiment.

26.6. QUANTUM ELECTRODYNAMICS (QED); QUANTIZATION OF GAUGE FIELD:

Having shown how to quantize the Spinor Field, it is now proper to consider Gauge Fields. The Quantum Field Theory of EM processes is called the QED, and is the simplest of the Gauge Theories. R.P. Feynman (1948), S. Tomonaga (1948) and J. Schwinger (1948) independently developed QED.

26.6.1 QUANTIZATION of the EM FIELD:

In both the complex scalar and the Dirac Fields, the free Lagrangian

L_{Free} = Invariant under U(1) transformation,

and one recognizes that *"globally"* charge, q is conserved: *i.e.*, $\sum q$ = constant. So each particle wave function ψ gives the effect

$$\psi \to \psi' = e^{-i\alpha q} \psi \qquad (26.6.1)$$

where α = a universal constant. $\alpha \sum q$ = constant phase. This (26.6.1) is a rotation (determined by the particle charge) in some INTERNAL space unconnected with space-time. The GAUGE PARTICLE in quantum mechanics consisted in evaluating the global *(i.e., z - t - independent)* U(1) phase invariance into a local *(i.e., z – t dependent)* U(1) phase invariance – the compensating fields being identified with the EM fields.

The Dirac Lagrangian,

$$L_D = \psi(i\gamma^\mu \partial_\mu - m_0)c, \qquad (26.6.2)$$

under LOCAL U(1) TRANSFORMATION

$$\psi(z,t) \to \psi'(z,t) = e^{-ie\chi(z,t)} \psi(z,t), \qquad q = e \qquad (26.6.3.)$$

provided that one replaces

$$\partial_\mu \to D^\mu = \partial^\mu + i e A^\mu, \qquad (26.6.4.)$$

(The EM Field, like any mass less field, possesses only two independent components, but is covariant- form invariance- described by a 4-vector potential, A_μ). Let the 4-vector potential, A_μ transforms as

$$A^\mu \to A'^\mu = A^\mu + \partial^\mu \chi, \qquad (26.6.5)$$

In this case, L_D becomes

$$L_D \to L_D + L_{int}. \qquad (26.6.6)$$

where $L_{int} = -e \amalg^\dagger \cdot \gamma^\mu \amalg A_\mu,$ (26.6.7)

But $\quad$ L=T-V

$$L_{int} \equiv V_D = e \amalg^\dagger \amalg A_0 - e \amalg^\dagger \alpha \amalg A = \rho A_0 - jA \qquad (26.6.8)$$

Thus the EM 4-vector current operator,

$$j^\mu_{EM} = e \amalg^\dagger \gamma^\mu \amalg = e \mathbb{N}^\mu. \qquad (26.6.9)$$

Conservation of j^μ_{EM} would follow 'global U(1) invariance' alone (*i.e.* χ = constant); The requirement for the force of the "local U(1) invariance" is that it has specified a unique form of the interaction, *i.e.* L_{int}. Indeed, this is just $j^\mu_{EM} A_\mu$, so that j^μ_{EM} is

(i) a SYMMETRY CURRENT,

(ii) it determines the precise way in which A^μ couples to the matter field ш.

$$A^\mu \equiv (\nabla, A)$$

A Gauge Transformation is

$$A^\mu \to A'^\mu = A^\mu + \partial^\mu \chi, \qquad (26.6.5)$$

Covariant form of the Maxwell's Equations is

$$j^\mu_{EM} = (\rho_{EM}, j^\mu)$$

and the Continuity Equation is

$$\partial_\mu j^\mu_{EM} = 0.$$

The Maxwell's Equations become

$$\partial_\mu F^{\mu\nu} = j^\nu_{EM}. \qquad (25.1.14)$$

with the field strength tensor (a 4-D curl).

$$F^{\mu\nu} = \partial^\mu A^\nu - \partial^\nu A^\mu. \qquad (25.1.15)\text{ b}$$

Since $\quad F^{\mu\nu}$ = invariant, obviously, under gauge transformation

$$A^\mu \to A'^\mu = A^\mu - \partial^\mu \chi,\qquad(26.6.5)$$

The Maxwell's Equations in this form are manifestly **gauge invariant**.

The Euler-Lagrange Equation (covariant field equations) satisfied by A^μ then is

$$\Box A^\mu - \partial^\nu(\partial_\mu A^\mu) = j^\nu{}_{EM}.\qquad(26.6.10)$$

$$\therefore\quad L_{EM} = -\frac{1}{4}F_{\mu\nu}F^{\mu\nu}.\qquad(26.6.11)$$

where $F_{\mu\nu} = \partial_\mu A_\nu - \partial_\nu A_\mu$. $\qquad(25.1.15)\,a$

is the quantized form of the field strength tensor.

i.e., $F_{\mu\nu}$ = invariant under Gauge Transformation.

In the same way the GLOBAL U(1) invariance of the complex scalar field may be generalized to a LOCAL invariance incorporating electromagnetism.

$$L_{KG} \to L_{KG} + L_{int}.\qquad(26.6.12)$$

where $L_{KG} = \partial_\mu \varphi^\dagger \, \partial^\mu \varphi - m_0^2 \varphi^\dagger \varphi$.

$$L_{int} = -ie[\varphi^\dagger \, \partial^\mu \varphi - (\partial^\mu \varphi^\dagger)\varphi]A_\mu + e^2 A^\mu A_\mu \varphi^\dagger \varphi.$$

$\therefore$ EM current, $j^\mu{}_{EM} = -\partial L_{int}/\partial A_\mu$, $\qquad(26.6.13)$

Substituting for L_{int},

$$j^\mu{}_{EM} = +ie[\varphi^\dagger \, \partial^\mu \varphi - (\partial^\mu \varphi^\dagger)\varphi]A_\mu - 2e^2 A^\mu \varphi^\dagger \varphi.\qquad(26.6.14)$$

The extra interaction Lagrangian, $\mathcal{L}_{int}$, gives rise to an interaction Hamiltonian, $\mathcal{H}_{int}$ that can be constructed

$$H_{int} = \int j^\mu{}_{EM} A_\mu d^3z.\qquad(26.6.15)$$

$$j^\mu{}_{EM} = e\, \text{Ш}^\dagger \gamma^\mu \, \text{Ш}\qquad(26.6.16)$$

The form of this interaction can be interpreted very simply in terms of the particle aspect of the quantum fields. It involves the product $\text{Ш}^\dagger$, Ш and A^μ, all at the same point.

To understand this considers the simpler fermion's case, for definiteness.

(i)   the field ш destroys electrons or creates positrons,
(ii)   The field ш† creates electrons and destroys positrons,
(iii)   The interaction is guaranteed to conserve charge and fermion number,
(iv)   The EM Field A^μ, creates and destroys photons.

Thus this single interaction term incorporates many processes.

Feynman diagrams illustrate a way of viewing EM processes. The diagrams are built from the interaction (EM coupling) vertex.

26.6.2 RADIATION GAUGE QUANTIZATION:

One has $L_{EM} = -\frac{1}{4} F_{\mu\nu} F^{\mu\nu}$. (26.6.11)

$F^{\mu\nu} = \partial^\mu A^\nu - \partial^\nu A^\mu$. (25.1.15) b

For a given EM Field, $A_\mu \neq$ unique; the Gauge Transformation leaves $F_{\mu\nu}$ = unchanged. This leads to the Lorentz Condition,

$\Box A(z) = -\partial_\mu A^\mu = 0$ (26.6.17)

A vector potential, A_μ obeying (26.6.17) is said to belong to the Lorentz Gauge.

Further if $\Lambda(z)$ satisfies such that

$\partial \Lambda(z)/\partial t = -\varphi$,

Since here $\varphi = 0$, and the potentials satisfy additional condition:

$\varphi = 0; \nabla . A = 0$. (26.6.18)

Such potentials are said to belong to the RADIATION GAUGE (or Coulomb gauge). In this gauge, A_μ has 2 components (In the LORENTZ GAUGE, A_μ has 3 components).

26.6.3 TO CALCULATE the Hamiltonian, H

Maxwell's Equations are

$\Box A^\mu - \partial^\nu (\partial_\mu A^\mu) = 0$ (26.6.19)

Applying (26.6.16), for free field case, one gets

$\Box A^\mu = 0$ (26.6.20)

In the radiation gauge, since $\varphi = 0; \nabla . A = 0$,

$\Box A = 0$ (26.6.21)

This is the K.G. equation for a mass less field, viz.

$A(z) = \int \{d^3k/(2\pi)^3 2k_0\} \sum_{\lambda=1,2} \hat{c}_\lambda(k)[\hat{a}_\lambda(k)e^{-ik\cdot z} + \hat{a}_\lambda^\dagger(k)e^{ik\cdot z}]$ (26.6.22)

$\hat{a}_{k\lambda} = \int d^3k \sqrt{\{(2\pi)^3 2k_0\}} f_k(z)^* i\partial_0 \varepsilon_{k\lambda} - A(z)$

$\hat{a}_{k\lambda}^\dagger = -\int d^3k \sqrt{\{(2\pi)^3 2k_0\}} f_k(z) i\partial_0 \varepsilon_{k\lambda} - A(z)$

where $\varepsilon_\lambda(k) = \varepsilon_{k\lambda} = \varepsilon^\mu$ the polarization vector of the wave,

$$A^\mu = N\varepsilon^\mu e^{-ikz},$$

Because of the COULOMB GAUGE, $\nabla \cdot A = 0$,

$$k \cdot \varepsilon_{k\lambda} = 0.$$

The Fourier expansion A_μ is given by

$$A_\mu(z) = \int d^3k / \{(2\pi)^3 2k_0\} \sum_{\lambda=0-3} \varepsilon_{\mu\lambda}(k)[\hat{a}_{k\lambda} e^{-ikz} + \hat{a}^\dagger_{k\lambda} e^{ikz}] \quad (26.6.23)$$

One should regard the components of the vector potential as independent quantities, thereby giving up the Lorentz Condition. One can have the Commutation Relations

$$[\hat{a}_{k\lambda}, \hat{a}^\dagger_{k'\lambda'}] = 0 \;;\text{ and } [\hat{a}^\dagger_{k\lambda}, \hat{a}^\dagger_{k'\lambda'}] = 0 \quad (26.6.24)$$

$\hat{a}^\dagger_\lambda$-s and $\hat{a}_\lambda$-s are creation and annihilation operators for four kind of photons (two transverse, one longitudinal and one time-like).

$$N_{k\lambda} = \hat{a}^\dagger_{k\lambda} \hat{a}_{k\lambda} \quad (26.6.25)$$

$$H_{EM} = \frac{1}{2}\int d^3z (E^2 + B^2) = \frac{1}{2}\int d^3z \,[(\partial A/\partial t)^2 + \nabla \Lambda A)^2]$$

$$= \sum_{\lambda=1,2} \int \{d^3k/(2\pi)^3 2k_0\}[\hat{a}^\dagger_{k\lambda}, \hat{a}_{k\lambda}] = ++\text{ve definite}. \quad (26.6.26)$$

Thus sacrificing the Lorentz invariance Condition, the total energy of a collection of photons, with transverse polarization, the physical degrees of freedom is quantized.

26.6.4 LORENTZ GAUGE QUANTIZATION:

For many purposes one requires the quantization on a Lorentz invariant manner. The canonical conjugate momentum $\pi(z)$ has corresponding

$$\pi^\mu = (\partial L_{int}/\partial t) A_\mu \quad (26.6.27)$$

Since $[A_0, \pi^0] = 0$,

and A_0 = a complex number, and not an operator, the covariance is lost. This and other conditions lead to, in general,

$$L_{int} = -\frac{1}{4} F_{\mu\nu} F^{\mu\nu} - \frac{1}{2}\lambda(\partial_\mu A^\mu)^2 \quad (26.6.28)$$

$$\Box A^\mu - (1-\lambda)\partial^\mu(\partial_\lambda A^\lambda) = 0 \quad (26.6.29)$$

$\lambda = 1$, is misleadingly known as the FEYNMAN GAUGE.

It can be obtained that

$$[\hat{a}_{k\lambda}, \hat{a}^{\dagger}_{k'\lambda'}] = -g^{\lambda\lambda'}(2\pi)^3 2k_0 \delta^3(k-k') \qquad (26.6.30)$$

which is <u>covariant</u>.

For scalar photons this becomes

$$[\hat{a}_{k0}, \hat{a}^{\dagger}_{k'0}] = -g^{\lambda\lambda'}(2\pi)^3 2k_0 \delta^3(k-k') \qquad (26.6.31)$$

The −ve sign on the right hand side of the equation gives problems for interpretation.

If $\quad |1\rangle = \int \{d^3k/(2\pi)^3 2k_0\} f(k) \hat{a}^{\dagger}_{k'0} |0\rangle \qquad (26.6.32)$

$$\langle 1||1\rangle = \int \{d^3k/(2\pi)^3 2k_0\} |f(k)|^2 \langle 0||0\rangle$$

Demanding $\hat{N}$ = integral of the number density operator, such that

$$\hat{N}|1\rangle = |1\rangle,$$

and inserting a negative sign,

$$\hat{N}|1\rangle = -\int \{d^3k/(2\pi)^3 2k_0\} | \hat{a}^{\dagger}_0(k)\hat{a}_0(k) \int \{d^3q/(2\pi)^3 2q_0\} f(q)|^2 \hat{a}^{\dagger}_0(q)|0\rangle$$

$$= \int \{d^3q/(2\pi)^3 2q_0\} f(q)|^2 \hat{a}^{\dagger}_0(q)|0\rangle = |1\rangle.$$

H_{EM} cannot give negative energy eigen values.

But H_{EM} can have negative energy expectation values.

26.6.5.   GUPTA-BLEULER Indefinite METRIC QUANTIZATION:

The negative values for the expectation $<H_{EM}>$ seen above can be imposing the condition that the physical state $|\psi\rangle$ should satisfy the subsidiary condition for all z;

$$(\partial_\mu A^\mu)|\psi\rangle = 0 \qquad (26.6.33)$$

for, decomposing into +ve and −ve frequency parts,

$$(\partial_\mu A^\mu)|\psi\rangle = [(\partial_\mu A^{+\mu}) + (\partial_\mu A^{-\mu})]|\psi\rangle = 0$$

But the negative frequency operator contains the creation operator, so not even the vacuum $|0\rangle$ could satisfy this identity. On the other hand, $A^{+\mu}$ contains the annihilation operator, and one may demand

$$(\partial_\mu A^{+\mu}(z))|\psi\rangle = 0 \qquad (26.6.34)$$

which is satisfied by the vacuum state.

$$\langle\psi|\partial_\mu A^{+\mu}(z)|\psi\rangle = \langle\psi|\partial_\mu A^{+\mu}(z) + \partial_\mu A^{-\mu}(z)|\psi\rangle = \langle\psi|\partial_\mu A^{-\mu}(z)|\psi\rangle$$

$$= \langle \psi | \partial_\mu A^{+\mu}(z) | \psi \rangle^* = 0.$$

This is the condition first formulated by Gupta and Bleuler. They proposed that the Lorentz Condition should be modified so that it holds good for the destruction part of the operator, i.e.

$$\partial_\mu A^{+\mu}(z) | \psi \rangle^* = 0.$$

Thai is, physical states are admixtures of longitudinal and time-like photons so that the above equation is valid.

i.e., in the momentum space this condition can be written as

$$[\hat{a}_{k3} - \hat{a}_{k0}] | \varphi \rangle = 0 \text{ , for all k.} \qquad (26.6.35).$$

In other words, the longitudinal and scalar photons are not observed as free particles. That means the contributions of the longitudinal and time-like photons to the H_{EM} cancel each other, leaving only the contribution of the (physical) transverse states.

This can be better presented by replacing the expectation value of an operator

$$<\hat{O}> = \langle \varphi | \hat{O} | \varphi \rangle \text{ by the definition}$$

$$<\hat{O}> = \langle \varphi | \hat{\eta} \hat{O} | \varphi \rangle$$

where $\hat{\eta}$ = a metric operator $\hat{\eta} = (-1)^m$.

The norm of the state vector $|\varphi\rangle$ is

$$\hat{N} | \varphi \rangle = \langle \varphi | \hat{\eta} \hat{O} | \varphi \rangle$$

The two definitions provide the $\square_{EM}$ with a positive expectation value and every state a positive norm.

26.7.   RENORMALIZATION:

The theory of electrons and photons that was described so far in particle interactions. But there are other particles and interactions: the proton, the neutron, the pion, the muon, the neutrino, *etc*. With the Local Quantum Field Theory seen above, when non-trivial calculations were attempted to higher order in α, the results aree infinite! These infinities arose precisely because of the local nature of the interaction. But the difficulties due to the said infinities became an asset with the development of *Renormalization* and QED. The cure of such difficulties is called 'renormalization'; and it involves redefinition of parameters like charge and mass of the electron. Renormalization is considered to be an essential property of any theory of elementary particle and their interactions.

Renormalizability is another constraint to impose in the theories of particles. The success of renormalizable QED, in the mid 20th Century, was spectacular. The electron magnetic moment (Lande g- factor) and other quantities were calculated to incredible accuracy and agreed well with increasingly precise experiments. QED became the paradigm of a successful physical theory.

The presentation of the subject of Field Theory has been unsatisfactory in at least two ways: (1) the relationship between the Green's Functions and quantities that can be measured in the Laboratory – Scattering Cross Sections and particle life times – has not been developed. This relationship is known as the 'reduction formulae', and it permits experimentally measured quantities to be found in terms of the various orders of Green's function, which in turn are expressed as combinations of field-free propagators. (2) the real Lagrangians have not been used., so real particles have not emerged from the calculations, without using any one of the various SYMMETRIES.

26.8 FEYNMAN DIAGRAMS:

Processes in which the absorption or emission of a *single photon* are spontaneous and stimulated ones. Both, in the fundamental theory of radiation processes as well as in many practical applications of non-linear optics, it will be necessary to study effects in which several or even *many photons* play a role. High order perturbation theory is useful and necessary for qualitative and quantitative discussions. The modern view of interactions of elementary charged particles and the EM Field is embodied in a Quantum Field Theory; QED. RP Feynman (1948), Tomonaga (1948) and Schwinger (1948) independently developed QED.

Earlier to Dirac's Quantum Mechanics looked difficult because of explanations with integral signs and complicated sum over states of particles. The introduction by Dirac of his bracket notation became acceptable. However, another simplification took place in the 1950s when physicists began to draw little squiggles and were still able to claim that they are doing calculations. In this second revolution in quantum mechanics, to get an overview of the possible processes, RP Feynman (1918 - 88) showed how squiggles could be used to portray mathematical expressions systematically. These diagrams go by his name. This technique of graphs is particularly useful to view EM processes.

26.8.1 S-MATRIX:

The PATH-INTEGRAL Method is used in the calculation of SCATTERING processes. When a particle is scattered on another, non-relativistically, by interaction through a potential $V(x)$ (which is small) then according to Perturbation Theory (Fig 26.1),

Propagator $\quad \Delta_F(q_f,t_f,q_i,t_i) = \langle q_f,t_f || q_i,t_i \rangle = K_0 + K_1 + K_2 + . + .$ (26.8.1)

where K_0 = Free Propagator,

K_1 = the 1st order correction to K_0, etc.,

for propagation with one interaction with the potential V. It can be shown that the solution to the perturbation series in K, called the BORN'S series, is

$$K(x_f,t_f,x_i,t_i) = K_0(x_f,t_f,x_i,t_i)$$
$$-\{(i/\hbar)\int K(x_f,t_f,x_i,t_i)V(x_i,t_i)K(x_f,t_f,x_i,t_i)dx_i dt_i - (i/\hbar)^2 \int - \quad (26.8.2)$$

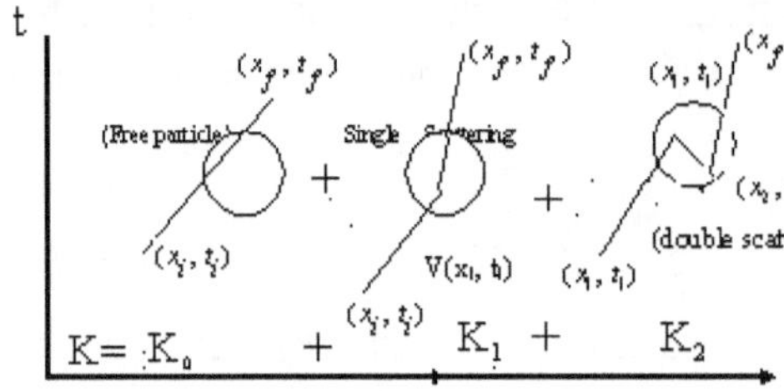

Fig. 26.1

It is known that $K_0(x_f,t_f,x_i,t_i)$ is the Green's function of the Schrodinger equation.

The SCATTERING AMPLITUDE, S = amplitude for detecting a final particle with definite momentum, ψ_{out}.

$$S = \int \psi_{out}^*(x_f,t_f)\psi^{(+)}(x_f,t_f)\, dx_f, \quad (26.8.3)$$

where the superscript (+) indicates that the $\psi^{(+)}$ corresponds to the wave which was free at $t = -\infty$, involving $K_0(x,t,x',t')$ which vanishes for $t' > t$. The scattering amplitude S is the overlap of the wave functions; and can be shown to be one of a matrix S, when

$$S_{fi} = \delta(k_i - k_f)$$
$$-\{(-i/\hbar)\int \psi_{out}^*(x_f,t_f)K_0(x_f,t_f,x,t)V(x,t)K_0(x,t,x_i,t_i)\psi_{in}(x_i,t_i)dx_f dx\, dx_i dt\}$$
(26.8.4)

This equation may be translated into a set of simple rules for the scattering amplitudes. These are called the Feynman Rules, which have been dealt with in Chapter 23.

R.P. Feynman developed the Propagator approach. The scattering process is described in terms of integral equations. The PROPAGATOR APPROACH to the NON-RELATIVISTIC Schrödinger Equation is as follows:

26.8.2 Feynman Rules (in Coordinate Space):

Let the first order approximation is taken and the diagram Fig 26.2

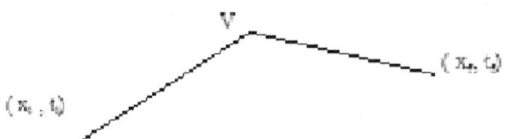

Fig. 26.2

$$A = \int \psi_o^*(x_f,t_f;x_i,t_i)\int \psi_o^+(x_f,t_f)dx_f \tag{26.8.5}$$

Making the correspondence may summarize the rules that translate this diagram into the mathematical expression for the scattering amplitude

$$S = K_0(x_2,t_2;x_1,t_1) \tag{26.8.6}$$

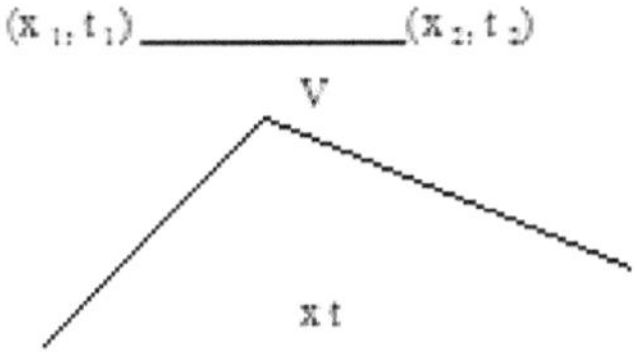

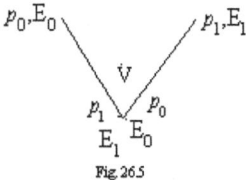

Fig. 26.3

$$S = -\{(-i/\hbar)\iint_{x\&t} V(x,t)\,dxdt \tag{26.8.7}$$

In addition one may mutiply by and Ψ_{out}^* at the ends of the diagram and integrate over the two relevant spatial variables.

26.8.3. Feynman Rules in momentum space.

p_0, E_0 \ / p_1, E_1
$\dot{V}$
p_1 \ / p_0
E_1 E_0

Fig. 26.5

Making the correspondence may summarize the rules that translate this diagram Fig. 26.5 into the mathematical expression for the scattering amplitude.

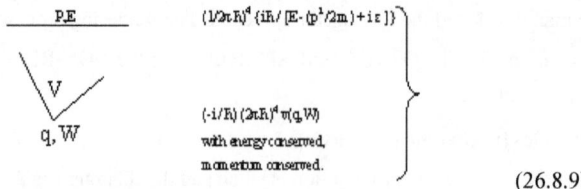

(26.8.9)

26.8.4 Feynman rules for SPONTANEOUS EMISSION of a photon:

The wave function, Φ(t)

$$\varphi(t) = \hat{a}_2^{\dagger}\phi_0 + \Sigma g_\lambda (1/\Delta\omega_2)[1-e^{i\Delta\omega t}]\hat{a}_1^{\dagger}\hat{b}_\lambda^{\dagger}\phi_0$$

$$= \hat{a}_2^{\dagger}\phi_0 + \Sigma \hat{c}_\lambda(t)\hat{a}_1^{\dagger}\hat{b}_\lambda^{\dagger}\phi_0 \qquad (26.8.10)$$

describes the spontaneous emission.

$\hat{a}_2^{\dagger}\phi_0$ = final state,

g_λ contains the external field.

The Feynman diagram for this expression is:

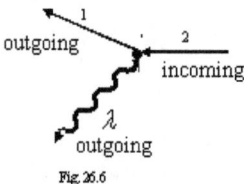

Fig. 26.6

This graph Fig. 26.6 is read from right to left; the solid line represents an electron, vertex is the interaction of the electron with the photon field, wavy line indicates a photon. In the diagram in which an incoming electron in the upper state 2 and no photon are present is transformed under the influence of the electron field interaction into a new state. In it the electron is now in state 1 while a photon with wave vector $k(=2\pi/\lambda)$ has been created. This can be interpreted by means of a graph. It is the task of the theory to calculate coefficients $\hat{c}_\lambda(t)$. The prescription for this calculation is read as follows:

Incoming electron wave:	$e^{i\varepsilon_2 t}$	→—○ 2
Outgoing " " " "	$e^{i\varepsilon_1 t}$	○—→
Outgoing photon:	$e^{i\Delta\omega_\lambda t}$	○∼∼∼
Vertex:	$-ig\lambda$	○
where	$\varepsilon_i = W_i/\hbar \quad i=1,2,3,...$	

The functions in the middle of the scheme must be multiplied with each other and finally one has to integrate the product from an initial time $t=0$ until the final time t.

This prescription yields

$$\hat{c}_\lambda(t) = -g_\lambda [e^{i\varepsilon_1 t} e^{-i\varepsilon_2 t} e^{i\Delta\omega t} - 1]/(\varepsilon_1 - \varepsilon_2 + \omega_\lambda) \tag{26.8.11}$$

26.8.5 Re-interpretation of STIMULATED EMISSION of Photon:

The diagram Fig. 26.7 shown indicates the induced emission of a photon:

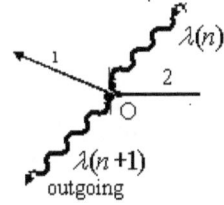

Fig 26.7

Incoming electron wave	$e^{-i\varepsilon_2 t}$	→—○ 2
Incoming Photon	$e^{-in\omega t}$	○∼∼∼ $\lambda(n)$
Outgoing electron wave	$e^{i\varepsilon_1 t}$	○—→ 1
$\lambda(n+1)$ outgoing Photon	$e^{i(n+1)\omega t}$	∼∼∼ $\lambda(n+1)$
Vertex	$ig_\lambda (n+1)^{1/2}$	○

One can calculate then the coefficients of the corresponding expansion of the wave functions in terms of photons and electron states. This is how the rules for calculating the time-dependent coefficients work.

26.8.6 Contribution to the SELF ENERGY of the electron:

Consider higher order perturbation theory and the action of two interaction Hamiltonians at time τ_1 and τ_2 can be represented by Feynman graph shown below (Fig. 26.8).

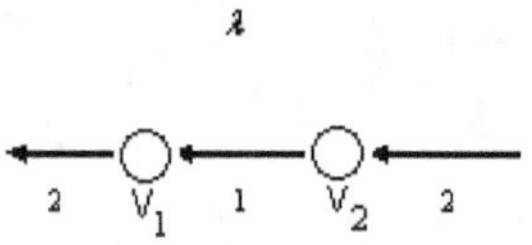

Fig 26.8

The vertices stem each time from an application of the interaction Hamiltonian, H_1.

Vertex 1

a) Incoming electron $\quad e^{i\varepsilon_2 \tau_1}$

b) Outgoing Photon $\quad -ig_\lambda$

c) Propagation of electron and photon from τ_1 to τ_2 $\quad e^{-i(\omega+\varepsilon_1)(\tau_2-\tau_1)}$

Vertex 2

Absorption of a photon $\quad -ig_\lambda^*$

Outgoing Free electron $\quad e^{i\varepsilon_2 \tau_2}$

$$\hat{c}_2(t) = \int_0^t d\tau_2 \, e^{i\varepsilon_2\tau_2} \sum (-ig_\lambda^*) \int_0^{\tau_2} e^{-i(\omega+\varepsilon_1)(\tau_2-\varepsilon_1)}(-ig_\lambda) \, e^{-i\varepsilon_2\tau_1} d\tau_1 \qquad (26.8.12)$$

where $0 \leq \tau_1 \leq \tau_2 \leq t$.

26.8.7 FEYNMAN GRAPH of a COMPLICATED PROCESS:

Virtual emission processes and change in the mass of electron:

After integration over τ_1,

$$\hat{c}_2(t) = \sum |g_\lambda|^2 \int_0^t [1-e^{-i(\varepsilon_2-\varepsilon_1-\omega)\tau_2}](1/i)/(\varepsilon_2-\varepsilon_1+\omega_\lambda)d\tau_2. \qquad (26.5.15)$$

Neglecting the oscillatory term (oscillating in time) the exponential function, for sufficiently long times, then $\quad \hat{c}_2(t) = -i\, t\, \Delta\varepsilon$ $\qquad (26.5.16)$

where $\Delta\varepsilon = \sum |g_\lambda|^2/(\varepsilon_2-\varepsilon_1+\omega_\lambda)$ $\qquad (26.5.17)$

In the I-picture, $|\Phi(t)\rangle_I = \hat{c}_2(t)\,\hat{a}_2^\dagger \varphi_0 + \hat{a}_2^\dagger \varphi_0 = (1-i\,t\,\Delta\varepsilon)\,\hat{a}_2^\dagger \varphi_0$. $\qquad (26.5.18)$

But $\quad |\Phi(t)\rangle_S = \hat{U}|\Phi(t)\rangle_I$,

$\hat{U} = e^{-i\hat{H}_0 t/\hbar}$.

$|\Phi(t)\rangle_S = e^{-i\Delta\varepsilon\, t} \hat{U}(\hat{a}_2^\dagger)\hat{U}^{-1}|\varphi_0\rangle = \hat{a}_2^\dagger e^{-i\Delta\varepsilon\, t}$.

$$= e^{-i(\varepsilon_2 + \Delta\varepsilon)t} |\varphi_0\rangle. \qquad (26.5.19)$$

Comparing with the stationary state of $\mathcal{H}(= \mathcal{H}_0 + \mathcal{H}_1)$,

$$|\Phi(t)\rangle = e^{-iWt} |\varphi_0\rangle,$$

one gets $W = \hbar(\varepsilon_2 + \Delta\varepsilon)$ \qquad (26.5.20)

i.e., $\hbar\Delta\varepsilon$ = energy shift of the electron in state 2,

$$\hbar \Delta\varepsilon = \hbar \sum |g_\lambda|^2 / (\varepsilon_2 - \varepsilon_1 + \omega_\lambda), \qquad (26.5.21)$$

from 2nd order Perturbation Theory

where (26.5.17) was used for $\Delta\varepsilon$.

$\Delta\varepsilon$ stems from the emission and subsequent absorption of a photon. Since the energy is not conserved during the intermediate state in which a photon is present the corresponding processes are called *virtual emission* of a photon. Thus the energy of an electron is shifted by the virtual emission and reabsorption of quanta (here photons).

This is the process involving the BOUND ELECTRON of an atom, which can occupy the states 1 and 2.

In the same way, for processes with FREE ELECTRONS with momentum, p the conservation of momentum requires $\quad p = p_{initial} + \hbar k.$ \qquad (26.5.22)

In this some more general case, the energy shift depends on p^2. This is usually called the SELF-ENERGY.

$$W_k^0 = W_0^0 + p^2/2m \qquad (26.5.23)$$

where $p = p_{initial}$, and one gets

$$\Delta W = \hbar\Delta\varepsilon = \Delta W_0 - Cp^2.$$

where $C = \delta/2m$, can be written; δ = small.

So $\quad W_k^0 = W_0^0 + p^2/2m^*$

with $\quad (1/m^*) = (1/m) - (\delta/m)$ \qquad (26.5.24)

showing VIRTUAL EMISSION PROCESSES INVOLVE CHANGE OF THE MASS OF THE ELECTRON.

Feynman graph of the Lamb Shift is shown in Fig. 26.9

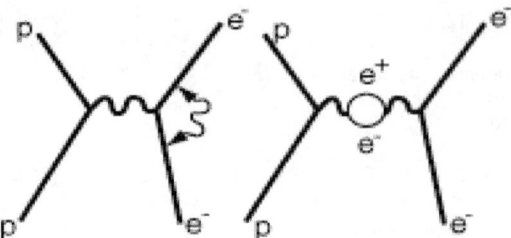

Fig. 26. 9 :   Feynman graph of the Lamb Shift

26.9. RENORMALIZATION OF THE MASS OF ELECTRON:
AND THEORY OF THE LAMB SHIFT:

It was seen that the EM SELF-ENERGY TERM has appeared for FREE ELECTRONS. Consider all levels of an atom. Let the initial and final states are $|n\rangle$ and $|n'\rangle$, respectively. The interaction Hamiltonian in the S-Picture for a H-atom (*i.e.* bound electron), $\hat{H}_{I,1}$,

$$\hat{H}_{I,1} = \hbar \sum_{\lambda,n,n'} \{g_{\lambda,n,n'} \hat{b}_\lambda^\dagger \hat{a}_n^\dagger \hat{a}_{n'} + g^*_{\lambda,n,n'} \hat{b}_\lambda \hat{a}_n^\dagger \hat{a}_{n'}\} \qquad (26.6.1)$$

Summing up over the intermediate states of both the virtually emitted photon and the electron, the

ENERGY SHIFT is $\qquad \Delta W_n / \hbar = \left(\sum_{\lambda,n'} \{|g_{\lambda,n,n'}|^2 / (\varepsilon_n - \varepsilon_{n'} - \omega_\lambda) \right) \qquad (26.6.2)$

where $W_n = \hbar \varepsilon_n$,

One can get (26.6.2) evaluated as

$$\Delta W_n / \hbar = C \int d^3k (1/\omega_k) \sum_{n',j} |\langle n|e_j p|n'\rangle|^2 / (\varepsilon_n - \varepsilon_{n'} - \omega_\lambda), \qquad (26.6.3)$$

where $\quad C = (1/2\pi)^3 \{e^2 / 2m^2 \hbar \varepsilon_0\}$

$(1/V) \sum_\lambda \Rightarrow \sum_{j=1,2} [d^3k/(2\pi)^3]$

$\omega_\lambda = \omega_k = ck$,

using $\int_{k\text{-space}} d^3k = \int_{\text{real space}} k^2 dk \int d\Omega$

$\int d\Omega = \sum_{n',j} |\langle n|e_j p|n'\rangle|^2 = 4\pi \frac{2}{3} |\langle n|p|n'\rangle|^2$,

$\therefore \Delta W_n / \hbar = (1/2\pi^2)\{2e^2 /3m^2 \hbar \varepsilon_0 c^3\} \int_0^\infty \omega d\omega \sum_{n'} |\langle n|p|n'\rangle|^2 /(\varepsilon_n - \varepsilon_{n'} - \omega_\lambda), \quad (26.6.4)$

Similarly for a FREE ELECTRON,

$$\Delta W_p / \hbar = (1/2\pi^2)\{2e^2/3m^2\hbar\varepsilon_0 c^3\} p^2 \int_0^\infty d\omega \qquad (26.6.5)$$

One can note that the EM SELF ENERGY for a FREE ELECTRON of momentum p is proportional to p^2.

For a FREE ELECTRON *without interaction*, with EM Field, BARE ELECTRON has "bare" mass, m_0.

$$W_p = p^2/2m_0. \qquad (26.6.6)$$

and $\quad \Delta W_p = -(1/2\pi^2)\{2e^2/3m^2\hbar\varepsilon_0 c^3\} p^2 \int_0^\infty d\omega \qquad (26.6.7)$

Total energy, $\quad W = W_p + \Delta W_p = p^2/2m \qquad (26.6.8)$

When the interaction is taken into consideration, the free electron gets mass, m.

$$\therefore \quad (1/m) = (1/m_0) - (1/2\pi)^2\{2e^2/3m^2\hbar\varepsilon_0 c^3\} p^2 \int_0^\infty d\omega$$

$$\boxed{(1/m) \equiv (1/m_0) - 2\hat{a}} \qquad (26.6.9)$$

This shift of the mass of an electron from its "bare" value, m_0 to its "observed" value, m, as result of the EM Self-energy, is called RENORMALIZATION of the mass. This is due to the zero-point a fluctuation of the quantized EM Field, a shift in the position of the electron occurs causing a perturbation similar in structure to the Darwin term. The argument used in the renormalization theory is as follows: In the case of a BOUND ELECTRON, $\Delta W_n / \hbar = \omega_n$ includes an infinite energy change that is already counted when one uses the observed mass in the $\hat{H}$ than the bare mass. In other words, in fact, one should start with the $\hat{H}$ of the H-atom in the presence of the radiation field

.i.e., $\quad \hat{H} = p^2/2m_0 - \{e^2/4\pi \varepsilon_0 r\} + \hat{H}_{int}. \qquad (26.6.10)$

Then using $\hat{H} = p^2/2m - \{e^2/4\pi \varepsilon_0 r\} + \{\hat{H}_{int} + \hat{a}p^2\} \qquad (26.6.11)$

thus if one use the observed free particle mass in the expression for the kinetic energy (which one always does) one should not count that part of $\hat{H}_{int}$ that produces the mass shift,.*i.e.*, one should regard

$$\{\hat{H}_{int} + \hat{a}p^2\} = \hat{H}_{eff} \text{ of an electron.} \qquad (26.6.12)$$

of a RENORMALIZED MASS, m with the radiation field.

Regarding the LAMB SHIFT, it is seen that, to first order in α $(=e^2/\hbar c)$, one must add $(<ap^2>$, the expectation value of the 2nd term in (26.6.12)) to $\Delta W_n/\hbar$ for the bound electron of equation (26.6.4). In order to avoid counting the EM interaction twice, once in m and once in $\hat{H}_{int}$,

More precisely the ENERGY SHIFT is given by

$$\Delta W_n/\hbar = \omega_n = a\int_0^\infty \omega d\omega \sum_{n'}\{|K\langle n|p|n'\rangle|^2/(\varepsilon_n - \varepsilon_{n'} - \omega)\} + [\langle n|p^2|n'\rangle/\omega] \quad (26.6.13)$$

where $a\int_0^\infty d\omega = (1/2\pi)^2\{2e^2/3m^2\hbar\varepsilon_0 c^3\}p^2\int_0^\infty d\omega$

One obtains $\quad \Delta W_n/\hbar = \omega_n$

$$= (1/2\pi)^2\{2e^2/3m^2\hbar\varepsilon_0 c^3\}\ell n|mc^2/(W_{n'}-W_n)|\{e^2\hbar^2/2\varepsilon_0\}|\varphi_n(0)|^2 \quad (26.6.14)$$

For hydrogen atom, $|\varphi_n(0)|^2 \neq 0$, only for the s-state. H. Bethe calculated the average value for the 2S level. Inserting all the numerical values,

$$\boxed{\omega_n = \Delta W_n/\hbar = 1040 \ MHz} \quad (26.6.15)$$

According to these considerations, a shift in the $2^2S_{1/2} \to 2^2P_{1/2}$ must be expected. Experimentally, W.E. Lamb and R.C Retherford (1947) observed (Fig 26.10) the shift in the case of Hydrogen spectrum.

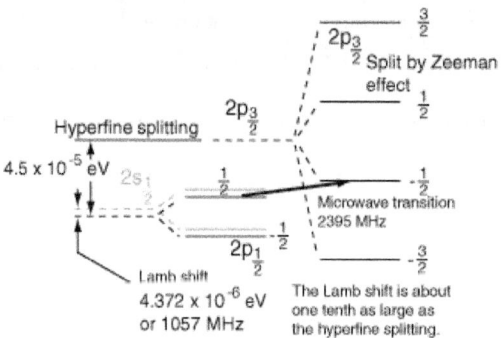

Fig. 26.10 : The Lamb Shift of the $2^2S_{1/2}$ level, of Hydrogen, lying higher than the $2^2P_{1/2}$ level, by 202.0 meV,

The discovery of the Lamb Shift (coming from vacuum polarization) of the $2^2S_{1/2}$ level, of hydrogen, lying higher than the $2^2P_{1/2}$ level, by 202.0 *meV*, initiated rapid and amazing progress in QED. The lines in Feynman Diagrams represent "bare" electrons and not the "real" (physical) electrons observed in the Laboratory. Real electrons are continuously emitting and reabsorbing virtual photons and CANNOT be detached from this accompanying cloud of virtual photons. Technically all the divergences in all Diagrams whatever can be absorbed into just two constants, which are the difference in mass (Δm) between the real (m) and bare (m_0) electron, and the corresponding charge difference (Δe):

$$m = m_0 + \Delta m \qquad (26.6.16)$$

$$e = e_0 + \Delta e. \qquad (26.6.17)$$

Then m_0, e_0, Δm and Δe are all infinite but unobservable while e and m remain finite. This procedure, called RENORMALIZATION, of the charge and mass, is a theory that gives finite reaction rates and observables at the cost of a restricted number of unobservable infinite constants. The property of Renormalizability coupled with the small size of structure constant means that it is possible to make predictions of high accuracy with QED whilst only calculating the contribution of relatively few low-order Feynman Diagrams. Thus

Effect	Energy contribution
Vacuum polarization	-27 MHz
Electron mass renormalization	+1017 MHz
Anomalous magnetic moment	+68 MHz
Total	+1058 MHz

Worked out Example 26.3.

Briefly describe the significance the Lamb Shift.

Solution:

When the Lamb Shift was experimentally determined, it provided a high precision verification of the theoretical calculations made with the Quantum Theory of Electro Dynamics (QED). These calculations predicted that electrons continually exchanged electrons, this being the mechanism by which electromagnetic force acted.

The effect of the continuous emission and absorption of photons on the electron-g factor could be calculated with great precision. The tiny Lamb Shift, measured with great precision, agreed to many decimal places, with the calculated result from QED. The measured precision gives the electron spin g-factor as

$$\boxed{g = 2.002319034386}. \qquad (26.6.18)$$

26.9.1. COLOUR AND QUANTUM CHROMO-DYNAMICS (QCD):

Careful examination of the wave functions of Baryons, in terms of Quark wave functions, required them to be completely Anti-symmetric. Accordingly, they are Symmetric rather than Anti-symmetric under the interchange of the quarks, although the quarks are Fermions. This requirement led O.W. Greenberg and others (1964) to introduce a new quantum number, known as COLOUR to achieve the necessary anti-symmetrization. With a new quantum number comes a new COLOUR SYMMETRY, which is recognized to be the symmetry underlying the Quantum Field Theory (QFT) of the Strong interaction. The QFT of the colour force is known as Quantum Chromo Dynamics (QCD).

26.10 CONSERVATION LAWS IN QUANTUM FIELD THEORIES:

Charge Conjugation replaces all particles with their ant-particles; Parity (P) Inversion inverts the coordinates of all particles: Time Reversal T runs all particle reactions backwards. It is a fundamental theorem in Quantum Field Theory that the Combined Symmetry operation of CPT applied to any particle interaction should produce another interaction process that is also a possible process Various experiments have shown that C, P, CP and T operations are all individually violated in certain particle decays; but no experiment has ever shown a violation of CPT invariance.

Table26.1

Conserved Quantity	Strong	Electro-magnetic	Weak
Parity, P	Yes	Yes	No
Charge Conjugation, C	Yes	Yes	No
Time reversal, T or CP	Yes	Yes	Usually, BUT: 1 in 500 violation in Kaon decays
CPT	Yes	Yes	Yes

26.10.1 QUANTIZATION SCHEMES IN FIELDS THEORY.

There is a number of different types of quantization schemes have been proposed over the decades, each with its own merits and drawbacks. Some of these were discussed in the earlier Chapters.

1).CANONICAL QUANTIZATION:

It is the most direct method. Here time is singled out as a special coordinate and Lorentz invariance is sacrificed.. Quantization is done only the physical modes. The system is manifest as unitary. In QED this method is not too difficult. But in non-Abelian gauge theories, this method is prohibitorily tedious.

2) COVARIANT QUANTIZATION (Gupta-Bluer)

It maintains full Lorentz symmetry – a great advantage. But unphysical states of negative norm are allowed to propagate as GHOSTS in the theory; and are eliminated only when constraints are applied to the state vectors.

3) PATH INTEGRAL approach

This is based on simple, intuitive principles that go into the very heart of the assumptions of quantum theory. Its drawback is that functional integration is a mathematically delicate operation that may not even exist in Minkowski space.

4) The BCCHI-ROUET-TYUPIN (BRST) APPROACH

It is one of the most convenient and practical covariant approaches for gauge theories. GHOSTS are allowed, but are eliminated by applying the BRST condition to the state vectors. The BRST approach can easily be expressed in terms of path integrals.

5) BATALIN – VILKOVISKY (BV) QUANTIZATION

This is a cumbersome programme, but it remains the only method that can quantize the most complicated actions.

6) STOCHASTIC QUANTIZATION:

It preserves gauge invariance. One has to postulate a fictitious 5th coordinate, so that the physical system settles down to the physical solution as the 5th coordinate evolves.

26.11. TOWARD A THEORY OF EVERYTHING (TOE):

All BASIC Processes are then known to be represented by the picture (Fig. 26.11) given below:

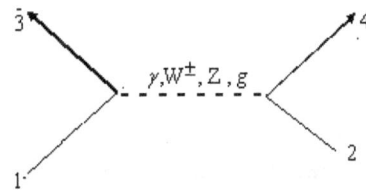

Fig.26.11: Feynmann diagram of all Basic Processes

Accordingly all the fundamental forces in nature, except gravitational interactions, seem to obey several common features:

(a)   These forces are all described by the same basic diagram above,

(b)   Each type of interaction posses a gauge symmetry group; SU(3) group for the strong interaction, and $SU(2)_W \otimes U(1)_\gamma$ for the electro-weak interaction. The HIGG'S FIELD breaks the electro-weak gauge group, while the colour and charge remain conserved. This is the STANDARD MODEL. This is shown in the chart; GIVING ONE THE POSSIBILITY OF UNIFYING ALL THE KNOWN FORCES via QFT.

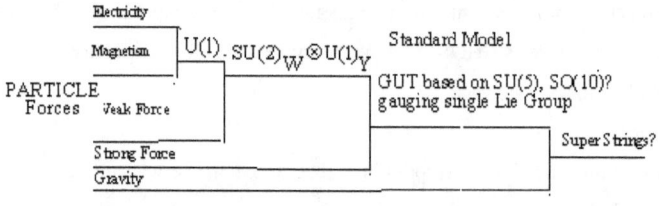

Fig. 26.12 GFT in Chart

According to GRAND UNIFICATION THEORY (GUT), the energy scale at which unification of all particle forces takes place is enormously large, about 10^{15} GeV, just below the Planck energy. Near the instant of the Big Bang, where such energies were found, the theory predicts that all three particle forces (excepting gravity) were unified by one GUT Symmetry. In this picture, as the Universe rapidly cooled down, the original GUT Symmetry was broken down successively into the present-day symmetries of the Standard Model. A typical breakdown scheme might be:

$$O(10) \to SU(5) \to SU(3) \otimes SU(2) \otimes U(1) \to SU(2) \otimes U(1)$$

Standard Model is thus a combination of Quantum Field Theories including QED, QCD and electro-weak unification that describes all the particles and their interactions. It is not a final theory as it contains around 20 constants that have to be put in form experimental measurements and as yet no satisfactory QUANTUM THEORY OF GRAVITY, which described separately in Einstein's General Theory of Relativistic Field Equations. .

Electrodynamics provides the pattern on which to model the Gauge Theory of the colour force. The starting point is to impose on the Dirac equation for the quarks.

$$\left(i\hat{\gamma}^{\mu}\partial_{\mu} - m_0\right)|\Psi\rangle = 0 \qquad (25.4.13)$$

In principle the amplitude for any two-body to n-body process can be obtained by drawing the Feynman diagrams of importance and adding up their contributions to the amplitude. There is a set of rules, similar to those for QED, which gives the factors associated with external lines and vertices. One important difference lies in the existence of the new types of vertices, which are only possible for a NON-ABELIAN symmetry (for Example, SU(3)).

The theory, which describes STRONG interactions in the Standard Model, is called QCD. QCD is similar to QED in that both describe interactions which are mediated by mass less spin-1 bosons coupling to conserved charges. Theories of this type are called GAUGE

THEORIES, because they have a characteristic symmetry called gauge invariance; and the spin -1 bosons are called GAUGE BOSONS. In the QED the bosons are PHOTONS; in QCD they are called GLUONS. Gluons have zero electric charge like photons, but couple to the colour charges, rather than to the electric charge. This leads immediately to the so-called FLAVOUR INDEPENDENCE of strong interactions; *i.e.* different quark flavours a = u, d, s, c, b and t must have identical strong interactions, because they exist in the same three COLOUR STATES r, g, b with the same possible colour charges.

A single theory that combines a GUT with quantum gravity is known as THEORY OF EVERYTHING (TOE). Some theorists think super-string theory will provide a TOE.

REVIEW QUESTIONS

R.Q. 26.1 Apply the Klein-Gordon Equation to a nucleon (point source) at a location independent of time interacting with a real scalar potential Φ, which describes a spin less particle, and obtain expression for the Yuckawa Potential.

R.Q. 26.2 (1) Give the Relativistically invariant form for the Euler-Lagrange Equation.

(2) Write an expression for the action S, if it has to be relativistically invariant.

(3) What does an invariant Lagrangian describe?

(4) Write down the expression for the Lagrangian which will give the K G Equation.

(Answers: 1) $(\partial L_S / \partial \phi) - \partial_\mu (\partial L_S / \partial_\mu \phi) = 0$, 2) $S[\varphi(z,t)] = \iint d^3z\, dt\, L_S(\varphi')$, 3) $L_S = $ invariant it describes a mass less particle, 4) $L_S = \frac{1}{2} \partial_\mu \varphi\, \partial^\mu \varphi - \frac{1}{2} m_0^2 \varphi^2$).

R.Q. 26.3 What is the theoretical basis behind the prediction that 'all particles have their anti-particle counterparts'?

R.Q. 26.4 Find the scattering amplitude fro the second order process shown in the diagram,

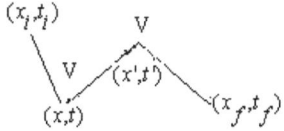

(Answer:

$$A^{(2)} = (-i/\hbar)^2 \int \psi_{out}^*(x_f,t_f;x_i,t_i) \int \psi_o^+(x_f,t_f) K_0(x_f,t_f;x',t') V(x',t') K_0(x',t';x,t)$$
$$V(x,t) K_0(x,t;x_i,t_i) dx_i dx dt dx_f$$

&&*&*&*&*&*&

APPENDIXES

VLUES OF PHYSICAL CONSTANTS

BIBLIOGRAPHY

APPENDIX .
Values of Physical Constants

[Ref: *CODATA*, 1998; *J. Phys. Chem. Ref. Data*, Vol. 28 (6) 1999; *Rev. Mod Phys.* 72.(2) 2000]

Planck constant	$h = 6.6256 \times 10^{-34} J-s$;
	$\hbar = (h/2\pi) = 1.054 \times 10^{-34} J-s$
Permittivity of free space	$\varepsilon_0 = 8.8542 \times 10^{-12} F\,m^{-1}$;
	$\varepsilon_0 = 8.8542 \times 10^{-12} C^2 N^{-1} m^{-2}$
The Coulomb constant	$k = (1/4\pi\,\varepsilon_0) = 8.9875 \times 10^9 Nm^2 C^{-2}$
	$k = (1/4\pi\,\varepsilon_0) = 8.9875 \times 10^9 F^{-1} m$
Permeability of free space	$\mu_0 = 4\pi \times 10^{-7} T\,m\,A^{-1}$,
	$\mu_0 = 4\pi \times 10^{-7} N\,A^{-2}$.
Gravitational constant	$G = 6.67 \times 10^{-11} Nm^2 kg^{-2}$
Speed of light in Free space	$c = 2.997925 \times 10^8 ms^{-1}$
Boltzmann constant	$k_B = 1.3805 \times 10^{-23} JK^{-1}$
Avogadro's constant	$N_A = 6.0225 \times 10^{23} mol^{-1}$
Stefan's constant	$\sigma = 5.67 \times 10^{-8} Wm^{-2} K^{-4}$,
Universal Gas constant	$R = 8.314\, J\, K^{-1} mol^{-1}$
Electronic Charge	$e = 1.6021 \times 10^{-19} C$
Electron Rest mass	$m_e = 9.1094 \times 10^{-31} kg$
	$m_e = 5.4858 \times 10^{-4} u = 0.5110\, MeV$
Electron Volt	$1\,eV = 1.6021 \times 10^{-19} J$
Proton Rest mass	$M_p = 1.6726 \times 10^{-27} kg$;
	$M_p = 1.007276\, u = 938.272\, MeV$
Mass of Neutron	$M_n = 1.67493 \times 10^{-27} kg$
	$M_n = 1.008665\, u = 939.57\, MeV$
Mass of Deuteron	$M_d = 2.013553\, u$
Mass of Alpha particle	$M_\alpha = 4.002602\, u$
Mass of H-atom	$M_H = 1.67493 \times 10^{-27} kg$;
	$M(_1^1 H) = 1.007825\, u$
Mass of Tritium	$M(_1^3 H) = 3.016049\, u$
Mass of He	$M(_2^4 He) = 4.00387\, u$
Mass of Nitrogen	$M(_7^{14} N) = 14.00752\, u$

BIBLIOGRAPHY

(References and books for Further Reading):

1) Schiff, L.I. (1968), *Quantum Mechanics* (MGH Kogakusha, Tokyo).
2) Bohm, David (1989), *Quantum Theory* (Dover Publication, NY).
3) Pauling, L. and Wilson, E.B. (1935), *Introduction to Quantum Mechanics* (MGH Kogakusha, Tokyo).
4) Merzbacher, E.(1970) *Quantum Mechanics*, II edn (John Wiley- Toppan Co., Tokyo).
5) David Park (1964), *Introduction to Quantum Theory* (MGH, NY).
6) Constantinescu,F. and Magyari,E. (1971), *Problems in Quantum Mechanics* (Pergamon, Oxford).
7) Schwabl, F. (1992), *Quantum Mechanics* (Narosa, New Delhi).
8) Gasiorowicz, S. (1996) *Quantum Physics* (John Wiley).
9) Bethe, H.A. and Jackiw (1968), *Intermediate Quantum Mechanics* (Benjamin, NY.).
10) Bethe, H.A. and Salpeter, EE, (1957), *Quantum Mechanics of One- and Two- Electron Atoms* (Springer-Verlag NY).
11) Dirac, PAM (1958) *The Principles of Quantum Mechanics* (Oxford Univ. Press, Oxford).
12) Feynman, R.P., Leighton, R.B. and Sands, M. (1965) *The Feynman Lectures on Physics, Vol. 3, Quantum Mechanics* (Addison-Wesley, Reading).
13) Landau, L.D. and Lifshitz, E.M. (1965), *Quantum Mechanics*, 2 nd Edn (Non-Relativistic Theory) (Addison-Wesley, Reading).
14) Messiah, A.(1999) *Quantum Mechanics* , 2 volumes, (Dover, NY).
15) Powell, J.L. and Craseman, B. (19 71), *Quantum Mechanics* (BI Publications, ND).
16) Rose, M.E. (1937), *Elementary Theory of Angular Momentum* (John Wiley, NY).
17) Sakurai, J.J. (1999) *Modern Quantum Mechanics* (Addison-Wesley, Reading).
18) Sakurai, J.J. (1999) *Advanced Quantum Mechanics* (Addison-Wesley, Reading)
19) Matthews, P.T. (1979) *Introduction to Quantum Mechanics* (TMH, New Delhi).
20) Wilezek, F. (2000) *What is Quantum Theory?* In Physics Today, June 2000.(AIP)
21) Born, Max and Wolf, E. (19 59) *Principles of Optics* (Pergamon, Oxford).
22) Fermi, E. (1995), *Notes on Quantum Mechanics* (Univ. Chicago Press, Chicago).
23) Fock, VA (1978), *Fundamentals of Quantum Mechanics* (MIR Pub, Moscow).
24) Reimes, Stanley (1990), *Many Electron Theory* (ELBS) (North-Holland, Amsterdam, 1972).
25) Davydov, AS (1976), *Quantum Mechanics* (Pergamon, Oxford).
26) Bjorken, JD and Drell, SD (1964), *Relativistic Quantum Mechanics* (MGH, NY).
27) Mattuck, RD (1976) *A Guide to Feynman Diagrams in the Many Body Problem* (MGH, NY).
28) Itzykson, C and Zuber, JB (1980), *Quantum Field Theory* (MDGH, NY).
29) Atchison, IJR and Hey, AJG (1989), *Gauge Theories in Particle Physics- A Practical Introduction* (Adam Hilger, Bristol).
30) Roman, Paul (1965) *Advanced Quantum Theory* (Addison Wesley, Reading).
31) Atkin, RH (1956) *Mathematics of Wave Mechanics* (William Heinemann, Melbourne).
32) Mandl, F (1960) Quantum Field Theory (Nescience,).
33) Siegfied Flugge (1971) *Practical Quantum Mechanics* II (Springer-Verlag, Berlin).
34) Mandl, F and Shaw (1984) *Quantum Field Theory* (Wiley, Chichester).
35) Power, EA , *Introductory Quantum Electrodynamics* (Longman, London).

36) Craig, DP and Thirunanchandran, T (1984) *Molecular QED* (AP, London)
37) Kaku, Michio (1993) *Quantum Field Theory -A Modern Introduction* (Oxford Univ Press, NY).
38) Davies, Paul (1989) *The New Physics* (Camb Univ Press, Cambridge, USA).
39) Akhiezer, AI and Berestetskii, VB (1962) *Elements of Quantum Electrodynamics* ((London Oldbairne Press).
40) March, NH, Young, W H and Sampather, S. (1967) *The Many Body Problem in Quantum Mechanics* (CUP).
41) Rojansky, V.(1958) *Introductory Quantum Mechanics* (Englewood Cliffs, NY).
42) Mott, NF (1952) *Elements of Wave Mechanics* (CUP, London).
43) Dicke RH and Wittke, JP (1960) *Introduction to Quantum Mechanics* (Addison Wesley, Reading).
44) Ryder, Lews H (1996) *Quantum Field Theory- Text* (CUP,).
45) Weinberg, S. (1996) *Quantum Theory of Fields* (Vol I and Vol II)(CUP).
46) Bohm, Arno (1989), *Quantum Mechanics, Foundation and Applications* 3rd Ed (Springer-Verlag, NY).
47) Lieb, Elliott H, (1997) *The Stability of Matter: From Atoms to Stars*, II Edn.(Springer-Verlag, Berlin).
48) Mavromatis, Harry A. (1992) *Exercises in Quantum Mechanics* (Kluwer Acad. Pub. Dordrecht).
49) Srinivasan, K. and Rajeswari, V (1993) *Quantum Theory of Angular Momentum* (Narosa, New Delhi).
50) Varshelovich, D A, Moskalev, A N and Rhersowiskii, V K (1989) *Quantum Theory of Angular Momentum* (World Sci. Singapore).
51) Arfken, G. (1985) *Mathematical Methods for Physicists*, 3rd Edn (AP).
52) Fano, U and Rau, ARD (1996) *Symmetries in Quantum Physics* (AP,).
53) Lawrie, Ian D. (1990) *A Unified Grand Tour of Theoretical Physics* (Adam Hilger, Bristol).
54) Kleinerst, H (1995) *Path Integrals in Quantum Mechanics, Statistics and Polymer Physics* (Plenum, NY).
55) Shankar, R (1997) *Principles of Quantum Mechanics* (Plenum, NY).
56) Ferry, David K. (1995) *Quantum Mechanics- n Introduction to Device Physicists and Electronic Engineers* (Inst. Physics Publ., Bristol).
57) Squires, GL (1995) *Problems in Quantum Mechanics with Solutions* (CUP,).
58) Pisani, C. (1996) (Ed) *Quantum Mechanical Ab-Initio Calculation of the Properties of Crystalline Materials*, Vol 67 of Lecture Notes on Chemistry (Springer,).
59) Griffiths, (1995) *Introduction to Quantum Mechanics* (PH, NJ).
60) Landau, RH (1996), *Quantum Mechanics* II (J W, NY).
61) Greiner (1994) *Quantum Mechanics- An Introduction* (3rd edn) (Springer, NY).
62) Greiner, Schafer Reinhardt (1994) *Quantum Chromodynamics* (Springer-Verlag, NY).
63) Greiner and Reihardt (1994), *Quantum Electrodynamics* (Springer-Verlag, NY).
64) Liboff, RL (1993), *Introductory Quantum Mechanics* (2nd Edn) (Addison Wesley, Reading).
65) Baggott, Jim (1992), *The Meaning of Quantum Theory – A Guide for students of Chemistry and Physics*, (Oxford University Press, Oxford).
66) Stevens, Charles F. (1995), *The Six Core Theories of Modern Physics* (The MIT Press, Cambridge).
67) Cohen – Tannoudji, C., Diu, B. & Laloe, F. (1977), *Quantum Mechanics*, Vol. 1 and 2 (John Wiley, New York).
68) Adams, Steve, (2000), *Frontiers- Twentieth Century Physics* (Taylor & Francis, London).

69) Devanarayanan, S. (2005), *Quantum Mechanics – Principles and Applications* (Sciech, Chennai)
70) Holzner, Steven (2009), *Quantum Physics for Dummies* (Wiley Publi, NJ).
71) Biedenharn, L.C., (1995), "Quantum Group Symmetry and q-tensor", (World Sci., NJ).
72) Annamma John & S. Devanarayanan, (1997) Indian J. Pure Appl. Phys., 35(1997) 387.

^&*&*&*&*&*^*

About the Author

Prof. S. DEVANARAYANAN, Ph.D. (IISc); D.Sc. (USA), Dip (Uppsala)

Dr. S. Devanarayanan was educated at the University College, Thiruvananthapuram, Indian Institute of Science, Bangalore, and Institute of Physics, Uppsala, Sweden. He had a brilliant academic career throughout. He was the Professor and Head of the Department of Physics, University of Kerala during 1993 – 2000; and has 37 years of teaching / research experience in Physics and materials science.. He has to his credit over 80 published research papers in standard scientific periodicals and 42 presentations in National and International Science events. A Monograph entitled THERMAL EXPANSION OF CRYSTALS (1979) and book "QUANTUM MECHANICS: Principles & Applications" (2005), "QUANTUM CHEMISTRY"(2013), "PHYSICS IN A NUTSHELL: Companion for Success in Competitive Tests" (2016), "A TEXT BOOK ON NUCLEAR PHYSICS" (2016) , and "FERROELECTRICITY IN CRYSTALS" (2016) were authored by him. He has served as a Professor in Physics at the University of Puerto Rico, USA, during 1989 –91. He was awarded the SIDA Fellowship and worked at The Institute of Physics, Uppsala, Sweden, during 1970 – 71. A Life Member of various academic bodies like the Indian Physics Association, American Physical Society, and Fellow of the Indian Cryogenic Council, his biography has found place a number of times in the publications of Marquis' Who's Who (USA), International Biographical Centre (UK), American Biographical Institute, Refacimento International, *etc*. The Govt. of Kerala appointed him as a member of the Commission of Enquiry on the working of the University of Kerala, in 2000.

As an experimental physicist / materials scientist his research specializations include Phase Transitions in crystals, Mössbauer Effect, Crystal growth, Vibrational spectroscopy and Atmospheric physics. Devanarayanan's early studies were on phase transitions in ferroelectric crystals at the Indian Institute of Science, where he had used Fizeau's optical interferometer and cryogenics. The work on single crystal sodium trihydro selenite is the outstanding among them. He continued the studies on magnetic transitions and magnetic structute in Fe_2P , down to liquid helium temperatures, at the Institute of Physics, Uppsala, using Mossbauer spectroscopy. This work and the extensive studies on ordering on the role of concentration of silicon in Fe-Si alloys form important contribution. Yet another investigation was on a series of phase transitions at cryogenic temperatures in crystalline lithium cesium sulphate using Raman Spectroscopy at University of Puerto Rico. He had encouraged his students to start work on in the field of biological crystals, vibrational spectroscopy and q-deformed oscillators in Solid State. He was responsible for analyses of rocket-sonde temperature data over a period of solar cycle in the middle atmosphere over Thumba, and at mid- and high- latitudes.

A number of students received Ph.D. and M. Phil. degrees under his research supervision. He had the special honour of being invited by the Royal Swedish Academy to submit proposals for the award of the Nobel Prize in Physics for 1995. Devanarayanan believes in Sir C.V. Raman's (NL) advice that one can become a good scientist only when one takes up research along with teaching at a University level. He has made academic visits in Sweden, Finland, Leningrad (USSR), The Netherlands, Germany, France, Australia, Czechoslovakia, Hungary, Austria, England, and USA. For more details, website: <researchgate.net>.

&&*&*&*&*&*^*&*&*&*&*

695

www.ingramcontent.com/pod-product-compliance
Lightning Source LLC
Chambersburg PA
CBHW080615190526
45169CB00009B/3186